Selected Papers of Abraham Robinson

Volume 2 Nonstandard Analysis and Philosophy

Abraham Robinson, 1970.

Selected Papers of Abraham Robinson

Edited by
H. J. Keisler, S. Körner, W. A. J. Luxemburg, and A. D. Young

Volume 2 Nonstandard Analysis and Philosophy
Edited and with introductions by
W. A. J. Luxemburg
and S. Körner

New Haven and London Yale University Press 1979

Printed in the United States of America by the Hamilton Printing Company, East Greenbush, New York.

Published in Great Britain, Europe, Africa, and Asia (except Japan) by Yale University Press, Ltd., London. Distributed in Australia and New Zealand by Book & Film Services, Artarmon, N.S.W., Australia; and in Japan by Harper & Row, Publishers, Tokyo Office.

Library of Congress Cataloging in Publication Data

Robinson, Abraham, 1918–1974.
Selected papers of Abraham Robinson.

Include bibliography.
CONTENTS: v. 1. Model theory and algebra.—v. 2. Nonstandard analysis and philosophy.—v. 3. Aeronautics.
1. Mathematics—Collected works. 2. Robinson, Abraham, 1918–1974. I. Keisler, H. Jerome. II. Title.
QA3.R66 1978 510'.8 77-92395
ISBN 0-300-02071-6 (v. 1)
0-300-02072-4 (v. 2)
0-300-02073-2 (v. 3)

Acknowledgment is gratefully made to the following for permission to reprint previously published articles by Abraham Robinson (and others):

Academic Press, Inc., for "Ordered Differential Fields," from J. Combinatorial Theory *Ser. A, 14 (1973).*

American Mathematical Society for the following:

From the Transactions *of the American Mathematical Society: "Syntactical Transforms" (with A. H. Lightstone), copyright © 1957, vol. 86, pp. 220–45; "Elementary Properties of Ordered Abelian Groups" (with Elias Zakon), copyright © 1960, vol. 96, pp. 222–36; "Local Differential Algebra," copyright © 1960, vol. 97, pp. 427–56; "Local Partial Differential Algebra" (with S. Halfin), copyright © 1963, vol. 109, pp. 165–80.*

From Proceedings of the International Congress of Mathematicians: *"On the Application of Symbolic Logic to Algebra," copyright © 1950, vol. 1, pp. 686–94.*

From Proceedings of the American Mathematical Society: *"Solution of a Problem by Erdos–Gillman–Henriksen," copyright © 1956, vol. 7, pp. 908–09.*

From the Proceedings of Symposia in Pure Mathematics: *"A Decision Method for Elementary Algebra and Geometry–Revisited," copyright © 1974, vol. 25, pp. 139–52.*

Association for Computing Machinery for "Random-Access Stored-Program Machines, an Approach to Programming Languages," from Journal of the Association for Computing Machinery, *vol. 11, no. 4 (1964). Copyright © 1964 by the Association for Computing Machinery, Inc.*

Association for Symbolic Logic for the following:

"On Predicates in Algebraically Closed Fields." Reprinted from the Journal of Symbolic Logic. *vol. 19, pp. 103–14, by permission of the Association for Symbolic Logic. Copyright 1954 by the Association for Symbolic Logic.*

"Note on a Problem of L. Henkin." Reprinted from the Journal of Symbolic Logic, *vol. 21, no. 1, pp. 33–35, by permission of the Association for Symbolic Logic. © 1956 by the Association for Symbolic Logic.*

"On the Representation of Herbrand Functions in Algebraically Closed Fields" (with A. H. Lightstone). Reprinted from the Journal of Symbolic Logic, *vol. 22, pp. 187–204, by permission of the Association for Symbolic Logic. © 1957 by the Association for Symbolic Logic.*

"On the Notion of Algebraic Closedness for Non-Commutative Groups and Fields." Reprinted from the Journal of Symbolic Logic, *vol. 36, pp. 441–44, by permission of the Association for Symbolic Logic. © 1971 by the Association for Symbolic Logic.*

Koninklijke Nederlandse Akademie van Wetenschappen for the following from Nedrl. Akad. Wetensch. Proc.: *"Non-standard Analysis," Ser. A 64, and Indag. Math. 23 (1961); "Non-standard Theory of Dedekind Rings," Ser. A 70, and Indag. Math. 29 (1967).*

London Mathematical Society for the following:

From J. London Math. Soc.: *"On the Integration of Hyperbolic Differential Equations," vol. 25 (1950).*

From Proc. London Math. Soc.: *"On Functional Transformations and Summability," (2) 52 (1950).*

Mathematical Association of America for "Function Theory on Some Nonarchimedean Fields," from Papers in the Foundations of Mathematics *(supplement to* American Mathematical Monthly, *June/July 1973), Slaught Memorial Papers No. 13, pp. 87–109 (1973).*

Noordhoff International Publishing Co. for "Some Thoughts on the History of Mathematics," from Compositio Math,. *vol. 20, pp. 188–193 (1968). Also in* Logic and the Foundations of Mathematics, *Groningen, Walters-Noordhoff, 1968.*

North-Holland Publishing Company for the following: "Formalism 64," from Proc. Internat. Congress for Logic, Methodology and Philos. Sci., *Jerusalem, 1964, pp. 228–246 (1965); "Topics in Non-Archimedean Mathematics," from the* Theory of Models Proc. 1963 Internat. Sympos., *Berkeley, California, pp. 285–298 (1965); "The Metaphysics of the Calculus," from* Problems in the Philosophy of Mathematics, *pp. 28–46 (1967); "Algebraic Function Fields and Nonstandard Arithmetic," from* Contributions to Nonstandard Analysis, *pp. 1–14 (1972); "Concerning Progress in the Philosophy of Mathematics," from* Proc. Logic, *Colloquium at Bristol, 1973, pp. 41–52 (1975).*

La Nuova Italia Editrice for "Model Theory," from Contemporary Philosophy: A Survey, Volume 1: Logic and the Foundations of Mathematics, *pp. 61–73 (1968).*

Pacific Journal of Mathematics for the following from Pacific J. Math.: *"On Generalized Limits and Linear Functionals," vol. 14 (1964). Copyright ©1964 by Pacific Journal of Mathematics; "Solution of an Invariant Subspace Problem of K. T. Smith and P. R. Halmos" (with Allen R. Bernstein), vol. 16 (1966). Copyright ©1966 by Pacific Journal of Mathematics.*

Philosophical Society of Finland for "On the Theory of Normal Families," from Acta Philosophica Fennica, *vol. 18 (1965).*

Rijksuniversiteit te Groningen Mathematisch Instituut for "Standard and Nonstandard Number Systems" (The Brouwer Memorial Lecture 1973), from Nieuw Archief voor Wiskunde *(3) 21 (1973).*

Springer-Verlag, New York, for "Enlarged Sheaves," from Proc. Victoria Symposium on Nonstandard Analysis 1972. Lecture Notes on Mathematics, *vol. 369, pp. 249–260 (1974).*

EDITORS' NOTE

The tragic loss caused by the premature death of Abraham Robinson brought about a deeply felt desire by his close friends, colleagues, and pupils to commemorate his exceptional gifts and personality. An appropriate way to accomplish this appeared to be the selection and editing of his writings on applied mathematics, mathematics, mathematical logic, philosophy, and nonstandard analysis, and the publication of these papers in these volumes.

This edition presents only an incomplete picture of Robinson's work and influence on mathematics and mathematical logic. It is impossible to do full justice to his many lectures, his philosophical and expository articles, and his many contributions to the work of his pupils. We do feel, however, that the present selection, together with his many important books and monographs, will show the mark his work has left on the development of mathematics and mathematical logic in this century.

It is with great pleasure and with deep gratitude that we acknowledge the help and advice of his widow, Renée Robinson. We also would like to extend our sincere thanks to Mrs. Jane Isay of the staff of Yale University Press for her care and professionalism in the production of these volumes.

H. J. Keisler
S. Körner
W. A. J. Luxemburg
A. D. Young

CONTENTS

[Numbers in brackets refer to numbers in the bibliography]

Nonstandard Analysis and Analysis

Philosophy

BIOGRAPHY OF ABRAHAM ROBINSON

George B. Seligman

Abraham Robinson was born in Waldenburg in Lower Silesia (then Germany) on October 6, 1918. He was the second of two children born to Abraham Robinsohn and Hedwig Lotte Robinsohn, born Bähr. His brother Saul was two years older. The apparent break with Jewish practice made in naming him after his father was justified on the grounds of the father's untimely death at the age of forty, shortly before young Abraham's birth.

The senior Abraham Robinsohn had studied literature and philosophy in Switzerland and had continued his study of philosophy in Oxford and London. He received a Ph.D. from the University of London for a dissertation on William James. He was a leader in Zionist movements from his student days in Switzerland, where he organized student groups for the study of modern Hebrew and the promotion of its use within Jewish society. For the rest of his life he served the Jewish communities of England and continental western Europe through his journalistic and organizational work, particularly in spreading the idea of a Jewish state among the younger people. He sought Jewish unity through language and drew up plans for organizing Jewish education around this linguistic core. His writing and editorial efforts were centered first in London and later in Berlin, where several periodicals owed to him much of their vitality, and perhaps even the very fact of their publication.

Leadership of the World Zionist Organization had passed, upon the death of Theodore Herzl in 1904, to David Wolffsohn, a Cologne businessman who had been a most enthusiastic supporter of the Organization and a personal friend of Herzl's. By 1912 Wolffsohn's health was failing and he was aware of his inability to do justice to all the responsibilities of his office. In particular, the extensive correspondence on behalf of the Organization, much of it necessarily in Hebrew, was a serious burden for a man who was primarily a fund-raiser. About the end of 1912, Wolffsohn engaged Abraham Robinsohn as his private secretary, with responsibility for this correspondence. Robinsohn also tutored Wolffsohn in Hebrew and served as his confidant and personal adviser. The latter capacity became more significant as Wolffsohn's health deteriorated and he shared with Robinsohn his visions and his concerns for the future of the Zionist movement.

David Wolffsohn died in the fall of 1914. His papers, and ultimately those of Herzl as well, became the responsibility of Robinsohn. The curator of these papers found himself in his later thirties, still unwed, but now with a hope for the financial security to undertake marriage. In early 1916 he married Hedwig Lotte Bähr, a teacher in Cologne who shared his dedication to Zionism. She was twenty-seven at the time. His duties with the Herzl and Wolffsohn archives took the couple to Berlin, where a son, Saul, was born in late 1916. They were only awaiting the end of the war to realize the dream of Jerusalem, where Robinsohn had been selected to head the Jewish national library on Mt. Scopus, when he died, suddenly, in May of 1918. He left unpublished a long biography of Wolffsohn and other writings. A shorter version of the biography, in German, was published by the Jüdischer Verlag in Berlin in 1921. The full manuscript lies on Mt. Scopus awaiting publication, not the least of the reasons for the delay being the accounts it is said to contain of conflicts within the Zionist organization, involving participants who long outlived Wolffsohn. A short volume of fables in poetry and prose in Hebrew, possibly written with his children in mind, was published by Eschkol in Berlin in 1924.

The expecting widow and her small son found shelter in her father's home in Waldenburg. It was there that the new Abraham was born in October, and there that he spent most of the first seven years of his life. The home seems to have been one lively with intellectual and social activity. In addition to carrying on with several of her late husband's projects, Frau Robinsohn kept up the poetic efforts she had pursued from her youth and maintained a vigorous interest in artistic and Zionist affairs. Her father, Jacob Bähr, bore the informal title *Prediger* (preacher). Born in Königsberg in East Prussia, he was a man of considerable learning and of forceful character, much admired in the community. He was primarily a teacher, an early and vocal proponent of Zionism. A lecturer and writer on Jewish educational and cultural matters, he held positions as head of the organization of Jewish teachers in Silesia and as cultural leader of the Jewish community. He fought for better conditions of employment for Jewish teachers, particularly by seeking contributions from wealthier communities to help support teachers in poorer ones. His house boasted a fine library and those close to the family credit his books and his example, as well as his teaching, as decisive influences on his grandsons.

When Jacob Bähr retired in 1925, the family moved to Breslau. For the next three years, until his death in 1928, the boys spent much of their time in the care and company of their grandfather, while the mother worked as secretary to the Keren Hajessod, the organization for raising money to purchase land and otherwise to aid settlement of Jews in Palestine.

Frau Robinsohn had extensive contacts in Zionist and literary circles, and her family had a history of involvement in social movements dating back at least to the participation of her maternal grandfather, from Westphalia, in the German revolution of 1848. She had returned to Berlin for a time after Abraham's birth in connection with some of her husband's projects. Letters from Zionist leaders attest to the value of these services. It is also known from letters that she received Chaim Weizmann on a visit to Breslau in 1927, and we may assume that this was only the high point in a series of stimulating visits at the home of Saul and Abraham. The boys were encouraged to read widely in the classics and current literature, as well as in Zionist writings, and acquaintances of the time report on discussions of their readings that the mother initiated in the home. Their spirit is recalled as that of liberal German humanism, while the family consciously maintained an identification with Jewish nationalism.

Abraham and Saul attended a private school in Breslau that was founded and headed by Rabbi Simonson. The school incorporated all levels from elementary school through high school, so the intellectual resources available to the boys stretched well beyond what children their age would normally require. Without display but very clearly, they showed their ability to take full advantage of the opportunity. The death of their father had deprived them of his passionate encouragement in Hebrew at home, but they made rapid progress there and in the Talmud under the direction of Rabbi Simonson. The rabbi's widow reports on his recollections: "The big boy was an extremely gifted child, but the little one was a genius." Eliezer Pinczover, a friend there and later in Jerusalem, remarks on Abby as a "flower blossoming in hiding," in that he avoided all show of what he knew and could do. Both acquaintances were struck by the seriousness and maturity of his bearing. Indeed, the picture they paint is almost distressingly serious, and not easily reconciled with the zestful curiousity and enthusiasm for life that those who met him in mature years cherished in him.

There was still another member of the family whose influence in these early years was significant. He afforded them a direct contact with the world of science and with a richer and more cosmopolitan society than was theirs in Waldenburg or Breslau. Dr. Isaac Robinsohn, the only brother of the their father, was head of the Rothschild Hospital in Vienna and had an international reputation for techniques and medical discoveries made in the developing science of radiology. He took on the responsibility of guardianship for the boys after their father's death, and they were regular summer guests at his villa on the outskirts of Vienna. In addition to his research and administrative duties, he acted on his deeply felt responsibility on the part of the medical profession, and on the part of society generally, to enable those who could not otherwise afford it to share in the benefits of medical advances. He promoted better care for the poor both in his hospital and through more broadly based programs of outpatient services. It is probably impossible to assign a specific weight to his influence in developing the principles of generosity that were so much a part of Abby's character. However, it seems fair to give ample credit to the impression his example must have made.

Meanwhile the evil disease that had been incubating in Germany was taking full possession of its host. The national elections of March 5, 1933, saw the National Socialist German Workers' Party, with its leader Hitler already established as chancellor, with its unofficial apparatus of terror augmented by a highly developed propaganda machine, poll 44 percent of the vote. Detention of opposition members of the new Reichstag and physical bullying of some waverers produced a majority of those present for the Enabling Act of March 23, which granted Hitler dictatorial powers. Frau Robinsohn made hurried plans to leave for Palestine.

Matters worsened quickly. On March 29 a manifesto of the Nazi party announced a boycott of Jewish shops. On March 30 the Robinsohn household was shaken by the report that the passports of Jews were to be called in and stamped invalid for foreign travel. Hasty arrangements were made, and Frau Robinsohn and Saul took the night train to Berlin to settle affairs there before leaving the country. The next morning Abraham was to accompany a friend directly to Brno, in Czechoslovakia. He kept a diary for this ten-day period, beginning with his waiting at the Breslau station for the friend and the train to Brno, only to discover that an error in their understanding as to its time of departure had brought him to the station too late. He thereupon followed his mother and brother to Berlin, where the family decided that their best chances lay by way of Munich, and from there through Italy.

His record of the trip is a sensitive chronicle of anxiety and hope, broken by episodes of eager sightseeing among the classical ruins of Rome and Athens, the natural beauty and teeming humanity of Naples, and even the grim hostility of Berlin, where the boycott was being proclaimed on every corner. (At this stage, Nazi control was inadequate to enforce the boycott and it was limited to one day.) Although twenty-four hours was more than the average stay at each of their stops, the boys and their mother managed to explore remarkably widely. This early manifestation of avid tourism in Abby will be not at all surprising to those who knew him later.

The Robinsohns' train left Berlin on the evening of April 1, crossed into Austria at noon the next day, and brought them through the Brenner Pass and into Italy by nightfall. By early morning on April 3 they were in Rome, leaving for Naples on the afternoon of April 4. On the evening of April 5 they boarded ship at Naples for Haifa via Athens, where the daylight hours of April 7 were spent on shore. Early in the morning of April 9 they arrived at Haifa.

The new home must have been spiritually exhilarating, but it was also necessary to make a living. Frau Robinsohn opened a *pension* on Rothschild Boulevard in Tel Aviv, where the boys, then fourteen and sixteen, entered the high school. Evidence of Abby's special interest in mathematics is available from this time, in the form of a carefully written set of notes in German on properties of conics and a somewhat later set, dated 1936 and in Hebrew, on the geometrical optics of lenses with surfaces generated by conics.

Within two years Saul had left Tel Aviv for the University in Jerusalem and the strain of running the *pension* had overcome the strength of Frau Robinsohn. The family decided to move to Jerusalem, taking an apartment in the house of friends. Abby spent his last school years there.

The time was one of great turmoil. Arab opposition to the influx of Jews and their acquisitions of property, financed largely from abroad, had hardened and was becoming more violent. Along with most of his companions, Abby was involved in Jewish defense activities. From at least as early as 1937, he was on duty in the principal defense organization, the Haganah. However, he does not seem to have been nearly so distracted from his studies by this excitement as were many of his classmates. One of them, Miriam Hermann of Jerusalem, recalls that the atmosphere was one of rebellion against the "irrelevant" academic requirements. It seems that the students were able to get hold of examination questions in advance and to set up a cooperative to produce answers for all. As one who knew him might expect, Abby shocked his classmates by quite pleasantly but firmly insisting that he would have nothing to do with the scheme. The lesson in integrity from a newcomer, probably somewhat younger than the norm for his class, must have been hard to take, but Mrs. Hermann reports that Abby's quiet commitment to principle under pressure had its effect on her character and on that of the others as well.

Abby took up his university studies in 1936. In addition to Haganah duty, which was unpaid, both he and Saul did tutoring in order to support themselves and their mother.

Although he was very well read in the humanities, mathematics had won his heart. The lectures of Abraham Fraenkel introduced him to research in logic and set theory. Professor Ernst Straus of UCLA, who was a younger student at the Hebrew University in those days, tells how Abby enjoyed the reputation among the other students of mathematics as being "the one to ask if you wanted to understand something." Straus adds that Abby fully lived up to that reputation and cites a lecture by Abby in the mathematics club as the start of his own interest in the Riemann zeta-function. By 1938 Fraenkel was proclaiming that there was nothing more he could teach Abby. Meanwhile Abby was taking on the most strenuous duties in the Haganah, with an uncomplaining acceptance and a quiet determination to do what had to be done. Companions in this service recall being impressed by his vast knowledge, particularly of German philosophy.

His first research papers date from these years, when he was not older than twenty. It is appropriate to his later career that his first paper was in logic, communicated by Fraenkel to Bernays for publication in the *Journal of Symbolic Logic*. His other teachers included Fekete and Levitzki, who was working on questions of nilpotency for rings all of whose elements are nilpotent. In 1938 Hopkins and Levitzki had independently published affirmative results for rings with minimum condition. Their proofs were somewhat labored, and even in the early 1950s this was counted as one of the more troublesome theorems in the theory of artinian rings. It was surprising to discover among Abby's files the galley proofs for a one-page proof, dated 1939 but never published, that yields nothing in the way of simplicity or incisiveness to the best proofs known today. The note is reproduced in these *Papers*.

In 1939 he won a scholarship to the Sorbonne and arrived in France at about the time the war started. Besides his studies, he managed to subscribe to a course of lessons in ballroom dancing and to learn something about skiing during the French winter.

The worsening condition of France left him little opportunity to develop his intersts. By June 14, 1940 the German army had entered Paris. With a fellow student from Jerusalem (now Professor Jacob Talmon of the Hebrew University), Abby joined the stream of refugees from the capital. Once again the flight was recorded in his notebooks, this time in Hebrew. At times they were in earshot of artillery, but they managed to reach the darkened station at Orleans in a freight car. From there they finally caught a train to Bordeaux, where he notes that the station was fully illuminated. The city was full of refugees; even the French government made Bordeaux its last station of withdrawal. There Paul Reynaud had to give up his desperate struggle to keep France alive. On June 16 he turned over the government to those committed to peace with Germany, and Marshal Pétain asked for an armistice the following day.

That was Monday, June 17. A notice was posted on the door of the British consulate instructing those British and British-protected citizens wishing to travel to England to come to the consulate with provisions. Abby and Talmon joined the line in the morning. In the afternoon they were brought by special train to the port of le Verdon and by launch to a ship standing offshore. Later in the day, bombs were dropped at le Verdon and ship's guns were engaged. The ship did not leave until the following day, when numerous officials and journalists joined the flight. Meanwhile Abby watched air battles overhead. Once under way that evening, they found craft of every sort sailing from ports in North Africa and western France along their course to Britain.

They were interned for three weeks at Crystal Palace, on the southern outskirts of London. Through the intercession of a friend from the days in the Haganah (now Professor Schimon Abramsky of Oxford and London), they were released. They were billeted, at government expense, with a working-class family in Brixton, in what Talmon describes as "a quarter of ill repute." The Jewish Agency for Palestine helped them with pocket money.

Intensive bombing of London had begun. Talmon recalls night after night, in those months as the nights lengthened, in air-raid shelters in the subways. The song and hilarity brought forth there to make the best of matters contrasted sharply with the depressing views of damaged London offered by the cold, drizzly days. One morning they returned from the shelter to find that their home had received a direct hit. For more than a week they wandered the streets homeless, lounging in railway and subway stations and spending their nights in the shelters.

Their friends and families in Jerusalem had been making desperate inquiries after them. Somehow the search had extended from the Hebrew University to Professor Norman Bentwich, a well-connected British Jew associated with the University of London. He managed to locate them and referred them to the International Student Service. Talmon, who had received an M.A. and had begun work on a doctoral dissertation, was given a scholarship at the London School of Economics and Political Science, which later evacuated to Cambridge. Abby had no degree whatever and would have had to be taken on as a beginning student at a British university with no means of support. They remained in close contact with each other and with the Abramskys even after Talmon had left London for Cambridge.

Although Abby tried to join the British armed forces, they were not yet ready to accept so new an arrival, especially a born German. In November 1940 he finally found a way to contribute to the war effort by joining one of the regiments being organized by the Free French under de Gaulle, who had flown to England from Bordeaux on the day Abby sailed. It is not clear why Abby was assigned to the air force, whether by choice or by chance. In either case, there could hardly have been much forethought. The affiliation shaped his career for many years to come.

We do not have a record of his activities with the Free French, except to know that he visited the Abramskys in London from time to time and that he held the rank of sergeant at the end of 1941. At one time, he volunteered for service in Africa. He underwent the full range of immunizations customary for military personnel heading for that sector, but the assignment failed to develop.

It was during this period that he submitted to the *Proceedings of the Royal Society of Edinburgh* his note on the distributive law in fields, published there in 1941. His scholarship and his talents gained recognition; among his souvenirs are several signed photographs from Free French officers, saluting him as "friend, collaborator, professor."

The occasion for the presentation of photographs was his parting from service with the Free French in January 1942 to join the British forces, to whom he had been on loan since December 1941. At that time he was appointed Scientific Officer in the Ministry of Aircraft Production, with assignment to the Royal Aircraft Establishment in Farnborough.

His first duties were in the department of structures, where he threw himself into the work with characteristic dedication and soon produced technical reports that caught the attention of workers in the department of aerodynamics. Later he was transferred to that section. Meanwhile he was familiarizing himself with the theoretical and practical tools of the aeronautical engineer. His companion at work, Hermann Jahn (now Professor Emeritus of Applied Mathematics at Southampton), recalls his reading aeronautics by flashlight in bed after the landlord had turned off the electricity at night. He carried his program of study to the stage of official qualification, by examination, as an aeronautical engineer. According to Jahn, Abby enjoyed a reputation for infallibility among his colleagues at Farnborough. Alec Young, who worked with him there and later, reports in an interview:

> It soon became apparent that Abraham Robinson was an exceptionally good applied mathematician and, more than that, he had an outstanding ability for understanding in physics and in the problems he dealt with, so that he could apply mathematics in a way that was extraordinarily stimulating as well as elegant.
>
> In a very short time he had mastered the subject of wing theory and in a number of extraordinarily brilliant papers expanded its frontiers, particularly in the field of supersonic aerodynamics. . . . By the end of the war I was aware of Abraham Robinson as one of the country's experts in the field of wing theory.

Much of his work was of course classified, so his mastery of aerodynamics became evident to scientists at large only with his postwar publications. The problems dealt with often had their origins in actual failures or other difficulties encountered in flight. Later on, as supersonic flight became an obvious development for the future, fundamental questions of design and performance were considered. His fellow workers report that Abby was quick to recognize oversimplifications that had led to error. In those cases he was most able in resolving the more complex problem, frequently seeing in it features of an important much more general class of problems and casting his solution in such a form as to contribute significantly to these. Some of the valuable unclassified work he did during this time, and later at Cranfield, was subsequently published by the Ministry of Supply through the Reports and Memoranda of the Aeronautics Research Council. They are listed individually in our bibliography.

He was not content with his scientific contributions to the war effort, but volunteered for service with the Home Guard, in anticipation of a possible invasion of Britain. As with the Haganah, he spent much of his off-duty time in maneuvers and hard training. Fortunately the threat of actual combat came to naught this time.

His spare time still allowed him to visit London regularly, often by bicycle, where he indulged his passion for sightseeing and took in galleries, concerts and theatrical performances. On one of these visits, on January 30, 1943, he met at the Abramskys' an artist, a young woman from Vienna, Renée Kopel, who was working in London as a fashion designer and acting with a group of exiled German actors who had established a German theater in London. Through the spring and summer months they spent more time together, discovering common pleasures not only in the cultural world of the metropolis but also in long quiet walks in the English countryside. As fall came on, he called for her one night after the theater with a present. She recalls that it was a calendar, and he urged her to examine it closely. She did not have to get past Januaary, where she found January 30, 1944, the anniversary of their meeting, marked as the date of their wedding.

Colleagues who have sat on committees with Abby will recall that he was patient and tolerant in listening to all sides of a question, and showed tact in presenting his own point of view. Still he made his point so well that he usually prevailed. His courtship went likewise, and the marriage took place as appointed. After a honeymoon in Cornwall, they found an apartment in a cottage in West Byfleet, between London and Farnborough, where there was a housekeeper who, according to Renée, was able to prepare delicious meals despite wartime shortages. Both continued to work daily in their usual jobs, although Renée had given up acting. The location offered splendid opportunities for walking, always one of their favorite recreations, and for cycling through the beautiful countryside.

After a year the cottage was sold and they took an apartment with a similar housekeeping arrangement (but with a cuisine of a decidedly lower order) in a house in Surbiton, closer to London but farther from Farnborough. It was clear by then that the end of the war was not far off. Abby worked hard at Farnborough, but yearned to carry on his work free of the obligation to serve military objectives above all others.

At the end of the war he joined Hermann Jahn and two other scientists on a mission to Germany to study the progress of German aeronautical research. Jahn recalls Abby's insisting on a side trip to Breselenz, where Riemann was born and, having arrived, questioning villagers (with little success) about the house of his birth. Their trip carried them to Frankfurt, to Hannover, and into upper Bavaria. Despite a cold and fever, Abby was eager to see everything.

When the College of Aeronautics at Cranfield was established after the war, Abby was offered a position in its faculty of mathematics. The appointment was somewhat irregular since his studies had been interrupted without his receiving a degree. However, it was agreed that his work fully qualified him to receive an M.Sc. degree from the Hebrew University, and he was sent to Jerusalem for the conferring of the degree. At the same time the Hebrew University offered him a faculty position. He declined on the grounds of his obligations in England, but made it clear that he hoped one day to return.

The Robinsons moved into a house belonging to the college, bought furniture, and established a household of their own. Renée reduced her schedule of work somewhat, staying overnight in London during midweek because of the difficulty of commuting such a distance. Meanwhile Abby was continuing his research on subsonic and supersonic wing theory. While Renée was off in London, he quietly took up a course of lessons in flying in order to be able to add another dimension of his understanding of flight. Besides his aerodynamical works of this period, he published papers on algebraic curves (with Motzkin), on nonassociative systems, on hyperbolic differential equations, and on functional equations and summability, all based on research done in or before 1948.

He had made contact with Professor Dienes, at Birkbeck College of the University of London, and visited London regularly in midweek to see Renée and to participate in seminars in logic, in which he was resuming activity. The immediate result of this activity was a thesis, "The Metamathematics of Algebraic Systems," for which he received the Ph.D. in London in 1949. An abstract of a portion of this work was sent to the organizers of the section on logic and foundations at the International Congress of Mathematicians in Cambridge (Mass.) in 1950, and it so impressed them that he was invited to address the congress. From this time on, logic dominates his research; publications in aeronautics continue, with diminishing frequency. He had been appointed Deputy Head of his department at Cranfield in 1950, and the project of writing a book on wing theory with his former student J. A. Laurmann was begun about then. When he was offered a position in 1951 as Associate Professor of Applied Mathematics at the University of Toronto, the chance to move to a first-class university with experts in all fields was an irresistible lure.

One of the drawbacks of moving to the New World was that it became harder to pursue in the Old his ardor for traveling and sight-seeing. Photographs show him and Renée in Italy, Switzerland, Scandinavia, Ireland, and the Lake District of England. These tours, together with his trips to Jerusalem in 1946 and to America in 1950, show that he made major excursions in every year from 1946 through 1951, more than one in some years. Of course he followed closely the establishement of the state of Israel and its battle against Arab efforts to strangle it at birth.

Some recollections of his early days in Toronto come from Professor J. A. Steketee, now at Delft Institute of Technology, who had come to Toronto the year before. He remembers teaching Abby to drive a small and temperamental English Ford that Abby had bought. By the end of the year the Ford had carried Abby and Renée to the Rockies and back. In a subsequent summer it made the long journey to Mexico, where Abby was so taken with the country, its relics from Aztec times and earlier, and its modern art and architecture that it took over first place on his tourist's list.

The Department of Applied Mathematics occupied an old private home on St. George Street in Toronto. Abby kept regular hours (about 8:30 to 5:30), often bringing sandwiches and eating at his desk. His self-discipline included an insistence on the completion of a daily quota of work on the book on wing theory. His teaching involved a graduate course on wing theory and less advanced courses on fluid dynamics and partial differential equations, as well as a share in the introductory mathematics courses taught as a service to engineering students.

Professor G. F. D. Duff of Toronto recalls with gratitude Abby's encouragement in his work on mixed initial boundary value problems for hyperbolic partial differential equations, and with admiration Abby's quickness in understanding and analyzing problems. Abby himself published a joint paper on the subject with his student L. L. Campbell in 1955. Duff emphasizes Abby's gift for going to fundamentals and for putting any problem in perspective against the broad background that his knowledge afforded.

In these days Abby was even busier with research in logic than with his work on applied mathematics. His thesis had been published in 1951. In 1955 there followed the *Théorie Metamathématique des Ideaux* (Gauthier-Villars), and in 1956 the *Complete Theories* (North-Holland). He also took up some metamathematical questions related to irreducibility of polynomials, work that was extended by his postdoctoral student P. C. Gilmore (now with IBM). Gilmore has reported admiringly on the excitement and enthusiasm with which Abby welcomed their collaboration, scouring the literature for leads to further applications of their results.

Model theory in general and its applications to the theory of algebraically closed fields and that of ordered fields and other ordered structures were particular concerns of his in Toronto. His first collaborations with his student A. H. Lightstone were generated there. This began a close relationship that lasted until their deaths, following close upon each other. Still he was active in research in applied mathematics, publishing in 1956 and 1957 three important papers on fluid dynamics, wave propagation in heterogeneous media, and transient stresses in beams. These were his last major contributions in traditional applied mathematics.

In 1956 he became Professor and Chairman of the Department of Applied Mathematics, and in that year W. A. J. Luxemburg came to Toronto. Professor Luxemburg, one of the editors of these papers and one of the first analysts to promote the idea of nonstandard analysis, has been his friend and admirer since then. Luxemburg was struck by Abby's enormous knowledge of mathematics and equally by his tact and caution in expressing judgment about the work of others.

The Robinsons had bought a small house in Toronto. His colleagues remember being entertained there quietly and graciously in the company of guests with diverse interests from the arts and the university.

From correspondence with Fraenkel we learn that he was regularly in touch with mathematical affairs in Israel. In 1952 he received, but was unable to accept, invitations to visit the Hebrew University and the Weizmann Institute. In the summer of 1954 he did manage to get to Israel for lectures in Jerusalem and Haifa. There soon followed an invitation from Fraenkel to join the faculty of the Hebrew University, but once again Abby declined.

He gave invited addresses to the National Research Council of Canada in 1951–52; to the Colloquium of Mathematical Logic at the Sorbonne, the National Aeronautical Establishment in Ottawa, and New York University in 1952–53; to the Canadian Symposium on Aerodynamics and the Association for Symbolic Logic in 1953–54; and in London, Zürich, Jerusalem, Haifa, Amsterdam, and Pittsburgh in 1954–55. The last of these was again an address to the Association for Symbolic Logic, and the Amsterdam address was delivered to the Symposium on the Mathematical Interpretation of Formal Systems, in association with the International Congress of 1954. This paper, "Ordered structures and related concepts," was published in the proceedings of that symposium by North-Holland in 1955.

It was appropriate both to the direction his interests had taken and to his commitment to Israel that he should be offered and accept the chair previously held by his teacher Fraenkel at the Hebrew University when it was offered him in 1957. His colleagues at Toronto were conscious of their great loss in his departure. Dinners and farewell ceremonies echoed with his praises and their expressions of regret. Yet it seemed destined to be so.

On the way to Israel he stopped in London to receive the D.Sc. in recognition of his cumulative achievements. Then the Robinsons set out by auto across Europe. Renée remembers driving from Zagreb to Belgrade, in both of which cities Abby gave lectures. The trip took much longer than expected. Finally they arrived at the university, and Abby left the car to learn where and when his lecture was to be held. As it turned out, he was already late and his listeners were waiting. He sampled opinion as to the most appropriate language and began at once in French. After a while, someone from the institute was finally sent out to the car to tell Renée what had happened and to bring her in.

The move to Jerusalem involved a considerable reduction in salary, as well as the loss of Renée's income from work with the Canadian Broadcasting Company, where she had performed in serials and in occasional longer shows. But it was all worth it to Abby. Renée tells how every stone in the city had its special appeal to him, how he loved to walk the streets of Jerusalem and to breathe in the historical air of its places. They traveled from one end of Israel to the other, visiting sites of archaeological interest and the shrines of all the religions. Renée recalls his insistence on attending midnight mass with the Franciscans on Mt. Zion their first Christmas.

Their first lodgings in Jerusalem were in a boarding house kept by a friend of his mother's from Breslau. It had a fine view of the city and counted among its residents a famous archaeologist with whom Abby could enjoy discussing ancient Israel and the places he was visiting. After three months they found a pleasant apartment with a good view. Abby's real preference would have been for something closer to the soil and the people, a house of the type the Arabs lived in.

Abby threw himself into the affairs of the university and into the educational development of the country with the full commitment that was his heritage. The custom of professors was to appear at their offices only at very limited hours; Abby spent every weekday there, his office door open to students and colleagues. His vision of Israel had room in it for the Arabs to enjoy dignity and equal opportunity. There were few Arab students at the Hebrew University, but Abby soon came to be known to them as a sympathetic listener and adviser. He did what he could to urge government action to supply schools in the Arab areas with more and better textbooks in Arabic, an effort that was less than completely successful.

Student activism at the university was high. Strikes of students were frequent. Abby's capacity for being a good listener and for wise mediation stood him and the university in good stead. Renée especially remembers one strike in the cafeteria that he was called in to settle. She also recalls his being displayed to the family and neighbors of an Arab student who had insisted that Abby stop by when returning from a visit to the high school and the holy places in Nazareth. In the rather primitive house, shared with the livestock and without electricity, they were served coffee and shown off with pride.

Still another student admirer was a senior Thai naval officer, who came for one year to study with him. His concept of an ideal relationship with his teacher was that he should live with the Robinsons and serve Abby while learning at his feet. Although frustrated in this plan, he later entertained Abby and Renée royally on their visit to Bangkok in 1961.

Abby was in the middle of university policymaking. His keen interest, his conscientiousness, his availability, and his wisdom made him a favorite for committee work and other responsibilities, and had him marked for more administrative duties in the future. A steady stream of distinguished visitors from abroad came to the university. Abby was a willing and valued guide, not only to that institution but to the city and surroundings as well. He was a principal organizer of the international symposium on logic in Jerusalem in 1961.

As measured by his rate of publication, his research was to some extent made to pay the price for his involvement in other matters. He published several papers on differential algebra, where differentially closed fields offered intriguing challenges to the model-theorist. Using Seidenberg's elimination theory, he found a model completion for the axioms of differential fields; then he could take the differentially closed fields to be models of the "closure" axioms associated with this completion. It would be appropriate to say that he *invented* differentiably closed fields. He wrote on mathematics in secondary education, on the mechanization of number theory and the theory of equations, and on non-archimedean complex function theory, one of his early efforts in nonstandard analysis. His most fundamental work from this period dates from 1960–61, an academic year he spent as Visiting Professor at Princeton University, and includes his first title in nonstandard analysis. He was also working on his *Introduction to Model Theory and the Metamathematics of Algebra*, eventually published in 1963. Only one paper in aeronautics dates from the Jerusalem period, the chapter on aerofoil theory in the *McGraw-Hill Handbook of Engineering Mechanics* of 1962.

Saul had been living in Haifa for some time. He was a history teacher and an official in the teachers' union, a position in which his opportunities for influence were limited largely to the union and to gatherings of affiliated groups in other countries. In 1959 he was invited to be educational director for UNESCO in Hamburg, a temporary position that was renewed after its three-year term had expired. In 1964 he was appointed director of education at the Max-Planck Institute in Berlin and professor of history at the Free University there. Leaving Israel was a serious rupture for him, particularly to return to the seat of the Nazi horrors. However, he saw in the new post an opportunity, through reform in the schools, to enable the youth of Germany and elsewhere to learn from the grisly past and to be guided toward a humane future. He pursued this goal vigorously until his death in Berlin in April 1972. At the time of his death he was recognized as one of the world's distinguished authorities on comparative education.

Travels from Israel included the International Congress of Mathematicians in Edinburgh in 1958, with longer stopovers in Belgium and on the Riviera. A trip to a symposium in Poland in 1959 offered chances to return to Waldenburg and Breslau (although now bearing different names, Walbrzych and Wroclaw), as well as side trips to Turkey, Switzerland, Paris, and Italy. In 1960 the trip to Princeton pursued a roundabout route touching Cyprus, Athens, southern France, the Pyrenees, and northern Spain; embarking by ship at Genoa for New York; and finally crossing the country to Los Angeles and returning.

Renée was not very happy in Jerusalem. Living on a border with a hostile neighbor often gave her a feeling of being cooped up in a garrison town. In London and Toronto she had felt a part of the world of fashion and of the arts, but found little similar stimulation in Jerusalem. Despite their frequent trips abroad, she still wanted to leave. Abby had visited the Department of Philosophy at the University of California at Los Angeles for some time in the spring of 1961. He had been quite taken with the University and with California. When an offer of a joint professorship in the departments of philosophy and mathematics came from UCLA in 1962, its attractions and Renée's unsettled state were too much for him, and he accepted. He continued to keep in close touch with developments in Israel and to visit whenever possible.

The return to Israel in 1961 and the move to Los Angeles in 1962 constituted a round-the-world trip for the Robinsons. From Princeton they had gone to the west coast of the United States, then to Hawaii and to Japan, where Abby lectured in numerous universities. Then followed lecturing and sight-seeing in Hong Kong, Thailand, Cambodia, India, and Nepal before the academic year in Jerusalem. The next year's final leg of the trip followed the more familiar transatlantic route.

After renting a house for the first five months in Los Angeles, Abby and Renée found a house, described by Renée as a "dream house," in Mandeville Canyon, with a splendid pool, surrounded by hills and geraniums. Several correspondents attest to the memorable beauty of the setting and to their enjoyment of visits there. It was an excellent base for launching trips to the national parks of the western states and to their favorite goal, Mexico, where they spent Christmas vacation in 1962.

The years in Los Angeles saw a sharper focusing of Abby's attention on applications of nonstandard models in arithmetic and analysis. His interest in mechanization of problems in arithmetic and algebra made it natural that he should be invited as a consultant to IBM, and several papers written in that capacity are included in this selection. One of the early triumphs of nonstandard analysis was his proof, together with his student A. R. Bernstein, of the existence of invariant subspaces for polynomially compact operators. He investigated the theory of normal families of complex functions from a nonstandard viewpoint and began the work on nonstandard methods in algebraic number theory that occupied him throughout his later years.

Angus Taylor, now Chancellor of the University of California at Santa Cruz, was chairman of the mathematics department at UCLA in Abby's first years there and later moved into the university's administration. We are indebted to him for an account of Abby's activities.

> We soon got well acquainted and he became a good friend and valued advisor in the conduct of departmental affairs. I had not previously known of his expertise in aerodynamics, wing-theory in particular. I soon learned that he was amazingly versatile and broad in his intellectual interests, which included a strong interest in the history of mathematics. Abby was very helpful in connection with my efforts

to develop applied mathematics as part of the departmental curriculum. He played a vigorous role in elevating every aspect of the departmental affairs.

After a while the demands of membership in two departments became larger than Abby wanted to bear, and we arranged for his transfer to full-time membership in Mathematics. This did not signify much change in Abby's interests, for he had been a logician in the Philosophy Department, and in the Mathematics Department he continued his work in logic, of course.

Non-standard analysis was taking a good deal of Abby's time then and he had graduate students who were working in this field under his direction. His book, *Non-Standard Analysis*, was completed while he was still at UCLA, in 1965.

It was quite natural that a man of Robinson's breadth of mind and qualities of academic statesmanship would be drawn into the functioning of the University's Academic Senate, which is the official faculty body established by the University's governing Board of Regents to authorize and supervise instruction and to advise the administration on matters of educational policy and faculty appointments and promotions. Robinson was chosen to serve on the Universitywide Committee on Educational Policy, and in the year 1964–65 was chairman of that committee. This made him an ex-officio member of the Academic Senate's representative Assembly and of the Academic Council. The Council is the steering committee of the Assembly as well as being the only body that meets monthly with the President of the University (*i.e.* of the nine-campus University of California system) to confer with him and advise him on a very broad range of matters of common concern to the faculty and the administration. I happened to be the chairman of the Assembly and the Council at the same time that Robinson was on these bodies in his ex-officio capacity as chairman of the Committee on Educational Policy.

It was an interesting time. The University was growing rapidly, at the rate of seven or eight thousand students per year. New campuses were developing at Irvine, San Diego, and Santa Cruz. We were helping the Divisions of the Senate at these new campuses to get organized, and we were reviewing the academic plans of the new campuses. It was a turbulent academic year, the central feature of which was the controversy at Berkeley that was sparked by the so-called Free Speech movement (FSM) and dramatized by student demonstrations and a massive sit-in in Sproul Hall. The Berkeley faculty was drawn into the controversy, there was a turn-over in the chancellorship at Berkeley, and there were many long meetings and special sessions of the Regents and the Academic Council as the issues were debated and investigative reports were discussed. The Academic Council was heavily involved in conferring with the University President and advising both him and the Regents in an effort (largely successful) to moderate the tensions and head off intemperate and unwise actions. Through all of this Abby was a source of wisdom and an effective leader of a very important Senate Committee, as well as a member of the influential Academic Council.

I left UCLA in February, 1966 to go to Berkeley as Vice President-Academic Affairs of the University of California system. By that time Abby was clearly recognized as the outstanding member of the UCLA Mathematics Department. We had attracted to UCLA many able young mathematicians, among them several who were logicians attracted by Robinson's presence. We had also, after many difficulties, established a program in applied mathematics. Abby had been an inspiration to me and had become a valuable friend both personally and intellectually. I was sorry to lose contact with him as the inevitable consequence of my move. I was still more sorry—indeed I was filled with the deepest regret when, in a later year, Abby decided to leave UCLA and go to Yale.

Ernst Straus remembers Abby's excitement over non-standard analysis in these days. Both Straus and Abby's later students at Yale recount his delight in quoting Fraenkel's view that mathematical creativity declines after age thirty and in having supplied a counterexample to that proposition. Straus also tells of his respect for tradition, and how he was distressed by the insults to academic courtesy and standards that he saw in student movements of the 1960s, perceiving in them "a general anti-intellectual tendency." This is not to say that he did not share the students' abhorrence of the war in Vietnam. Straus describes him as "the spirit of" a committee of professors organized to protest vigorously, with an emphasis on rational persuasion and dignified example.

In the summers of 1962, 1963, and 1964, the Robinsons drove across the country to spend four to six weeks at Yorktown Heights, where Abby served as a consultant at the IBM Watson Research Center. When time allowed, they included as much sightseeing as was available on these trips. National parks were especially sought out. The conference in Bristol in 1964 offered a chance for a sea voyage to Europe and to Israel, including visits to Holland, Berlin, Vienna, and Switzerland and a cruise through the Greek islands. In 1965 they traveled to Oregon, and in 1966 Abby visited Tübingen, Locarno, and Finland on the way to and from the International Congress in Moscow.

In 1965–66, the year following the hectic year on the Academic Council, Abby found a chance to get away to St. Catharine's College at Oxford, where he spent the fall term. He also lectured in Paris at the Sorbonne and in Rome, in each of the latter cases taking advantage of the chance to exercise another of the languages he had mastered. On his return trip to Los Angeles, he stopped to talk in New Haven and to discuss with officers of the University and of the departments of mathematics and philosophy the future of logic at Yale. It had been decided that the mathematics department should have one of the two chairs in logic previously held in philosophy, and we were anxious to have Abby consider an offer. He had declined other offers in the United States and Canada, some of them prestigious professorships not requiring teaching. Moreover, his visit to New Haven came in the depressing days of winter, after a similar season in northern Europe. Renée made it quite clear that she preferred southern California, and his declining of Yale's offer was not long in coming.

On their return to Los Angeles, Abby was made freshly aware of the size that the academic enterprise had taken on. He already knew well its vulnerability to changes in political climate. A smaller private university offered him close contact with all members of a distinguished senior faculty in mathematics, a regular flow of talented postdoctoral researchers and graduate students, and the opportunity for easy contact with scholars in fields other than his own. No doubt his awareness of tradition lent appeal to the case of a university nearly twice as old as the oldest of his previous affiliations. He persuaded Renée that it was worth sacrificing the southern California climate and let Nathan Jacobson, then Chairman of the Yale Mathematics Department, know that he was now receptive to an offer.

The proposal for his appointment met with enthusiasm from all sides among the Yale faculty. Outside mathematics and philosophy, members of the Engineering and Applied Science faculty were particularly quick to recognize a splendid opportunity for Yale. He came to New Haven as Professor of Mathematics in the fall of 1967.

After first renting a suburban house, the Robinsons purchased at the beginning of 1968 a spacious modern house on a dead-end road on the slopes of the hill constituting Sleeping Giant State Park. The surroundings were wooded, eliminating the broad vistas they had known in California. On the other hand, extensive opportunities for walks in the park were right at their doorstep, and they were taken with pleasure and regularity.

He was soon the mentor of a lively group of younger logicians. Promising recent Ph.D.'s, among them K. J. Barwise, P. Eklof, E. Fisher, P. Kelemen, M. Lerman, J. Schmerl, S. Simpson, D. Saracino, and V. Weispfenning took postdoctoral positions at Yale and profited from his keen questioning and the generous sharing of his time and wisdom. Established logicians, among them A. Levy, G. Sacks, and G. Sabbagh, were happy to make visits of a semester or longer. Graduate students were charmed and excited by the promise held out for model theory and nonstandard methods in analysis and arithmetic by his courses, and many sought him out as adviser. He refused none, and gave generously of his attention to all. It was soon evident, though not from complaints on Abby's part, that a second senior logician was needed to share these burdens. After abortive efforts to appoint someone whose status rivaled his, Abby advised building for the future by bringing Angus Macintyre from Aberdeen. Macintyre had only one abbreviated and agonizing year as Abby's colleague, during much of which Abby was incapable of generating the kind of excitement Macintyre had every reason to expect. Nevertheless, he is cognizant of this brief experience as a priceless formative step in his own development. He had the chance to work closely with Abby's research students, and both he and other colleagues at Yale owe to the luster that Abby shed on them the fact that outstanding young logicians and students of logic have continued to value the opportunity to join the Yale Mathematics Department.

Although he was quickly involved in the administration of the Department of Mathematics as Director of Graduate Studies and in the councils of the university as a member of the Advisory Committee for the Physical Sciences, his involvement with the central problems and policies of the university was less than it had been in Jerusalem or in Los Angeles. We fully expected he would be our next departmental chairman until it became clear that he would not survive to take on those duties. In the meantime, he was chosen President of the Association for Symbolic Logic. His influence among his fellow scholars reached its peak, not only through his presidency but because of the volume and power of his publications, his lectures in all parts of the world, and his following among younger logicians and mathematical converts to nonstandard methods.

About forty articles and books testify to his activity at Yale, several of them joint work with scholars who brought different interests and found in Abby and his nonstandard methods a fecund collaborator and exciting challenges. Among these joint works we may cite those with Lightstone and with Luxemburg in pure nonstandard analysis; with Kelemen on applications in physics; with D. J. Brown on applications in economics; with Weispfenning in algebra; and the very striking work, published posthumously, with P. Roquette on diophantine equations. The collaboration with Roquette began late, and Roquette was able to make only one of his scheduled two visits to New Haven while Abby was still alive. His remarks, both in the obituary article in the *Bulletin of the London Mathematical Society* (1976) and in letters to Renée, make it clear that he valued their brief association as an opportunity of exceptional scientific and human dimensions.

From students and young colleagues we have some of the warmest reminiscences of his helpfulness and his charm. They recall his delight in the successes of his students, always eager to have one of them "crack a hard nut," in his words. The awkward moment when he proposed to them that further discourse be on a first-name basis seems to be a treasured memory, as was the flattering one when he would ask if they "could spare the time" to go out for a cup of coffee. In his office or in the hall it seemed he always could spare the time—the time to listen, to dredge up associations out of his rich memory, to ask the gentle but pointed questions that both raised hopes and pricked consciences.

Honors now showered down on him. In 1967 he had been the center of attention at the First International Meeting on Nonstandard Analysis, organized by Luxemburg at Caltech. In 1970 he was invited to speak at the International Congress of Mathematicians at Nice, and he combined the trip to this engagement with the Shearman Lectures at University College in London and an invited address to the British Mathematical Colloquium in York. In 1971 he was appointed to the prestigious Sterling Professorship at Yale and was honored by an invitation to give the Earle Raymond Hedrick Lectures at the summer meeting of the Mathematical Association of America at Pennsylvania State University. The cafeterias there buzzed with excited discussion of nonstandard analysis after each of his talks. In that year he also gave a series of five lectures on nonstandard analysis with applications to complex function theory, Hilbert spaces, and algebraic curves at the University of Quebec in Montreal, besides speaking at the Tarski Symposium in Berkeley and the International Congress on the Logic, Methodology and Philosophy of Science in Bucharest. The following year he was elected a Fellow of the American Academy of Arts and Sciences and gave the Delong Lectures at the University of Colorado.

In the spring of 1973 he was a visiting member of the Institute for Advanced Study, working with Kurt Gödel, the logician he most admired. Letters and comments by Gödel show that he in turn held Abby in highest esteem. Later that spring he went to Leiden to be the second recipient of the Brouwer Medal, awarded by the Dutch Mathematical Society. The award was made by Professor A. Heyting, and his presentation is to be found in the *Nieuw Archiv voor Wiskunde* for 1973, immediately after the text of Abby's lecture on "Standard and nonstandard number systems." Then he went to Warsaw, where he lectured on various aspects of his work over a period of about ten days. During that spring he also gave lectures in Amsterdam and Utrecht. For a vacation the Robinsons traveled to Jerusalem, Vienna, and Switzerland. In the summer Abby participated in the Logic Colloquium at Bristol and worked at Heidelberg with Roquette in their exciting, but too brief, collaboration. In 1974 he was elected to membership in the National Academy of Sciences of the United States.

His passion for tourism was only partially satisfied by the opportunities connected with the occasions listed above. He and Renée found time for winter vacations in Florida, the Caribbean, Yucatán, and Portugal. His schedule was so full, however, that it was practically impossible for him to take a trip devoted purely to sight-seeing and relaxation. Thus when he went to Lake Como, it was to the summer institute at Varenna. When he went to Chile in 1970, it was to organize a Congress on Symbolic Logic, and a similar situation prevailed two years later in Brasïlia. For this occasion he had diligently worked his way through a teach yourself Portuguese course and did give the first of his series of lectures in Portuguese. At the request of the Spainish-speaking members of his audience, he relented and gave the subsequent lectures in English. Linguistic skills were one of his sources of quiet pride. To the ear of a native speaker of standard American English, reasonably attuned to detect unusual influences, he gave few indications indeed of his origin outside this continent. This achievement is the more remarkable if we remember that he was thirty-three before he spent any extended time in the New World, and forty-two before he stayed in the United States for a prolonged period. His lecturing career thus included talks in English, French, German, Hebrew, Italian, Portuguese and Spanish, to the writer's knowledge. His reading knowledge extended further, including at least Russian and Greek.

His scientific contributions during this period are treated in some detail in the discussions of his work. Besides nonstandard analysis and its applications, and his work with Roquette, he seems to have been particularly pleased by his success in adapting the methods of forcing, introduced by Paul Cohen in his striking work on the continuum hypothesis, to extend the scope of his achievements on complete theories.

He was an exciting and stimulating colleague. He would inquire as to what one was working on, presenting the problem of choosing a level of communication. We soon learned that it was not at all necessary to try to oversimplify matters for him. His interest was genuine in several respects: his human interest in people, his desire to keep abreast of mathematics on all fronts, and his constant exploration in new fields of application of model-theoretic methods. He was the best listener in the department, both in his willingness to listen and in the likelihood that he would have a penetrating comment. Outside mathematics and in the context of the University at large, he was a prized conversationalist. His comrades in the Fellowship of Davenport College admired his ability to discuss almost any subject knowledgeably, but with modesty. He did feel a mission to communicate to nonmathematicians some of his enthusiasm for mathematics and volunteered to give a seminar with the title "Mathematics is Beautiful" to the students in his college. In the faculty–student committee that reviewed such matters, there was some reluctance to accept this profession of sentiment as the title for a course, out of fear that a precedent might be set by accepting a declarative sentence for a course title. The seminar subsequently became "The Beauty of Mathematics."

Saul died of a heart attack in April 1972. Heart disease had been the cause of their father's early death, too, and Abby had reason to believe in the family's special susceptibility to the dangerous ailment. He had led a vigorous life remarkably free of medical problems, but now he was aware that it could be snuffed out in an instant. No signs of heart trouble appeared to justify his concern, however. In the fall of 1973 he resumed lecturing at Yale, also joining with Professor Stephan Körner in directing a new interdepartmental seminar in mathematics and philosophy. He began to be troubled by abdominal pains. We do not know how long he bore these without complaining, but by November he had to acknowledge them to Renée. He continued his regular schedule until Thanksgiving recess at the end of November, when he entered the hospital for a series of tests. These gave abundant grounds for suspecting cancer of the pancreas, and an exploratory operation revealed that the disease was beyond surgical remedy.

To recuperate from the operation, he was moved to the Yale Infirmary, across the street from the Mathematics Building. His students and colleagues could drop in frequently, and he felt closer to their activities. Renée was constantly at hand, doing her best to help him conserve time and energy for the tasks he felt were most important. He was under no illusions. He insisted that no extraordinary measures to prolong life be undertaken. He would do the best he could, for as long as he could. Carol Wood, his former student, asked him in early 1974 how she should answer inquiries about his health. He instructed her to tell them frankly that he was dying.

When the effects of the operation were sufficiently healed, he was allowed to go home. I recall driving him and Renée home. Along the way we had to stop at a pharmacy to pick up a prescription. When she brought it out after some time, Renée commented on the elaborate bookkeeping that had had to be done. Abby replied that this was only to be expected in the case of such medicines, with an aside to me that showed he was being given strong narcotics.

He had been scheduled to go to Australia, one continent he had missed. That trip had to be scrapped, but he insisted on beginning his graduate course in model theory in January. It was soon evident that he could not stand to lecture, and the classroom was awkward for a seated lecture. The class was removed to his modest office, where a dozen or so hearers crowded in. The disease and the drugs forced him to struggle to concentrate, but his wit still could flash out, and his listeners' laughter would then fill the narrow corridor outside his office. He was still able to discuss the genesis of some of the central ideas, and he gave insights into the psychology of his own mathematical invention that he might have been too self-effacing for under other circumstances.

Soon he had to give up the class and return to the infirmary. His periods of alertness grew rarer and shorter. Nevertheless, he could brighten at the visits of old friends and former students, who came to see him one more time. Out of his very limited store of energy, he could still draw together enough to concentrate a few minutes as a graduate student (in this case, Peter Winkler) reported the latest progress of his thesis. In a very few brief questions he was able to show what kinds of problems lay beyond the result and to encourage Winkler to get on with them.

By the beginning of April, it seemed the end must be very near. Visitors often found Abby asleep under the influence of the drugs and Renée in a chair asleep with fatigue. Death came on Thursday afternoon, April 11.

It is not hard to find lessons in Abby's life and in his character. Tributes sent to Renée and tributes read at his memorial service in Dwight Chapel, on September 15, 1974, listed these in abundance. I shall not resist entirely the temptation to cite a very few here, especially those that Renée found paramount.

He could laugh. Almost always the object of his laughter was one aspect or another of the human condition or of human institutions that struck him as paradoxical. His own behavior was by no means excluded from what he found laughable. All who recall him find no instance of cruel laughter at the expense of anyone. His targets were either harmless foibles or paradoxes, or else conditions that, in his steady optimism, he regarded as capable of improvement. Cartoons in *Punch* would provoke peals of laughter from him.

He was generous. He was generous of his time, his attention, his advice, his praise, and in his withholding of scorn when others would have been all too ready to unleash it. He had considerable pride, but it did not find expression in priority fights or a defensive need to advertise his own work. He was less reticent about his exploits in a beer-drinking contest or a walking contest, where a bit of boasting could harm no one. He was generous with his money; he was ready to make sacrifices in salary to help graduate students continue their work over the summer. Above all, he was generous in spirit, seeing in each human being a creature of the stuff he was made of and thereby deserving the respect and tolerance he felt was his own due.

He was a man of principle. Whatever theological dogmas had been part of his upbringing were shed except for their ethical significance, incorporated by his mind into a firm rational humanism. Whatever the actual course of events, he tried to conduct himself as would a person of good will in a world destined to be inherited by the fair and generous. When it was a matter of playing the game according to rules laid down for the good of society generally, he insisted on following the rules to the letter. When some part of society strayed and needed setting right, he was ready to bear more than his share of the sacrifice and danger involved in restoring reason. His adherence to principle extended to his working habits; he refused to break off a spell of work, once begun, until three pages were done.

He loved life and saw beauty in it. He found pleasure in food, theater, music, art, and books, ranging from rare old classics of mathematics to mystery thrillers. At the time they discussed marriage, he told Renée that what he most hoped they would find in their life together was "much love and much beauty." In his travels, in his scholarship, in his homes and the things that surrounded him there, and especially in his human contacts one senses the conviction that whatever "I" is, it is fortunate to find itself in a world that has such a capacity for delighting it.

He is buried on a hillside in Har Menuchot Cemetery outside Jerusalem. From the site of his grave one looks down into a green valley. Away on the other side of the valley roll the hills of the land he loved best.

INTRODUCTION TO PAPERS ON NONSTANDARD ANALYSIS AND ANALYSIS

W. A. J. Luxemburg

Every student of mathematics is aware of the fact that infinitesimals played a major role in the early history of the calculus. Although the basic idea of an infinitesimal quantity goes back to antiquity, it was Leibniz who first apparently conceived of the idea to adjoin to the finite numbers new ideal elements which could serve as infinitely small as well as infinitely large numbers in such a way that they could be combined arithmetically with the finite numbers in the same manner as the finite numbers could. In particular, he thought of his differentials as infinitesimals although in his original definition of this concept he does not allude to infinitesimals.

Newton, on the other hand, rejected the idea of considering differentials as infinitesimals. He expressed his preference for postulating the existence of quantities which can be diminished without end to form a basis for his theory of fluxions.

Because of the logical inconsistencies in the rules that Leibniz and Newton gave to govern differentials, their ideas were severely attacked not only from a technical but also from a philosophical point of view. When Leibniz and his followers failed to clarify their ideas on how to construct a number system which would, in addition to the finite numbers, include infinitely small as well as infinitely large numbers, the acceptance of the concept of an infinitesimal to form a basis for the foundation of the calculus gradually lost ground. Instead the limit concept emerged to take its place. Through the work of d'Alembert, Cauchy, and Weierstrass it became apparent that the limit concept had won the contest and it was readily accepted as the basic concept on which to found the calculus.

Infinitesimals survived, however, not only as a figure of speech but also as a tool of the intuition to help the art of invention as visualized by Leibniz. Strong reservations as to the usefulness of this concept versus the limit concept were still voiced. Thus at the end of the nineteenth century the eminent mathematician C. S. Peirce declares in "The Law of Mind" (*Monist* 2 [1891/92]: 543–45): "The idea of an infinitesimal involves no contradiction—As a mathematician, I prefer the method of infinitesimals to that of limits as far easier and less infested with snares."

G. Cantor, the discoverer of the theory of transfinite cardinal and ordinal numbers, favored the limit concept and went so far as to assert that on the basis of his new theory he could prove the impossibility of the existence of infinitesimals. But whatever opinions were expressed about the existence of infinitesimals it was generally felt that the creation of a consistent theory of infinitesimals for the calculus, although highly desirable, remained a hopeless task.

The world's foremost set theoretician, A. H. Fraenkel, discussed the problem of the infinitesimal in the first edition of his book *Einleitung in die Mengenlehre*, which appeared in 1928. From the outset Fraenkel favors the generally held opinion that the real test for the efficacy of a theory of infinitesimals lies in its applicability to the calculus. He goes on to mention that until that time all the proposed theories of infinitesimals had failed to provide even a new proof of the mean-value theorem. Despite all the failures, however, Fraenkel believes that it is not inconceivable that someday a second Cantor may appear who will create a consistent theory of infinitesimals for the calculus. On the basis of his judgment of the state of development of the foundations of mathematics he was compelled to conclude, however, that it was highly unlikely that such a new development would take place in the foreseeable future.

About forty years later Fraenkel returns to the problem of the infinitesimal in his fascinating autobiography *Lebenskreise,* which was posthumously published in 1967. In it he tells the reader in a number of revealing passages of his experiences at the University of Marburg. Particularly, he reveals how deeply disturbed he was when he came into contact with the ideas concerning the infinitely small expressed by the Marburger school of philosophers. Those ideas originated with Hermann Cohen in his *Prinzip der Infinitesimalmethode* (1883) and were further developed by Paul Natorp in his *Logischen Grundlagen der exakten Wisschenschaften* (1910). Apparently the Marburger philosophers were of the opinion that the existence of infinitesimals could be established on the basis of Cantor's theory by arguing that, if in a ratio $1 : \omega = x : 1$, ω is an infinite ordinal number in the sense of Cantor, then x must be considered to be an infinitesimal.

Fraenkel then informs his readers that recently (in 1960) his former pupil Abraham Robinson succeeded in an entirely different and novel way to create a rigorous and highly surprising theory of infinitesimals for the calculus, which once and for all vindicates Leibniz and returns the concept of an infinitesimal to the honorable position it once occupied in the foundation of the calculus.

In an invited address in January 1961 to a joint meeting of the American Mathematical Society and the Mathematical Association of America, Abraham Robinson first revealed to the general mathematical community his solution to the age-old problem of giving a rigorous foundation for the calculus based on a theory of infinitesimals. He baptized his new approach to the foundations of analysis "nonstandard analysis." This somewhat unusual name was chosen by Robinson because nonstandard analysis is nothing but the detailed study of certain models of the real number system which are called nonstandard models by logicians. The text of this invited address, if it ever existed, is not available. But it is fair to say that the substance of it is contained in Robinson's first publication on this subject entitled "Non-standard Analysis," which appeared in 1961 in the December issue of the *Proceedings of the Royal Dutch Academy of Sciences of Amsterdam.* In this paper Robinson summarizes all the important aspects of the new method. It is clear from the outset that this new approach gives more than a calculus developed within the framework of a non-archimedean totally ordered field, which by nature contains elements that may be called infinitely small. Indeed, the surprising new element in Robinson's approach is that by considering a nonstandard model of the reals he defines a proper extension of the reals in the form of a non-archimedean totally ordered field, which is directly related to the reals as expressed by the logical transfer principle that the new extended field is, in fact, an elementary extension of the reals. This leads immediately to the possibility of extending all the properties of the reals as well as all the elementary functions of the calculus to the extended field, and of studying their behavior within the framework of a new number system which now contains infinitely small as well as infinitely large numbers. This paper presents an overwhelming amount of evidence on the effectiveness of the new method. The important operation of taking the standard part of a finite number is introduced. This fundamental operation of the new method may also be viewed as giving a precise meaning to Leibniz's rule that a differential *dx* can be neglected at the end of a calculation in an expression containing only finite quantities but not during the calculation. The new proofs which are sketched for classical results taken from areas such as the theory of limits, differential and integral calculus, differential geometry, potential theory and classical applied mathematics are marvels of elegance and also provide new insights into the old problems. A very detailed version of all the new ideas that are touched upon in this paper formed the basis for a monograph entitled *Non-Standard Analysis,* which appeared in 1966 in the series *Studies in Logic and the Foundations of Mathematics*.

At the same time, Robinson made a careful investigation into the history of the theory of infinitesimals. His findings are presented in a brilliant essay contained in his book on nonstandard analysis.

In the intervening years, while preparing the manuscript of his book, Robinson wrote, besides several other articles, three important fundamental papers on subjects dealing with nonstandard analysis. The first of these, entitled "On Generalized Limits and Linear Functionals," appeared in 1964 in the *Pacific Journal of Mathematics*. In this paper Robinson shows, by using nonstandard methods, the existence of the generalized limits in the sense of Banach and Mazur for bounded sequences of complex numbers. This is accomplished by expressing such limits in the form of the standard part of a simple expression in terms of the nonstandard elements of the sequence.

This surprising result is then used to give a new and simple representation of the dual space of the classical Banach space of bounded sequences. We may mention here that Robinson's interest in summability theory, to which we shall return later in more detail, stems from the early 1950s.

This important paper was to a large extent responsible for the introduction, by the present author, of the nonstandard hull of a normed linear space in order to extend the representation theory to other Banach spaces, a notion which proved to be such a useful tool in the hands of L. Moore Jr. and W. Henson.

In the Nevanlinna anniversary volume, which appeared in 1965, Robinson shows how effective nonstandard methods can be applied to the theory of normal families of analytic functions. Not only are new proofs given for old results but also new insights are gained by this new approach. A novelty of this paper ("On the Theory of Normal Families") is his theory of polynomials whose degree may be a nonstandard integer, and its applications to the theory of the location of the roots of polynomials and analytic functions. A number of striking new results concerning the location of roots are given at the end of the paper.

Among all the papers Robinson wrote on nonstandard analysis dealing with problems in analysis and functional analysis the most famous is his paper with A. R. Bernstein entitled "Solution of an Invariant Subspace Problem of K. T. Smith and P. R. Halmos," which appeared in 1966 in the *Pacific Journal of Mathematics*. We remind the reader that one of the main problems in the spectral theory or reduction theory of bounded linear operators is to determine the nontrivial closed invariant subspaces. The first general result, that compact operators on a Hilbert space have nontrivial closed invariant subspaces, was proved by the late Professor J. von Neumann. The proof and its generalizations to compact operators on a Banach space appeared in 1954 in a paper by N. Aronszajn and K. T. Smith. During the next ten years no further progress was made with the general problem. In 1963, however, in the series Lectures on Modern Mathematics, P. R. Halmos in his article "A Glimpse into Hilbert Space" popularized the problem again and joined K. T. Smith to conjecture that perhaps every operator whose square is compact may have a nontrivial closed invariant subspace. It was this article by Halmos which inspired Robinson to show, by means of his newly developed method, that in fact this conjecture is true. Roughly speaking, the main idea of the proof was the observation that the result obviously holds in the finite dimensional case, and so, by the transfer principle, it holds in nonstandard Hilbert spaces whose dimension is a nonstandard integer. By selecting such a space in which the original Hilbert space can be embedded, Robinson was able to show by means of an ingenious reduction method that the original operator also has a nontrivial closed invariant subspace.

This method was extended and refined by A. R. Bernstein to obtain a proof of the result that if an operator on a Banach space is polynomially compact, then it has already a nontrivial closed invariant subspace. These results brought about a renewed interest in the very difficult, if not hopeless, problem of finding invariant subspaces. During the last decade many important contributions to the invariant subspace problem have been made. In 1973 the Russian mathematician V. I. Lomonosov settled in the affirmative Robinson's conjecture that every operator which commutes with a non-zero compact operator has a nontrivial closed invariant subspace. Just recently the Swedish mathematician Per Enflo showed, by means of an ingenious construction, that not every bounded operator on a Banach space has a nontrivial invariant subspace. The same problem for Hilbert spaces remains open.

From this period also stems an interesting article by Robinson on languages which are based on nonstandard arithmetic that appeared in 1963 in the *Nagoya Mathematical Journal*. In the article Robinson studies a logical calculus of infinitary nature in which the ordinary natural numbers are replaced by the numbers of a nonstandard model of arithmetic. This allows one to include formulas whose length may be a nonstandard integer. In doing so, Robinson emphasized the fact that the notion of arithmetic is relative even at the metamathematical level.

During this fruitful period Robinson also started to give form to his increased interest in applying his new method to algebraic number theory. In 1966 two papers appeared in the *Atti della Accademia Nazionale dei Lincei Rendiconti* entitled "A New Approach to the Theory of Algebraic Numbers." An invited address given in

1967 at the First International Conference on Nonstandard Analysis, which was held in Pasadena at the California Institute of Technology under the auspices of the Office of Naval Research, was also devoted to nonstandard algebraic number theory. All this preparatory work led to his fundamental contributions to class field theory and diophantine geometry.

In his book *Non-Standard Analysis* Robinson had already shown that in order to deal more effectively with nonstandard models in mathematics one should consider so-called enlargements, which are nonstandard models in which every standard concurrent binary relation has a bound. Robinson's work in algebraic number theory consists of introducing and analyzing the properties of "enlargements" of algebraic number fields. This idea was conceived to present a most universal "completion" of such a field in which Hensel's p-adic and Chevalley's idèles are contained and can be analyzed by means of the transfer principle. In this way Robinson presented new insights into the classical ideas of Kronecker, Hensel, Hasse, Chevalley, and others. Articles that represent these ideas are: an invited address at a meeting of the American Mathematical Society in 1967 which appeared in the *Bulletin of the American Mathematical Society* (vol. 73) under the title "Nonstandard Arithmetic"; the paper "Algebraic Function Fields and Nonstandard Arithmetic" presented at the "Tagung" on Nonstandard Analysis held in 1970 at the Oberwolfach Mathematical Research Institute and published in the Proceedings of the meeting under the title "Contributions to Nonstandard Analysis"; the fundamental paper "Nonstandard Points on Algebraic Curves" published in 1973 in the *Journal of Number Theory*; and his posthumously published joint paper "On the Finiteness Theorem of Siegel and Mahler Concerning Diophantine Equations," which the coauthor P. Roquette prepared for publication in the *Journal of Number Theory*. All these papers are included in this volume.

The fundamental importance of Robinson's creation of nonstandard analysis was soon acknowledged by large segments of the mathematical community. In 1973 Robinson was awarded for this outstanding achievement the L. E. J. Brouwer medal by the Dutch Mathematical Society "Het Wiskundig Genootschap." In his acceptance speech, published in 1973 in volume 21 of the *Nieuw Archief voor Wiskunde* under the title "Standard and Nonstandard Number Systems," Robinson discusses the philosophical question why, throughout the history of mathematics, certain algebraic or arithmetical structures were honored by the name *number systems*. The address concludes with the observation that "situations exist in mathematics today whose final solution may come again in the form of a system of numbers that is unimaginable today and will be commonplace tomorrow."

A few years before his untimely death Robinson came into contact with the mathematical economist Dr. D. J. Brown of Yale University. From their discussions on the problems concerning exchange economies in which the set of traders is allowed to grow without bound, Robinson immediately realized that a more natural approach would be to consider exchange economies in which the set of traders is a nonstandard integer. This nonstandard approach led to the resolution of Edgeworth's conjecture that, if the set of traders increases, the core approaches the set of competitive equilibrium. This was accomplished by showing that in a nonstandard exchange economy the concept of the core and competitive equilibrium are the same. These and related results are contained in two joint papers with D. J. Brown entitled "The Cores of Large Exchange Economies" and "Nonstandard Exchange Economies," respectively. This work has recently been carried forward in a very essential way by H. J. Keisler. Some of the surprising results of Keisler's work, which is still in preparation, were announced by Keisler at the Abraham Robinson Memorial Conference held at Yale University in May 1975.

A very important aspect of nonstandard analysis hinted at earlier is that the new method does not consist merely in adding in a consistent way infinitesimals to the reals. This had been done successfully. In the 1890s, Tullio Levi-Civita, responding to a question of Veronese concerning geometries, constructed a non-archimedean totally ordered field whose elements are formal power series.

At a very early stage in the development of nonstandard analysis Robinson felt that a definite relation between his nonstandard number systems and those of Levi-Civita and others had to exist. Also, from the point of view of the theory of asymptotics he was particularly interested in the formal power series fields of Levi-Civita. In the latter part of the 1960s and early 1970s, Robinson occupied himself very intensely with this problem. A solution was found in showing that, by using a positive infinitesimal as a new scale, one can define a new non-archimedean valuation field in which the formal power series fields of Levi-Civita can be embedded. This led also to a completely new approach to certain questions in asymptotics and a new definition of the so-called Popken norm. Part of this work was reported on in an invited address at a Conference on the Foundations of Mathematics held in 1968 at the Naval Academy in Annapolis. The Proceedings of the Conference appear in volume 13 of the Slaught Memorial Papers published by the Mathematical Association of America. A draft of an extended version of this paper was modified by the late Professor A. H. Lightstone, a former student of Abraham Robinson, into a text for a book which was published shortly after Robinson's death as volume 13 of the North-Holland Mathematical Library under the title *Non-Archimedean Fields and Asymptotic Expansions*. Recently, in continuing the study of these new valuation fields introduced by Robinson, I have shown that they are maximal in the sense of Kaplansky and are all isomorphic algebraically. These and related results have appeared (1976) in volume 25 of the *Israel Journal of Mathematics,* which was dedicated to the memory of Abraham Robinson.

Finally I would like to say a few words about the question of whether the new nonstandard number systems may form an adequate framework in which to analyze problems arising in theoretical physics dealing with infinities. In his first paper Robinson hints at such a possibility: "the question whether or not a scale of nonstandard analysis is appropriate to the physical world really amounts to asking whether or not such a system provides a better explanation of certain observable phenomena than the standard system of real numbers . . . . The possibility that this is the case should be borne in mind." In two papers written jointly with Peter J. Kelemen which appeared in volume 13 of the *Journal of Mathematical Physics* the first steps in this direction were made. Subsequently, Professor Max Dresden and some of his pupils have successfully continued this line of research. Representative of their work is a series of papers entitled "A 'Nonstandard Approach' to the Thermodynamic Limit," which appear in volume A13 of the *Physical Review* and subsequent issues. There is no question that nonstandard number systems may find their rightful place here, as foreseen by Abraham Robinson.

The picture of Abraham Robinson's mathematical oeuvre would not be complete without the inclusion of the papers on analysis stemming from the early 1950s.

Of the two papers dealing with the theory of summability, one is coauthored with G. G. Lorentz. The other paper deals with a question of the integration of a certain type of hyperbolic partial differential equation.

As I indicated earlier, Robinson's interest in summability goes back to the days when he was in contact with the mathematicians of Birkbeck College, London University. At that time Robinson came into contact with Professor R. G. Cooke, who was actively engaged in research on summability. Summability theory attracted him immediately and he contributed extensively to the subject. A number of his results appeared for the first time in R. G. Cooke's book entitled *Infinite Matrices and Sequence Spaces,* which appeared in 1950. Because these results are of great interest, I will detail them here.

If A and B are two T-matrices which are absolutely equivalent for all bounded sequences and if the two T-matrices C and D have the same property, then the matrices AC and BD also are absolutely equivalent for all bounded sequences.

We remind the reader that a T-matrix A is in general an infinite matrix with complex entries which satisfy the Silverman–Toeplitz conditions for defining a regular and consistent method of summability.

Two T-matrices A and B are called absolutely equivalent for a class Z of sequences whenever for each sequence $z \in Z$ we have $\lim_{n\to\infty}[(Az)_n - (Bz)_n] = 0$.

An important concept in the theory of summability is that of a core of a sequence introduced by K. Knopp. It is in fact the closed convex hull of the set of the limit points of a sequence. Robinson was quick to realize the significance of this concept in summability theory and contributed the following results to the above-mentioned book of R. G. Cooke's.

In order for a T-matrix $A = (a_{nk})$ to have the property that the core of every bounded sequence contains the core of its A-transform, it is necessary and sufficient that $\lim_{n\to\infty} \sum_{k=1}^{\infty} |a_{nk}| = 1$.

In order for the core of the A-transform of any bounded sequence to be contained in its core it is necessary and sufficient that the T-matrix A be absolutely equivalent, with respect to the family of all bounded sequences, to a non-negative T-matrix.

The latter result was generalized by Robinson in the following sense. Let $A = (a_{nk})$ be a real T-matrix and let $\{\theta_k\}$ be a sequence of positive numbers such that $\lim_{k\to\infty} \theta_k = \infty$ and $\Sigma_1^\infty |a_{nk}|\, \theta_k$ is finite for all n. Then the core of the A-transform of any sequence $\{s_k\}$ satisfying $|s_k| \leq \theta_k$ for all k is contained in the core of $\{s_k\}$ if and only if there exists a non-negative matrix $B = (b_{nk})$ such that $\sum_{k=1}^{\infty} b_{nk}\theta_k$ and $\Sigma_1^\infty |a_{nk} - b_{nk}|\theta_k$

are finite for all n, and A is absolutely equivalent to B for the family of all sequences $\{s_k\}$ with $|s_k| \leq \theta_k$ for all k.

The famous Pringsheim–Vivanti theorem, which asserts that $z = 1$ is a singularity of every analytic function in the unit disk whose coefficients of Taylor expansion are non-negative, was derived by Robinson from the following interesting theorem he discovered.

If $f(z) = \sum_{n=0}^{\infty} c_n z^n$ is analytic in $|z| < R$ and if f can be continued analytically in its principal star domain S to define a single-valued analytic function, say f, then for each $w \in S$ the value $f(w)$ is contained in the core of the sequence of partial sums of the power series $\sum_{n=0}^{\infty} c_n w^n$.

The most important results concerning the core, however, are contained in his paper with G. G. Lorentz entitled "Core-consistency and Total Inclusion for Methods of Summability," which was published in the *Canadian Journal of Mathematics* in 1953.

In this paper the following interesting question is examined. Let A and B be two regular summability methods such that A is stronger than B in the sense that each complex bounded sequence has the core of its A-transform contained in the core of its B-transform. Does it necessarily follow that A is divisible by B on the right? The answer is in general no. But the authors show that if the elements of B are non-negative, it is almost true in the sense that there exists a regular summability method C with non-negative entries such that CB-A is a matrix $D = (d_{nk})$ satisfying $\lim_{n \to \infty} \sum_0^{\infty} |d_{nk}| = 0$.

In the paper "On Functional Transformations and Summability" (*Proc. London Math. Soc.* [2] 52 [1950]: 132–60), Robinson generalizes the classical Silverman–Toeplitz theorem, which gives conditions on a matrix $A = (A_{nk})$ with complex entries such that the sequence $t_k = \sum A_{nk} s_n$ converges to s whenever the sequence $\{s_n\}$ converges to s. In Robinson's paper the s_n are now allowed to be elements of a general Banach space S and the elements A_{nk} are assumed to be bounded linear operators on S.

This kind of generalization was suggested to him by studying some of the classical summability methods, such as that of Borel. By introducing a new concept of the norm of a sequence of bounded linear operators Robinson showed that the Silverman–Toeplitz conditions, suitably modified for the general case, are again necessary and sufficient for the regularity and consistency of the method.

It is most likely that the concept of the norm of a sequence of bounded linear operators newly arrived at in this paper will prove fruitful in other questions concerning summability methods in Banach spaces.

Out of his work on Applied Mathematics grew his paper "On the Integration of Hyperbolic Differential Equations," which appeared in the *Journal of the London Mathematics Society* (25 [1950]: 209–17). In it Robinson responds to the challenge of solving Cauchy's problem for a linear hyperbolic equation of the form

$$\sum_{k=0}^{n} \sum_{\ell=0}^{n} \alpha_{k,\ell}(x, y) \frac{\delta^2 z}{\delta x^k \, \delta y^\ell} = \alpha_0(x,y),$$

where $\alpha_{n,0} \neq 0$ and the Cauchy data for the solution $z = z(x,y)$ are described for $x = 0$.

The main gist of the paper is to show that under the mild condition that the characteristic equation $\sum_{i=0}^{n} (-1)^i \alpha_{ni} \gamma^{n-i} = 0$ has distinct real roots, the problem can be reduced to the solution of a Cauchy problem for a system of first-order quasi-linear equations of the form $\frac{\delta f_i}{\delta x} + c_i \frac{\delta f}{\delta y} = F_i(x,y,f, \ldots, f_m), i = 1,2, \ldots, m.$

These equations can be solved by integrating along the curves $\frac{dy}{dx} = c_i$. The paper is historically significant because of the numerical aspects of the method.

It is hoped that the above words of introduction to Robinson's important oeuvre in nonstandard analysis and analysis will give the reader a few glimpses into the wealth of significant new ideas which are contained in Robinson's work. We are certain that his ideas will remain a vital source of inspiration for many generations of mathematicians.

INTRODUCTION TO PAPERS ON PHILOSOPHY

S. Körner

A mathematician who is concerned both with applying mathematics to the physical world and with inquiring into its logical structure and foundations cannot fail to become interested in philosophy and—if he is blessed with the gifts of an Abraham Robinson—to become a creative philosopher and, in particular, a creative philosopher of mathematics. He was convinced that, in the words of Immanuel Kant, "mathematics in its application to natural science" indispensably requires some metaphysics and that "since mathematics must here necessarily borrow from metaphysics, it need not be ashamed to let itself be seen in the company of the latter." His conviction that the application or, more precisely, the applicability of mathematics in the natural sciences poses inescapable problems to philosophy inspired a great deal of his historical scholarship, his study of logic, and, above all, his work in the philosophy of mathematics. It also influenced his ideas on the teaching of mathematics. Abraham Robinson's contribution to the philosophy of mathematics can, I believe, be best understood if one sees him as a successor of Leibniz and Hilbert. He was their successor not just in the sense of finding their ideas congenial, but in the deeper sense of developing, and adding to, their mathematical insights; of reflecting on their and his own mathematical discoveries and developing a philosophy of mathematics in the light of these reflections.

The purpose of the following introductory remarks to Abraham Robinson's philosophical papers is not to restate what he himself has expressed with admirable clarity, precision, and elegance, but to consider his philosophical position in the light of some classical and recent theories which he regarded as important and to which he reacted in developing his own philosophy of mathematics. The introduction falls naturally into three parts. The first contains a brief description and comparison of the views held by Leibniz and Kant on the nature of mathematical infinity because these views influenced all subsequent thought on the foundations of mathematics. The second contains a sketch of some relevant features of Hilbert's formalism, Gödel's platonism, constructivism, and logical positivism, as understood by Robinson. In the last part Robinson's own formalist conception of mathematics, especially infinitary mathematics, will be characterized and briefly illustrated.

Both Leibniz and Kant distinguish an absolute, theological concept of infinity, possessed by a perfect God in whose existence both of them believe, from mathematical concepts of infinity. As regards the latter both of them in their different ways reject as empirically empty the notions of an actually infinite set, of an infinitely large and of an infinitely small quantity. Robinson shares this general view with them and their formalist and constructivist successors. Whereas his philosophical and his logico-mathematical writings show the deep and direct influence of Leibniz, of which

Robinson was fully aware and which he gratefully acknowledged, Kant's influence on him is less direct and seems to have been mainly mediated by Hilbert.

Leibniz's philosophical views on the nature of the infinitely great and the infinitely small are on the one hand connected with his metaphysical system, in which the principle of continuity plays an important role, and on the other hand with his development of the infinitesimal calculus. In a letter to Des Bosses (Erdmann, p. 436) he states, as he also does elsewhere, that he regards the infinitely large and the infinitely small as fictions of the mind which enable one to adopt a succinct manner of speaking (*modum loquendi compendiosum*) and which can be eliminated after having been put to use in mathematical reasoning. For the infinitely small, in particular, everybody can substitute as small a quantity as he wishes (*tam parvum quam volet*).

In a brief memoir (C.I. Gerhardt, p. 350), which Robinson quotes twice in his *Non-Standard Analysis*, Leibniz characterizes his use of infinitesimals by comparing it with the method of Archimedes. His own method, he says, "differs from the style of Archimedes only in its linguistic expressions," which "are more direct and more suitable to serve the art of invention" (*plus conformes à l'art d'inventer*). What is striking about this quotation (and what struck Robinson about it) is that it foreshadows an analogous relation between the Cauchy–Weierstrass version of analysis on the one hand and nonstandard analysis on the other.

In spite of the mathematical differences between Leibniz's infinitesimal calculus and Robinson's nonstandard analysis, they both presuppose or suggest a similar philosophical conception of infinity, namely, that the infinitely large and the infinitely small are not only fictions, but fictions which are well founded (*fictiones bene fundatae*) in the sense that their use advances our knowledge about the real world and that it is possible to formulate clear rules for their introduction and use as well as for their eventual elimination. In other words, while the fictitious concepts may be treated *as if* they were nonfictitious, i.e., applicable, concepts, the context of their identifiability or equivalence (as opposed to their identity) can be precisely delimited. Because Leibniz's approach was considered irremediably inconsistent, hardly any efforts were made to improve this delimitation. An exception is Vaihinger's general theory of fictions, in which he tries to justify the use of infinitesimals by a "method of opposite mistakes" which neutralize each other—a method which, however, was far too imprecise to attract the notice of mathematicians (see *Die Philosophie des Als Ob* [Berlin, 1913], pp. 511 ff.).

For Kant the notion of an infinitely large or infinitely small set or quantity is an "Ideal," i.e., a notion which is neither abstracted from, nor applicable to, sense-experience and which, therefore, is theoretically empty, in particular mathematically and scientifically empty. Moreover the attempt to apply the Ideas to sense-experience and physical phenomena leads to contradictions. Yet, in spite of their theoretical emptiness the Ideas are not practically empty, but acquire a nontheoretical or practical meaning from man's moral experience. From the apparent antinomy of an Idea's being theoretically meaningless and practically meaningful there arises a task, which is one of the sources of Hilbert's program: the task of proving that no inconsistency results from adjoining a theoretically empty, but practically significant, Idea, such as the Idea of an actual infinity, to a system of theoretically, in particular mathematically and scientifically, significant concepts and statements. The conception of this task is still recognizably present in Robinson's version of formalism.

In developing his own philosophy of mathematics, Robinson reacted to some recent philosophical views which he considered noteworthy, though unacceptable. Of these he found Hilbert's formalism most congenial and Gödel's platonism most alien. As regards constructivism and logical positivism, he admitted the possibility that under the pressure of new mathematical developments he might undergo a conversion to one of them—a possibility which in the case of platonism he wholly excluded (see "From a Formalist's Point of View," [104] in bibliography).

Hilbert's distinction between a theoretically meaningful finitary and a theoretically meaningless infinitary mathematics corresponds to the Kantian distinction between mathematical concepts, which are applicable to perceptual or imaginable objects, and mathematical Ideas, which are not. Hilbert's formalism is further based on the observation that much of mathematics can be regarded as a purely formal, rule-governed manipulation of symbols which allows one to introduce purely formal notions of well-formed formula, theorem, consistency etc., that is to say, notions which are independent of any theoretical meaning which the symbols may have. His program was to formalize all mathematics, including infinitary mathematics, completely and *using only finite mathematical reasoning* to prove that the so formalized mathematics is formally consistent. When as a result of Gödel's incompleteness theorems it became clear that Hilbert's program could not be implemented because his finite mathematics was not strong enough for the purpose, it became necessary either to abandon or to reform formalism. Robinson's own mathematical work and his philosophical reflection on it, as well as his view of the history of mathematics and of its philosophy, clearly pointed toward the reformation of formalism and was to become his central philosophical interest.

In order to understand Robinson's objection to platonism, one must be aware that he—like most other mathematicians—uses this term in a very broad sense which covers not only the doctrine of Plato and his school that the objects of mathematics are part of a mind-independent reality, but also the so-called conceptualist doctrine that the objects of mathematics, although they are for their existence dependent on the activity of a human mind, are the same for everybody. Since Robinson's opposition to platonism is mainly directed against the kind of conceptualist epistemology represented by Gödel, it is proper to mention some of its salient points.

Gödel holds in particular that "we do have something like a perception" of mathematical objects, including the objects of classical set-theory, and bases this thesis on the alleged fact that the axioms of set-theory "force themselves upon us as being true." Like Kant, he holds that there is a mathematical intuition, but unlike Kant, Hilbert, and Robinson, he regards this intuition as extending also to infinite objects. He further holds that mathematical intuition need not be conceived "as giving an immediate knowledge of the objects concerned," but that in mathematics "as in the case of physical experience we *form* our ideas" of mathematical objects "on the basis of something else which *is* immediately given'' (see ''What is Cantor's Continuum Problem,'' pp. 258–73, in *Philosophy of Mathematics*, edited by P. Benacerraf and H. Putnam, Englewood Cliffs, N.J. [1964]). The platonist doctrine which Robinson finds particularly unacceptable is that there exist—independently of our minds or as products of their activity—actually infinite sets or infinitesimal quantities about which we can make statements which are true and, hence, meaningful in the same sense as are our statements about the objects which we discern in the physical world.

Robinson's program of reforming Hilbert's formalism implies his agreement with Hilbert's doctrine that infinitary mathematics, though theoretically meaningless, is nevertheless for a variety of theoretical and practical reasons worth pursuing. It also explains on the one hand his interest in, and on the other his rejection of, Brouwer's intuitionism and other more recent versions of the constructivist philosophy of mathematics. For constructivism considers infinitary mathematics not only as theoretically meaningless but also as not being mathematics at all. Brouwer, like Hilbert, claims to have derived his general conception of the nature of mathematics from Kant, but he would not agree that Kant's proof of the consistency of the Ideas with our beliefs about the phenomenal world of our experience has anything to do with pure or applied mathematics, whatever relevance the proof may have to morality or religion.

The philosophical views so far discussed differ radically in their delimitation of a mathematical core, containing statements which are true of an objectively or intersubjectively existing reality and consequently do not admit of genuine alternatives. They do, however, agree that there is such a mathematical core. This common assumption is opposed to the conventionalist view—associated with various kinds of pragmatism and logical positivism—according to which all mathematical statements are based on linguistic conventions and hence admit of alternatives. Robinson, though in no way convinced by it, considers as worthy of consideration the positivist view that all mathematics, and even ''all logic as part of language, is ultimately arbitrary and is regulated only by its usefulness in coping with the empirical world'' (see ''From a Formalist's Point of View,'' p. 48). In particular, he does not regard it as wholly inconceivable that quantum mechanics may require, or suggest, the adoption of a nonclassical logic. (For such a proposal see, e.g., G. Birkhoff and J. von Neumann, ''The Logic of Quantum Mechanics,'' *Annals of Mathematics* 37 [1963]: 823–43.)

Robinson's philosophy of mathematics emerges naturally from his reflective awareness of the interaction of philosophical and mathematical ideas in his own thinking as well as in the minds of the other philosophers and mathematicians whose problems he shared, even if he could not accept their solutions. Foremost among them is Leibniz, whose doctrine of possible worlds and whose infinitesimal calculus can be regarded as anticipations of modern model theory and of nonstandard analysis and whose account of mathematical fictions can be regarded as anticipating Robinson's formalist philosophy of mathematics.

Although this philosophy underwent some minor changes during his lifetime, it rested throughout on the following three principles:

1. the principle of the *theoretical* meaninglessness of the notion of an infinite

totality, i.e., "that any mention, or purported mention, of infinite totalities is, literally, meaningless";

2. the principle of the usefulness of the notion of infinity as a *well-founded* fiction, i.e., that mathematicians should "continue to act *as if* infinite totalities existed" (see "Formalism 64," p. 230); and

3. the principle that there exists an *intuitive, nonconventional core* of mathematics and logic which is presupposed in all mathematical thinking (see e.g., "From a Formalist Point of View," p. 48).

The precise delimitation of these principles depended for Robinson not so much on philosophical argument and speculation as on mathematical practice. Thus while he rejected the creed of the extreme platonist "who sees the world of the actual infinite spread out before him," he does hold that "any universal sentence of elementary arithmetic" has a "direct intuitive meaning" although it does refer to an unlimited domain of individuals. And, as has been pointed out earlier, he admits in general the possibility of redrawing the borders of the nonconventional core of intuitive mathematics and logic (see "From a Formalist's Point of View," pp. 48–49).

One of Robinson's most important contributions to the philosophy of mathematics was his clarification and application of the Leibnizian notion of well-founded mathematical fictions. Thus, although he agrees on the whole with Leibniz's introduction of fictions into the calculus, he does not consider them sufficiently well founded. More precisely, while he accepts the extension by Leibniz and his successors of the real number system by infinitesimals, he objects that they were not able "to state with *sufficient precision* just what rules were supposed to govern their extended number system," e.g., to state precisely why this system was not Archimedian (see e.g., *Non-Standard Analysis*, p. 266).

The precision required is that of model theory, in particular of nonstandard analysis. This theory replaces, for example, the vague claim that x and $x+dx$ can be treated as if they were equal by the demonstration that they are equivalent in a precise sense which allows their substitution for each other in some well-defined relations but not in others. And it replaces the vague claim that finite and infinite quantities can be treated as if they had the same properties by the precise demonstration that the system of real numbers and its enlargement by the infinitesimals satisfy the same set of sentences of the lower predicate calculus. Robinson's formalist program, as based on his three principles, is capable of further applications, for example, in providing more precise and more detailed elucidations of the relations between pure and applied mathematics (see "Concerning Progress in the Philosophy of Mathematics"). His premature death has deprived us of the light which his thought would have thrown on this and many other philosophical and mathematical problems.

Nonstandard Analysis and Analysis

NON-STANDARD ANALYSIS

BY

ABRAHAM ROBINSON

(Communicated by Prof. A. HEYTING at the meeting of April 29, 1961)

1. *Introduction.*

It has been known since the publication of a classical paper by SKOLEM [9] that there exist proper extensions of the system of natural numbers $N(n \geqslant 0)$ which possess all properties of N that are formulated in the Lower predicate calculus in terms of some given set of number-theoretic relations or functions, e.g. addition, multiplication, and equality. Such an extension of the natural numbers is known as a (strong) non-standard model of arithmetic.

Now let R_0 be the set of all real numbers. Let K_0 be the set of sentences formulated in the Lower predicate calculus in terms of (individual constants for) all elements of R_0 and in terms of (distinct symbols for) all relations that are definable in R_0, including singulary relations. In a well-defined sense all elementary statements about *functions* in R_0 can be expressed within K_0. Thus if the real-valued function $f(x)$ is defined on the subset of S of R_0, and not elsewhere, then there exists a binary relation $F(x, y)$ in the vocabulary of K_0 such that $F(a, b)$ holds in R_0 if and only if $a \in S$ and $b = f(a)$. The fact that $F(x, y)$ denotes a function with domain of definition S is expressed by the sentence

$$[(x)(y)[F(x,y) \supset T(x)]] \wedge [(x)[T(x) \supset [(\exists y)(z)[F(x,y) \wedge [F(x,z) \supset E(y,z)]]]]],$$

where $T(x)$ is the singulary relation which defines S and $E(y, z)$ stands for $y = z$. Note that a different relation, $F'(x, y)$ corresponds to "the same" $f(x)$ if it is obtained the restricting the domain of definition of $f(x)$ to a proper subset of S. However, for ease of understanding we shall in the sequel use a less formal notation and include expressions like $x = y$, $y = f(x)$, $xy = z$ among our sentences. It is not difficult to translate these sentences into the strict formalism of K_0.

Let R^* be a model of K_0 which is a proper extension of R_0 with respect to all the relations and individual constants contained in R_0. R^* will be called a *non-standard model of analysis*. The existence of non-standard models of analysis follows from a familiar application of the extended completeness theorem of the Lower predicate calculus (*e.g.* [7]). Such models may also be constructed in the form of ultra-powers (*e.g.* [6]). The latter method affords us an insight into the structure of non-standard models of analysis and enables us to discuss the question to what extent we can single out certain distinguished models of this kind. Considerable progress can be made in this direction, but for the work of the present paper any one non-standard model of analysis will do as well as another.

It is our main purpose to show that these models provide a natural approach to the age old problem of producing a calculus involving infinitesimal (infinitely small) and infinitely large quantities. As is well known, the use of infinitesimals, strongly advocated by Leibnitz and unhesitatingly accepted by Euler fell into disrepute after the advent of Cauchy's methods which put Mathematical Analysis on a firm foundation. Accepting Cauchy's standards of rigor, later workers in the domain of non-archimedean quantities concerned themselves only with fragments of the edifice of Mathematical Analysis. We mention only du Bois–Reymond's Calculus of infinities [2] and Hahn's work on non-archimedean fields [4] which in turn were followed by the theories of Artin–Schreier [1] and, returning to analysis, of Hewitt [5] and Erdös, Gillman, and Henriksen [3]. Finally, a recent and rather successful effort of developing a calculus of infinitesimals is due to Schmieden and Laugwitz [8] whose number system consists of infinite sequences of rational numbers. The drawback of this system is that it includes zero-divisors and that it is only partially ordered. In consequence, many classical results of the Differential and Integral calculus have to be modified to meet the changed circumstances.

Our present approach yields a proper extension of classical Analysis. That is to say, the standard properties of specific functions (e.g. the trigonometric functions, the Bessel functions) and relations, in a sense made precise within the framework of the Lower predicate calculus, still hold in the wider system. However, the new system contains also infinitely small and infinitely large quantities and so we may reformulate the classical definitions of the Infinitesimal calculus within a Calculus of infinitesimals and at the same time add certain new notions and results.

There are various non-trivial interconnections between the theories mentioned in this introduction. For example (as is not generally realized) the ultra-power construction coincides, in certain special cases which are relevant here, with the construction of residue fields in the theory of rings of continuous functions. Similarly, there are various connections between these theories and the work of the present paper. We have no space to deal with them here. Details and proofs of the results described in the present paper will be given in the volume "Introduction to Model theory and to the Metamathematics of Algebra" which is due to be published in the series "Studies in Logic and the Foundations of Mathematics".

2. *Non-standard analysis and non-archimedean fields.*

Let R^* be any non-standard model of analysis. Then $R^* \supseteq R_0$ but $R \neq R_0$. It follows that R^* is non-archimedean. The elements of $R_0 \subseteq R^*$

will be referred to as the *standard* elements of R^* or standard numbers. Since R^* is non-archimedean the following two subsets of R^* are not empty.

M_0, the set of all $a \in R^*$ such that $|a| < r$ for some $r \in R_0$. M_0 is a ring. The elements of M_0 will be said to be *finite*.

M_1, the set of all $a \in R^*$ such that $|a| < r$ for all positive $r \in R_0$. The elements of M_1 will be said to be *infinitesimal*. M_1 is a prime ideal in M_0 and M_0/M_1 is isomorphic to R_0.

Let $S = (a_1, a_2, \ldots)$ be any subset of R^*. We write $O(a_1, a_2, \ldots)$ for the module $M_0 a_1 + M_0 a_2 + \ldots \subseteq R^*$ (weak sum) and we write $o(a_1, a_2, \ldots)$ for the module $M_1 a_1 + M_1 a_2 + \ldots \subseteq R^*$. In particular $M_0 = O(1)$ and $M_1 = o(1)$. We write $a =_1 b$ if $a - b \in M_1$, i.e. if the difference between a and b is infinitesimal and we say in that case that a is *infinitely close* to b. Every finite number (i.e. every element of M_0) is infinitely close to some standard number. We write

$$a = b \text{ mod. } O(a_1, a_2, \ldots)$$

if $a - b \in O(a_1, a_2, \ldots)$, with a similar notation for o.

So far we have formulated only a number of obvious, and in part well-known, notions and facts concerning all ordered fields which are extensions of the field of real numbers. We now make use of the fact that R^* is a non-standard model of analysis. Let $N'(x)$ be the singulary relation which defines the natural numbers in R_0. Then $N'(x)$ defines a set N^* in R^*. N^* turns out to be a non-standard model of the natural numbers with respect to all relations definable in N. Similarly, we obtain a non-standard model of the rational numbers, R_1^* as a subset of R^*. It can be shown that every standard transcendental number is infinitely close to some element of R_1^*.

Syntactically, or linguistically, our method depends on the fact that we may enrich our vocabulary by the introduction of new relations, such as $R_0'(x)$, $M_0'(x)$, $M_1'(x)$ which define R_0, M_0, M_1, in R^*. (Note that the singulary relations just mentioned are, provably, not definable in terms of the vocabulary of K_0). We are therefore in a position to reformulate the notions and procedures of classical analysis in non-archimedean language. Since all the "standard" results of analysis still hold we may make use of them as much as we please and we may therefore carry out our reformulation either at the level of the fundamental definitions (of a limit, of an integral, ...) or at the level of the proof or, finally, by introducing non-standard notions into a result obtained by classical methods.

We consider in the first instance functions, relations, sets, etc. which are defined already in R_0, so that appropriate symbols are available for them in the original vocabulary. Such concepts will be called *standard* (functions, relations, sets, etc.). For example the interval $a < x < b$, will be called a standard interval in R^* if a and b are standard numbers (elements of R_0). The interval $\eta < x < b$, where b is standard and η infinitesimal positive, is not a standard interval. The function $\sin x$ is a standard function. Strictly speaking we ought to refer here also to the interval of definition of the function but, as in ordinary analysis, this will frequently be taken for granted.

At a more advanced stage it turns out to be nexessary to go beyond standard functions, etc. Consider a standard function of $n+1$ variables, $y=f(x_1, \ldots, x_n, t)$, $n \geqslant 1$. Regard t as a parameter and define $g(x_1, \ldots, x_n)$ by $g(x_1, \ldots, x_n)=f(x_1, \ldots, x_n, \tau)$ where τ is not (necessarily) standard. Then $g(x_1, \ldots, x_n)$ will be called a *quasi-standard* function. (Note that need not be a continuous function of its arguments.) For example, the function $f(x, n)=\sqrt{n/\pi}\, e^{-nx^2}$ is a standard function of two variables. The function $g(x)=f(f, \omega)=\sqrt{\omega/\pi}\, e^{-\omega x^2}$, where ω is an infinite (non-finite) positive number, is quasi-standard.

Quasi-standard relations, etc. are introduced in a similar way. For example, the interval $\eta<x<b$ mentioned above is quasi-standard.

3. *Examples in non-standard Analysis.*

Let s_n be a standard sequence, i.e. a function defined in the first instance on the natural numbers N and taking values in R_0. Then the definition of s_n extends automatically to the elements of N^* (the non-standard positive integers). Let s be a standard number. We define –

3.1. s is called the *limit* of s_n iff s_ω is infinitely close to s for all infinitely large positive integers ω. In symbols –

$$(\omega)[N'(\omega) \wedge \sim R_0(\omega) \supset |s-s_\omega| \in M_1].$$

This compares with the classical definition (s is the limit of s_n if for every $\delta>0$ there exists a positive integer ω_0 such that, etc.). It can be shown that the two definitions are equivalent under the stated conditions (s_n and s standard). That is to say $\lim_{n\to\infty} s_n=s$ in R_0 if and only if 3.1 holds in R^*. The proof involves the formalization of the classical definition as a sentence in K_0. Similarly, the following is an equivalent definition of the concept of a limit point (accumulation point) of a sequence for standard s_n and s.

3.2. s is a limit point of s_n iff $s_\omega =_1 s$ for some infinite positive integer ω.

Similarly –

3.3. A standard sequence s_n is bounded if and only if s_n is finite for all infinitely large n.

The theorem of Bolzano–Weierstrass for standard bounded sequences can of course still be proved by classical methods. Using non-standard analysis we obtain an alternative proof along the following lines.

Let $\langle a, b\rangle$ be a closed interval containing all elements of the standard sequence s_n. Then there exists a sentence X of K_0 which states that for every positive integer m, the partition of $\langle a, b\rangle$ into m sub-intervals of equal length $(b-a)/m$ yields at least one sub-interval I that contains an unbounded number of elements of s_n. X holds also in R^* and so, taking m infinite, we obtain a sub-interval I^* of $\langle a, b\rangle$, of infinitesimal length, that contains an unbounded number of elements of s_n. Both end points of I^* are infinitely close to a single standard number s. This is the required limit point.

We have the following version of Cauchy's necessary and sufficient condition for convergence, which may either be proved directly or transcribed from the standard version.

3.4. The standard sequence s_n converges iff $s_\omega =_1 s_{\omega'}$ for all infinitely large ω and ω'.

The theory of infinite series may be reduced to that of infinite sequences in the usual way.

Coming next to functions of a real variable, suppose that the standard function $f(x)$ is defined in a standard open interval $a<x<b$. Then the standard number l is the limit of $f(x)$ as x tends to be from the left iff $f(b-\eta)=_1 l$ for all positive infinitesimal η. $f(x)$ is continuous at the standard point x_0, $a<x_0<b$ iff $f(x_0+\eta)=_1 f(x_0)$ for all infinitesimal η. Again these conditions are equivalent to the classical definitions. Accordingly, $f(x)$ is continuous in (a, b) if $f(x_0+\eta)=_1 f(x_0)$ for all *standard* x_0 in the open interval and for all infinitesimal η. The natural question now arises what non-standard condition corresponds to *uniform* continuity in the interval $a<x<b$. The answer is

3.5. $f(x)$ is uniformly continuous in (a, b) – a, b, and $f(x)$ standard – if $f(x_0+\eta)=_1 f(x_0)$ for all infinitesimal η and for *all* x_0 in the open interval (a, b).

In a similar way, we may distinguish between ordinary continuity and uniform continuity of a standard sequence of functions $s_n(x)$.

3.5. $f(x)$ has the derivative f_0 at the standard point x_0 (f_0 a standard number) if for all infinitesimal η

$$\frac{f(x_0+\eta)-f(x_0)}{\eta} =_1 f_0$$

a formula which may be used in practice. Various "standard" results of the Differential calculus, including Rolle's theorem can in fact be established readily by means of non-standard Analysis.

We touch only briefly upon integration and remark that, up to infinitesimal quantities, Cauchy's integral and Riemann's integral can be defined by means of a partition of a given standard interval into an infinite number of subintervals of infinitesimal length combined with the formation of the usual sums such as $\Sigma(x_n-x_{n-1})y_n$. The non-standard definition of the Lebesgue integral appears to be more intricate and has not been carried out in detail so far.

Next, we discuss differential notation in connection with functions of several variables. We do not select a specific element of M_1 as *the* differential but regard any infinitesimal increments as differentials of the independent variables. Thus, the theorem on the existence of the total differential may be stated as follows.

3.6. Let (x_0, y_0) be a standard point (a point with standard numbers as coordinates), and let S be the standard plane set given by

$$(x-x_0)^2+(y-y_0)^2<r^2$$

where r is a standard positive number. Suppose that the standard function $f(x, y)$ possesses continuous first derivatives in S. Let dx, dy be any pair of infinitesimal numbers and let $df=f(x_0+dx, y_0+dy)-f(x_0, y_0)$. Then

3.7. $$df=\frac{\partial f}{\partial x_0}\,dx+\frac{\partial f}{\partial y_0}\,dy \quad \text{mod. } o(dx, dy).$$

Although this formula does not yield ordinary equality between the left and right hand side it can be applied without difficulty, e.g. to the calculation of the derivative of a function defined implicitly by $f(x, y)=0$.

More generally, it is true that much of the classical work in Differential Geometry has been done in terms of a vague notion of infinitesimals, and the same applies to Analytical Mechanics. It is a matter of general belief that all this work could, if necessary, be rewritten to conform to the rigor of contemporary Mathematics but nobody would think of carrying out this task. It is therefore not without interest that we may now justify the use of infinitesimals in all these problems directly. As an example, the case of the osculating plane of a skew curve has been considered in detail. Thus, let C be a standard skew curve in three dimensions (the equations for C are expressed in terms of standard functions) and let P be a standard point on C. Let Π be the set of all planes drawn through P and any two neighboring points P_1, P_2 such that P, P_1, P_2, are not collinear. Then the osculating plane of C at P may be defined as a standard plane p through P such that p is *infinitely near* to all elements of Π in a sense which can be made precise without difficulty. This definition leads to the usual equation for the osculating plane.

Going on to an example which is of greater contemporary interest, let G be a *standard* n-parametric Lie group, n finite. Thus G is defined by analytic functions in R_0, but the passage to R^* extends it automatically to a wider group G^*. Let $(e_1, \ldots, e_n)$ be the set of parameters for the identity in both G and G^*. Then the "infinitely small" transformations in G^* are given by the sets $(e_1+\eta_1, \ldots, e_n+\eta_n)$, where $\eta_1, \ldots, \eta_n$ are infinitesimal. These transformations now constitute a genuine subgroup G' of G^*, which may be analyzed further.

We pass on to the consideration of quasi-standard functions. Let $f(x, t)$ be a standard function defined in a standard set S_1, and let $g(x)=f(x, \omega)$, where ω is non-standard, e.g. infinite or infinitesimal.

Then it can be shown that all functionals or operators which apply to the standard functions $g_t(x) = f(x, t)$, t standard, are extended in a natural and unique way to the function $g(x)$. For example, suppose that the integrals

$$\int_a^b g_t(x)dx = \int_a^b f(x, t)dx = F(t)$$

are defined, in R_0, in some definite sense (e.g. as Riemann integrals) for the range of t under consideration. Then $F(t)$ is a standard function, and we shall regard $F(\omega)$ as the value of the integral $\int_a^b g(x)dx$. It is not difficult to see that if the function $g(x)$ is obtained from different families $f_1(x, t)$, $f_2(x, t)$, so that $g(x) = f_1(x, \omega_1) = f_2(x, \omega_2)$ then the use of either of these families leads to the same value for the integral $\int_a^b g(x)dx$. Moreover, the definition preserves the properties of an integral to the extent to which they can be expressed in the Lower predicate calculus, e.g. approximation by sums of the form $\Sigma(x_n - x_{n-1})y_n$.

The same argument applies to functional operators. Thus, if the derivatives $\dfrac{\partial f(x, t)}{\partial x} = h(x, t)$ exist then we define $\dfrac{dg}{dt} = h(x, \omega)$. It may be mentioned that all these definitions take on a rather more concrete form it we consider non-standard models which are in the form of ultraproducts.

Quasi-standard functions yield a natural realization of generalized functions. Thus, a Dirac delta-function on an interval I may be defined as a quasi-standard function $\delta(x)$ such that for a given standard x_0 in I, $\delta(x)$ is infinitesimal for all standard $x \neq x_0$ in I, and $\int_I \delta(x)dx = 1$. For instance, $\sqrt{\omega/\pi}\, e^{-\omega(x-x_0)^2}$, where ω is an infinite natural number, is a delta function for $I = R^*$ and $(1 + \cos(x - x_0))^\omega / \int_{-\pi}^{\pi} (1 + \cos t)^\omega dt$ is a delta function in the interval $(-\pi, \pi)$. For given I and $x_0 \in I$, there are many delta functions, as opposed to the situation in the theory of distributions. Quasi-functions can be added, subtracted, multiplied, and divided by one another provided the divisor does not vanish. It is natural to consider such functions in connection with a concrete application.

For any finite number a we write $b = st\{a\}$ (read 'b is the standard part of a') for the standard number b which is infinitely close to a. Consider now a standard function $\phi = \phi(x, y, z)$ which is harmonic in a region V bounded by a standard surface S. Let $P = (x, y, z)$ be a standard point in the interior of V, so that (Green's formula)

$$\phi(x, y, z) = \frac{1}{4\pi} \int_S \left(\frac{1}{r} \frac{\partial \phi}{\partial n} - \frac{\partial}{\partial n} \left(\frac{1}{r} \right) \phi \right) dS.$$

The formula is usually obtained by applying Green's identity to the pair of functions ϕ and $\psi = 1/r$. The singularity of ψ at P is taken into account by a familiar procedure.

Instead of taking $\psi = 1/r$ we may introduce the potential ψ_ϱ of a homogeneous sphere of infinitesimal radius ϱ round P. In this case there is no singularity at P and after first checking that Green's identity applies, we obtain the formula

$$\phi(x, y, z) = \frac{1}{4\pi}\, st\left\{\int\limits_S \left(\psi_\varrho \frac{\partial\phi}{\partial n} - \frac{\partial\psi_\varrho}{\partial n}\phi\right) dS\right\}.$$

An interesting result is obtained if we apply similar considerations to Volterra's formula for the solution of the two-dimensional wave equation.

Coming next to classical Applied Mathematics, it would of course be natural to reword the usual statements about particles of fluid and about infinitesimal surfaces and volumes in terms of the present theory. However, we pass over this possibility and consider instead a particular point in Fluid mechanics that gives rise to certain conceptual difficulties.

It is the assumption of boundary layer theory, e.g. for flow round a body or through a pipe, that the equations of inviscid flow are valid everywhere except in a narrow layer along the wall. It is found that the thickness of the layer, δ, is proportional to $R^{-1/2}$ where R is the Reynolds number, supposed large. Within this boundary layer, the flow is determined by means of the boundary layer equations which are obtained by simplifying the Navier–Stokes equations of viscous flow. However, when solving these equations, it proves natural to suppose that the boundary layer is *infinitely thick*. For example, for the case of a straight wall along the x-axis, the boundary layer equations are solved for boundary conditions at $y = 0$ and $y \to \infty$ a procedure which is clearly incompatible with the previous assumption on the smallness of δ. This conceptual difficulty can be resolved by supposing that the inviscid fluid equations hold for all positive standard values $y > 0$ while the influence of viscosity is confined to values of y that belong to $O(\delta)$, δ infinitesimal (so that the Reynolds number R is infinite). Introducing $y' = y/\delta$ we may then derive and solve the boundary layer equations for $0 < y' < \infty$ which is a region in which the flow has not been defined previously. There are other problems in continuous media mechanics that should be amenable to a similar analysis.

In reality it is of course not true that the region in which viscosity is effective may be regarded as infinitely thin. It can in fact be seen with the naked eye both in certain laboratory experiments and in every day life. Thus, the above model is intended only as a conceptually clear picture within which it should be easier to discuss some of the more intricate theoretical questions of the subject such as the conditions near the edge of the layer.

For phenomena on a different scale, such as are considered in Modern

Physics, the dimensions of a particular body or process may not be observable directly. Accordingly. the question whether or not a scale of non-standard analysis is appropriate to the physical world really amounts to asking whether or not such a system provides a better explanation of certain observable phenomena than the standard system of real numbers. The possibility that this is the case should be borne in mind.

Fine Hall,
Princeton University

REFERENCES

1. ARTIN, E. and O. SCHREIER, Algebraische Konstruktion reeller Körper, Abhandlungen, Math. Seminar, Hamburg, **5**, 85–99 (1926).
2. DU BOIS-REYMOND, Über asymptotische Werthe, infinitäre Approximationen und infinitäre Auflösung von Gleichungen, Mathematische Annalen, **8**, 363–414 (1875).
3. ERDÖS, P., L. GILLMAN and M. HENRIKSEN, An isomorphism theorem for real-closed fields, Annals of Mathematics **61**, 542–554 (1955).
4. HAHN, H., Über die nichtarchimedischen Grössensysteme, Sitzungsberichte der kaiserlichen Akademie der Wissenschaften (Vienna), **116**, section IIa, 601–655 (1907).
5. HEWITT, E., Rings of real-valued continuous functions, Transactions of the American Mathematical Society, **64**, 65–99 (1948).
6. KOCHEN, S. B., Ultraproducts in the theory of models, to be published in the Annals of Mathematics.
7. ROBINSON, A., On the metamathematics of algebra, Studies in Logic and the Foundations of Mathematics, Amsterdam 1951.
8. SCHMIEDEN, C. and D. LAUGWITZ, Eine Erweiterung der Infinitesimalrechnung, Mathematische Zeitschrift, **69**, 1–39 (1958).
9. SKOLEM, T., Über die Nichtcharakterisierbarkeit der Zahlenreihe mittels endlich oder abzählbar unendlich vieler Aussagen mit ausschliesslich Zahlenvariablen, Fundamenta Mathematicae **23**, 150–161 (1934).

ON LANGUAGES WHICH ARE BASED ON NON-STANDARD ARITHMETIC[1]

ABRAHAM ROBINSON

1. Introduction. The natural numbers play a part in the formulation of logical syntax inasmuch as they are used to count the symbols in a sentence, or the sentences in a proof, etc. In the present paper, we shall study an infinitary logical calculus which is based on replacing the ordinary natural numbers in the capacity just mentioned, by a non-standard model of arithmetic. (Compare refs. 3, 5, 6 for some other logical calculi of infinitary nature). Thus, our calculus will include formulae of length n for any natural number n, finite or infinite, in the chosen non-standard model of arithmetic. Evidently, the study of such formulae can be of interest only if we introduce concepts which are beyond the power of expression of the notions borrowed from the standard case. It turns out that the concept of truth in a model, when defined by means of Skolem functions has this character and involves a curious phenomenon which is analogous to one first pointed out by H. Steinhaus and J. Mycielski for another kind of infinitary language. Thus, while in the standard predicate calculus the negation of a sentence in prenex normal form is reduced to prenex normal form by changing the type of the quantifiers and by shifting the sign of negation, this procedure is not legitimate when truth is defined in this way in our calculus.

The plan of this paper is as follows. In the second section we detail two auxiliary results which are used later. One of them is a generalization, due to Craig (ref. 1), of Beth's theorem on definitions. The second result is Tarski's well known theorem on the non-definability of the set of elementary sentences that are true in arithmetic. The proof of this theorem by means of a non-

Received May 1, 1962.

[1] The work on the present paper was carried out in part while the author held a grant from the National Science Foundation (No. G 14006) at the University of California in Berkeley.

standard model of arithmetic leads up to the main subject of the paper. In the third section, we introduce the non-standard language which forms the main subject of this paper, and discuss its syntactical properties. Finally, sections 4-6 deal with two types of truth definition for our non-standard language, and their interrelations.

The present paper may thus be regarded as a study of one of the implications of the existence of non-standard models of arithmetic. However, it is not intended as a mere exercise in the use of this concept. In order to appreciate this, notice that in the usual investigations of problems of incompleteness and non-categoricity these phenomena are regarded as properties of certain axiomatic systems, which are detected when the systems in question are investigated from a metamathematical point of view in which arithmetic is supposedly absolute. By introducing non-standard arithmetic already at the stage of the construction of the sentences of the object language, we emphasize the fact that the notion of arithmetic may be relative even at the metamathematical level. However, we do not escape the convenient assumption that at the back of all our investigations there is some system of arithmetic which is absolute.

In the sequel, we shall refer to the Lower predicate calculus without going into the details of the particular version to be used here. In fact, our general arguments apply equally to the various standard versions of the calculus. However, for the sake of definiteness we mention here that we shall envisage a language in which the relation or function symbols do not determine the number of their places. Thus, the same relation symbol may be used, in different connections, to denote both binary and ternary relations. Round brackets are used to enclose the arguments of a relation or of a function, or to enclose a quantifier, while square brackets serve to determine the syntactical grouping. Finally, no quantifier shall contain a variable that is already quantified in its scope.

2. Auxiliary results. Let L be an elementary language (a language of the Lower predicate calculus) whose set of extralogical constants comprises (symbols for) relations, individual constants, and functions. Let X be a sentence in L which contains an n-place relation R, $n>0$, and in addition a number of other extralogical constants which constitute a set S. Suppose that S is divided into two subsets S_1 and S_2 such that S_1 contains at least one n-place relation, $m>0$. Let X' be the sentence obtained from X by replacing the elements of S_2, as well as R, by distinct extralogical constants of the same kind which were not contained in X. Thus, R is replaced by an n-place relation R' and S_2 is replaced by a set which will be called S_2'. We may indicate the connection between X and X' by writing

$$X = F(S_1, S_2, R), \qquad X' = F(S_1, S_2', R')$$

The following result, due to W. Craig (ref. 1) is a generalization of Beth's theorem on definitions

2.1. THEOREM. *Suppose that*

2.2. $$\vdash X \wedge X' \supset [(\forall x_1) \cdots (\forall x_n)[R(x_1, \ldots, x_n) \equiv R'(x_1, \ldots, x_n)]]$$

Then there exists a predicate (*wff*) $Q(x_1, \ldots, x_n)$ *whose set of extralogical constants belongs to* S_1 *such that*

2.3. $$\vdash X \supset [(\forall x_1) \cdots (\forall x_n)[R \equiv Q]].$$

Next, consider an elementary language N whose extralogical constants are the relation of identity, $I(x, y)$ — i.e. $x = y$ —, the individual constant 0, and the functions $\sigma(x, y)$, $\pi(x, y)$ — i.e. $x+y$, xy —, as well as the successor function, which will be denoted by $\phi(x)$. A familiar procedure of arithmetization associates with every atomic symbol, term, wff, and sentence of N a Gödel number such that the set S of Gödel numbers of sentences of N is definable in N i.e. is given by a predicate of N (and is, moreover, recursive). Let V be the set of Gödel numbers of sentences of N which are true for the natural numbers. Then

2.4. THEOREM (Tarski). *V is not definable in N.*

Proof. Let x, y and z be three arbitrary but definite variables in N. Let P be the set of predicates $R(x, y)$ in N which contain no free variables other than x and y and which do not contain z. Let Q be the set of predicates $S(x, y)$ which are of the form

2.5. $$R(x, y) \wedge (\forall z)[[(\exists w) I(\sigma(z, w), y)] \supset \sim R(x, z)]$$

where $R(x, y)$ belongs to P. Thus $S(x, y)$ states, in a different notation, that

$y = \mu z R(x, z)$. The relation between the Gödel number of any element of P and the corresponding element of Q is recursive.

By the *length* of a formula $R(x, y)$ we mean the number of its atomic symbols, including brackets, connectives, quantifiers, variables, and extralogical constants, separate occurrences being counted separately. The predicate $A(w, t)$, "w is the Gödel number of some formula and t is the length of that formula", is recursive and may be represented as a predicate in N. The same applies to the predicate $B(w, t)$, "w is the Gödel number of some formula whose length does not exceed t". Finally, let $C(u, v, w, t)$ be the predicate "w is the Gödel number of some element $R(x, y)$ of P and u and v are natural numbers, and t is the Gödel number of the formula obtained by substituting the numerals of u and v in the predicate $S(x, y)$ which is given by 2.5." Then $C(u, v, w, t)$ is again recursive and may be taken to be formalized as a predicate in N.

Let M_0 be the system of (non-negative) natural numbers, and suppose, contrary to the assertion of Theorem 2.4 that the set of Gödel numbers of sentences of N which are true in M_0 is given by a predicate $T(z)$ in N. Then the following sentence is true in M_0,

2.6. $$(\forall u)(\forall s)(\exists v)(\forall w)(\forall t)[B(w, s) \wedge C(u, v, w, t) \supset \sim T(t)]$$

This sentence states, informally, that for any given natural numbers u and s there exists a natural number v such that for all elements $R(x, y)$ of P whose length does not exceed s, $z = v$ is not the smallest number which satisfies $R(u, z)$. The truth of this sentence in M_0 follows immediately from the fact that the number of elements of P of length not greater than a given natural number, is finite.

Let M be any non-standard strong model of arithmetic, i.e. a model of V which is a proper extension of M_0. Let a be an infinite (non-standard) element of M. Then the intersection of all elementary submodels of M which include a is known to be a strong non-standard model of arithmetic, which will be denoted by $M_0(a)$ (ref. 4). For every element b of $M_0(a)$ there exists a predicate $R_b(x, y)$ of P such that for the corresponding element $S_b(x, y)$ of Q (see 2.5 above), the sentence $S_b(\mathbf{a}, \mathbf{b})$ holds in $M_0(a)$. (Note that the passage from a number to the corresponding numeral is indicated by bold face print.)

Now let c be any infinite element of $M_0(a)$. Since 2.6 holds in $M_0(a)$ it follows that the sentence

2.7. $$(\exists v)(\forall w)(\forall t)[B(w, \mathbf{c}) \wedge C(\mathbf{a}, v, w, t) \supset \sim T(t)]$$

also holds in that structure. Hence, for some element b of $M_0(a)$,

2.8. $$(\forall w)(\forall t)[B(w, \mathbf{c}) \wedge C(\mathbf{a}, \mathbf{b}, w, t) \supset \sim T(t)]$$

Let $R_b(x, y)$ be the predicate introduced above. The length of $R_b(x, y)$ is a natural number in the ordinary sense, d say, which is smaller than c in $M_0(a)$. Let n be the Gödel number of $R_b(x, y)$ then $B(\mathbf{n}, \mathbf{c})$ is true in $M_0(a)$, and the same applies to the sentence

$$(\forall t)[B(\mathbf{n}, \mathbf{c}) \wedge C(\mathbf{a}, \mathbf{b}, \mathbf{n}, t) \supset \sim T(t)],$$

in view of 2.7, and hence to

2.9. $$(\forall t)[C(\mathbf{a}, \mathbf{b}, \mathbf{n}, t) \supset \sim T(t)]$$

On the other hand, let m be the Gödel number of $S_b(\mathbf{a}, \mathbf{b})$ then the sentence $[C(\mathbf{a}, \mathbf{b}, \mathbf{n}, \mathbf{m}) \wedge T(\mathbf{m})]$ holds in $M_0(a)$, by the definition of $R_b(x, y)$. It follows that

2.10. $$(\exists t)[C(\mathbf{a}, \mathbf{b}, \mathbf{n}, t) \wedge T(t)]$$

holds in $M_0(a)$ and this contradicts 2.9. The proof of 2.4 is now complete.

While the above method involves as much arithmetization as any other proof, we were able to avoid the introduction of diagonal arguments by the use of non-standard models. Nevertheless, it may be said that there is a certain relation between our method and the Berry paradox. Essentially, the scope of the method includes Gödel's incompleteness theorem since this is a consequence of Tarski's theorem, given certain ancillary considerations.

Let $W(x)$ be a predicate of N which defines the set of Gödel numbers of sentences of N, and let M be any non-standard strong model of arithmetic. Since $W(x)$ is satisfied in M_0 by an infinite and hence unbounded set of elements, the same is true in relation to M. It follows that $W(x)$ holds in M for certain infinite elements. Now there exists a recursive predicate $S(x, y, t)$, "t is the Gödel number of the yth element of the wff whose Gödel number is x", which may be taken to be formalized within N. The following sentence then holds in M_0 and hence in M.

2.11. $$(\forall x)(\forall y)(\forall z)(\exists w)[A(x, y) \wedge Q(z, y) \supset S(x, z, w)].$$

In this sentence, $A(x, y)$ signifies "y is the length of the formula whose Gödel number is x", as before, and $Q(z, y)$ has been introduced to denote the order relation $(\exists v)E(\sigma(z, v), y)$ or, briefly, $z \leq y$. Since there is no uniform bound to the length of the wff of N there exist infinite elements a and b of M such that the sentence $A(\mathbf{a}, \mathbf{b})$ holds in M. This sentence states, formally, that a is the Gödel number of a wff of length b. This is something more than a purely formal statement since we can actually recover the ordered set of "Gödel numbers of atomic symbols" which constitute the wff defined by a. Indeed, 2.11 shows that for any element c of M which does not exceed b there exists an element d of M which is the "Gödel number of the cth element of a". If d is finite then it corresponds to one of the original atomic symbols and there is no need to use quotation marks. This will certainly be the case when c is the Gödel number of a connective, or of a bracket, or of a quantifier. However, if d is infinite then we have to regard it as the Gödel number of a new atomic symbol. In order to realize these ideas systematically, it seems advantageous not to limit oneself to the language N, which describes arithmetic but to carry out the extension of the original finitary language within a more general framework. This will be done in the next section.

3. Construction of a non-standard language. Let U be a set of individuals, of a cardinal which is greater than or equal to $2^{\aleph_0}$, and otherwise arbitrary (unless and until it is defined more closely for a particular purpose). Suppose that distinct symbols have been assigned to all individuals of U, to all relations on U, and to all functions defined on U. Without fear of confusion we denote the corresponding structure again by U. In particular, the symbol $I(x, y)$ will be taken to correspond to the relation of identity in U. Let K be the set of all sentences of the Lower predicate calculus which are formulated in terms of these extralogical constants and which hold in U. Select a countable subset U_0 of U, and let $R_0(x)$ be the relation which determines U_0, i.e. which holds in U precisely for the elements of U_0. Let a be an individual constant outside the vocabulary of K. Consider the set $K' = K \cup R_0(a) \cup \{\sim I(a, b_\nu)\}$ where b_ν varies over all elements of U. Then a familiar argument shows that K' is consistent and hence posesses a model, U' say. U' is a proper extesion of U since it contains $\boldsymbol{a}$ (or an element denoted by $\boldsymbol{a}$), and a is different from the elements of M_0 since it satisfies $\{\sim I(\boldsymbol{a}, b_\nu)\}$ and from the elements of $U-U_0$ since it satisfies $R_0(\boldsymbol{a})$. It is not difficult to see that $I(x, y)$ may be supposed to denote the identity also in U'.

U' is a model of K. To every set and relation in U there corresponds a set or relation in U' which satisfies the same predicates in the language of K. Let $R(x)$ be a predicate of this kind and let S and S' be the sets determined by $R(x)$ in U and U' respectively. If S is empty then S' also is empty since the sentence $(\forall x)[\sim R(x)]$ belongs to K. Suppose next that S contains a finite and positive number of elements, $n>0$. Then the sentence $(\exists x_1)\cdots(\exists x_n)(\forall y)$ $[R(x_1)\wedge\cdots\wedge R(x_n)\wedge R(y)\supset I(x_1, y)\vee\cdots\vee I(x_n, y)]$ holds in U and hence in U'. Thus S' cannot contain more than n different elements. Since S' is an extension of S it follows that S' coincides with S.

Suppose next that S is precisely countable. Then there exists a relation $J(x, y)$ which defines a one-to-one correspondence between S and U_0 in U. Let U_0' be the set defined by $R_0(x)$ in U'. Then $J(x, y)$ defines a one-to-one correspondence between S' and U_0' in U'. In particular, for some element b of S', $J(b, \boldsymbol{a})$ holds in U'. But a does not belong to U_0 and so b cannot belong to S'. Thus, S' is a proper extension of S.

Finally, if S is an arbitrary infinite subset of U then it contains a countable subset, S_0 say. If S' and S_0' are the corresponding subsets of U then $S'\supseteq S_0'$ and $S_0'-S_0$ is not empty. But $S_0'\cap S=S_0$ and so $S'-S\supseteq S_0'-S_0$, i.e. S' is a proper extension of S.

Summing up we find that infinite sets are extended on passing from U to U' while finite subsets remain unchanged. Now let M be any structure whose domain of individuals, D, is a subset of U. The relations and functions of M are not in themselves relations and functions of U since they are defined only on D. However any function, e.g. $\phi(x, y)$ which is defined on D can be extended — usually in many different ways — to a function $\phi_1(x, y)$ which is defined on U. A similar statement applies to the relations which are defined on D. Suppose then that for every relation and function of M we have selected an extension to the whole of U. Let $R(x)$ be the relation which defines D in U. For every sentence X which is defined in M we may then construct the sentence X_1 which is obtained from X by replacing every (symbol for a) relation or function in M by the (symbol for) the corresponding extension to U and by relativizing the result with respect to $R(x)$ (for the operation of relativization, see e.g. ref 7). Then X_1 holds in U if and only if X holds in M.

Let H be the set of all sentences X which are defined and hold in M and let H_1 be the set of corresponding sentences X_1. Then $H_1 \subseteq K$ and so the sentences of H_1 hold in U'. Let D' be the subset of U' which is determined by $R(x)$ in U'. By restricting the relations and functions of H_1 in U' to D' we then obtain a structure M' which is an extension of M and a model of H. M' is thus an elementary extension of M. It is actually independent of the particular choice of the extended relations and functions.

Select a countable subset D_0 of U and choose functions $\sigma(x, y)$ and $\pi(x, y)$ (i.e. $x+y$ and xy) on D_0 in such a way that D_0 is turned into a standard model of arithmetic M_0 ("the" natural numbers). Let M_0' be the corresponding structure in U'. Then M_0' also is a strong model of arithmetic and since the underlying domain of individuals D_0' must be a proper extension of D_0, M_0' is a non-standard strong model of arithmetic.

We now suppose that certain subsets of U are regarded as the constituents of a language L of the first order predicate calculus. That is to say, there are certain disjoint sets of individuals of U, of adequate cardinal numbers, which serve as brackets, commas, connectives ($\sim$, $\wedge$, and $\vee$), quantifiers ($\forall$ and $\exists$), variables, individual constants, relations, and functions of L. Since they are individuals of U, the relations of L will, in general, be distinct from the symbols which denote relations of U. Moreover, the relations of L will certainly be different from the relations of U since the latter are sets of n-tuples of elements of U. Accordingly, we shall call these individuals of U, *L-relations*, and we shall similarly refer to L-variables, L-connectives, etc.

Going farther, we suppose that the terms and well-ordered formulae (wff) of L also constitute subsets of U ("L-terms", L-wff"). To regard terms and wff, including sentences, as individuals in a certain domain is entirely in keeping with the axiomatic approach to the syntax of a formal language. Thus, we have in U one-place relations $Q_v(x)$, "x is an L-variable", $Q_c(x)$, "x is an L-connective", $Q_f(x)$, "x is an L-wff", $Q_s(x)$, "x is an L-sentence", etc. Various connections which exist between these relations are then expressed by sentences of K, for example

$$(\forall x)[Q_s(x) \supset Q_f(x)].$$

Again, there exists a relation $S(x, y, z)$ in U which states that x is an L-wff and y is a natural number (an element of D_0) and z is the yth atomic symbol (L-bracket, L-relation, L-variable, etc.) in X. Then the following sentences belong to K.

3.1. $$(\forall z)(\forall w)[S(x, y, z) \wedge S(x, y, w) \supset I(z, w)],$$

3.2. $$(\forall x)(\forall y)[[(\forall u)(\forall v)[S(x, u, v) \equiv S(y, u, v)]] \supset I(x, y)].$$

The second sentence involves the tacit assumption, which is usually taken for granted, that a sentence is determined completely by its sequence of atomic symbols. We note that any discussion of the question how to relate different occurrences of the same symbol is entirely redundant in the present approach.

Passing to U', we see that the relations which in U define the various sets of L-symbols and L-formulae, define corresponding sets in U'. If the original sets are finite, as is the case for the connectives and for the quantifiers, then they remain unchanged on passing to U'. If the original set is infinite, as is the case for the sentences, then the corresponding set in U' is a proper extension of it. The extended language will be denoted by L' and, accordingly, we shall refer to its variables, connectives, and so on, as L'-variables, L'-connectives, etc. Now let F and F' be the sets of L-wff and of L'-wff, respectively. Then $F \subseteq U$, $F' \subseteq U'$ and F is a proper subset of F'. Every element a of F' determines a certain ordered set of atomic symbols in the sense that for every element n of D' (i.e. for every natural number in U') which does not exceed a particular $m \in D'$ — the *length* of a — there is a unique L'-atomic symbol b such that $S(a, n, b)$ holds in U', where $S(x, y, z)$ is the relation introduced above ("b is the nth atomic symbol of a"). We know that such m and b exist because this fact is expressed by an elementary sentence which holds in U.

The set of L'-wff and more particularly, of L'-sentences, is quite varied. Thus, for every non-standard natural number l in U' there exists an L'-sentence whose length exceeds l. That this is so can be deduced immediately from the existence of sentences of unbounded length in U. Again, if L includes relations of n places for all natural numbers n, then L' will include relations whose number of places, n, is infinite, more precisely n is an infinite element of M_0'.

On the other hand, we cannot assign the elements of an infinitely long wff in the present sense arbitrarily even in cases where this would appear to yield a formula which is intuitively meaningful, as will be shown presently.

We recall that the order type of any non-standard model of arithmetic can be expressed in the form $\omega + ({}^*\omega + \omega)\theta$ where θ is some dense order type without first or last element (compare ref. 2). It follows that if m is an infinite element of M' then the ordered set of elements of M' which are smaller than or equal to m is of order type $\omega + ({}^*\omega + \omega)\theta' + {}^*\omega$ where θ' is again dense without first or last element. Accordingly the ordered sequence of any L'-wff which is of infinite length possesses an order type of this kind.

To continue, we require some simple facts concerning finite partial sequences. By a finite partial sequence σ we mean any function whose domain is a finite set of natural numbers and whose range is included in a given set, G. By the length of σ we mean the greatest element in the domain of σ *plus* 1. The length of the empty sequence (sequence with empty domain) shall be zero, by definition.

The number of non-empty finite partial sequences with range in U is equal to the cardinal of U. Accordingly these sequences can be indexed in U in the following sense. There exists a subset J of U and a three place relation $N(x, y, z)$ on U such that x is an element of J and y is a natural number (an element of M_0) and such that the following conditions are satisfied

3.3. $(\forall x)(\forall y)(\forall z)(\forall w)[N(x, y, z) \wedge N(x, y, w) \supset I(z, w)]$.

3.4. $(\forall x)(\exists y)(\forall z)(\forall w)[N(x, z, w) \supset Q(z, y)]$.

3.5. $(\forall x)(\forall y)[[(\forall z)(\forall w)[N(x, z, w) \equiv N(y, z, w)]] \supset I(x, y)]$.

Of these, 3.3 states that, for given x, $N(x, y, z)$ defines z as a function of y, 3.4 affirms that the domain of any such function is bounded by a natural number y (where $Q(z, y)$ denotes the relation $z \leq y$ between natural numbers, as before) and 3.5 ensures that no two different elements of J define the same function.

Morever, we suppose that as x varies over the elements of J *all* finite partial sequences with values in U occur as functions defined by $N(x, y, z)$. It is evident that this condition can be satisfied (in many ways) by a suitable choice of J and N, but it does not correspond to any sentence in K.

Suppose that J is determined by a relation $Q_U(x)$ in U. Let G be any subset of U then we may obtain all finite partial sequences with values in G by restricting J to a suitable subset, J_G say. Let $Q_G(x)$ be the relation which determines J_G then we may define a corresponding three place relation $N_G(x, y, z)$ by

3.6. $$N_G(x, y, z) = Q_G(x) \wedge N(x, y, z).$$

We note that as a consequence of our stated assumptions the sentence

3.7. $$(\forall x)[[(\exists y)(\exists z)N(x, y, z)] \equiv Q_U(x)]$$

belongs to K. (We have excluded the empty sequence from the indexing, otherwise the equivalence in 3.7 would have to be replaced by an implication.) Accodingly, J is determined completely by $N(x, y, z)$. Similarly, for any $G \subseteq U$, J_G is determined by N_G.

Passing to U', we find that $Q_U(x)$ determines a subset J' of U' which contains J as a proper subset. Any element a of J' then defines a partial sequence with domain in M_0' by means of the predicate $N(a, y, z)$. Indeed, this predicate defines z as a function of y in view of 3.3, such that the arguments are in M_0' by virtue of an earlier condition. Moreover, in view of 3.4, the domain of the sequence is bounded by some element of M_0'. Any function which is defined in this way by some $N(a, y, z)$ will be called a pseudo-finite partial sequence. Notice that the set of pseudo-finite partial sequences in U' does not depend on the particular choice of $N(x, y, z)$ (which implies a definite choice for J). Indeed, suppose that we have chosen two different relations on U, $N_1(x, y, z)$ and $N_2(x, y, z)$, with corresponding index sets J_1 and J_2 in such a way that all the conditions laid down for N and J are satisfied by N_1 and J_1 and by N_2 and J_2, respectively. Let $Q_1(x)$ and $Q_2(x)$ determine the sets J_1 and J_2, respectively. Then if a belongs to J_1, $N_1(a, y, z)$ defines a non-empty finite partial sequence in U. But this sequence must occur also for some value of the first argument in $N_2(x, y, t)$. Accordingly, the sentences

$$(\forall x)[Q_1(x) \supset [(\exists w)[(\forall y)(\forall z)[N_1(x, y, z) \equiv N_2(w, y, z)]]]]$$

and

$$(\forall x)[Q_2(x) \supset [(\exists w)[(\forall y)(\forall z)[N_2(x, y, z) \equiv N_1(w, y, z)]]]]$$

hold in U and hence belong to K. Thus, the two sentences hold also in U' and this shows that the sets of pseudo-finite partial sequences defined by N_1 and N_2 coincide. The set of non-empty finite partial sequences in U will be denoted by Σ and the set of non-empty pseudo-finite partial sequences in U' by Σ'. We shall imagine that the index set J has been fixed once and for all. A sequence will be said to be total if its domain contains with every number also all smaller numbers. Such sequences form subsets Σ_0 and Σ_0' of Σ and Σ' respectively.

Now let T be the set of L-sentences, $T \subseteq U$, and let Σ_T be the subset of Σ_0 whose elements take values in T, i.e. which are finite partial sequences of L-sentences. Let $J_T \subseteq J$ be the corresponding index set. Then the rule which assigns to every non-empty finite total sequence of L-sentences, $\sigma = (p_0, \ldots, p_n)$ the L-sentence

3.8. $$q = [p_0 \wedge [p_1 \wedge [p_2 \wedge \cdots \wedge p_n] \cdots]$$

defines a function ψ on Σ_T, where $q = \psi(t)$ if t is the index of σ in J. ψ can be represented by a relation $Q\psi(x, y)$ which holds if x is the index of the sequence $\sigma = (p_0, \ldots, p_n)$ and y is given by 3.8. Passing to U' we see that $Q\psi(x, y)$ assigns a sentence also to every non-empty pseudo-finite total sequence.

We note that if in 3.8 we count the symbols from left to right, beginning with 0, then the places with index $\equiv 0(3)$ are filled with left brackets, those with index $\equiv 1(3)$ by sentences, and those with index $\equiv 2(3)$ by the connective of conjunction, up to the place with index $3n-1$. The place with index $3n$ is filled by the last sentence, p_n. This is followed by n right brackets. Thus, the length of q is $4n+1$.

Let n be an infinite natural number in M_0' and let a and b be two distinct L-sentences. It would then appear to be intuitively possible to define a particular L'-sentence r of length $4n+1$ which is a repeated conjunction of the sentences a and b simply by substituting in s and q (see 3.8) the sentence a for p_i, i finite and the sentence b for p_i, i infinite, $i \leq n$. This is indeed a well-defined ordered set of atomic symbols, of length $4n+1$ as we can see from the detailed description given in the preceding paragraph. However, it is not an L'-sentence. For if r were an L'-sentence then the sentences which fill the places with index $3k+1$ in r, $0 \leq k \leq n-1$ would constitute a pseudo-finite total sequence ρ of length n such that the elements of the sequence are equal to a for all finite subscripts and to b for all infinite subscripts. Such a sequence ρ cannot belong to Σ_0'. For, using an argument which can easily be formalized, ρ satisfies the condition that its 0th element is a and that whenever an element of ρ equals a the next element, if any, is also equal to a. Hence by the axiom of induction which holds in U', if $\rho \in \Sigma_0'$ then all elements of ρ are equal to a, contrary to construction. Thus, there is no L'-sentence of the kind described above.

The deductive calculus of the language L' will be discussed only briefly. We take for granted the deductive calculus of the language L. More particularly, we take it that there exists a subset T_0 of the set of sentences T which is constituted by the *L-axioms*. An L-proof is a non-empty finite total sequence of L-sentences such that every element of the sequence is either an L-axiom of it is deducible from one or two earlier elements of the sequence as an ***immediate consequence*** in accordance with the usual rules of the lower predicate calculus. Thus, the proofs constitute a subset Σ_P of Σ_T. Let J_P be the subset of J whose elements induce the elements of Σ_P and let $Q_P(x)$ be the relation on U which determines J_P.

An L-sentence is provable if it is an element of some L-proof. Thus, "z is provable" corresponds to the predicate $(\exists x)(\exists y)[Q_P(x) \wedge N(x, y, z)]$. The set of provable sentences in U will be denoted by T_P.

More generally, if A is a set of L-sentences in U, which is given by a relation $Q_A(x)$, then we define ***an A-proof*** as a non-empty finite total sequence of L-sentences such that every element of the sequence is either an L-axiom or an element of A or it is deducible from one or two earlier elements of the sequence as an immediate consequence. For given A, the A-proofs constitute a subset Σ_A of Σ_T. Let J_A be the subset of J whose elements index the elements of Σ_A and let $Q_A(x)$ be the relation on U which determines J_A. An L-sentence is A-provable if it is an element of some A-proof. Thus, "z is A-provable" is given by $(\exists x)(\exists y)[Q_A(x) \wedge N(x, y, z)]$. The set of A-provable sentences will be denoted by T_A.

The predicate $Q_P(x)$ defines in U' a set of L'-sentences, T_P' which is an extension of T_P. The elements of T_P' will be said to be L'-provable. An L-sentence which is not L provable is not L'-provable either. On the other hand if T_0' is the set of L'-axioms (corresponding to the set T_0 in U) then it can be shown that not all sentences which are L'-provable are deducible from

T'_0 by the rules of the lower predicate calculus. Indeed, since for any natural number n there exist L-provable sentences which cannot be proved in less than n steps, or more precisely which do not occur in proofs of length $<n$, it follows that there exist L-provable sentences which do not occur in any proof of finite length.

More generally, let A' be any set of L'-sentences. Then we define the set T'_A of L'-sentences which are A'-provable in the following way. Among the subsets of B' of A' there are some which are defined by relations that occur already in U, i.e. which correspond to sets of L-sentences B in U. Among these subsets B' of A' there are, for example, all finite subsets of A'. For any B which corresponds to such a B' there exists a relation $R_B(z)$ which determines the B-provable sentences in U. $R_B(z)$ determines a set T'_B in v'. Then $T'_{A'}$ is by definition the union of all such sets. In particular, if the original A' is itself a set which corresponds to a set of L-sentences, A, then $T'_{A'}$ corresponds, in U' to the set of A-provable sentences in U.

4. Semantics of a non-standard language. In the remainder of this paper, we shall be concerned with the semantics of the non-standard language L' in an extension U' of U. For this purpose, we first have to discuss the notion of a truth definition in the standard language L within the framework of the set U.

Let M be a structure whose domain of individuals, D, is a subset of U, with a set of relations $P(M)$ and a set of functions $F(M)$. Let $R(x)$ be the relation which determines D in U. As discussed in detail at the beginning of the preceding section, the elements of $P(M)$ and of $F(M)$ are not, in general, defined on U, but the sentences about M which involve the elements of $P(M)$ and $F(M)$ can be related to sentences about U by means of the operation of relativization with respect to $R(x)$.

We now wish to use the language L in order to describe the structure M. For this purpose, we first introduce correspondences between the individuals of D and some L-individuals; between the relations of $P(M)$ and some L-relations; and between the functions of $F(M)$ and some L-functions. It will be assumed that a sufficient number of L-symbols is available for this purpose.

Let $Q_{ic}(x)$ be the relation which determines the set of L-individuals in U, and suppose that a correspondence between the elements of D and a set of

L-individuals is provided by a relation $C(x, y)$ on U which satisfies the following conditions.

4.1. $(\forall x)(\forall y)[C(x, y) \supset R(x) \wedge Q_{ic}(y)]$,

4.2. $(\forall x)(\forall y)(\forall z)[C(x, y) \wedge C(x, z) \supset I(y, z)]$,

4.3. $(\forall x)(\forall y)(\forall z)[C(y, x) \wedge C(z, x) \supset I(y, z)]$,

4.4. $(\forall x)(\exists y)[R(x) \supset C(x, y)]$.

The sentences 4.1-4.4 state, briefly, that $C(x, y)$ establishes a one-to-one correspondence between D and a set of L-individuals. The set of L-individuals which corresponds to individuals of M will be denoted by G. It is given by the predicate $(\exists x)C(x, y)$.

The correspondence between the elements of $R(M)$ and a set of L-relations is expressed in a way which is rather different. We suppose that for every element $R(x_1, \ldots, x_n)$ of $P(M)$, $n \geq 0$, we are given an L-relation r and an $(n+1)$-place relation on U, $S_R(y_0, y_1, \ldots, y_n)$ such that different L-relations correspond to different elements of $R(M)$ and such that the sentences

4.5. $$(\forall x_1)\cdots(\forall x_n)(\forall y_1)\cdots(\forall y_n)[C(x_1, y_1) \wedge \cdots \wedge C(x_n, y_n) \supset [R(x_1, \ldots, x_n) \equiv S_R(r, y_1, \ldots, y_n)]]$$

all belong to K. Let G_P be the set of L-relations which are employed in this way.

Similarly, we suppose that there is a one-to-one correspondence between the elements of $F(M)$ and a certain subset G_F of the set of L-functions such that for every function $\phi(x_1, \ldots, x_n) \in F(M)$ and the corresponding $g \in G_F$ there exists an $(n+2)$-place relation $S_\phi(y_0, y_1, \ldots, y_{n+1})$ on U for which

4.6. $$(\forall x_1)\cdots(\forall x_n)(\forall x_{n+1})(\forall y_1)\cdots(\forall y_n)(\forall y_{n+1})[C(x_1, y_1) \wedge \cdots \wedge C(x_n, y_n) \wedge C(x_{n+1}, y_{n+1}) \supset [I(\phi(x_1, \ldots, x_n), x_{n+1}) \equiv S_\phi(g, y_1, \ldots, y_n, y_{n+1})]]$$

belongs to K.

We do not exclude the possibility that the set of L-individuals, G, coincides with the set of individuals of M, D, and that at the same time, the relation $C(x, y)$ coincides with the restriction of the relation of identity $I(x, y)$ to D. Even if this is not the case from the outset we may, if we wish, define a structure M^* isomorphic to M which possesses that property. Thus, if R is a relation of M then we choose the corresponding relation of M^* on the set of individuals G as the relation which satisfies

$$4.7. \quad (\forall x_1)\cdots(\forall x_n)(\forall y_1)\cdots(\forall y_n)[C(x_1, y_1)\wedge\cdots\wedge C(x_n, y_n) \equiv [R(x_1, \ldots, x_n) \supset R^*(y_1, \ldots, y_n)]].$$

In other words, $R^*(y_1, \ldots, y_n)$ is coextensive with the predicate $S_R(r, y_1, \ldots, y_n)$. The corresponding functions of M^* may be defined in a similar way.

We also observe that while the correspondences between the elements of D, $R(M)$, and $F(M)$, and G, G_R and G_F, respectively, have to be fixed in order to make the semantical interpretation of an L-sentence in M definite, the particular choice of S_R and S_ϕ is irrelevant.

Let G_T be the set of all L-sentences whose atomic symbols include no extralogical constants other than those belonging to $G \cup G_R \cup G_F$. Since we are at the moment dealing with the language L which is based on standard arithmetic it is not difficult to reformulate the usual procedure for the determination of the truth or falsehood of a sentence of G_T with respect to the structure M within the present framework.

Employing the usual terminology, we call a sentence atomic if it does not involve any connectives or quantifiers and hence does not involve any variables. Disregarding the brackets and commas, an atomic sentence consists of a relation which is followed by a sequence of terms. The terms themselves are not atomic symbols but are given by sequences of functions and of individuals combined according to certain definite rules. Let the atomic L-sentence $a \in G_T$ be indicated by the expression $r(t_1, \ldots, t_n)$ where r is an L-relation and $t_1, \ldots, t_n$ are L-terms. By virtue of 4.6 every term t_i corresponds to an individual b_i of D and, again by means of the correspondence r itself corresponds to a relation R. We now ask whether $R(b_1, \ldots, b_n)$ holds in M. As it stands this question does not refer directly to the structure U, but as explained earlier all sentences about M can be reduced to sentences about U. Thus, finally, we say that a holds in M, or is true in M if and only if $R(b_1, \ldots, b_n)$ holds in M. The truth values of all other sentences of G_T in M are now determined by means of the conditions that a negation is true in M iff the original L-sentence is not true in M, a disjunction is true in M iff at least one of its disjuncts is true, etc. together with the usual conditions for the quantifications. Note that the connection between L-sentence and its negation, or of two L-sentences and their disjunction is given by two and three place relations defined on U, e.g. $N(x, y)$ "x, y are L-sentences and y is the negation of x". The relation between x and y can be analyzed further by means of the strings of atomic symbols which define any $\bar{L}$-sentence but this is not necessary for our present purpose. Let the set of all L-sentences which hold in M according to this definition be denoted by G_M and let $Q_T(x)$ and $Q_M(x)$ be the relations of U which determine G_T and G_M, respectively.

On passing to U', the structure M is transformed into a structure M' which is an extension of M and that all sentences (in the ordinary sense) which are defined and hold in M, hold also in M' (i.e. M' is an elementary extension of M). We now *define* that the set of L'-sentences which hold in M' *according to the internal truth definition* (or briefly, *internally*) is the set G'_M which is determined by the relation $Q_M(x)$ in U'. Among the elements of G'_M are the L-sentences of G_M which correspond in a natural way to the sentences (in the ordinary sense) that hold in M. Beyond that, G'_M contains also a wide variety of additional sentences including sentences of infinite length. All the ordinary rules of semantics which were mentioned above still hold in the present case. Thus, the negation of an L'-sentence holds in M' iff the sentence does not hold in M', a conjunction holds in M' iff both conjuncts hold in that structure, a disjunction holds in M' iff at least one of the disjuncts holds in M'. If $q(x)$ is an L'-wff with a single free variable and with extralogical constants in C' C'_P C'_F then $(\exists x)q(x)$ (or, more precisely, the L'-sentence which is obtained from $q(x)$ by existential quantification) holds in M' iff $q(a)$ holds in M' for at least one $a \in b'$ and $(\forall x)q(x)$ holds in M' iff $q(a)$ holds in M' for all $a \in b'$.

Now let G_Q be the set of L-predicates (L-wff with *at least* one free variable) whose extralogical constants belong to $G \cup G_P \cup G_F$. Such an L-predicate is said to be *defined* in M. Let $(x_0, x_1, \ldots, x_n)$ be the finite total sequence of variables of an element q of G_Q, taken in their order of first occurrence from left to right. Let $\Sigma_0(M)$ be the set of all finite total sequences in U whose elements belong to M and let H_M be the subset of J which indexes the elements of $\Sigma_0(M)$. Let $R_Q(x)$ and $H(x)$ be the relations which determine the sets G_Q and H_M in U.

Then the set G'_Q which is determined by $R_Q(x)$ in U' is precisely the set of all L'-predicates which are *defined* in M', i.e. whose set of extralogical constants belongs to $G' \cup G'_P \cup G'_F$. Similarly, the set $H'_{M'}$ defined by $H(x)$ in U' indexes the set $\Sigma'_0(M')$ of all pseudo-finite total sequences in U' whose elements belong to M'.

The condition "x is an L-predicate which is defined in M and y is the index of an element σ of $\Sigma_0(M)$ whose length equals the number of free variables of x, and z is the L-sentence resulting from the substitution of the elements of σ in their given order for the free variables of x" can be expressed by a relation $W(x, y, z)$ in U. Then $W(x, y, z)$ has a corresponding meaning in U', where the number of free variables in a given predicate may now be infinite. In either U or U', if x is an L-(or L'-) predicate which is defined in $M(M')$ and y is the index of an element σ of $\Sigma_0(M)$ (or $\Sigma'_0(M')$) then the corresponding z is an L-(or L'-) sentence which is defined in M (or M'). In particular, if q is an element of G'_Q and the number of variables in q is infinite, then these variables constitute a pseudo-finite total sequence $\xi = (x_0, x_1, \ldots, x_n)$. The substitution of an element $\sigma = (a_0, a_1, \ldots, a_n)$ of $\Sigma'_0(M')$ for ξ yields an L'-sentence which is defined in M'. Conversely, it is not difficult to see that if σ is an ordered set of elements of M' whose order type is the same as that of ξ, and if the substitution of the elements of σ for the corresponding elements of ξ in q yields an L'-sentence then σ belongs to $\Sigma'_0(M')$. For given n, the elements of $\Sigma'_0(M')$ of length $n+1$ are divided into two classes according as the result of the substitution does or does not hold in M'. Accordingly, q defines a kind of infinitary pseudo-relation in M' which is defined for some but not all ordered sets of elements of M' which have the order type of ξ.

5. Skolem operators and the external truth definition. To continue we require some auxiliary considerations which yield a modified version of Henkin's notion of generalized quantifiers (ref. 3). We shall be concerned only with linearly ordered quantification.

Let S be an ordered set which is partitioned into two subsets, S_1 and S_2. (Thus, $S_1 \cup S_2 = S$, $S_1 \cap S_2 = 0$). Let Φ_1 and Φ_2 be two sets of functions which are defined in S_1 and S_2 respectively and which take values in a given set A. By a *Skolem operator* we mean a function ψ whose domain is Φ_1 and whose values belong to Φ_2 (i.e. a mapping from the functions of Φ_1 to functions of Φ_2)

in such a way that if $g = \phi(f)$ and $k = \phi(h)$ and $f(s_1) = h(s_1)$ for all elements s_1 of S_1 which precede any particular $s_2 \in S_2$ in the given ordering of S, then $g(s_2) = k(s_2)$.

Thus, if X is a sentence in prenex normal form in the Lower predicate calculus,

$$X = q_0 q_1 \cdots q_{n-1} Z(x_0, x_1, \ldots, x_{n-1})$$

where, for $i = 0, \ldots, n-1$ the quantifier q_i quantifies the variable x_i, and where Z does not contain any quantifiers, let S be the ordered set $(0, 1, \ldots, n-1)$, and let S_1 and S_2 be the subsets of S whose elements are, respectively, the subscripts of the universal and of the existential quantifiers of X. Suppose that X holds in a structure M with domain of individuals A. Then, corresponding to every existential quantifier in X, we may define a Skolem (or Herbrand) function of the preceding universally quantified variables in such a way that the introduction of these functions into Z yields a true sentence for all values of the remaining arguments of Z. For example, if $X = (\forall x_0)(\exists x_1)(\forall x_2)(\exists x_3) Z(x_0, x_1, x_2, x_3)$ holds in M then there are Skolem function $\phi(x_0)$, $\phi_3(x_0, x_2)$, such that $Z(x_0, \phi_1(x_0), x_2, \phi_3(x_0, x_2))$ holds in M for all values of the arguments x_0, x_2 in A. Defining $S = (0, 1, 2, 3)$, $S_1 = (0, 2)$, $S_2 = (1, 3)$, Φ_1 as the set of all functions on 0, 2 and taking values in A, Φ_2 as the set of all functions on 1, 3 and taking values in A, we may regard the mapping $(x_0, x_2) \to (\phi_1(x_0), \phi_3(x_0, x_2))$ as a Skolem operator. The same is true for general X, as above. Conversely (at first disregarding the question whether the substitution of the Skolem functions makes the matrix true universally) any Skolem operator from the set of functions of elements of A defined on S_1 into the set of functions of elements of A defined on S_2 yields a set of Skolem functions. Accordingly, we may say that a sentence X as above holds in a structure M if and only if there exists a Skolem operator which satisfies the condition just described.

We now consider these notions in connection with the language L. Let J be an index set for all partial finite sequences in U as before, except that in the present section we shall for convenience suppose that J includes an index for the empty sequence. Let a_1 and a_2 be indexes for two disjoint finite total sequences of natural numbers (elements of M_0) σ_1 and σ_2. The number of such pairs, (σ_1, σ_2), is countable. Let Φ_i be the set of all finite partial sequences whose domain is σ_i and whose values are L-individuals, $i=1, 2$. Then the cardinal of both Φ_1 and Φ_2 is not greater than (and, except for empty σ_i is equal to) the cardinal of the set of all L-individuals, λ say. We shall now suppose that the cardinal of U is not less than 2^λ. Since the number of Skolem operators from Φ_1 to Φ_2 does not exceed 2^λ, it follows that there exists a set J_ψ in U which indexes the Skolem operators. Various relations may now be associated with J_ψ. For example, there exists a relation $A_\psi(x, y, z, u, v, w)$ which signifies "x and y are indexes of disjoint finite total sequences of natural numbers as above, σ_1 and σ_2 say, z is the index of a Skolem operator as defined, from sequences with domain σ_1 and values in the set of L-individuals to sequences with domain σ_2 and values in the set of L-individuals, u is the index of a particular partial sequence of L-individuals with domain σ_1, v is a natural number which is an element of σ_2, and w is the value of $\psi(t)$ for the argument v."

Let X be any sentence. By a ***prefix*** of X we mean any sequence of quantifiers Q such that $X=QZ$ where Z may contain further quantifiers. We may consider the same notion also for L-sentences. Accordingly, there exists a relation $Q_{pr}(t, x, y)$ which has the following significance. "t is an L-sentence which possesses a prefix π such that x is the index of the sequence of (finite) ordinal numbers which correspond to universal quantifiers in π and y is the index of the sequence of ordinal numbers which correspond to existential quantifiers in π." Thus if t begins with four quantifiers of which the 0th and 2nd are universal and the 1st and 3rd are existential as in the particular example given above then x is the index of the sequence $(0, 2)$ and y is the index of the sequence $(1, 3)$. Then the following relation is closely connected with A_ψ. "t is an L-sentence and $Q_{pr}(t, x, y)$ holds, and z is the index of a Skolem operator ψ from sequences with domain σ_1 to sequences with domain σ_2. Moreover, u is the index of a particular finite partial sequence of L-individuals τ, and v is the L-sentence resulting from the substitution of the elements of τ for the variables corresponding to σ_1 and of the elements of $\psi(\tau)$ for the variables which correspond to σ_2, in the L-wff which results from the deletion of π." We denote this relation by $B_\psi(t, x, y, z, u, v)$. Observe that in the present context we find it convenient to introduce the Skolem operator as an operation between sequences of L-symbols, and not of individuals of any particular model.

Let G be any sentence of L-individuals. We shall presently make use of a relativized version of B_ψ, to be denoted by $B_{\psi G}(t, x, y, z, u, v)$ which is to hold if $B_\psi(t, x, y, z, u, v)$ holds and if moreover u is the index of a sequence of L-individuals which belong to G, and ψ (with index z) transforms all sequences with elements in G into sequences with elements in G. Let $Q_G(x)$ be the relation which determines G in U, and let $R_G(x, u)$ signify that the values of τ (determined by u) are in G and the domain of τ is σ_1 (given by x).

Let G_T be the set of L-sentences which are defined in a structure M according to some correspondence between L-symbols and individuals, relations, and functors of M as explained in the previous chapter. Suppose that G is the set of L-individuals which correspond to individuals of M, determined by $Q_G(x)$, and let G_M be the set of all L-sentences which hold in M, determined by $Q_M(x)$, as before. Then the following sentence holds in U

5.1. $$(\forall t)(\forall x)(\forall y)[Q_{pr}(t, x, y) \supset [Q_M(t) \equiv [(\exists z)(\forall u)(\forall v)[[R_G(x, u) \wedge B_{\psi G}(t, x, y, z, u, v)] \supset Q_M(v)]]]].$$

5.1 may be regarded as a formal statement of the fact that the truth of a sentence in a given model can be expressed in terms of the existence of appropriate Skolem functions as described in detail at the beginning of this section for the case of a sentence in prenex normal form. Passing to U', we see that 5.1 still expresses a certain connection between a structure M' in U' as introduced in section 3 and the set of L'-sentences which are defined in U' according to given correspondences between L'-symbols and individuals, relations, or functors of M'. However, we have to bear in mind that the sets of sequences which occur in the definition of a Skolem operator in U' are confined to sequences of pseudo-finite character. On the other hand, the notion of substitution retains its meaning.

So far we have relied on the passage from U to U' not only in the formulation of the non-standard language L' but also in its model—theoretic interpretation. We shall now introduce an alternative truth definition which relies directly on the notion of Skolem functions. It will be convenient to formulate this new concept only for a certain subset of the set of L'-sentences whose elements have a relatively simple structure. While it would not be hard to widen the scope of our definition somewhat, it might be difficult to extend it to the set of all L'-sentences.

For given U and U', and a language L' in U', we define subsets of L'-terms, L'-atomic formulae, etc. as follows.

An L'-term will be called *simple* if it represents the result of a finite number of applications of functors with a finite number of arguments. An L'-atomic formula is *simple* if it consists of an L'-relation with a finite number of arguments filled by simple L'-terms. In this connection, the word finite is to be interpreted in the absolute sense so that simple terms and relations are constructed from their atomic symbols exactly as in the standard lower predicate calculus.

Simple L'-wff are defined inductively as follows. An L'-atomic formula is a simple L'-wff. The negation of a simple L'-wff is simple. An L'-wff which is obtained from a set of simple L'-wff by repeated conjunction alone is simple. An L'-wff which is obtained from a set of simple L'-wff by repeated disjunction alone is simple. Finally, an L'-wff which is obtained from a simple L'-wff by repeated quantification is simple. Note that the last three operations may involve an infinite number of conjunctions, or disjunctions, or quantifications, and that both universal and existential quantification may appear in the same operation of repeated quantification.

Every simple L'-wff which is obtained by negation determines uniquely the L'-wff from which it is obtained. Every simple L'-wff which is obtained by conjunction determines uniquely a sequence of simple L'-wff from which it is obtained by repeated conjunction and which are not themselves conjunctions. A corresponding fact holds for disjunctions. Every simple L'-wff which is obtained by quantification determines uniquely a simple L'-wff from which it is obtained by repeated quantification and which is not itself obtained by quantification. The truth of all these statements follows from the fact that they hold for the sentences of L in U and can be transferred to the sentences of L' in U'.

The rank of a simple L'-wff which is an atomic L'-wff is, by definition, 0. The rank of a simple L'-wff which is obtained by negation from a simple L'-wff of rank n is, by definition, $n+1$. The rank of a simple L'-wff which is obtained by repeated conjunction from the particular sequence of L'-wff mentioned in the preceding sequence is, by definition of rank $n+1$ if all the elements of the sequence are of rank not exceeding n and at least one of them is of rank n. A similar definition is introduced for disjunctions. Finally, if a simple L'-wff is obtained by repeated quantification from a simple L'-wff which is not itself obtained by quantification, and which is of rank n then the rank of the quantified L'-wff is again defined to be $n+1$. These definitions assign unique ranks to the elements of a subset Ω of the set of simple L'-wff. The elements of Ω will be said to be simple L'-wff *of finite rank.*

Now let M be a structure in the standard sense of the lower predicate calculus, i.e. with finitary relations and functors (if any) but not necessarily contained in either U or U'. A one-to-one correspondence between the individuals, relations, and functors of M and some L'-individuals, L'-relations, and L'-functors (constituting sets G', G'_R, G'_F) then determines a set of elements of Ω which are defined in M, i.e. those simple L'-atomic formulae whose L'-individuals, etc., belong to the correspondence, and all other simple L'-wff of finite rank that are constructed from them by a finite number of negations and of repeated conjunctlons, disjunctions, and quantifications. The set of these elements of Ω will be denoted by Ω_W.

The ***external truth definition*** for sentences of Ω_W with respect to the structure M (for the given correspondence of atomic symbols) will now be introduced in the following way.

For elements of Ω_W which are of rank 0 we use the standard truth definition. That is to say if a sentence of Ω_W which is of rank 0 is of the form $r(a_1, \ldots, a_n)$ where r is an L'-relation and $a_1, \ldots, a_n$ are L'-individuals then $r(a_1, \ldots, a_n)$ is true (or holds) in M externally if and only if the corresponding n-tuple of individuals of M belongs to the relation of M that corresponds to r. If the sentence is of the form $r(t_1, \ldots, t_n)$ where $t_1, \ldots, t_n$ are L'-terms without variables then we first ***evaluate*** these terms, i.e. we replace them by L'-individuals which correspond to the functional values of $t_1, \ldots, t_n$ in M and then proceed as before.

Suppose now that we have already assigned external truth values to all sentences of Ω_W of rank $<n$ where n is a positive integer. Let a be any sentence of Ω_W which is of rank n. If n is a conjunction then, as explained above, a is obtained by repeated conjunction from a specific sequence of Ω_W-

sentences which are not conjunctions and which are of rank $\leq n-1$. If and only if all these are true in M externally, then we define that a is true in M externally. Similarly, if a is a disjunction then we define that a is true externally if and only if at least one of the elements of the corresponding sequence is true externally. If a is a negation, of an Ω_W-sentence, b say, then a is true externally if and only if b is not true externally.

Finally, suppose that a is obtained by repeated quantification from an element of Ω_W which is not itself obtained by quantification, b say, where b is a predicate not a sentence. We may symbolize the connection between a and b briefly by $a = qb$ where q indicates a finite or infinite sequence of quantifiers. The quantified variables in q constitute an ordered set of length n, where n is a finite or infinite natural number. The natural numbers whose places in q are occupied by universal and existential quantifiers respectively (beginning with 0) constitute two disjoint sets of natural numbers in M'_0, σ_1 and σ_2 say. Let Φ_i be the set of all partial pseudo-finite sequences of elements of G' with domain σ_i, $i = 1, 2$. Let Ψ be the set of *all* Skolem operators from Φ_1 to Φ_2, where the underlying set $\sigma_1 \cup \sigma_2$ is ordered as in M'_0. Then we shall say that a is true in M externally if there exists a Skolem operator $\psi \in \Psi$ such that if we substitute the individuals of the set $(\phi_1, \psi(\phi_1))$ for the corresponding free variables of b then we obtain an Ω_W-sentence (of rank $n-1$) which is true in M externally. In this connection, we recall that the set of pseudo-finite sequences is defined by means of an index set J' in U'. On the other hand, we have not restricted Ψ by the stipulation that its elements be given by some index set. It might appear at first sight that this procedure is somewhat arbitrary. However, in actual fact we only introduce the restriction on the elements of Φ_1 and Φ_2 (i.e. to pseudo-finite sequences) in order to ensure that the result of substituting $(\phi_1, \psi(\phi_1))$ for the free variables of b yields an L'-sentence at all. In fact, it can be shown without difficulty that this last condition is entirely equivalent to the above restriction on the elements of Φ_1 and Φ_2. We shall see presently that it implies no corresponding restriction on the elements of Ψ.

This completes our definition of external truth. It is not difficult to verify that if we substitute sentences of Ω_W for the propositional variables in any tautology of the standard propositional calculus then we obtain a sentence which holds in M externally. Similarly if we substitute an element of Ω_W with a simple free variable for the predicate variable F in one of the formulae $[(\forall x)F(x)] \supset F(a)$ or $F(a) \supset [(\exists x)F(x)]$ then we obtain a sentence of Ω_W which holds in M externally (provided we obtain a wff at all as usual). Finally, it is not difficult to verify that the modus ponens and the two rules of deduction of the predicate calculus also lead from sentences of Ω_W which are true in M externally to sentences of the same kind. Accordingly, we may say that the external truth definition is a reasonable one. We might further suspect that it actually coincides with the interior truth definition in cases where both definitions are applicable. Thus, let M' be a structure in U' which corresponds to some structure in U and let G_T be the set of L'-sentences defined in M according to some given correspondences between L'-symbols and individuals, relations, and functions of M', as in section 3. By the same correspondences, the subset G_T^* of G_T which consists of the simple L'-sentences of finite rank in G_T is defined in M' according to the rules of the present section. It is not difficult to see that if a sentence of G_T^* *does not involve any quantifiers* then it holds in M' according to the external truth definition if and only if it holds in M' according to the internal truth definition.

Indeed, for a simple L'-sentence of rank 0, a, the external truth definition is the same as for the standard predicate calculus and, since 4.5 and 4.6 hold also in U', the internal truth definition for a (whose relations and functions are finitary) also is the same as for the standard predicate calculus. It follows that the two definitions coincide in this case.

Supposing that we have already proved our assertion for sentences of rank smaller than n, where $n>0$, let a be a simple L'-sentence of rank n which does not involve any quantifiers. If a is obtained by conjunction, then (see above) it is a repeated conjunction of a set of L'-sentences of rank $<n$ which are not themselves conjunctions.

According to the external truth definition, a is true in M' if and only if all the elements of the set of conjuncts just mentioned are true in M'. Now this is a condition which holds in U and can be formulated as a sentence of K and, accordingly, holds also in U'. It follows that the external and internal truth definitions for a coincide. The argument is similar for disjunctions and rather simpler for negations.

Now suppose that the L'-sentence a has been obtained from a simple L'-sentence b of finite rank by (possibly, infinitely repeated) quantification, where b is free of quantifiers. Within U, if an L-sentence in prenex normal form holds in a structure M then there exists a corresponding Skolem operator ψ. This fact can be formulated as a sentence of K and, accordingly, holds also in U'. We conclude that if a holds in M' according to the internal truth definition, then it holds in M' also according to the external truth definition. The converse does not follow in this way since we have admitted Skolem operators in U' which are not necessarily indexed in U'. More precisely, we shall show in the next section that the internal and external truth definitions do not coincide in all cases.

6. Discrepancy between internal and external truth definitions. Let U be the structure considered throughout this paper and let M_0^* be the selected standard model of the natural numbers and L the standard Lower predicate calculus within U, as before. K being again the set of all sentences which are defined and hold in U, let H be the set of sentences of K which do not include any individual constants. Let W be any model of H. We claim that W contains a partial structure, W_0, which is isomorphic to U.

Indeed, for any element a of U, K includes a relation $R_a(x)$ which holds only for a. Thus the sentences

6.1. $R_a(a)$,

6.2. $(\exists x)R_a(x)$,

6.3. $(\forall x)(\forall y)[R_a(x)\wedge R_a(y)\supset I(x, y)]$

belong to K. 6.2 and 6.3 belong also to H and, accordingly hold in W. It is easy to see that we may impose the condition that $I(x, y)$ is the relation of identity also in W without limiting the validity of our argument. Let W_0 be the partial structure of W which consists of all elements b in W for which there exists an a in U such that $R_a(b)$ holds in W. This establishes a one-to-one correspondence C: $a\leftrightarrow b$ between the elements of U and the elements of W_0. We are going to show that C establishes an isomorphism between U and W_0.

Let $R(x_1, \ldots, x_n)$ be any relation in U and let $a_1, \ldots, a_n$ be a set of elements of U. Then $R(a_1, \ldots, a_n)$ holds in U if and only if the sentence

6.4. $$(\forall x_1)\cdots(\forall x_n)[R_{a_1}(x_1)\wedge\cdots\wedge R_{a_n}(x_n)\supset R(x_1,\ldots,x_n)]$$

belongs to K and hence to H. Let $a_i \leftrightarrow b_i$ under the correspondence C then $R_{a_i}(b_i)$ holds in W. Thus if $R(a_1, \ldots, a_n)$ holds in U, then 6.4 belongs to H and so $R(b_1, \ldots, b_n)$ holds in W_0. Conversely, if $R(b_1, \ldots, b_n)$ holds in W_0 then 6.4 holds in H and so $R(a_1, \ldots, a_n)$ holds in U. This shows that C is an isomorphic correspondence with respect to the relations of the two structures and a similar argument applies with respect to the functions. It follows that if we identify the b_i with the corresponding a_i then we may regard W as an extension of U, and a model of K.

Let $S_0^*(x, y, z)$ and $P_0^*(x, y, z)$ be two relations which determine addition and multiplication within the set of natural numbers in U, M_0^*. More precisely, we shall suppose that $S_0^*(a, b, c)$ holds if and only if a, b, c belong to M_0^* and $a+b=c$, and $P_0^*(a, b, c)$ holds if and only if a, b, c belong to M_0^* and $ab=c$. In particular, $S_0^*(a, b, c)$ and $P_0^*(a, b, c)$ do not hold if at least one of a, b, c is outside M_0. Also, let $I_0^*(x, y)$ coincide with the relation of identity $I^*(x, y)$ on M_0^*, while $I_0^*(a, b)$ does not hold if at least one of the constants a, b is outside M_0^*. Thus $I_0^*(a, a)$ holds only for element a of M_0^*.

In section 2 above, we considered a model of the natural numbers, M_0, which included the relation of identity $I(x, y)$ and the functions $\sigma(x, y)$, $\pi(x, y)$ and $\phi(x)$, i.e. $x+y$, xy and $x'=x+1$. It is not difficult to show, and is usually taken for granted, that we may replace the three functions just mentioned by two three-place predicates, $S(x, y, z)$ and $P(x, y, z)$, which stand for $x+y=z$ and $xy=z$. Thus, every relation $Q(x_1, \ldots, x_n)$ within the domain of natural numbers which is definable in the Lower predicate calculus in terms of the relation of identity, $I(x, y)$ and the functions σ, π and ϕ, is definable also in terms of I, S and P. This applies to the model of the natural numbers, M_0, as introduced in section 2, i.e. without reference to any structure U which includes M_0. Suppose however, that M_0 coincides with M_0^* as above. In that case the relations I, S, P are obtained by restricting I_0^*, S_0^*, P_0^* to M_0^*.

Let $Q(x_1, \ldots, x_n)$, $n \geq 1$, be any relation which constitutes a subset of $M_0 = M_0^*$ (i.e. which holds for the n-tuples of a certain set of natural numbers). Then we claim that Q is definable in terms of I, S, and P if and only if it is definable in terms of I_0, P_0 and S_0 (with reference to U).

To prove our assertion we denote by $R_0^*(x)$ the relation which defines M_0^* within U. Thus, $R_0^*(a)$ holds in U precisely when a belongs to M_0^*. Let $Q(x_1, \ldots, x_n)$ be a predicate which is defined in $M_0 = M_0^*$ in terms of the relations I, S, and P (without reference to U), and let $Q(x_1, \ldots, x_n)$ be obtained from Q by relativizing that predicate with respect to R_0^* and by replacing I, S, P everywhere by I_0^*, S_0^*, P_0^*, respectively. Let $a_1, \ldots, a_n$ be elements of M_0. Then $Q(a_1, \ldots, a_n)$ holds in M_0 if and only if $Q_0(a_1, \ldots, a_n)$ holds in U. It follows that the predicate 6.5 $T(x_1, \ldots, x_n) = R(x_1) \wedge \cdots \wedge R(x_n) \wedge Q^*(x_1, \ldots, x_n)$ determines the same set as Q, where T is interpreted with reference to U. Now $(\forall x)[R_0^*(x) \equiv I_0^*(x, x)]$ holds in U. It follows that if $T^*(x_1, \ldots, x_n)$ is obtained from $T(x_1, \ldots, x_n)$ by substituting $I_0^*(x, x)$ everywhere for $R_0^*(x)$ then T^* defines the same set as T. Thus T^* is defined in terms of I_0^*, S_0^*, P_0^* and determines the same set within U as Q determines in M_0, without reference to U.

To prove the converse, we introduce a relation of order 0, F^*("False") which does not hold in U, by definition. Let F be the restriction of F^* to M_0. Thus, F is defined but does not hold in M_0. Let $Q^*(x_1, \ldots, x_n)$, $n \geq 1$, be a predicate which is formulated in terms of I_0^*, S_0^*, P_0^* and which defines a subset of M_0^n (the space of points $(a_1, \ldots, a_n)$ with $a_i \in M_0$) when $M_0 = M_0^*$ is regarded as a partial structure of U. It will be sufficient to consider only Q^* which are in prenex normal form,

6.6. $$Q^*(x_1, \ldots, x_n) = q_1 q_2 \cdots q_k\ Z(x_1, \ldots, x_n, y_1, \ldots, y_k)$$

where Z does not contain any quantifiers and where q_i quantifies y_i, $i = 1, \ldots, k$.

We define the predicates $Q_0(x_1, \ldots, x_n, y_1, \ldots, y_k)$, $Q_1(x_1, \ldots, x_n, y_1, \ldots, y_{k-1}), \ldots, Q_k(x_1, \ldots, x_n)$ as follows.

6.7. $$Q_0(x_1, \ldots, x_n, y_1, \ldots, y_k) = Z(x_1, \ldots, x_n, y_1, \ldots, y_k).$$

Supposing that $Q_j\ (x_1, \ldots, x_n, y_1, \ldots, y_{k-j})$ has been defined already, $0 \leq j < k$, we distinguish two cases.

If q_{k-j} is an existential quantifier, $q_{k-j} = (\exists y_{k-j})$ then we put

6.8. $Q_{j+1}(x_1, \ldots, x_n, y_1, \ldots, y_{k-j-1}) = (\exists y_{k-j})[R_0(y_{k-j})$
$\wedge [Q_j(x_1, \ldots, x_n, y_1, \ldots, y_{k-j}) \wedge Q_j^*(x_1, \ldots, x_n, y_1, \ldots, y_{k-j-1})]]$

where Q_j^* is obtained from Q_j by replacing every atomic formula in Q_j which contains y_{k-j} by F_0^*. If q_{k-j} is a universal quantifier, $q_{k-j} = (\forall y_{k-j})$, we put

6.9. $Q_{j+1}(x_1, \ldots, x_n, y_1, \ldots, y_{k-j-1}) = (\forall y_{k-j})[R_0^*(y_{k-j}) \supset$
$[Q_j(x_1, \ldots, x_n, y_1, \ldots, y_{k-j}) \wedge Q_j^*(x_1, \ldots, x_n, y_1, \ldots, y_{k-j-1})]]$

where Q_j^* is defined as before.

Taking the former case, we observe that the sentence

6.10. $(\forall x_1) \cdots (\forall x_n) \cdots (\forall y_{k-j-1})[(\exists y_{k-j})Q_j(x_1, \ldots, x_n, y_1, \ldots, y_{k-j}) \equiv$
$Q_{j+1}(x_1, \ldots, x_n, y_1, \ldots, y_{k-j-1})]$

holds in U. Indeed, the sentence

6.11. $(\forall x_1) \cdots (\forall x_n) \cdots (\forall y_{k-j-1})[[(\exists y_{k-j})Q_j(x_1, \ldots, x_n, y_1, \ldots, y_{k-j}) \equiv$
$[(\exists y_{k-j})[[R_0^*(y_{k-j}) \wedge Q_j(x_1, \ldots, x_n, y_1, \ldots, y_{k-j})]] \vee [(\exists y_{k-j})$
$[\sim R_0^*(y_{k-j}) \wedge Q_j(x_1, \ldots, x_n, y_1, \ldots, y_{k-j})]]]$

is a theorem of the Lower predicate calculus. But

$(\forall x_1) \cdots (\forall x_n) \cdots (\forall y_{k-j})[[\sim R_0^*(y_{k-j}) \wedge Q_j(x_1, \ldots, x_n, y_1, \ldots, y_{k-j})] \equiv$
$Q_j^*(x_1, \ldots, x_n, y_1, \ldots, y_{k-j})]$

holds in U since any atomic sentence of Q_j that contains an individual a which is in U but outside M_0^*, cannot hold in U. It follows, by the rules of the Lower predicate calculus that the sentences

$(\forall x_1) \cdots (\forall x_n) \cdots (\forall y_{k-j-1})(\exists y_{k-j})[[\sim R_0^*(y_{k-j}) \wedge$
$Q_j(x_1, \ldots, x_n, y_1, \ldots, y_{k-j})] \equiv Q_j^*(x_1, \ldots, x_n, y_1, \ldots, y_{k-j-1})]$

and

6.12. $(\forall x_1) \ldots (\forall x_n) \ldots (\forall y_{k-j-1})[[(\exists y_{k-j})[\sim R_0^*(y_{k-j}) \wedge$
$Q_j(x_1, \ldots, x_n, y_1, \ldots, y_{k-j})]] \equiv Q_j^*(x_1, \ldots, x_n, y_1, \ldots, y_{k-j-1})]$

also hold in U. Combining 6.12 with 6.11, we obtain 6.10.

Suppose now that q_{k-j} is a universal quantifier, $q_{k-j} = (\forall y_{k-j})$. In that case, similar considerations show that

6.13. $(\forall x_1) \cdots (\forall x_n) \cdots (\forall y_{k-j-1})[[(\forall y_{k-j})Q_j(x_1, \ldots, x_n, y_1, \ldots, y_{k-j})]$
$\equiv Q_{j+1}(x_1, \ldots, x_n, y_1, \ldots, y_{k-j-1})]$

holds in U. Thus, in either case,

6.14. $(\forall x_1)\cdots(\forall x_n)\cdots(\forall y_{k-j-1})[[q_{k-j}Q_j(x_1, \ldots, x_n, y_1, \ldots, y_{k-j})] \equiv Q_{j+1}(x_1, \ldots, x_n, y_1, \ldots, y_{k-j-1})]$ holds in U. Applying 6.14 for $j=0, 1, \ldots, k-1$ and taking into account 6.7, we conclude that $(\forall x_1)\cdots(\forall x_n)[q_1, \ldots, q_k Q_0(x_1, \ldots, x_n, y_1, \ldots, y_k) \equiv Q_k(x_1, \ldots, x_n)]$

i.e. 6.15. $(\forall x_1)\cdots(\forall x_n)[Q^*(x_1, \ldots, x_n) \equiv Q_k(x_1, \ldots, x_n)]$ holds in U.

Next, we define another sequence of predicates, $T_0(x_1, \ldots, x_n, y_1, \ldots, y_k)$, $T_1(x_1, \ldots, x_n, y_1, \ldots, y_{k-1}), \ldots, T_k(x_1, \ldots, x_n)$ by

6.16. $T_0(x_1, \ldots, x_n, y_1, \ldots, y_k) = Z(x_1, \ldots, x_n, y_1, \ldots, y_k)$ and by

6.17. $T_{j+1}(x_1, \ldots, x_n, y_1, \ldots, y_{k-j-1}) = q_{k-j}[T_j(x_1, \ldots, x_n, y_1, \ldots, y_{k-j}) \wedge T_j^*(x_1, \ldots, x_n, y_1, \ldots, y_{k-j-1})]$

for $0 \leq j < k-1$, where T_j^* is obtained from T_j by replacing every atomic formula in T_j which contains y_{k-j} by F_0^*.

By considering successive j, $j=0, 1, \ldots, k$, we see that $Q_j(x_1, \ldots, x_n, y_1, \ldots, y_{k-j-1})$ is obtained by relativizing the predicate $T_j(x_1, \ldots, x_n, y_1, \ldots, y_{k-j-1})$ with respect to $R_0^*(x)$. It follows that the predicate $T(x_1, \ldots, x_n)$ which is obtained from $T_k(x_1, \ldots, x_n)$ by replacing I_0^*, S_0^*, P_0^*, F_0^* everywhere by I, S, P, F, respectively, holds in M_0 for any $a_1, \ldots, a_n$ in M_0 if and only if $Q_k(x_1, \ldots, x_n)$ holds in U, i.e. by 6.15, if and only if $Q^*(a_1, \ldots, a_n)$ holds in U. $T(x_1, \ldots, x_n)$ is not as yet the required predicate since it contains, in addition to I, S, P, the relation F. However, replacing F everywhere in $T(x_1, \ldots, x_n)$ by $(\forall x)[I(x, x) \wedge \sim I(x, x)]$ we obtain a predicate $Q(x_1, \ldots, x_n)$ which is obtained in terms of I, S, and P and which holds in M_0 without reference to U if and only if $Q^*(x_1, \ldots, x_n)$ holds in U. This completes the proof of our assertion.

With U and $M_0 = M_0^*$ as before, let L be the Lower predicate calculus within U as introduced in seetion 3. Suppose that a set G of L-individuals is in one-to-one correspondence with the elements of M_0 and let j, s, and p be L-relations which correspond to the relations $I(x, y)$, $S(x, y, z)$ and $P(x, y, z)$ in M_0, respectively. Thus, in the notation of section 4 with $M = M_0, G_R = \{j, s, p\}$ while G_F is empty. Putting it in a less formal way, we may say that j, s, and p, *denote* the respective relations in M_0. Let G_W be the set of L-wff formulated in terms of these L-symbols and let G_T be the set of L-sentences which belong to G_W. Let G_P be the set of elements of G_T which are in prenex normal form and such that the matrix (the quantifier-free part of the sentence) is in Boolean normal form as a conjunction of disjunctions of atomic sentences and of negations of such sentences. Let G_T^* be the set of elements of G_T which hold in M_0 and let G_P^* be the corresponding subset of G_P.

We may map the wff of G_W on a subset S_W of $M_0 = M_0^*$ by the familiar process of arithmetization. Then S_W is recursive. Let S_T, S_P, S_T^*, S_P^* be the sets of natural numbers which correspond to G_T, G_P, G_T^*, and G_P^* respectively in this mapping. Then S_T and J_P also are recursive. Moreover, since there is a recursive procedure for transforming any given sentence in prenex normal form with a matrix in Boolean normal form, there exists a recusive number-theoretic relation $K(x, y)$ such that the first argument determines the second argument — i.e. $(\forall x)(\forall y)(\forall z)[K(x, y) \wedge K(x, z) \supset I(y, z)]$ — and such that for every element a of S_T there exists an element b of S_P for which $K(a, b)$ holds. Moreover, $K(a, b)$ holds only if a is in S_T and b is in S_P and a and b are the Gödel numbers of equivalent sentences. Since $K(x, y)$ is recursive, it is definable in terms of the relations I, S, and P.

Let $V_T(x)$ and $V_P(x)$ be the relations which determine the sets S_T^* and S_P^* in M_0, without reference to U. Then

6.18. $$(\forall x)[V_T(x) \equiv (\exists y)[K(x, y) \wedge V_P(y)]]$$

holds in M_0. By Tarski's theorem (2.4 above), $V_T(x)$ cannot be expressed in terms of the operations of addition, multiplication and succession and as remarked earlier, this is equivalent to the fact that $V_T(x)$ cannot be expressed in terms of I, S and P. On the other hand, $K(x, y)$ can be so expressed. Hence, by 6.18, $V_P(x)$ cannot be expressed in terms of the relations I, S and P. Combining this conclusion with the result of the previous argument we obtain

6.19. **Theorem.** S_P^*, *regarded as a subset of* U, *is not defined by any predicate which is formulated in terms of the relations* $I_0^*(x, y)$, $S_0^*(x, y, z)$, $P_0^*(x, y, z)$ *alone.*

As before, let H be the set of all sentences which are defined in terms of the relations of U but without individual constants, and which hold in U. In particular, H contains sentences which involve the relation $I_0^*(x, y)$, $S_0^*(x, y, z)$, $P_0^*(x, y, z)$ and $V_P(x)$. Let H' be obtained from H by replacing all relations of H *except* I_0^*, S_0^*, P_0^*, by distinct new relations. In particular, let $V_P'(x)$ be the relation which replaces $V_P(x)$. We are going to prove

6.20. THEOREM. *The set* $H = H \cup H' \cup \{(\exists x)[\sim[V_P(x) \equiv V_P'(x)]]\}$ *is consistent.*

Proof. Suppose on the contrary that H^* is contradictory. Then

6.21. $$H \cup H' \vdash (\forall x)[V_P(x) \equiv V_P'(x)].$$

In 6.21, we may replace H and H' by finite subsets H_1 and H_2 of H and H' respectively. Moreover—if necessary by adding a number of sentences—we may suppose that H_2 is obtained by replacing the relations of H, except I_0^*, S_0^*, P_0^* by the new relations of H'. Let X be the conjunction of the sentences of H_1 then the conjunction of the sentences of H_2, H' say is obtained from X in the same manner. Thus, $X \in H$, $X' \in H'$ and

6.22. $$X \cap X' \vdash (\forall x)[V_P(x) \equiv V_P'(x)].$$

Theorem 2.1 now implies that there exists a predicate $Q(x)$ which is defined in terms of I_0^*, S_0^* and P_0^* alone, such that

6.23. $$X \vdash (\forall x)[V_P(X) \equiv Q(x)].$$

But $X \in H$ holds in U and so 6.23 entails that $V_P(x)$ is expressible in terms of I_0^*, S_0^*, P_0^*. This contradicts 6.19 and proves 6.20.

Let U^* be a model of H^*. U^* contains a model of arithmetic, $\overline{M}_0$, which is defined in terms of I_0^*, S_0^*, P_0^*. $\overline{M}_0$ cannot be the standard model since in that model $(\forall x)[V_P(x) \equiv V_P'(x)]$. We denote by U_1 the structure which is obtained from U^* by taking into account only the relations of H, and by U_1' the corresponding structure which takes into account only the relations of H'. Then U_1 and U_1' are models of H and H' respectively. They have in common the non-standard model of arithmetic, $\overline{M}_0$, which is a proper extension of the standard model of arithmetic M_0 in U. Thus the theory of sections 2–5 applies to both U_1 and U_1'. The sets which constitute the Lower predicate calculus within U_1 and U_1', respectively, need not coincide. However, if G_{1W} and G_{1W}' are the sets which correspond to G_W in U_1 and U_1' respectively then there is a natural one-to-one correspondence between G_{1W} and G_{1W}' in the Gödel numbering of their elements. That is to say if $a \in G_{1W}$ and $a' \in G_{1W}'$ then we define $a \leftrightarrow a'$ if a and a' have the same Gödel numbers in $\overline{M}_0$. This correspondence is an isomorphism in the sense that if $a \leftrightarrow a'$ then a and a' possess the same length l (where l is an element of $\overline{M}_0$). Moreover, the nth atomic symbols of a and a' either are the "same" logical constants other than variables, or variables with the same Gödel numbers, or they denote the same extralogical constants (relations, natural numbers). The sentences of G_{1W} are mapped on the sentences of G'_{1W} under this correspondence, $G_{1T} \leftrightarrow G'_{1T}$, and the same applies to the sentences in prenex normal form, $G_{1P} \leftrightarrow G'_{1P}$.

Now let G^*_{1P} and $G^{*\prime}_{1P}$ be the sets of sentences in prenex normal form within the Lower predicate calculi of U and U', respectively which hold in $\overline{M}_0$. Let A and A' be the sets of Gödel numbers of G^*_{1P} and $G^{*\prime}_{1P}$, then A and A' consist of the elements of $\overline{M}_0$ which satisfy $V_P(x)$ and $V'_P(x)$, respectively. U^* is a model of H^* and so the sentence $(\exists x)[\sim[V_P(x) \equiv V'_P(x)]]$ holds in U^*. Thus, there exists an element m of U^* such that $\sim[V_P(m) \equiv V'_P(m)]$ holds in U^*. It follows that either $V_P(m)$ holds in U^* and $V'_P(m)$ does not hold in U^* or V_P does not hold in U^* and $V'_P(m)$ holds in U^*. Clearly, it does not restrict the generality of our considerations to suppose that the former case applies. m is the Gödel number of sentences a and a' within the Lower predicate calculi of U_1 and U'_1 respectively. Then $a \leftrightarrow a'$, and so a and a' are "isomorphic" in the sense detailed above. The sentence a holds in $\overline{M}_0$, since $V_P(m)$ holds in U^*. a belongs to G_{1P}, which includes G^*_{1P} as a subset and so a' belongs to G'_{1P}. But a' does not belong to $G^{*\prime}_{1P}$ since $V'_P(m)$ does not hold in U^*, and so a' does not belong to $G^{*\prime}_{1T}$ either.

Let b and b' be the negations of a and a' within the Lower predicate calculi of U and U', respectively, then $b \leftrightarrow b'$. Let c be obtained from b by the standard transformation of the negation of a sentence in prenex normal form into a sentence in prenex normal form, i.e. by changing the type of the quantifier of b, existential to universal and vice versa, while the sign of negation is shifted to the front of the matrix of b. We shall say, briefly, that c is obtained from a by negation and normalization. Thus, if c' is similarly obtained from a' by negation and normalization then $c \leftrightarrow c'$.

a holds in $\overline{M}_0$ and so b and c are defined but do not hold in $\overline{M}_0$, all according to the internal truth definition. On the other hand, a' is defined but does not hold in $\overline{M}_0$ and so b' and c' hold in $\overline{M}_0$, again according to the internal truth definition.

Now all sentences in prenex normal form with matrices in Boolean normal form are simple formulae in the sense introduced in section 5, of ranks not exceeding 4. Accordingly, we may apply the external truth definition to the sentences of G_{1P} and G'_{1P}. Since corresponding sentences of G_{1P} and G'_{1P} are given by the same Gödel numbers and are isomorphic in the sense explained above, it will be seen that corresponding elements of G_{1P} and G'_{1P} either both hold or both do not hold in $\overline{M}_0$. But a holds in $\overline{M}_0$ according to the external truth definition. Similarly, c' holds in $\overline{M}_0$ according to the internal truth definition and so both c and c' hold in $\overline{M}_0$ according to the external truth definition. b and b' are the negations of a and a' respectively, and so these sentences do not hold in $\overline{M}_0$ according to the external truth definition either.

Disregarding the relations which were introduced on passing from H to H' and recalling that our structure U^* is also a model of K, we thus obtain the following theorem.

6.24. THEOREM. *There exists a model U^* of K including a model of arithmetic $\overline{M}_0$, and three sentences a, b, c, in the Lower predicate calculus L^* within U^* such that the following conditions are satisfied. a is an L^*-sentence in prenex normal form with a matrix in Boolean normal form; b is obtained from a by negation and c is obtained from b by normalization; accordingly, c is again in prenex normal form. The sentence a holds in $\overline{M}_0$ according to the internal truth definition, and the sentences b and c do not hold in $\overline{M}_0$ according to that definition. The sentences a and c hold in $\overline{M}_0$ according to the external truth definition and the sentence b does not hold in $\overline{M}_0$ according to that definition.*

Our argument was based on the assumption that $V(a)$ holds in U^* while $V'(a)$ does not hold in that structure. If the opposite is true we only have to interchange the roles of U_1 and U'_1.

Theorem 6.24 shows that a particular sentence (i.e. c) may well be true in $\overline{M}_0$ according to the external truth definition but false according to the internal truth definition. Moreover, whereas our intuition which is derived from the finite case, would indicate that b and c (which is obtained from b by normalization) must be true simultaneously, the opposite applies if we adopt the external truth definition. As mentioned in the introduction, an analogus phenomenon was noticed by H. Steinhaus and J. Micielski for certain sequences of quantifiers of order type ω. No such sequences are possible in our calculus and it is not clear whether the connection between the two phenomena is more than superficial.

We may make our result somewhat more precise by imposing certain conditions on the form of the prefix in a. Thus, every sentence in prenex normal form is equivalent to another sentence in prenex normal form in which the quantifiers alternate. For example in ordinary notation, $(\forall x)(\forall y)(\exists z)(\exists w)$ $Q(x, y, z, w)$, where Q is free of quantifiers, is equivalent to $(\forall x)(\exists u)(\forall y)(\exists z)$ $(\forall v)(\exists u)[Q(x, y, z, w) \wedge I(u, u) \wedge I(v, v)]$ where $I(u, u)$ and $I(v, v)$ need be included only if we wish to avoid empty quantification. Accordingly, we may restrict the sets G_P, G_P^* which were employed in the proof of 6.24 to sentences in whose prefix the quantifier alternate. This leads to conclusion that the sentences a and c in 6.24 may also be supposed to be of this type.

In the external truth definition we combine laws of formation which are based on non-standard arithmetic with a standard semantical approach. As a result we may employ a non-standard language L' even in order to describe the standard system of natural numbers, M_0. This contrasts with previous efforts to penetrate into the world of non-standard arithmetic by employing the standard Lower predicate calculus in order to discuss a non-standard model of arithmetic.

References

[1] W. Craig, Three uses of the Herbrand-Gentzen theorem in relating model theory and proof theory, Journal of Symbolic Logic, vol. **22** (1957), pp. 269-285.

[2] L. Henkin, Completeness in the theory of types, Journal of Symbolic Logic, vol. **15** (1950), pp. 81–91.

[3] L. Henkin, Some remarks on infinitely long formulas, *Infinitistic Methods*—Proceedings of the Symposium on Foundations of Mathematics, Warsaw 1959—pub. 1961, pp. 167–183.

[4] A. Robinson, Model theory and non-standard arithmetic, *Infinitistic Methods*—Proceedings of the Symposium on Foundations of Mathematics, Warsaw 1959—pub. 1961, pp. 265–302.

[5] D. Scott and A. Tarski, The sentential calculus with infinitely long expressions, Colloquium Mathematicum, vol. **6** (1958), pp. 165-170.

[6] A. Tarski, Remarks on predicate logic with infinitely long expressions, Colloquium Mathematicum, vol. **6** (1958), pp. 171–176.

[7] A. Tarski, A. Mostowski and R. M. Robinson, Undecidable Theories, Studies in Logic and the Foundations of Mathemctics, Amsterdam 1953.

ON GENERALIZED LIMITS AND LINEAR FUNCTIONALS

ABRAHAM ROBINSON

1. Introduction. Let m be the space of bounded sequences of real numbers, $\sigma = \{s_n\}$, $n = 1, 2, 3, \cdots$ under the norm $\|\sigma\| = \sup |s_n|$. Then m includes the set of all convergent sequences of real numbers. Let $A = (a_{kn})$, k, $n = 1, 2, 3, \cdots$ be a real Toeplitz-matrix, i.e. an infinite matrix of real numbers which satisfies the conditions

$$\lim_{k\to\infty} a_{kn} = 0\ , \qquad n = 1, 2, \cdots\ ; \tag{1.1}$$

$$\sum_{n=1}^{\infty} |a_{kn}| < M\ , \tag{1.2}$$

where M is a positive number which is independent of k;

$$\lim_{k\to\infty} \sum_{n=1}^{\infty} a_{kn} = 1\ . \tag{1.3}$$

It is well-known that A defines a regular method of summation. That is to say, if $\{s_n\}$ is a convergent sequence in M, with limit s, then the transform $\{t_k\}$ of $\{s_n\}$ by A exists, where

$$t_k = \sum_{n=1}^{\infty} a_{kn} s_n \tag{1.4}$$

and $\{t_k\}$ converges to the same limit as $\{s_n\}$,

$$\lim_{k\to\infty} t_k = \lim_{n\to\infty} s_n = s\ . \tag{1.5}$$

If $\{s_n\}$ is an arbitrary element of m then $\{t_k\}$ exists anyhow and may (or may not) converge to a limit, which is then called the A-limit of $\{s_n\}$. However, it has been shown by Steinhaus that for every Toeplitz-matrix A there exists a sequence $\{s_n\}$ in m which is not summed by A, i.e. such that $\{t_k\}$ does not converge (compare e.g. Ref. 2). Nevertheless the method of summation by infinite matrices is very useful in various branches of real and complex Function Theory. We shall call the limits obtained in this way *Toeplitz-limits*, to distinguish them from other types of generalized limits such as those discussed below.

A straightforward application of the Hahn-Banach extension theorem shows that there exist continuous linear functionals $F(x)$ defined *on* m such that for every convergent sequence $\sigma = \{s_n\}$ in m, $F(\sigma) = \lim_{n\to\infty} s_n$. Such a functional will be called a *Hahn-Banach-limit.*

Received March 21, 1963.

In view of Steinhaus' theorem, mentioned above, a Hahn-Banach-limit cannot be a Toeplitz-limit, although every Toeplitz-limit can be extended to a Hahn-Banach-limit, again by applying the Hahn-Banach extension theorem.

If $F(x)$ is a Hahn-Banach-limit such that

1.6. (positivity) $s_n \geqq 0$ for $n = 1, 2, \cdots$ entails $F(\{s_n\}) \geqq 0$ and such that

1.7. (shift invariance) $F(\{s_n\}) = F(\{s_{n+1}\})$ for all $\{s_n\} \in m$— then $F(x)$ will be called a *Mazur-Banach limit*. It is known that there exist Mazur-Banach limits (compare ref. 1 or ref. 10).

In spite of their greater efficiency, the generalized limits which are defined by means of the methods of Functional Analysis (Hahn-Banach limits) cannot in general be used to replace the more concrete matrix methods of summation (Toeplitz-limits). In the present paper, we propose to use the methods of Non-standard Analysis in order to bridge the gap between the Hahn-Banach-limits on one hand and Toeplits-limits on the other hand. Thus, we shall derive generalized limits which, while clearly related to the matrix methods, are efficient for *all* bounded sequences. These generalized limits are given by certain linear forms with coefficients in a nonstandard model of analysis, and some of them satisfy also conditions 1.6 and 1.7. More generally, it will be shown that there exist non-standard models of analysis in which all continuous linear functionals on m can be represented by linear forms in a sense which will be made precise in due course.

The foundations of Non-standard Analysis are sketched in Ref. 7 and developed in greater detail in Refs. 8 and 9. The scope of the theory of Ref. 8 is more comprehensive inasmuch as it is based on a higher order language, L. We shall adopt L as the basis of the present paper, but in order to follow it, it will be sufficient to suppose that L is some formal language which is appropriate to the predicate calculus of order ω.

As mentioned in Ref. 7, particular non-standard models of analysis are provided by the ultrapowers of the real numbers, R_0, which constitute proper extensions of R_0. The ultrapower technique applies both to the lower predicate calculus and to the higher order language which will be considered here. For the notions which can be expressd in the lower predicate calculus, W. A. J. Luxemburg has given a detailed and expert development of Non-standard Analysis in terms of ultrapowers (Ref. 5, compare also Ref. 6). When specific ultrapowers are used, the explicit use of a formal language can be avoided and the fact that a non-standard model of analysis has in a definite formal sense the *same* properties as the standard model remains in the background, as a heuristic principle. This has the advantage of making the subject comprehensible to analysts who are not familiar with the formal languages of Mathematical Logic and also reveals certain aspects of the procedure which are not apparent if the structure of the non-standard model remains unspecified. On the other hand, by using a formal language we may establish the truth of a vast number of useful assertions about the non-standard model simply but rigorously by transfer from the standard case. If no formal language is used we have to prove all these results ab initio.

2. Q-sequences. Let R_0 be the field of real numbers, and let K_0 be the set of all sentences which are formulated in a higher order language L, as in Ref. 8 (compare § 1, above) and which hold in R_0. Thus, L contains (symbols for) all real numbers, for all n-ary relations on real numbers, including unary relations, or *sets*, and also for concepts of higher type such as relations between sets, relations between relations, etc. In such a language, an infinite sequence $\{s_n\}$, $n = 1, 2, 3, \cdots$ is represented by a binary relation $\sigma(x, y)$ in which the domain of the first variable is the set of positive integers, such that for every positive integer n there is a unique real number a for which $\sigma(n, a)$ holds in R_0 (and hence, is a sentence of K_0). The set of all these relations then determines a unary relation in L, $\theta(x)$, say, i.e. $\theta(\sigma)$ holds in R_0 if and only if σ represents an infinite sequence as indicated.

Now let $^*R$ be a non-standard model of K_0, i.e. a model of K_0 which is different from R_0 and hence, is a proper extension of R_0. Then the unary relation of L which determines the set of natural numbers, N, in R_0 determines a non-standard model of the natural numbers, $^*N$, in $^*R$, and the predicate $\theta(x)$ determines a set of non-standard infinite sequences in $^*R$, i.e. a set of functions whose domain is the set of positive integers in $^*R$, $^*N$-$\{0\}$, and which take values in $^*R$. The sequences which are determined in this way are called Q-sequences (Q for quasi-standard). In particular, if $\{s_n\}$ is an infinite sequence in R_0, which corresponds to a binary relation $\sigma(x, y)$ as above, then $\sigma(x, y)$ determines in $^*R$ a Q-sequence which coincides with $\{s_n\}$ for all finite values of the subscript n, and which will be denoted by $^*\{s_n\}$. A Q-sequence which is obtained in this way by the extension of a sequence in R_0 is called an S-sequence (S for standard). For example, the sequence which is given by $s_n = 1/n$ for all positive integers, finite or infinite, is an S-sequence. There exist Q-sequences which are not S-sequences. For example, if ω is a fixed infinite natural number then the sequence which is given by

(2.1)
$$a_n = 1/\omega \quad \text{for} \quad 1 \leqq n \leqq \omega$$
$$a_n = 0 \quad \text{for} \quad n > \omega$$

is not an S-sequence, since every S-sequence takes standard real values for standard (finite) subscripts, but it is a Q-sequence, for the assertion,

"For every positive integer ω there exists an infinite sequence which is given by 2.1,"

holds in R_0, and therefore, can be formulated as a sentence of K_0 and holds also in $^*R$.

More generally, if (a_{kn}) is a Toeplits-matrix, and ω is a fixed infinite natural number then the sequence

2.2. $a_n = a_{\omega n}$ for all positive integers n, finite or infinite,

is a Q-sequence. It cannot be an S-sequence, for 1.1 implies that for every finite positive integer n, $a_{\omega n}$ is infinitesimal. If the sequence were an S-sequence then $a_{\omega n} = a_n$ would at the same time have to be a standard number and so $a_n = 0$ for all finite n. It is easy to see that the only S-sequence which has this property is the zero sequence, $a_n = 0$ for *all* n, and this contradicts 1.3, which implies that $\sum_{n=1}^{\infty} a_{\omega n}$ is infinitely close to 1. In this connection, $\sum_{n=1}^{\infty} a_{\omega n}$ is to be understood in the sense of the classical (Weierstrass) definition.

The equation 2.1 is a special case of 2.2. It is obtained by taking for (a_{kn}) the matrix of arithmetic means, $a_{kn} = 1/k$ for $n \leqq k$, $a_{kn} = 0$ for $n > k$.

On the other hand it is not difficult to define sequence which are not even Q-sequences. For example, the sequence which is given by

(2.3)
$$a_n = 0 \quad \text{for all finite positive integers } n\,,$$
$$a_n = 1 \quad \text{for all infinite positive } n\,,$$

cannot be a Q-sequence, for if it were, then the set of all infinite natural numbers would be definable in $^*R$ as the set of all natural numbers for which $a_n = 1$. Every nonempty set of natural numbers which is definable in $^*R$ must have a first element, for this is a property of subsets of natural numbers which can be formulated as a sentence of K_0. The set of infinite natural numbers in $^*R$ is not empty but does not possess a first element. This shows that 2.3 cannot be a Q-sequence.

A more general property of Q-sequences, which will be made use of in the sequel is as follows.

2.3. THEOREM. *Let $\{A_n\}$ be a Q-sequence such that A_n is infinitesimal for all finite n. Then there exists an infinite natural number ω such that A_n is infinitesimal for all $n < \omega$.*

Proof. A *Q-set* is a set which exists in $^*R$ as a model of K_0 (just as a *Q-sequence* is a sequence which exists in $^*R$). Supposing that $\{A_n\}$ satisfies the assumptions of the theorem, let B be the set of all positive integers in $^*R$ such that $n|A_n| \geqq 1$. Then B is a Q-set (since it is defined in terms of a Q-sequence). If B is empty then $n|A_n| < 1$ and so $|A_n| < 1/n$ for all positive integers n. This shows in particular that A_n is infinitesimal for all infinite n since $1/n$ is then infinitesimal. We conclude that in this case the conclusion of the theorem is satisfied by all infinite positive integers ω. If B is not empty then it includes a smallest element, ω, for it is a property of every nonempty subset of the natural numbers in R_0, and hence also in $^*R$, to possess a smallest element. Moreover, ω must be infinite, otherwise $\omega|A_\omega|$ would be infinitesimal and hence, smaller than 1. For $n < \omega$ we have again $n|A_n| < 1$, $|A_n| < 1/n$, A_n is infinitesimal also for all infinite n less than ω. This completes the proof of 2.3.

Let $\{a_n\}$ be a Q-sequence. Then the infinite sum $\sum_{n=1}^{\infty}|a_n|$ may or may not exist in the sense of the classical (Weierstrass) definition, as applied to $^*R$. Thus, $\sum_{n=1}^{\infty}|a_n|$ exists if and only if the partial sums $\sum_{n=1}^{k}|a_n|$ is uniformly bounded in $^*R$ as k varies over all finite or infinite positive integers. In particular, if $\sum_{n=1}^{\infty}|a_n|$ exists and equals a *finite* number in $^*R$ (i.e. a number of $^*R$ which is smaller than some number of R_0) then $\{a_n\}$ will be called an *S-bounded form*. The reason for this terminology will become apparent presently. A Q-sequence $\{a_n\}$ is an S-bounded form if and only if there exists a standard real number A and that $\sum_{n=1}^{k}|a_n| \leqq A$ for all k, finite or infinite.

Let $^*m$ be the extension of the space m to $^*R$. Thus, $^*m$ consists of all Q-sequences which are bounded in $^*R$. In particular, $^*m$ contains all the S-sequences $^*\sigma$ which are the extensions to $^*R$ of bounded sequences $\sigma = \{s_n\}$ in R_0.

Let $\alpha = \{a_n\}$ be an S-bounded from, such that $\sum_{n=1}^{\infty}|a| = A$, where A is a finite number. We shall associate with α a functional $F_\alpha(x)$ with domain m and range in R_0, in the following way. For every $\sigma = \{s_n\}$ in m, the sum $\sum_{n=1}^{\infty} a_n s_n$ exists in $^*R$ since, for every finite or infinite positive integer k,

$$\left|\sum_{n=1}^{k} a_n s_n\right| \leqq \sum_{n=1}^{k}|a_n||s_n| \leqq ||\sigma\sum_{n=1}^{k}|a_n| \leqq ||\sigma||A\ , \tag{2.4}$$

taking into account that $||\sigma||$ which is a bound for the absolute values of the element of σ must be a bound also for the absolute values of the elements of $^*\sigma$. Moreover, (2.4) shows that $\sum_{n=1}^{\infty} a_n s_n$ is actually finite, since it cannot exceed the finite number $||\sigma||A$, and accordingly possesses a standard part. Denoting by ${}^0a$ the standard part of any finite a, we now put

$$F_\alpha(\sigma) = {}^0\Big(\sum_{n=1}^{\infty} a_n s_n\Big) \tag{2.5}$$

and we claim that $F_\alpha(x)$ is a continuous linear functional on m.

Indeed, if $\sigma = \{s_n\}$ and $\sigma' = \{s'_n\}$ are two sequences in m and λ is a real number (in R_0) then

$$\begin{aligned} F_\alpha(\sigma + \sigma') &= {}^0\Big(\sum_{n=1}^{\infty} a_n(s_n + s'_n)\Big) = {}^0\Big(\sum_{n=1}^{\infty} a_n s_n + \sum_{n=1}^{\infty} a_n s'_n\Big) \\ &= {}^0\Big(\sum_{n=1}^{\infty} a_n s_n\Big) + {}^0\Big(\sum_{n=1}^{\infty} a_n s'_n\Big) = F_\alpha(\sigma) + F_\alpha(\sigma') \end{aligned}$$

and

$$F_\alpha(\lambda\sigma) = {}^0\Big(\sum_{n=1}^{\infty} a_n \lambda s_n\Big) = {}^0\Big(\lambda \sum_{n=1}^{\infty} a_n s_n\Big) = \lambda\,{}^0\Big(\sum_{n=1}^{\infty} a_n s_n\Big) = \lambda F_\alpha(\sigma)\,,$$

showing that the functional is linear. Also, by 2.4, $| F_\alpha(\sigma) | \leqq A \| \sigma \|$ showing that $F_\alpha(x)$ is continuous and, moreover that the norm of $F_\alpha(x)$ does not exceed A, $\| F_\alpha(x) \| \leqq A$.

3. **Generalized limits.** Let α be an S-bounded form, and let $F_\alpha(x)$ be the linear functional associated with α, as introduced in § 2 above. Then

3.1. THEOREM. *In order that $F_\alpha(x)$ be a Hahn-Banach limit for a given S-bounded form $\alpha = \{a_n\}$ is necessary and sufficient that*

3.2. *a_n is infinitesimal for all finite n, $a_n \simeq 0$, and*

3.3. *$\sum_{n=1}^{\infty} a_n$ is infinitely close to 1, $\sum_{n=1}^{\infty} a_n \simeq 1$.*

Proof. The conditions are necessary. For any finite positive integer n, let σ be the sequence (in m) which is defined by

$$s_k = 0 \quad \text{for} \quad k \neq n, \qquad s_n = 1\,. \tag{3.4}$$

Then 3.4 holds also for the extension of σ to ${}^*R$, and so

$$F_\alpha(\sigma) = {}^0\Big(\sum_{k=1}^{\infty} a_k s_k\Big) = {}^0a_n$$

where ${}^0a_n$ is the standard part of a_n. On the other hand, since $F_\alpha(x)$ is a Hahn-Banach limit, we must have $F_\alpha(\sigma) = 0$, which is the limit of σ. Hence ${}^0a_n = 0$, i.e. $a_n \simeq 0$, a_n is infinitesimal.

Again, let σ be the sequence in which is defined by

$$(3.5) \qquad s_n = 1 \qquad \text{for all } n\,.$$

Then the limit of σ is 1, and $s_n = 1$ also for all infinite values of n. Hence, $F_\alpha(\sigma) = {}^0(\sum_{n=1}^{\infty} s_n) = 1$, and so $\sum_{n=1}^{\infty} \simeq 1$, proving 3.3.

The conditions are also sufficient. For any positive integer n, put $A_n = \sum_{k=1}^{n} |a_k|$. Then 3.2 implies that A_n is infinitesimal for all finite n. Hence, by 2.3, there exists an infinite integer ω such that $A_{\omega-1} = \sum_{k=1}^{\omega-1} |a_k| = \eta$, say, is infinitesimal. On the other hand, since the sum $\sum_{n=1}^{\infty} |a_n|$ exists, there is an infinite integer $\Omega > \omega$, such that $\sum_{n=\Omega}^{\infty} |a_n| \leqq \eta$. Referring to 3.3 we may then conclude that

$$\sum_{n=\omega}^{\Omega} a_n \simeq 1$$

or, which is the same, that

$$\sum_{n=\omega}^{\Omega} a_n = 1 - \varepsilon$$

where ε is infinitesimal (positive, or negative, or zero).

Now let $\sigma = \{s_n\}$ be any convergent sequence in m (i.e. in R_0), with limit s. Then $s - s_n$ is infinitesimal for all infinite n. Let θ be the lowest upper bound of the elements of the set T of all numbers $|s - s_n|$ for $n \geqq \omega$. θ exists since T is a bounded Q-set, and the fact that every bounded set has a lowest upper bound transfers from R_0 to $^*R$ since it can be expressed as a sentence of K_0. But the elements of T are all infinitesimal, and so θ also must be infinitesimal (possibly zero). For if τ is an upper bound for T, and τ is not infinitesimal then $\frac{1}{2}\tau$ also is an upper bound for T and so τ cannot be the lowest upper bound of T.

We have to show that

$$s = F_\alpha(\sigma) = {}^0\!\left(\sum_{n=1}^{\infty} a_n s_n\right)$$

or, which is the same that $|s - \sum_{n=1}^{\infty} a_n s_n|$ is infinitesimal.

Now, computing in $^*R$,

$$\begin{aligned}
\left|s - \sum_{n=1}^{\infty} a_n s_n\right| &= \left|\left(\sum_{n=\omega}^{\Omega} a_n + \varepsilon\right)s - \sum_{n=1}^{\infty} a_n s_n\right| \\
&= \left|\varepsilon s - \sum_{n=1}^{\omega-1} a_n s_n + \sum_{n=\omega}^{\Omega-1} a_n(s - s_n) - \sum_{n=\Omega}^{\infty} a_n s_n\right| \\
&\leqq |\varepsilon|\,|s| + \sum_{n=1}^{\omega-1} |a_n|\,|s_n| + \sum_{n=\omega}^{\Omega-1} |a_n|\,|s - s_n| + \sum_{n=\Omega}^{\infty} |a_n|\,|s_n| \\
&\leqq |\varepsilon|\,|s| + \eta\|\sigma\| + A\theta + \eta\|\sigma\| \leqq (|\varepsilon| + 2\eta)\|\sigma\| + A\theta
\end{aligned}$$

where the sum on the right hand side is indeed infinitesimal. This completes the proof of 3.1.

Let (a_{kn}) be any Toeplitz-matrix in the ordinary sense, i.e. in R_0, and let ω be an infinite positive integer. Then the sequence $\alpha = \{a_n\}$ which is given by 2.2 is an S-bounded form since, by 1.2,

$$\sum_{n=1}^{\infty} | a_n | = \sum_{n=1}^{\infty} | a_{\omega n} | < M .$$

At the same time, α satisfies 3.2 and 3.3 by 1.1 and 1.3 respectively. It follows that $F_\alpha(x)$ is a generalized limit.

If we wish to ensure that an S-bounded form defines a Banach-Mazur limit we require an additional condition.

3.6. THEOREM. *Let $\alpha = \{a_n\}$ be an S-bounded form which satisfies 3.2 and 3.3 such that $a_n \geqq 0$ for all n and such that*

$$\sum_{n=1}^{\infty} | a_{n+1} - a_n | \tag{3.7}$$

is infinitesimal. Then $F_\alpha(x)$ is a Banach-Mazur limit.

Proof. We know from 3.1 that $F_\alpha(x)$ is at any rate a Hahn-Banach limit. Thus, it only remains for us to show that $F_\alpha(x)$ satisfies condition 1.7, for any convergent sequence $\sigma = \{s_n\}$ in m. Computing in $^*R$, we have in fact

$$\left| \sum_{n=1}^{\infty} a_n s_n - \sum_{n=1}^{\infty} a_n s_{n+1} \right| = \left| a_1 s_1 + \sum_{n=1}^{\infty} (a_{n+1} - a_n) s_{n+1} \right|$$
$$\leqq | a_1 | \, \| \sigma \| + \left(\sum_{n=1}^{\infty} | a_{n+1} - a_n | \right) \| \sigma \| ,$$

where the right hand side is infinitesimal, by 3.2 and 3.7. Hence,

$$\sum_{n=1}^{\infty} a_n s_n \simeq \sum_{n=1}^{\infty} a_n s_{n+1}$$

and so

$$F_\alpha(\{s_n\}) = {}^0\!\left(\sum_{n=1}^{\infty} a_n s_n \right) = {}^0\!\left(\sum_{n=1}^{\infty} a_n s_{n+1} \right) = F_\alpha(\{s_{n+1}\}) ,$$

showing that 1.7 is satisfied. The truth of 1.6 is obvious, proving 3.6.

If $a_n = a_{\omega n}$, for infinite ω, where (a_{kn}) is a Toeplitz-matrix as above then 3.7 is satisfied provided

$$\lim_{k \to \infty} | a_{k,\, n+1} - a_{kn} | = 0 \tag{3.8}$$

in the standard sense, i.e. in R_0, and $a_n \geqq 0$ for all n, provided the elements of the matrix are nonnegative.

A particular example of a sequence $\{a_n\}$ which provides a Banach-Mazur limit is given by 2.1.

4. Representation of continuous linear functionals on m in non-standard analysis. We have seen that every S-bounded form in a non-standard model of analysis gives rise to a continuous linear functional on m. Conversely, we may ask whether every continuous linear functional on m can be represented in this way. In this direction, we have the following rather strong result, whose proof is the main purpose of this section.

4.1. THEOREM. *There exists a non-standard model of analysis, $^*R$, such that for every continuous linear functional $F(x)$ on m, there is an S-bounded form α in $^*R$ such that $F(x) \equiv F_\alpha(x)$ identically on m.*

In order to prove 4.1, we require some auxiliary considerations which refer to the standard case (i.e. to R_0).

Let $\sigma^1, \cdots, \sigma^\nu$ be any finite sequence of elements of m, where

$$\sigma^i = \{s_1^i, s_2^i, \cdots, s_n^i, \cdots\}\,, \qquad i = 1, \cdots, \nu$$

and let ε be positive, otherwise arbitrary. Then (compare Ref. 1, p. 69) there exists a positive integer μ, such that for every set of real numbers $\lambda_1, \cdots, \lambda_\nu$,

$$\| \lambda_1\sigma^1 + \lambda_2\sigma^2 + \cdots + \lambda_\nu\sigma^\nu \| \leqq \max_{1\leqq i\leqq\mu} | \lambda_1 s_i^1 + \lambda_2 s_i^2 + \cdots + \lambda_\nu s_i^\nu |(1+\varepsilon)\,.$$

It follows that if we define the μ-dimensional vectors τ^i, $i = 1, \cdots, \nu$, by

$$\tau^i = (s_1^i, s_2^i, \cdots, s_\mu^i)$$

and use the sup (lowest upper bound) norm in the space m' spanned by these vectors, then for every set of real numbers $\lambda_1, \cdots, \lambda_\nu$

$$\| \lambda_1\sigma^1 + \lambda_2\sigma^2 + \cdots + \lambda_\nu\sigma^\nu \| \leqq (1+\varepsilon)\| \lambda_1\tau^1 + \lambda_2\tau^2 + \cdots + \lambda_v\tau^\nu \|\,.$$

Hence, for any continuous linear functional $F(x)$ on m,

$$\begin{aligned} | F(\lambda_1\sigma^1 + \cdots + \lambda_\nu\sigma^\nu) | &\leqq \| F(x) \| \, \| \lambda_1\sigma^1 + \cdots + \lambda_\nu\sigma^\nu \| \\ &\leqq (1+\varepsilon)\| F(x) \| \, \| \lambda_1\tau^1 + \cdots + \lambda_\nu\tau^\nu \| \end{aligned}$$

which is equivalent to

$$(4.2) \quad | \lambda_1 F(\sigma^1) + \cdots + \lambda_2 F(\sigma^\nu) | \leqq (1+\varepsilon)\| F(x) \| \, \| \lambda_1\tau^1 + \cdots + \lambda_\nu\tau^\nu \|\,.$$

In m', define a functional $G(x)$ by

$$G(\lambda_1\tau^1 + \cdots + \lambda_\nu\tau^\nu) = \lambda_1 F(\sigma^1) + \cdots + \lambda_\nu F(\sigma^\nu) . \tag{4.3}$$

This definition is unique, for 4.2 shows that if two representations of the form $\lambda_1\tau^1 + \cdots + \lambda_\nu\tau^\nu$ coincide, then the corresponding expressions on the right hand side of 4.3 coincide. Moreover, by 4.3, $G(x)$ is clearly additive, and by 4.2 it is also continuous, with norm $\|G(x)\| \leqq (1+\varepsilon)\|F(x)\|$ (where the norm of F and G refer to the spaces m and m' respectively).

Now let m_μ be the full μ-dimensional real space under the sup norm, so that m' is a subspace of m_μ. Extend $G(x)$ to all of m_μ without increasing its norm and, without fear of confusion, call the result again $G(x)$. Let $\xi^i = (\delta_{i1}, \delta_{i2}, \cdots, \delta_{i\mu})$, $i = 1, \cdots, \mu$, where δ_{ik} is the Kronecker delta, and put $a_i = G(\xi^i)$. Let

$$\xi^0 = \xi^1 \, sg\, a_1 + \xi^2 \, sg\, a_2 + \cdots + \xi^\mu \, sg\, a_\mu$$

(where $sg\, a$ is -1, 0, or 1 according a is negative, zero, or positive) then $\|\xi^0\| = 1$ provided at least one of the a_i is different from zero. Also,

$$\begin{aligned} G(\xi^0) = |a_1| + |a_2| + \cdots + |a_\mu| &\leqq \|G(x)\|\,\|\xi^0\| \\ &\leqq (1+\varepsilon)\|F(x)\|\,|\xi^0\| \end{aligned}$$

and so, in any case

$$|a_1| + |a_2| + \cdots + |a_3| \leqq (1+\varepsilon)\|F(x)\| .$$

Now

$$\tau^i = s_1^i\xi^1 + s_2^i\xi^2 + \cdots + s_\mu^i\xi^\mu$$

and so

$$F(\sigma^i) = G(\tau^i) = a_1 s_1^i + a_2 s_2^i + \cdots + a_\mu s_\mu^i .$$

Summing up, we have established (compare Ref. 1, p. 70).

4.4. THEOREM. *Let $F(x)$ be a continuous linear functional on m, let $\varepsilon > 0$, and let $\sigma^i = \{s_n^i\}$ be arbitrary elements of m, $\varepsilon = 1, \cdots, \nu$. Then there exist real numbers $a_1, \cdots, a_\mu$, such that*

$$a_1 s_1^i + a_2 s_2^i + \cdots + a_\mu s_\mu^i = F(\sigma_i) , \qquad i = 1, \cdots, \nu \tag{4.5}$$

while

$$|a_1| + |a_2| + \cdots + |a_\mu| \leqq (1+\varepsilon)\|F(x)\| . \tag{4.6}$$

We are now in a position to prove 4.1.

Consider the statement:

"x is an absolutely convergent sequence, $x = \{x_1, x_2, \cdots\}$ such that $\sum_{n=1}^{\infty} |x_n|$ does not exceed y, and z is a bounded sequence, $z = \{z_1, z_2, z_3, \cdots\}$, such that $\sum_{n=1}^{\infty} x_n z_n$ is equal to w."

This may be formulated in L as a predicate $Q(x, y, z, w)$ whose extralogical constants occur in K_0. We now extend our vocabulary by introducing for each bounded linear functional $F(x)$ on m a constant a_F, different a_F being used for different functionals. For any continuous linear functional $F(x)$ on m, and for any σ in M, we define a sequence of formal sentences $Y_n(F, \sigma)$ by

$$(4.7) \qquad Y_n(F, \sigma) = Q\left(a_F, \left(1 + \frac{1}{n}\right) \| F(x) \|, \sigma, F(\sigma)\right) \qquad n = 1, 2, \cdots .$$

Let K_1 be the set of all sentences which are obtained in this way. We claim that the set $H = K_0 \cup K_1$ is consistent. In order to establish this fact it is sufficient to show that $H' = K_0 \cup K_1'$ is consistent where K_1' is an arbitrary finite subset of K_1. We may limit the class of K_1' to be considered somewhat by observing that for any F and σ,

$$Y_n(F, \sigma) \supset Y_m(F, \sigma)$$

is deducible from K_0 provided $n > m$. This implies that we may suppose the subscripts of the sentences which belong to the given K_1' to be all equal. Indeed, if this is not the case from the outset we replace them all by the greatest subscripts which occurs in K_1' to begin with.

Suppose in accordance with this remark that K_1' consists of the sentences

$$(4.8) \qquad \begin{array}{l} Y_n(F_1, \sigma^1),\ Y_n(F_1, \sigma^2),\ \cdots,\ Y_n(F_1, \sigma^{\nu_1}) , \\ Y_n(F_2, \sigma^{\nu_1+1}),\ Y_n(F_2, \sigma^{\nu_1+2}),\ \cdots,\ Y_n(F_2, \sigma^{\nu_2}) , \\ \cdots\cdots\cdots\cdots\cdots\cdots\cdots\cdots\cdots\cdots \\ Y_n(F_k, \sigma^{\nu_{k-1}+1}),\ Y_n(F_k, \sigma^{\nu_{k-1}+2}),\ \cdots,\ Y_n(F_k, \sigma^{\nu_k}) . \end{array}$$

In order to prove that H' is consistent we shall show that, with a suitable interpretation of the constants $a_{F_1}, \cdots, a_{F_k}$, R_0 becomes a model of H'. It is in fact evident that R_0 is a model of K_0. Interpreting a_{F_1} as the infinite sequence

$$\{a_1, \cdots, a_\mu, 0, 0, 0, \cdots\}$$

for a suitable set of real numbers $(a_1, \cdots, a_\mu)$ such as exists according to Theorem 4.4, if we put $\varepsilon = 1/n$ and identify F with F_1, and $\sigma^1, \cdots, \sigma^\nu$ with the present $\sigma^1, \cdots, \sigma^{\nu_1}$ respectively, we find that the sentences in the first line of 4.8 also holds in R_0. A similar procedure shows that, with the appropriate interpretation, the remaining sentences of 4.8 also hold in R_0. This shows that H' is consistent and hence, that H is consistent.

Let $^*R$ be a model of H. For any continuous linear functional on m, $F(x)$, the constant a_F denotes an infinite sequence

$$\alpha = \{a_1, a_2, a_3, \cdots, a_k, \cdots\}$$

(where the subscript varies over the positive integers in $^*R$) such that for any $\sigma = \{s_k\}$ in m,

$$\sum_{k=1}^{\infty} a_k s_k = F(\sigma) . \tag{4.9}$$

Moreover, for any finite positive integer n,

$$\sum_{k=1}^{\infty} | a_k | \leqq \left(1 + \frac{1}{n}\right) || F(x) || . \tag{4.10}$$

Indeed, the validity of 4.9 and 4.10 is asserted by $Y_n(F, \sigma)$, and this sentence holds in $^*R$. Since n is a positive and finite integer, but otherwise arbitrary, we may conclude from 4.10 that

$${}^0\left(\sum_{k=1}^{\infty} | a_k |\right) \leqq || F(x) || \tag{4.11}$$

so that a is an S-bounded form. Equation 4.9 shows that $F_\alpha(x) \equiv F(x)$ identically on m since it implies the weaker relation ${}^0(\sum_{k=1}^{\infty} a_k s_k) = F(\sigma)$. $^*R$ is a model of K_0 but it cannot be the standard model, R_0, for it is not true that every continuous linear functional in ${}^0R$ can be represented as in 4.9. This completes the proof of 4.1.

Suppose now that we have strict inequality in 4.11, then there exists a standard number M, $0 < M < || F(x) ||$ such that

$$\sum_{k=1}^{\infty} | a_k | \leqq M .$$

Hence, by 4.9,

$$| F(\sigma) | = \left| \sum_{k=1}^{\infty} a_k s_k \right| \leqq || \sigma || \sum_{k=1}^{\infty} | a_k | \leqq || \sigma || M$$

for all σ in m. But this contradicts the definition of $|| F(x) ||$ and shows that 4.11 may be replaced by

$${}^0\left(\sum_{k=1}^{\infty} | a_k |\right) = || F(x) || . \tag{4.12}$$

We also observe that the sequences which correspond to the a_{F_i} in the consistency proof for H' are all finite, i.e. they take the value zero for sufficiently large subscripts. We may therefore add this as a requirement for the sequences $\alpha = \{a_1, a_2, \cdots, a_k, \cdots\}$ and we may attain in this way that every α is Q-finite i.e. that a_k is equal to zero for subscripts greater than some ν which depends on α and which may be infinite. Our remarks are summed up in the following corollary.

4.13. COROLLARY TO 4.1. *In* 4.1, *we may add the three conditions that the sequences* α *are finite, that* $\sum_{k=1}^{\infty} a_k s_k = F(\sigma)$ *for every* σ *in* m—*and not only* ${}^0(\sum_{k=1}^{\infty} a_k s_k) = F(\sigma)$—*and that* ${}^0(\sum_{k=1}^{\infty} |a_k|) = \| F(x) \|$.

5. Representation of continuous linear functionals on m by ultrafilter limits. In order to be able to follow the contents of this section, the reader should have some familiarity with the theory of ultrapowers. He will then see that the non-standard model of analysis constructed in the preceding section may be obtained as an ultrapower of R_0. Let $J = \{r\}$ be the index set of the ultrapower and let D be the ultrafilter on J which determines $^*R$. An infinite sequence in $^*R$ then is an equivalence class, with respect to D, of sets of infinite sequences $\sigma^\nu = \{s_1^\nu, s_2^\nu, \cdots\}$ which are indexed in J.

A set of real numbers a_ν which is indexed in J tends to the limit a in the ultrafilter D, and we write

$$\lim_J a_\nu = a(D) ,$$

if for every positive ε, the set $\{\nu \mid |a - a_\nu| < \varepsilon\}$ belongs to D. With this notation, we have the following "translation" of 4.1, taking into account 4.13.

5.1. THEOREM. *There exists an index set* $J = \{\nu\}$ *and an ultrafilter* D *on* J *such that—*

for every continuous linear functional $F(x)$ *on the space* m *there exists a set of sequences* $a^\nu = \{a_1^\nu, a_2^\nu, \cdots\}$ *which is indexed by* J *and which satisfies the following conditions.*

5.2. *The sequences* a^ν *are finite in the sense that* $a_n^\nu = 0$ *for all* ν *greater than some positive integer* k_ν,

$$\lim \sum_{n=1}^{k_\nu} | a_n | = \| F(x) \| \qquad (D) , \tag{5.3}$$

and

5.4. *for any* $\sigma = \{s_1, s_2, \cdots\}$ *in* m,

$$\lim_{J} \sum_{n=1}^{k_\nu} a_n s_n = F(\sigma)(D) .$$

Moreover, in view of 4.9 *we may replace* 5.4 *by the condition*

5.5. *for any* σ *in* m, *the set*

$$\left\{\nu \,\middle|\, \sum_{n=1}^{k_\nu} a_n s_n = F(\sigma)\right\}$$

belongs to D—

which is somewhat stronger than 5.4.

Theorem 5.1 is not only an immediate consequence of 4.1 and 4.13, but conversely it implies those results as we may see by taking as the required non-standard model of analysis the ultrapower $(R_0)^J_D$, i.e. the direct product of a set of copies of R_0 which are indexed in J, reduced with respect to the ultrafilter D. Theorem 5.1 has the "advantage" that it does not involve the notion of a non-standard model of analysis but it lacks the intuitive significance of the preceeding theory.

Except for the requirement of positivity in the definition of the Banach-Mazur limit all our concepts and results carry over to the complex case. The proofs also remain applicable with minor modifications. In particular, in § 4.4 $sg\, a$ has to be replaced by $e^{-i \arg a}$ for $a \neq 0$.

The reader may find it interesting to compare 5.1 above with the representation of continuous linear functionals on m by Moore-Smith limits, which is due to Hildebrandt (Ref. 4) and with the matrix representation due to Mazur (compare Ref. 1, p. 72) which, however, applies only to separable subspaces of m.

In conclusion the author is pleased to acknowledge that his thinking on the subject of the present paper has been stimulated by conversations with P. Katz, W. A. J. Luxemburg, A. Meir, and D. Scott.

References

1. S. Banach, *Théorie des opérations linéaires*, Warsaw, 1932.
2. R. G. Cooke, *Infinite matrices and sequence spaces*, London, 1950.
3. N. Dunford and J. T. Schwartz, *Linear operators, part I: general theory*, New York, 1958.
4. T. H. Hildebrandt, *On bounded functional operations*, Trans. Amer. Math. Soc., **36** (1934), 868–875.
5. W. A. J. Luxemburg, *Non-standard Analysis*, lecture notes, California Institute of Technology, 1962.
6. ———, *Two applications of the method of construction by ultrapowers to Analysis*, Bull. Amer. Math. Soc., **68** (1962), 316–419.

7. A. Robinson, *Non-standard Analysis*, Proc. Royal Acad. Sc., Amsterdam, series A, **64** (1961), 423–440.

8. ———, *Complex function theory over non-archimedean fields*, Technical (scientific) note no. 30, U.S. Air Force contract No. **61** (052)-187, Jerusalem, 1962.

9. ———, *Introduction to model theory and to the metamathematics of algebra*, Amsterdam, 1963.

10. R. Sikorski, *On the existence of the generalized limit*, Studia Mathematicae **12** (1951), 117–124.

UNIVERSITY OF CALIFORNIA, LOS ANGELES

On the Theory of Normal Families

Abraham Robinson [1]

1. *Introduction*

For about one hundred and fifty years after its inception in the seventeenth century, Mathematical Analysis developed vigorously on inadequate foundations. And despite this inadequacy the precise formal or quantitative results produced by the leading mathematicians of that period have stood the test of time. In the first half of the nineteenth century, the concept of the limit, which had been advocated previously by Newton and d'Alembert, gained ascendancy. Cauchy, whose influence was instrumental in bringing about this change, still based his arguments on the intuitive concept of an infinitely small number as a variable tending to zero. At the same time (disregarding a well-known exception), he handled this concept with remarkable efficiency and set the stage for the formally more satisfactory theory of Bolzano and Weierstrass, now known briefly as the ε, δ -method. It was this precise approach which paved the way for the formulation of more general and more abstract concepts. Among these, we shall be concerned particularly with the notion of compactness. As applied to functions of a complex variable, this notion has led to the beautiful theory of normal families developed largely by Paul Montel. Once the basic results of that theory have been obtained, it becomes easy to give a qualitative exposition of some of the most important chapters of complex variable theory such as the Picard theory and the theory of univalent and multivalent functions. This, in turn provides a natural background for more precise quantitative theories. The leading role played by Rolf Nevanlinna in the development of the latter may justify the inclusion of the present paper in a volume dedicated to him.

[1] The author acknowledges with thanks the support received from the National Science Foundation (Grant No GP — 4038).

After the victory of the limit concept, the previously accepted approach, which was based on the notion of infinitely small and infinitely large quantities, fell into disuse. However, recent developments in Mathematical Logic have shown that this approach can be carried out within a framework which is just as rigorous as the ε, δ -method of classical Analysis. This framework permits the development of the theory of functions of a complex variable, including the theory of normal families. It involves the introduction of certain generalized or ideal functions which may have infinitely small and infinitely large values, both in their range and in their domain. We shall show how the theory of these generalized functions can be correlated with the classical theory of normal families and how it can be used to obtain results in standard complex variable theory.

The plan of the paper is as follows: The logical formulations of our theory are explained in section 2; some facts about real and complex variable theory are developed on this basis in section 3; and normal families are considered in section 4. Sections 5 and 6 contain applications to univalent or multivalent functions, to algebraic or algebroid functions, and to lacunary polynomials.

The theory presented here belongs to the area that has come to be known as Non-standard Analysis. The elementary parts of Non-standard Analysis are presented in references 1, 5 and 6. The elements of the non-standard theory of functions of a complex variable are detailed in reference 7, which also contains applications to Picard's theory. A comprehensive account of Non-standard Analysis is due to be published in book form (reference 8.) Since limitations of space force us to confine the description of the logical foundations of the theory to a sketch, the reader is advised to consult the references just mentioned for further details.

The main purpose of this paper is the presentation of a new way of looking at certain topics in complex variable theory. Nevertheless, it is believed that some of the results given here (in sections 5 and 6) have not been stated elsewhere, although they are formulated entirely in classical terms.

2. *Non-standard Models of Analysis*

Let R be the field of real numbers. We introduce a formal language, L, which enables us to make statements about R. L shall include the following symbols:

Connectives

$\sim$ (negation, "not"), $\vee$ (disjunction, inclusive "or"), $\wedge$ (conjunction, "and"), $\supset$ (implication, "if . . . then"), $\equiv$ (equivalence or biconditional, "if and only if.")

Quantifiers

$(\exists)$ (existential, "there exists") and $(\forall)$ (universal, "for all")

Parentheses

[,], left and right.

Variables

x, y, z . . . a countable set.

Individual Constants

(briefly, *constants*) which denote real numbers. We assume that for every real number, there is a constant which denotes it.

Similarly, we assume that for every set of real numbers, for every relation between real numbers, for every set of sets, every set of relations, etc. there is a symbol (a *relation symbol*) which denotes it. For example, $Q(x,y)$ may denote the relation of order $x < y$, $E(x,y)$ the equality, $S(x,y,z)$ may denote the three-place relation which holds whenever $x + y = z$, $F(x,y)$ the relation which holds whenever $y = e^x$, etc.

The last two examples show that we do not require any symbols for functions, since these may be represented by relations (i.e., by their "graphs.")

From the above symbols, we may form sentences which express facts about the real numbers, for example, the sentence

$$(\forall x)(\forall y)(\forall)\ [S(x,y,z) \supset S(y,z,x)]$$

which expresses the commutativity of addition. We shall permit quantification also with respect to higher order entities i.e., sets, relations, sets of relations, etc. For example, the sentence

$$(\forall x)(\forall y)(\exists t)\ [Q(x,y) \equiv t(y)]$$

expresses the fact that for every number x there exists the set of the numbers wich are greater than x.

Within this framework, we can express all statements which are likely to arise in classical Analysis. Replacing this vague claim by a more precise assertion, the language L permits us to formulate within it all statements which will be required by our present theory.

The precise delimitation of the set of all meaningful sentences in L is to some extent a matter of convention. In particular, it is quite customary to introduce certain *type restrictions* by which, for example, an expression such as $A(A)$ — which asserts that A satisfies itself — is neither true nor false, but *meaningless.* Although the need for ruling out such expressions does not arise directly in our theory, we can certainly do without them.

Let K be the set of all sentences which are formulated by means of the above-mentioned symbols and which are true for the field of real numbers, R. A familiar procedure of Model Theory shows that there exists a proper extension of R, to be denoted by $*R$, such that all sentences of K are true also for $*R$, in the following special sense. The individual constants which occur in K and which denote numbers in R, denote the same numbers in $*R$. The connectives have their usual interpretation. The symbols of K which denote sets, relations, sets of relations, etc. in R denote sets, relations, sets of relations, etc., respectively, in $*R$. However, the totality of subsets of $*R$ (i.e., sets of elements of $*R$) which are to be taken into account in interpreting the appropriate quantifications, "there exists a set (of elements of $*R$)" "for all sets (of elements of $*R$)" do not refer to the set of all subsets of $*R$ but out to a certain subset of the set of subsets of $*R$ to be called *internal* sets of elements of $*R$. Similarly, the expressions "there exists a relation (between elements of $*R$)" "for all relations", refer *not* to all relations between elements of $*R$ but out to a certain subset of these, to be called *internal* relations. Corresponding qualifications apply to the interpretation of quantification for other higher order entities. Thus, there are internal sets, internal relations, internal sets of relations, etc., and also internal functions, i.e., functions which are represented by internal relations. For our purposes, it is not necessary to know precisely which sets, relations, etc., are internal, although we shall see presently that we can decide this issue in certain cases. The essential point is that there exists some such system of internal entities for which $*R$ becomes a model of K. $*R$ will be called a non-standard model of analysis. We add that there are many extensions of R which may serve as non-standard models of analysis. We choose one of them and keep it fixed throughout the discussion.

As stated already, the individual constants which denote (individual) elements of R shall denote the same elements within $*R$. Regarded as elements of $*R$, we call these the *standard elements* of

$^*R$ or, more specifically, the *standard numbers* of $^*R$, regarding the other elements of $^*R$ from now on also as *real numbers*. Similarly, we call any set of elements of $^*R$, any relation between elements of $^*R$, any set of sets etc. *standard* if the entity in question is denoted by a relation symbol of L that denotes also some corresponding entity in R. However, in this case the entities of R and $^*R$ which are denoted by the same symbol in K, do not, in general, coincide.

If we wish to express, within L, statements about arbitrary elements of $^*R$, or arbitrary subsets, etc. of $^*R$, then we have to suppose that L contains in addition to the individual constants and relation symbols enumerated at the beginning of this section also constants and relation symbols to denote the remaining entities of $^*R$. If a sentence X in L contains only symbols of the former kind, i.e., symbols that denote entities in R, then we call X *standard*. It follows from the definition of K that, for any standard X, either X or $\sim X$ belongs to K. We conclude that any standard sentence X holds (i.e., is true) in $^*R$ if and only if it holds in R. For if X holds in R then $X \epsilon K$, and so X holds in $^*R$ since all sentences of X hold in $^*R$. While if X does not hold in R then $\sim X$ holds in R and belongs to K and so $\sim X$ holds also in $^*R$. It follows that X cannot hold in $^*R$. This simple observation will be made use of repeatedly in the sequel.

$^*R$ is an ordered field. For R is an ordered field and so K contains a set of axioms for the concept of an ordered field. $^*R$ is non-archimedean, for any ordered field which is a proper extension of the field of real numbers is non-archimedean.

Let N be the set of natural numbers regarded as a subset of R. Let $^*N$ be the subset of $^*R$ which is denoted in L by the same symbol as N. Then $^*N$ is an extension of N. The properties of $^*N$ to the extent in which they can be expressed by standard sentences are the same as the properties of N. $^*N$ is a proper extension of N. For the statement "For every real number r there exists a natural number greater than r" can be expressed as a sentence X of K and, accordingly, must be true also in $^*R$. Now, let a be a particular real number in $^*R$ which is greater than all standard real numbers. There exists such an a since $^*R$ is non-archimedean. Since X is true in $^*R$ there exists within $^*R$ a "natural number" (i.e., an element of $^*N$) which is greater than a and hence is greater than all standard natural numbers. $^*N$ is a non-standard model of arithmetic of the kind first considered by Skolem for a first order language.

It is easy to show that any element of $*N - N$ is greater than all elements of N. The elements of $*N - N$ will be called *infinite natural numbers*. As we have just seen, $*N - N$ is not empty.

$*N - N$ does not have a smallest element. For suppose that n is the smallest number in $*N - N$. Then $n \neq 0$. It follows that there exists an $m \, \varepsilon \, *N - N$ such that $n = m + 1$. But m cannot be standard for then n would be standard, $n \, \varepsilon \, N$. Hence $m \, \varepsilon \, *N - N$, contrary to the definition of n.

We conclude that the set of infinite natural numbers, $*N - N$, is not an internal set. For the fact that every non-empty set of natural numbers includes a smallest element can be expressed as a statement of K and, accordingly, is true also for $*R$, where we have to interpret *set* as *internal set*. On the other hand, any set of numbers in $*R$ which can be defined by means of the constants and relation symbols of K is an internal set; for example, the set of all prime numbers of $*N$ (numbers greater than 1 and divisible only by themselves and by 1) within $*N$ is an internal set.

Let $y = f(x)$ be a function from a subset of R into R. This function is represented with L by a two-place relation $(F(x,y)$. The same relation denotes a function also in $*R$, which coincides with the original function for standard value of the argument. We shall denote this extended function also by $f(x)$ (just as, for example, we use the same notation, *sin* x, for the function in question, both as a function in the field of real numbers, and as a function in the field of complex numbers.) Similar remarks apply to functions of several variables. However, if A is any set (of any type) in R, we shall denote the corresponding set in $*R$ by $*A$.

A number $a \, \varepsilon \, *R$ is called *finite* if $|a| < r$ for some standard r, otherwise a is *infinite*. This is consistent with the notion of an infinite natural number introduced above. a is called *infinitesimal* if $|a| < r$ for all positive standard r. We write $a \simeq b$ for a and b in $*R$ if $a - b$ is infinitesimal. This is an equivalence relation in $*R$. The corresponding equivalence classes are called *monads*. Using a simple Dedekind cut argument, we may show that for any finite number a there exists a unique standard number called the *standard part* of a, and denoted by ${}^0a$ such that $a \simeq {}^0a$. Thus, any monad which contains a finite number contains a unique standard number. Such a monad is called a *finite monad*. The references mentioned in section 1 may be consulted for a systematic exposition of the properties of the relation $\simeq$. In the present paper, we shall make use of some simple properties of this relation without proof. A number of less obvious facts involving it will be considered in the next section.

3. *Continuity in real and complex function theory*

Let a and b be two standard real numbers and let $f(x)$ be a standard function defined for $a < x < b$. Then $f(x)$ is defined in the first instance within R but, as we have pointed out, this definition extends automatically to $^*R$. Let x_0 be a standard real number such that $a < x_0 < b$. The proof of the following theorem provides a typical example of an elementary argument in Non-standard Analysis.

3.1. *Theorem*

$f(x)$ is continuous at x_0 (in the classical sense) if and only if $f(x_1) \simeq f(x_0)$ for all x_1 such that $x_1 \simeq x_0$, i.e., if and only if $f(x_1)$ is in the same monad as $f(x_0)$ for all x_1 that are in the same monad as x_0.

Proof

Suppose that $f(x)$ is continuous at x_0 in the classical sense, and let $x_1 \simeq x_0$. Then we have to prove $f(x_1) \simeq f(x_0)$. That is to say, we have to show that $|f(x_1) - f(x_0)| < \varepsilon$ for any standard positive ε. Given such an ε, we know that, with R, there exists a positive δ such that "for all x, $|x - x_0| < \delta$ implies $|f(x) - f(x_0)| < \varepsilon$." The statement in quotation marks can be expressed as a standard sentence, so the fact that it is true in R shows that it is true also in $^*R$. But $x_1 \simeq x_0$ implies that $|x_1 - x_0| < \delta$, and so $|f(x_1) - f(x_0)| < \varepsilon$, as asserted.

We have shown that the condition of the theorem is necessary. To prove that it is also sufficient, we suppose that it is satisfied and consider any standard x_0 and ε such that $a < x < b$ and $\varepsilon > 0$. Then it is true in $^*R$ that "there exists a positive y such that for all x, $|x - x_0| < y$ implies $|f(x) - f(x_0)| < \varepsilon$." Indeed, such a y exists within $^*R$ since any positive infinitesimal $y = \delta$ will do for the purpose. But the statement in quotation marks can be expressed as a standard sentence and so it must be true also in R. Thus, we find that, in R, there is a positive $y = \delta$ (where δ must now be standard) such that $|f(x_1) - f(x_0)| < \varepsilon$ provided $|x_1 - x_0| < \delta$. This shows that $f(x)$ is continuous at x_0 in the classical sence and completes the proof of the theorem.

We prove in a similar way

3.2. *Theorem*

Let $f(x)$ be defined in the interval $a < x < b$, a and b standard, as before. In order that $f(x)$ be uniformly continuous in the interval, it is necessary and sufficient that $f(x_1) \simeq f(x_0)$ for all x_0 and x_1 in $^*R$ such that $a < x_0 < b$, $a < x_1 < b$, and $x_1 \simeq x_0$.

Now let a and b be two numbers in $*R$, $a < b$, and let $f(x)$ be an internal function which is defined in the interval $a < x < b$ in $*R$. The concept of continuity for such a function now splits in a natural way into two. Thus, consider the function $f(x) = \sin \omega x$ where ω is an infinite positive number. This function is defined as an internal function in $*R$ in the following sense. Within R, we may define sin cx for any positive a as the unique solution of the equation $y'' + c^2 y = 0$, which has a continuous second derivative and satisfies the initial conditions $y = 0$, $y' = c$ for $x = 0$. Since the fact that such a function exists and is unique can be expressed by a standard sentence, it must be true also in $*R$. Accordingly, there is a unique function which has a continuous second derivative and satisfies the above differential equation with the specified boundary conditions, also for $c = \omega$, where ω is positive infinite. This is the function which we denote by *sin* ωx. It is defined for all real x, and more particularly, for $0 < x < 1$. Moreover, *sin* ωx is continuous at any point $x = x_0$ in that interval in the sense of the formal ε, δ-definition, since this is true for all the functions *sin* cx within R. That is to say, for any positive ε *in* $*R$ there exists a positive δ *in* $*R$ such that $| x_1 - x_0 | < \delta$ implies $| \sin \omega x_1 - \sin \omega x_0 | < \varepsilon$.

On the other hand, we may also require that for any standard positive ε (i.e., for $\varepsilon > 0$ in R) there exist a standard positive δ such that $| x_1 - x_0 | < \delta$ implies $| f(x_1) - f(x_0) | < \varepsilon$ for any x_1 in the domain of definition of the function $f(x)$. If this condition is satisfied, then we shall say that $f(x)$ is S-continuous at x_0. The function *sin* ωx is not S-continuous at any point x_0.

3.3 *Theorem*

Let $f(x)$ be an internal function with domain of definition $a < x < b$. In order that $f(x)$ be S-continuous at a point x_0, $a < x_0 < b$, it is necessary and sufficient that $x_1 \simeq x_0$ imply $f(x_1) \simeq f(x_0)$ for all x_1 such that $a < x_1 < b$.

If a, b, x_0, and $f(x)$ are all standard, then 3.3 reduces to 3.1.

Proof of 3.3

Necessity is proved similarly as in 3.1 To prove sufficiency, let ε be any standard positive number. Then for any x_1 such that $a < x_1 < b$, $| f(x_1) - f(x_0) | < \varepsilon$ since $f(x_1) - f(x_0)$ is actually infinitesimal. It is therefore true for all infinite natural numbers n that "for all x_1 such that $a < x_1 < b$, $| x_1 - x_0 | < 1/n$ implies $| f(x_1) - f(x_0) | < \varepsilon$."

Let A be the set of natural numbers n which satisfy the condition in quotation marks. Then A is internal. As we have just seen, $A \supset {}^*N - N$. But ${}^*N - N$ is not an internal set, so A must contain also a finite positive integer, *e.g.*, $n = \nu$. Putting $\delta = 1/\nu$, we see that $| f(x_1) - f(x_0) | < \varepsilon$ provided $a < x_1 < b$ and $| x_1 - x_0 | < \delta$. This shows that $f(x)$ is *S*-continuous at x_0 and completes the proof of 3.3.

We may regard a complex number as an ordered pair of real numbers and, accordingly, we may express the theory of functions of a complex variable also within L. Functions of a complex variable may then be represented as two-place relations between ordered pairs of real numbers or, alternatively as four-place relations between real numbers. Which of these (or similar) representations we choose is irrelevant for our arguments. Accordingly, we shall from now on talk of the complex plane, $C = R \times R$, and of the corresponding extended complex plane, ${}^*C = {}^*R \times {}^*R$. Again, if we wish to discuss complex variable theory on the Riemann sphere, we may realize the latter by the points of a sphere in three-dimensional space, e.g., $x^2 + y^2 + (z - 1)^2 = 1$.

A "complex number" $z \,\varepsilon\, {}^*C$ is said to be *finite* if $| z |$ is finite, otherwise z is *infinite*. z is *infinitesimal*, if $| z |$ is infinitesimal. For $z_1 \,\varepsilon\, {}^*C$, $z_2 \,\varepsilon\, {}^*C$ we write $z_1 \simeq z_2$ if $z_1 - z_2$ is infinitesimal. The equivalence classes determined by this equivalence relation are again called *monads*. A monad which contains a finite point, contains a unique standard plint. In this case, monad is called *finite*. Thus, if $z = x + iy$ is finite, then there exists a unique standard point, written ${}^0z$ and called the standard part of such that $z \simeq {}^0z$. Moreover, in that case ${}^0z = {}^0x + i{}^0y$.

Similar notions can be defined on the Riemann sphere if we use the chordal metric. In this case, any point is either finite, and belongs to the monad of some standard point, or it is infinite and belongs to the monad of the point at infinity.

Let D be an internal set of points in ${}^*C$, and let μ be a monad which is a subset of D while z_0 is a point which belongs to μ.

3.4 *Theorem*

There exists a standard positive r such that all points of the disk $\triangle = \{z \mid | z - z_0 < r\}$ belong to D.

Proof

For all infinite n, the disk $\triangle$ is a subset of μ and hence, a subset of D. Thus, the set A of natural numbers n such that $\{z \mid | z - z_0 | < 1/n\} \subset D$ contains all infinite natural numbers. But then (compare the argument used in the proof of 3.3). A contains also some finite positive ν. Putting $r = 1/\nu$, we obtain

$$\triangle = \{z \mid |z - z_0| < r\} \subset D,$$

which proves the assertion.

The union of all monads which are contained entirely in D is called the *S-interior* of D. The union of all finite monads which are contained entirely in D is called the *finite S-interior* of D. The various notions of continuity are obtained by obvious adaptation from the corresponding definitions for real functions. Thus, $f(z)$ is S-continuous at z_0 if for every standard positive ε there exists a standard positive δ such that $|z_1 - z_0| < \delta$ implies $|f(z_1) - f(z_0)| < \varepsilon$.

3.5 *Theorem*

Let $f(z)$ be an internal complex function of a complex variable which is defined in an internal set D, and let z_0 be a point which belongs to the S-interior of D. Then $f(z)$ is S-continuous at z_0 if and only if $f(z_1) \simeq f(z_0)$ for all z_1 such that $z_1 \simeq z_0$, i.e., if and only if $f(z_1)$ belongs to the same monad as $f(z_0)$ whenever z_1 belongs to the same monad as z_0.

The proof is similar to that of Theorem 3.3.

Observe that if $f(z)$ is S-continuous at some point which belongs to the S-interior of D then it is S-continuous at any other point of the same monad,

4. *Normal functions and normal families*

Let D be an internal subset of $^*C$ which is a *domain*. That is to say, if A is the set of all domains (open and connected sets of points) in C, then D belongs to the corresponding set $^*A$ in the non-standard model.

Let $f(z)$ be an internal function which is holomorphic in D. That is to say, if B is the set of ordered pairs (X, y) such that X is a domain in C and y is a function which is holomorphic in X, in the ordinary sense, then $(D, f(z))$ belongs to the corresponding set $^*B$ in the non-standard model.

Let μ be a monad which belongs entirely to D, $\mu \subset D$. We say that *$f(z)$ is normal in μ* if $f(z)$ is S-continuous at some point, and hence at all points, of μ and also if $f(z)$ is infinite for all points of μ. Moreover, in these cases, we shall say also that *$f(z)$ is normal* at z_0 for any $z_0 \varepsilon \mu$. If $f(z)$ is not normal in a monad $\mu \subset D$ (or at a point $z_0 \varepsilon \mu$) then we say that $f(z)$ is irregular in μ (or, at z_0).

4.1 *Theorem*

In order that $f(z)$ be irregular in the monad $\mu \subset D$, it is necessary and sufficient that there exist two points $z_1 \varepsilon \mu$ and $z_2 \varepsilon \mu$ such that $f(z_1)$ is finite and $f(z_2)$ is infinite.

Proof

The condition is sufficient. For if $f(z)$ is normal in μ then it is either infinite throughout μ or it is S-continuous at some point $z_0 \varepsilon \mu$. The first possibility is evidently ruled out by the condition of the theorem. But, if $f(z)$ were S-continuous at $z_0 \varepsilon \mu$ then, by 3.5, $f(z_1) \simeq f(z_0)$ $f(z_2) \simeq f(z_0)$ and so $f(z_1) \simeq f(z_2)$ for all z_1, z_2 in μ, i.e., $f(z_1) - f(z_2)$ is infinitesimal. This is incompatible with the condition that $f(z_1)$ be finite and $f(z_2)$ infinite, for in that case $f(z_2) - f(z_1)$ also is infinite.

The condition of the theorem is also necessary. For suppose $f(z)$ is irregular in μ. Then $f(z)$ must be finite for some point of μ e.g., z_1. Now, if $f(z) \simeq f(z_1)$ for all $z \varepsilon \mu$ then $f(z)$ would be normal in μ by 3.5. It follows that there exists a $z_0 \varepsilon \mu$ such that $f(z_0) - f(z_1)$ is not infinitesimal. Put $| z_0 - z_1 | = r$ and let γ be the circle $| z - z_1 | = \sqrt{r}$. Then $\gamma \subset \mu$ since $\sqrt{r}$, like r, is infinitesimal. Moreover, since $f(z)$ is holomorphic in D there exists a $z_2 \varepsilon \gamma$ such that $| f(z) \leq | f(z_2) |$ for all z in the closed disk $| z - z_1 | \leq \sqrt{r}$. Hence, by the lemma of Schwarz

$$4.2 \qquad | f(z_0) - f(z_1) | \leq \frac{| z_0 - z_1 |}{\sqrt{r}} | f(z_2) | = \sqrt{r} \, | f(z_2) |$$

But $f(z_0) - f(z_1)$ is not infinitesimal so there exists a standard positive number a such that $| f(z_0) - f(z_1) | \geq a$. Hence, by 4.2

$$4.3 \qquad | f(z_2) | \geq \frac{a}{\sqrt{r}}$$

where the right hand side of the inequality is infinite. This shows that $f(z_2)$ is infinite and completes the proof of the theorem.

Now suppose more particularly that the domain D is standard. That is to say, there exists a domain $\triangle$ in the ordinary complex plane such that $D = {}^*\triangle$. Then it is not difficult to see that the *finite* S-interior D_i of D is connected.

Let $f(z)$ be an internal function which is defined and holomorphic in the standard domain $D = {}^*\triangle$. Suppose that $f(z)$ is finite for all points of the finite S-interior, D_i, of D. It then follows from 4.1 that $f(z)$ is normal, more precisely S-continuous, at all points of D_i. We define a standard complex function ${}^0f(z)$ called the *standard part* of

$f(z)$, in D in the following way. Let z_0 be any point of $\triangle$, and let μ be the monad that contains z_0. Then $\mu \subset D_i$ and so the values of $f(z)$ for $z \varepsilon \mu$ all have the same standard part, ${}^0(f(z_0))$. We set ${}^0f(z_0) = {}^0(f(z_0))$. This defines ${}^0f(z)$ in $\triangle$ and hence, on passing from R to $^*R$, in $^*\triangle$. It is not difficult to see that, by virtue of the S-continuity of $f(z)$, the function ${}^0f(z)$ is continuous in the classical sense in $\triangle$ and hence in $^*\triangle$.

Moreover, ${}^0f(z)$ is holomorphic in $\triangle$, and hence in $^*\triangle$. For let $z_0 \varepsilon \triangle$ and let r be standard positive and such that the closed circular disk $|z - z_0| \leq r$ is entirely in $\triangle$. Let $z_1 \varepsilon \triangle$ and such that $|z_1 - z_0| < r$. Then, by Cauchy's formula, which applies also to $f(z)$

4.4 $$f(z_1) = \frac{1}{2\pi i}\int_\lambda \frac{f(z)}{z - z_1}\,dz$$

where the integral on the right hand side is taken round the circle $\gamma : |z - z_0| = r$. Now the expression $|{}^0f(z) - f(z)|$ has a maximum on γ, since both $f(z)$ and ${}^0f(z)$ are continuous. This maximum must be infinitesimal, by the definition of ${}^0f(z)$. From this fact, we deduce without difficulty that $\frac{1}{2\pi i}\int_\lambda \frac{{}^0f(z)}{z - z_1}\,dz - \frac{1}{2\pi i}\int_\lambda \frac{f(z)}{z - z_1}\,dz$ is infinitesimal. At the same time, the difference ${}^0f(z_1) - f(z_1)$ must be infinitesimal. Hence

$$ {}^0f(z_1) \simeq f(z_1) = \frac{1}{2\pi i}\int_\gamma \frac{f(z)}{z - z_1}\,dz \simeq \frac{1}{2\pi i}\int_\gamma \frac{{}^0f(z)}{z - z_1}\,dz$$

4.5 $${}^0f(z_1) \simeq \frac{1}{2\pi i}\int_\gamma \frac{{}^0f(z)}{z - z_1}\,dz$$

But the complex numbers on the left and right hand sides of 4.5 are both standard and so we may replace $\simeq$ by $=$,

$${}^0f(z)_1) = \frac{1}{2\pi i}\int_\gamma \frac{{}^0f(z)}{z - z_1}\,dz$$

This shows that ${}^0f(z)$ satisfies Cauchy's formula for $|z - z_0| < r$ in $\triangle$ and, accordingly, is holomorphic at z_0 and hence, throughout $\triangle$.

4.6 *Theorem*

Let $f(z)$ be an internal function which is holomorphic for $|z| < r$, r standard positive. Suppose that $f(z)$ is infinitesimal but different from zero for all infinitesimal z. Then there exists a standard positive $\varrho < r$ such that $f(z)$ is infinitesimal for $|z| < \varrho$.

Proof

Since $f(z)$ is infinitesimal for infinitesimal z, it is S-continuous at the origin. Hence (as can also be seen directly) there is a standard positive $r' \leq r$ such that $|f(z)| \leq 1$ for $|z| < r'$. Accordingly, the function $F(z) = {}^0f(z)$ exists for $|z| < r'$ and is holomorphic for such z. Moreover, $F(0) = {}^0f(0)) = 0$. Suppose that $F(z)$ is not identically zero for $|z| < r'$. Then there exists a standard positive $q < r'$ such that $F(z) \neq 0$ for $0 < |z| \leq q$. On the other hand, there must be a number z_1 such that $0 < {}^0|z_1| \leq q$ and such that $f(z_1)$ is infinitesimal. If $f(z)$ vanishes for some z_1 for which $|z_1| \leq q$ then this is an appropriate z_1 since ${}^0|z_1| \neq 0$ by one of the assumptions of the theorem. If $f(z)$ does not vanish for $|z| \leq q$ then it takes its minimum in this closed disk for some z_1 such that $|z_1| = q$ and hence ${}^0|z_1| = q$. But then $f(z_1)$ is infinitesimal since $|f(z_1)| \leq |f(0)|$ and $f(0)$ is infinitesimal. Put $z_2 = {}^0z_1$. Then $|z_2| = |{}^0z_1| = {}^0|z_1|$ and so $0 < |z_2| \leq q$. Also, $F(z_2) = {}^0(f(z_2)) = {}^0(f(z_1)) = 0$ and this contradicts the fact that $F(z) \neq 0$ for $0 < |z| \leq q$. We conclude that $F(z)$ must be identically zero for $|z| < r'$. It follows that $f(z)$ is infinitesimal for $|z| < \frac{1}{2} r'$ since $F(z) \simeq f(z)$ for such z. Thus, the conclusion of the theorem is satisfied for $\varrho = \frac{1}{2} r'$.

4.7 *Theorem*

Let $f(z)$ be an internal function which is holomorphic for $|z| < r$, r standard positive. Suppose that $f(z)$ is infinite for all infinitesimal z. Then there exists a standard positive $\varrho < r$ such that $f(z)$ is infinite for $|z| < \varrho$.

Proof

It follows from the assumptions of the theorem that $f(z) \neq 0$ for $|z| < 1/n$ where n is any infinite natural number. Hence, by an argument used repeatedly already, there exists a finite natural number ν such that $f(z) \neq 0$ for $|z| < r' = 1/\nu$. Accordingly the function $g(z) = 1/f(z)$ is holomorphic for $|z| < r'$. Moreover, $g(z)$ is infinitesimal for infinitesimal z and different from zero for such z (and in fact for all z such that $|z| < r'$). Hence, there exists a standard

$\varrho > 0$ such that $g(z)$ is infinitesimal for $|z| < \varrho$. But then $f(z)$ is infinite for $|z| < \varrho$, as required.

4.8 *Theorem*

Let $f(z)$ be holomorphic in the standard domain $D = {}^*\triangle$ and normal at all points of the finite S-interior of D, D_i. Then $f(z)$ is either finite at all points of D_i or infinite at all points of D_i.

Proof

Suppose that the assumptions of the theorem are satisfied. Then $\triangle$ is open. Let $\triangle_1$ be the set of all points z in $\triangle$ such that $f(z)$ is finite. We claim that $\triangle_1$ is open. For suppose that $f(z_0)$ is finite, $z_0 \in \triangle$, and let μ be the monad that contains z_0 so that $\mu \subset {}^*\triangle = D$. Since $f(z_0)$ is finite, $f(z)$ is S-continuous at z_0, so there exists a standard $\delta > 0$ such that $|f(z_1) - f(z_0)| < 1$ for $|z_1 - z_0| < \delta$. Then $|f(z_1))| \leq |f(z_1) - f(z_0)| + |f(z_0) \leq 1 + |f(z_0)|$ so that $f(z_1)$ is finite. This shows that $\triangle_1$ is open.

Let $\triangle_2$ be the set of points z of $\triangle$ such that $f(z)$ is infinite. We claim that $\triangle_2$ is open. For suppose $z_0 \varepsilon \triangle_2$ and let μ be the monad that contains z_0. Then $f(z)$ is infinite for all $z \varepsilon \mu$. Applying Theorem 4.7 to the function $f(z) = f(z_0 + z)$ we find that there exists a standard positive ϱ such that $f(z)$ is infinite for $|z - z_0| < \varrho$ This shows that $\triangle_2$ is open.

Now $\triangle = \triangle_1 \cup \triangle_2$, while $\triangle$ is connected. This shows that either $\triangle_1$ is empty or $\triangle_2$ is empty, and proves the theorem.

We shall now formulate and prove a theorem which correlates our present theory with the classical theory of normal families.

4.9 *Theorem*

Let $\triangle$ be a domain in the ordinary complex plan and let $F = \{f(z)\}$ be a family of functions which are holomorphic in $\triangle$, in the classical sence. Then, F is normal in $\triangle$ if and only if the functions $f(z) \varepsilon {}^*F$ are normal at all standard points $z \varepsilon \triangle$.

Proof

Suppose that the condition of the theorem is satisfied and let $\{f_n(z)\}$ be a sequence of elements of F. We have to show that we can extract from $\{f_n(z)\}$ a subsequence which converges either to a holomorphic function $g(z)$ or to infinity, uniformly on every compact subset of $\triangle$.

In the non-standard model, the subscripts of $\{f_n(z)\}$ range over all elements of ${}^*N$. Let ω be an infinite natural number and consider the function $f_\omega(z)$. By assumption, $f_\omega(z)$ is normal at all points of D_i. Hence, by 4.8, $f_\omega(z)$ is either finite at all points of D_i or infinite at all points of D_i.

Suppose first that $f_\omega(z)$ is finite at all points of D_i. Then $g(z) = {}^0f_\omega(z)$ exists and is holomorphic in $\triangle$ and $D = \triangle^*$. Let E be any compact subset of $\triangle$. It can be shown that $^*E \subset D_i$, although we omit the proof of this assertion. It follows that if n and m are any finite positive integers then it is true in $^*R$ that

"there exists a natural number $k > m$ such that $| g(z) - f_k(z) | < 1/n$ for all $z \varepsilon {}^*E$."

Indeed, the number $k = \omega$ is appropriate since $| g(z) - f_\omega(z) |$ is actually infinitesimal for all $z \varepsilon {}^*E$ and so, a fortiori $| g(z) - f_\omega(z) | < 1/n$. Formulating the statement in quotation marks as a standard sentence and reinterpreting that sentence in the standard model, we find that there exists a finite natural number $k > m$ such that $| g(z) - f_k(z) | < 1/n$ for all $z \in E$. Using this fact for $n = 1, 2, 3, \ldots$, we may choose a sequence of compact subsets of $\triangle$, $E_1 \subset E_2 \subset E_3 \subset \ldots$ and a sequence of natural numbers $k_1 < k_2 < k_3 \ldots$, all in the standard model, such that $\bigcup_n E_n = \triangle$ and

$| g(z) - f_{k_n}(z) | < 1/n$ for all $z \varepsilon E_n$.

This inequality shows that $\lim_{n\to\infty} f_{k_n}(z) \sim g(z)$ uniformly in any E_j.

Suppose next that $f_\omega(z)$ is infinite throughout D_i. A similar method then shows that, for a given sequence of sets E_n as above, we may find $k_1 < k_2 < k_3 < \ldots$ such that

$$| f_{k_n}(z) | > 1/n \text{ for all } z \varepsilon E_n$$

The two conditions together show that F is normal.

Conversely, suppose that F is normal and let $f(z) \varepsilon {}^*F$, and $z_0 \varepsilon \triangle$. Then we have to show that $f(z)$ is normal at z_0. Suppose that $f(z)$ is not normal at z_0. Then there exist points z_1 and z_2 in the monad of z_0, $z_1 \simeq z_0$, $z_2 \simeq z_0$ such that $f(z_1)$ is finite but $f(z_2)$ is infinite. Accordingly, we may choose a standard positive number b such that $| f(z_1) | \leq b$. For any finite natural number n it is then true in the non-standard model that "there exist an element of $^*F$ (*i.e.*, $f(z)$) and complex numbers z' and z'' (*i.e.*, z_1 and z_2) such that $| z' - z_0 | < 1/n$, $| z'' - z_0 | < 1/n$, $| f(z') | \leq b$, $| f(z'') | > n$." Formulating the statement in quotation marks as a standard sentence and reinterpreting in the standard model, we find that there exists a sequence of elements of F, $\{f_\omega(z)\}$ and sequences $z_n' \to z_0$, $z_n'' \to z_0$ such that $| f(z_n') | \leq b$, $| f(z_n'') | > n$, $n = 1, 2, \ldots$ This shows that we cannot extract a convergent subsequence from $\{f_n(z)\}$ in any neighborhood of z_0, and hence that $\{f_n(z)\}$ is not normal in $\triangle$. The proof of 4.9. is now complete.

Although the present paper is concerned with normal families of holomorphic functions, we may quote in passing the following result for meromorphic functions which is parallel to 4.9.

4.10 *Theorem*

Let $\triangle$ be a domain on the ordinary Riemann sphere. Then a family F of functions which are meromorphic in $\triangle$ is normal if and only if all elements of $*F$ are S-continuous in the chordal metric at all points of $\triangle$.

Returning to families of holomorphic functions, we observe that in consequence of 4.9, a condition which yields a normal family of functions in the classical sense, will frequently lead to internal functions which are normal in the sense of the present theory. We may also proceed in the opposite direction as exemplified by the following proof of a classical theorem (4.13.)

4.11 *Theorem*

Suppose that the internal function $f(z)$ is holomorphic in the standard domain $D = *\triangle$, and let $z_0 \varepsilon \triangle$, while $\mu \subset D$ is the monad of z_0. Then if $f(z) \neq 0$ in μ and $f(z)$ is normal in μ, the function $g(z) = 1/f(z)$ also is normal in μ.

Proof. A familiar argument shows that there exists a standard positive $r > 0$ such that $f(z)$ is holomorphic and different from zero for $|z - z_0| < r$. Hence, $g(z)$ is holomorphic for $|z - z_0| < r$. Now if $g(z)$ were not holomorphic at z_0 then it vould take both finite and infinite values in μ. But then $f(z)$ would take in μ both infinitesimal values and values that are not infinitesimal. This in turn would imply that $f(z)$ is not normal at z_0, a contradiction which proves 4.11.

4.12. *Theorem*

Let the internal function $f(z)$ be holomorphic in a standard domain which includes a disk $|z - z_0| < r$ where r is standard and positive. Suppose that there exist a standard complex w_0 and a standard positive ϱ such that $w = f(z)$ does not take any values in the disk $|w - w_0| < \varrho$ for z in the monad of z_0. Then $f(z)$ is normal at z_0.

Proof

Put $g(z) = \dfrac{1}{f(z) - w_0}$, then $|g(z)| \leq 1/\varrho$ for $z \simeq z_0$ and so $g(z)$ is normal, more precisely, S-continuous at z_0 by 4.1. It now follows from 4.11 that $f(z) - w_0 = 1/g(z)$ is normal at z_0 and so the same applies to $f(z)$.

4.13 *Theorem*

Let F be a family of functions which are holomorphic in a domain $\triangle$ in the ordinary complex plane C. Suppose that all elements $f(z)$ of F omit the values which are contained in a disk $|w - w_0| < \varrho$, $\varrho > 0$ in the ordinary complex plane. Then F is normal in $\triangle$.

Proof

Let $f(z) \varepsilon {}^*F$. Since the condition that the elements of F omit the values w such that $|w - w_0| < \varrho$ can be formulated as a standard sentence, this is a property of $f(z)$ also. Let z_0 be a standard point which belongs to $\triangle$. Then the conditions of 4.12 apply and so $f(z)$ is normal at z_0. It It now follows from 4.9 that F is normal in $\triangle$, proving the theorem.

4.14 *Theorem*

Let the internal function $f(z)$ be holomorphlc in a domain which includes a disk $|z - z_0| < r$, r standard and positive. Suppose that there exist two finite complex numbers a and b which do not both belong to the same monad such that $f(z) \neq a$ and $f(z) \neq b$ for all z in the monad of z_0. Then $f(z)$ is normal at z_0.

Proof

Suppose that $f(z)$ is not normal in the monad of z_0, μ. Then $f(z)$ is finite somewhere in μ and we may suppose, for simplicity and without loss of generality, that $f(z)$ is finite for $z = z_0$. Put

$$g(z) = \frac{1}{b - z} f(z_0 + z) - \frac{a}{b - a}$$

then $g(z)$ is holomorphic for $|z| < r$, and $g\,(0)$ is finite and $g(z) \neq 0$, $g(z) \neq 1$ for $z \simeq 0$. It then follows from a (by now) familiar argument that there exists a standard positive $\varrho \leq r$ such that $g(z) \neq 0$ and $g(z) \neq 1$ for all $|z| < \varrho$. Moreover, $|g\,(0)| < \alpha$ for some standard positive α. Hence, by Schottky's theorem, there exists a standard m such that $|g(z)| \leq m$ for $|z| \leq \frac{1}{3}\varrho$. But then $g(z)$ is normal, more precisely S-continuous at the origin, and so $f(z)$ is normal at z_0. This contradiction proves 4.14.

From 4.14 we may deduce without difficulty the so-called *fundamental criterion* (critère fondamental) for normal families. However, we shall prefer to use 4.14 directly.

To conclude this section, we consider quasi-normal families of holomorphic functions. The basic result, corresponding to 4.9 is as follows:

4.15 *Theorem*

Let $\triangle$ be a domain in the ordinary complex plane and let $F = \{f(z)\}$ be a family of functions which are holomorphic in $\triangle$ in the classical sence. Then F is quasi-normal in $\triangle$, of order $q \geq 0$ if and only if (i) every function $f(z) \varepsilon {}^*F$ is irregular for at most q points $z \varepsilon \triangle$ and (ii) there exists a function $f(z) \varepsilon \triangle$ such that $f(z)$ is irregular at precisely q distinct points $z \varepsilon \triangle$.

To prove 4.15, we proceed as follows: We first suppose that (i) is satisfied and show, similarly as in the proof of the "sufficiency" part of 4.9 that F is quasi-normal of order q at most. Next, using the method of proof employed for the "necessity part" of 4.9, we show that (ii) implies the existence of a sequence of functions extracted from F which possesses q effectively irregular points. Thus, (i) and (ii) together imply that F is quasi-normal of order q in $\triangle$. Conversely, if F is quasi-normal of order q in $\triangle$, then it also follows from the arguments mentioned already that no $f(z) \varepsilon {}^*F$ can be irregular at more than q distinct $z \varepsilon \triangle$ (for then there would exist a sequence of functions extracted from F which possesses at least $q + 1$ effectively irregular points); and that there must be an $f(z) \varepsilon {}^*F$ which is irregular at q distinct $z \varepsilon \triangle$ (for otherwise F would be quasi-normal of order $q - 1$ at most.) The reader will have no difficulty in filling in the details of the proof.

5. *Applications*

It is now easy to apply the theory of normal families to various branches of the theory of functions of a complex variable by considering the "ideal" as well as the "ordinary" elements of a family of functions (*i.e.*, the functions $f(z) \varepsilon {}^*F$) and not by the extraction of sequences and subsequences of functions as in the classical approach. Rather than illustrate our method by the proof of well-established results, we shall give some examples of theorems which apparently have not been stated before but which suggest themselves naturally within the context of our present theory.

We begin with some preliminary remarks. Suppose the internal function $f(z)$ is holomorphic for $| z - z_0 | < r$, where z_0 is standard and r is standard positive; and that $f(z)$ is normal for $0 <^0 | z - z_0| < r$, but irregular at the point z_0. Let ϱ be standard positive and smaller than r. Then $f(z)$ is either finite for all z such that ${}^0 | z | = \varrho$, or infinite for all such z, by 4.8. But in the former case, $|f(z)|$ would then be bounded by a finite number in the monad of z_0, by the maximum principle, and hence would be normal and S-continuous at z_0, by 4.1. We conclude that $f(z)$ is infinite for all z such that $0 <^0 |z - z_0| < r$.

On the other hand, if a is any finite complex number, then $f(z) = a$ for some z in the monad of z_0. For if this is not the case then $f(z) - a \neq 0$ for $|z - z_0| < q$ where q is some standard positive number, and so $g(z) = \frac{1}{f(z) - a}$ is holomorphic for $|z - z_0| < q$. But then $g(z) \simeq 0$ for $0 <^0 |z - z_0| < q$ and so $g(z) \simeq 0$ for $^0|z - z_0| = 0$ also, i.e., for $z \simeq z_0$. This in turn implies that $f(z)$ is infinite for $z \simeq z_0$, contrary to the assumption that $f(z)$ is irregular at z_0.

The word "standard" preceding the statement of the next theorem indicates that we are dealing with classical function theory. However, the methods of proof are non-standard.

5.1. *Theorem* (*standard*)

Let a_0 be a fixed complex number, $a_0 \neq 0$, and let q be a fixed positive integer. Let F be the family of the functions

5.2 $f(z) = a_0 + a_1 z + a_2 z^2 + \ldots$

which are holomorphic and (at most) q-valent in the unit circle and let r and δ be two positive numbers, $r < 1$. Then there exists a positive $\alpha = \alpha(a_0, r, \delta)$ such that for every $f(z) \varepsilon F$ there are q points $z_1, \ldots, z_q$, $|z_j| \leq r$, $j = 1, \ldots, q$, such that

$|f(z)| \geq \alpha$ provided $|z| \leq r$ and $|z - z_j| \geq \delta$, $j = 1, \ldots, g$.

Proof

Suppose that there exist a_0, r, δ which satisfy the assumptions of the theorem, although its conclusion does not hold. That is to say, "for every $\alpha > 0$ there exists an $f(z) \varepsilon F$ such that for all q-tuples of complex numbers $z_1, \ldots, z_q$, $|z_j| \leq r$, $j = 1, \ldots, g$ there exists a z_0 for which $|f(z_0)| < \alpha$, and $|z_0| \leq r$, $|z_0 - z_j| \geq \delta$, $j = 1, \ldots, q$." Now, the statement in quotation marks can be expressed by a standard sentence. Reinterpreting that sentenc ein the non-standard model and choosing α positive but infinitesimal, we find that there exists a function $f(z) \varepsilon {}^*F$ such that for all q-tuples $z_1, \ldots, z_q$, $|z_j| \leq r$, $j = 1, \ldots, q$, there exists a z_0 for which $f(z_0)$ is infinitesimal where $|z_0| \leq r$, $|z_0 - z_j| \geq \delta$, $j = 1, \ldots, q$.

Consider such a function $f(z)$. $f(z)$ must be q-valent, at most, in the unit circle since this is a property of the elements of F and hence, of the elements of ${}^*F$. Suppose there exist $q + 1$ distinct standard points z'_j, $|z'_j| < 1$, $j = 1, \ldots, q + 1$ at which $f(z)$ is irregular.

Now, by 4.14, the finite values omitted by $f(z)$ in the monad of any one of these z'_j all belong to a single monad μ_j. It follows that every finite value which does not belong $\mu_1 \cup \mu_2 \cup \ldots \cup \mu_{q+1}$ is taken in the monads of $z'_1, z'_2, \ldots, z'_{q+1}$, i.e., at least $q + 1$ times. This contradiction shows that there cannot be more than q standard points in the interior of the unit circle at which $f(z)$ is irregular.

Suppose that $f(z)$ is irregular at one standard point z'_1 at least, $|z'_1| < 1$. Then $f(z)$ is infinite for any z in the S-interior of the unit circle which does not belong to a monad in which $f(z)$ is irregular. It follows that if we choose a set of standard points z_j, $|z_j| \leq r$, $j = 1, \ldots, q$ which includes all standard points z'_k, $|z'_k| \leq r$ at which $f(z)$ is irregular then $f(z_0)$ is infinite for all z_0 for which $|z_0| \leq r$ and $|z_0 - z_j| \geq \delta$, $j = 1, \ldots, q$ and hence certainly $|f(z_0)| \geq \alpha$. This contradiction shows that $f(z)$ is normal everywhere in the S-interior D_i of the unit circle. Moreover, $f(0) = a_0$ is finite and so $f(z)$ is finite at all points of D_i. It follows that the function $g(z) = {}^0f(z)$ exists and is holomorphic for $|z| < 1$. Since the points $|z| \leq r$ for which $f(z)$ is infinitesimal cannot be enclosed in q circular disks $|z - z_j| < \delta$, $|z_j| \leq r$ it follows that there are at least $q + 1$ standard points $\bar{z}_j$, $|\bar{z}_j| \leq r$ such that $f(z)$ is infinitesimal for all z in the monads of $\bar{z}_j$, $j = 1, \ldots, q + 1$. This implies that $g(\bar{z}_j) = {}^0f(\bar{z}_j) = 0$, $j = 1, \ldots, q + 1$. Now $g(0) = a_0 \neq 0$ so $g(z)$ is not constant. Accordingly, we may surround the points $\bar{z}_j$ with sufficiently small non-intersecting circles C_j: $|z - \bar{z}_j| = \varrho$, such that $g(z) \neq 0$ on all C_j, $j = 1, \ldots, q + 1$. But $f(z) = g(z) + (f(z) - g(z))$ where $f(z) - g(z)$ is infinitesimal. Hence, by Rouché's theorem, $f(z)$ vanishes at some point of the open disk bounded by C_j, $j = 1, \ldots, q + 1$, and this again contradicts the fact that $f(z)$ is q-valent. This completes the proof of 5.1.

We may vary the conditions which define the family F, *e.g.*, by replacing $a_0 \neq 0$ by $a_0 = 0$, $a_1 = 1$.

Our next example will be concerned with algebroid functions. For a specified $n \geq 1$, we consider a family F of n-valued functions $w = f(z)$ which are defined by equations

5.2 $G(w, z) \equiv w^n + P_1(z)w^{n-1} + \ldots + P_n(z) = 0$ where $P_1(z), \ldots, P_n(z)$ are holomorphic in the unit circle. We limit the family F further by supposing (i) that the values of $P_i(0), \ldots, P_n(0)$ are specified, $P_j(0) = a_j$, $j = 1, \ldots, n$; (ii) that there is a specified positive integer q, the same for all functions of the family, such that $w = f(z)$ possesses no more than q branch points in the unit circle;

(iii) that there exists a positive $\varrho < 1$, the same for all functions of the family, such that $f(z)$ possesses not more than one branch point for which $|z| < \varrho$; and (iv) that $f(z)$ omits the values 0 and 1, i.e., $G(0, z) \neq 0$, $G(1, z) \neq 0$ for $|z| < 1$.

Observe that condition (ii) is satisfied, for example, if the functions $G(w, z)$ are algebraic of bounded degree.

5.3 *Theorem (standard)*

Suppose that a family of functions, F, satisfies the stated conditions for fixed n, q and ϱ, and let r be positive and smaller than 1. Then there exists a positive $m = m(r)$ such that $|f(z)| \leq m$ for all $f(z) \varepsilon F$, $|z| \leq r$.

Proof

Suppose that, for some particular F and r, the theorem is not true. Then the following statement can be formulated as a standard sentence: "For every $m > 0$, there exist a function $f(z) \varepsilon F$ and a point z_0, $|z_0| \leq r$ such that $|f(z_0)| > m$." Reinterpreting this sentence in the non-standard model and choosing a positive infinite m, we conclude that there exist a function $f(z) \varepsilon {}^*F$ and a point z_0, $|z_0| \leq r$ such that $f(z_0)$ is infinite. We propose to show that this leads to a contradiction. More precisely, we shall show that any $f(z) \varepsilon {}^*F$ is finite for z in the S-interior of the unit circle, ${}^0|z| < 1$.

As a first step in this direction, we establish the existence of a standard point z_1, $|z_1| < 1$ such that there is no branch point of $f(z)$ in the monad of z_1 and such that all determinations of $f(z)$ are finite in the monad of z_1. If there is no branch point of z_1 in the monad of zero, then $z_1 = 0$ will do for this purpose. For in that case, there is a standard $\sigma > 0$ such that $f(z)$ consists of n (distinct or coinciding) functions $f_k(z)$ which omit 0 and 1 and, hence, are normal in the monad of zero. Moreover, these functions reduce to the roots of the equation

5.4 $w^n + a_0 w^{n-1} + \ldots + a_n = 0$

for $z = 0$ and so the $f_k(z)$ are actually finite and S-continuous in the monad of zero.

Now, suppose that the point $\xi \simeq 0$ is a branch point. In that case, there can be no other branch point z with $|z| < \varrho$ by (iii). We introduce variables $t_k = (z - \xi)^{1/p_k}$, p_k a positive integer, such that the function $\varphi_k(t_k) = f_k(z(t_k))$ is a holomorphic function of t_k for sufficiently small t_k, $|t_k| < \tau_k$ where τ_k is standard positive, $k = 1, \ldots, n$. Since $\varphi_k(t_k)$ omits 0 and 1, it is normal for ${}^0|t_k| < \tau_k$. Moreover, $\varphi_k(t_k)$ takes a finite value (i.e., a root of 5.4) for some value of t_k, corresponding to $z = 0$. Hence, $\varphi_k\ (t_k)$ is finite for $^0 \mid t_k \mid < \tau_k$, by 4.8. We may therefore choose z_1 as any standard number different from zero such that the corresponding t_k satisfy $^0 \mid t_k \mid < \tau_k$.

Now, let z_0 be any standard point which satisfies $\mid z_0 \mid < \mid$ and such that $f(z)$ has no branch point in the monad of z_0. We may choose a simply connected standard domain D which contains both z_0 and z_1. Then $f(z)$ is represented on D by a set of single-valued analytic functions which are normal at all standard points of D and are finite at z_1 and hence by 4.8 throughout D and, in particular, in the monad of z_0.

Suppose finally that $\mid z_0 \mid < 1$, z_0 standard and that $f(z)$ possesses one or more branch points in the monad of z_0. Surround z_0 with a small circle C of radius λ, λ standard, and such that C is in the interior of the unit circle and all branch points of $f(z)$ which are not in the monad of z_0 are at a distance of at least 2λ from z_0. Then the coefficients $P_j(z)$ of the function $G(w,z)$ which corresponds to $f(z)$ (see 5.2) are, up to sign, the fundamental symmetrical functions of the different determinations $f_k(z)$ of $f(z)$ and since these are finite on C, the $P_j(z)$ also must be finite on C. But then, by the maximum principle, the $P_j(z)$ must be finite also for $\mid z - z_0 \mid < \lambda$ and, in particular, must be finite in the monad of z_0. This in turn implies that the several determinations of $f(z)$ which are given by 5.2, are finite in the monad of z_0. Accordingly, $f(z)$ is finite for all z such that $^0 \mid z \mid < 1$. This proves the theorem.

The proof did not contain any direct reference to the Riemann surface of $f(z)$ since we wanted to avoid a detailed discussion of this notion within the framework of the present theory. However, we may mention that the ideas of Non-standard Analysis can be extended so as to include general topological notions.

6. *Lacunary polynomials*

A lacunary polynomial of rank $r > 1$ is a polynomial with complex coefficients,

$$6.1 \quad p(z) = a_1 z^{n_1} + a_2\ z^{n_2} + \ldots + a\ z^{n_r}$$

$$a_j \neq 0,\ j = 1, \ldots, r,\ n_1 < n_2 \ldots < n_r$$

(The commonly accepted name "lacunary polynomial", which is somewhat misleading, refers to the fact that, for a given r there are likely to be many gaps in the degrees of the monomials which appear in 6.1)

6.2 *Theorem (standard)*

Let r be a positive integer and let F be the family of all lacunary polynomials of the form

$$6.3\quad p(z) = z + a_2 z^{n_2} + \ldots + a_r z^{n_r}\ .$$

$$a_j \neq 0,\ j = 2, \ldots, r,\ 1 < n_2 < \ldots < n_r$$

which have no zero $z \neq 0$ in the unit circle. Then there exists a positive ϱ such that all elements of F are univalent for $|z| < \varrho$.

In order to prove 6.2, we require some auxiliary considerations. Let az^m and bz^n be two monomials in the non-standard model, so that m and n may be finite or infinite natural numbers and suppose $a \neq 0$, $b \neq 0$ and $m \neq n$. We shall say that az^m *dominates* bz^m *at the origin*, $az^m \succ bz^n$ or $bz^n \prec az^m$ if there exists an infinitesimal $\varrho_0 > 0$ such that $|a|\varrho^m > |b|\varrho^n$ for all infinitesimal $\varrho > \varrho_0$ i.e., if $|a/b|\varrho^{m-n} > 1$ for such ϱ.

Consider the function $g(\varrho) = |a/b|\varrho^{m-n}$. If $m > n$ then $g(\varrho)$ increases strictly from 0 to ∞ as ϱ increases from 0 to ∞ in $^*R$. Accordingly there is a unique ϱ' such that $g(\varrho') = 1$. If this ϱ' is infinitesimal, then $az^m \succ bz^n$; if ϱ' is not infinitesimal, $az^m \prec bz^n$. If $m < n$, $g(\varrho)$ decreases strictly from ∞ in $^*R$, and again $g(\varrho') = 1$ for a unique ϱ'. In this case, $az^m \prec bz^n$ if ϱ' is infinitesimal, and $az^m \succ bz^n$ if ϱ' is not infinitesimal. Thus, either az^m dominates bz^n or bz^n dominates az^m. We also see without difficulty that a necessary and sufficient condition for az^m to dominate bz^n is that there exist a positive ϱ_1 which is not infinitesimal such that $|a|\varrho^m > |b|\varrho^n$ for all positive $\varrho < \varrho_1$ which are not infinitesimal.

If az^m, bz^n, cz^p are three monomials of different degrees, then $az^m \succ bz^n$ and $bz^n \succ cz^p$ together imply $az^m \succ cz^p$.

Now, let $p(z)$ be a lacunary polynomial of rank r in the non-standard model and suppose that $p(z)$ is given by 6.1. From what has been said already, it follows that one of the monomials $a_j z^{n_j}$ in 6.1 dominates all others. We call the exponent of this particular monomial, n_j, the *order of* $p(z)$ *at the origin*, $n_j = \text{ordo } p(z)$.

6.4 *Theorem*

If ordo $p(z) = m$, where m is a finite natural number then the number of roots of $p(z)$ in the monad of zero, counting multiplicities, is m. If ordo $p(z)$ is an infinite natural number, then for any finite positive integer k there exist infinitesimal complex numbers $z_1, \ldots, z_k$ such that $p(z) = (z - z_1) \ldots (z - z_k)\ q(z)$ where $q(z)$ is an internal polynomial.

Proof

Let az^m, bz^n be two monomials such that $az^m \succ bz^n$ and let q be any finite positive integer. We claim that $|\ az^m\ | \geq q\ |\ bz^n\ |$ for sufficiently large infinitesimal z. Indeed, if $m > n$ so that $g(\varrho') = 1$ for an infinitesimal ϱ, then if $\varrho_0 = q\varrho'$ we have, for $|\ z\ | > \varrho_0$,

$$\left|\frac{a}{b}\right| \left| z^{m-n} \right| \geq \left|\frac{a}{b}\right| \varrho_0{}^{m-n} \geq \frac{\varrho_0}{\varrho'} \left|\frac{a}{b}\right| \varrho_0{}^{m-n} \geq q$$

i.e., $|\ az^m\ | \geq q\ |\ bz^n$, proving our assertion in this case. If $m < n$ so that $g(\varrho') = 1$ for a ϱ' which is not infinitesimal, then we have, for *all* infinitesimal z

$$\left|\frac{a}{b}\right| \left| z^{m-n} \right| \geq \left|\frac{z}{\varrho'}\right|^{m-n} \left|\frac{a}{b}\right| \varrho'^{m-n} \geq \left|\frac{\varrho'}{z}\right| \left|\frac{a}{b}\right| \varrho'^{m-n} \geq q$$

since z/ϱ' is infinitesimal. Thus, $|\ az^m\ | \geq q\ |\ bz^n\ |$ in this case also.

Now, let $p(z)$ be given by 6.1 and let $a_j z^{n_j}$ be the monomial of $p(z)$ which dominates all others. Putting $g = r$ we then have, for sufficiently large infinitesimal z, $\frac{1}{\mathrm{r}} \left| a_j z^{n_j} \right| > a_\nu z^{n_\nu}$,

$$\frac{r}{r-1} \left| a_j z^{n_j} \right| \geq \sum_{\nu \neq j} \left| a_\nu z^n \right|$$

This in turn implies that there exists an infinitesimal $\sigma > 0$ such that

6.5 $|\ a_j\ z^{n_j}\ | > |\ p(z) - a_j\ z^{n_j}\ |$

for $|\ z\ | \geq \sigma$, z infinitesimal. Hence, by Rouché's theorem, the number of zeros inside any circle $|\ z\ | < \sigma'$ for σ' infinitesimal but not less than σ is the same as the number of zeros of $a_j z^{n_j}$. If n_j is finite, ordo $p(z) = n_j = m$, then this number is precisely m, in agreement with the conclusion of the theorem in this case. If ordo $p(z) = n_j = m$ is infinite, then $p(z)$ has infinitesimal roots $z_1 \ldots z_k$ for any finite k, and

$p(z)$ is then divisible by the product $(z - z_1) \ldots (z - z_k)$. This is the conclusion stated in 6.4.

Proof of 6.2.

Referring to the family of polynomials F which is described in the statement of 6.2, we observe that any $p(z)$ in $^*F$ also can have only one root in the monad of zero, and so ordo $p(z) = 1$, by 6.4. Thus, $|z| \geq |p(z) - z|$ for sufficiently large infinitesimal $|z|$ and hence also for sufficiently small $|z| = \varrho$ which is not infinitesimal. For such z

6.6 $|p(z)| \leq |p(z) - z| + |z| \leq 2|z|$

Hence, choosing a particular non-infinitesimal $\varrho > 0$ such that 6.6 applies for $|z| = \varrho$ we have, for $|z| < \varrho$, by Schwarz' lemma

$$|p(z)| \leq \frac{|z|}{\varrho} \max_{|z'|=\varrho} |p(z')| \leq |z| \frac{2\varrho}{\varrho} = 2|z|$$

This shows that $p(z)$ is infinitesimal for infinitesimal z.

Now suppose that the conclusion of 6.2 is not true for some particular r. Then "for any positive ϱ there exists a $p(z) \varepsilon F$ such that $p(z_1) = p(z_2)$ for some z_2, z_2 for which $|z_1| < \varrho$, $|z_2| < \varrho$, $z_1 \neq z_2$." Formulating the statement in quotation marks as a standard sentence, reinterpreting in the non-standard model and choosing ϱ positive infinitesimal, we can find a $p(z) \varepsilon {}^*F$ such that $p(z_1) = p(z_2) = \alpha$ for some infinitesimal z_1 and z_2, $z_1 \neq z_2$. Since we have just shown that $p(z)$ is infinitesimal for infinitesimal z, α must be infinitesimal. By a previous argument, $|z| \geq r|a_j z^{n_j}|$, $j = 2, \ldots, r$ for sufficiently large infinitesimal z and so for such z

$$6.7 \quad |p(z)| \geq |z| - |z - p(z)| \geq \frac{1}{r}|z|$$

Since 6.7. holds for large infinitesimal z it must hold also for small z which are not infinitesimal. If follows that there exists a standard positive σ such that $|p(z)| \geq \frac{\sigma}{r}$ for $|z| = \sigma$. Hence, by Rouché's sheorem, the number of roots of $p(z)$ for $|z| < \sigma$, *i.e.*, 1, must be the tame as the number of roots of $p(z) - \alpha$ for such z. This contradicts the fact that $p(z)$ takes the value α twice for $|z| < \sigma$ and hence, proves the theorem.

The reader may wonder whether it is not always possible to find conventional proofs in place of the "non-standard proofs" of standard theorems exemplified in the present paper. The answer is that, though this may complicate matters, it is always possible to "translate" a proof of the kind given here into a standard proof by the use of ultrapowers, more particularly hyperreal fields (references 1,5.) Nevertheless, we venture to suggest that our approach has a certain natural appeal, as shown by the fact that it was preceded in history by a long line of attempts to introduce infinitely small and infinitely large numbers into Analysis.

References

1. W. A. J. LUXEMBURG. **Non-standard Analysis (lectures on A. Robinson's theory of infinitesimals and infinitely large numbers).** California Institute of Technology 1962.

2. P. MONTEL. **Leçons sur les familles normales** (Collection Borel). Paris 1927.

3. P. MONTEL. **Leçons sur les fonctions univalentes et multivalentes** (Collection Borel). Paris 1933.

4. A. OSTROWSKI. *Über Folgen analytischer Funktionen und einige Verschärfungen des Picardschen Satzes,* **Mathematische Zeitschrift,** vol. 24 (1926), pp. 215—258.

5. A. ROBINSON. *Non-standard Analysis,* **Proceedings of the Royal Academy of Sciences, Amsterdam, Sec. A,** vol. 64 (1961), pp. 432—440.

6. A. ROBINSON. *Introduction to model theory and to the metamathematics of Algebra.* **Studies in Logic and the Foundations of Mathematics,** Amsterdam 1963.

7. A. ROBINSON. *Complex function theory over non-archimedean fields, Technical* **(Scientific note No. 30, USAF Contract** No. 61 (052) — 187), Jerusalem, 1962.

8. A. ROBINSON. *Non-standard Analysis.* **Studies in Logic and the Foundations of Mathematics.** Amsterdam. (To be published.)

University of California, Los Angeles

SOLUTION OF AN INVARIANT SUBSPACE PROBLEM OF K. T. SMITH AND P. R. HALMOS

ALLEN R. BERNSTEIN AND ABRAHAM ROBINSON

The following theorem is proved.

Let T be a bounded linear operator on an infinite-dimensional Hilbert space H over the complex numbers and let $p(z) \neq 0$ be a polynomial with complex coefficients such that $p(T)$ is completely continuous (compact). Then T leaves invariant at least one closed linear subspace of H other than H or $\{0\}$.

For $p(z) = z^2$ this settles a problem raised by P. R. Halmos and K. T. Smith.

The proof is within the framework of Nonstandard Analysis. That is to say, we associate with the Hilbert space H (which, ruling out trivial cases, may be supposed separable) a larger space, $^*H$, which has the same formal properties within a language L. L is a higher order language but $^*H$ still exists if we interpret the sentences of L in the sense of Henkin. The system of *natural numbers* which is associated with $^*H$ is a nonstandard model of arithmetic, i.e., it contains elements other than the standard natural numbers. The problem is solved by reducing it to the consideration of invariant subspaces in a subspace of $^*H$ the number of whose dimensions is a nonstandard positive integer.

1. Introduction.

We shall prove:

MAIN THEOREM 1.1. *Let T be a bounded linear operator on an infinite-dimensional Hilbert space H over the complex numbers and let $p(z) \neq 0$ be a polynomial with complex coefficients such that $p(T)$ is completely continuous* (*compact*). *Then T leaves invariant at least one closed subspace of H other than H or* $\{0\}$.

For $p(z) = z^2$ this settles Problem No. 9 raised by Halmos in [2] and there credited to K. T. Smith. For this case, a first proof was given by one of us (A.R.) while the other (A.R.B.) provided an alternative proof which extends to the case considered in 1.1. The argument given below combines the two proofs, both of which are based on Nonstandard Analysis. The Nonstandard Analysis of Hilbert space was developed previously by A.R. as far as the spectral analysis of completely continuous self-adjoint operators (compare [7]) while A.R.B. has disposed of the spectral theorem for bounded self-adjoint operators

---

Received July 5, 1964, and in revised form December 10, 1964. The authors acknowledge with thanks the support received from the National Science Foundation (Grant No. GP-1812).

by the same method. The general theory will be sketched here only as far as it is required for the proof of our main theorem.

Some of our arguments are adapted from the proofs of the theorem for $p(z) = z$, i.e., when T is itself completely continuous, which are due to von Neumann and Aronszajn for Hilbert space, as above, and to Aronszajn and K. T. Smith for general Banach spaces [1].

The particular version of Nonstandard Analysis which is convenient here relies on a higher order predicate language, L, which includes symbols for all complex numbers, all sets and relations of such numbers, all sets of such sets and relations, all relations of relations, etc. Quantification with respect to variables of all these types is permitted. Within this framework, a sequence of complex numbers, $y = s_n$, $n = 1, 2, 3, \cdots$, is given by a many-one relation $S(n, y)$ when n varies over the set of positive integers, P. The separable Hilbert space, H, may then be represented as a set of such sequences (i.e., as l_2) while a particular operator on H is identified with a relation of relations.

Let K be the set of sentences formulated in L which hold in the field of complex numbers, C. K includes sentences about, or involving, the sets of real numbers and of natural numbers, since these may be regarded as subsets of the complex numbers which are named in L. It also includes sentences about Hilbert space as represented above.

Nonstandard Analysis is based on the fact that, in addition to C, K possesses other models, which are proper extensions of C. We single out any one of them, $^*C$, calling it *the nonstandard model*, as opposed to the *standard model*, C. However, $^*C$ is a model of K only if the notions of set, relations, etc. are interpreted in $^*C$ in the sense of the higher order model theory of Henkin [3]. That is to say, the sets of sets, relations, etc., which are taken into account in the interpretation of a sentence in $^*C$ may (and will) be proper subsets of the corresponding sets over $^*C$ in the absolute sense. The sets, relations, etc. which are taken into account in the interpretation in $^*C$ will be called *admissible*.

The basic properties and notions of Nonstandard Analysis which are expounded in [4] and [5] are applicable here. Thus, an individual of $^*C$ (which will still be called a complex number) may or may not be an element of C, i.e., a complex number in the ordinary sense or *standard* number, briefly an *S-number*. Every *finite* complex number a is infinitely close to a unique standard complex number, ${}^0a$. That is to say, if $|a|$ is smaller than some real S-number, then there exists a complex S-number, ${}^0a$, the *standard part* of a, such that $|a - {}^0a|$ is smaller than all positive S-numbers. A number which is infinitely close to 0 is *infinitely small* or *infinitesimal*. In particular, 0 is the only S-number which is infinitesimal. A complex number a which is not finite, i.e., which is such that $|a|$ is greater than any S-number, is *infinite*. There exist elements of $^*C$ which are infinite.

Every set, relation, etc. in C possesses a natural extension to $^*C$. This is simply the set, relation, $\cdots$, in $^*C$ which is denoted by the same symbol in L. At our convenience, we may, or may not, denote it by the same symbol also in our notation (which is not necessarily part of L). Thus, we shall denote the extension of the set of positive integers, P, to $^*C$ by $^*P$ but if $\sigma = \{a_n\}$ is a sequence of complex numbers in C then we shall denote its extension to $^*C$ still by $\sigma = \{a_n\}$. According to the definition of an infinite number which was given above, the infinite positive integers in $^*C$ are just the elements of $^*P - P$.

The following results are basic (for the proofs see [5] and [6]).

THEOREM 1.2. *The sequence* $\{a_n\}$ *in* C *converges to a limit* a *(*a *an S-number) if and only if the extension of* $\{a_n\}$ *in* $^*C$ *satisfies the condition that* $|a - a_n|$ *is infinitesimal for all infinite* n.

THEOREM 1.3. *Let* $\{a_n\}$ *be an admissible sequence in* $^*C$ *such that* a_n *is infinitesimal for all finite* n. *Then there exists an infinite positive integer* ω *(i.e.,* $\omega \in {}^*P - P$*) such that* a_n *is infinitesimal for all* n *smaller than* ω.

$\{a_n\}$ is called *admissible* in $^*C$ if the relation representing $\{a_n\}$ belongs to the set of relations which are admissible in the sense explained above. Admissible operators, etc., are defined in a similar way. 1.3. shows that the sequence $\{a_n\}$ which is defined by $a_n = 0$ for finite n and by $a_n = 1$ for infinite n is not admissible in $^*C$.

2. Nonstandard Hilbert space. The selected representation of the Hilbert space H consists of all sequences $\{s_n\}$ of complex numbers such that $\|\sigma\|^2 = \sum_{n=1}^{\infty} |s_n|^2$ converges. The corresponding space $^*H$ over $^*C$ consists of all admissible sequences $\{s_n\}$ in $^*C$ such that $\|\sigma\|^2 = \sum_{n=1}^{\infty} |s_n|^2$ converges, i.e., such that it satisfies the formal (classical) definition of convergence in L.

Among the points of $^*H$ are the extensions of points of H (as sequences). We identify the points of H with their extension in $^*H$ and may then regard H as a subset (though not an admissible subset) of $^*H$.

A point σ of $^*H$ is called *norm-finite* if $\|\sigma\|$ is a finite real number in the sense explained in section 1. σ is *near-standard* if $\|\sigma - \sigma^0\|$ is infinitesimal for some ${}^0\sigma \in H$. If such a ${}^0\sigma$ exists then it is determined uniquely by σ. It is called the *standard part* of σ.

Applying 1.2. to the partial sums of any point $\sigma = \{s_n\}$ in H, we obtain:

THEOREM 2.1. *For any $\sigma = \{s_n\}$ in H and any infinite positive integer ω, the sum $\sum_{n=\omega}^{\infty} |s_n|^2$ is infinitesimal.*

Next, we sketch the proof of:

THEOREM 2.2. *A point $\sigma = \{s_n\}$ in $^*H$ is near-standard if and only if it is norm-finite and if at the same time $\sum_{n=\omega}^{\infty} |s_n|^2$ is infinitesimal for all infinite ω.*

Suppose that $\| \sigma - {}^0\sigma \|$ is infinitesimal for some ${}^0\sigma$ in H. Then $\| \sigma \| = \| \sigma - {}^0\sigma + {}^0\sigma \| \leqq \| \sigma - {}^0\sigma \| + \| {}^0\sigma \| < 1 + \| {}^0\sigma \|$ so that σ is norm-finite. Also, let ${}^0\sigma = \{s'_n\}$, then $\sum_{n=\omega}^{\infty} |s'_n|^2$ is infinitesimal for infinite ω, by 2.1. Also, $\sum_{n=\omega}^{\infty} |s_n - s'_n|^2$ is infinitesimal since this sum cannot exceed $\| \sigma - {}^0\sigma \|^2$. But

$$\sum_{n=\omega}^{\infty} |s_n|^2 \leqq \left(\left(\sum_{n=\omega}^{\infty} |s_n - s'_n|^2\right)^{1/2} + \left(\sum_{n=\omega}^{\infty} |s'_n|^2\right)^{1/2}\right)^2,$$

showing that the conditions of 2.2 are necessary.

Supposing that they are satisfied, $\| \sigma \|$ is finite, hence $|s_n|$ is finite for any n and s_n possesses a standard part, ${}^0s_n$. Consider the sequence $\{{}^0s_n\}$ in C. It can be shown that $\sum_{n=1}^{\infty} |{}^0s_n|^2$ converges in C and hence, represents a point σ' in H and $^*H$. Thus, if $\sigma' = \{s'_n\}$ then $s'_n = {}^0s_n$ for finite n but not necessarily for infinite n. Since, for all finite k, $\sum_{n=1}^{k} |s_n - s'_n|^2 = \sum_{n=1}^{k} |s_n - {}^0s_n|^2$ is infinitesimal, it follows from 1.3 that $\sum_{n=1}^{k} |s_n - s'_n|^2$ is still infinitesimal for some infinite k, $k = \omega - 1$, say. On the other hand, $\sum_{n=\omega}^{\infty} |s_n|^2$ is infinitesimal by assumption, and $\sum_{n=\omega}^{\infty} |s'_n|^2$ is infinitesimal, by 2.1. The inequality

$$\| \sigma - \sigma' \|^2 = \sum_{n=1}^{\infty} |s_n - s'_n|^2 \leqq \sum_{n=1}^{\omega-1} |s_n - s'_n|^2 + \left(\left(\sum_{n=\omega}^{\infty} |s_n|^2\right)^{1/2} + \left(\sum_{n=\omega}^{\infty} |s'_n|^2\right)^{1/2}\right)^2,$$

then shows that $\| \sigma - \sigma' \|$ is infinitesimal, σ is near-standard with standard part ${}^0\sigma = \sigma'$.

The following theorem is proved in [7] for general topological spaces but under somewhat different conditions.

THEOREM 2.3. *Let A be a compact set of points in H. Then all points of $^*A$ (i.e., of the set which corresponds to A in $^*H$) are near-standard.*

Indeed, suppose that A is compact but that $\sigma \in {}^*A$ is not near-standard. Then there exists a standard positive r such that $\| \sigma - \tau \| > r$ for all $\tau \in H$. This is trivial if σ is not norm-finite. If σ is norm-finite, then by 2.2, there exists an infinite positive integer ω such that $\sum_{n=\omega}^{\infty} |s_n|^2 > 2r^2$ for some standard positive number r. For any $\tau = \{t_n\}$ in H, $\sum_{n=\omega}^{\infty} |t_n|^2$ is infinitesimal. Hence

$$\|\sigma - \tau\| = \left(\sum_{n=1}^{\infty} |s_n - t_n|^2\right)^{1/2} \geqq \left(\sum_{n=\omega}^{\infty} |s_n - t_n|^2\right)^{1/2}$$
$$\geqq \left(\sum_{n=\omega}^{\infty} |s_n|^2\right)^{1/2} - \left(\sum_{n=\omega}^{\infty} |t_n|^2\right)^{1/2} > r .$$

On the other hand, since A is compact it possesses an r-net, i.e., for some finite number of points in A, $\tau_1, \cdots, \tau_m$, and for all ξ in A, $\|\xi - \tau_i\| < r$ for some i, $1 \leqq i \leqq m$. But, for the specified $\tau_1, \cdots, \tau_m$, this is a property of H which can be formulated as a sentence of K. It follows that for all points ξ of $^*A$ also $\|\xi - \tau_i\| < r$ for some i, $1 \leqq i \leqq m$. This contradiction proves the theorem.

3. **Operators in nonstandard Hilbert space.** An operator from H into H may be regarded as a relation between elements of H, i.e., between sequences of elements of C (which are themselves relations). The corresponding operator in $^*H$, which is denoted by the same symbol in L, will be denoted here also by T. This cannot give rise to any confusion. For if $\tau = T\sigma$ in H then $\tau = T\sigma$ also in $^*H$ since $\tau = T\sigma$ can be expressed by a sentence of K.

In particular, let T be a bounded linear operator defined on all of H. For the assumed representation of H by sequences, T has a matrix representation, $T = (a_{jk})$, $j, k = 1, 2, 3, \cdots$. The coefficients of this matrix satisfy the conditions:

3.1. $$\sum_{k=1}^{\infty} |a_{jk}|^2 < \infty \qquad j = 1, 2, 3, \cdots$$
$$\sum_{j=1}^{\infty} |a_{jk}|^2 < \infty \qquad k = 1, 2, 3, \cdots$$

In $^*H$ these subscripts of (a_{jk}) vary also over the infinite positive integers. By 3.1 and 2.1., $\sum_{k=\omega}^{\infty} |a_{jk}|^2$ is infinitesimal for infinite ω, provided j is finite. This is not necessarily true for infinite j as shown by the matrix for the identity operator.

THEOREM 3.2. *Let T be a completely continuous (compact) linear operator on H. Then T maps every norm-finite point in $^*H$ on a near-standard point.*

Proof. If σ is norm-finite then $\|\sigma\| < r$ for some positive S-number r. The sphere $B = \{\xi \mid \|\xi\| < r\}$ is bounded in H and is mapped by T on a set whose closure, A, is compact. If the corresponding sets in

$^*H$ are $^*B$ and $^*A$ respectively then $^*B$ contains σ (since σ satisfies the defining condition of B) and so $^*A$ contains $T\sigma$. But $^*A$ contains only near-standard points, by 2.3, so $T\sigma$ is near-standard, proving 3.2.

In a somewhat different setting [7] the converse of 3.2 is also true.

THEOREM 3.3. *If $T = (a_{jk})$ is a completely continuous linear operator on H, then a_{jk} is infinitesimal for all infinite k (j finite or infinite).*

Proof. For finite j, this follows from the fact that $\sum_{n=k}^{\infty} | a_{jk} |^2$ is then infinitesimal. For infinite j, define $\sigma = \{s_n\}$ by $s_n = 0$ for $n \neq k$ and by $s_k = 1$. Then $\| \sigma \| = 1$, so $\tau = \{t_j\} = T\sigma$ must be near-standard, by 3.2, where $t_j = \sum_{n=1}^{\infty} a_{jn} s_n = a_{jk}$. But then $t_j = a_{jk}$ must be infinitesimal for infinite j, by 2.2.

An operator $T = (a_{jk})$ will be called *almost superdiagonal* if $a_{jk} = 0$ for $j > k + 1$, $k = 1, 2, 3, \cdots$. This definition depends on the specified basis of H.

THEOREM 3.4. *Let T be a bounded linear operator on H which is almost superdiagonal. Let*

3.5. $$p(z) = c_0 + c_1 z + \cdots + c_m z^m,\ c_m \neq 0,\ m \geqq 1$$

be a polynomial with standard complex coefficients such that $p(T)$ is completely continuous. Then there exists an infinite positive integer ω such that $a_{\omega+1,\omega}$ is infinitesimal.

Proof. Put $Q = (b_{jk}) = p(T)$. We show by direct computation that, for any $h \geqq 1$,

3.6. $$b_{h+m,h} = c_m a_{h+1,h} a_{h+2,h+1} a_{h+3,h+2} \cdots a_{h+m,h+m-1} .$$

By 3.3, $b_{h+m,h}$ is infinitesimal for all infinite h. Since c_m is not infinitesimal, one of the remaining factors on the right hand side of 3.6 must be infinitesimal, e.g, $a_{h+j+1,h+j}$, $0 \leqq j < m$. Setting $\omega = h + j$, we obtain the theorem.

4. Projection operators. Let E be any admissible closed linear subspace of $^*H$ within the nonstandard model under consideration. The corresponding projection operator, which reduces to the identity on E, will be denoted by P_E. Given E, we define a subset $^\circ E$ of H as follows. For any $\sigma \in H$, $\sigma \in {}^\circ E$ if and only if $\| \sigma - \sigma' \|$ is infinitesimal for some $\sigma' \in E$. Since, by a familiar property of projection operators, $\| \sigma - \sigma' \| \geqq \| \sigma - P_E \sigma \|$, it follows that $\sigma \in {}^\circ E$ if and only if $\| \sigma - P_E \sigma \|$ is infinitesimal. In that case, $\sigma = {}^\circ(P_E \sigma)$. More generally, if τ is a near-standard element of E then ${}^\circ\tau \in {}^\circ E$.

The tools developed so far suffice to establish the following theorem, 4.1, as well as the subsequent theorems, 4.2 and 4.3.

THEOREM 4.1. *Given E as above, the set ${}^\circ E$ is a closed linear subspace of H.*

Proof. Let σ_1, σ_2 be elements of ${}^\circ E$. There exist elements τ_1, τ_2 of E such that $\|\sigma_1 - \tau_1\|$ and $\|\sigma_2 - \tau_2\|$ are infinitesimal. Then $\tau_1 + \tau_2$ belongs to E and

$$\|(\sigma_1 + \sigma_2) - (\tau_1 + \tau_2)\| \leqq \|\sigma_1 - \tau_1\| + \|\sigma_2 - \tau_2\|$$

so that the left hand side of this inequality also is infinitesimal. Hence, $\sigma_1 + \sigma_2$ belongs to ${}^\circ E$. Again for $\sigma \in {}^\circ E$ and λ standard complex, there exists $\tau \in E$ such that $\|\sigma - \tau\|$ is infinitesimal. Then $\lambda\tau \in E$ and $\|\lambda\sigma - \lambda\tau\| = |\lambda|\,\|\sigma - \tau\|$ is infinitesimal and so $\lambda\sigma \in {}^\circ E$. This shows that ${}^\circ E$ is linear in the algebraic sense.

Now let $\sigma_n \to \sigma$, where the σ_n are defined for standard natural n and belong to ${}^\circ E$, and σ belongs to H. In order to prove that ${}^\circ E$ is closed we have to show that σ belongs to ${}^\circ E$. By assumption, the distances $\|\sigma_n - P_E\sigma_n\|$ are infinitesimal for all $n \in N$. Hence, by Theorem 1.3 there exists an infinite natural number ω such that $\|\sigma_n - P_E\sigma_n\|$ is infinitesimal for all $n < \omega$. The sequence of points $\{\sigma_n\}$ in ${}^\circ E \subseteq H$ extends, in ${}^*H$, to a sequence of points defined for all $n \in {}^*N$. Moreover, by 1.2 above, the fact that $\sigma_n \to \sigma$ in H implies that $\|\sigma_n - \sigma\|$ is infinitesimal for all infinite n. Hence, for all infinite n less than ω, $\|\sigma - P_E\sigma_n\|$, which does not exceed

$$\|\sigma - \sigma_n\| + \|\sigma_n - P_E\sigma_n\| ,$$

also must be infinitesimal. But $P_E\sigma_n \in E$ and so $\sigma \in {}^\circ E$, as required. This completes the proof of 4.1.

Let ω be an infinite natural number. The closed linear subspace of ${}^*H$ which consists of all points $\sigma = \{s_n\}$ such that $s_n = 0$ for $n > \omega$ will be denoted by H_ω. The corresponding projection operator, which will be denoted by P maps any $\sigma = \{s_n\}$ in ${}^*H$ into the point $\sigma' = \{s'_n\}$, where $s'_n = s_n$ for $n \leqq \omega$ and $s'_n = 0$ for $n > \omega$. For any point $\sigma \in H$, $\|\sigma - P\sigma\| = (\sum_{n=\omega+1}^{\infty} |s_n|^2)^{1/2}$ is infinitesimal, by 2.1.

For any bounded linear operator T on H let $T' = PTP$, and let T_ω be the restriction of T' to H_ω. Then $\|T'\| \leqq \|P\|^2\,\|T\| \leqq \|T\|$ and so $\|T_\omega\| \leqq \|T\|$.

THEOREM 4.2. *Let E be an admissible closed linear subspace of H_ω which is invariant for T_ω, i.e., $T_\omega E \subseteq E$. Then ${}^\circ E$ is* *invariant for* T, $T{}^\circ E \subseteq {}^\circ E$.

Proof. Let $\sigma \in {}^\circ E$, then we have to show that $T\sigma \in {}^\circ E$. By assumption, there exists a $\tau \in E$ such that $\|\sigma - \tau\|$ is infinitesimal. Then $T_\omega \tau \in E$, i.e., $PT\tau \in E$. Thus, in order to show that $T\sigma$ is infinitely close to E, we only have to establish that the quantity $a = \|T\sigma - PT\tau\|$ is infinitesimal. Now

$$\begin{aligned} a = \|T\sigma - PT\tau\| &= \|T\sigma - PT\sigma + PT(\sigma - \tau)\| \\ &\leqq \|T\sigma - PT\sigma\| + \|P\|\,\|T\|\,\|\sigma - \tau\| \end{aligned}$$

and $\|T\|$ is a standard real number, while $\|P\| \leqq 1$ and $\|\sigma - \tau\|$ is infinitesimal. At the same time $T\sigma$ is a point of H and so the difference $T\sigma - PT\sigma$ is infinitesimal, as shown above. It follows that a is infinitesimal, and this is sufficient for the proof of 4.2.

The number of dimensions of H_ω as defined within the language L is ω, $d(H_\omega) = \omega$. In this sense, H_ω is "finite-dimensional". Similarly, with every admissible closed linear subspace E of H_ω, there is associated a natural number $d(E)$ in $^*C$, which may be finite or infinite, and which has the properties of a dimension to the extent to which these can be expressed as sentences of K.

THEOREM 4.3. *Let E and E_1 be two admissible closed linear subspaces of H_ω such that $E \subseteq E_1$ and $d(E_1) = d(E) + 1$. Then ${}^\circ E \subseteq {}^\circ E_1$ and any two points of ${}^\circ E_1$ are linearly dependent modulo ${}^\circ E$.*

Proof. Since $E \subseteq E_1$, it is trivial that ${}^\circ E \subseteq {}^\circ E_1$. Now suppose that ${}^\circ E_1$ contains two points σ_1 and σ_2 which are linearly independent modulo ${}^\circ E$. Then σ_1 and σ_2 are infinitely close to points τ_1, τ_2 of E_1, respectively. Since the dimension of E_1 exceeds that of E only by one, there must be a representation

4.4. $$\tau_2 = \lambda\tau_1 + \tau$$

or vice versa, where $\tau \in E$ and λ is an element of $^*C$. Now if λ were infinitesimal (including $\lambda = 0$) τ_2 would be infinitely close to E, and so σ_2 would be infinitely close to E and would belong to ${}^\circ E$. This is contrary to the assumption that σ_1 and σ_2 are linearly independent modulo ${}^\circ E$. If λ were infinite, then the relation

$$\tau_1 = \lambda^{-1}\tau_2 - \lambda^{-1}\tau$$

(in which λ^{-1} is infinitesimal and $\lambda^{-1}\tau$ belongs to E) would show that σ_1 belongs to ${}^\circ E$. Note that both τ_1 and τ_2 are norm-finite since they are infinitely close to the standard points σ_1 and σ_2, respectively.

We conclude that λ possesses a standard part, ${}^\circ\lambda$, and that ${}^\circ\lambda \neq 0$.

Also, $\tau = \tau_2 - \lambda\tau_1$ is infinitely close to $\sigma = \sigma_2 - {}^\circ\lambda\sigma_1$, since

$$\begin{aligned} \| \tau - \sigma \| &= \| \tau_2 - \lambda\tau_1 - (\sigma_2 - {}^\circ\lambda\sigma_1) \| \\ &\leqq \| \tau_2 - \sigma_2 \| + |\lambda| \, \| \tau_1 - \sigma_1 \| + |\lambda - {}^\circ\lambda| \, \| \sigma_1 \| \end{aligned}$$

so that $\| \tau - \sigma \|$ is infinitesimal. It follows that σ belongs to ${}^\circ E$ and that σ_1 and σ_2 are linearly dependent modulo ${}^\circ E$. This contradiction proves the theorem.

5. Proof of the main theorem. We are now ready to prove 1.1. To begin with, we work in the standard model, i.e., in an ordinary Hilbert space H over the complex numbers, C. Our method, like that of [1] is based on the fact that in a *finite-dimensional* space, of dimension μ say, any linear operator possesses a chain of invariant subspaces

5.1. $$E_0 \subseteqq E_1 \subseteqq E_2 \subseteqq \cdots \subseteqq E_\mu$$

where $d(E_j) = j$, $0 \leqq j \leqq \mu$, so that $E_0 = \{0\}$.

The proof of 1.1. is trivial [1] unless for every $\sigma \neq 0$ in H, the set $A = \{\sigma, T\sigma, T^2\sigma, \cdots, T^n\sigma, \cdots\}$ is linearly independent algebraically and generates the entire space. Assuming from now on that this is the case, we choose σ such that $\| \sigma \| = 1$, and we replace A by an equivalent orthonormal set $B = \{\sigma = \eta_1, \eta_2, \eta_3, \cdots \eta_n, \cdots\}$ by the Gram-Schmidt method. Then $\{\sigma, T\sigma, \cdots, T^{n-1}\sigma\}$ and $\{\eta_1, \eta_2, \cdots \eta_n\}$ are linearly dependent upon each other. We deduce without difficulty that T is almost superdiagonal with respect to the basis B. Representing any $\tau \in H$ by the sequence $\{t_n\}$, where $t_n = (\tau, \eta_n)$, we may then identify H with the sequence space considered in the preceding sections. Thus, if $T = (a_{jk})$ in this representation, then $a_{jk} = 0$ for $j > k + 1$, $k = 1, 2, 3, \cdots$ and, passing to ${}^*C$ and ${}^*H$, there exists an infinite positive integer ω such that $a_{\omega+1,\omega}$ is infinitesimal, by 3.4. ω will be kept fixed from now on, and for it we consider the space H_ω and the operators P and $T' = PTP$ introduced in Section 4 above.

Let $\xi = \{x_i\}$ be any norm-finite element of ${}^*H$. Consider the difference

$$(TP - T')\xi = (I - P)TP\,\xi = \zeta = \{z_n\}\,.$$

We obtain by direct computation that $z_{\omega+1} = a_{\omega+1,\omega}x_\omega$, and $z_n = 0$ for $n \neq \omega + 1$. Hence $\| \zeta \| \leqq |a_{\omega+1,\omega}| \, \| \xi \|$, so that ζ is infinitesimal. Using the equivalence relation $\tau_1 \sim \tau_2$ for points of ${}^*H$ such that $\| \tau_1 - \tau_2 \|$ is infinitesimal, we have shown that $TP\xi \sim T'\xi$, where the points on both sides of this equivalence are norm-finite. We then prove by induction that:

5.2. $T^r P\xi \sim (T')^r \xi$ for norm-finite ξ, $r = 1, 2, 3, \cdots$.

The case $r = 1$ has just been disposed of. Suppose 5.2 proved for $r - 1$, $r \geqq 2$. Then

$$T^r P\xi \sim T(T')^{r-1}\xi = TP(T')^{r-1}\xi \sim T'(T')^{r-1}\xi = (T')^r\xi$$

where we have made use of the first equivalence for $(T')^{r-1}\xi$ in place of ξ. Applying 5.2 to the monomials of $p(T)$, and taking into account that $P\xi \sim \xi$ for $\xi \in H_\omega$, we obtain

5.3. $p(T)\xi \sim p(T')\xi$ for norm-finite ξ in H_ω.

Let T_ω be the restriction of T' to H_ω, as in Section 4. Since H_ω is "finite" more precisely ω-dimensional in the sense of Nonstandard Analysis, there exists a chain of subspaces as in 5.1 with $\mu = \omega$, such that $T_\omega E_j \subseteqq E_j$, $j = 0, 1, 2, \cdots, \omega$. The E_j are also linear subspaces of $^*H$. They are finite-dimensional, hence closed, in the sense of Nonstandard Analysis, i.e., they satisfy the formal condition of closedness as expressed within the language L. Let P_j be the projection operator from $^*H$ onto E_j, $j = 0, 1, 2, \cdots, \omega$, so that $P_\omega = P$.

Suppose $p(z)$ is given by 3.5. For any $\xi \neq 0$ in H, $p(T)\xi$ must be different from 0 otherwise $\xi, T\xi, \cdots, T^n\xi$ would be linearly dependent, contrary to assumption. Choose ξ in H with $\|\xi\| = 1$. Since $\xi \sim P\xi$, $p(T)\xi \sim p(T)P\xi$, so $p(T)P\xi$ is not infinitesimal and by 5.3, $p(T)P\xi$ and hence $p(T')\xi$ is not infinitesimal. Thus, $\|p(T')\xi\| > r$ for some standard positive r. Consider the expressions

5.4. $r_j = \|p(T')\xi - p(T')P_j\xi\|$, $j = 0, 1, 2, \cdots, \omega$,

and note that $r_j \leqq \|p(T')\|\,\|\xi - P_j\xi\|$. We have $r_0 = \|p(T')\xi\|$ so $r_0 > r$. Also $\|\xi - P_\omega\xi\| = \|\xi - P\xi\|$ is infinitesimal, hence $r_\omega < r/2$. It follows that there exists a smallest positive integer λ with may be finite or infinite, such that $r_\lambda < r/2$ but $r_{\lambda-1} \geqq r/2$.

With every E_j, we associate the closed linear subspace $^\circ E_j$ of H which was defined in Section 4. Now $^\circ E_{\lambda-1}$ cannot coincide with H, more particularly, it cannot include ξ. For if it did, then $\|\xi - P_{\lambda-1}\xi\|$ would be infinitesimal, so $r_{\lambda-1}$, which is bounded by $\|p(T')\|\,\|\xi - P_{\lambda-1}\xi\|$ would be infinitesimal, contrary to the choice of λ.

On the other hand $^\circ E_\lambda$ cannot reduce to $\{0\}$. Consider the point $\eta = p(T')P_\lambda\xi$. $\eta \in E_\lambda$ since $P_\lambda\xi \in E_\lambda$ and E_λ is invariant under $p(T_\omega)$ and, equivalently, under $p(T')$. Also, since $P_\lambda\xi \in H_\omega$,

$$\eta = p(T')P_\lambda\xi \sim p(T)P_\lambda\xi\ ,$$

where the right-hand side is near-standard, by 3.2, since $P_\lambda\xi$ is norm-finite and $p(T)$ is completely continuous. It follows that η possesses a standard part, $°\eta$, and that $°\eta$ belongs to $°E_\lambda$. Again, $°\eta = 0$ would imply that η is infinitesimal. Hence, by 5.4

$$r_\lambda \geqq \| p(T')\xi \| - \| p(T')P_\lambda \xi \| > r - \zeta$$

where ζ is infinitesimal. Hence $r_\lambda > r/2$, contrary to the choice of λ. We conclude that $°E_\lambda$ contains a point different from 0, i.e., $°\eta$.

Both $°E_{\lambda-1}$ and $°E_\lambda$ are invariant for T, by 4.2. If neither were a proper invariant subspace of H for T we should have $°E_{\lambda-1} = \{0\}$, $°E_\lambda = H$. But this contradicts 4.3, proving 1.1.

REFERENCES

1. N. Aronszajn and K. T. Smith, *Invariant subspaces of completely continuous operators*, Annals of Math. **60** (1954), 345-350.
2. P. R. Halmos, *A glimpse into Hilbert space*, Lectures on Modern Mathematics, Vol. I, New York-London (1963), 1-22.
3. L. Henkin, *Completeness in the theory of types*, J. Symbolic Logic **15** (1950), 81-91.
4. A. Robinson, *Non-standard analysis*, Proc. Royal Acad. Sci, Amsterdam, Ser. A **64**, 432-440.
5. ———, *Introduction to model theory and to the metamathematics of Algebra*, Studies in Logic and the Foundations of Mathematics, Amsterdam, 1963.
6. ———, *On generalized limits and linear functionals*, Pacific J. Math. **14** (1964), 269-283.
7. ———, *Topics in non-archimedean Mathematics*, Proceedings of the Symposium on Model Theory, Berkeley, 1963. (To be published.)

UNIVERSITY OF WISCONSIN
AND
UNIVERSITY OF CALIFORNIA, LOS ANGELES

TOPICS IN NON-ARCHIMEDEAN MATHEMATICS

ABRAHAM ROBINSON

University of California, Los Angeles, California, U.S.A.

1. Introduction. Students of the history of mathematics know that Leibniz wished to base the differential and integral calculus on a number system that includes infinitely small and infinitely large quantities. More precisely, he regarded the new numbers as ideal elements which were supposed to have the same properties as the familiar real numbers and stated that their introduction was useful for the *art of invention.* Neither Leibniz nor his immediate successors were able to establish a rational framework within which this claim could be substantiated and (with rare exceptions) the attempt was abandoned by later generations, who followed the lines laid down by Cauchy and Weierstrass. Even so, there were individual mathematicians such as du Bois–Reymond, Levi–Civita, and quite recently, C. Schmieden and D. Laugwitz, who tried to revive Leibniz' ideas and who obtained certain interesting though partial results in this direction.

In the fall of 1960, it occurred to the present speaker that suitable extensions of the real number system – including infinitely small and infinitely large quantities as required by Leibniz, and yet possessing in a definite sense *the same properties* as the real numbers – can be obtained without difficulty by model-theoretic arguments. Such extensions (which are by no means unique) are analogous and related to the non-standard models of arithmetic first envisaged by Skolem. Accordingly, they have been called non-standard models of analysis. Some early results in this field were given in A. Robinson [61a] and developed in greater detail in A. Robinson [63]. Since then the scope of the theory has been enlarged in successive stages. In the present paper I propose to show by means of a number of suitably chosen examples how this has been done and how it has resulted in developments which should be of interest to mathematicians and logicians alike. On the mathematical side, I shall touch upon the theory of functions of a complex variable, topological spaces, and linear operators in Hilbert space.

2. Non-standard analysis of real numbers. Let $^\circ\mathfrak{R}$ be the field of real numbers. In a first-order language which contains distinct symbols for *all* individuals (numbers) *and* relations of $^\circ\mathfrak{R}$ (including the relations $E(x, y)$, $Q(x, y)$, $S(x, y, z)$, $P(x, y, z)$ for $x=y$, $x<y$, $x+y=z$ and $xy=z$ respectively), let K_0 be the set of all sentences which hold in $^\circ\mathfrak{R}$. By the usual finiteness (or compactness) argument, K_0 is consistent with the set of sentences $H=\{Q(0, a) \wedge Q(a, b_\nu)\}$ where 0 denotes the zero of $^\circ\mathfrak{R}$, as usual, b_ν varies over (symbols for) all positive elements of $^\circ\mathfrak{R}$, and a is an additional individual constant. Let $^*\mathfrak{R}$ be a model of $K_0 \cup H$. In this model the element (denoted by) a is positive and *infinitesimal*, i.e., such that the repeated addition of $|a|$ to itself any finite number of times yields a result smaller than 1. $^*\mathfrak{R}$ is thus a non-archimedean field, as we may conclude also from the fact that it is a proper extension of $^\circ\mathfrak{R}$. More particularly, $^*\mathfrak{R}$ is a proper elementary extension of $^\circ\mathfrak{R}$. It will be said to be a non-standard model of analysis.

Let $N(x)$ be the one-place predicate (one-place relation) which characterizes the set of natural numbers $^\circ\mathfrak{N}$ in $^\circ\mathfrak{R}$. Then the subset $^*\mathfrak{N}$ of $^*\mathfrak{R}$ which consists of all elements $^*\mathfrak{R}$ that satisfy $N(x)$ constitutes a proper elementary extension of $^\circ\mathfrak{N}$, and hence a non-standard model of arithmetic.

The infinitesimal elements a of $^*\mathfrak{R}$ can be characterized also by the property that $|a|<b$ for all positive elements b of $^\circ\mathfrak{R}$. If $|a|<b$ for *some* positive element of $^\circ\mathfrak{R}$ then a is said to be *finite*. For any finite element a of $^*\mathfrak{R}$, there exists a uniquely determined element of $^\circ\mathfrak{R}$, denoted by $^\circ a$ or st(a) and called the *standard part* of a, such that the difference $^\circ a-a$ is infinitesimal.

Within this framework we may reconstruct the elementary parts of the differential and integral calculus, as shown in some detail in the closing sections of A. Robinson [63]. A typical example is as follows.

Any infinite sequence of real numbers $\{s_n\}$ is represented in the language of K_0 by a two-place relation $F(x, y)$ whose domain of the first place coincides with $^\circ\mathfrak{N}$, and such that the argument of the first place determines the argument of the second place uniquely in $^\circ\mathfrak{R}$. These facts about F in relation to $^\circ\mathfrak{N}$ can be stated as sentences of K_0, and, accordingly, hold also in $^*\mathfrak{R}$. In ordinary language, $F(x, y)$ represents in $^*\mathfrak{R}$ a sequence $\{^*s_n\}$, where $^*s_n$ belongs to $^*\mathfrak{R}$ for all n in $^*\mathfrak{N}$, and $^*s_n=s_n$ for finite n.

Suppose $\{s_n\}$ is bounded in $^\circ\mathfrak{R}$. Thus, for some m in $^\circ\mathfrak{R}$ and for all n in $^\circ\mathfrak{N}$, $|s_n|<m$. Since this is a fact which can be expressed as a sentence

of $^{\circ}K$, $|{}^*s_n|<m$ also in $^*\mathfrak{R}$, i.e., also for arbitrary *infinite* $n=\omega$, where an element ω of $^*\mathfrak{R}$ is *infinite* if it belongs to $^*\mathfrak{R}-{}^{\circ}\mathfrak{R}$. Since $|{}^*s_\omega|<m$ it is finite and accordingly (see above) possesses a standard part $\sigma=\mathrm{st}({}^*s_\omega)$. We claim that σ is a limit point (accumulation point) of $\{s_n\}$, according to the standard definition in $^{\circ}\mathfrak{R}$. In order to verify this claim we have to show that for any given $\varepsilon>0$ in $^{\circ}\mathfrak{R}$ and for any $n \in {}^{\circ}\mathfrak{R}$, there exists an $P \in {}^{\circ}\mathfrak{R}$ which is greater than n such that $|s_P-\sigma|<\varepsilon$. Now the property of x, "x is a natural number greater then n such that $|s_x-\sigma|<\varepsilon$", can be formulated as a predicate $Q(x)$ in the language of K_0, and we have to show that the sentence $Y=(\exists x)Q(x)$ holds in $^{\circ}\mathfrak{R}$, i.e., that it belongs to K_0. For this purpose, we only have to establish that Y holds in $^*\mathfrak{R}$ (for either Y or $\sim Y$ holds in $^{\circ}\mathfrak{R}$, and if $\sim Y$ holds in $^{\circ}\mathfrak{R}$ then it belongs to K_0 and hence holds also in $^*\mathfrak{R}$). But $Q(\omega)$ holds in $^*\mathfrak{R}$ since $|{}^*s_\omega-\sigma|$ is infinitesimal, and so Y holds in $^*\mathfrak{R}$, as required. We have proved:

Every bounded sequence (in $^{\circ}\mathfrak{R}$) possesses an accumulation point; in other words, we have proved the theorem of Bolzano and Weierstrass.

3. Higher-order non-standard analysis. Within the framework described so far, we can prove theorems on *individual* functions and relations, i.e., theorems which do not involve any higher order quantification. A survey shows that this suffices for the more elementary parts of the theories of real and complex variables. Moreover, while we do not enlarge the domain of functions and other higher order concepts directly on passing from $^{\circ}\mathfrak{R}$ to $^*\mathfrak{R}$, this can be done in a roundabout way. For example, if $g(x, y)$ is a function of two variables in $^{\circ}\mathfrak{R}$, we may consider the function $f(x)=g(x, \omega)$ for some value of ω in $^*\mathfrak{R}-{}^{\circ}\mathfrak{R}$, and we may then obtain a function of one variable in $^*\mathfrak{R}$ which is not obtained by the extension of any function of one variable in $^{\circ}\mathfrak{R}$. Thus, taking

$$g(x, y)=(y/\pi)^{1/2}\exp(-yx^2)$$

and putting $y=\omega$, an infinite natural number, we obtain a function $f(x)=g(x, \omega)$ which is infinitesimal for all standard x (i.e., all x in $^{\circ}\mathfrak{R}$) except $x=0$. Now consider the function

$$J(y)=\int_{-\infty}^{\infty} g(x, y)dx$$

which is defined for all y in $^{\circ}\mathfrak{R}$ (and is, in the present case, known to be equal to 1 for all such y). Then we may *regard* $J(y)$, for all y in $^*\mathfrak{R}$ also, as the integral $\int_{-\infty}^{\infty} g(x, y)dx$. In this sense

$$\int_{-\infty}^{\infty} f(x)dx = 1$$

where $f(x) = (\omega/\pi)^{1/2} \exp(-\omega x^2)$ as described above. This shows that $f(x)$ may be thought of as a realization of Dirac's delta function.

However, if one wishes to handle problems involving higher order concepts more freely, it is appropriate to extend the general framework of the analysis. There are various ways for doing this, either along set-theoretical or along type-theoretical lines. The author has carried out the procedure in detail for a particular version of type theory (A. Robinson [62]). It is based on the assumption that our formal language contains, in addition to symbols for all individuals in the set of real numbers (or equivalently, in any given set of cardinal $\mathfrak{c}$), also symbols for all sets, relations, sets of sets, sets of relations, etc., and that quantification is permitted with respect to all of these. By $^\circ\mathfrak{R}$ we shall now mean the higher-order structure which can be described in this language, and by K_0 the set of all sentences which hold in this structure. Following Henkin [50a], we know that the finiteness principle of the lower predicate calculus can be extended to the kind of language now under consideration provided we reinterpret the notion of the set of all subsets, or of all relations, sets of relations, etc., in a given structure so as to include (possibly) only a subset of the totality of all entities of the given type. In this sense, the set of sentences $K_0 \cup H$, with $H = \{Q(0, a) \wedge Q(a, b_\nu)\}$ as before, and with the new definition of K_0, possesses a model $^*\mathfrak{R}$ which is a proper extension of $^\circ\mathfrak{R}$. The reinterpretation of higher-order concepts which is necessary if we wish to insure that $^*\mathfrak{R}$ satisfies the sentences of K_0 is illustrated by the following example. Let α be the symbol which denotes the set $^\circ A$ of all subsets of $^\circ\mathfrak{R}$ when the sentences of K_0 are interpreted in $^\circ\mathfrak{R}$ and consider the interpretation of α with reference to $^*\mathfrak{R}$. It is not difficult to see that α still denotes a set $^*A$ of sets of individuals of $^*\mathfrak{R}$. However, we are going to show that $^*A$ cannot now include *all* sets of individuals of $^*\mathfrak{R}$. More particularly, let $^\circ\mathfrak{N}$ and $^*\mathfrak{N}$ be the set of natural numbers in $^\circ\mathfrak{R}$ and the corresponding set in $^*\mathfrak{R}$, as before. Then $^\circ\mathfrak{N}$ cannot be included in A. For $^*\mathfrak{N}$ is included in $^*A$ since $^\circ\mathfrak{N}$ is included in $^\circ A$. Also, it is a property of the set $^\circ A$ which can be expressed as a sentence of K_0 and, accordingly, applies also to $^*A$, that the difference of any two elements of the set in question ($^\circ A$ or $^*A$) also belongs to the set. Thus, if $^\circ\mathfrak{N}$ belongs to $^*A$ then

$^*\mathfrak{N} - {}^\circ\mathfrak{N}$ also belongs to $^*A$. On the other hand, every non-empty set of natural numbers in $^\circ\mathfrak{N}$ possesses a smallest element and this must be true also in $^*\mathfrak{N}$. But $^*\mathfrak{N} - {}^\circ\mathfrak{N}$ is a non-empty set of natural numbers which does not possess a smallest element for if $\omega \in {}^*\mathfrak{N} - {}^\circ\mathfrak{N}$ (i.e., if ω is an infinite natural number) then $\omega - 1 \in {}^*\mathfrak{N} - {}^\circ\mathfrak{N}$ as well. This proves our assertion.

The theory of functions of a complex variable can be brought within the scope of the theory of $^\circ\mathfrak{R}$ by regarding complex numbers as pairs of real numbers. A complex function of a complex variable $z = f(w)$, with $z = x + iy$, $w = u + iv$, can then be expressed by a four-place relation $F(x, y, u, v)$ such that $F(a, b, c, d)$ holds for real numbers a, b, c, d if and only if $a + ib = f(c + id)$. Thus, if $\Phi(t)$ is a one-place relation which characterizes a family of complex functions of a complex variable – i.e., which is satisfied by certain four-place relations – in $^\circ\mathfrak{R}$, then $\Phi(t)$ will define a corresponding family of functions in $^*\mathfrak{R}$. These "generalized" functions can be used in order to investigate the properties of the "standard" complex functions in $^\circ\mathfrak{R}$, as shown in detail in A. Robinson [62]. Thus, the employment of functions of this kind can replace the familiar normal families method in the proof of Riemann's mapping theorem and in the theory of exceptional values of entire functions (Picard's theorem, Julia's directions).

I shall now give a concrete illustration of the methods which become available in this way. It is chosen from the so-called analytic theory of polynomials which is concerned with the location of the zeros of complex polynomials in the complex plane.

Let $^\circ Z$ be the complex plane in the ordinary sense. We may regard it as a set of pairs of elements of $^\circ\mathfrak{R}$. Let $^\circ\Pi$ be the set of all complex polynomials with complex coefficients. As we pass from $^\circ\mathfrak{R}$ to $^*\mathfrak{R}$, $^\circ Z$ is converted into an extension $^*Z$ of $^\circ Z$, and $^\circ\Pi$ is converted into a set of functions $^*\Pi$ which have the properties of the elements of $^\circ\Pi$ to the extent in which they can be expressed as sentences of K_0. In particular, $^*\Pi$ contains the extensions to $^*Z$ of all the standard polynomials $p(z)$ with coefficients in $^\circ Z$.

With every polynomial $p(z)$ in $^\circ\Pi$ and with every natural number $k \in {}^\circ\mathfrak{N}$, there is associated a complex number $a_k \in {}^\circ\mathfrak{Z}$ which is the coefficient of z^k in the MacLaurin expression for $p(z)$. For $p(z) \neq 0$, the least k for which $a_k \neq 0$ is the *degree* of $p(z)$. Also, the number of non-zero coefficients of any $p(z) \in {}^\circ\Pi$ will be called the *rank* of $p(z)$. Finally, given any positive number ϱ in $^\circ\mathfrak{R}$ we define $\Omega[p(z), \varrho]$ as the number of zeros z of $p(z) \neq 0$, such that $|z| < \varrho$, taking into account multiplicities.

All these functionals extend to $^*\Pi$ within $^*\mathfrak{R}$. However, it must be borne in mind that the degree and rank of a polynomial (element of $^*\Pi$) may now be infinite natural numbers. Some thought is required in order to see how $\nu = \Omega[p(z), \varrho]$ can be expressed within our formal language, but this can be done by formalizing

"there exists a finite set $\{z_1, \ldots, z_\nu\}$ such that $|z_i| < \varrho$ for $i = 1, 2, \ldots, \nu$ and $p(z) = (z - z_1) \ldots (z - z_\nu)q(z)$ where $q(z)$ is a polynomial which does not vanish for any z, $|z| < \varrho$."

Here again, ν may be infinite for $p(z)$ and ϱ in $^*\mathfrak{R}$.

A *monomial* in $^\circ\Pi$ or $^*\Pi$ is, as usual, a polynomial of rank 1. The general form of a monomial in $^*\Pi$ is az^n where $a \in {}^*Z$ and $n \in {}^*\mathfrak{N}$, and where a and n may or may not belong to $^\circ Z$ and $^\circ\mathfrak{N}$, respectively. Any polynomial of *finite* positive rank may be written as a *finite* sum (*finite* in the ordinary sense, in both cases) of monomials of different degrees, so

$$p(z) = c_1 z^{n_1} + c_2 z^{n_2} + \ldots + c_r z^{n_r}, \quad r \geqslant 1 \tag{3.1}$$

where $n_1 < n_2 < \ldots < n_r$ and $c_i \neq 0$, $1 \leqslant i \leqslant r$, and where the n_i may be finite or infinite.

Let az^m and bz^n be two monomials, of different degrees, finite or infinite. We say that bz^n *dominates* az^m, $bz^n \gg az^m$, if there exists a *finite* positive number ϱ in $^\circ\mathfrak{R}$ and that $|bz^n| > |az^m|$ for all z such that $|z|$ is *finite* and greater than ϱ. In order to appreciate the significance of this relation notice that for a, b, n, m in $^\circ\mathfrak{R}$, the monomial of higher degree dominates the other. This is not necessarily true in $^*\mathfrak{R}$ but it can still be shown that either $az^m \gg bz^n$ or $bz^n \gg az^m$ (and the two cases are mutually exclusive). It can also be shown that if $bz^n \gg az^m$ and $cz^k \gg bz^n$, and $k \neq m$, then $cz^k \gg az^m$. This shows that among the monomials which constitute a polynomial of finite rank as written out in 3.1 there is one that dominates all the others. The degree of this monomial will be called the *order* of the polynomial, ord $p(z)$. It turns out that, for sufficiently large finite $|z|$ the modulus of the dominant monomial becomes larger than the sum of the remaining monomials of $p(z)$. On this basis, an adaptation of classical arguments (use of Rouché's theorem) provides the following result.

Theorem 3.2. *Let $p(z)$ be a non-zero polynomial of finite rank in*

*II. *If* ord $p(z)=\mu$ *is finite,* $\mu \in {}^\circ\mathfrak{N}$, *then the number of finite zeros of* $p(z)$ (*i.e., with finite* $|z|$) *is* μ. *If* ord $p(z)$ *is infinite then the number of finite zeros of* $p(z)$ *is infinite.*

This theorem leads to a far-reaching generalization of the following classical result which is due to Kakeya.

Let $p(z)$ be a polynomial of degree not exceeding n, $n \geqslant 1$, such that at least $k \geqslant 1$ roots of $p(z)$ are included in the circle $|z| \leqslant \varrho$. There exists a positive number $\theta=\theta(n, k)$—which does not depend on the particular coefficients of $p(z)$—such that $p'(z)$ has at least $k-1$ roots in the circle $|z| \leqslant \varrho\theta$.

Equivalently, we may state in the conclusion that $zp'(z)$ has at least k roots in the circle $|z| \leqslant \varrho\theta$. Thus, Kakeya's theorem is a special case of the following:

Theorem 3.3. *There exists a function* $\phi(r, k)$ *whose arguments range over the positive integers (classically, i.e., in* ${}^\circ\mathfrak{N}$) *and which takes real positive values (i.e., in* ${}^\circ\mathfrak{R}$) *and which possesses the following property: For any positive real* ϱ *and for any polynomial of rank not exceeding* r,

$$p(z)=c_1 z^{n_1}+c_2 z^{n_2}+ \dots c_r z^{n_r},\ 0 \leqslant n_1 < n_2 < \dots < n_r,$$

and for any set of complex numbers λ_i *such that* $0 \leqslant |\lambda_1| \leqslant |\lambda_2| \leqslant \dots \leqslant |\lambda_r|$ —*if* $p(z)$ *possesses at least* k *roots in the circle* $|z| \leqslant \varrho$ *then the polynomial*

$$g(z)=\lambda_1 c_1 z^{n_1}+\lambda_2 c_2 z^{n_2}+ \dots +\lambda_r c_r z^{n_r}$$

possesses at least k *roots in the circle* $|z| \leqslant \varrho\phi(r, k)$.

Note that this is a classical theorem, without reference to any non-standard extension of the real or complex numbers. The substitution $z=\varrho z'$ shows that, for the proof, it is sufficient to suppose $\varrho=1$.

Given any positive integers r and k in ${}^\circ\mathfrak{N}$, suppose that no $\phi(r, k)$ as asserted by the theorem exists. Then the following statement holds in ${}^\circ\mathfrak{R}$.

"For every $\sigma > 0$, there exist a polynomial $p(z)$ and constants λ_i that satisfy the assumptions of the theorem, with $\varrho=1$, such that the corresponding $g(z)$ possesses not more than $k-1$ roots in the circle $|z| \leqslant \sigma$."

The statement in quotation marks can be expressed as a sentence X which then belongs to K_0, and hence holds also in ${}^*\mathfrak{R}$. Choosing for σ an *infinite* positive number (an element of ${}^*\mathfrak{R}$ greater than any number of ${}^\circ\mathfrak{R}$), we then obtain $p(z)$ and λ_i in ${}^*\mathfrak{R}$ which satisfy the assumptions of the theorem, with $\varrho=1$, while $g(z)$ has less then k zeros in the circle $|z|\leqslant\sigma$ and, a fortiori, has less than k finite zeros (and hence does not vanish identically). However, bearing in mind Theorem 3.2 above, we shall refute this possibility by showing that $g(z)$ must be of order k at least.

Indeed, since $g(z)$ does not vanish identically, it includes a non-vanishing monomial, $\lambda_j c_j z^{n_j}$ say, which dominates all the other monomials of $g(z)$, so that $\operatorname{ord} g(z)=n_j$. Similarly, $p(z)$ includes a dominant polynomial, $c_i z^{n_i}$ say, and since $p(z)$ has at least k finite zeros we have $n_i=\operatorname{ord} p(z)\geqslant k$. Now suppose that $j<i$, $n_j<n_i$. At the same time, $c_i z^{n_i}$ dominates $c_j z^{n_j}$ so that

$$|c_i z^{n_i}/c_j z^{n_j}|>1$$

for sufficiently large finite $|z|$. But $|\lambda_i|\geqslant|\lambda_j|$, by assumption, and so, for such $|z|$,

$$|\lambda_i c_i z^{n_i}/\lambda_j c_j z^{n_j}|=|\lambda_i/\lambda_j|\,|c_i z^{n_i}/c_j z^{n_j}|>1,$$

which contradicts the fact that $\lambda_j c_j z^{n_j}$ is the dominant monomial in $g(z)$. We conclude that $j\geqslant i$, $n_j\geqslant n_i\geqslant k$, so that $g(z)$ possesses at least k finite zeros. This contradiction proves 3.3.

Kakeya's theorem is obtained from 3.3 by taking $\lambda_i=n_i=i-1$. Other applications of our method to the analytic theory of polynomials are given in A. Robinson [62].

4. Application to topological spaces. As we pass from classical analysis to the consideration of other structures, e.g., of topological spaces, we have to decide how to select the appropriate extensions that correspond to the non-archimedean fields used in the preceding sections. We are going to describe a scheme which is adequate for this purpose and which is suitable for both first- and higher-order languages.

Let ${}^\circ\mathfrak{M}$ be a structure which corresponds either to a first-order language, as in Section 2, or to a higher-order language, as in Section 3. Let K_0 be the set of all sentences which hold in ${}^\circ\mathfrak{M}$, within the language under consideration, in terms of a given vocabulary. For the purpose of the following definition it is not necessary that this vocabulary include symbols for *all* individuals and relations (of first and higher order, as the case may be) in ${}^\circ\mathfrak{M}$.

Let $Q(x, y)$ be any two-place predicate (well-formed formula with two free variables) which is definable in the vocabulary of K_0. We shall say that $Q(x, y)$ is *concurrent* in $°\mathfrak{M}$ if for every finite set of entities (individuals or relations) $a_1, \ldots, a_n, n \geqslant 1$, which belong to the domain of the first place of Q in $°\mathfrak{M}$, there exists an entity b in $°\mathfrak{M}$ such that $Q(a_1, b), \ldots, Q(a_n, b)$ all hold in $°\mathfrak{M}$. In this connection, we say that a_i belongs to the domain of the first place of Q if $Q(a_i, b)$ holds in $°\mathfrak{M}$ for *some* b.

For every concurrent predicate Q we have introduced a distinct symbol b_Q, and we define K_Q as the set of sentences $\{Q(a_\nu, b_Q)\}$ where a_ν varies over all (symbols for) elements of the domain of the first place of Q. Let

$$H = K_0 \cup \bigcup_Q K_Q$$

where Q varies over all concurrent predicates in $°\mathfrak{M}$ which are defined in the given vocabulary. A familiar argument shows that H is consistent. The models $^*\mathfrak{M}$ of H will take over the part played previously by the non-standard models of analysis $^*\mathfrak{R}$. At the same time we may verify without difficulty that the more general procedure sketched above also yields non-standard models of analysis, as in Sections 2 and 3, for the appropriate vocabularies. Thus, taking the first-order case considered in Section 2, with a vocabulary which includes (symbols for) all individuals and relations in the field of real numbers, $°\mathfrak{M} = °\mathfrak{R}$, we claim that $^*\mathfrak{M}$ as introduced above is a non-standard model of analysis in the sense of Section 2. Indeed, consider the relation which is in ordinary language

"y is positive and is smaller than x."

Evidently, this can be expressed as a predicate in the language of K_0, more particularly, like all other predicates that range over individuals it is actually given by an atomic relation in the language of K_0, $Q(x, y)$ say. The domain of the first place of Q consists of all positive real numbers. It follows that in any model $^*\mathfrak{M}$ of H, b_Q denotes an element which is positive and smaller than any ordinary (standard) real number. This shows that $^*\mathfrak{M}$ is a non-standard model of analysis. Similar arguments apply to the higher order theory.

Adopting the higher order approach, let $°\mathfrak{M}$ be an arbitrary topological space, and suppose that the vocabulary of the language includes (symbols for) all individuals, sets, relations, sets of relations, etc. Let $^*\mathfrak{M}$ be an extension of $°\mathfrak{M}$ as introduced above. The points (individuals) of $^*\mathfrak{M}$ include, in particular, the points of $°\mathfrak{M}$, which will be called *standard* *points*. Similarly, if $°U$ is the set of open sets in $°\mathfrak{M}$, and $^*U$ is the corresponding set in $^*\mathfrak{M}$, then $^*U$ contains in particular the extensions of the open sets of $°\mathfrak{M}$. These extensions will be called the *standard open sets* of $^*\mathfrak{M}$. Notice that in general, a standard open set is a proper extension of the corresponding set in $°\mathfrak{M}$. For example, if $°\mathfrak{M}=°\mathfrak{R}$ as in Section 3, with the usual topology and $°A$ is the open interval between 0 and 1 then the corresponding $^*A$ contains all elements of $^*\mathfrak{R}$ between 0 and 1, standard or not, but at the same time is a *standard* open set according to our definition.

For any point a in $^*\mathfrak{M}$, we define the *monad* of a as the intersection of all standard open sets which include a. If $°\mathfrak{M}=°\mathfrak{R}$ and $^*\mathfrak{M}=^*\mathfrak{R}$, and a is a standard point (ordinary real number) then the monad of a consists of all points b in $^*\mathfrak{R}$ such that $b-a$ is infinitesimal. In general, the monad of a point a need not be a set within the theory of $^*\mathfrak{M}$ (just as $°\mathfrak{R}$ is not a set within the theory of $^*\mathfrak{R}$). A point of $^*\mathfrak{M}$ which is included in the monad of a standard point is called *near-standard*. We are going to prove

Theorem 4.1. *A space* $°\mathfrak{M}$ *is compact if and only if all points of* $^*\mathfrak{M}$ *are near-standard.*

In this theorem, $^*\mathfrak{M}$ is *any* extension of $°\mathfrak{M}$ of the kind introduced above. A space is called compact if it satisfies the Heine–Borel condition, without reference to any separation axiom.

Proof of 4.1. Suppose that $°\mathfrak{M}$ is not compact. Then there exists an open covering of $°\mathfrak{M}$, $°B=\{°A_\nu\}$ such that for every finite subset $\{°A_1, \ldots, °A_n\}$ of $°B$ there is a point in $°\mathfrak{M}$ which does not belong to $°A_1 \cup °A_2 \cup \ldots \cup °A_n$. Consider the relation $Q(x, y)$ which is expressed informally by

"The complement of x belongs to $°B$ and x contains y."

Then $Q(x, y)$ is concurrent, according to the definition given above. It follows that $^*\mathfrak{M}$ contains a point b_Q such that b_Q is contained in the complements of all $^*A_\nu$ (which are the extensions of the corresponding $°A_\nu$ from $°\mathfrak{M}$ to $^*\mathfrak{M}$). Thus, for every standard point in $^*\mathfrak{M}$, b_Q belongs to the complement of some standard open set that contains the point in question, and hence does not belong to its monad. This shows that b_Q is not near-standard and establishes the sufficiency of the condition.

Conversely, suppose that the point b in $^*\mathfrak{M}$ does not belong to the monad of any standard point. It is then possible to choose for every standard point a_ν a standard open set $^*A_\nu$ which includes a_ν and which does not include b. The set of corresponding open sets in $^\circ\mathfrak{M}$, $^\circ B=\{^\circ A_\nu\}$, is then an open covering of $^\circ\mathfrak{M}$. We claim that no finite subset of $^\circ B$ covers $^\circ\mathfrak{M}$. For if the subset $\{^\circ A_1, \ldots, {}^\circ A_k\}$ of $^\circ B$ were a covering of $^\circ\mathfrak{M}$, then the formal equivalent of the statement

"For all points x, $x \in {}^\circ A_1 \cup {}^\circ A_2 \cup \ldots \cup {}^\circ A_k$"

would hold in $^\circ\mathfrak{M}$ and so

"For all points x, $x \in {}^*A_1 \cup {}^*A_2 \cup \ldots \cup {}^*A_k$"

would hold in $^\circ\mathfrak{M}$. In particular, then $b \in {}^*A_1 \cup {}^*A_2 \cup \ldots \cup {}^*A_k$, which is contrary to the definition of the $^*A_\nu$. This completes the proof of 4.1.

Theorem 4.1 provides a simple proof of the theorem of Heine–Borel that a closed and bounded set $^\circ A$ in (ordinary) n-dimensional Euclidean space $^\circ\mathfrak{R}^n$, $n \geqslant 1$, is compact. Indeed, by 4.1 we only have to show that all points of $^*A$ are near-standard if $^\circ A$ is a closed and bounded set. Let $a=(a_1, \ldots a_n)$ be any point in $^*A$. Then the coordinates $a_1, \ldots, a_n$ of a are all finite, and hence possess standard parts, $^\circ a_1, \ldots, {}^\circ a_n$. Then it can be shown without difficulty that the point $^\circ a=(^\circ a_1, \ldots, {}^\circ a_n)$ belongs to $^\circ A$ and that a belongs to the monad of $^\circ a$.

5. Application to normed linear spaces. Let $^\circ\mathfrak{M}$ be a metric space. In the usual topology induced by the metric of $^\circ\mathfrak{M}$, a point a in $^*\mathfrak{M}$ is near-standard if and only if the distance of a from some standard point in $^*\mathfrak{M}$ is infinitesimal. This is true in particular if $^\circ\mathfrak{M}$ is a normed linear space. In that case, we call a point a in $^*\mathfrak{M}$ *norm-finite* if the norm of a is a finite real number as defined in Section 2 above.

Now let $^\circ\mathfrak{M}$ be a Banach space, and let $^\circ T$ be a linear operator on $^\circ\mathfrak{M}$ into $^\circ\mathfrak{M}$. $^\circ T$ is said to be compact if it transforms every bounded set in $^\circ\mathfrak{M}$ into a relatively compact set, i.e., into a set which is compact in $^\circ\mathfrak{M}$. This important class of operators can be characterized very simply in terms of the corresponding $^*T$ in $^*\mathfrak{M}$. It turns out that $^\circ T$ is compact if and only if $^*T$ maps every norm-finite point on a near-standard point. This definition can be used among other things in order to give a direct proof of the fact that an integral operator with continuous kernel is compact in the space of continuous functions.

Finally, let us consider the application of our method to (infinite-dimensional) separable Hilbert space. Such a space can be represented by the infinite sequences $\{s_n\}$ such that $\sum s_n^2$ is finite or such that $\sum s_n \bar{s}_n$ is finite (for real or complex Hilbert space, as the case may be). Accordingly we may consider its theory already within the framework of Section 3, although the more general approach of Section 4 appears more natural. Within this framework let ${}^{\circ}\mathfrak{M}$ be the given Hilbert space, together with its sets, relations, sets of relations, etc., as before, and together with its ordinary field of coefficients, ${}^{\circ}\mathfrak{R}$ say (for the real case), and let ${}^{*}\mathfrak{M}$ be an extension, as introduced in Section 4, incorporating a non-standard model of analysis, ${}^{*}\mathfrak{R}$, as the appropriate extension of ${}^{\circ}\mathfrak{R}$.

Let $\{e_1, e_2, e_3, \ldots\}$ be an orthogonal basis in ${}^{\circ}\mathfrak{M}$ so that an arbitrary point of the space can be represented by a sequence of coordinates $\{\lambda_1, \lambda_2, \lambda_3, \ldots\}$ with respect to this basis. On passing to ${}^{*}\mathfrak{M}$, $\{e_i\}$ is extended to a set $\{{}^{*}e_i\}$, i.e., a set of points in the extended space, with subscripts ranging over the appropriate extension of the natural numbers, ${}^{*}\mathfrak{N}$, such that $e_i = {}^{*}e_i$ for all finite i. Any point in the extended space is given by a sequence $\{\lambda_i\}$ with i ranging over ${}^{*}\mathfrak{N}$ and λ_i in ${}^{*}\mathfrak{R}$, such that the infinite sum $\sum_{i=1}^{\infty} \lambda_i^2$ converges in the sense of ${}^{*}\mathfrak{M}$. That is to say, $\sum_{i=1}^{\infty} \lambda_i^2$ may be a finite real number in ${}^{*}\mathfrak{R}$, or an infinite real number in ${}^{*}\mathfrak{R}$, provided it exists at all. It can be shown that the point represented by $\{\lambda_i\}$ is *norm-finite* in the sense obtained above precisely when $\sum_{i=1}^{\infty} \lambda_i^2$ is finite (i.e., not greater than some element of ${}^{\circ}\mathfrak{R}$) and that it is *near-standard* if and only if $\sum_{i=\omega}^{\infty} \lambda_i^2$ is infinitesimal for all infinite natural numbers ω.

We shall now indicate how this set-up can be used for the development of spectral theory. Let ${}^{\circ}T$ be a linear operator which is defined and hence bounded, on the space ${}^{\circ}\mathfrak{M}$. Let ${}^{*}T$ be the extension of ${}^{\circ}T$ to the space ${}^{*}\mathfrak{M}$. Now choose an infinite natural number $\omega \in {}^{*}\mathfrak{N}$ and let ${}^{\omega}\mathfrak{M}$ be the subspace of ${}^{*}\mathfrak{M}$ which consists of the point of ${}^{*}\mathfrak{M}$ whose coordinates λ_i vanish for $i > \omega$. Then ${}^{\omega}\mathfrak{M}$ is a "finite-dimensional" space in a non-standard sense, i.e., its dimension is the "natural" number ω.

Let ${}^{\omega}P$ be the projection operator from ${}^{*}\mathfrak{M}$ onto ${}^{\omega}\mathfrak{M}$, so

$${}^{\omega}P\colon \{\lambda_1, \lambda_2, \ldots, \lambda_{\omega}, \lambda_{\omega+1}, \lambda_{\omega+2}, \ldots\} \to \{\lambda_1, \lambda_2, \ldots, \lambda_{\omega}, 0, 0, \ldots\}.$$

Define the operator ${}^{\omega}T$ as the restriction of the product ${}^{\omega}PT{}^{\omega}P$ to

${}^{\omega}\mathfrak{M}$. Then ${}^{\omega}T$ is a linear operator on the "finite-dimensional" space ${}^{\omega}\mathfrak{M}$ and, accordingly, can be represented by an $\omega \times \omega$ matrix in the usual way. In particular, if ${}^{\circ}T$ is self-adjoint, so are ${}^{*}T$ and ${}^{\omega}T$. Applying the standard theory of matrices in finite-dimensional spaces, we find that ${}^{\omega}T$ possesses ω eigenvalues, equal or different, and to each eigenvalue there corresponds an eigenspace such that ${}^{\omega}\mathfrak{M}$ is the direct sum of these spaces. The question now arises how to reduce these results about ${}^{\omega}T$ to results about ${}^{\circ}T$ in ${}^{\circ}\mathfrak{M}$. This reduction is not trivial, but we have carried it out in detail for compact operators ${}^{\circ}T$. In that case the eigenvalues λ_i of ${}^{\omega}T$ are finite and hence, possess standard parts ${}^{\circ}\lambda_i$. It can then be shown that any ${}^{\circ}\lambda_i$ which is different from 0 is an eigenvalue of ${}^{\circ}T$. Moreover, for λ_i for which ${}^{\circ}\lambda_i \neq 0$, i.e., which are not infinitesimal, any corresponding eigenvector turns out to be near-standard, owing to the compactness of ${}^{\circ}T$.

If ${}^{\circ}T$ is not compact, we can at least appreciate in a general way how it may happen that certain elements of the spectrum are left without eigenvectors. Such elements may be expected to correspond to eigenvalues λ_i of ${}^{\omega}T$ in ${}^{\omega}\mathfrak{M}$ whose eigenvectors, though chosen to be of finite norm, are not near-standard, and hence do not correspond to any eigenvectors in ${}^{\circ}\mathfrak{M}$.

6. Methodological considerations. As mentioned in A. Robinson [61a] and [62], certain non-standard models of analysis can be realized as ultraproducts. This has the advantage of making the theory more concrete in appearance. Luxemburg (in his lecture notes, California Institute of Technology, 1962) and Takeuti [62] have in fact based their contributions to non-standard analysis on structures of this type. Using a theorem due to Frayne one can show that the models ${}^{*}\mathfrak{M}$ introduced in Section 4 above can be embedded in ultrapowers of ${}^{\circ}\mathfrak{M}$ and hence, that for any given ${}^{\circ}\mathfrak{M}$, ${}^{*}\mathfrak{M}$ may be taken as an ultrapower of ${}^{\circ}\mathfrak{M}$. There are other structures of special types which would serve our purpose. However, for all the applications of the theory made so far, any one of the structures ${}^{*}\mathfrak{M}$ introduced in Section 4 is as good as another. Accordingly, we preferred to base our considerations on the formal properties of the models ${}^{*}\mathfrak{M}$ rather than on their mode of construction.

It may be said that in spite of the model-theoretic garb of our method, it really represents an extension of the *deductive procedures* used in various branches of mathematics. And, if we wish to pass to a more philosophical level of discussion, we may well argue that the operation of the existence of standard models as opposed to non-standard models, e.g., of arithmetic, has a meaning only inasmuch as it implies the adoption of certain rules of deduction which distinguish the standard model from any non-standard model that is under consideration at the same time.

Logica matematica. — *A new approach to the theory of algebraic numbers.* Nota di ABRAHAM ROBINSON, presentata (*) dal Socio B. SEGRE.

RIASSUNTO. — Nella teoria degli ideli e adeli, gli ideali di un anello di Dedekind D (per esempio di numeri algebrici o di funzioni algebriche) sono moltiplicativamente isomorfi, a meno di elementi associati, ad un sotto–insieme di un anello, Δ, che è un'estensione di D. L'idea fondamentale della teoria suaccennata fu concepita da Prüfer e sviluppata poi da von Neumann, Chevalley e altri. In questa Nota mostriamo come una teoria di questo tipo può venire rielaborata adoperando modelli non–standard definiti per mezzo di un linguaggio formalizzato.

1. In the theory of idèles and adèles (e.g. refs. [1], [3]), the ideals of a given Dedekind domain D, e.g. within the realm of algebraic numbers or of algebraic functions of one variable are shown to be multiplicatively isomorphic, up to associated elements, to a subset of a ring, Δ, which is an extension of D. In the special cases mentioned, the idea was first realized by Prüfer and, following him, by von Neumann, while the case of general Dedekind domains was treated by Krull (refs. [2], [4], [5]). We propose to show that the theory can be developed conveniently by the use of the notion of a non-standard enlargement of a given mathematical structure which has, in recent years, been applied extensively to Analysis (refs. [6]–[9]). In the present Note, we shall describe our method independently of the above mentioned approaches, leaving the study of the connection with them for a later paper. We also propose to show subsequently that our tools are sufficiently powerful to cope with infinite algebraic extensions of the rational number field and of rational function fields of one variable.

2. Let M be a mathematical structure. Let K be the set of sentences which are formulated in a higher order predicate language which comprises all finite types (or alternatively, which are formulated within a suitable set-theoretic framework) and which are true in M. Consider any binary relation $R(x, y)$ between individuals of M or between entities of other types (e.g. between individuals and sets). An entity will be said to be *in the domain of the first variable* of R if there exists an entity b such that $R(a, b)$ holds in M. $R(x, y)$ will be called *concurrent* if for any finite set of entities, $a_1, \cdots, a_n$ which are in the domain of the first variable of R, there exists an entity b such that $R(a_1, b), \cdots, R(a_n, b)$ all hold in M (and hence, belong to K). It is not difficult to establish the existence of *enlargements* of M; i.e. of extensions *M of M such that (i), M satisfies all sentences of K and (ii) for any concurrent

(*) Nella seduta del 12 febbraio 1966.

relation $R(x,y)$ in M there exists an entity b_R in *M such that $R(a, b_R)$ holds in *M for all entities a which belong to the domain of the first variable of R *in* M. However, referring to (i), it is understood that we interpret the sentences of K in *M in non-standard fashion. That is to say, assertions concerning entities of D other than individuals (e.g. sets, relations between individuals, relations between relations, relations between sets) are to be interpreted (in general) *not* with regard to the totality of such entities in D but with regard to an appropriate sub-class of these, called *internal* (or *admissible*) entities. For example, if N is the structure of the natural numbers and *N is an enlargement of N, then the relation $x<y$, is concurrent. It follows that *N contains numbers which are *infinite*, i.e., larger that all numbers of N. However, it is easy to see that the set of all infinite numbers of *N (still to be called *natural*) does not contain a smallest element. It follows that this set is not internal. For any non empty set of natural numbers in N does possess a smallest element, and this is a fact which can be expressed within K and, accordingly, must be true in *N for all non-empty *internal* sets. On the other hand, the set of prime numbers in *N is an internal set in *N as can be seen by interpreting, in *N, a sentence of K which asserts the existence of this set (for N).

When dealing with a given mathematical theory, it is necessary to *enlarge* all the mathematical structures involved in the theory, simultaneously. This can be done by supposing that these structures have first been embedded in a single structure, M. Thus, when dealing with a ring R, we shall require not only an enlargement *R of R but also a simultaneous enlargement *N of the natural numbers N; for a discussion in ring theory may involve the powers of a ring element, which are given by a mapping from $R \times N$ into R.

3. Let D be a Dedekind domain. By a *proper ideal* we shall mean any ideal other than the zero-ideal, 0, or the entire ring. We shall suppose that D possesses at least one proper ideal, i.e., that it is not a field. By a *prime ideal* we shall mean a *proper* prime ideal. Every proper ideal in D possesses a unique representation as the product of powers of distinct prime ideals.

We embed D and the natural numbers, N, simultaneously in a structure M and consider an enlargement *M of M. M contains enlargements *D and *N of D and N respectively. *D is an integral domain, as can be seen by expressing the fact that D is an integral domain within K and reinterpreting in M. (In the particular case when D is a ring of algebraic numbers, N may be regarded as a substructure of D; and M may then be taken to coincide with D. Accordingly, *M will coincide with *D). For any $a \in D$, $a \neq 0$, and for any proper ideal J in D we write $n = \text{ord}_J(a)$ for the uniquely determined natural number n such that $a \in J^n$ but $a \notin J^{n+1}$.

Let Ω be the set of ideals in D, and let $^*\Omega$ be the corresponding entity in *M. That is to say, $^*\Omega$ is denoted by the same symbol as Ω in the vocabulary of K. The elements of $^*\Omega$ are the *internal* ideals of *D. An internal ideal $A \in {}^*\Omega$ is *standard*, by definition, if there exists an ideal B in D, such t $A = {}^*B$, i.e. such that A is denoted by the same symbol as B in the vocabulary of K. The notation $n = \mathrm{ord}_J(a)$ still has a meaning for all proper internal ideals in *D and for all $a \in {}^*D$, $a \neq 0$, but n may now be a finite *or infinite* natural number.

We define the monad μ of *D as the intersection of all proper standard ideals in *D. Thus $\mu = \bigcap_\nu {}^*J_\nu$, where J_ν varies over the proper ideals of D and $^*J_\nu$ is the corresponding ideal in *D in each case. μ is an ideal in D, since it is an intersection of ideals; but it can be shown that μ is not internal. An equivalent definition for μ is $\mu = \bigcap_\nu {}^*P_\nu^n$, where P_ν varies over the prime ideals of D and n varies over the finite natural numbers. The following facts can now be established without difficulty.

(i) μ does not contain any element of D other than zero. For if $a \in D$, $a \neq 0$, then there exists a prime ideal P in D such that $a \notin P^n$ for a sufficiently large finite n. Hence $a \notin {}^*P^n$, $a \notin \mu$.

(ii) Let B be any ideal in D. Then $a \in {}^*B \cap D$ if and only if $a \in B$. For the relation $a \in B$ can be expressed within K for any $a \in D$ and so, for such $a \in {}^*B$ if and only if $a \in B$.

(iii) There exists an internal proper ideal J in *D (i.e., $J \in {}^*\Omega$, $J \neq 0$, $J \neq {}^*D$) such that $J \subset \mu$.

For consider the relation $R(x, y)$ which is defined by the following expression: "x is a proper ideal, and y is a proper ideal which is included in x". One verifies that $R(x,y)$ is concurrent in D (and in M). Accordingly, by one of the basic properties of M, there exists an internal proper ideal J in *D such that $J \subset {}^*J_\nu$, for any proper ideal J_ν in D. Hence $J \subset \mu = \bigcap_\nu {}^*J_\nu$, as asserted.

4. Let $\Delta = {}^*D/\mu$, so that Δ is the quotient ring of *D with respect to μ and, for any internal proper ideal $J \subset \mu$ in *D, such as exists according to 2. (iii) above, let $\Delta_J = {}^*D/J$. Denote by φ, φ_J the canonical (homomorphic) mapping from *D onto Δ, Δ_J, respectively. For J as considered there are mappings ψ_J from Δ_J onto Δ with kernel $\varphi_J(\mu)$ and we have the commutative diagrams

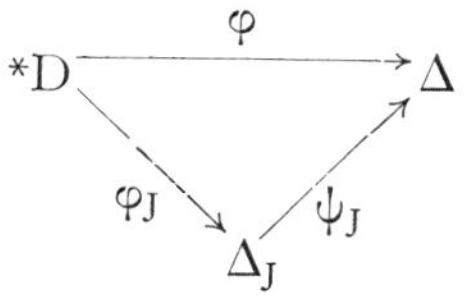

By a standard results on Dedekind domains, the quotient ring of D with respect to a proper ideal is a principal ideal ring. Formulating this fact within K and reinterpreting in *M, we conclude that the *internal* ideals of any quotient ring of *D with respect to an *internal* ideal are principal. The internal ideals of *D/J are just the images $\varphi_J(A)$ of the internal ideals A of *D. Moreover, a homomorphic image of a principal ideal is principal. Hence, the images $\varphi(A) = \psi_J(\varphi_J(A))$ are principal ideals in Δ for all internal ideals A in *D. Conversely, if F is a principal ideal in Δ, $F = (a)$, say, let $a = \varphi(b)$, $b \in {}^*D$.

Then F = φ ((*b*)), so that F is the image of an internal ideal. We conclude that an ideal in Δ is principal if and only if it is the image of an internal ideal in *D under φ.

Moreover, if A and B are ideals in D, then φ ((A , B)) = (φ (A) , φ (B)). It follows that the domain of principal ideals in Δ is closed under the operation of taking the greatest common divisor. We conclude that, if an ideal F in Δ has a finite base $F = (a_1, \cdots, a_n)$, then F is actually principal since $F = (\cdots(a_1), (a_2)), (a_3)), \cdots, (a_n))$. We conclude further that for any *a* and *b* in Δ, $a \neq 0$, $b \neq 0$, there exists a greatest common divisor *d*; i.e., *d* divides *a* and *b*, and any *d'* which divides *a* and *b* divides also *d*.

The basis of a principal ideal in Δ, J = (*a*), where *a* is not a divisor of zero, is uniquely defined up to associated elements of Δ, i.e., up to multiplication by a unit of Δ. If A_1 and A_2 are distinct ideals of D, $A_1 \neq A_2$, then $^*A_1 \neq {}^*A_2$ and $\varphi(^*A_1) \neq \varphi(^*A_2)$ in view of 3, (ii) above. Thus, φ induces a one-to-one multiplicative mapping Φ from the proper ideals of D into the classes of associated elements of Δ.

The mapping φ is one-to-one on D: for, if $a \neq b$ for elements *a* and *b* of D, then $a - b \neq 0$ and so $\varphi(a - b) \neq 0$ by 2, (i) above, and hence $\varphi(a) - \varphi(b) \neq 0$, $\varphi(a) \neq \varphi(b)$. Thus, φ provides an injection of D into Δ; in other words, Δ may be regarded as an extension of D. Accordingly, Δ satisfies the requirements mentioned at the beginning of this Note. A more detailed description of the structure of Δ will be given in a subsequent paper (1).

BIBLIOGRAPHY.

[1] C. CHEVALLEY, *Généralisation de la théorie du corps de classes pour les extensions infinies* « Journal de Mathématiques pure et appliquées », ser. 9, *15*, 359–371 (1936).

[2] W. KRULL, *Idealtheorie*, « Ergebnisse der Mathematik und ihrer Grenzgebiete », *4* (1935).

[3] S. LANG, *Algebraic Numbers*, 1964.

[4] J. VON NEUMANN, *Zur Prüjerschen Theorie der idealen Zahlen*, « Acta Szeged », *2*, 193–227 (1926).

[5] H. PRÜFER, *Neue Begründung der algebraischen Zahlentheorie*, « Mathematische Annalen », *94*, 198–243 (1925).

[6] A. ROBINSON, *On generalised limits and linear functionals*, « Pacific Journal of Mathematics », *14*, 269–283 (1964).

[7] A. ROBINSON, *On the theory of normal families*, « Acta Philosophica Fennica », fasc. 18 (Rolf Nevanlinna Anniversary volume), 159–184 (1965).

[8] A. ROBINSON, *Introduction to Model theory and to the Metamathematics of Algebra*, Studies in Logic and the Foundations of Mathematics (1963).

[9] A. ROBINSON, *Non-standard Analysis*, Studies in Logic and the Foundations of Mathematics (to be published).

(1) The author acknowledges with thanks that the research which led to the present Note was supported, in part, by the National Science Foundation (Grant No. GP–4038).

Logica matematica. — *A new approach to the theory of algebraic numbers.* Nota II di ABRAHAM ROBINSON (*) presentata (**) dal Socio B. SEGRE.

RIASSUNTO. — Questa Nota II continua lo sviluppo (iniziato nella Nota I) di una teoria degli anelli di Dedekind, D, usufruendo dei metodi dell' « analisi non-standard ». In particolare si mostra che ogni elemento di D possiede una rappresentazione come prodotto di elementi primi di *D. Questa rappresentazione è unica, a prescindere dall'ordine dei fattori e dalla loro eventuale sostituzione con elementi ad essi associati.

1. In a previous paper (ref. [1]), we considered a Dedekind domain D containing at least one proper ideal (i.e., ideal different from D and zero), and we introduced an *enlargement* *D of D. Defining the *monad* μ of *D by $\mu = \bigcap_{\nu}^{*} J_{\nu}$, where J_{ν} varies over the proper ideals of D, we then investigated the quotient ring $\Delta = {}^*D/\mu$. We showed that the canonical mapping φ from *D to Δ maps all *internal* ideals in *D on principal ideals in Δ; and that any finite set of non-zero elements in Δ possesses a greatest common divisor. Continuing our investigation, we shall arrive at a detailed understanding of the laws of divisibility and factorization in Δ. The reader is referred to ref. [1] for some definitions and results used in the present paper.

2. Let J be an internal proper ideal in *D.

THEOREM.—$\varphi(J) = \Delta$ *if and only if all the internal prime divisors of* J *are non-standard.*

Proof.—The condition is necessary. For suppose *P divides J where P is a prime ideal in D. Then $P \neq D$ so there exists an element $a \in D - P$. Now if $\varphi(J) = \Delta$ then $\varphi({}^*P) = \Delta$ and so $\varphi(a) \in \varphi({}^*P)$. But then $a \in ({}^*P, \mu) = {}^*P$, contrary to the fact that $a \notin {}^*P$, by transfer from D to *D.

The condition is also sufficient. We established in ref., section 3, (iii), that there exists an internal proper ideal in *D, to be called here J_0, such that $J_0 \subset \mu$. Let $J_1 = (J, J_0)$; then J_1 does not possess any standard prime divisors. Let $J_2 = J : J_1$. Then $J_2 \subset \mu$ since J_2 is divisible by all ${}^*P_{\nu}^{n}$ where P_{ν} is any prime ideal in D and n is any finite natural number. Also, $(J, J_2) = {}^*D$, and so $(J, \mu) = {}^*D$, $\varphi(J) = \Delta$, as asserted.

3. Let J be an internal proper ideal in *D. We know that $\varphi(J) = 0$ if and only if J is divisible by all ${}^*P_{\nu}^{n}$, P_{ν} any prime ideal in D, n any finite natural number. $\varphi(J)$ will be called a zero divisor if $\varphi(J) \neq 0$ and if there exists an element $a \neq 0$ in Δ such that $\varphi(J)\, a = 0$.

(*) Supported, in part by the National Science Foundation (Grant. No. GP-4038).
(**) Nella seduta del 16 aprile 1966.

THEOREM.—*In order that* φ (J) *be a zero divisor, it is necessary and sufficient that* $\varphi(J) \neq 0$ *and that there exist a prime ideal* P *in* D *such that* J *is divisible by all finite powers of* *P.

Proof.—The conditions are necessary. For suppose that φ (J) is a zero divisor. Then $\varphi(J) \neq 0$, by definition. Suppose that J is not divisible by any infinite power of any $^*P_\nu$, where P_ν varies over the prime ideals of D, but that φ (J) is a zero divisor. Then $\varphi(J)\,a = 0$ for some $a \neq 0$ in Δ, $a = \varphi(b)$ say, where $b \in {}^*D$, $b \neq 0$. Let $J_1 = (b)$, so that J_1 is an internal ideal in *D. Then $\varphi(J)\,\varphi(J_1) = \varphi(J J_1) = 0$. It follows that for any prime ideal P_ν in D, $J J_1$ is divisible by all finite powers of $^*P_\nu$, and hence is divisible by some infinite power of $^*P_\nu$. Writing $\mathrm{ord}_Q(A)$ for the exponent of the highest power of an ideal Q, by which an ideal $A \neq 0$ is divisible, we then have $\mathrm{ord}_{^*P_\nu}(J J_1) = m_\nu$ where m_ν is an infinite natural number and $\mathrm{ord}_{^*P_\nu}(J) = n_\nu$, where n_ν is finite. But $\mathrm{ord}_Q(AB) = \mathrm{ord}_Q(A) + \mathrm{ord}_Q(B)$ for any ideals Q, $A \neq 0$, $B \neq 0$, and so $\mathrm{ord}_{^*P_\nu}(J_1) = \mathrm{ord}_{^*P_\nu}(J J_1) - \mathrm{ord}_{^*P_\nu}(J) = m_\nu - n_\nu$ is infinite. This implies $\varphi(J_1) = 0$, $a = 0$, contrary to assumption.

The conditions (taken together) are also sufficient. Suppose that $\varphi(J) \neq 0$ and that there exists a prime ideal P in D such that $\mathrm{ord}_{^*P}(J) = n$ is infinite. Let $J_0 \subset \mu$ be the ideal which was introduced in Section 2 above and let $\mathrm{ord}_{^*P}(J_0) = m$, so that m is infinite. Then the ideal $J_1 = J_0 : {}^*P^m$ is not divisible by *P. Hence, J_1 is not a subset of μ although $J J_1 \subset \mu$, $\varphi(J J_1) = 0$. Choose $b \in J_1 - \mu$, so that $a = \varphi(b) \neq 0$. Then $\varphi(J)\,a \subset \varphi(J J_1)$ and so $\varphi(J)\,a = 0$, φ (J) is a zero divisor.

4. Let J be an internal proper ideal in *D, such that φ (J) is different from the zero ideal and is not a zero divisor.

THEOREM.—φ (J) *is a prime ideal in* Δ *if and only if* $\mathrm{ord}_{^*P_\nu}(J) = 0$ *for all prime ideals* P_ν *in* D *except one*, P, *for which* $\mathrm{ord}_{^*P} = 1$.

Proof.—The condition is sufficient. Suppose that $J = {}^*P J_1$ where J_1 does not have any standard prime divisors (so that $J_1 = {}^*D$ or J_1 is the product of non standard prime ideals). Then $\varphi(J) = \varphi({}^*P)\,\varphi(J_1) = \varphi({}^*P)$. Let $ab \in \varphi({}^*P)$ where $a = \varphi(a_0)$, $b = \varphi(b_0)$. Then $\varphi(a_0)\,\varphi(b_0) = \varphi(a_0 b_0) \in \varphi({}^*P)$ and so $a_0 b_0 \in ({}^*P, \mu) = {}^*P$. Since *P is prime it follows that one of the factors a_0 or b_0 belongs to *P, e.g. $a_0 \in {}^*P$. Hence $\varphi(a_0) \in \varphi({}^*P)$, $a \in \varphi({}^*P)$. On the other hand, $\varphi({}^*P) \neq \Delta$, by Section 2, and $\varphi({}^*P) \neq 0$. This shows that $\varphi(J) = \varphi({}^*P)$ is prime.

The condition is also necessary. For suppose that $J = {}^*P_1^m\,{}^*P_2^n Q$ where P_1 and P_2 are prime ideals in D, equal or different, and Q is not divisible by either P_1 or P_2. Since φ (J) is not a zero divisor m and n must be finite. Now $\varphi(J) = (\varphi({}^*P_1))^m (\varphi({}^*P_2))^n \varphi(Q)$. If φ (J) were prime we should then conclude that either $\varphi({}^*P_1) = \varphi(J)$ or $\varphi({}^*P_2) = \varphi(J)$ or $\varphi(Q) = \varphi(J)$. In the first case $\varphi({}^*P_1) \subset \varphi({}^*P_1\,{}^*P_2)$ and so ${}^*P_1 \subset ({}^*P_1\,{}^*P_2, \mu) = {}^*P_1\,{}^*P_2$, which is impossible. A similar argument applies to the remaining two cases. This completes the proof.

As shown in the course of the proof, $\varphi(J) = \varphi({}^*P)$ if *P is the unique prime divisor of J. Thus, as P ranges over all the prime ideals of D, $\varphi({}^*P)$

ranges over all the prime ideals φ (J) in Δ which are mentioned in the theorem. Moreover, if P_1 and P_2 are distinct prime ideals in D, then $\varphi(^*P_1) \neq \varphi(^*P_2)$. For if $a \in P_1 - P_2$ then $a \in {}^*P_1 - {}^*P_2$ and so $\varphi(a) \in \varphi(^*P_1)$. However, $\varphi(a) \in \varphi(^*P_2)$ would imply $a \in (^*P_2, \mu) = {}^*P_2$, contrary to assumption. Thus, φ provides a multiplicative bijection between the standard prime ideals of *D and the prime ideals of Δ which are not zero divisors and which are images of internal ideals in *D.

5. The monad μ is not an internal set in the sense of Non-standard Analysis. For suppose it were. Then $\text{ord}_{^*P}(\mu) = n_P$ would be infinite for all prime ideals P in D. Choose one such $P = P_0$ and define $\mu_0 = \mu : P_0$. Then $\mu_0 \supset \mu$, $\mu_0 \neq \mu$. On the other hand, μ_0 would still be divisible by all finite powers of *P for any prime ideal P in D, and so $\mu_0 \subset \mu$, $\mu_0 = \mu$. This contradiction shows that μ is not internal.

It follows that if we regard $\Delta = {}^*D/\mu$ as a set of subsets of *D then Δ cannot be an internal entity; for in that case, μ also would be an internal entity.

Let J be a principal ideal in Δ, $J = (a)$. We know that an ideal in Δ is principal if and only if it is the image under φ of an internal ideal in *D. If $J = \Delta$, then a is a unit in Δ and, conversely, if a is a unit in Δ then $J = \Delta$. If a is a zero divisor in Δ, so that $ab = 0$ in Δ for $b \neq 0$ then J is a zero divisor for in that case $Jb = 0$. Conversely, if J is a zero divisor, then a is a zero divisor.

Suppose now that $J = (a)$, $a \neq 0$, where a is neither a unit nor a zero divisor. As usual, a is said to be *prime* if a is divisible only by units (invertible elements) and by elements associated with a.

THEOREM.—*J is prime if and only if a is prime.*

Proof.—Let J be prime. If $a = bc$ then $bc \in J$, hence either $b \in J$ or $c \in J$. If $b \in J$, $b = da$ for some $d \in \Delta$, then $dac = a$, $a(dc - 1) = 0$. Since a is not a zero divisor, we conclude that $dc - 1 = 0$, c is a unit, b is associated with a. A similar argument applies if $c \in J$. Hence, a is prime. On the other hand, suppose that a is prime and that $bc \in J$ but $b \notin J$, $c \notin J$. Let $(a, b) = (d)$ so that d is a greatest common divisor of a and b. Since d divides a but is not in J it must be a unit, $(d) = \Delta$. Similarly $(a, c) = \Delta$, and so $(a, b)(a, c) = \Delta$. But $(a, b)(a, c) = (a^2, ac, ba, bc) = J$ and so $J = \Delta$, contrary to assumption. This proves our assertion.

6. Let $a \in \Delta$, $a \neq 0$, and let p be a prime element of Δ. If $p^n \mid a$ but $p^{n+1}\, a$ for some finite natural number $n \geq 0$ then we set $\text{ord}_p(a) = n$. If $p^n \mid a$ for all finite natural numbers n then we set $\text{ord}_p(a) = \infty$. Notice that p^n is not defined for infinite n.

For a and p as above, set $(a) = A$, $(p) = P$ in Δ. Then there exists an internal ideal J in *D and a prime ideal Q in D such that $A = \varphi(J)$, $P = \varphi(^*Q)$. Suppose that $p^n \mid a$ for some finite n. Then $P^n \mid A$ and so $\varphi(^*Q^n) \mid \varphi(J)$, $\varphi(J) \subset \varphi \subset (^*Q^n)$, $J \subset (^*Q^n, \mu) = {}^*Q^n$, $^*Q^n \mid J$. Conversely, if $^*Q^n \mid J$ then $P^n \mid A$ and so $p^n \mid a$. We conclude that if $\text{ord}_{^*Q}(J)$ is an infinite natural number then ord_p

$(a) = \infty$, and conversely; while if $\text{ord}_{*Q}(J)$ is finite then $\text{ord}_p(a)$ is finite and conversely, and in this case $\text{ord}_p(a) = \text{ord}_{*Q}(J)$.

THEOREM.—*For* $a, b, p \in \Delta$, $a \neq 0$, $b \neq 0$, *p prime*:

$$\text{ord}_p(ab) = \text{ord}_p(a) + \text{ord}_p(b)$$

In this connection, the sum on the right hand side is defined to be ∞ if at least one of $\text{ord}_p(a)$, $\text{ord}_p(b)$ is ∞.

For, introducing A, P, J, Q as above and setting $(b) = B$, $B = \varphi(K)$, where K is an internal ideal in *D, suppose next that $\text{ord}_p(a)$ and $\text{ord}_p(b)$ are finite. Then $\text{ord}_{*Q}(J)$ and $\text{ord}_{*Q}(K)$ also are finite and

$$\text{ord}_{*Q}(JK) = \text{ord}_{*Q}(J) + \text{ord}_{*Q}(K)$$

applies in *D. But

$$(ab) = AB = \varphi(J)\,\varphi(K) = \varphi(JK)$$

and so $\text{ord}_p(ab) = \text{ord}_{*Q}(JK)$. This establishes the assertion for finite $\text{ord}_p(a)$, $\text{ord}_p(b)$. The remaining cases can be disposed of in a similar way.

The fact that $\text{ord}_p(a) = \infty$ if and only if $\text{ord}_{*Q}(J)$ is infinite also shows that $a \in \Delta$, $a \neq 0$, is a zero divisor if and only if $\text{ord}_p(a) = \infty$ for at least one prime element p of Δ and a is a unit if and only if $\text{ord}_p(a) = 0$ for all prime elements p of Δ.

7. If p and q are prime elements of Δ then $\text{ord}_p(q) = 1$ if q is associated with p, and $\text{ord}_p(q) = 0$ in the alternative case. From every class of associated primes select one, π, calling it a representative prime. Now let $a \in \Delta$, $a \neq 0$, a not a zero divisor and possessing only a finite number of distinct representative primes $\pi_1, \cdots, \pi_j$, $j \geq 0$ as divisors. Thus, $\text{ord}_{\pi_i}(a) = n_i$ is a positive integer for $\pi_1, \cdots, \pi_i$ and $\text{ord}_p(a) = 0$ for any prime p not associated with one of these.

Consider the product $b = \Pi_{i=1}^{j} \pi_i^{n_i}$. b is a divisor of a. For since $\pi_1^{n_1} | a$, we have $a = \pi_1^{n_1} a_1$ where $\text{ord}_{\pi_1}(a_1) = 0$, $\text{ord}_{\pi_i}(a_1) = n_i$, $i = 2, \cdots, j$, by the theorem of Section 6. Continuing in this way we show that $b | a$, $a = \varepsilon b$. Then $\text{ord}_{\pi_i}(\varepsilon) = 0$, $i = 1, \cdots, j$ and more generally $\text{ord}_p(a) = 0$ for all prime elements p of Δ. This shows that ε is a unit. Thus, we have represented a as a product of powers of distinct representative primes multiplied by a unit. It is not difficult to see that this representation is unique.

Now suppose in addition that a is an element of D regarded as a subset of Δ. Let $(a) = P_1^{n_1} \cdots P_j^{n_j}$ be the representation of (a) as a product of powers of distinct prime ideals in D, so that $*(a) = *P_1^{n_1} \cdots *P_j^{n_j}$ in *D. By Section 4 and 5 there exist representative primes $\pi_1, \cdots, \pi_j$ of Δ such that $\varphi(*P_i) = (\pi_i)$, $i = 1, \cdots, j$. Then $\text{ord}_{\pi_i}(a) = n_i$, $i = 1, \cdots, j$ in Δ and $\text{ord}_p(a) = 0$ for prime elements of Δ not associated with $\pi_1, \cdots, \pi_j$. Hence, $a = \varepsilon \pi_1^{n_1} \cdots \pi_j^{n_j}$, where ε is a unit. Thus, the multiplicative system H generated by the units and prime elements of Δ contains all non-zero elements of D. For every element a of H there exists an internal ideal J in *D such that $\varphi(J) = (a)$. For if $a = \varepsilon p_1^{n_1} \cdots p_j^{n_j}$ where ε is a unit and $p_1, \cdots, p_j$ are prime elements of Δ, then there exist prime ideals $P_1, \cdots, P_j$ in D such that $\varphi(^*P_i) = (P_i)$, $i = 1, \cdots, j$. Then the ideal J defined by $J = {}^*P_1^{n_1} \cdots {}^*P_j^{n_j}$ satisfies $\varphi(J) = \varphi(^*P_1^{n_1}) \cdots \varphi(^*P_j^{n_j}) = (a)$, as required. The elements of H will be called *Prüfer-finite*.

THEOREM.—*Let* $a \in \Delta$, $a \neq 0$. *Then a is Prüfer-finite if and only if there exists an element* $b \in D$, $b \neq 0$ *such that* $a \mid b$ *in* Δ.

Proof.—The condition is sufficient. For let $b \in D$, $b \neq 0$, so that b can be written as $b = \varepsilon \pi_1^{n_1} \cdots \pi_j^{n_j}$ where $\pi_1, \cdots, \pi_j$ are representative primes and ε is a unit. Then $a \mid b$ implies that a is not a zero-divisor and that it can be written in the form $a = \eta \pi_1^{m_1} \cdots \pi_j^{m_j}$ where η is a unit and $0 \leq m_i \leq n_i$, $i = 1, \cdots, j$. Hence $a \in H$.

The condition is also necessary. For suppose $a \in H$, $a = \varepsilon p_1^{n_1} \cdots p_j^{n_j}$ where ε is a unit and $p_1, \cdots, p_j$ are prime elements of Δ. Then $(p_i) = \varphi(^*P_i)$, $i = 1, \ldots, j$ where $P_1, \cdots, P_j$ are prime ideals in D, and so $(a) = \varphi(^*P_1^{n_1} \cdots {}^*P_j^{n_j})$. Choose $b_i \in P_i$, $b_i \neq 0$, then $b = b_1^{n_1} \cdots b_j^{n_j}$ is different from zero and is contained in ${}^*P_1^{n_1} \cdots {}^*P_j^{n_j}$ and hence in (a). Thus, $a \mid b$, as required.

8. In addition to the elements of H, we may consider also elements of Δ which, while not zero divisors, are divisible by an infinite number of distinct representative primes. Let a be such an element of Δ. Then the function $\mathrm{ord}_P(a)$ takes finite values only. Moreover, if b is a second element of this kind and $\mathrm{ord}_\pi(a) = \mathrm{ord}_\pi(b)$ for all representative primes π then a and b must be associated. For let J and K be internal ideals in *D such that $(a) = \varphi(J)$, $(b) = \varphi(K)$. Let R be the set of internal prime ideals P in *D such that $\mathrm{ord}_P(J) = \mathrm{ord}_P(K) = n_P > 0$. Let $Q = \prod_{P \in R} P^{n_P}$. Then $Q \mid J$, $Q \mid K$, $J = QJ'$, $K = QK'$. Also, R includes all standard prime divisors of J and K and so $\varphi(J') = \varphi(K') = \Delta$. Hence, $\varphi(J) = \varphi(Q) = \varphi(K)$, $(a) = (b)$, a and b are associated.

Now let $f(\pi)$ be any function from the representative primes into the finite natural numbers. We claim that there exists an element $a \in \Delta$ such that $\mathrm{ord}_\pi(a) = f(\pi)$ for all representative primes π. Indeed, for any prime ideal P in D there is a unique representative prime π such that $\varphi(^*P) = (\pi)$, and we write $\pi = g(P)$, $f(\pi) = f(g(P)) = h(P)$. Consider the relation $R(x, y)$ which holds if x is a prime ideal in D and y is an ideal in D, $y \neq 0$, such that $\mathrm{ord}_x(y) = h(x)$. Then R is concurrent. It follows that there exists an internal ideal $J \neq 0$ in *D such that $\mathrm{ord}_{^*P}(J) = h(P)$ for all prime ideals P in D. Let $a \in \Delta$ such that $(a) = \varphi(J)$. Then $\mathrm{ord}_\pi(a) = f(\pi)$ for any representative prime element π—as required.

BIBLIOGRAPHY.

[1] A. ROBINSON, *A new approach to the theory of algebraic numbers*, « Rend. Acc. Naz. Lincei », *40*, 227–225 (1966).

NON-STANDARD THEORY OF DEDEKIND RINGS

BY

ABRAHAM ROBINSON [1])

(Communicated by Prof. A. HEIJTING at the meeting of March 18, 1967)

1. *Introduction.* Let D be a Dedekind ring and let $^*D$ be an enlargement of D in the sense of reference 7 (see section 2, below). It has been shown (refs. 8, 9) that there exists an ideal μ in $^*D$ such that $\mu \cap D = \{0\}$ and such that the ideals of D correspond to *principal* ideals in the ring $\Delta = {}^*D/\mu$. This leads to a theory of unique decomposition (disregarding multiplication by units) of the elements of D in Δ. In the present paper, we relate Δ to the standard theory of P-adic numbers and of adèles.

When D is the ring of rational integers, Δ is included in a class of structures introduced by MACDOWELL and SPECKER (ref. 6). However, in that case the main advantage of Δ – i.e., that it provides a unique prime decomposition for the elements of D – is not apparent, since this property is possessed already by D itself.

2. *Enlargements which are ultrapowers.* We recall the definition of an enlargement (ref. 7). Let M be a higher order structure and let K be the set of all sentences which hold in M for a specified assignment of symbols to denote the entities (individuals and relations) of M. A relation $R(x, y)$ of any type in M is said to be *concurrent* if it possesses the following property. Let $a_1, \ldots, a_k$, $k \geqslant 1$ be entities in the domain of the first argument of R, i.e. such that $R(a_i, b_i)$, $i = 1, \ldots, k$ holds for certain entities b_i of M. Then there exists a b such that $R(a_i, b)$, $i = 1, \ldots, k$ holds in M. An extension $^*M$ of M is called an *enlargement* of M if $^*M$ is a model of K and if for every concurrent relation $R(x, y)$ in M, there exists a b_R such that $R(a, b_R)$ holds in $^*M$ for every a in the domain of the first argument of R in M.

For a given M, the existence of an enlargement $^*M$ of M can be established by several methods, and the results mentioned in the introduction and detailed in references 8 and 9, apply to *any* enlargement. However, the particular property required in order to relate Δ to the ring of adèles over D is implied by the special assumption that $^*D$ has been constructed as an ultrapower. Accordingly, we proceed to show that for any given structure M there exists an enlargement $^*M$ which is an ultrapower of M. The reader may consult references 1, 3, 4, 5, 7, for the general notion of an ultrapower.

[1]) Work supported in part by Grant No. GP-5600 from the National Science Foundation.

The entities of $^*M$ are to be functions $f(x)$ from an "index set" I, to be specified presently, into entities of the structure M. More particularly, the set of individuals of $^*M$ is the set of all functions from I into the set of individuals of M; the n-ary relations between individuals of $^*M$ are given by the set of functions from I into n-ary relations between individuals of M; and so on, for relations of higher type.

Let F be any ultrafilter in I, arbitrary but fixed. Then the set theoretic interpretation of an n-ary relation $R=\varrho(x)$ in $^*M$, introduced as above, is as follows. R shall hold between individuals $a_i=f_i(x)$, $i=1, \ldots, n$ of $^*M$ if and only if the set $\{x|\varrho(x)$ holds between $f_1(x), \ldots, f_n(x)\}$ is an element of F. With this definition, and without any further assumptions on I, $^*M$ satisfies all sentences that hold in M, "in Henkin's sense" (ref. 2, compare ref. 7).

We now make a particular choice for I. I shall be the set of all functions $g(R)$ which assign to every concurrent relation R in M a subset of the domain of the first argument of R. It is easy to verify that there exist many concurrent relations so that the domain of such a function $g(R)$ is not empty. Since there exist concurrent relations of many types, the definition of $g(R)$ violates standard type restrictions. This can be avoided, but only at the cost of more complicated definitions, a precaution which seems unnecessary in the present context.

We introduce a partial ordering into I by stipulating that $h\prec g$ if $h(R)\subset g(R)$ for all concurrent relations R. Furthermore, we define $h\cup g$ as the function k which is given by $h(R)\cup g(R)=k(R)$ for all concurrent R.

Let F_0 be set of subsets α_g of I which are given by $\alpha_g=\{h|g\prec h\}$. Any such α_g is not empty since $g\in\alpha_g$. Also, if $\alpha_g\in F_0$ and $\alpha_h\in F_0$ then $\alpha_g\cap\alpha_h\in F_0$ since $\alpha_g\cap\alpha_h$ coincides with $\alpha_{g\cup h}$. Thus, F_0 constitutes a filter basis in I. Let F_1 be the filter generated by F_0, i.e. the set of all subsets of I that contain elements of F_0, and let F be an ultrafilter which is an extension of F_1. We claim that with this definition of F, the structure $^*M$, constructed as above, becomes an enlargement of M.

Let $R(x, y)$ be a concurrent relation in M, of arbitrary type. Recall that an entity (individual or relation) in $^*M$ is a function $f(g)$ from I into the entities of appropriate type in M. Define an entity $b_R=f(g)$ in $^*M$ as follows. For every $g=g(x)$ in I select $f(g)$ as an entity b of M such that $R(a, b)$ holds in M for all elements a of the finite set $A=g(R)$. This is possible, since R is concurrent.

M is injected into (embeddded in) $^*M$ by identifying any entity a in M with the constant function $a\equiv\varphi(g)$ on I. For any such a which belongs to the domain of the first argument of R consider the set $S=\{g|R(a, f(g))$ holds in $M\}$. By the choice of $f(g)$, the condition that $R(a, f(g))$ holds in M will certainly by satisfied if $a\in g(R)$. Hence, $S\supset T$ where $T=\{g|a\in g(R)\}$. Thus, in order to show that $R(\varphi(g), f(g))$ holds in $^*M$, we only have to verify that T, and hence S, belongs to the ultrafilter F. But T consists precisely of those elements g of I such that $g_a\prec g$

where the function $g_a(x)$ is defined by $g_a(x)=\phi$ for $x \neq R$ and by $g_a(R)=\{a\}$. This shows that $R(a, b_R)$ holds in $^*M$ and completes the proof that $^*M$ is an enlargement of M.

The particular property of ultrapower enlargements which will be required in the sequel (section 4 below) is given by the following definition.

2.1. The enlargement $^*M$ of a given M will be called *comprehensive* if for any mapping $f(y)$ from a set A (of any type) in M into *the extension* $^*B$ of a set B in M, there exists an internal mapping $\varphi(y)$ which coincides with $f(y)$ on A.

We are going to show that *if* $^*M$ *is an ultrapower of* M *as constructed above then* $^*M$ *is comprehensive.*

Let A, B and $f(y)$ be as specified in 2.1. The elements of $^*B$ are functions $g(x)$ defined on I and such that the sets $J_g=\{x|g(x) \in B\} \in F$ or, as we shall say also, in this case such that $g(x) \in B$ almost everywhere. (More generally, one says that an internal entity of the enlargement, $g(x)$, possesses a property p *almost everywhere* if the set $\{x|g(x)$ has property $p\}$ belongs to F.) However, we may even suppose that $g(x) \in B$ is satisfied for all $x \in I$, i.e. that $J_g=I$ for if this is not the case from the outset then, by the definition of the relation of equality in $^*M$ we obtain an entity $g'(x)$ equal to $g(x)$ by putting $g'(x)=g(x)$ for $x \in J_g$, and $g'(x)=b$ for an arbitrary $b \in B$ if $x \notin J_g$.

The elements a of A are represented in $^*M$ by the constant functions $h(x)=a$ for all $x \in I$ and the given function $f(y)$ maps any $a \in A$ on a function $g_a(x) \in {}^*B$. The required function $\varphi(y)$ will now be determined as an entity in the ultrapower by specifying its component functions $\varphi_x(y)$ for any $x \in I$ and $y \in A$, thus $-\varphi_x(y)=g_a(x)$. For any $h(x)\equiv a$ in $^*M$, we then have $\varphi_x(h)=g_a(x)$, so that $\varphi(y)$ coincides with $f(y)$ on A. For other values of y in $^*A$, $\varphi(y)$ is at any rate an element of $^*B$ since $g_y(x)$ belongs to B for all $x \in I$ and $y \in A$. Thus, $\varphi(y)$ possesses the properties required by 2.1, $^*M$ is comprehensive.

Let S be a finite or countable (not necessarily internal) set of infinite natural numbers in a comprehensive enlargement $^*M$. We propose to show that *there exists an infinite natural number in* $^*M$ *such that* $m<n$ *for all* $n \in S$.

The assertion is obviously true if S is finite, for in that case it contains a smallest element n_0 and $m=n_0-1$ is then infinite. Suppose then that S is countable and has been ranged as infinite sequence $\{n_0, n_1, n_2, \ldots\}$. Putting $A=B=N$ (the set of natural numbers) in 2.1, we see that there exists an internal function $\varphi(y)$ from $^*N$ into $^*N$ such that $\varphi(k)=n_k$ for all finite natural k. Consider the internal subset of $^*N$ which is given by

$$\{k|\varphi(h)>k \text{ for all } h\leqslant k\}.$$

Evidently this set includes all finite k since $\varphi(0), \varphi(1), \ldots, \varphi(k)$ are then all infinite. Accordingly, the set must include also some infinite natural number $k=m$. Since $m<\varphi(h)$ for all $h\leqslant m$ we have, in particular, $m<n_h=\varphi(h)$ for all finite h. Thus $m<n$ for all $n \in S$, as asserted.

3. *P-adic integers over Dedekind rings.* Let D be a Dedekind ring which is not a field and let P be a prime ideal in D other than 0 or D. Let ${}^*D$ be an enlargement of D. At this stage, ${}^*D$ need not be given by an ultrapower. As usual, ${}^*P$ is the extension of P to ${}^*D$.

Let μ_P and λ_P be the intersection of all finite powers of ${}^*P$, and the union of all infinite powers of ${}^*P$, respectively. Evidently $\mu_P \supset \lambda_P$. Moreover, if $b \in \mu_P$ so that $b \in {}^*P^n$ for all finite n, then the application of the axiom of induction to ${}^*D$ shows that $b \in {}^*P^n$ also for some infinite n, for otherwise the set of n for which $b \notin {}^*P^n$ would be neither empty, nor would it contain a smallest element. Hence $b \in \lambda_P$, $\lambda_P=\mu_P$.

μ_P *is prime.* Indeed, suppose $ab \in \mu_P$ then $ab \in {}^*P^n$ for some infinite n. On the other hand, if $a \notin \mu_P$, $b \notin \mu_P$ then there exist finite natural numbers k and l such that $a \notin {}^*P^k$, $b \notin {}^*P^l$. This shows that μ_P is prime.

We conclude that $\Delta_P={}^*D/\mu_P$ is an integral domain. Moreover, $D \cap \mu_P=\phi$ since any $a\neq 0$ in D is not contained in ${}^*P^n$ for sufficiently high finite n. This shows that the mapping $D \to \Delta_P$ which is the restriction of the canonical map ${}^*D \to \Delta_P$ to D is a bijection and Δ_P may be regarded as an extension of D.

Now let B_P be the set of all (ordinary) infinite sequences $\{d_n\}$, $d_n \in D$, which are Cauchy convergent in the P-adic valuation of D, and let $J_P \subset B_P$ be the ideal of null-sequences in B_P. Then the quotient ring B_P/J_P is the ring of P-adic integers over D. Passing to ${}^*D$, we see that if $\{d_n\}$ is a standard sequence in ${}^*J_P$, then for any infinite n, d_n is contained in some infinite power of ${}^*P$ and hence, is contained in μ_P. On the other hand, let $\{d_n\}$ be any standard sequence in ${}^*B_P$. By the Cauchy convergence of $\{d_n\}$, the difference d_m-d_n belongs to μ_P for all infinite m and n. It follows that if $d_n \in \mu_P$ for some infinite n then $d_n \in \mu_P$ for all infinite n.

We are going to show that D_P can be injected into Δ_P. Denote by ψ and ψ the canonical mappings ${}^*D \to \Delta_P={}^*D/\mu_P$ and $B_P \to D_P=B_P/J_P$, respectively. Choose an infinite natural number w and map B_P into ${}^*D$ by $\varphi(\{d_n\})=d_w$.

Let $\{d_n\}$ be any element of B_P and put $b=\psi(d_w)$ and $c=\chi(\{d_n\})$. Suppose that, at the same time, $c=\chi(\{d'_n\})$ for some other $\{d'_n\} \in B_P$ and put $b'=\psi(d'_w)$. Then $\chi(\{d_n-d'_n\})=0$ and so $\{d_n-d'_n\}$ belongs to the kernel of χ, $\{d_n-d'_n\} \in J_P$. But then $d_w-d'_w \in \mu_P$, and so $b-b'=\psi(d_w-d'_w)=0$, $b=b'$. We conclude that if we assign to any given $c \in D_P$ for any choice of $\{d_n\} \in B_P$ such that $c=\chi(\{d_n\})$, the element $b=\psi(\varphi(\{d_n\}))$ of Δ_P then this provides a mapping from D_P into Δ_P. Denoting this mapping by π, we see without difficulty that π is a homomorphism. The kernel of this mapping consists of the χ-images of sequences $\{d_n\} \in B_P$ such that $d_w \in \mu_P$, and this is the case only if $\{d_n\}$ is a null-sequence, $\{d_n\} \in J_P$, $\chi(\{d_n\})=0$. Thus, π is an injection of D_P into Δ_P.

It is natural to ask whether $\pi(D_P)=\Delta_P$. We are going to show that this is indeed the case *provided the quotient field D/P is finite.*

Assuming that D/P is finite let b be any element of $^*D$. In order to prove our assertion, it is sufficient to show that there exists a $a\{d_n\}\in B_P$ such that $b\equiv d_w(\mu_P)$. For in that case $\pi\chi(\{d_n\})=\psi d_w=\psi b$, and so the range of π is the range of ψ, which is Δ_P.

Given $b\in {}^*D$, we have the following alternative. Either there exists, for every finite natural number n, a $d_n\in D$ such that $b\equiv d_n(^*P^n)$ or there exists a finite natural number n such that $b\equiv d(^*P^n)$ for some $d\in D$ but $b\not\equiv d(^*P^{n+1})$ for all $d\in D$. In the former case we only need consider a standard sequence $\{d_n\}$ such that $b\equiv d_n(^*P^n)$ for all finite n. Then $d_m-d_n\in P^n$ for finite $m\geqslant n$ and so $\{d_n\}\in B_P$, and $d_m-d_w\in\mu_P$ for all infinite natural m. Also, since $b-d_n\in {}^*P^n$ for all finite n, $b-d_m\in {}^*P^m$ also for sufficiently low infinite m, and hence $b-d_m\in\mu_P$. This shows that $b-d_w=(b-d_m)+(d_m-d_w)\in\mu_P$, as required.

The alternative case cannot arrise. For if $b\equiv d(^*P^n)$ for finite n, consider $b-d$. The quotient ring D/P^{n+1} is finite since D/P is finite. Accordingly, there exist elements $a_1,\ldots,a_k\in D$ such that is true in D that "for every x in D, at least one of $a_1-x^2, a_2-x,\ldots,a_k-x$ is contained in P^{n+1}". Reinterpreting the statement in quotation marks in $^*D$, we conclude that for one of these a_i, $a_i-(b-d)\in {}^*P^{n+1}$. Putting $d'=d+a_i$, we see that $b\equiv d'(^*P^{n+1})$. Accordingly, the second case cannot arise and our assertion is proved.

The interrelation of the various mappings is expressed by the commutative diagram

$$\begin{array}{ccc} B_P & \xrightarrow{\ \varphi\ } & ^*D \\ \chi\downarrow & & \downarrow\psi \\ D_P & \xrightarrow{\ \pi\ } & \Delta_P \end{array}$$

Observe that the condition that D/P be finite is satisfied if D is the ring of rational integers or the ring of integers in an algebraic number field (a *finite* extension of the rationals). On the other hand, there are cases when πD_P is a proper subset of Δ_P as shown by the following example.

Let $D=C[x]$ where C is the field of complex numbers, and let the prime ideal P be given by $P=(x)$. Then D/P is isomorphic to C and $\Delta_P={}^*D/\mu_P$ is isomorphic to a ring of power series in the standard sense

$$a_0+a_1x+a_2x^2+\ldots \qquad (3.1)$$

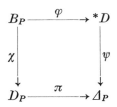

where the a_n are elements of $^*D$ such that $\{a_n\}$ constitutes an inital segment of a (Q-finite or infinite) sequence in $^*D$. But πD_P contains only power series 3.1. with standard coefficients, and two elements of Δ_P are different as soon as the corresponding power series 3.1 differ for at least one subscript. It follows that if a_0 is any non-standard element of $^*D$ then $\pi c \neq \psi a_0$ for all $c \in D_P$. Hence $\pi D_P \neq \Delta_P$.

4. *The ring Δ.* In references 8 and 9 we considered the ring Δ which is obtained from a ring $^*D$ as above as a homomorphic image whose kernel is the ideal μ. μ is the intersection of all powers $^*P^n$, where P ranges over the prime ideals of D and n ranges over the finite integers and is thus also the intersection of all the ideals μ_P where P is prime and non-trivial (different from 0, D). We shall discuss the connection between Δ and the rings Δ_P of the preceding section.

For any non-trivial prime ideal P in D, define $\nu_P = \bigcap_{Q \neq P} \mu_Q$, where Q ranges over the non-trivial prime ideals other than P. We claim that $(\mu_P, \nu_P) = {}^*D$. In this connection, (μ_P, ν_P), the greatest common divisor of μ_P and ν_P is defined, as usual, as the set $\{x \mid$ There exist $y \in \mu_P$, $z \in \nu_P$ such that $x = y + z\}$.

To begin with, we observe that there exists an internal ideal $J_P \neq 0$ in $^*D$ such that $J_P \subset \nu_P$. For we know (ref. 8) that there exists a non-zero internal ideal which is contained in μ, and $\mu \subset \nu_P$. Let n be the greatest natural number such that $^*P^n \supset J_P$, where n may be finite or infinite and let $K_P = J_P : {}^*P^n$. Then $K_P \subset \nu_P$. Indeed, for any non-trivial prime ideal $Q \neq P$, K_P is divisible by the same powers of $^*Q$ as J_P and, hence, is divisible also by an infinite power of Q. Hence, for $Q = P$, $K_P \subset \mu_Q$ and hence, $K_P \subset \nu_P$.

Let w be any infinite natural number, then $^*P^w$ and K_P are relatively prime and so $(^*P^w, K_P) = {}^*D$. But $^*P^w \subset \mu_P$ and $K_P \subset \nu_P$ and so $(^*P^w, K_P) \subset (\mu_P, \nu_P)$. Hence, $(\mu_P, \nu_P) = {}^*D$, as asserted.

By one of the isomorphism theorems (which holds for all ideals in $^*D$, internal or not), $\Delta_P = {}^*D/\mu_P = (\mu_P, \nu_P)/\mu_P$ is isomorphic to $\nu_P/\nu_P \cap \mu_P = \nu_P/\mu$. Let $\Pi_P(\nu_P/\mu) = \Lambda$ be the strong direct sum of the rings ν_P/μ. We map $^*D$ into Λ in the following way.

Since $(\mu_P, \nu_P) = {}^*D$, there exist elements $a_P \in \mu_P$, $b_P = \nu_P$ such that $a_P + b_P = 1$. Hence, for any $c \in {}^*D$, the element $c_P = cb_P$ belongs to ν_P, while $c_P - c = ca_P$ belongs to μ_P, so

$$c_P \in \nu_P \quad , \quad c_P - c \in \mu_P. \tag{4.1}$$

Suppose now that some other $c'_P \in {}^*D$ satisfies the same conditions, i.e.

$$c'_P \in \nu_P \quad , \quad c'_P - c \in \mu_P.$$

Then $c_P - c'_P \in \nu_P$, $c_P - c'_P \in \mu_P$ and so $c_P - c'_P \in \mu_P \cap \nu_P = \mu$. Let ψ_P be the canonical mapping of ν_P onto ν_P/μ, then we define $\varphi_P c$ for any $c \in {}^*D$ by $\varphi_P c = \psi_P \gamma$ for any γ which satisfies

$$\gamma \equiv 0(\nu_P) \quad \gamma \equiv c(\mu_P).$$

This definition is unique for, as we have just seen, there are such γ, e.g.

$\gamma = c_P$, and they all have the same ψ_P-image, for given c. Taking the vector of all $\varphi_P c$ determined in this way, we obtain the required mapping of $^*D$ into Λ, to be denoted by λ.

λ is a homomorphism. Let $c, d \in {}^*D$ and

$$\gamma_P \equiv 0(\nu_P) \quad , \quad \gamma_P \equiv c(\mu_P)$$

$$\delta_P \equiv 0(\nu_P) \quad , \quad \delta_P \equiv d(\mu_P).$$

Then

$$\gamma_P + \delta_P \equiv 0(\nu_P) \quad , \quad \gamma_P + \delta_P \equiv c + d(\mu_P)$$

$$\gamma_P \delta_P \equiv 0(\nu_P) \quad , \quad \gamma_P \delta_P \equiv 0(\mu_P)$$

and so

$$\lambda(c+d) = \lambda c + \lambda d \quad , \quad \lambda(cd) = \lambda c \cdot \lambda d.$$

Suppose now that c belongs to the kernel of this homomorphism, $\lambda c = 0$. Thus, for all non-trivial prime ideals P, $\psi_P \gamma_P = 0$ where

$$\gamma_P \equiv 0(\nu_P) \quad , \quad \gamma_P \equiv c(\mu_P).$$

But $\psi_P \gamma_P = 0$ implies $\gamma_P \in \mu$ and, therefore, $\gamma_P \in \mu_P$ and $c \in \mu_P$; and so, $\lambda c = 0$. This shows that the kernel of λ is precisely μ and hence, that $\Lambda = {}^*D/\mu$ is isomorphic to $\lambda {}^*D$. At the same time, as we have seen, Λ is isomorphic to the strong direct sum of the Λ_P and so we have arrived at the conclusion that Λ can be injected into the strong direct sum $\Pi_P \Lambda_P = H$.

It is natural to ask under what condition this injection Λ is actually onto. This will be the case if to any set $\{\gamma_P\}$ which is indexed by the set Π of all non-trivial prime ideals P of D and such that $\gamma_P \in \nu_P$ for all $P \in \Pi$ there exists a $c \in {}^*D$ such that $\gamma_P \equiv c(\mu_P)$ for all $P \in \Pi$. If Π is finite, we only need to set $c = \Sigma_{P \in \pi} \gamma_P$ in order to obtain an element of the required nature. We shall now show that an element c of this kind can also be found if Π is countable *provided the enlargement* $^*D$ *is comprehensive* (see 2.1).

On this assumption, suppose that $\{\gamma_P\}$ is a set as mentioned. We range the elements of Π in an infinite sequence $P_0, P_1, \ldots, P_n, \ldots$ $(n \in N)$. Defining $\delta(y) = \gamma_{P_y}$ for $y \in N$, we see from 2.1 that there exists an internal function $\varphi(y)$ from $^*N$ into $^*D$ such that $\varphi(n) = \gamma_{P_n}$ for all finite n.

Let m be any finite natural number. Since $\gamma_{P_n} \in \nu_{P_n}$ for all finite n there exists, for every finite natural number $n \neq m$ an *infinite* natural number k_n such that $\varphi(n) \in ({}^*P_m)^{k_n}$. Using the last result of section 2, we may strengthen this conclusion by affirming that there exists an infinite natural number $h = h_m$ such that $\varphi(n) \in ({}^*P_m)^{h_m}$ for $n \neq m$, where h_m is now independent of n. And, using the same result for the second time, we may conclude further that there exists an infinite natural number w such that $\varphi(n) \in ({}^*P_m)^w$ for $n \neq m$, n and m finite.

For a fixed finite m, $\varphi(n) \in ({}^*P_m)^w$ for all finite $n \neq m$. Accordingly, there exists an infinite natural number λ_m such that $\varphi(n) \in ({}^*P_m)^w$ for all $n < \lambda_m$, $n \neq m$. But then, again using the last result of section 2, there exists an infinite natural number λ such that $\lambda < \lambda_m$ for $m = 0, 1, 2, 3, \ldots$ ($m \in N$). We conclude that $\varphi(n) \in ({}^*P_m)^w$ for all finite n and all $n \leqslant \lambda$, $n \neq m$, and hence, in particular, for all infinite $n \leqslant \lambda$.

Put.

$$c = \sum_{n=0}^{\lambda} \varphi(n).$$

Then, for any finite m

$$c - \gamma_{P_m} = \sum_{\substack{n=0 \\ n \neq m}}^{\lambda} \varphi(n) \in ({}^*P_m)^w$$

and so

$$\gamma_{P_m} \equiv c(\mu_{P_m}), \quad m = 0, 1, 2, 3, \ldots$$

This shows that the specified injections of Δ into Λ and H are onto, Δ is isomorphic to Λ and to H.

If D is countable then Π is finite or countable. Also, as shown in section 3, if D/P is finite for all $P \in \Pi$ then Δ_P is isomorphic to the P-adic completion of D. Combining these assumptions, we therefore obtain the following result.

4.2. Theorem. Let D be a countable Dedekind ring which is not a field such that the quotient rings D/P are finite for all non-trivial prime ideal P in D. Let ${}^*D$ be a comprehensive enlargement of D and let μ be the intersection of all extensions ${}^*Q$ to ${}^*D$ of non-zero standard ideals Q in D. Then the ring $\Delta = {}^*D/\mu$ is isomorphic to the strong direct sum of the P-adic completions of D.

Although we have not introduced any topological notions in the present paper we may mention that μ and the μ_P are in fact the monads of zero in the appropriate topologies of ${}^*D$. Also, if we were to use valuations explicitly, it would be more natural to develop the theory for the quotient fields of D and ${}^*D$ and, where appropriate, to consider also Archimedean valuations.

The most interesting special cases covered by 4.2 are those for which D is the ring of integers in the field of rational numbers or in a finite algebraic extension of the field of rational numbers. For these cases, Δ is isomorphic to the ring of adèles with entire p-adic or P-adic components but without components (or with zero components) in the Archimedean completions.

The reader may also find it interesting to apply the results of references 7 and 8 to these cases. For example, section 2 of reference shows that if D is the ring of rational integers then the units of Δ are the canonical images of those elements of ${}^*D$ which possess infinite prime divisors only, and this can also be seen more directly by means of Wilson's theorem.

Suppose now that D is the ring of integers in a finite algebraic extension of the rationals. Then we claim that an element $a \in {}^*D$ belongs to the monad μ if and only if a is divisible by all standard positive integers. For suppose $a \in \mu$ and let $(m) = P_1^{\alpha_1} \dots P_k^{\alpha_k}$ be the representation of the ideal (m) as a product of powers of prime ideals in D, where m is an arbitrary standard positive integer. Then a is contained in ${}^*P_1^{\alpha_1}, \dots, {}^*P_k^{\alpha_k}$ by assumption and hence, is contained also in ${}^*(m)$, $m|a$. Conversely, suppose that a is divisible by all standard positive integers and let P be any non-trivial prime ideal in D. Then P contains a (standard) prime p and $p^n|a$ for any finite positive n. Hence $a \in {}^*(p^n) \in {}^*P^n$, showing that $a \in \mu$.

Now let D_0 be the ring of *all* algebraic integers and let μ_0 be the set of elements of ${}^*D_0$ which are divisible by all finite positive natural numbers. Then μ_0 is an ideal. Let $\Delta_0 = {}^*D_0/\mu_0$ and let D_1 be the ring of integers in any finite algebraic number field, $D_1 \subset D_0$. Then the preceding observation shows that the monad of ${}^*D_1$, μ, is given by $\mu_1 = {}^*D_1 \cap \mu_0$ and so $\Delta_1 = {}^*D_1/\mu_1$ is (isomorphic to) the restriction of Δ_0 to the canonical image of ${}^*D_1$ in Δ_0. Thus, the rings Δ_1 which correspond to the various algebraic number fields are embedded in a natural way in one and the same ring Δ.

University of California
Los Angeles

REFERENCES

1. Frayne, T., A. C. Morel and D. S. Scott, Reduced direct products. Fundamenta Mathematicae, **51**, 195–227 (1962).
2. Henkin, L., Completeness in the theory of types. Journal of Symbolic Logic, **15**, 81–91 (1950).
3. Keisler, H. J., Ultraproducts and elementary classes. Proceedings of the Royal Academy of Sciences, Amsterdam, Ser. **A, 64**, 477–495 (1962).
4. Kochen, S., Ultraproducts in the theory of models. Annals of Mathematics, Ser. 2, **79**, 221–261 (1961).
5. Luxemburg, W. A. J., Non-standard Analysis (mimeographed). Pasadena 1962.
6. MacDowell, R. and E. Specker, Modelle der Arithmetik. Infinitistic Methods (Symposium on Foundations of Mathematics, Warsaw 1959) Warsaw, 1961, pp. 257–263.
7. Robinson, A., Non-standard Analysis. Studies in Logic and the Foundations of Mathematics, Amsterdam, 1966.
8. ———, A new approach to the theory of algebraic numbers. Accademia Nazionale dei Lincei, Rendiconti della Classe di Scienze, etc., Ser. VIII, **40**, 222–225 (1966).
9. ———, A new approach to the theory of algebraic numbers, Nota II. Accademia Nazionale dei Lincei, Rendiconti della Classe di Scienze, etc., Ser. VIII, **40**, 770–774 (1966).

NONSTANDARD ARITHMETIC[1]

ABRAHAM ROBINSON

1. Introduction. In 1934 it was pointed out by Thoralf Skolem [23] that there exist proper extensions of the natural number system which have, in some sense, "the same properties" as the natural numbers. As the title of his paper indicates, Skolem was interested only in showing that no axiomatic system specified in a formal language (in his case the Lower Predicate Calculus) can characterize the natural numbers categorically; and he did not concern himself further with the properties of the structures whose existence he had established. In due course these and similar structures became known as *nonstandard models of arithmetic* and papers concerned with them, wholly or in part, including certain applications to other fields, appeared in the literature (e.g. [7], [9], [11], [14], [15], [16], [17]). Beginning in the fall of 1960, the application of similar ideas to analysis led to a rapid development in which nonstandard models of arithmetic played an auxiliary but vital part. It turned out that these ideas provide a firm foundation for the nonarchimedean approach to the Differential and Integral Calculus which predominated until the middle of the nineteenth century when it was discarded as unsound and replaced by the ϵ, δ method of Weierstrass. Going beyond this area, which is particularly interesting from a historical point of view, the new method (which has come to be known as *Nonstandard Analysis*) can be presented in a form which is sufficiently general to make it applicable also to mathematical theories which do not involve any metric concept, e.g., to general topological spaces [18].

In the present paper we shall show how the experience gained with this more general approach can be used in order to throw new light also on arithmetic or more precisely, on the classical arithmetical theories which have grown out of elementary arithmetic, such as the theory of ideals, the theory of p-adic numbers, and class field theory. Thus we shall provide new foundations for infinite Galois theory and for the theory of idèles. Beyond that, we shall develop a theory of *ideals* for the case of infinite abelian extensions in class field theory. This is remarkable, for Chevalley introduced idèles [2] precisely in

[1] An address delivered before the San Jose meeting of the Society on April 22, 1967 by invitation of the Committee to Select Hour Speakers for Far-Western Sectional Meetings. Research supported in part by the National Science Foundation, Grant No. GP-5600; received by the editors June 17, 1967.

order to deal with infinite extensions since, classically, the ideals in the ground field cannot cope with this case.

§5 below is related to the theory of p-adic completions and of idèles in Dedekind rings which is developed in [**17**], [**20**].

I acknowledge with thanks several stimulating conversations with W. A. J. Luxemburg, A. M. MacBeath and O. Todd while participating in a program on Nonstandard Analysis sponsored by the Office of Naval Research at the California Institute of Technology. Among others from whose knowledge I have benefited in connection with the problems considered here, I wish to mention particularly P. Roquette, E. G. Straus, and H. Zassenhaus.

2. **Enlargements and ultrapowers.** In this section we give an informal description of the framework which is required for our subsequent arguments. The reader may consult [**18**] for a formal development.

Let M be a mathematical structure of any kind and let $R(x, y)$ be a binary relation of arbitrary type in M. Thus, R may be a relation between individuals of M or between individuals and functions, or between sets and binary relations, etc. By the *first domain* of R, D_R, we mean the set of all entities (individuals, relations, functions, $\cdots$) a for which there exists a b such that $R(a, b)$ holds (is satisfied) in M. We shall say that R is *concurrent* if, for every finite subset $\{a_1, \cdots, a_n\}$ of D_R, $n \geq 1$, there exists an entity b in M such that $R(a_1, b)$, $R(a_2, b)$, $\cdots$, $R(a_n, b)$ all hold in M. Now let M' be an extension of M. We shall say that the relation $R(x, y)$ in M is *bounded in* M' if there exists an entity b_R in M' such that $R(a, b_R)$ holds in M' for all $a \in D_R$, i.e., for all elements of the first domain of R *in* M. b_R will be called a *bound* for R.

An extension $^*M$ of M is called an *enlargement* of M if all concurrent relations of M are bounded in $^*M$ and if, moreover, all statements which hold in M hold also in $^*M$ in a sense which will now be explained.

Let K be the set of all statements which hold in M. We may imagine that these statements are expressed in a formal language L which includes symbols for all individuals of M, for all sets of M, and for all functions, relations, sets of relations, etc. of all (finite) types. In addition, L is supposed to include the usual connectives, $\neg$ (*not*), $\vee$ (*or*), $\wedge$ (*and*), $\supset$ (*if* $\cdots$ *then*), and also variables and quantifiers (*for all* and *there exists*). Quantification is permitted with respect to entities of all types (e.g., "for all functions of two variables," "there exists a ternary relation between sets").

To every entity (individual, function, relation, $\cdots$) R in M, there corresponds an entity $^*R$ in $^*M$, which is denoted by the same symbol in L. A relation R holds between entities $S_1, \cdots, S_n$ in M if and only if $^*R$ holds between $^*S_1, \cdots, {}^*S_n$ in $^*M$. On this basis, any statement X of K can be reinterpreted in $^*M$ where we assign their usual meaning to the connectives and to quantification with respect to individuals. However, when interpreting quantification with respect to entities of higher type in $^*M$, we shall (in general) not refer to the totality of entities of the type in question but to a certain subclass of such entities called *internal*. Thus, the phrase which in M signifies "there exists a function of three variables" is to be interpreted in $^*M$ as "there exists an internal function of three variables," and, similarly, "for all binary relations in M" corresponds to "for all internal binary relations in $^*M$." It is in this sense that the statements of K are required to hold also in $^*M$ (for some fixed determination of the class of internal entities) where the bounds of concurrent relations introduced above also must be internal.

It is a simple consequence of the compactness theorem (finiteness principle) that every structure M possesses an enlargement $^*M$. $^*M$ is a proper extension of M if and only if the number of individuals of M is infinite and, in this case, there are many nonisomorphic enlargements for the given M. In particular $^*M$ can be constructed as a suitable ultrapower of M [4], [8], [10]. However, in many cases the mode of construction of $^*M$ is irrelevant and all necessary information concerning it can be extracted from the defining properties of an enlargement as laid down above. There are exceptions to this and some of them will be discussed in due course.

The following example, which is fundamental, will show how we may be able to decide that a particular entity is not internal.

Let N be the system of natural numbers and let $^*N$ be an enlargement of N. The operations of additions and multiplications extend automatically from N to $^*N$. The relation of order, $x<y$, is concurrent in N. It follows that it possesses a bound in $^*N$, to be denoted by b. Then $0<b$, $1<b$, and, quite generally, $n<b$ in $^*N$ for all natural numbers $n\in N$. This shows that $^*N$ is a proper extension of N, in agreement with the general statement made above on enlargements of infinite structures. From now on all individuals of $^*N$ will be called natural numbers, the numbers of N being *standard* and *finite* while the remaining numbers of $^*N$ are *nonstandard* and *infinite*. It is not difficult to show that any infinite natural number is greater than every finite natural number.

The set of infinite natural numbers, $^*N-N$, cannot be internal.

For it is a fact of N that "every nonempty set of natural numbers includes a smallest element." Reinterpreting the statement in quotes for $^*N$, we conclude that every *internal* set of numbers of $^*N$ includes a smallest element. For if a is a finite number, $a+1$ also is finite; so if $a \in {}^*N - N$, i.e. if a is infinite, then $a-1$ also is infinite. This shows that $^*N - N$ cannot be internal.

For an example of an internal set, consider the set A of all infinite natural numbers greater than some infinite natural number a, $A = \{y \mid y > a\}$. It is true in N that "for every natural number x there exists a set z which consists of all natural numbers y such that $y > x$." But the statement in quotes must be true also in $^*N$ in the sense that for every number x in $^*N$ there exists an *internal* set z such that $z = \{y \mid y > x\}$. It follows in particular that $A = \{y \mid y > a\}$ is an internal set.

When applying nonstandard analysis to other mathematical structures, e.g., to a topological space T, it is essential to consider not only an enlargement $^*T$ of T but to enlarge simultaneously all other mathematical structures which occur in the argument, e.g., the natural numbers, N. This can be done by taking for M some structure (e.g., a model of Set Theory) which includes both T and N. We then work in an enlargement $^*M$ of M which contains simultaneous enlargements $^*T$ and $^*N$ of T and N.

Although we have assumed for the definition of an enlargement that *all* concurrent binary relations in M possess bounds in $^*M$, only a small proportion of these will be required in practice. Thus, in retrospect, it is then possible to weaken the definitions of an enlargement by supposing that only the concurrent relations that are involved in the argument possess bounds in $^*M$. We shall then say that $^*M$ is an enlargement *for the relations* in question.

In particular, let M be a structure which includes the natural numbers N and let $^*M$ be an enlargement of M for the concurrent relation $x < y$ between natural numbers. Then $^*M$ contains an extension $^*N$ of N which is an enlargement of N for the relation $x < y$. We claim that if $R(x, y)$ is a concurrent relation in M, of any type, with countable first domain, then $R(x, y)$ possesses a bound in $^*M$. Indeed, let $A = \{a_0, a_1, a_2, \cdots\}$ enumerate the first domain of such a relation. By assumption, there exists a sequence of entities $B = \{b_0, b_1, b_2, \cdots\}$ in M such that $R(a_j, b_k)$ holds in M for $j \leq k$, $k = 0, 1, 2, \cdots$. The "sequences" $^*A$ and $^*B$ which correspond to A and B in M then have subscripts ranging over all numbers $n \in {}^*N$. The statement "for every natural number x and for every natural number $y < x$, $R(a_y, b_x)$ holds in M" must then be true also in $^*M$, where it applies, more precisely, to the extension $^*R$ of R. In particular, it is therefore true in $^*M$ that, for an arbitrary *infinite* natural number ω, $R(a_y, b_\omega)$ holds in $^*M$ for all $y \leqq \omega$ and, hence, for all finite y. This shows that b_ω is a bound for $R(x, y)$.

Suppose in particular that $^*M$ is an ultrapower M_D^N of M where the natural numbers N are included in M and also serve as index set for copies of M, and D is a free ultrafilter on N. Then the internal entities of $^*M$ are simply all sequences $a = \{a_n\}$, $b = \{b_n\}$, $S = \{S_n\}$, $\cdots$ of entities of corresponding types in M. Two entities of $^*M$ are regarded as equal if the set of subscripts on which they coincide belongs to D. And if, for example, a and b are individuals in $^*M$ and S is a binary relation, then $S(a, b)$ holds in $^*M$ by definition if the set

$$\{n \mid S_n(a_n, b_n) \text{ holds in } M\}$$

belongs to D. Then $^*M$ is an enlargement of M for the concurrent relation $<$ between natural numbers, for $\{0, 1, 2, 3, \cdots\}$ is a number of $^*N$ which is greater than any element of N. Accordingly, all concurrent relations of M with countable first domain possess bounds in $^*M$.

3. **Infinite Galois theory.** Let F be a commutative algebraic field and let Φ be a separable and normal algebraic extension of F. That is to say, Φ is an algebraic extension of F and if a polynomial $f(x) \in F[x]$ which is irreducible in F possesses a root in Φ, then $f(x)$ splits into distinct linear factors in Φ.

If Φ is of finite degree over F, we have the standard Galois theory for Φ/F which establishes a bijection between the subgroups of the group of automorphisms G of Φ/F and the subfields of Φ which are extensions of F. Dedekind pointed out that this correspondence breaks down if Φ is of infinite degree over F, and Krull showed [11] that the situation can be saved by restricting consideration to subgroups of G which are closed in a certain topology. Here we shall give an independent approach to the problem and shall then establish its connection with Krull's theory. We shall suppose from now on that Φ is an infinite extension of F.

Let $^*\Phi$ be an enlargement of Φ and let $^*F$ be the corresponding enlargement of F, $^*F \subset {}^*\Phi$. Let G be the Galois group of Φ/F so that $^*G$ is the corresponding group for $^*\Phi/{}^*F$. Consider the binary relation $R(x, y)$ in Φ which is defined as follows:

"x and y are finite normal algebraic extensions of F and subfields of Φ and $x \subset y$."

It is not hard to see that $R(x, y)$ is concurrent and hence possesses a bound Ψ in $^*\Phi$. By the definition of a bound, Ψ is a subfield of $^*\Phi$ and an extension of $^*F$, and moreover, $\Psi \supset {}^*A$, where A is any subfield of Φ which is a finite normal extension of F. But the union of such fields A is equal to Φ, and $^*A \supset A$, and so $\Psi \supset \Phi$. At the same time, reinterpreting the defining properties of $R(x, y)$ in $^*\Phi$, we see that Ψ is a "finite" extension of $^*F$ in the sense of the enlargement. That is to say, there exists a natural number $n \in {}^*N$, which may and will be infinite, such that n is the degree of Ψ over $^*F$. Following a suggestion of M. Machover, we shall say that Ψ is of *starfinite* degree over $^*F$ (in place of *Q-finite* for *quasi-finite* as in [18]). There exists a "polynomial" $f(x)$ of degree n with coefficients in* F such that Ψ is the splitting field of $f(x)$. That is to say, all general statements which can be made about splitting fields of polynomials in the ordinary case can be made also about Ψ and $f(x)$, with the appropriate interpretation in the enlargement. In particular, there exists the Galois group H of Ψ over $^*F$. H consists of all internal automorphisms of Ψ which leave the elements of $^*F$ invariant, where the word *internal* will be used throughout in the sense introduced in §2 above. We have the usual Galois correspondence between the *internal* subgroups of H and the *internal* subfields of Ψ which are extensions of $^*F$.

Let $\sigma \in H$. By a standard result on extensions of isomorphisms, σ is the restriction of some element of $^*G$ to Ψ. Moreover, since Φ is a union of finite normal extensions of F, $\sigma\Phi \subset \Phi$ and $\sigma^{-1}\Phi \subset \Phi$ and so $\sigma\Phi = \Phi$. Thus, the restriction $^\circ\sigma$ of σ to Φ is an automorphism of Φ which leaves the elements of F invariant. Then $^\circ\sigma \in G$, and $\sigma \rightarrow {}^\circ\sigma$ determines a homomorphism from H onto G, whose kernel consists of the elements of H that leave all elements of Φ invariant. We write $H \rightarrow {}^\circ H = G$.

Let Θ be a subfield of Φ and an extension of F, $F \subset \Theta \subset \Phi$, and put $(^*\Theta)_\Psi = {}^*\Theta \cap \Psi$. Let H_Θ be the subgroup of H which corresponds to $(^*\Theta)_\Psi$ under the Galois correspondence. Define $^\circ H_\Theta$ by

$$^\circ H_\Theta = \{\tau \mid \tau = {}^\circ\sigma \text{ for some } \sigma \in H_\Theta\}.$$

As shown, the elements of $^\circ H_\Theta$ are automorphisms of Φ/F so that $^\circ H_\Theta \subset G$. More precisely, $^\circ H_\Theta$ is a subgroup of G.

We claim that Θ *is the set of invariants of* Φ *under* $^\circ H_\Theta$.

In fact, $\Theta \subset (^*\Theta)_\Psi$, and the elements of $(^*\Theta)_\Psi$ are invariant under the automorphisms of H_Θ. On the other hand, if $a \in \Phi$, but $a \notin \Theta$, then $a \notin {}^*\Theta$ and so $a \notin (^*\Theta)_\Psi$. It follows that there exists a $\sigma \in H_\Theta$ such that $\sigma a \neq a$ and so $^\circ\sigma a \neq a$. This proves our assertion.

Conversely, $^{\circ}H_{\Theta}$ *is the set of automorphisms of G which leave the elements of* Θ *invariant.*

Suppose that $\sigma \in G$ leaves the elements of Θ invariant. Then $^*\sigma$ leaves the elements of $^*\Theta$ invariant. Thus $(^*\sigma)_{\Psi}$, the restriction of $^*\sigma$ to Ψ, leaves the elements of $(^*\Theta)_{\Psi}$ invariant. Hence, $(^*\sigma)_{\Psi} \in H_{\Theta}$ and, further, $\sigma = {}^{\circ}((^*\sigma)_{\Psi}) \in {}^{\circ}H_{\Theta}$. This completes the argument.

Accordingly, we have a mapping $\gamma: \Theta \rightarrow {}^{\circ}H_{\Theta}$ from the subfields of Φ which are extensions of F into the set of subgroups of G. The main question is how to characterize the subgroups of G which belong to the image of γ.

For any $\sigma \in {}^*G$, we define $^{\circ}\sigma$ as the restriction of σ from $^*\Phi$ to Φ. $^{\circ}\sigma$ is then identical with $^{\circ}(\sigma_{\Psi})$ where σ_{Ψ} is the restriction of σ from $^*\Phi$ to Ψ and $^{\circ}(\sigma_{\Psi})$ is the further restriction of σ_{Ψ} from Ψ to Φ as introduced previously. For any subset S of G, which may be internal or external, we define $^{\circ}S$ by

$$^{\circ}S = \{\tau \mid \tau = {}^{\circ}\sigma \text{ for some } \sigma \in S\}.$$

3.1 THEOREM. *A subgroup J of G belongs to the image of* γ *if and only if* $^{\circ}(^*J) = J$.

Observe that $^{\circ}(^*J) \supset J$ for all subsets J of G. Accordingly, the condition of the theorem may be replaced by $^{\circ}(^*J) \subset J$.

The condition is necessary. For suppose J belongs to the image of γ, so that $J = {}^{\circ}H_{\Theta}$ for some field Θ, $F \subset \Theta \subset \Phi$, as above. Then J is the set of $\sigma \in G$ under which the elements of Θ are invariant; and so, by one of the basic properties of enlargements, $^*J$ is the set of $\sigma \in {}^*G$ under which the elements of $^*\Theta$ are invariant. It follows that all elements of Θ are invariant under the automorphisms which belong to $^{\circ}(^*J)$ and so $^{\circ}(^*J) \subset J$, as asserted.

Conversely, suppose that $^{\circ}(^*J) = J$. Let $(^*J)_{\Psi}$ be the group which consists of the restrictions of the elements of $^*J$ to Ψ. Then $(^*J)_{\Psi}$ is internal. Let Λ be the subfield of Ψ which corresponds to $(^*J)_{\Psi}$ under the Galois correspondence for Ψ. Then the elements of Λ are invariant under the automorphisms of $(^*J)_{\Psi}$ and so the elements of $\Theta = \Lambda \cap \Phi$ are invariant under the elements of $^{\circ}((^*J)_{\Psi}) = {}^{\circ}(^*J) = J$. Moreover, if $a \in \Phi - \Theta$, then $a \in \Psi - \Lambda$ and so $\sigma a \neq a$ for some $\sigma \in (^*J)_{\Psi}$. Hence $^{\circ}\sigma a \neq a$ where $^{\circ}\sigma \in {}^{\circ}((^*J)_{\Psi}) = J$. Thus, Θ consists of all elements of Θ which are invariant under J, and further, $^*\Theta$ consists of all elements of $^*\Theta$ which are invariant under $^*J$. But this shows that $(^*\Theta)_{\Psi} = {}^*\Theta \cap \Psi$ consists of all elements of Ψ which are invariant under the automorphisms of $(^*J)_{\Psi}$; and so, $\Theta_{\Psi} = \Lambda$. Thus, in our previous notation, $(^*J)_{\Psi} = H_{\Theta}$ and $J = {}^{\circ}H_{\Theta}$. Hence $\gamma: \Theta \rightarrow J$, as required. This completes the proof of 3.1.

3.1 can be used for the proof of standard results, for example of the well known

3.2 THEOREM. *$J=\gamma(\Theta)$ is a normal subgroup of G if and only if Θ is a normal extension of F.*

PROOF. If $\Theta\subset\Psi$ is a normal extension of F then $^*\Theta$ is a normal extension of $^*F$. Hence, $(^*\Theta)_\Psi$ is a normal extension of $^*F$. Conversely, if $(^*\Theta)_\Psi$ is normal over $^*F$ then $\Theta=(^*\Theta)_\Psi\cap\Phi$ is normal over F. On the other hand, if J is a normal subgroup of G then $^*J$ is a normal subgroup of $^*G$ and so $\sigma{}^*J\sigma^{-1}\subset{}^*J$ for any $\sigma\in{}^*G$. Hence, if $J'=(^*J)_\Psi$, then $\sigma_\Psi J'\sigma_\Psi{}^{-1}\subset J'$, so J' is normal in the Galois group H of Ψ over $^*F$. Also, $J={}^\circ(J')$ so J, which is the image of J' in the mapping $H\to{}^\circ H=G$, also is normal. Thus, if $^\circ(^*J)=J$ then J is normal if and only if $(^*J)_\Psi$ is normal; and for any subfield Θ of Φ over F, Θ is normal if and only if $(^*\Theta)_\Psi$ is normal. But $(^*J)_\Psi$ and $(^*\Theta)_\Psi$ correspond in the Galois correspondence of $\Psi/{}^*F$ if J and Θ correspond in the Galois correspondence γ of Φ/F. Hence J is normal if and only if Θ is normal. This proves 3.2.

The connection of our condition $^\circ(^*J)=J$ with Krull's theory is provided by the following observations.

If $^\circ(^*J)=J$ and $\sigma\in G-J$, then there exist $a_1,\cdots,a_n\in\Phi$ such that all $\sigma'\in G$ which coincide with σ on $a_1,\cdots,a_n$, $\sigma'a_i=\sigma a_i$ $(i=1,\cdots,n)$ do not belong to J either.

In this condition, we might replace $a_1,\cdots,a_n$ by a single a such that $F(a)$ includes $a_1,\cdots,a_n$.

PROOF. Given $\sigma\in G-J$, suppose on the contrary that for every $a_1,\cdots,a_n\in\Phi$ there exists $\sigma'\in G$ which coincides with σ on $a_1,\cdots,a_n$ and such that $\sigma'\in J$. Then the relation $R(x,y)$ which is defined by "$x\in\Phi$ and $y\in J$ and $\sigma x=yx$" is concurrent. Thus, there exists a $y=\tau$ in $^*J$ such that $\sigma a=\tau a$ for all $a\in\Phi$. But then $\sigma={}^\circ\tau\in{}^\circ(^*J)=J$, a contradiction.

Conversely, given a subgroup $J\subset G$, suppose that for every $\sigma\in G-J$ there exists a finite set $\{a_1,\cdots,a_n\}\subset\Phi$ such that $\sigma'\in G-J$ for all $\sigma'\in G$ which coincide with σ at $a_1,\cdots,a_n$, i.e., $\sigma a_i=\sigma' a_i$, $(i=1,\cdots,n)$. Then $^\circ(^*J)=J$.

For suppose $\sigma\in G-J$, but $\sigma={}^\circ\tau$ for $\tau={}^*J$. Then $\sigma a=\tau a$ for all $a\in\Phi$ and, in particular, $\sigma a_i={}^*\sigma a_i=\tau a_i$ for $i=1,\cdots,n$. Hence, applying the condition of the theorem to $^*\Phi$, $\tau\in{}^*G-{}^*J$. This contradiction proves the assertion.

In the Krull topology, a fundamental system of neighborhoods of the identity in G is provided by the subgroups which leave finite extensions of F invariant. We have just shown that $^\circ(^*J)=J$ if and only if J is closed in that topology.

If Φ is an algebraic number field, the Krull topology satisfies the first (and also the second) axiom of countability. It is then true (compare Theorem 9.3.12 of [**18**]) that the standard part $^\circ A$ of any internal subset A of G is closed. It follows that, in that case, a subgroup J of G is closed in the Krull topology and, equivalently, belongs to the Galois correspondence for Φ over F if and only if J is the standard part of an internal subgroup H of $^*G$, i.e., $J = {}^\circ H$. W. A. J. Luxemburg showed recently [**13**] that even without assuming the validity of an axiom of countability, the standard part of any internal subset of an enlargement $^*T$ of a topological space T must be closed provided $^*T$ is a particular kind of enlargement, a so-called *saturated model.* For this type of enlargement, the above conclusion concerning closed subgroups of G applies for uncountable Φ also.

Coming back to the general case, we observe, for future reference, that the Krull topology of G can be defined without mention of Φ, since a fundamental system of neighborhoods of the identity in G is given by the set of subgroups of G which are of finite index in G.

As we have seen, G is a homomorphic image of the starfinite group H, $H \to {}^\circ H = G$. G is also *pro-finite* ([**22**], i.e., it is a projective limit of finite groups. We are going to show that every pro-finite group is a homomorphic image of a starfinite group. To see this, let the group G be the projective limit of a set of finite groups $\{G_\alpha\}$ which is indexed on a preordered set I, filtered to the right, with a specified system of homomorphic maps $f_{\alpha\beta}\colon G_\beta \to G_\alpha$. Passing to an enlargement, we see that there exist elements $\omega \in {}^*I$ which majorize all elements of I, $\alpha \leqq \omega$ for all $\alpha \in I$. Then G_ω is a starfinite group since the G_α are finite for standard α. In order to map G_ω homomorphically on G, we observe that the elements of G are points $g = \{\gamma_\alpha\}$ of the cartesian product $\prod G_\alpha$ where $\gamma_\alpha \in G_\alpha$ for each $\alpha \in I$. We define a mapping $\phi\colon G_\omega \to G$ by

$$\gamma \to \{\gamma_\alpha\} = \{f_{\alpha\omega}\gamma\} = g$$

for any $\gamma \in G$. Then $\gamma_\alpha = f_{\alpha\beta}\gamma_\beta$ for all α and β in I so that $g = \phi$ belongs to G. Evidently, ϕ is a homomorphism. It only remains to be shown that it is onto.

Let $g = \{\gamma_\alpha\}$ be an arbitrary element of I and set $\gamma = \gamma_\omega$. Then $\phi\gamma = \{f_{\alpha\omega}\gamma_\omega\} = \{\gamma_\alpha\} = g$. This completes the argument.

4. Absolutely algebraic fields of prime characteristic. Let p be a standard prime number. p will remain fixed throughout this section. Let F be the prime field of characteristic p. For every positive integer n there exists a field Φ_n which contains just p^n elements. Φ_n is unique up to isomorphism, and all finite extensions of F are obtained in this way. If $m|n$ then Φ_n contains just one field (isomorphic to) Φ_m. The elements of Φ_n constitute the set of roots of the equation $x^{p^n}-x=0$.

Suppose now that Φ is any algebraic extension of F, possibly infinite. We consider an enlargement $^*\Phi$ of Φ. $^*\Phi$ contains an enlargement $^*F$ of F but in the present case, $^*F=F$, since F is finite. An argument similar to that given at the beginning of §3 above shows that there exists a natural number n (which may now be infinite) such that $\Phi\subset\Phi_n\subset{}^*\Phi$. To continue, we introduce Steinitz' g-numbers, now known also as *surnatural* or supernatural numbers ([5], [12], [21]). A *surnatural* number is a symbolic expression $g=2^{\nu_0}3^{\nu_1}\cdots p_k^{\nu_k}\cdots$ where p_k ranges over all (standard) primes and the ν_k are natural numbers, $\nu_k\in N$, or else $\nu_k=\infty$ where the "symbol" ∞ is taken to be greater than any $\nu\in N$. Surnatural numbers are multiplied by adding exponents with the convention that for all natural ν, $\infty+\nu=\nu+\infty=\infty+\infty=\infty$. The surnatural number $g=2^{\nu_0}3^{\nu_1}\cdots p_k^{\nu_k}\cdots$ divides the surnatural number $h=2^{\mu_0}3^{\mu_1}\cdots p_k^{\mu_k}\cdots$, and we write $g|h$, if $\nu_k\leqq\mu_k$ for all k. The g.c.d. and l.c.m. of a finite or infinite set of surnatural numbers are defined in the usual way.

Observe that so far the notions concerning natural numbers, including the introduction of the "symbol" ∞, were constructed entirely within a framework M regarded as *standard*. Passing to an enlargement $^*M$ of M, we now define the *surnatural part*, $[n]$, of any finite or infinite number $n\in{}^*N$ by $[n]=2^{\nu_0}3^{\nu_1}\cdots p_k^{\nu_k}\cdots$ where ν_k is the exponent of p_k in the prime power decomposition of n if that exponent is finite; and $\nu_k=\infty$ if the exponent in question is infinite, with p_k ranging over the finite primes. Thus $[n]$ is a standard surnatural number, and the mapping $n\rightarrow[n]$ is external (not internal).

For $\Phi\subset\Phi_n$, as above, Φ_n contains exactly the same absolutely algebraic elements as Φ. For if $\bar{F}$ is the algebraic closure of F (unique up to isomorphism) and $a\in\bar{F}-\Phi$, then $a\in{}^*\bar{F}-{}^*\Phi$ in the enlargement and so $a\notin{}^*\Phi$ and, a fortiori, $a\notin\Phi_n$.

Now let m be any finite natural number which divides n. This will be the case if and only if m divides $[n]$. Then Φ_n contains (a field isomorphic to) Φ_m as in the standard case. Since all elements of Φ_m are absolutely algebraic, it follows that Φ contains Φ_m. Conversely, if $\Phi_m\subset\Phi$ for finite m then $\Phi_m\subset\Phi_n$ and so m divides both n and $[n]$. Thus $[n]$ (but not n) depends only on Φ and is the l.c.m. of all finite m such that $\Phi_m\subset\Phi$. We call $[n]$ *the Steinitz number of* Φ, $g(\Phi)$.

Within this framework, Steinitz' result, that to every surnatural number g there exists one and (up to isomorphism) only one absolutely algebraic field of characteristic p whose Steinitz number is g, can be proved as follows. Given g, it is easy to show that there exists a finite or infinite natural number n such that $[n]=g$. Let Φ_n be the corresponding field which contains p^n elements and let Φ be the subfield of Φ_n which consists of the absolutely algebraic elements of Φ_n. Then the argument of the preceding paragraph shows that g is the l.c.m. of all finite natural numbers m such that $\Phi_m \subset \Phi$ and hence, $g=\gamma(\Phi)$. On the other hand, let Φ and Ψ be two absolutely algebraic fields of characteristic p with the same Steinitz number g. In order to prove that Φ and Ψ are isomorphic, we may suppose more particularly that they are contained in the same algebraic closure of F, $\overline{F}$, and we shall then show that Φ and Ψ actually coincide. Choose a field $\Phi_n \supset \Phi$ as in the preceding paragraphs and choose a field $\Phi_m \supset \Psi$ in the same way for Ψ. Then $\Phi_n \subset {}^*\Phi \subset {}^*\overline{F}$ and $\Phi_m \subset {}^*\Psi \subset {}^*\overline{F}$ and $[n]=[m]=g$. Let k be the g.c.d. of n and m so that again $[k]=g$. Let Φ_k be the subfield of ${}^*\overline{F}$ which contains just p^k elements. Then $\Phi_k \subset \Phi_n \cap \Phi_m$. Let Θ be the field which consists of the absolutely algebraic elements of Φ_k, then we claim that $\Theta=\Phi$. For let $\alpha \in \Phi$, then the field Λ generated by α over F contains just p^l elements for some finite l. Then $l|n$ and hence $l|g$ and $l|k$. This implies $\Lambda \subset \Phi_k$, and further, $\alpha \in \Phi_k$, $\alpha \in \Theta$. Accordingly, $\Theta=\Phi$ and, similarly, $\Theta=\Psi$, and hence $\Phi=\Psi$, as asserted.

5. *P*-**adic numbers and valuation theory.** Let F be a finite algebraic extension of the field of rational numbers Q (e.g., Q itself), and let ${}^*F$ be an enlargement of F. For a given *archimedean* valuation V of F and hence, of ${}^*F$, we denote by F_0 the set of elements a of ${}^*F$ such that $|a|$ is *finite* in the given valuation, i.e., such that $|a| \leqq r$ for some standard real number r. Also, we denote by F_1 the set of elements a of ${}^*F$ such that $|a|$ is *infinitesimal*, i.e., such that $|a| < r$ for all standard positive r. Then it is not difficult to see that F_0 is a valuation ring and F_1 is a valuation ideal. The valuation V' which is induced in ${}^*F$ by the ring F_0 is *nonarchimedean* although the original valuation V was archimedean. The valuation group Γ' of V' is (isomorphic to) the multiplicative quotient group of ${}^*F-\{0\}$ with respect to the group F_0-F_1. If V is real, F_0/F_1 is isomorphic to the field of real numbers R as can be seen most directly by injecting F into R and hence ${}^*F$ into ${}^*R$. The corresponding homomorphism ϕ: $F_0 \to R$ consists of taking the standard part of any $a \in F_0$ in the metric induced by V. That is to say, for any $a \in F_0$, $\phi(a)={}^\circ a$ is the uniquely determined real number such that $\phi(a)-a \in F_1$. The mapping ϕ is surjective; for if r is any standard real number, then there exists a standard sequence $\{s_n\}$, $s_n \in F$, $n \in N$ such that $\lim_{n\to\infty} s_n=r$. Passing to the enlargement, we then have $s_\omega - r \in F_1$ for any infinite

natural ω and so ${}^\circ s_n = r$. Similarly, if V is complex then F_0/F_1 is isomorphic to the field of complex numbers.

Now let P be a prime ideal in F. For any $a \in F$, $a \neq 0$, we define $\text{ord}_P(a)$ as usual as the exponent of P in the prime power decomposition of the ideal (a), $\text{ord}_P(a) \lesseqgtr 0$; and we set $\text{ord}_P(a) = \infty$ for $a = 0$, by convention. The definition of $\text{ord}_P(a)$ extends to ${}^*F$ and then ranges over the finite and infinite rational integers and ∞. We define F_0 (for the given P) as the set of elements a of ${}^*F$ such that $\text{ord}_P(a)$ is greater than some finite negative integer, and we define $F_1 \subset F_0$ as the set of $a \in {}^*F$ such that $\text{ord}_P(a)$ is an infinite natural number or ∞. As before, F_0 is a valuation ring in ${}^*F$ and F_1 is its valuation ideal; the corresponding valuation group is the multiplicative quotient group of ${}^*F - \{0\}$ with respect to $F_0 - F_1$, and the valuation V_P' thus obtained is nonarchimedean. However, V_P' is still different from the P-adic valuation V_P of F or ${}^*F$.

It is not difficult to show that F_0/F_1 is (isomorphic to) the P-adic completion of F. In particular, if F is the field of rational numbers and $P = \{p\}$ where p is a standard prime number, then F_0/F_1 coincides with the field of p-adic numbers. Thus, we obtain the archimedean and nonarchimedean completions of F within the framework of Nonstandard Arithmetic by introducing appropriate valuation rings and ideals in all cases. The uniformity of the procedure becomes even more apparent if we put $\text{ord}(a) = -|a|$, for we then have $a \in F_1$ if $\text{ord}(a)$ is infinite or equal to ∞, just as in the non-archimedean case. The analogy can be pursued further, but here we observe only that it motivates the notation used in the sequel for archimedean divisors.

From now on, we shall use the notation $\text{ord}_P(a)$ for both $a \in F$ and $a \in {}^*F$, and both for prime ideals in F and for "infinite primes" P where the word "infinite" is used here in the sense of valuation theory, not in the sense of nonstandard arithmetic. In order to minimize confusion, we shall call such symbolic primes (places) from now on only *archimedean*, while the primes (places) which correspond to prime ideals will be called *nonarchimedean* replacing the terms *infinite* and *finite* of valuation theory in this context. (Observe that in nonstandard arithmetic it is just the finite primes that have a good claim to being called archimedean!) As for the "symbol" ∞, it is neither finite nor infinite in our sense but is a standard entity which is greater than any finite or infinite natural number in the enlargement since it was taken to be greater than any finite natural number in the standard framework.

By a *surdivisor* g in F, we mean a formal infinite product $g = \prod P_j^{\nu_j}$

where P_j ranges over the standard archimedean and nonarchimedean primes and ν_j may be any standard natural number or ∞. However, if P_j is archimedean, then we admit for ν_j only the values 0 and ∞. ν_j is the *exponent* of P_j in g and P_j *occurs* in g if $\nu_j>0$. For $F=Q$ we may replace P_j by the corresponding rational prime number p_j so that the surdivisors of Q in which the archimedean prime of Q does not occur may be regarded also as surnatural numbers. A surdivisor will be called a *divisor* if the number of primes which occur in it is finite, and if $\nu_j\neq\infty$ for all nonarchimedean P_j. Surdivisors are multiplied by adding exponents. Divisibility, g.c.d., and l.c.m. are defined in the usual way.

Surdivisors are defined as standard entities which have a meaning relative to both F and $^*F$. Only such surdivisors will be considered. We observe, however, that the notion of divisibility and related notions which are defined in the next paragraph are not standard and not even internal.

Let $a\in{}^*F$. The surdivisor $g=\prod P_j^{\nu_j}$ *divides* a if $\mathrm{ord}_{P_j}(a)\geqq\nu_j$ in case ν_j is a natural number $\nu_j\in N$, and $\mathrm{ord}_{P_j}(a)$ is an infinite natural number in case $\nu_j=\infty$. a is said to be *entire* for g if $\mathrm{ord}_{P_j}(a)\geqq 0$ for $\nu_j\in N$ and $\mathrm{ord}_{P_j}(a)$ is greater than some standard negative integer for $\nu_j=\infty$. For the given g, the ring $F_g\subset{}^*F$ is then defined as the set of $a\in{}^*F$ which are entire for g; and $J_g\subset F_g$ is defined as the set of $a\in{}^*F$ which are divisible by g. Then J_g is an ideal in F_g. Let K_g be the quotient ring F_g/J_g.

The following special cases are basic.

5.1. $g=1$, i.e., $\nu_j=0$ for all P_j. Then $F_g=J_g={}^*F$ and so $K_g=\{0\}$.

5.2. $g=P_j$ where P_j is a nonarchimedean prime, i.e., $\nu_j=1$ and $\nu_i=0$ for all other P_i. Let $^\circ F_g$ and $^\circ J_g$ be, respectively, the restrictions of F_g and J_g to F. Then $^\circ F_g$ is the valuation ring for the P_j-adic valuation V_{P_j} and $^\circ J_g$ is the corresponding valuation ideal. It follows that $^\circ F_g/^\circ J_g$ is a finite field of characteristic p where p is the rational prime number contained in P_j. But finite sets are not extended on passing from F to $^*F$ and so

$$^*(^\circ F_g/^\circ J_g) = {}^*(^\circ F_g)/^*(^\circ J_g) = F_g/J_g = K_g$$

is the same finite field of characteristic p.

5.3. $g=P_j^{\nu_j}$ where P_j is a nonarchimedean prime and ν_j is a natural number greater than 1 ($\nu_i=0$ for all other P_i). We see, similarly as in 5.2, that K_g is now a finite ring of prime characteristic.

5.4. $g=P_j^\infty$ where P_j is an archimedean or nonarchimedean prime (and $\nu_i=0$ for all other P_i). In this case, F_g and J_g reduce to the ring F_0 and the ideal F_1 introduced at the beginning of this section. It follows that K_g is the P_j-adic completion of F if P_j is nonarchimedean, or the field of real or of complex numbers if P_j is real or complex archimedean, respectively.

Now let $g=\prod P_j^{\nu_j}$ be any surdivisor, $g\neq 1$. For any P_j which occurs in g, we call $P_j^{\nu_j}$ a *primary factor* of g and we denote it by g_j. We regard g_j as a surdivisor in which $\nu_i=0$ for $i\neq j$ and we write $g=\prod g_j$ where j ranges only over the subscripts for which $\nu_j>0$.

We are going to prove

5.5 THEOREM. *For any ultrapower enlargement, K_g is isomorphic to $\prod K_{g_j}$ where $\prod$ indicates the strong direct product (strong direct sum).*

PROOF. We may suppose that g possesses at least two distinct primary factors. For any primary factor g_j of g, we construct a homomorphism α_j: $K_g\to K_{g_j}$ as follows. Let $h_j=\prod_{i\neq j} g_i$ so that $g=g_jh_j$. Then $F_g=F_{g_j}\cap F_{h_j}$ and $J_g=J_{g_j}\cap J_{h_j}$. Put $J'_{g_j}=J_{g_j}\cap F_g$ and $J_{h_j}'=J_{h_j}\cap F_g$ so that J'_{g_j} and J'_{h_j} are ideals in F_g.

We claim that $(J'_{g_j}, J'_{h_j})=F_g$. That is to say, we assert the existence of a, $b\in{}^*F$ such that $a+b=1$ where a and b are entire for g and, moreover, $a\in J_{g_i}$ and $b\in J_{h_j}$. In order to find an appropriate a, we have to satisfy the conditions

5.6 $$\text{ord}_{P_j}(x) \geqq \nu_j,$$

5.7 $$\text{ord}_{P_i}(x-1) \geqq \nu_i \quad \text{for any other } P_i \text{ occurring in } g.$$

If $\nu_j=\infty$ or $\nu_i=\infty$ we now replace 5.6 and 5.7 by sequences of conditions,

5.8 $$\text{ord}_{P_j}(x) \geqq 0, \text{ord}_{P_j}(x) \geqq 1, \cdots, \text{ord}_{P_j}(x) \geqq n, \cdots$$

or

5.9 $$\text{ord}_{P_i}(x-1) \geqq 0, \text{ord}_{P_i}(x-1) \geqq 1, \cdots, \text{ord}_{P_i}(x-1) \geqq n, \cdots,$$

respectively, where n ranges over all finite natural numbers. The approximation theorem of valuation theory shows that any finite number of conditions as in 5.8 and 5.9 can be satisfied already by some x in F. Using an appropriate concurrent relation, we may conclude that all conditions 5.6, 5.7, or if ν_j, $\nu_i=\infty$, 5.8, 5.9 can be satisfied simultaneously by some $x=a$ in ${}^*F$. Since $\text{ord}_{P_i}(x-1)\geqq\nu_i>0$ implies $\text{ord}_{P_i}(x)\geqq 0$, we conclude that $a\in F_g$, while $\text{ord}_{P_j}(a)\geqq\nu_j$ shows that $a\in J_{g_i}$. Hence $a\in J'_{g_j}$.

Next we determine $b'\in{}^*F$ such that $x=b'$ satisfies the conditions $\text{ord}_{P_j}(x-1)\geqq\nu_j$ while $\text{ord}_{P_i}(x)\geqq\nu_i$ for all other P_i. This can again be done by means of an appropriate concurrent relation, and yields an $x=b'$ which belongs to J'_{h_j}. Furthermore, if P_j is nonarchimedean,

$$\text{ord}_{P_j}(a + b' - 1) \geqq \min(\text{ord}_{P_j}(a), \text{ord}_{P_j}(b' - 1)) \geqq \nu_j,$$

while if P_j is archimedean, so that $\nu_j = \infty$, then at any rate $\text{ord}_{P_j}(a+b'-1) = \infty$ also. Similarly, for $i \neq j$, $\text{ord}_{P_i}(a+b'-1) = \text{ord}_{P_i}((a-1)+b') \geqq \nu_i$ in all cases, and so $a+b'-1 \in J_g$. But $J_g \subset J'_{h_j}$ and so $b = b' - (a+b'-1) = 1 - a \in J'_{h_j}$ where $a+b=1$. This shows that $(J'_{g_j}, J'_{h_j}) = F_g$, i.e., J'_{g_j} and J'_{h_j} are comaximal in F_g.

Now let c be any element of F_g. With a and b as above, put $c_j = cb$. Then $c_j \in J'_{h_j}$ and $c - c_j = c(a+b) - cb = ca \in J'_{g_j}$. Suppose that some element $c' \in F_g$ satisfies the same conditions as c_j, i.e.,

5.10 $$c' \in J'_{h_j}, \qquad c - c' \in J'_{g_j}.$$

Then $c' - c_j \in J'_{h_j}$, $c' - c_j \in J'_{g_j}$ and so $c' - c_j \in J_g$. Hence, denoting by ϕ_j the canonical map from J'_{h_j} onto J'_{h_j}/J_g, we see that $c \to \phi_j(c')$, for c' satisfying 5.10, defines a mapping ψ_j from F_g into J'_{h_j}/J_g. It is not difficult to show that ψ_j is a homomorphism.

On the other hand,

$$J'_{h_j}/J_g = J_{h_j} \cap F_g / J_{h_j} \cap J_{g_j} = J_{h_j} \cap F_{g_j} / J_{h_j} \cap J_{g_j},$$

since $J_{h_j} \cap F_g = J_{h_j} \cap F_{h_j} \cap F_{g_j} = J_{h_j} \cap F_{g_j}$. We claim that $J_{h_j} \cap F_{g_j} / J_{h_j} \cap F_{g_j}$ is isomorphic to $F_{g_j}/J_{g_j} = K_{g_j}$. Indeed, the cosets of $J_{h_j} \cap F_{g_j}$ with respect to $J_{h_j} \cap J_{g_j}$ are subsets of the cosets of F_{g_j} with respect to J_{g_j} and every coset of the latter class contains at most one of the former. Accordingly, it only remains to be shown that every coset of the second class contains at least one coset of the first class. Thus, given any $f \in F_{g_j}$ we are required to find an $f' \in J_{g_i}$ such that $f - f' \in J_{h_j}$. In other words, we have to show that there exists an $x = f'$ which satisfies the conditions $\text{ord}_{P_j}(x) \geqq \nu_j$, $\text{ord}_{P_i}(f-x) \geqq \nu_i$ for any other P_i, and this can again be done by combining an application of the approximation theorem with the introduction of a suitable concurrent relation. Hence, if we map every coset of $J_{h_j} \cap F_{g_j}$ on the coset of F_{g_j}, in which it is contained, we obtain an isomorphic mapping χ_j from J'_{h_j}/J_g onto K_{g_j}. It follows that $\lambda_j = \chi_j \psi_j$ is a homomorphic mapping from F_g into K_{g_j} and

$$\lambda: c \to (\lambda_1(c), \lambda_2(c), \cdots, \lambda_j(c), \cdots),$$

where λ_j is included only for P_j which occur in g, is a homomorphic mapping from F_g into $\prod K_{g_j}$. The kernel of λ is the set $\{c \mid \psi_j(c) = 0$ for all P_j in $g\}$. For such P_j the corresponding c' (see 5.10) belongs to J_g. Hence $c \in J_{g_j}$ for all appropriate g·, and so $c \in J_g$.

Accordingly, λ induces an injection σ of $K_g = F_g/J_g$ into $\prod K_{g_j}$. It remains to be shown that λ is a surjection. For this purpose we employ, once again, the approximation theorem. Let $(c_0, c_1, \cdots, c_j, \cdots)$ be an arbitrary element of $\prod K_{g_j}$. We have to find a $c \in F_g$ such that $\lambda_j(c) = c_j$, i.e., we have to find a $c \in {}^*F$ such that c is entire for g and $\lambda_j(c) = c_j$. Thus, for arbitrary $d_j = \chi_j^{-1}(c_j) \in J'_{h_j}/J_g$, we have to find $c \in {}^*F$, which is entire for g such that $\phi_j(c) = d_j$. Or, again, we may give $c'_j \in J'_{h_j}$ arbitrarily (where $d_j = \phi_j(c'_j)$) and we then have to find $c \in {}^*F$ such that (compare 5.10)

5.11 $$c - c'_j \in J'_{g_j} \quad \text{for all } P_j \text{ which occur in } g$$

where c is entire for g. However, the last condition is now redundant since $c'_j \in J'_{h_j} \subset F_g$ and $J'_{g_j} \subset F_g$, so any c which satisfies 5.11 must be entire for g and this is true even if we relax the condition $c'_j \in J'_{g_j}$ and require only

5.12 $$c'_j \in F_g.$$

In order to satisfy 5.11 subject to the condition 5.12 it is sufficient to find $x = c$, which satisfies

5.13 $$\operatorname{ord}_{P_j}(x - c'_j) = \infty \quad \text{for all } P_j \text{ in } g;$$

for since $c'_j \in F_g$, we then have $c - c'_j \in F_g$ automatically and we certainly have $\operatorname{ord}_{P_j}(c - c'_j) \geqq \nu_j$. Furthermore, we may replace 5.13 by sequences of conditions,

5.14 $$\operatorname{ord}_{P_j}(x - c'_j) \geqq 0, \operatorname{ord}_{P_j}(x - c'_j) \geqq 1, \cdots, \operatorname{ord}_{P_j}(x - c'_j) \geqq k, \cdots$$

where k ranges over the finite natural numbers.

At this point, we make use of our assumption that $^*F$ is an ultrapower. It can then be shown [20] that for any sequence of internal entities $\{S_n\}$ of the enlargement, with subscripts ranging over N, there exists an *internal* sequence T_n in $^*F$, n ranging over $^*N$ such that $S_n = T_n$ for all $n \in N$.

We range the conditions 5.14 in a simply infinite sequence with subscripts in N and we denote the P_j, c'_j, and k which occur in the nth condition by $P^{(n)}$, $c^{(n)}$, and $k^{(n)}$, respectively. Evidently each P_j, c_j, and k will appear repeatedly in the sequence. We now consider the sequence of ordered triples $S_n = (P^{(n)}, c^{(n)}, k^{(n)})$, $n \in N$, and we extend it to an internal sequence $\{T_n\}$ as above, $n \in {}^*N$. The following sentence then holds in $^*F$, by virtue of the approximation theorem, for every finite or infinite m.

"There exists a $\xi \in {}^*F$ such that $\operatorname{ord}_{P^{(n)}}(\xi - c^{(n)}) \geqq k^{(n)}$ for all $n \leqq m$."

For any infinite m, a corresponding $x = \xi$ then satisfies all the conditions of 5.14. This completes the proof of Theorem 5.5.

Notice that we introduced the assumption that $^*F$ is an ultrapower for the last part of our proof because the argument from concurrent relations applies in our present framework only to *standard* binary relations.

Consider in particular the surdivisors $\gamma = \prod P_j^\infty$ where P_j ranges over all archimedean and nonarchimedean primes in F and $\delta = \prod P_j^\infty$ where P_j ranges over the nonarchimedean primes only. We call these the adelic (the restricted adelic) surdivisors, respectively. Still supposing that we are dealing with an ultrapower enlargement, we know that K_γ (K_δ) is isomorphic to $\prod K_{g_j}$ where $g_j = P_j^\infty$ and P_j ranges over all primes (over all nonarchimedean primes) in F. Writing ϕ for the isomorphism in question, $\phi(K_\gamma) = \prod K_{g_j}$ or $\phi(K_\delta) = \prod K_{g_j}$, as the case may be, we then have for any $c \in K_\gamma$ (for any $c \in K_\delta$)

$$\phi(c) = (c_0, c_1, \cdots, c_j, \cdots)$$

where c_j ranges over the P-adic and archimedean completions of F (over the P-adic completions only).

The adèle ring A_γ (the restricted adèle ring A_δ) is defined classically as the ring of elements $(c_0, c_1, \cdots, c_j, \cdots) \in \prod K_{g_j}$ such that $\text{ord}_{P_j}(c_j) \geqq 0$ for all but a finite number of j. Similarly, the idèle group I_γ (the restricted idèle group I_δ) is defined as the multiplicative group of elements of $\prod K_{g_j}$ such that $c_j \neq 0$ for all j and $\text{ord}_{P_j}(c_j) = 0$ for all but a finite number of j. Recalling that $K_\gamma = F_\gamma / J_\gamma$ ($K_\delta = F_\delta / J_\delta$), we write μ_γ (μ_δ) for the homomorphism $F_\gamma \to \prod K_{g_j}$ ($F_\delta \to \prod K_{g_j}$) with kernel J_γ (J_δ). For $a \in F_\gamma$ ($a \in F_\delta$) we then have $\text{ord}_{P_j}(a) \geqq 0$ if and only if $\text{ord}_{P_j}(\mu_\gamma(a)) \geqq 0$ ($\text{ord}_{P_j}(\mu_\delta(a)) \geqq 0$) for nonarchimedean P_j and $^\circ(\text{ord}_{P_j}(a)) \geqq 0$ if and only if $\text{ord}_{P_j}(\mu_\gamma(a)) \geqq 0$ for archimedean P_j.

The properties of adèles and idèles are reflected in the properties of their inverse images in F_γ and F_δ by μ_γ and μ_δ (compare [20] where the corresponding question is discussed for Dedekind rings). Let a be any element of F_γ (F_δ) such that $a \neq 0$. Then the entire or fractional ideal (a) can be written as a product of finite or infinite powers $\text{ord}_{Q_j}(a)$ of a finite or infinite number of prime ideals Q_j in $^*F$ where the Q_j may be standard or nonstandard. However, for standard Q_j the exponent of Q_j cannot be negative infinite, since a belongs to F_γ (belongs If to F_δ). If $\text{ord}_{Q_j}(a)$ is finite for all standard Q_j and is zero for almost all such Q_j, then we call a Prüfer-finite. Then $\mu_\gamma(a)$ ($\mu_\delta(a)$) is an idèle (a restricted idèle) if and only if a is Prüfer-finite and $\mu_\gamma(a)$ ($\mu_\delta(a)$) is an idèle unit if and only if $\text{ord}_{Q_j}(a) = 0$ for all standard Q_j, i.e., if a has nonstandard prime ideal factors only. Thus, for standard a, $\mu_\gamma(a)$ and $\mu_\delta(a)$ are idèle units if and only if a is a unit (invertible algebraic integer) in $^*F$.

Now let A be a standard entire ideal in F, so that $^*A$ is the corresponding ideal in $*F$. Then $*A$ has a two element basis, $*A=(\alpha, \beta)$ where α and β are algebraic integers in $*F$ and where β may be any arbitrary nonzero element of $*A$. In particular, we may choose a $\beta \neq 0$ which is divisible by the restricted adelic surdivisor δ. Then $\mu_\delta(\beta)=0$, while $\alpha \in F_\delta$ since α is an integer. Hence, $\mu_\delta(*A)=\mu_\delta(\alpha)$, and so $*A$ corresponds to a principal ideal in $\prod K_{\mathfrak{g}_j}$ (as is also evident from the theory of idèles). It follows that $\mathrm{ord}_{P_j}(A)=\mathrm{ord}_{P_j}(\alpha)$ for all standard prime ideals P_j and there exists an entire ideal B in $*F$ such that $*AB=(\alpha)$ where B is divisible only by nonstandard prime ideals. This shows, incidentally, that to every class $*C$ of ideals (multiplicative coset of principal ideals) in $*F$ and to every preassigned finite set of standard prime ideals S, there exist entire ideals in $*F$ which are not divisible by any ideal in S. Indeed, we only have to take the standard ideal $*A$ as a representative of the class $*C^{-1}$, then the above B is an ideal of the required type. And since our conclusion holds for $*F$, it holds also in F, by the usual argument for transferring conclusions from F to $*F$ and vice versa. In this case, however, only the argument is of interest, since the result is a simple consequence of the classical theory.

To sum up the results of the present section, we have seen that the fields and rings which are commonly associated by the theory of valuations with a given algebraic number field can all be obtained by a uniform procedure as homomorphic images of internal or external (noninternal) subrings of $*F$.

6. **Class field theory.** Up to this point, we have used enlargements, in a sense, only as auxiliary concepts, that is to say, we have shown how they can be employed in the construction and investigation of classical theories. We shall now consider a situation in which a classical theory is known to fail in the standard model but can still be carried out in an enlargement. This situation occurs in the class field theory of infinite abelian extensions and was in fact the occasion for the introduction of idèles by Chevalley [2], although not their only justification. To quote the example pointed out by Chevalley, let $F=Q$ be the field of rational numbers and, for a given prime number $p \geqq 3$, let $F_n=F(\zeta_n)$ where ζ_n is a primitive p^nth root of unity. Also, let $F_\infty=\bigcup_n F_n$. In order to define the class group for F_n in accordance with the standard precept, we introduce Q_0 as the multiplicative group of rational numbers prime to p and entire for p, and I_n as the subgroup of Q_0 which is given by $I_n=\{q \mid q\equiv 1\,(p^n)\}$. Then the multiplicative quotient group Q_0/I_n is the class group of F_n. By analogy with the finite case, we might now expect $Q_0/\bigcap_n I_n$ to be the class group of $\bigcup_n F_n$, but this conclusion is spurious since $\bigcap_n I_n=\{1\}$.

For any finite algebraic extension of the rationals Q, as in §5, let

$g = \prod P_j^{\nu_j}$ be a surdivisor in F. An entire or fractional internal ideal A in $^*F$ is called *entire for* g if $A \neq 0$ and if $\text{ord}_{P_j} A \geqq 0$ for all nonarchimedean primes P_j which occur in g. Notice that if ν_j is infinite for some nonarchimedean P_j in g, then $a \in {}^*F$ may be entire for g according to the definition of the previous section which was introduced relative to the valuation ring $F_{P_j^\infty}$, although $A = (a)$ is not entire for g according to our present definition. A is called *prime for* g if $A \neq 0$, and if for any nonarchimedean P_j which occurs in g, $\text{ord}_{P_j} A \leqq 0$.

We define the ray modulo g, R_g, as the subset of $^*F$ whose elements a are defined by the following conditions. If $P_j^{\nu_j}$ is a primary factor of g and P_j is nonarchimedean, then $a - 1$ is divisible by $P_j^{\nu_j}$; while if P_j is real archimedean then $a > 0$ in the corresponding embedding of $^*F$ in the enlargement of the real numbers, $^*R$; and (to provide for a trivial case) $a \neq 0$. R_g is a multiplicative group. We denote by I_g the corresponding *ideal ray group*, i.e., the group of principal ideals (a) in $^*F$ such that $a \in R_g$.

Now let m be a divisor in F, where we recall that a surdivisor is called a divisor if the number of its primary factors $P_j^{\nu_j}$ is finite and if at the same time $\nu_j = \infty$ only for archimedean P_j. To m, there corresponds an ideal ray group J_m in F, in the classical sense, where J_m consists of the ideals $A \neq 0$ in F such that $A = (a)$ for some $a \in F$ that satisfies $\text{ord}_{P_j}(a-1) \geqq \nu_j$ for all nonarchimedean P_j occurring in m, and $a > 0$ for the order corresponding to any real archimedean P_j in m. It is not difficult to see that $I_m = {}^*J_m$. Standard class field theory assigns to J_m an abelian normal extension F_m of F as its class field. F_m is unique up to isomorphism and hence is determined uniquely if we require $F_m \subset \bar{F}$ where $\bar{F}$ is a fixed algebraic closure of F. For any surdivisor g we now define the class field for g, F_g, as the compositum of all fields $F_m \subset \bar{F}$ as m ranges over the divisors of F such that $m \mid g$. (For $m = g$, the notations F_m and F_g are consistent.)

Let h be any internal divisor in $^*F$, $h = \prod Q_j^{\mu_j}$ where, by the definition of a divisor in $^*F$, Q_j ranges over the primes of $^*F$, and the set of Q_j for which $\mu_j \neq 0$ is starfinite. Moreover, μ_j may now be a finite or infinite natural number or ∞, but the latter case is possible only if Q_j is archimedean. In general, Q_j may be standard or nonstandard internal, but if Q_j is archimedean then it must be standard since the set of archimedean primes in F is finite and, accordingly, is not enlarged on passing to $^*F$. If Q_j is standard then we may write $Q_j = {}^*P_j$ for some prime P_j in F and there is no essential limitation in assuming that the subscript (j) is the same on both sides. Let $^*g$ be the injection of a surdivisor g into the enlargement (where we append the star in order to avoid misunderstandings). For example, if $g=P_j^\infty$ then $*g={}^*P_j^\infty$. Suppose that P_j is nonarchimedean and let h be the divisor $h={}^*P_j^\omega$ where ω is an infinite natural number. Then the standard definition of divisibility, when transferred to the enlargement, forces us to conclude that $h|{}^*g$ but not ${}^*g|h$ in ${}^*F$. On the other hand, consider any $a\in{}^*F$ such that $\text{ord}_{P_j}(a)=\omega-1$. By our definition of divisibility by surdivisors, $g|a$ but not $h|a$. In order to cope with this discrepancy we need a new notion of divisibility of a divisor by a surdivisor. We shall say that the surdivisor $g=\prod P_j^{\nu_j}$ *surdivides* the divisor $h=\prod Q_j^{\mu_j}$ if for any finite natural ν_j and $Q_j={}^*P_j$ we have $\mu_j\geqq\nu_j$ (including the possibility that $\mu_j=\infty$); while if $\nu_j=\infty$, then μ_j is an infinite natural number, or $\mu_j=\infty$. We denote this relation by $g\|h$. It holds in the special case considered above although ${}^*g|h$ does not hold in that case. If, however, h is a standard divisor then $g\|h$ only if ${}^*g|h$.

For any surdivisor in F there exists a divisor h in ${}^*F$ which is surdivisible by g. In order to see this, range all standard P_j in a sequence with subscripts in N, beginning with the archimedean primes. Put

$$6.1 \qquad g_k = P_0^{\nu_0(k)}P_1^{\nu_1(k)}\cdots P_k^{\nu_k(k)}, \qquad k=0,1,2,\cdots$$

where $\nu_j(k)=\nu_j$ if ν_j is a natural number, $\nu_j(k)=\infty$ if $\nu_j=\infty$ and P_j is archimedean and $\nu_j(k)=k$ if $\nu_j=\infty$ and P_j is nonarchimedean. Then the g_k are divisors in F. Passing to the enlargement and putting $h=g_\omega$ for arbitrary infinite ω, we see that h is a divisor in ${}^*F$. It is not difficult to verify that $g\|h$, for if $\nu_j=\infty$ and P_j is nonarchimedean then $\nu_j(\omega)=\omega$.

Applying the standard theory to a divisor h in ${}^*F$, we obtain an ideal ray group J_h and a class field F_h of J_h where ${}^*F\subset F_h$ and where we may suppose that $F_h\subset{}^*\overline{F}$. Let ${}^\circ F_h=F_h\cap\overline{F}$, then ${}^\circ F_h$ is the *standard part* of F_h, i.e. ${}^\circ F_h$ contains just the elements of F_h which are standard.

If g is a surdivisor and ${}^*g\|h$ then $J_h\subset I_g$. In that case, also, ${}^\circ F_h\supset F_g$ where F_g is the class field of g, as before. For let $\alpha\in F_g$, then α belongs to the compositum of a finite number of fields $F_{m_1},\cdots,F_{m_k}$ corresponding to divisors $m_1|g,\cdots,m_k|g$. It follows that $\alpha\in F_m$ where m is the l.c.m. of $m_1,\cdots,m_k$. But $m|g$ with $g\|h$ entails ${}^*m|h$ in the enlargement and so $F_m\subset{}^*F_m=F_{*m}\subset F_h$. Hence $\alpha\in{}^\circ F_h$; $F_g\subset{}^\circ F_h$ as asserted.

Let H be the intersection of the fields ${}^\circ F_h$ as h ranges over the divisors which are surdivided by g. *We claim that* $H=F_g$.

Since we have already shown that $F_h \supset F_g$ for $g \| h$, it follows that $H \supset F_g$. It only remains for us to establish inclusion in the opposite direction. Let $\alpha \in H$, then $\alpha \in F_h$ for any h such that $g \| h$. Also, in the notation of 6.1 above, $g \| g_k$ for all infinite natural numbers k and so $\alpha \in F_{g_k}$ for all infinite k. But $\{F_{g_k}\}$ is an internal sequence and so we may conclude that $\alpha \in F_{g_k}$ also for sufficiently large finite k. Now, for all finite k, $g_k | g$ and so $F_{g_k} \subset F_g$. Hence $\alpha \in F_g$, as asserted.

Let Γ_g be the Galois group of F_g/F. For any divisor h which is surdivided by g as before, let Γ_h be the Galois group of $F_h/{}^*F$. Let C_h be the group of ideals in ${}^*F$ which are entire and prime for h, then $J_h \subset C_h$. Moreover, there is a homomorphism from C_h onto Γ_h, with kernel J_h, which is given by the Artin symbol

$$Q \to \sigma = \left(\frac{F_h/{}^*F}{Q}\right), \quad Q \in C_h, \quad \sigma \in \Gamma_h.$$

Let ϕ be the canonical mapping from Γ_h to Γ_g with kernel K where K consists of the $\sigma \in \Gamma_h$ that leave the elements of $F_g \subset F_h$ invariant. K is not necessarily internal. The mapping ϕ is onto, since every automorphism of F_g over F can be extended to an automorphism of ${}^\circ F_h$ over F which can then be extended to an automorphism of $\overline{F}$ over F. On passing to the enlargement, this in turn can be extended to an internal automorphism of ${}^*\overline{F}$ over ${}^*F$, and then restricted to an automorphism of F_h over ${}^*F$. Further restriction to ${}^\circ F_h$ and then to F_g leads us back to the original automorphism, which therefore belongs to the range of ϕ.

Let C_g be the group of ideals in ${}^*F$ which are entire and prime for g. We define a *generalized Artin symbol* to indicate a mapping from C_g to Γ_g by

$$Q \to \sigma = \left(\frac{F_g/F}{Q}\right) = \phi\left(\frac{F_h/{}^*F}{Q}\right)$$

for all ideals $Q \in C_g$. The mapping is onto, since ϕ is onto.

We claim that this definition is independent of our particular choice of h (provided $g \| h$, as above). Indeed, let k be any other divisor such that $g \| k$. We may suppose that $h | k$, for if this is not the case from the outset we may then prove that we obtain the same interpretation of the generalized Artin symbol by taking the g.c.d. of h and k as we do by taking h or k.

Suppose that $g \| h$ and $h | k$ and hence $g \| k$. If C_k is the group of ideals of F which are entire and prime for k, we then have $I_g \supset J_h \supset J_k$ and $C_g \supset C_h \supset C_k$ and $F_g \subset F_h \subset F_k$. Also, if Γ_k is the Galois group of

$F_k/{}^*F$ and ϕ' is the canonical mapping from Γ_k to Γ_g (corresponding to ϕ for h) and ψ is the mapping from Γ_k to Γ_h whose kernel consists of the automorphisms in Γ_k that leave the elements of F_h invariant, then $\phi' = \phi\psi$. Now let Q be any prime ideal in C_k with norm NQ. Then if

$$\sigma = \left(\frac{F_k/{}^*F}{Q}\right)$$

we have, for all $\alpha \in F_h$,

6.2 $$\sigma\alpha \equiv \alpha^{NQ}(Q).$$

But the same condition is satisfied by the restriction of σ to F_h, i.e., by $\psi\sigma$, and so

$$\left(\frac{F_h/{}^*F}{Q}\right) = \psi\left(\frac{F_k/{}^*F}{Q}\right).$$

Hence

$$\phi\left(\frac{F_h/{}^*F}{Q}\right) = \phi\psi\left(\frac{F_k/{}^*F}{Q}\right) = \phi'\left(\frac{F_k/{}^*F}{Q}\right),$$

which shows that our definition of the generalized Artin symbol is independent of the particular choice of h, if Q is a prime ideal. The general result now follows from the multiplicativity of the symbol.

Let e be the identity in Γ_g, while I_g is the ideal ray group of g as introduced previously. We claim that *for any* $Q \in C_g$, $Q \in I_g$ *if and only if*

$$\left(\frac{F_g/F}{Q}\right) = e.$$

To see this, observe that

$$\left(\frac{F_h/{}^*F}{Q}\right)$$

is the identity in Γ_γ if and only if $Q \in J_h$. Now suppose $Q \in I_g$ so that $Q = (q)$ where $q \in R_g$. Take $h = \prod Q_i^{\mu_i}$ where $\mu_i = \text{ord}_{Q_i}(q-1)$ for any nonarchimedean Q_i and $\mu_i = \infty$ for any archimedean Q_i which occurs in g. Then $g \| h$ and $Q \in J_h$, and hence

$$\left(\frac{F_h/{}^*F}{Q}\right) = e$$

and so

$$\left(\frac{F_g/F}{Q}\right) = e.$$

Conversely, suppose that

$$\left(\frac{F_g/F}{Q}\right) = e$$

but that $Q \notin I_g$. Let $\{g_k\}$ be the sequence defined by 6.1. $\{g_k\}$ is internal and $g \| g_k$ for all infinite k. It follows that, for such k, $Q \notin I_{g_k}$. Hence, $Q \notin I_m$ where $m = {}^*g_k$ for sufficiently large finite k, and so

$$\left(\frac{F_m/F}{Q}\right) \neq e.$$

On the other hand, $m | g$ and so another argument involving 6.2, applied this time to $\alpha \in F_m$, shows that

$$\left(\frac{F_m/F}{Q}\right) = e$$

in Γ_m, a contradiction which proves our assertion.

We have now shown that I_g is the kernel of the homomorphism provided by the generalized Artin symbol. Thus, the symbol induces an isomorphism between C_g/I_g and Γ_g, $C_g/I_g \simeq \Gamma_g$, as one would like to expect of a class field. The isomorphism also provides a correspondence between the subfields of F_g on one hand and certain subgroups of C_g/I_g on the other hand via the Galois group Γ_g. Thus the subgroups of C_g/I_g which appear in this correspondence are just those that are closed in the Krull topology.

Suppose in particular that $g = \gamma$, where γ was defined in the preceding section, although the complex archimedean factors of γ are now irrelevant. Then F_γ is the compositum of all abelian extensions of F, i.e., it is the maximal abelian extension of F over A. The group I_γ now consists of all principal ideals (a) such that $a \in {}^*F$ is totally real and $a-1$ is divisible by $\omega!$ for some infinite natural number ω. The group C_γ/I_γ, which is isomorphic to the Galois group of F_γ/F can also be expressed in terms of the idèles of ${}^*F$. However, the discussion of this and other topics which are evidently still required in order to complete the picture must be left for another occasion (compare [21]).

While in the present section we have made effective use of infinite prime numbers or nonstandard prime ideals as elements of fields or of multiplicative groups, they have not, so far, occurred as divisors or as characteristics of fields. A simple application of infinite prime numbers in this direction is as follows (see [16]). Suppose that a sentence X which is formulated in the Lower Predicate Calculus in terms of equality, addition, and multiplication is false for fields of arbitrarily high characteristic. Thus, the statement "for every natural number n there exists a prime number $p>n$ such that X is false in a field of characteristic p" is true for the standard natural numbers and hence is true also in an enlargement. Choosing n infinite, we see that there is a field F of infinite characteristic p such that X is false in F. But when looked at from the outside, F is actually of characteristic 0 since it is not of any finite characteristic. We have proved the well-known result that if X is true for all fields of characteristic 0 then it is also true for all fields of characteristic $p>p_0$ where p_0 depends on X.

A much deeper result which can be stated readily in terms of infinite primes is the famous theorem of Ax and Kochen. Let F_p be the prime (minimal) field of infinite prime characteristic p and let $F_p\{x\}$ be the field of formal power series of x adjoined to F_p within an enlargement. Thus, the subscripts of a series $\sum_{n=-q}^{\infty} a_n x^n \in F_p\{x\}$ range over the nonstandard integers of the enlargement. Let $\overline{F}_p$ be the field of p-adic numbers in the same enlargement. Then $F_p\{x\}$ is elementarily equivalent to $\overline{F}_p$ with respect to the *standard language* of the Lower Predicate Calculus with a vocabulary for the field operations and for valuation in an ordered group. This is a rather elegant reformulation of the Ax-Kochen result, but the available methods of proof are based either on model theoretic methods [1] or on the elimination of quantifiers [3]. It would be interesting to handle the problem effectively by nonstandard methods.

7. Concluding remarks. As we have seen, our methods offer, in many cases, alternatives to familiar infinitary constructions and passages to the limit. It is quite likely that at some future date a deeper understanding of the structure of definitions and proofs will enable us to provide systematic translations from one framework to the other. And we may recall here that already Pascal and Leibniz maintained that the respective infinitesimal methods employed by them differed from the Greek method of exhaustion only in the manner of speaking. Coming next to the mathematician's desire for obtaining an intuitive picture of his universe of discourse, the use of ultrapowers as representations of enlargements is entirely appropriate, although even this does not lead to a categorical (unique) enlargement except by means of artificial restrictions. Beyond that, the use of ultrapowers (see Theorem 5.5) or of other special models (see Luxemburg's result quoted in §3 above) may actually be required in order to prove particular propositions.

As far as the results of the present paper on algebraic number fields are concerned, the argument at the end of §3 shows that they all remain true in an ultrapower on a countable index set for a free ultrafilter. Within this framework, the infinite numbers and ideals may be regarded as just another kind of limit. Moreover, the procedure by which we obtain an ultrapower F_D^N from a countable direct product of fields F^N can be combined into a single step with the further homomorphisms $F_0 \to F_0/F_1$ onto various completions of F. In particular, we may thus obtain the real numbers R by taking a countable direct power of the rational numbers Q^N and a free ultrafilter D on N. An element $q = \{q_n\} \in Q^N$ will be called *finite* if there exists a rational number r such that $\{n \mid |q_n| < r\} \in D$. Let Q_0 be the set of finite elements of Q^N; then Q_0 is a subring of Q^N. Let Q_1 be the set of infinitesimal elements of Q^N, i.e., of elements $q = \{q_n\}$, such that $\{n \mid |q_n| < r\} \in D$ for the positive rational numbers r. Then $Q_1 \subset Q_0$ and Q_1 is an ideal in Q_0. The quotient ring Q_0/Q_1 is isomorphic to the field of real numbers. In this way we obtain a procedure which bears a general similarity to the method of completion by Cauchy sequences but is quite different from it in detail.

On the other hand, it would seem wasteful to give up the logical basis of our method altogether, for it alone provides the setting within which we may deduce the validity of statements in $*M$ quite generally from their validity in M and vice versa. Without this setting, any property which is known to apply to M has to be established for $*M$ separately in each case, and while this is certainly possible it has been contrary to good mathematical practice ever since the days of Theaetetus.

It may be too early to say whether the methods of Nonstandard Analysis will ever become accepted (or "standard") tools of mathematics. At any rate, it is remarkable that an idea which once formed the basis for most of the work in the Differential and Integral Calculus and which was declared bankrupt one hundred years ago (after a long but admittedly fraudulent career) has, after all, enough vitality to make a meaningful contribution to a subject as far removed from its origins as the theory of algebraic number fields.

Bibliography

1. J. Ax and S. Kochen, *Diophantine problems over local fields*, Amer. J. Math. **87** (1965), 605–630, 631–698.

2. C. Chevalley, *Généralisation de la théorie du corps de classes pour les extensions infinies*, J. Math. Pures Appl. (9) **15** (1936), 359–371.

3. P. J. Cohen, *Decision procedures for real and p-adic fields*, Unpublished.

4. T. Frayne, D. C. Morel and D. S. Scott, *Reduced direct products*, Fund. Math. **51** (1962), 195–227.

5. D. K. Harrison, *Finite and infinite primes for rings and fields*, Mem. Amer. Math. Soc. No. 68 (1966).

6. H. Hasse, *Bericht über neuere Untersuchungen und Probleme aus der Theorie der algebraischen Zahlkörper*, Teil I, Klassenkörpertheorie, Jber. Deutsch. Math.-Verein. **35** (1926), 1–66; **36** (1927), 233–311.

7. L. Henkin, *The completeness of the first order functional calculus*, J. Symbolic Logic **14** (1949), 159–166.

8. H. J. Keisler, *Ultraproducts and elementary classes*, Proc. Roy. Acad. Sci. Amsterdam Ser. A **64** (1962), 477–495.

9. J. G. Kemeny, *Undecidable problems of elementary number theory*, Math. Ann. **135** (1958), 160–169.

10. S. Kochen, *Ultraproducts in the theory of models*, Ann. of Math. (2) **79** (1961), 221–261.

11. W. Krull, *Galoissche Theorie unendlicher algebraischer Erweiterungen*, Math. Ann. **100** (1928), 678–698.

12. S. Lang, *Rapport sur la cohomologie des groupes*, Benjamin, New York, 1966.

13. W. A. J. Luxemburg, *A new approach to the theory of monads*, Symposium on Applications of Model Theory to Analysis and Algebra, Pasadena, May 1967, (to appear).

14. R. MacDowell and E. Specker, *Modelle der Arithmetik*, Infinitistic Methods (Symposium on Foundations of Mathematics, Warsaw, 1959), Państwowe Wydawnictwo Naukowe, Warsaw, 1961, pp. 257–263.

15. E. Mendelson, *On non-standard models for number theory*, Essays on the Foundations of Mathematics (Fraenkel anniversary volume), Magnes Press, Jerusalem, 1961, pp. 259–268.

16. A. Robinson, *Les rapports entre le calcul déductif et l'interprétation sémantique d'un système axiomatique*, Colloques Internationaux du Centre National de la Recherche Scientifique, No. **36**, Paris, 1953, pp. 35–52.

17. ———, *Model theory and non-standard arithmetic*, Infinitistic Methods (Symposium on Foundations of Mathematics, Warsaw, 1959), Państwowe Wydawnictwo Naukowe, Warsaw, 1961, pp. 265–302.

18. ———, *Non-standard analysis*, Studies in Logic and the Foundations of Mathematics, North-Holland, Amsterdam, 1966.

19. ———, *A new approach to the theory of algebraic numbers*, Accad. Naz. Lincei Rend. (8) **40** (1966), 222–225, 770–774.

20. ———, *Non-standard theory of Dedekind rings*, Proc. Roy. Acad. Sci., Amsterdam, (to appear).

21. ———, *Non-standard algebraic number theory*, Symposium on Applications of Model Theory to Analysis and Algebra, Pasadena, May 1967, (to appear).

22. J.-P. Serre, *Cohomologie Galoisienne*. Lecture Notes in Mathematics No. 5, Springer, Berlin, 1964.

23. T. Skolem, *Über die Nichtcharakterisierbarkeit der Zahlenreihe mittels endlich oder unendlich vieler Aussagen mit ausschliesslich Zahlenvariablen*, Fund. Math. **23** (1934), 150–161.

UNIVERSITY OF CALIFORNIA, LOS ANGELES, AND
YALE UNIVERSITY

On some applications of model theory to algebra and analysis

by **ABRAHAM ROBINSON** (Los Angeles)

1. Introduction.

In the winter of 1966 the author had the honor of being invited to give a series of lectures at the Mathematical Institute « Guido Castelnuovo », University of Rome. The present paper has been written in connection with that occasion and deals with problems and methods of the kind described in the lectures. While it is the purpose of the paper to provide an introduction to the subject, several of the results detailed below are believed to be new (1).

The plan of this paper is as follows. In section 2 and 3 we delineate the basic notions of Model theory and describe some of the directions in which it has been developed. In sections 4 and 5 we discuss applications to the theories of commutative and of ordered fields. The latter are related to Hilbert's seventeenth problem. In section 6 we introduce Non-standard Analysis. This is a relatively new field of application of Mathematical Logic. It is based on the discovery that the notions of Model theory can clear up the inconsistencies which disturbed and ultimately blocked the development of Mathematical Analysis by means of infinitely small and infinitely large quantities. In section 7 we apply Non-standard Analysis to the theory of lattices in the Geometry of Numbers.

(1) The writer wishes to express his thanks to Professors B. Segre, L. Lombardo-Radice, and E. Casari for having made his visit to the University of Rome possible. He is also happy to acknowledge the support of the National Science Foundation (Grant No. GP-4038) in connection with the work described here.

2. Model theory.

At the basis of Model theory is the conceptual dichotomy between mathematical structures, on one hand, and the formal language or languages which are employed in order to describe these structures, on the other hand. Here we shall not be concerned with the degree of reality which can be ascribed to mathematical structures or the extent to which such ontological questions are at all meaningful; nor shall we discuss the philosophical problems relating to the very nature of a formal language and of the components which constitute it.

We shall choose our formal language L, to begin with, within the framework of the Lower Predicate Calculus. This means that the formulae of the language are built up from individual constant symbols $a, b, c, \dots$, from individual variables, $x, y, z, \dots$, and from n ary relation symbols, $R(x_1, \dots, x_n), \dots$; together with propositional connectives, $\neg$ (negation), $\vee$ (disjunction), $\wedge$ (conjunction), $\supset$ (implication), $\equiv$ (equivalence or biconditional); the two quantifiers $(\exists)$ (there exists-) and $(\forall)$ (for all-); and parentheses, [, and]. We may, if we so desire, delete from this list the connectives of conjunction, implication, and equivalence, and the universal quantifier, $(\forall)$, since these can be *defined* in terms of the remaining symbols. We may also omit the parentheses and transfer their function to the mode of construction of well formed formulae. We shall not follow up these possibilities here.

We shall suppose that the reader is familiar with the notion of a well formed formula or, at least, that he is familiar with the customary interpretation of the above symbols and can discern whether a string (finite sequence) of such symbols « makes sense ». Observe that within the Lower Predicate Calculus, quantification is permitted with respect to individual variables only, $(\exists x)$, $(\forall y), \dots$. A *sentence* is a well formed formula in which all variables are quantified (i. e., are within the scope of a corresponding quantifier).

A structure Σ consists of a set of individuals, A, and a set P of relations defined on A, and we may take Σ as the ordered pair of these two sets, $\Sigma = \langle A, P \rangle$. We may regard any relation R in P, set theoretically, as a set of ordered n-tuples of A.

The interpretation of a sentence X with reference to a structure Σ presupposes the assignment of a meaning, within Σ, to the relation symbols and individual constant symbols of X. That is to say, there must be given a correspondence which assigns to every individual constant symbol of X an individual in A, and to every relation symbol of X a relation in P. Moreover, if we are concerned with the interpretation of a set of sentences K then the correspondence in question is to assign meanings to the symbols which occur in the sentences of K consistently, i. e., an individual constant or relation symbol is supposed to have the same meaning for all the sentences of K in which it occurs. Finally, we may if we so wish, assume that the correspondence is one-to one, i. e., that two distinct symbols cannot denote the same entity of the structure, and in order to make matters more definite we shall assume that this is the case.

Let $R(a_1, \dots, a_n)$ be an *atomic sentence*, i. e., a sentence which is obtained by attaching n individual constant symbols to a relation symbol of order n, and suppose that, order a specified correspondence R, $a_1, \dots, a_n$ denote a relation R', and individual constants $a'_1, \dots, a'_n$ in a structure M, respectively. Then we say that $R(a_1, \dots, a_n)$ *is true in* M, or *holds in* M, it and only if the n-tuple $\langle a'_1, \dots, a'_n \rangle$ belongs to the relation R in M. Furthermore, having determined whether the sentences X and Y hold or do not hold in M, we say that $\neg X$ holds in M if and only if X does not hold in M; that $X \vee Y$ holds in M if and only if at least one of the two sentences X or Y holds in M; that $X \wedge Y$ holds in M it and only if both X and Y hold in M; that $X \supset Y$ does not hold in M if X holds in M but Y does not hold in M, but that $X \supset Y$ holds in M in all other cases; and that $X \equiv Y$ holds in M if both X and Y hold in M and if neither X nor Y holds in M but that $X \equiv Y$ does not hold in M in the remaining cases.

Now let Y be a sentence which has been obtained by existential quantification. Thus, Y is of the form $Y = (\exists x) Z(x)$. Let $Z(a)$ be the sentence obtained from the formula $Z(x)$ by substituting the individual constant symbol a everywhere for the variable x. Then Y shall hold in M if and only if $Z(a)$ holds in M for some individual constant symbol denoting an individual of A. And if $Y = (\forall x) Z(x)$, then Y shall hold in M if and only if $Z(a)$ holds in M for a denoting any individual of A. If all sentences of a set of sentences K hold in M then we say that M is a model ot K. The reader may consult ref. 15 for a more detailed discussion of these matters.

A set of sentences which possesses a model is said to be (semantically) *consistent*. A set of sentences which is not consistent is *contradictory.* If a sentence X is true in any model of a set of sentences K — for a consistent interpretation of the symbols of the set $K \cup \{X\}$, see above — then we say that X is a (semantical) consequence of K, $K \Vdash X$. The basic result of the subject from a model-theoretic point of view is the *finiteness principle*, first stated by A. I. Malcev.

2.1. THEOREM. *If any finite subset of a set of sentences K is consistent, then K is consistent.*

An immediate consequence of 2.1 is

2.2. THEOREM. *If the sentence X is a consequence of a set of sentences K then X is a consequence also of some finite subset of K.*

Observe that we have not introduced any notion of formal (syntactical) deducibility. Although the introduction of such notions within the present framework is rather natural it is not essential for our purposes. However, Gödel's completeness theorem shows that for the accepted deductive systems of the Lower Predicate Calculus the deductive (syntactical) and semantical notions of consistency and of consequence are equivalent, respectively.

The following principle, though formulated here in model-theoretic terms, has its origin in a rule of inference.

2.3. THEOREM. *Let K be a set of sentences, X a single sentence, and Y a sentence of the form $Y = Z(a)$ where the symbol a occurs neither in X nor in the sentences of K. Suppose that*

$$K \Vdash Y \supset X.$$

Then

$$K \Vdash [(\exists y)\, Z(y)] \supset X$$

where y is any variable which does not occur in Y.

The reader may consult ref. 15 for the proofs of Theorems 2 1-2.3.

3. Examples from Model theory.

We are now ready to discuss a typical problem of « pure Model Theory ». It forms a suitable introduction to the more « applied » problems which will be discussed subsequently.

A sentence X is said to be in *prenex normal form* if, when read from left to right, none of its quantifiers is preceded by a connective. Equivalently, a sentence is in prenex normal form if, in its successive construction from atomic formulae, the introduction of all the connectives precedes the introduction of any quantifier. X is said to be an ***existential*** sentence if it is in prenex normal form and contains existential quantifiers only (if any). It is not difficult to see that if an existential sentence X holds in a structure M then X holds also in all structures which are extensions of M. Conversely, it has been shown by Tarski (ref. 17) that if a sentence X has the property that its truth in any structure M entails its truth also in any extension of M then there exists an existential sentence Y such that $X \equiv Y$ is true universally, i.e. is true in all structures in which the symbols of X and Y are interpreted. The following generalization of this result was proved, independently, by Łoś and the present author (refs. 9,11)

3.1. THEOREM. *Let K be a consistent set of sentences and let X be a sentence whose relation and individual constant symbols occur in the sentences of K. Suppose that whenever X holds in a structure M then X holds also in all extensions of M which are models of K. Then there exists an existential sentence Y such that $K \# X \equiv Y$.*

The converse of this theorem is again obvious.

In order to prove 3.1. we shall employ a simple but useful artifice konow as the *method of diagrams*. Let M be a structure and let K be a set of sentences. Suppose we wish to concern ourselves exclusively with models of K which are isomorphic to extensions of M. In order to characterize such structures within our model-theoretic framework we introduce the notion of a diagram. Assuming that we have assigned to all the relations and individual constants corresponding symbols in the formal language L, let D be the set of all atomic sentences (e. g. $R(a_1, \dots, a_n)$) which hold in M, together with the negatives of any such sentence, provided the sentence does not hold in M (e. g. $\neg R(a_1, \dots, a_n)$). For example, let M be the field of rational numbers formulated in terms of the relation of equality $E(x, y)$ (read $x = y$) and the relations of addition, $S(x, y, z)$ (read $x + y = z$), and multiplication, $P(x, y, z)$ (read $xy = z$). Then, with the usual denotation of the numerals, the sentences $E(1, 1)$ and $S(7, 6, 13)$ belong to the diagram D of M; and $P(7, 6, 40)$ does not belong to D, so that the negation of this sentence $\neg P(7, 6, 40)$ belongs to D. Coming back to the general case, we see without difficulty that every model of D is an extension of a structure which is isomorphic to M and, conversely, every structure which is an extension of an isomorph of M is a model of D for an appropriate interpretation of the symbols of D. The assumption that a sentence X holds in all models of K which are extensions of a given structure M now turns out to be equivalent to $K \cup D \Vdash X$ where D is the diagram of M.

We are now ready to prove 3.1. For given K and X as detailed in the theorem, let H be the set of all existential sentences Y formulated by means of relation and individual constant symbols which occur in K and such that $K \Vdash Y \supset X$. The set H is not empty and has the property of being *(quasi-)disjunctive.* That is to say, if Y_1 and Y_2 are elements of H, e. g.

$$Y_1 = (\exists z_1) \dots (\exists z_j)\, Q_1(z_1, \dots, z_j),$$
$$Y_1 = (\exists v_1) \dots (\exists v_l)\, Q_2(v_1, \dots, v_l),$$

then there exists an element Y_3 of H, i. e.

$$Y_3 = (\exists z_1) \dots (\exists z_j)(\exists v_1) \dots (\exists v_l)\,[Q_1(z_1, \dots, z_j) \vee Q_2(v_1, \dots, v_l)]$$

such that $[Y_1 \vee Y_2] \equiv Y_3$ is true universally.

Consider the set $J = K \cup \{X\} \cup \{\neg Y_v\}$ where Y_v varies over the element of H. We propose to show that J is contradictory. Suppose on the contrary that J is consistent and let M be a model of J and D its diagram. Since X holds in M, it must hold also in all extensions of M which are models of K and, accordingly, must be a consequence of $K \cup D$, $K \cup D \Vdash X$. It now follows from 2.2. that there is a finite subset D' of D such that X is a consequence of $K \cup D'$. Let $Z(a_1, \dots, a_m)$ be the conjunction of the elements of D', taken in any order, where we have displayed the individual constant symbols which do not occur in K, if any. Since X is a consequence of $K \cup D'$ we then have

$$K \Vdash Z(a_1, \dots, a_m) \supset X$$

and hence, by the repeated application of 2.3,

3.2 $$K \Vdash [\exists y_1) \dots (\mathrm{E} y_m)\, Z(y_1, \dots, y_m)] \supset X.$$

Put

$$W = (\exists y_1) \dots (\exists y_m)\, Z(y_1, \dots, y_m).$$

Comparing 3,2 with the definition of H we see that W belongs to H and hence that $\neg W$ holds in M. On the other hand, W must hold in M since $Z(a_1, \dots, a_m)$ holds in M, being a conjunction of elements of D. This contradiction shows that M cannot exist, J must be contradictory.

Since J is contradictory, it follows that a finite subset of J must be contradictory. Hence, there must be a finite number of elements of H, to be called $Y_1, \dots, Y_k$ such that

$$K \Vdash \neg [X \wedge \neg Y_1 \wedge \dots \wedge \neg Y_k]$$

i. e. such that

3.3 $$K \Vdash X \supset [Y_1 \vee \dots \vee Y_k].$$

Now H has been shown to be disjunctive so there exists a sentence $Y \in H$ such that $[Y_1 \vee Y_2 \vee \dots \vee Y_k] \equiv Y$ is true universally. Hence, by 3.3,

3.4 $$K \Vdash X \supset Y$$

while

3.5 $$K \Vdash Y \supset X$$

by the definition of H. Combining 3.4 end 3.5 we see that

$$K \Vdash X \equiv Y.$$

This completes the proof of 3.1.

Theorem 3.1 belongs to a group of problems which call for the establishment of a connection between the syntactical properties of a sentence or set of sentences on one hand, the set-theoretic properties of its system of models on the other hand.

Two structures, M and M', are said to be *elementarily equivalent* — with respect to a given set of relation and individual constant symbols — if every sentence which is formulated in terms of these symbols and which hold in M holds also M' and vice versa. In the next two sections we shall consider several examples of the following problem. Let M and M' be two structures such that M' is an extension of M. Does there exist an existential sentence X such that X holds in M' although $\neg X$ holds in M? We observe that this problem is equivalent to the following. Does there exist an extension M'' of M' which is elementarily equivalent to M (with respect to set of relation and individual constant symbols for all relations and individuals of M)? If such a structure M'' exists then, clearly, every existential sentence which holds in M' must hold in M'' and hence also in M. Conversely, suppose that every existential sentence which is formulated by means of the above mentioned set of symbols and which holds in any extension M' of a structure M, holds already in M. Let K be the set of all sentences which are true in M. Then K contains the diagram D of M. Let M' a particular extension of M with diagram D', $D' \supset D$. In order to see that there exists an extension M'' of M' which is elementarily equivalent to M we only have to verify that $K \cup D'$ is consistent; for any model of $K \cup D'$ is then isomorphic to a model of K which is an extension of M'.

Now, if $K \cup D'$ were contradictory then there would exist sentences $Y_1, Y_2, \dots, Y_n$ in D' such that $K \cup \{Y_1, \dots, Y_n\}$ is contradictory. Put $Z(a_1, \dots, a_m) = Y_1 \wedge \dots \wedge Y_n$, where $a_1, \dots, a_m$ are just those individual constant symbols of $Y_1 \wedge \dots \wedge Y_n$ which do not occur in K. Then $K \cup \{Z(a_1, \dots, a_m)]$ is contradictory. A simple argument involving 2.3 then shows that $K \cup \{(\exists x_1) \dots (\exists x_m) Z(x_1, \dots, x_m)\}$ is contradictory. But the sentence $W = (\exists x_1) \dots (\exists x_n) Z(x_1, \dots, x_n)$ holds in M' and so W must hold also in M. It follows that W actually belongs to K, while K, being a set of sentences which hold in a given structure, must be consistent. We conclude that $K \cup D'$ is consistent and hence, possesses a model M, as required.

Given a consistent set of sentences K, suppose that any two models of K are elementarily equivalent with respect to the individual constant and relation symbols which occur in K. This will be the case if and only if for any sentence X which involves only relation and individual constant symbols which occur in K, either X is a consequence of K or $\neg X$ is a consequence of K. If this condition is satisfied then K is called *complete*. We shall also say that a *concept* is complete if some set of axioms (and hence, any set of axioms) which characterizes this concept is complete. It has been shown by Tarski (ref. 18), as a by-product of his decision procedure for the elementary theory of real numbers, that the concept of a real-closed field is complete. That is to say that if K is a set of sentences which consists of axioms for the notion of an ordered field, formulated in terms of the relations of equality, addition, and multiplication, as above, and in terms of a relation $Q(x, y)$ for order (read $x < y$), together with sentences which state that every positive number has a square root and every polynomial of odd degree has a root — then K is complete. Many other important algebraic concepts have been shown to be complete, e. g. the concept of an algebraically closed field of specified characteristic (see ref. 14). More recent works by Ax and Kochen (ref. 3) and by Yershov (ref. 19) on p-adic fields deserves particular mention in this connection.

We have described only a few topics of the Theory of Models, which developed over the last twenty years and has by now become a very extensive subject. The reader may consult ref. 15 and 20 for further information. While relying on the finiteness principle of section 2, we have not indicated the mode of the construction of the structure or structures whose existence is asserted there. In this connection, a special method of construction known as the method of reduced direct products has gained prominence in recent years (see refs. 6 and 7).

4. Field extensions.

Let M and M' be commutative fields, M' an extension of M. We assume as before that M and M' are specified in terms of the relations E, S, and P (equality, addition, and multiplication). Let A be a set of individual constant symbols which denote individuals of M (by means of given one-to-one correspondence with the individuals of M). Let K be the set of existential sentences which are formulated in terms of E, S and P and in terms of symbols of A. We shall concern ourselves with the question whether any sentence which belongs to K and which holds in M' must hold also in M. In some cases, the answer to this question is trivially negative, e.g. if M is the field of rational numbers an M' is obtained from M by the adjunction of an irrational algebraic number. In other cases, the answer is trivially positive in the sense that if follows directly from known results. Such is the case if M is an algebraically closed field. A situation which has some interest of its own is disposed of by the following theorem.

4.1. Theorem. *With M and K as detailed above, let $M' = M(t_1, \dots, t_n)$ $n \geq 1$, where the set $\{t_1, \dots, t_n\}$ is algebraically independent over M. Then every sentence $X \in K$ which holds in M' will hold already in M.*

Proof. Let

$$X = (\exists\, y_1) \dots (\exists\, y_m)\; Q(y_1, \dots, y_m)$$

where Q does not contain any further quantifier. For the purpose of the present proof we shall call a sentence Y *equivalent* to X if $X \equiv Y$ is true in all fields which are extensions of M.

We first transform X into a equivalent sentence $Y = (\exists y_1) \dots \dots (\exists y_m) R(y_1, \dots, y_m)$ in which R is free of quantifiers and in a conjunction of disjunctions of atomic sentences and (or) of negations of such sentences. Thus, R is in one of the boolean normal forms and can be determined from Q by a series of applications of the rules of the propositional calculus. Next, we eliminate the negations in the following way. We replace the formulae of the form $\neg S(\alpha, \beta, \gamma)$ — where α, β, and γ are either variables or individual constant symbols — successively by $(\exists z)[S(\alpha, \beta, z) \wedge \neg E(z, \gamma)]$, z being a variable which did not occur previously; similarly, we replace formulae $\neg P(\alpha, \beta, \gamma)$ by $(\exists w)[P(\alpha, \beta, w) \wedge \neg E(w, \gamma)]$, where w is any variable which did not occur previously; and we then replace formulae $\neg E(\alpha, \beta)$ by $(\exists x)(\exists y)[S(\alpha, x, \beta) \wedge \wedge P(x, y, 1)]$ where x and y did not occur previously and $\wedge$ has its usual denotation. Finally, we transform the result into prenex normal form by shifting the new quantifiers to the beginning of the sentence, in any order. It is not difficult to verify that the result is an existential sentence Z which is equivalent to Y. For example, it will be evident that $(\exists x)(\exists y)[S(\alpha, x, \beta) \wedge P(x, y, 1)]$ is true precisely if $x = \beta - \alpha$ is invertible, in other words if $\alpha \neq \beta$, i. e. if $\neg E(\alpha, \beta)$ is true. Once again, using the distributive laws or the propositional calculus, we may replace Z by an equivalent existential sentence of the form

$$W = (\exists y_1) \dots (\exists y_k) N(y_1, \dots, y_k)$$

where N is a disjunction of conjunctions of atomic formulae. According to our assumption, there are k individuals in M', to be denoted by $b_1, \dots, b_k$, such that $N(b_1, \dots, b_k)$ holds in M'. But if so, then one of the conjunctions which constitute $N(b_1, \dots, b_k)$, say $N'(b_{i_1}, \dots, b_{i_l})$, hold in M'. Each one of the b_1 denotes a rational function with coefficients in M, $\dfrac{p_i(t_1, \dots, t_n)}{q_i(t_1, \dots, t_n)}$. Now N' is a conjunction of atomic formulae of one of the forms $E(\alpha, \beta)$, $S(\alpha, \beta, \gamma)$, $P(\alpha, \beta, \gamma)$, where α, β, γ denote either individuals of M or rational functions of the type just indicated. It follows that we shall obtain formulae *true in* M if we substitute for $t_1, \dots, t_n$ arbitrary individuals $c_1, \dots, c_n$ of M, such that $q_i(c_1, \dots, c_n) \neq 0$, $i = i_1, \dots, i_l$. Since M is infinite and the number of q_i is finite it is an elementary fact of field theory that there are (many) n tuples $c_1, \dots, c_n$ which satisfy this condition. We conclude that W holds in M, and the same therefore applies to the sentence X, which is equivalent to W. This completes the proof 4.1.

It will be seen that the above result, as well as the arguments involved in its proof, are only a few steps removed from common Algebra. Although our next result will be a little less simple, both the problems and the methods described here may in fact be regarded as part of a natural development of Algebra, the « Metamathematics of Algebra ».

5. Extension of ordered fields.

An ordered field may be given in terms of the relations E, S, and P, as before, and in terms of an additional relation $Q(x, y)$ (for $x < y$). Let M and M' be commutative ordered fields, M' an extension of M, and let A be a set of individuals constant symbols denoting the individuals of M. Finally, let K be the set of existential sentences formulated in terms of E, S, P, and the symbols of A. As before, we ask whether every sentence of K which holds in M', holds also in M. An affirmative answer can be obtained immediately from known results if M is real closed.

The following theorem is analogous to 4.1.

5.1. THEOREM. *Let M be an archimedean ordered field. Let the field M' be given by an ordering of $M(t_1, \dots, t_n)$ which is an extension of the ordering of M, where the set $\{t_1, \dots, t_n\}$ is algebraically independent over M, $n \geq 1$. Then every sentence $X \in K$ that holds in M', will hold already in M.*

PROOF. Let M_0 be the real closure of M and let M_0' be the real closure of M'. Then $M \subset M_0 \subset M_0'$ and $M \subset M' \subset M_0'$ and M_0' still satisfies X. We interpolate between M_0 and M_0' a finite sequence of ordered fields $M_{0,i}$, $i = 1, \dots, n$, where $M_{0,0} = M_0$, and $M_{0,i+1}$ is the real closure of $M_{0.i}(t_{i+1})$, $i = 0, \dots, n-1$. Then $M_{0,n} = M_0'$. Now X holds in $M_{0,n}$ since $M_{0.n}$ includes M', and if X holds in $M_{0,i}$ then it holds also in every $M_{0,j}$ for $i < j \leq n$. If follows that there exists a smallest natusal number $k \leq n$ such that X holds in $M_{0,k}$. We propose to show that $k = 0$. Suppose on the contrary that $k > 0$ and put $M_1 = M_{0,k-1}$. Then X does not hold in M_1. Let D_1 be the diagram of M_1, using for the individual of $M \subset M_1$ the symbols of

A with their previous interpretation. Let K_0 be a set of axioms, in the Lower Predicate Calculus, for the notation of a real closed ordered field (compare the end of Section 3 above). Then every model M_2 of $K_0 \cup D_1$ is a real closed ordered field which is isomosphic to an extension of M_1. Now let $B = \{b_\nu\}$ be the set of individual constant symbols which occur in D_1, b_ν denoting b'_ν in M_1 and let c be an individual constant symbol which does not occur in D_1. Let H be the set of sentences $Q(c, b_\nu)$ or $Q(b_\nu, c)$ where we include the former if $t_k < b'_\nu$, and the latter if $b_\nu < t_k$, in $M_1(t_k)$. If we assign to c the interpretation t_k, $M_1(t_k)$ then becomes a model of $K_0 \cup D_1 \cup H$. On the other hand, every model M_2 of $K_0 \cup D_1 \cup H$ contains a field isomorphic to M_1, whose individuals are denoted by the corresponding symbols of D_1. For our present purposes, we may actually identify that field with M_1, so that M_2 becomes an extension of M_1. Let t be the individual of M_2 which is denoted by c, then $t < b'_\nu$ or $b'_\nu < t$ in M_2 according as $t_k < b'_\nu$ or $b'_\nu < t_k$ in M_1. Consider the field $M_1(t)$, which is a subfield of M_2. Every element p of $M_1(t)$ is a rational function of t with coefficients in M_1. Since M_1 is real closed, we may write p as a ratio of products,

5.2 $$p = a \frac{\Pi[(t - \alpha_i)^2 + \beta_i^2]\,\Pi(t - \gamma_i)}{\Pi[(t - \alpha'_i)^2 + \beta_i'^2]\,\Pi(t - \gamma'_i)}$$

where $a, \alpha_i, \beta_i, \gamma_i, \alpha'_i, \beta'_i, \gamma'_i$ belong to M_1 and where one or more of the products on the right hand side may be empty. At any rate, the square factors are by necessity positive while the sign of the linear factors depend on whether $t < \gamma_i$ or $t > \gamma_i$. The sign and number of these factors together with the sign of a determine the sign of p. It follows that $M_1(t)$ as an *ordered field* is isomorphic to $M_1(t_x) \subset M_{0,k}$ under a correspondence which maps p on

5.3 $$q = a \frac{\Pi[(t_k - \alpha_i)^2 + \beta_i^2]\,\Pi(t_k - \gamma_i)}{\Pi[t_k - \alpha'_i)^2 + \beta_i'^2]\,\Pi(t_k - \gamma'_i)}\,.$$

Moreover, if $\overline{M_1(t)}$ is the real closure of $M_1(t)$ within M_2, we may then conclude that $\overline{M_1(t)}$ is isomorphic as an ordered field to $M_{0,k}$. This skows that X holds in $\overline{M_1(t)}$ and, hence, in M_2. But if so, then X must be a consequence of $K_0 \cup D_1 \cup H$, and hence must be a consequence of $K_0 \cup D_1 \cup H_1$ for some finite subset of H_1 of H. Now M_1 is certainly a model of $K_0 \cup D_1$. We claim that it is also a model of H_1 for an appropriate interpretation of the symbol c.

Indeed, H_1 consists of a finite number of sentences of the form $Q(c, b_\nu)$ or $Q(b_\nu, c)$ where b_ν denotes an element of M_1. If b_1' is such an element of M_1 for which the corresponding b_ν satisfies $Q(c, b_\nu)$, and b_2' is an element of M_1 for which the b_ν which denotes it satisfies $Q(b_\nu, c)$ then $t < b_1'$ and $b_2' < t$ in $M_1(t)$ and so $b_2' < b_1'$. Since the ordering of M_1 is dense, it follows that we can find already in M_1 an element c' such that $c' < b_\nu'$ or $c' > b_\nu'$ according as $Q(c, b_\nu) \in H_1$ or $Q(b_\nu, c) \in H_1$ where b_ν denotes b_ν'. Thus, M_1 is a model of H_1 provided we take c to denote c'. We conclude that X, which is a consequence of $K_0 \cup D_1 \cup H_1$ holds in $M_1 = M_{0,k-1}$, a contradition which establishes our claim that $k = 0$.

So far, we have shown that X holds in the real closure of M, M_0. To continue, we replace X successively by equivalent sentences of simpler form (compare Section 4) where Y is now *equivalent* to X if $X \equiv Y$ holds in all *ordered* fields which are extensions of M. Thus, we first replace X by a sentence $(\exists y_1) \dots (\exists y_m) R(y_1, \dots, y_m)$ in which R is a conjunction of disjunctions of atomic sentences and (or) of negations of such sentences. We then eliminate negations $S(\alpha, \beta, \gamma)$ and $P(\alpha, \beta, \gamma)$ as in Section 4, and we eliminate negations of the form $E(\alpha, \beta)$, $Q(\alpha, \beta)$ by $Q(\alpha, \beta) \vee Q(\beta, \alpha)$ and $Q(\beta, \alpha) \vee E(\alpha, \beta)$, respectively. Shifting quantifiers and using the distributive laws, if necessary, we again arrive at a sentence

5.4 $$W = (\exists y_1) \dots (\exists y_k) N(y_1, \dots, y_k)$$

which is equivalent to X, where N is a disjunction of conjunctions of atomic formulae. And, observing that a sentence $N(b_1, \dots, b_k)$ can hold in M' only if one of the conjunctions which make up N (when taken in disjunction) holds in M', we see that we may assume immediately that $N(y_1, \dots, y_k)$ is a conjunction of atomic formulae. Our previous argument shows that W holds not only in M' but also in the real closure of M, M_0.

Translated into ordinary algebraic language, 5.4 asserts the existence of a solution $y_1, \dots, y_k$ for a set of equations and inequalities of one of the forms

5.5 $$\alpha_i + \beta_i = \gamma_i$$

$$\alpha_i \beta_i = \gamma_i$$

$$\alpha_i = \beta_i$$

5.6 $$\alpha_i < \beta_i$$

where the α_i, β_i, γ_i stand either for elements of M or for the unknowns y_i. We may replace 5.5 by a single polynomial equation by considering instead

5.7 $$\Sigma (\alpha_i + \beta_i - \gamma_i)^2 + \Sigma (\alpha_i \beta_i - \gamma_i)^2 + \Sigma (\alpha_i - \beta_i)^2 = 0.$$

Thus our problem can be reduced to the following.

Given a polynomial equation

5.8 $$p(y_1, \quad , y_k) = 0$$

with coefficients in an archimedean ordered field M, show that 5.8 has a solution in M, subject to the additional conditions 5.6. if it is known that such a solution exists in an ordering M' of $M(t_1, \dots, t_n)$.

A solution in M' such as exists by assumption may be taken to be of the form

5.9 $$y_j = \frac{p_j(t_1, \dots, t_n)}{q_j(t_1, \dots, t_n)}, \quad q_j(t_1, \dots, t_n) > 0, \quad j = 1, \dots, k$$

where the p_j and q_j are polynomials with coefficients in M (which may reduce to constants). Since the validity of 5.8. is preserved on substituting for $t_1, \dots, t_n$ elements $z_1, \dots, z_n$ of M which make $q_j(z_1, \dots, z_n)$ positive, $j = 1, \dots, n$, it will be sufficient to show that there exist $z_1, \dots, z_n \in M$ which satisfy this condition as well as certain additional conditions arising from 5.6. For example, if an inequality which comes under the heading of 5.6. is of the form $y_l < y_m$ then the corresponding condition on $z_1, \dots, z_n$ is

$$p_m(z_1, \dots, z_n) q_l(z_1, \dots, z_n) - p_l(z_1, \dots, z_n) q_m(z_1, \dots, z_n) > 0.$$

Thus a set

$$y_j = \frac{p_j(z_1, \dots, z_n)}{q_j(z_1, \dots, z_n)}$$

will satisfy the conditions of our problem, provided the z_i satisfy a certain finite set of polynomial inequalities

5.10 $$P_j(z_1, \dots, z_n) > 0, \qquad j = 1, \dots, s$$

with coefficients in M. We have to show that such z_i exist in M, knowing that in M' a solution is provided by $z_i = t_i$, $i = 1, \dots, n$.

Now the assertion that such a solution exists, i. e., in a semi-formal, but easily understood notation,

5.11 $$(\exists z_1)\dots(\exists z_n)[P_1(z_1,\dots,z_n)>0 \wedge P_2(z_1,\dots,z_n)>0 \wedge\dots\wedge P_s(z_1,\dots,z_n)>0]$$

evidently can be expressed by a sentence $Z \in K$. Since Z holds in M' the first part of our proof shows that Z holds in the real closure of M, M_0. But M was supposed archimedean, so both M and M_0 may be regarded as subfields of the field of real numbers, R. Let S be the set of points $\langle z_1,\dots,z_n\rangle$ in n dimensional real Euclidean space R^n, which satisfy the inequalities 5.10. S is not empty since it includes points with coordinates in M_0. But S is open, so it must include also points with rational coordinates. This completes the proof of 5.1. since the rational numbers are included in M.

Observe that we have not assumed that M' is (necessarily) archimedean, otherwise the result would be trivial. On the other hand, the assumption that M is archimedean is essential, as shown by the following counter-example.

We put $M = F(r)$ where F is the field of rational numbers and r is transcendental over F, and we order M by the prescription that $r > a$ for all $a \in F$. We introduce at the same time an auxiliary field $M^* = F(x)$ where x is transcendental over F and M^* is ordered in the same way as M_1 i. e. by making $x > a$ for all $a \in F$.

Let $M' = M(t) = F(r, t)$ where t is transcendental over M. Then the elements of M' are rational functions of r and t with coefficients in M. In particular, let $p(r, t) \in M'$ be a non-zero polynomial of r and t (with coefficients in M). Define $p^*(x)$ by $\pi^*(x) = p(x^2, x)$. We define an ordering of M' in the following way. If $p^*(x) \neq 0$ then $p(r, t)$ shall be *positive* if $p^*(x) = p(x^2, x)$ is positive in M. Now suppose that $p^*(x) = 0$. Then the remainder theorem shows that $p(r, t)$ can be written in the form

$$p(r, t) = p_1(r, t)(r - t^2)^m$$

where m is a positive integer and $p_1(r, t)$ is a polynomial of r and t with coefficients in F such that $p_1(x^2, x) \equiv 0$. We now define that $p(r, t)$ is positive if $p_1(r, t)$ is positive according to the definition given previously. Finally, we define that a rational function of r and t which belongs to M' is positive if it can be written as a ratio of positive functions. Let P be the set of elements of M' which are positive according to these definitions. Then P is additive and multiplicative and does not contain the zero polynomial. Moreover, if a is any non-zero element of M' then either $a \in P$ or $-a \in P$, and any element of $M \subset M'$ belongs to P if and only if it is positive in M. This shows that P defines an ordering of M', in which P is precisely the set of positive elements, and this ordering is an extension of the ordering of M. Regarding M' from now on as an ordered field, we consider the pair of inequalities

5.12 $$r - 1 < y^2 < r.$$

Evidently, the assertion that there exists a y which satisfies 5.12 can be formulated as an existential sentence X which belongs to the set K defined at the beginning of this section. Then X holds in M' since $y = t \in M'$ satisfies

$$t^2 - (r - 1) > 0, \qquad r - t^2 > 0.$$

On the other hand, X does not hold in M. For if it did then there would exist polynomials of r, $p(r)$ and $q(r)$, with rational coefficients, such that

5.13 $$r - 1 < \left(\frac{p(r)}{q(r)}\right)^2 < r.$$

This implies

5.14 $$(p(r))^2 > (q(r))^2 (r - 1), \qquad (q(r))^2 r > (p(r))^2.$$

But the coefficients of the leading terms of $(p(r))^2$ and $(q(r))^2$ are positive. Hence, writing $\deg p(r)$ and $\deg q(r)$ for the degrees of $p(r)$ and $q(r)$, we conclude from 5.14. that

$$2 \deg p(r) \geq 2 \deg q(r) + 1, \qquad 2 \deg q(r) + 1 \geq 2 \deg p(r).$$

But then $\deg p(r) > \deg q(r)$ according to the first of these inequalities, and so

$$2 \deg p(r) > 2 \deg q(r) + 1 \geq 2 \deg p(r).$$

This contradiction shows that X does not hold in M. Thus, the assumption that M is archimedean is indeed essential.

Theorem 5.1. leads to an immediate proof of Artin's theorem on positive definite functions which solved Hilbert's seventeenth problem, (ref. 1). Thus, suppose that a certain polynomial $p(t_1, \ldots, t_n)$ with rational coefficients cannot be written as a sum of squares of rational functions with rational coefficients. A relatively simple argument of Artin and Schreier (ref. 2) then shows that there is an ordering of the field $M' = M(t_1, \ldots, t_n)$ in which $p(t_1, \ldots, t_n) \in M'$ is negative. Now the assertion « there exist $x_1, \ldots, x_n$ such that $p(x_1, \ldots, x_n) < 0$ » can be formulated without difficulty as an existential sentence X within the set K introduced at the beginning of this section. As we have just seen, X holds in M' in the specified ordering. We may therefore conclude by means of 5.1. that X holds also in M, i. e. that there exist rational numbers $a_1, \ldots, a_n$ such that $p(a_1, \ldots, a_n) < 0$. In other words, we have established Artin's result that a polynomial $p(x_1, \ldots, x_n)$ with rational coefficients which is non-negative for all rational arguments, can be written as a sum of squares of rational functions with rational coefficients.

It should be mentioned that 5.1. and the method used for proving it are related to, though different from, earlier metamathematical works on Hilbert's seventeenth problem (refs. 5, 8, 12, 13).

6. Non-standard Analysis.

We now turn from Algebra to Analysis. The subject with which we shall deal appears at first sight to be far removed from the algebraic applications considered previously. But even apart from the obvious fact that we shall still employ the tools from Logic introduced earlier, there is another interesting connection, which we shall discuss briefly in the closing sentences of this paper.

The historical antecedents of the subject with which we shall be concerned from now on go back to the beginnings of the infinitesimal Calculus and beyond. The time-honored and, until recently unsuccessful, idea which we shall realize here is that of developing mathematical Analysis by embedding the domain of real numbers R in a proper extension $^*R$ which possesses *the same properties as* R. Such an extension $^*R$ of R must be non archimedean and this is the fact which enables us to find in $^*R$ the infinitely small (infinitesimal) and infinitely large elements whose existence was taken for granted in the early stages of development of the Calculus. But is it at all permissible to assume that such a system $^*R$ exists? The answer to this question would appear to be negative in view of the fact that there are systems of axioms for the real numbers which are *categorical* i. e. which determine a unique mathematical structure, up to isomorphism. Though not formulated explicitly at the time it was this sort of difficulty which embarrassed and, finally, discredited, the theory of infinitely small quantities as a tool in the development of Analysis. However, the difficulty vanishes as soon as we confine the notion « the same properties » to a specified formal language, with the appropriate interpretation in R and $^*R$. It turns out that there are several possibilities of this kind. Thus, let K be the set of all sentences which hold in R and which are formulated in the Lower Predicate Calculus in terms of individual constant and relation symbols for all individuals of R (i. e. the real numbers) and for all n-ary relations which exist in R, $n = 1, 2, \ldots$. Among the latter, there are the relations of equality, addition, and multiplication introduced earlier and denoted by $E(x, y)$, $S(x, y, z)$, and $P(x, y, z)$; but also a relation, to be denoted by $N(x)$, which characterizes the set of natural numbers, i. e. such that $N(a)$ holds in R if and only if a denotes a natural number; a binary relation, to be denoted by $F(x, y)$, say, which expresses the functional connection provided by any given function of a single variable, $y = f(x)$; and corresponding relations for functions of several variables.

We claim that there exists a proper extension $^*R$ of R which is a model of K. In order to see this, we introduce an individual constant symbol, b, which does not occur in K, and we define a set of sentences H by $H = \{\urcorner E(a_\nu, b)\}$ where a_ν varies over all (symbols for) real numbers. If K' and H' are finite subsets of K and H respectively then R becomes a model of $K' \cup H'$ if we interpret b in R by any number which is different from all numbers denoted by symbols a_ν which occur in H' or in K'. It follows that $K \cup H$ is consistent, by 2.1., and posses a model $^*R$. $^*R$ may be chosen as an extension of R since K contains the diagram of R; and $^*R$ is an ordered commutative field since the axioms for a commutative field are contained in K. Moreover, $^*R$ is a proper extension of R for since $^*R$ is a model of H, the individual of $^*R$ which is denoted by b must be different from all number of R. It follows that $^*R$ is non-archimedean. We call $^*R$ a *non-standard model of Analysis.*

Let A be any set of individuals of R (i.e., less formally, any any subset of R, any set of real numbers). Then A is characterized by a singulary relation symbol $S(x)$ in K, i.e. $S(a) \in K$ if and only if a denotes an element of R. It is not difficult to see that the set of individuals of $^*R$ which satisfy $S(x)$ (i.e. are denoted by a symbol b such that $S(b)$ holds in $^*R$) is an extension $^*A$ of A. Similarly for any n-ary relation A in R there is a corresponding extension $^*A$ in $^*R$, i.e., the relation which is denoted by the same relation symbol of K. And the properties of $^*A$ are just the same as the properties of A to the extent to which these can be expressed by sentences of K. In particular, to the set of natural numbers N in R there corresponds an extension $^*N$ in $^*R$ *which has the same properties* as N in the sense just indicated. It is easy to see that $^*N$ must be a proper extension of N (a so-called *non-standard model of Arithmetic*). For it is a property of R which can be expressed within K that « to every real number r there is a natural number n which is greater than r ». It follows that the statement in quotation marks holds also when reinterpreted in $^*R$, there exists an n in $^*N$ which is greater than r. But $^*R$ is non-archimedean and, accordingly, contains an element greater than all elements of N in the specified ordering of $^*R$. We conclude that it contains also an element of $^*N$ greater than all numbers of N.

From now on we shall extend the name (*real or natural*) *numbers* to the individuals of $^*R$ and $^*N$, reserving the qualification *standard* to the elements of R and N. The numbers of N will also be called *finite* while those of $^*N - N$ will be termed *infinite.* An infinite natural number must be greater than any finite natural number. There is no smallest infinite natural number, for if n is infinite, hence different from 0, it must have a predecessor, $n - 1$, which is then also infinite.

A real number r is called *finite* if $|r|$ is smaller than some standard real number (where we may define $|r|$ by max $(r_1 - r)$). If r is not finite it is *infinite.* These definitions are consistent with the terminology just introduced for natural numbers. A real number r is *infinitesimal* or *infinitely small* if $|r|$ is smaller than all standard positive real numbers. In particular, 0 is regarded as infinitesimal. For any finite real number r there exists a unique standard real number ^0r, called the *standard part* of r such that $r - {^0r}$ is infinitesimal. Given any standard real number r, the set of real numbers for which r is the standard part is called the *monad of* r, $\mu(r)$. The monad of 0 consists of all infinitesimal numbers.

Let $\{t_n\}$ be any infinite sequence of real numbers in R in the standard sense. $\{t_n\}$ is represented in the language of K be a binary relation $T(x, y)$, which stands for $y = t_x$. The same relation gives rise in $^*R$ to an infinite sequence in the sense of $^*R$, i. e. to a mapping from the natural numbers of $^*N$ into the real numbers of $^*R$. This is an extension of the original sequence. We shall find it convenient to denote it by the same symbol, $\{t_n\}$, so that the range of the subscript will become apparent only from the context.

6.1. THEOREM. *Let $\{t_n\}$ be an infinite sequence in R (a standard sequence) and let t be a standard real number. In order that* $\lim_{n \to \infty} t_n = t$ *in the classical sense, it is necessary and sufficient that $t_w - t$ be infinitesimal for all infinite natural numbers w.*

PROOF. Suppose that $\lim_{n \to \infty} t_n = t$ in the classical sense, and let w be any infinite natural number. Let r be standard real and positive. In order to prove that the condition of the theorem is necessary, we only have to establish that $|t_w - t| < r$. Now it follows from the classical definition of a limit that there exists a finite natural number n_0 such that « for all natural numbers n greater than n_0, $|t_n - t| < r$. » But the statement in quotation marks can be formulated as a sentence of K and accordingly holds also in $^*R$. Moreover, w, which is an infinite natural number, must be greater than n_0. Accordingly, $|t_w - t| < r$, which proves that the condition of the theorem is necessary.

The condition is also sufficient. For suppose that it is satisfied. Then we have to prove that $\lim_{n \to \infty} t_n = t$ in the classical sense. Now the following statement in quotation marks is true in $^*R$ for any given standard positive ε.

« There exists a natural number n_0 such that for any natural number $n > n_0$, $|t_n - t| < \varepsilon$ ».

This statement can be expressed by a sentence X in the Lower Predicate Calculus using only symbols of K. In that sentence, n_0 does not appear explicitly but corresponds to a variable, quantified existentially. To see that the statement holds in $^*R$ we only have to choose for n_0 any infinite natural number. Then any m greater than n_0 is also infinite and so $|t_n - t|$ is even infinitesimal, and certainly smaller than ε.

Now if X dit not hold in R then $\neg X$ would hold in R and hence, would belong to K and hold in $^*R$. But X holds in $^*R$, and so we may conclude that it holds also in R. This shows that there existe *in* R a natural number n_0 such that $|t_n - t| < \varepsilon$ for $n > n_0$, and proves $\lim_{n \to \infty} t_n = t$, the condition of the theorem is sufficient.

Following these lines, the classical Differential and Integral Calculus can be supplement with, and to a large extent superseded by, arguments in terms of a language of infinitesimals as called for by Leibniz and his disciples. However much of Analysis is expressed in the setting of a language of higher order or, alternatively, within the framework of axiomatic set theory, and involves the introduction of higher entities such as sets of relations, relations between relations, etc., and quantification with respect to sets, relations, sets of relations, etc. But even if K is the set of sentences which hold in R and which are formulated in terms of higher order language, it is still true that there exists a proper extension $^*R$ of R which satisfies all sentences of K; provided we agree to a special interpretation of the sentences of K in $^*R$, as follows. Instead of supposing (as would be natural) that statement concerning sets, relations, etc., refer to the totality of such entities in $^*R$, we interpret the sentences of K on the assuption that they refer only to certain subclasses of these entities called *internal*. Thus, if a sentence $X \in K$ affirms « there exists a set ... » or « for all relations ... » we interpret these phrases as « there exists an *internal* set ... », « for all internal *relations* ... ». The argument which leads to the conclusion that such a structure $^*R$ exists is not constructive and does not enable us to decide directly whether a particular entity is internal or not. However, in many cases we may answer such questions precisely by making use of the fact that $^*R$ is a model of K. For example, let $^*N$ be the set of natural numbers, finite or infinite, in $^*R$, as before. Then every non-empty *internal* set of natural numbers (bubset of $^*N$) must posses a first element; for this is a property of sets of natural numbers which can be expressed within K. But the set $^*N - N$, i.e. the set of infinite natural numbers does not include a smallest element, as shown previously. It follows that $^*N - N$ is not internal.

Just as there are, in $^*R$, internal sets of relations, etc., so we may talk of internal functions of higher order (e.g. set functions) i.e. functions which are expressed by internal relations.

The reader may consult ref. 10 or ref. 15 for more detailed expositions of Non-standard Analysis within the framework of the Lower Predicate Calculus. The higher order theory is developed in detail in ref. 16. In that book we have shown also how the theory can be generalized so as to apply not only to the real numbers, and to metric structures closely related to the real numbers, but also to any other kind of mathematical structure, e.g. to topological spaces.

In the next section we shall apply these ideas to certain topics in the Geometry of Numbers.

7. Applications to the Geometry of Numbers.

We shall be concerned with sets of points in n-dimensional real Euclidean space R^n. Statements about R^n can be expressed in terms of n-tuples of real numbers. Having chosen a non-standard model of Analysis, $*R$, we may consider the corresponding n-dimensional space $(*R)^n$. To every set of points A in R^n there corresponds a set $*A$ of elements of $(*R)^n$ and these also will from now on be called *points*. For the applications which we have in mind, we shall require a non-standard model $*R$ which satisfies all sentences that are true in R and are formulated within the framework of a higher order language as sketched at the end of the preceding section. Let $x = (x_1, \dots, x_n)$ be a point in $(*R)^n$ then we call x *finite* if the norm of x, $\| x \| = \sqrt{\{x_1^2 + \dots + x_n^2\}}$ is finite in the sense of section 6. If x is not finite then it is called *infinite*. A point is called *standard* if it belong to R^n or which is the same, if all its coordinates are standard. For any finite point x there exists a unique standard point ${}^0x$ such that $\|x - {}^0x\|$ is infinitesimal. ${}^0x$ is called the standard part of x. The coordinates of ${}^0x$ are just the standard parts of the coordinates of $x = (x_1, \dots, x_n)$, ${}^0x = ({}^0x_1, \dots, {}^0x_n)$.

Let B be any set of points in $(*R)^n$. Then the *standard part* ${}^0B$ of B is defined as the subset of R^n consisting of all points that are standard parts of points of B.

An (n-dimensional) lattice Λ is defined as the totality of points $\lambda_1 a_1 + \dots + \lambda_n a_n$ where $\lambda_1, \dots, \lambda_n$ vary over the integers, while $\{a_1, \dots, a_n\}$ is a set of linearly independent points in R^n, i.e. such that the determinant $|a_{ik}|$ is different from 0 for $a_i = (a_{i1}, \dots, a_{in})$, $i = 1, \dots, n$. $(a_1, \dots, a_n)$ is a *basis* of Λ. Let L_n be the set of n-dimensional lattices. Corresponding to L_n there exists an internal set $*L_n$ of internal sets of points in $({}^*R)^n$-still to be called lattices. Any $\Lambda \in {}^*L_n$ consists of the totality of points $\lambda_1 a_1 + \dots + \lambda_n a_n$, $\lambda_i \in {}^*N$, $i = 1, \dots, n$ for a given set of points $a_i = (a_{i1}, \dots, a_{in})$, $i = 1, \dots, n$, with non-vanishing determinant $|a_{ik}|$. For such a Λ, consider its standard part ${}^0\Lambda$. ${}^0\Lambda$ is not empty since it contains the origin $0 = (0, \dots, 0)$. Moreover, ${}^0\Lambda$ constitutes a group with respect to addition. For if $a \in {}^0\Lambda$, $b \in {}^0\Lambda$, then there exist points $a_1 \in \Lambda$, $b_1 \in \Lambda$ such that $a = {}^0a_1$, $b = {}^0b_1$. Then $a + b = {}^0(a_1 + b_1)$ and since $a_1 + b_1$ belong to Λ, we may conclude that $a + b$ belong to ${}^0\Lambda$.

We shall now study the question under what conditions ${}^0\Lambda$ is a lattice in R^n. For any lattice Λ we denote by $d(\Lambda)$ the *determinant* of Λ, i.e. the smallest positive value of the determinants $|a_{ik}| = \det(a_1, \dots, a_n)$ for $a_1, \dots, a_n \in \Lambda$. ($d(\Lambda)$ is the absolute value of the determinant of any basis of Λ). Moreover, we write

$$|\Lambda| = \inf_{a \in \Lambda - \{0\}} \|a\|$$

so that $|\Lambda|$ is the greatest lower bound of the norms of non-zero points of Λ.

7.1. THEOREM. *In order that ${}^0\Lambda$ be a lattice it is necessary and sufficient that $d(\Lambda)$ is infinite and $|\Lambda|$ is not infinitesimal.*

PROOF. The conditions are necessary. For suppose ${}^0\Lambda$ is a lattice. Let $b_i = (b_{i1}, \dots, b_{in})$, $i = 1, \dots, n$, be points of ${}^0\Lambda$ such that $d({}^0\Lambda) = |b_{ik}| \neq 0$. By assumption, there exist points of Λ, $a_i = (a_{i1}, \dots, a_{in})$, such that $b_i = ({}^0a_{i1}, \dots, {}^0a_{in})$. Then ${}^0|a_{ik}| = |{}^0a_{ik}| = |b_{ik}|$ and so $|a_{ik}| \neq 0$. It follows that $d(\Lambda)$ cannot exceed the absolute value of $|a_{ik}|$. But this must be finite, again because of the relation $|{}^0a_{ik}| = |b_{ik}|$.

Now suppose $|\Lambda|$ is infinitesimal. Then $\|a\|$ is infinitesimal for some $a \in \Lambda$, $a \neq 0$. We claim that in this case ${}^0\Lambda$ contains an entire straight line through the origin, so that it cannot be a lattice. Consider the point $\alpha = (1/\|a\|)\, a$. Clearly $\|\alpha\| = 1$, so α possesses a standard part, ${}^0\alpha$, and $\|{}^0\alpha\| = 1$. We may then prove that the entire straight line $x = {}^0\alpha t$ belongs to ${}^0\Lambda$. Take any standard t, then we have to show that the distance of the point $x = {}^0\alpha t$ from some point of Λ is ifinitely small. Let $n = [(1/\|a\|)\, t]$, i.e. n is the greatest integer not exceeding $(1/\|a\|)\, t]$. Put $x' = \alpha t$, $x'' = na$. Then

$x'' \in \Lambda$ and

$$\| x - x'' \| \leq \| x - x' \| + \| x' - x'' \| = \| ({}^0\alpha - \alpha)\, t \| + \left\| a \left(\frac{t}{\|a\|} - n \right) \right\| \leq \| {}^0\alpha - \alpha \| \, |t| + \| a \| .$$

But $\| {}^0\alpha - \alpha \|$ is infinitesimal by the definition of the standard part and $\| a \|$ is infinitesimal by assumption. It follows that $\| x - x'' \|$ is infinitesimal, x is the standard part of x'', and so $x \in {}^0\Lambda$. But a lattice is discrete and cannot contain a straight line. This proves the first half of the theorem.

To show that the condition is also sufficient, we apply Minkowski's convex body theorem. This shows that, for any standard $\tau > 0$ we may choose a standard $\sigma > 0$ such that the box

$$| x_i | \leq \sigma \, | \Lambda |, \quad | x_j | \leq \tau \, | \Lambda |, \quad j \neq i$$

contains at least one lattice point a_i of Λ other than 0. For the volume of the base is $\tau^{n-1} \sigma \, | \Lambda |^{\iota}$ and this can be made greater than $d(\Lambda)$ by an appropriate choice of the ***standard*** σ since $| \Lambda |$ is not infinitesimal. It follows that a_i is finite and if $a_i = (a_{i1}, \dots, a_{in})$ then $| a_{ii} | \geq | \Lambda |$, $| a_{ij} | \leq \tau \, | \Lambda |$ for $i \neq j$. Hence ${}^0a_i = ({}^0a_{i1}, \dots, {}^0a_{in})$ is finite and different from 0. By choosing τ sufficiently small, we may then make sure that the determinant $\det(a_1, \dots, a_n)$ differs by as small a standard number as we please from the product $\Pi_{i=1}^{n} a_{ii}$, and hence, that it is not infinitesimal. We conclude that $\det({}^0a_1, \dots$ $\dots, {}^0a_n) = {}^0(\det(a_1, \dots, a_n)) \neq 0$ so that ${}^0\Lambda$ includes n linearly independent points.

Moreover, ${}^0\Lambda$ must be discrete, for suppose that it contains a point a such that $\| a \| < \frac{1}{2} \, {}^0| \Lambda |$. Then $a = {}^0b$ for some $b \in \Lambda$ and $\| b \| < | \Lambda |$. This contradicts the definition of $| \Lambda |$ and completes the proof of the theorem.

Now suppose that the conditions of 7.1. are satisfied so that ${}^0\Lambda$ is a lattice. Let $\{a_1, \dots, a_n\}$ be a basis for ${}^0\Lambda$, and let $b_1, \dots, b_n$ be the uniquely determined points of Λ such that $a_i = {}^0b_i$, $i = 1, \dots, n$ Then

7.2. THEOREM. *$\{b_1, \dots, b_n\}$ is a basis for Λ.*

PROOF. Since $\det(a_1, \dots, a_n)$ is a standard number different from 0 it follows that $\det(b_1, \dots, b_n) \neq 0$. Thus the points $b_1, \dots, b_n$

are linearly independent. It follows that if b is any point in $({}^*R)^n$ then there are real numbers $\lambda_1, \ldots, \lambda_n$ in ${}^*R$ such that $b = \lambda_1 b_1 + \ldots + \lambda_n b_n$. Now suppose $b \in \Lambda$ and let $\mu_i = [\lambda_i]$, the greatest integer smaller than λ_i, $i = 1, \ldots, n$. Then $0 \leq \lambda_i - \mu_i < 1$. We have to show that $\lambda_i - \mu_i = 0$. Consider

$$b' = {}^0(\lambda_1 - \mu_1)\,{}^0b_1 + \ldots + {}^0(\lambda_n - \mu_n)\,{}^0b_n$$

$$= {}^0((\lambda_1 - \mu_1)\,b_1 + \ldots + (\lambda_n - \mu_n)\,b_n) = {}^0(b - \mu_1 b_1 - \ldots - \mu_n b_n).$$

Since the expression in parentheses on the right hand side belongs to Λ, b' is in ${}^0\Lambda$. But then ${}^0(\lambda_i - \mu_i)$, $i = 1, \ldots,$ must be integers, i.e. either 0 or 1. Thus, either $\lambda_i - \mu_i$ or $\lambda_i - \mu_i - 1$ is infinitely small, and we put $\nu_i = \mu_i$ in the former case and $\nu_i = \mu_i + 1$ in the latter. Consider the point $c \in \Lambda$ which is given by

$$c = b - \nu_1 b_1 - \ldots - \nu_n b_n = (\lambda_1 - \nu_1)\,b_1 + \ldots + (\lambda_n - \nu_n)\,b_n\,.$$

The right hand side shows that $\| c \|$ is infinitesimal, since the numbers $| \lambda_i - \nu_i |$ are infinitesimal and the points b_i are finite. But $| \Lambda |$ is not infinitesimal, and so $c = 0$, $\lambda_i = \nu_i$, $i = 1, \ldots, n$. This shows that $\lambda_i = \mu_i + 1$ cannot occur and $\lambda_i = \mu_i$ throughout, proving the theorem.

Next we shall prove some classical theorems by means of the results obtained above. A lattice Λ is said to be ***admissible*** for a set of points S (in R^n) if $\Lambda \cap S$ contains at most the origin. The number $\Delta(S)$ is defined by

$$\Delta(S) = \inf \mathbf{d}(\Lambda) \quad (\Lambda \text{ is admissible for } S).$$

If there is no lattice admissible for S, we put $\Delta(S) = \infty$.

7.3. Theorem. *Let $\{S_r\}$, $r = 1, 2, 3, \ldots$, be a sequence of open sets of points in R^n, such that*

$$\{0\} \subset S_1 \subset S_2 \subset S_3 \subset \ldots$$

let $S = U_r S_r$. Then

$$\Delta(S) = \lim_{r \to \infty} \Delta(S_r)$$

Proof. Observe that the theorem refers to ordinary space. In order to prove it, we pass to $({}^*R)^n$. Then $S_r \subset ({}^*R)^n$ also for infinite r.

Suppose to begin with that there exist an admissible lattice for S so that $\Delta(S)$ is « finite » (i.e. not ∞). Then $\Delta(S_r)$ is a standard real number for finite r and a number of $^*R$ for infinite R. Moreover

$$\Delta(S_1) \leq (S_2) \leq \ldots \leq \Delta(S_r) \leq \ldots \leq \Delta(S)$$

and this is again true for both finite and infinite r. It follows from 6.1. that in order to prove the assertion of the theorem, we only have to show that $\Delta(S) - \Delta(S_r)$ is infinitesimal for all infinite r.

Let r be infinite and suppose that, contrary to the desired conclusion $^0(\Delta(S_r)) < \Delta(S)$. Then there exists a standard real number d such that $\Delta(S_r) < d < \Delta(S)$. It now follows from the definition of $\Delta(S_r)$ that there exists (in $(^*R)^n$) a lattice Λ which is admissible for S_r such that $d(\Lambda) \leq d$. Moreover, since S_1 is an open set there exists a standard number $\varrho > 0$ such that $a \in S_1$ for $\| a \| \leq \varrho$ and hence $a \in S_r$ for $\| a \| \leq \varrho$. Hence $a \notin \Lambda$ for $\| a \| \leq \varrho$, $a \neq 0$, and so $| \Lambda | \geq \varrho$. Theorem 7.1. now shows that $^0\Lambda$ is a lattice in R^n.

Let $\{a_1, \ldots, a_n\}$ be a basis for $^0\Lambda$, $a_i = {}^0b_i$ where $b_i \in \Lambda$, $i = 1, \ldots, n$. Then $\{b_1, \ldots, b_n\}$ is a basis for Λ, by 7.2. Also, $d(^0\Lambda) = |\det(a_1, \ldots, a_n)|$ differs only by an infinitely small number from $d(\Lambda) = |\det(b_1, \ldots, b_n|$, which is smaller than or equal to d. Hence $d(^0\Lambda) < \frac{1}{2}(d + \Delta(S))$. On the other hand, we are going to show that $^0\Lambda$ is admissible for S. By the definition of $\Delta(S)$, this implies $d(^0\Lambda) \geq \Delta(S)$, a contradiction which proves the theorem.

Let $a \in S$, $a \neq 0$. Then $a \in S_k$ for sufficiently high finite k. Since the S_k are open and $S_k \subset S_m$ for $k < m$, there exists a standard $\mu > 0$ such that all points x in the sphere $\| x - a \| \leq \mu$ belong to S_k for $k > k_0$, k_0 a finite integer. But then $\| x - a \| \leq \mu$ implies $x \in S_r$, also for all $x \in (^*R)^n$. Now if a belonged to $^0\Lambda$ then $a = {}^0b$ for some $b \in \Lambda$, which would then satisfy $\| b - a \| \leq \mu$ and hence would belong to S_r. This contradicts the fact that Λ is admissible for S_r and completes the proof of the theorem in this case.

There is no difficulty in disposing of the case $\Delta(S) = \infty$ in a similar way. In that case, we only have to show that the sequence $\Delta(S_r)$ cannot consist of real numbers which are uniformly bounded in R. The details are left to the reader.

Classically, 7.3. is preved by means of Mahler's compactness principle for lattices (see ref- 4 where further references are given). Our Theorem 7.1. is related to that principle and replaces it, here and in other applications. However, we are now going to show that we can also derive Mahler's compactness principle from 7.1.

7.4. THEOREM (Mahler). *Let $\{\Lambda_k\}$ be an infinite sequence of lattices in R^n. Suppose that there exists two positive numbers ϱ and σ such that*

$$d(\Lambda_k) \leq \varrho, \quad |\Lambda_k| \geq \sigma \qquad r = 1, 2, 3, \dots .$$

Then there exist a sequence of natural numbers $k_1 < k_2 < k_3 < \dots < k_j < \dots$ and a lattice P such that the sequence of lattices $\{\Lambda_{k_j}\}$ converges to P. That is to say, there are bases $\{a_1, \dots, a_n\}$ and $\{a_1^{(k_j)}, \dots, a_n^{(k_j)}\}$ for P and the Λ_{k_j} respectively such that

$$\text{7.5} \qquad \lim_{n\to\infty} a_i^{(k_j)} = a_i, \qquad i = 1, \dots, n.$$

PROOF. Again the theorem refers to ordinary n-dimensional space but in order to prove it we pass to $({}^*R)^n$. Let w be any infinite natural number in ${}^*R$. Since the inequalities $d(\Lambda_k) \leq \varrho$, $|\Lambda_k| \geq \sigma$ hold for all finite k, they must be true also for $k = w$. Thus, the conditions of Theorem 7.1. apply to $\Lambda = \Lambda_w$ and the set $P = {}^0\Lambda_w$ is a lattice in R^n. Let $\{a_1, \dots, a_n\}$ be a basis for P, then $a_i = {}^0a_i^{(w)}$, $i = 1, \dots, n$, for certain points $a_i^{(w)}$ in Λ_w, and $\{a_1^{(w)}, \dots, a_n^{(w)}\}$ is a basis for Λ_w. Evidently

$$\text{7.6} \qquad |a_i - a_i^{(w)}| < \frac{1}{j}, \qquad i = 1, \dots, n$$

for all finite natural numbers j.

Consider the following statement, for given natural j.

« There exists a natural number k and points $a_1^{(k)}, \dots, a_n^{(k)}$ which constitute a basis for Λ_k such that

$$\text{7.7} \qquad |a_i - a_i^{(k)}| < \frac{1}{j}, \qquad i = 1, \dots, n \text{ ».}$$

This statement can be expressed as a formal sentence X_j within the specified higher order language for R and, accordingly, is either true or false in R. Observe that the number k and the coordinates of the points $a_1^{(k)}, \dots, a_n^{(k)}$ are represented in X_j not by individual constants but as quantified variables.

We show first that X_1 holds in R. For in the alternative case, $\neg X_1$ would hold in R, hence in ${}^*R$. But X_1 actually holds in ${}^*R$ as we see by setting $k = w$, $a^{(k)} = a^{(w)}$. Since, therefore, X_1 holds in R there exist a positive integer $k = k_1$ and a basis $\{a_1^{(k_1)}, \dots, a_n^{(k_1)}\}$ for A_{k_1} such that

$$| a_i - a_i^{(k_1)} | < 1, \quad i = 1, \dots, n.$$

For $j = 2$, we replace X_2 by a sentence X_2' which expresses X_2 in conjunction with the condition $k > k_1$. Then X_2' holds in $^*R$, being realised there by $k = w$, $a_i^{(k)} = a_i^{(w)}$ - i. e. by the same system which we used earlier in order to establish the existence of a suitable k_1. Hence X_2' holds also in R, there exist a positive integer $k = k_2$ and a basis $\{a_1^{(k_2)}, \dots, a_n^{(k_n)}\}$ for A_{k_2} such that $k_2 > k$, and

$$| a_i - a_i^{(k_2)} | < \frac{1}{2}, \quad i = 1, \dots, n.$$

Next we introduce a sentence X_3' which expresses X_3 in conjunction with the condition $k > k_2$ and we establish the existence of a standard $k_3 > k_2$ such that

$$| a_i - a_i^{(k_3)} | < \frac{1}{3}, \quad i = 1, \dots, n.$$

Continuing in this way, we obtain a sequence of positive integers $k_1 < k_2 < k_3 < \dots$ and bases $\{a_1^{(k_j)}, \dots, a_n^{(k_j)}\}$ for A_{k_j} such that

$$| a_i - a_i^{(k_j)} | < \frac{1}{j}, \quad i = 1, \dots, n.$$

Thus 7.5. is satisfied and the theorem is proved.

There are many other fields in Mathematics where compactness arguments can be replaced by the use of Non-standard Analysis, and the reader may consult ref. 16 for further examples. It is a matter of taste whether we wish to regard our present method as a remote reformulation of such argument or whether we wish to assert rather that compactness arguments (e. g. selection principles) were introduced into Analysis in order to fill a gap due to the historical breakdown of the method of infinitesimals.

To conclude, we may point out an interesting methodological connection between the application to algebra detailed in Section 5 above, which leads to a proof Artin's theorem, and the theory of the present section. In both cases a salient point of the proof consists in establishing the existence of certain « ideal (weak) solutions » satisfying specified conditions, from which we then deduce the existence of « real (strong) solutions ». Thus, in Section 5 one shows that, for a certain ordering of $M' = M(t_1, \dots, t_n)$ the *transcendental* elements $t_1, \dots, t_n$ make $p(t_1, \dots, t_n)$ negative, and one deduces from this fact that there exist *rational* numbers $a_1, \dots, a_n$ for which $p(a_1, \dots, a_n) < 0$; while in the proof of 7.3. we first introduce the lattice Λ as an ideal construct satisfying certain conditions, and we deduce from it the existence of the « real » lattice ${}^0\Lambda$. Our formulation will remind the reader of procedures current in contemporary Analysis, particularly in the theory of partial differential equations, and this similarity is by no means a matter of accident.

LIST OF REFERENCES

[1] E. ARTIN, *Über die Zerlegung definiter Funktionen in Quadrate*, Abhandlungen des mathematischen Seminars der hamburgischen Universität, vol. 5, 1927, pp. 100-115.

[2] E. ARTIN and O. SCHREIER, *Algebraische Konstruktion reeller Körper*, Abhandlungen des mathematischen Seminars der hamburgischen Universität, vol. 5, 1927, pp. 85-99.

[3] J. AX and S. KOCHEN, *Diophantine problems over local fields*, American Journal of Mathematics, vol. 87, 1965, pp. 605-630, 631-648.

[4] J. W. S. CASSELS, *An Introduction to the Geometry of Numbers*, Grundlehren der Mathematischen Wissenschaften, vol. 99. Berlin-Göttingen-Heidelberg 1959.

[5] L. HENKIN, *Sums of squares*, Summaries of talks presented at the Summer Institute of Symbolic Logic at Cornell University, 1957, pp. 284-291.

[6] H. J. KEISLER, *Ultraproducs and elementary classes*, Proceedings of the Royal Academy of Sciences, Amsterdam, Ser. A, vol. 64, 1961, pp. 477-495.

[7] S. KOCHEN, *Ultraproducts in the theory of models*, Annals of Mathematics, ser. 2, vol. 74, 1961, pp. 221-261.

[8] G. KREISEL, *The mathematical significance of consistency proofs*, Journal of Symbolic Logic, vol. 23, 1958, pp. 155-182.

[9] J. ŁOŚ, *On the extending of models, I*, Fundamenta Mathematicae, vol. 42, 1955, pp. 38-54.

[10] W. A. J. LUXEMBURG, *Non-standard Analysis*, Lectures on A. Robinson's theory of infinitesimals and infinitely large numbers, Pasadena, 1962.

[11] A. ROBINSON, *On a problem of L. Henkin*, Journal of Symbolic Logic, vol. 21, 1956, pp. 33-35.

[12] A. ROBINSON, *On ordered fields and definite functions*, Mathematische Annalen, vol. 130, 1955/56, pp. 257-271.

[13] A. ROBINSON, *Further remarks on ordered fields and definite functions*, Mathematische Annalen, vol. 130, 1955/56, pp. 405-409.

[14] A. ROBINSON, *Complete Theories*, Studies in Logic and the Foundations of Mathematics, Amsterdam, 1956.

[15] A. ROBINSON, *Introduction to Model Theory and to the Metamathematics of Algebra*, Studies in Logic and the Foundations of Mathematics, Amsterdam, 1963.

[16] A. ROBINSON, *Non-Standard Analysis*, Studies in Logic and the Foundations of Mathematics, Amsterdam, 1966.

[17] A. TARSKI, *Contributions to the theory of models*, Proceedings of the Royal Academy of Sciences, Amsterdam, ser. A, vol. 57, 1954, pp. 572-581, 582-588, vol. 58, 1955 pp. 56-64.

[18] A. TARSKI and J. C. C. MCKINSEY, *A decision method for elementary Algebra and Geometry*, 1948, 2nd ed. Berkely and Los Angeles, 1951.

[19] Y. L. YERSHOV, *Ob elementarnikh teoriakh lokalnikh polei*, Algebra i Logika, vol. 4, fasc. 2, pp. 5-30.

[20] J. W. ADDISON, L. HENKIN, and A. TARSKI (ed.), *The Theory of Models*, Proceedings of the 1963 International Symposium at Berkeley, Studies in Logic and the Foundations of Mathematics, Amsterdam, 1965.

[*Entrato in Redazione il 4 giugno 1966*]

Topics in Nonstandard Algebraic Number Theory

by ABRAHAM ROBINSON[1]

1. Introduction

Nonstandard analysis was put forward originally as a consistent framework for the development of the calculus and of other branches of analysis in terms of infinitely small and infinitely large quantities. Later, the basic ideas of this method were extended to other branches of mathematics, in particular, to general topology (see [6]). The present paper is one of a series (compare [7–9]) in which the nonstandard approach is applied to algebraic number theory and to related topics. The paper falls into two parts. In Sections 2–4 we consider the theory of entire ideals in an infinite algebraic number field; that is, in an infinite algebraic extension of the field of rational numbers. One of the difficulties that arises in the classical multiplicative theory of these ideals is that the cancellation rule no longer applies to it. We shall show how this situation can be alleviated by our nonstandard approach.

In Sections 5 and 6 we continue the development of class field theory for infinite algebraic number fields which was begun in [9]. We show how the main result obtained there can be related to the classical theory of Chevalley and Weil.

We shall now give an informal introduction to the basic ideas of nonstandard analysis as applied to the problems studied in the present paper. The reader may consult [6] and [8] for details.

[1] The author acknowledges with thanks that the research leading to the present paper was supported in part by the National Science Foundation. He also wishes to express his appreciation of the support given by the Office of Naval Research to a program in nonstandard analysis which was conducted at the California Institute of Technology during the academic year 1967–1968 and in which he participated during the winter term. Finally, the author thanks David Cantor for an instructive discussion in connection with the problems considered here.

Let F be an infinite algebraic extension of the field of rational numbers Q. We consider the properties of F in a higher-order language L, which includes symbols for all individuals (that is, numbers) of F, all subsets of F, all relations of two, three, $\cdots$ variables between individuals of F, all functions from individuals to individuals of F, and, more generally, all relations and functions of finite type (for example, functions from sets of numbers into relations between numbers) that can be defined beginning with the individuals of F. Let K be the set of all sentences formulated in L which hold (are true) in F. Then there exists a structure $^*F$ with the following properties.

1.1. $^*F$ is a model of K "in Henkin's sense." That is, all sentences of K hold in $^*F$, provided, however, that we interpret all quantifiers other than those referring to individuals in a nonstandard fashion, as follows. Within the class of all entities of any given type other than 0 (the type of individuals), there is distinguished a certain subclass of entities called *internal*. And, for such a type, the quantifiers "for all x" and "there exists an x" are to be interpreted as "for all *internal* x," "there exists an *internal* x" (of the given type). F can be injected into $^*F$, and so $^*F$ may, and will, be regarded as an extension of F.

1.2. Every concurrent binary relation $R(x, y)$ in F possesses a bound in $^*F$.

The relation $R(x, y)$, of any type, called *concurrent* if for any $a_1, \cdots, a_n$, $n \geq 1$, for which there exists $b_1, \cdots, b_n$ such that $R(a_1, b_1)$, $\cdots$, $R(a_n, b_n)$ hold in F, there also exists a b such that $R(a_1, b), \cdots,$ $R(a_n, b)$ hold in F. And the entity b_R in $^*F$ is called a *bound* for $R(x, y)$ if $R(a, b_R)$ holds in $^*F$ for any a for which there exists a b such that $R(a, b)$ holds in F.

Any structure $^*F$ which satisfies 1.1 and 1.2 is called an *enlargement* of F. In particular, $^*F$ can be constructed as an ultrapower of F. (See [3] and [4] for the notions of an ultrapower.) Enlargements which are ultrapowers have special properties that will be used implicitly in Sections 5 and 6 on the theory of idèles. On the other hand, the arguments of Sections 2–4 on ideal theory in infinite number fields apply to enlargements in general. Moreover, the reader will check that we shall have to rely on the existence of bounds only for a few concurrent relations. For all these cases, the existence of bounds is ensured already for any ultrapower of F that is based on a free ultrafilter in a countable index set. Thus *all* results of the present paper hold for such ultrapowers (compare [9]).

We stated earlier that there is an embedding of F into $^*F$. For the case of an ultrapower, this is obtained by identifying any entity a of F with the entity of $^*F$ which corresponds to the constant function $f(x) \equiv a$ on the index set of $^*F$. In terms of the language L, an entity b of $^*F$ corresponds to an entity a of F if it is denoted by the same symbol in L. The entities of $^*F$ which occur in this correspondence are called *standard* entities. If no confusion is likely to arise, we shall refer to entities of F also as standard. Notice, however, that if entities of F and $^*F$ correspond as indicated, then they are not, in general, extensionally the same. For example, if N is the set of natural numbers in F, then the corresponding set in $^*F$, to be denoted by $^*N$, contains, in addition to the standard natural numbers $0, 1, 2, \cdots$ also certain nonstandard, or infinite, natural numbers. To see that there must be such infinite numbers in $^*N$, consider the relation $R(x, y)$, which is defined by the condition "x and y are natural numbers and $x < y$." It is easy to verify that $R(x, y)$ is concurrent. Let b_R be a bound for R in $^*F$. Then b_R is a natural number in $^*F$; that is, an element of $^*N$ and, moreover, $0 < b_R$, $1 < b_R$, $2 < b_R$, and so on. This proves that b_R is a nonstandard, "infinite" natural number in $^*F$. More generally, it is not difficult to see that if A and $^*A$ are corresponding sets in F and $^*F$, then $^*A$ is a proper extension of A if and only if A is infinite.

Let B be the set of all finite sets of a given type in F. On passing to $^*F$, B is extended to a set $^*B$ whose elements will be called *star-finite*. An internal set A in $^*F$ is star-finite if and only if there is an internal one-to-one correspondence between the elements of A and a set of natural numbers $\{x \mid 0 \leq x \leq n\}$, where n is a finite or infinite element of $^*N$.

2. Ideals in Infinite Algebraic Number Fields

Let the fields F and Q be defined as in Section 1. On passing from F to an enlargement $^*F$, Q is extended to a subfield $^*Q$ of $^*F$. Let Q^i and F^i be the rings of integers in Q and F so that $Q^i = Z$, the ring of rational integers, and let $^*Q^i = {}^*Z$ and $^*F^i$ be the corresponding rings in the enlargement. Let $\Phi = \{F_n\}$ be a tower of subfields of F,

$$Q = F_0 \subset F_1 \subset F_2 \subset \cdots$$

such that all the F_n are finite extensions of Q and such that $\bigcup_n F_n = F$. Then the corresponding entity in $^*F$, $^*\Phi$, is a mapping from $^*N$ into the set of subfields of $^*F$, $^*\Phi = \{H_n\}$, say. For finite n, $H_n = {}^*F_n$, while for any infinite $n = \omega$, we have $F_m \subset H_\omega$ for any finite m. We conclude that $F = \bigcup_n F_n \subset H_\omega$, where n ranges only over finite subscripts. Thus $F \subset H_\omega \subset {}^*F$. Put $H_\omega = H$.

H is a *star-finite* extension of $^*Q$. That is to say, there exists an $\alpha \in {}^*F$

such that $H = {}^*Q(\alpha)$, where α is algebraic over ${}^*Q$ and where the symbol ${}^*Q(\alpha)$ and the term *algebraic* both have to be interpreted in the nonstandard sense. Thus the degree ω of α may well be (and actually is, for infinite F, as assumed) infinite; that is, n is an infinite natural number.

Let $S = \{J\}$ be the set of ideals in F^i, and let ${}^*S$ be the corresponding set in ${}^*F^i$. Thus ${}^*S$ consists of all internal sets of numbers of ${}^*F$ which are entire ideals in ${}^*F^i$. Let S_ω be the set of internal and entire ideals in H^i, where $H^i = H \cap {}^*F_i$. For any $J \in S$ we have the canonical mapping $J \to {}^*J$ into ${}^*S$. On the other hand, for any ideal $J \in {}^*S$, we define $\varphi(J) = J_H$ by $J_H = J \cap H_\omega$. Then $\varphi({}^*J) \cap F = J$. The ring of integers in H will be denoted by H^i, so that $H^i = {}^*F^i \cap H$. Let S_H be the set of internal ideals in H^i. Then the ideals of S_H are either principal or they have two-element bases, since this is true for the ideal of any finite algebraic number field.

Two different ideals $J_1, J_2 \in S_H$ may have the same intersection with F^i. However, to any $J \in S$ there corresponds a unique ideal $J_H \in S_H$ by the mapping $\psi(J)$ which is defined by $\psi(J) = \varphi({}^*J) = J_H$. The inverse of this map is simply $\psi^{-1}(J_H) = J_H \cap F^i$.

If $J \in S$ is prime, so is ${}^*J$, since the property of being prime can be expressed within the language L. Also, if any $J \in {}^*S$ is prime, so are the restrictions of J to H and to F. It follows that if $J \in S$ is prime, so is $J_H = \psi(J)$. At the same time, if any $J \in S_H$ is prime, so is its restriction to F.

The mapping ψ is not, in general, multiplicative. To see this, let F be the field of all algebraic numbers and let $J \in S$ be given by

$$J = (2^{2^{-1}}, 2^{2^{-2}}, \cdots, 2^{2^{-k}}, \cdots),$$

where $k = 1, 2, \cdots$ ranges over the finite positive integers. Then $J^2 = J$. But $1 \notin J$, so that J is not the unit ideal in F, and so $1 \notin J_H = \psi(J)$, J_H is not the unit ideal in H. It follows that $J_H{}^2 \neq J_H$, in other words, $(\psi(J))^2 \neq \psi(J)$. For, since the cancellation rule applies in H, $J_H{}^2 = J_H$ would imply that J_H is the unit ideal in H.

For any subset A of (the set of individuals of) ${}^*F$, we define the *standard part* of A, ${}^0A$, by ${}^0A = A \cap F$, and we write $\sigma(A) = {}^0A$ or

$$A \xrightarrow{\sigma} {}^0A.$$

Now let J_1 and J_2 be ideals in S_H and let

$$J_1 \xrightarrow{\sigma} {}^0J_1, \qquad J_2 \xrightarrow{\sigma} {}^0J_2, \qquad J_1J_2 \xrightarrow{\sigma} J,$$

so that $J = {}^0(J_1J_2)$. Then any $c \in {}^0J_1{}^0J_2$ can be written in the form

$$c = a_1b_1 + \cdots + a_kb_k,$$

where k is a finite positive integer and $a_i \in {}^0J_1$, $b_1 \in {}^0J_2$, $i = 1, \cdots, k$. Hence ${}^0J_1{}^0J_2 \subset J$. In general, the sign of inclusion in this relation cannot be replaced by $=$. However, if we add the assumption that there exist ideals K_1, K_2 in S such that $J_1 = \psi(K_1) = K_{1H}$, $J_2 = \psi(K_2) = K_{2H}$, then

$$*K_1*K_2 = *(K_1K_2)$$

Hence $$K_{1H}K_{2H} \subset (*(K_1K_2))$$

and so $${}^0(K_{1H}K_{2H}) \subset {}^0(*(K_1K_2)) = K_1K_2;$$

that is, $$J = {}^0(J_1J_2) \subset {}^0J_1{}^0J_2.$$

Thus ${}^0J_1{}^0J_2 = {}^0(J_1J_2)$ in this case. A corresponding conclusion applies for any finite number of ideals $J_i \in S_H$ such that $J_i = \psi(K_i)$, $K_i \in S$. We have proved

Theorem 2.1. *Let $K_1, \cdots, K_m$ be any finite number of ideals in S, $m \in N$. Then*

$${}^0(\psi(K_1)){}^0(\psi(K_2)) \cdots {}^0(\psi(K_m)) = {}^0(\psi(K_1)\psi(K_2) \cdots \psi(K_m)).$$

By a familiar result of commutative ring theory, any maximal ideal in S is prime (excluding the unit ideal, as usual). Conversely, let P be any prime ideal in S, $P \neq 0$, $P \neq F^i$; then the following nonstandard argument shows that P *is maximal.*

Since P is prime and since the property that an ideal is prime can be expressed in L, $*P$ also must be prime. Hence $P_H = \psi(P)$ is prime. If P is not maximal, then it is included in a maximal ideal Q in S, $P \subset Q \subset F^i$, $P \neq Q$, $Q \neq F^i$. Since $F^i - Q$ is not empty, $H^i - Q_H$ cannot be empty either, and so Q_H is a proper ideal which is a proper extension of P_H. But this is impossible, since any prime ideal in H must be maximal, as is the case for all finite algebraic number fields. Hence P is maximal, as asserted.

Let J be any proper ideal in S; that is, $J \neq 0$, $J \neq F^i$. A familiar argument shows that J possesses at least one prime divisor. We shall call J *primary* if it possesses exactly one prime divisor, P. Then $J_H \subset P_H$ and so there exists an ideal $Q \in S_H$ such that $J_H = P_HQ$. Suppose now that J_H possesses, in addition to P_H, another prime divisor P'. Then $1 \notin P'$ and so $1 \notin {}^0P'$ and at the same time $J \subset {}^0P'$. Assuming that J is primary, we conclude that either ${}^0P' = F^i$ or ${}^0P' = P$. But the former possibility is ruled out by $1 \notin {}^0P'$ and so ${}^0P' = P$. It follows that all prime divisors of J_H have the same standard part.

Now let J be any proper ideal in S, so that $J_H = \psi(J)$ is a proper ideal in S_H. Then J_H can be decomposed into a product of powers of distinct prime ideals,

$$J_H = P_1^{n_1} \cdots P_k^{n_k}, \qquad k \geq 1, \quad n_j \geq 1 \qquad \text{for } j = 1, \cdots, k,$$

where k and the n_j may be finite or infinite. Suppose that all the P_j have the same standard part, P. Then

$$J_H = \bigcap_j P_j^{n_j}$$

and so

$$J = \bigcap_j {}^0(P_j^{n_j}).$$

Now let Q be any proper prime ideal in S which divides J. Then $\psi(Q) = Q_H$ is a proper prime ideal which divides J_H. Accordingly, Q_H coincides with one of the P_j, and so $Q = {}^0Q_H = {}^0P_j = P$. We have proved

Theorem 2.2. *A proper ideal $J \in S$ is primary if and only if the prime divisors of $\psi(J)$ all have the same standard part.*

3. The Ring Δ_H

Let μ be the subset of $^*F^i$ which is defined by

3:1. $\mu = \{x \mid x \in {}^*F^i$ *any x is divisible by all nonzero standard rational integers*$\}$.

Then μ is an external ideal in $^*F^i$. We shall study the quotient ring $\Delta = {}^*F^i/\mu$. A corresponding investigation for the case that F is a finite algebraic number field or, more generally, a Dedekind ring has been carried out in [7–9].

The elements of μ are characterized also by the property that they are divisible by all nonzero elements of F^i. For if $a \in F^i$, $a \neq 0$, and Na is the norm of a, then $a \mid Na$, a divides Na, and so $Na \mid x$ implies $a \mid x$. Yet another equivalent condition is that for every standard rational prime p there exists an infinite natural number n such that $p^n \mid x$.

Let δ be the canonical mapping

$$\delta\colon {}^*F^i \to \Delta.$$

Then δ injects $F^i \subset {}^*F^i$ into Δ, since μ does not contain any standard elements other than 0. Let $\mu_H = \mu \cap H$; then we may identify $H^i/\mu_H = \Delta_H$ with a subring of Δ. Let δ_H be the restriction of δ to H, so that δ_H maps H on Δ_H and injects F^i into Δ_H.

Theorem 3.2. *Let $a \in F^i$, $a \neq 0$. Then $\delta_H(a)$ is invertible in Δ_H if and only if all prime ideals $P_j \in S_H$ which divide the ideal $(a)_H$ generated by a in H have norms NP_j that are powers of nonstandard primes.*

Remark. If F is a finite algebraic number field, then any proper internal prime ideal P in $^*F^i$ is standard if and only if its norm is a power of a standard rational prime. Even if P is not standard, its norm must be a *finite* power f of a rational prime (which is then infinite), since f is bounded by the degree of F over Q. In the case under consideration here, where F is an infinite extension of Q, we first pass from $^*F^i$ to H^i to be able to apply the results of standard ideal theory. Even in H^i, the norm of a proper internal prime ideal must be a power of a rational prime.

Proof of Theorem 3.2. For any $a \in H$, $N_H a$ shall denote throughout the norm of a in H over $^*Q$.

The conclusion of the theorem is satisfied if and only if Na, which is a rational integer, is divisible only by *infinite* prime numbers. For let

$$(a) = P_1^{k_1} \cdots P_l^{k_l}$$

be the prime power decomposition of the ideal $(a) \in S_H$. Then

$$|N_H a| = (N_H P_1)^{k_1} \cdots (N_H P_l)^{k_l},$$

where $N_H P_j$, $j = 1, \cdots, l$, is a finite or infinite power of the unique prime number contained in it.

Now let $b \neq 0$ be a rational integer such that all prime divisors of b are nonstandard. Let p be the smallest positive prime divisor of b so that p is infinite. Then b and $(p-1)!$ are coprime, and so there exist rational integers k and l such that

$$k(p-1)! + lb = 1.$$

Thus

3.3. $$lb \equiv 1 \bmod (p-1)!$$

Suppose now that a satisfies the hypothesis of the theorem and put $b = N_H a = aa'$, where $a' \in H^i$. Then

$$aa'l \equiv 1 \bmod (p-1)!$$

by 3.3. But $(p-1)! \in \mu_H$, and so

$$\delta_H(a)\delta(a'l) = 1,$$

which shows that $\delta_H(a)$ is invertible in Δ_H.

Conversely, suppose that δ_H is invertible in Δ_H. Then there exists a number $a' \in H^i$ such that $aa' - 1 \in \mu_H$. Thus $aa' - 1$ is divisible by $n!$ for all finite natural numbers n. A basic argument of nonstandard analysis now shows that $aa' - 1$ must be divisible also by ω for some infinite natural number ω. To sketch the argument briefly, let $A \subset {}^*N$

be the set of natural numbers n, finite or infinite, such that $n!$ does not divide $aa' - 1$. A is internal. If A is empty, then we have finished. Now it is a fact concerning standard natural numbers, which can be expressed as a sentence of K, that every nonempty set of natural numbers possesses a smallest element, and the same must therefore be true in $^*N$. Hence, if A is not empty, then it has a smallest element, $\omega + 1$, say, where ω is infinite, and $aa' - 1$ is then divisible by ω. Thus

$$aa' = 1 + k\omega!$$

where $k \in H^i$, and so

3.4. $$N_H a N_H a' = 1 + l\omega!$$

where l must now be a rational integer. Suppose that the standard prime number q divides $N_H a$. Since q also divides $\omega!$, it would then follow from 3.4 that q also divides 1, which is impossible. This completes the proof of Theorem 3.2.

Theorem 3.5. *Let $J \in S_H$. Then δ_H maps J on a principal ideal J' in Δ_H.*

Proof. Let $J \in S_H$. If $J = (0)$, then $J' = (0)$ and we have finished. If $J \neq (0)$ and $a \in J$, $a \neq 0$, then for any infinite natural number ω. J contains also the number $a\omega!$ By a standard result, which applies also in H, there exists a $b \in J$ such that $J = (b, a\omega!)$. Hence $J' = \delta_H(J) = (\delta_H(b), \delta_H(a\omega!))$. But $\delta_H(a\omega!) \in \mu_H$, and so $J' = (\delta_H(b))$, as asserted.

Theorem 3.6. *Let $J \in S_H$ and suppose that J contains a standard number $a \neq 0$. If there exist numbers $b \in H^i$, $c \in H^i$ such that*

$$J' = \delta_H(J) = (\delta_H(b)) = (\delta_H(c)),$$

then $\delta_H(b)$ and $\delta_H(c)$ are associated elements in Δ_H. That is, there exist invertible *elements m' and n' of Δ_H such that $\delta_H(b) = m'\delta_H(c)$ and $\delta_H(c) = n'\delta_H(b)$.*

Proof. Suppose that b and c satisfy the assumptions of the theorem. Then there exists an $m' \in \Delta_H$ such that $\delta_H(b) = m'\delta_H(c)$. Choose $m \in H^i$ such that $m' = \delta_H(m)$. Then $\delta_H(b - mc) = 0$, and so

$$b - mc \in \mu_H.$$

Similarly, there exists an $n \in H^i$ such that

$$c - nb \in \mu_H.$$

Hence $$nb - nmc \in \mu_H.$$

3.7. $$(1 - nm)c \in \mu_H.$$

By assumption, there exists a standard number $a \in J$, $a \neq 0$. Then $\delta(a) \in J'$ and so $\delta(a) = k'\delta(c)$ for some $k' \in \Delta_H$. But $k' = \delta_H(k)$ for some $k \in H^i$ and, for such k, $\delta_H(a) - \delta_H(k)\delta_H(c) = 0$ and hence

$$a - kc \in \mu_H,$$
$$(1 - nm)a - (1 - nm)kc \in \mu_H,$$

and, taking into account 3.7,

3.8. $$(1 - nm)a \in \mu_H.$$

Now it is easy to see that if $q \in \mu_H$ and γ is a standard positive integer, then $q/\gamma \in \mu_H$. But $a \mid Na$ (where Na is now the standard norm of the standard number a) and so $q/a = (q/Na)(Na/a)$ also belongs to μ_H. Hence, dividing $(1 - nm)a$ by a, we obtain, from 3.8,

$$1 - nm \in \mu_H$$

and so $$\delta_H(n)\delta_H(m) = 1.$$

Hence $m' = \delta_H(m)$ and $n' = \delta_H(n)$ are invertible, proving the theorem.

Theorem 3.9. *Let $J \in S_H$ and suppose that J contains a standard number $a \neq 0$. Suppose further that $J = (b, m)$, where $m \in \mu_H$. Then there exists an ideal $D \in S_H$ such that $JD = (b)$, where all prime divisors of $N_H D$ are nonstandard.*

Proof. Given J, a, and b with the specified properties there exists an ideal $D \in S_H$ such that $JD = (b)$, by standard ideal theory in finite algebraic number fields. Also, $\mu_H \subset J$, since any number of H_i which is divisible by all standard positive integers is divisible also by a.

Suppose now that D has a prime divisor $P \in S_H$ whose norm $N_H P$ is a power of a standard prime number p; p is the unique rational prime number contained in P. Then $ap \in JP$ and so, as above, $\mu_H \subset JP$ and, in particular, $m \in JP$. Also, $JP \supset JD = (b)$ and so $b \in JP$ and $J = (b, m) \subset JP$. But $J \supset JP$, and so $J = JP$. But this implies $P = H^i$, by standard ideal theory—which contradicts our assumptions and proves the theorem.

In particular, the conditions of 3.4 apply if the ideal J considered in Theorem 3.5 contains a standard element different from 0.

4. Topological Considerations

Let $T \subset S$ be the set of proper prime ideals in F^i. The *Krull topology* of T [5] is defined as follows. A subset A of T is *open* if for any $J \in A$ there exists a subfield F' of F which is finite (that is, a finite algebraic extension of Q) such that all $J' \in T$ for which $J' \cap F' = J \cap F$ also belong to A. In other words, for any $J \in T$, a fundamental system of neighborhoods of J is defined by the set of ideals of T which coincide with J on some finite subfield of F.

Passing to the enlargement $^*F$, we recall that the *monad* $\mu(J)$ of any $J \in T$ in the sense of nonstandard topology [6] is defined as the set $\bigcap_\nu {}^*A_\nu$ where A_ν ranges over the open neighborhoods of J in T. To define $\mu(J)$ more directly in terms of the algebraic situation we introduce $\lambda(F)$ as the set $\bigcup_\nu {}^*F_\nu$ where F_ν ranges over all finite subfields of F. It is not difficult to see that $\lambda(F)$ is a field that contains H. However, $\lambda(F)$ must be a proper subfield of $^*F$. For, on the one hand, all elements of $\lambda(F)$ are algebraic over $^*Q$ in the standard sense; that is, their degrees over $^*Q$ are finite natural numbers. On the other hand, since F is infinite over Q, the degrees of its elements are unbounded. It follows that $^*F$ contains elements of infinite degree over $^*Q$ and so $\lambda(F) \neq {}^*F$.

Theorem 4.1. *Let $J \in T$. Then $\mu(J)$ consists of all $J' \in {}^*T$ such that $J' \cap \lambda(F) = {}^*J \cap \lambda(F)$.*

Proof. Let A be any open set that includes J and suppose that $J' \cap \lambda(F) = {}^*J \cap \lambda(F)$. By assumption, there exists a finite field $F' \subset F$ such that A contains all ideals of T with the same restrictions to F' as J. It follows that $^*A$ contains all ideals of $^*T$ that have the same restrictions to $^*F'$ as $^*J$. This is true of J', since $\lambda(F)$ includes $^*F'$.

Conversely, suppose that $J' \in \mu(J)$, but that $J' \cap \lambda(F) \neq {}^*J \cap \lambda(F)$. It follows that $((J' - {}^*J) \cup ({}^*J - J'))$ has a nonempty intersection with $\lambda(F)$ and hence has a nonempty intersection with $^*F'$ for some finite subfield F' of F. Define A as the set of all ideals of T that coincide with J on F'. Then $J \in A$ and A is open and so $^*A \supset \mu(J)$. But $J' \in {}^*A$, since J' does not coincide with $^*J$ on $^*F'$. This completes the proof of 4.1.

As usual in nonstandard analysis, we may derive certain standard properties of T from results about $^*T$. For example, 4.1 shows immediately that an ideal of $^*T$ cannot belong to the monads of two distinct ideals of T. It follows [6] that T is a Hausdorff space.

5. Theory of Idèles

From now on, let F be a finite algebraic extension of the field of rational numbers, Q, and let F_{P_ν} be the completions of F, where P_ν ranges over the finite and infinite (or Archimedean) primes in F. Let $M = \Pi_\nu F_{P_\nu}$ be the strong direct product of the completions of F and let $I \subset H$ be the group of idèles over F. Thus I consists of the elements of M whose components are different from zero and are nonunits for a finite number of finite primes P_ν only. We may regard the F_{P_ν} as extensions of F (given a particular injection of F into F_{P_ν} for each ν). The diagonal injection of F into M then yields an embedding of F', the multiplicative group of F, into I. The idèle class group C is defined as the quotient group $C = I/F'$.

It has been shown in [9] (compare [8]) that M is isomorphic to the quotient ring F_0/μ, where F_0 and μ consist of all elements of ${}^*F$ whose absolute values are finite or infinitesimal, respectively, in the valuations associated with all the standard P_ν. Equivalently, F_0 consists of the elements of ${}^*F$ that are finite in all Archimedean valuations and whose orders are nonnegative, or negative but finite, for all standard finite primes; and μ consists of the elements of ${}^*F$ that are infinitesimal in all Archimedean valuations and whose orders are positive and infinite for the standard finite primes.

Let H be the set of entire or functional internal ideals J in ${}^*F$ such that $J \neq (0)$ and $\text{ord}_{P_\nu} J = 0$ for all standard finite primes P_ν. In other words, H consists of the nonzero ideals whose prime power decomposition does not contain any standard prime ideals. Let H_0 be the set of principal ideals in ${}^*F$, which can be written $J = (a)$, where a is positive in all real Archimedean valuations and $\text{ord}_{P_\nu} (a - 1)$—the order of $a - 1$ for the prime P_ν—is positive infinite for all standard finite primes. H_0 is a subset of H, as can be seen from the equation

$$\begin{aligned}\text{ord}_{P_\nu} J = \text{ord}_{P_\nu} a &= \text{ord}_{P_\nu} (a - 1 + 1)\\ &= \min(\text{ord}_{P_\nu} (a - 1), \text{ord}_{P_\nu} 1) = 0.\end{aligned}$$

Moreover, H_0 is multiplicatively closed, since

$$\begin{aligned}\text{ord}_{P_\nu} (ab - 1) &= \text{ord}_{P_\nu} ((a - 1)b + (b - 1))\\ &\geq \min(\text{ord}_{P_\nu} (a - 1)b, \text{ord}_{P_\nu} (b - 1)).\end{aligned}$$

This shows that if $\text{ord}_{P_\nu} (a - 1)$ and $\text{ord}_{P_\nu} (b - 1)$ are infinite, and hence $\text{ord}_{P_\nu} b = 0$, then $\text{ord}_{P_\nu} (ab - 1)$ also is infinite. Finally, H_0 is closed with respect to inversion, since $J^{-1} = (1/a)$ for $J = (a)$ and

$$\text{ord}_{P_\nu} \left(\frac{1}{a} - 1\right) = \text{ord}_{P_\nu} (a - 1) - \text{ord}_{P_\nu} a,$$

which is infinite if $\text{ord}_{P_\nu} (a - 1)$ is infinite and $\text{ord}_{P_\nu} a = 0$.

Having verified that H_0 is a multiplicative subgroup of H, we consider next the quotient group $D = H/H_0$. *We propose to show that D is a homomorphic image of the idèle class group C.*

For this purpose, we have to construct an epimorphism from C to

D. Let β be the canonical epimorphism from I to C, and let c be any element of C. Let b and b' be corresponding elements of I, so that $c = \beta b = \beta b'$. Then $b' = fb$ for some $f \in F'$ (and, conversely the existence of an $f \in F'$ such that $b' = fb$ shows that $\beta b = \beta b'$).

As mentioned earlier, there is an epimorphism, to be called here α, from the ring F_0 onto $M = \Pi_\nu F_{P_\nu}$, where the kernel of α is the set μ introduced above. We may assume that the embedding of F into M has been determined in such a way that for any $f \in F$, F being regarded as a subset of F_0, $\alpha f = f$. Moreover (see [9], sec. 5), if α_ν is the P_ν-component of αa for $a \in F_0$, then a is infinitesimal for any given Archimedean P_ν if and only if $a_\nu = 0$; for standard finite P_ν, $\text{ord}_{P_\nu}\, a$ is infinite if and only if $a_\nu = 0$, and if $\text{ord}_{P_\nu}\, a$ is finite, then it is equal to $\text{ord}_{P_\nu}\, a_\nu$. It follows that $a \in F_0$ belongs to $\alpha^{-1}I$ if and only if a is not infinitesimal for any Archimedean valuation while the number of standard prime divisors of (a) and the (positive or negative) powers in which they appear in (a) are all finite. This condition shows immediately that $G = \alpha^{-1}I$ is a multiplicative group.

For any $b \in I$ and for any $a \in b$ such that $b = \alpha a$, we may have at the same time $b = \alpha a'$ if and only if $a' = a + m$ for some $m \in \mu$. Thus, if $a \xrightarrow{\alpha} b$ and $f \in F'$, $m \in \mu$, then

$$fa + m \xrightarrow{\alpha} fb.$$

And if at the same time $b \xrightarrow{\beta} c$ and so $a \xrightarrow{\beta\alpha} c$, then

$$fa + m \xrightarrow{\beta\alpha} c.$$

Conversely, if $a \in G$ and $a \xrightarrow{\beta\alpha} c$ and at the same time $a' \xrightarrow{\beta\alpha} c$ for some $a' \in G$, then we claim that there exist $f \in F'$ and $m \in \mu$ such that $a' = fa + m$. Indeed, let $b = \alpha a$, $b' = \alpha a'$ so that $c = \beta b = \beta b'$. Then we know already that there exists an $f \in F'$ such that $b' = fb$. It follows that $b' = \alpha(fa)$ and so $fa - a' \in \mu$; that is, $a' = fa + m$ for some $m \in \mu$, as asserted.

Next, we introduce a homomorphism γ from G into H in the following way. For any $a \in G$ consider the representation of (a) as a product of powers of prime ideals in $^*F$.

5.1. $$(a) = P_1^{n_1} \cdots P_l^{n_l}$$

where l and the n_j may be finite or infinite and the n_j may be positive or negative. By the definition of G, the n_j must be finite for standard P_j, and only a finite number of such P_j appear in the product on the right side of 5.1. Hence, by deleting all such $P_j^{n_j}$ we obtain an *internal* ideal

$J = \gamma a$, which is an element of H, and we may verify immediately that γ is a homomorphism. The kernel of γ is just the multiplicative group of principal ideals (a) for $a \in F'$, which is isomorphic to the quotient group of F' by its group of units.

Finally, there is a canonical homomorphism from H onto $D = H/H_0$, which will be denoted by δ. Combining γ with δ, we obtain a homomorphism δ_γ from G into D. *We claim that δ_γ is an epimorphism into D.*

To see this, let $d \in D$ and $J \in H$ such that $d = \delta J$. Choose a standard ideal $^*J' \neq s$ (where J' is an ideal in F) such that $J \cdot {}^*J'$ is principal. The existence of such a J' follows immediately from the fact that the number of ideal classes in F is finite and hence does not increase on passing to $^*F$. Thus $J \cdot {}^*J' = (a')$ for some $a' \neq 0$ in $^*F$. Now choose an $a'' \in {}^*F$ that is positive in all real Archimedean valuations and is such that $\text{ord}_{P_\nu}(a'' - 1)$ is positive infinite for all standard finite primes and $a = a'a'' \in G$. Equivalently, the last condition can be replaced by the requirement that, for all Archimedean valuations, a is finite but not infinitesimal. A suitable a'' may be found as follows.

Range all standard *finite* primes in a sequence $\{P(n)\}$ and let ω be an infinite natural number. $\{P(n)\}$ is continued automatically from F to $^*F$ and the set $^*\{P(n)\}_{n \leq \omega}$ contains all standard finite primes and is *star-finite*; that is, finite in the sense of the enlargement. Given a' as above, choose a'' so as to satisfy the conditions

5.2. $$\text{ord}_{P(n)}(a'' - 1) > \omega \qquad \text{for } n \leq \omega$$

and

5.3. $$\left| a'' - \frac{e^{i\theta_\nu}}{a'} \right| < \frac{1}{\omega |a'|}$$

for all Archimedean valuations, where $\theta_\nu = \arg a'$ for the several valuations. Since their number is star-finite, conditions 5.2 and 5.3 can be satisfied simultaneously, by virtue of the approximation theorem. Also, from 5.3, and equivalent to it, we obtain for all Archimedean valuations,

5.4. $$| \, |a'| a'' - 1 | < \frac{1}{\omega}.$$

This shows that a'' is positive for all real valuations. More generally, 5.4. shows that, for all Archimedean valuations, $a = a'a''$ must be infinitely close to the unit circle and so $a \in G$.

Now let $(a'') = Q_1^{k_1} \cdots Q_m^{k_m}$ be the prime power decomposition of the ideal (a''). Then none of the Q_i is standard and so $\gamma a = \gamma(a'a'') =$

$J \cdot (a'')$. Moreover, by the conditions imposed on a'', $(a'') \in H_0$ and so $\delta(a'') = 1$ and $\delta\gamma a = \delta J \delta(a'') = d$. This shows that $\delta\gamma$ is onto D.

We shall now show that for any two elements a and a' of G, $\beta\alpha a = \beta\alpha a'$ implies $\delta\gamma a = \delta\gamma a'$.

We know already that $\beta\alpha a = \beta\alpha a'$ implies the existence of $f \in F'$, $m \in \mu$, such that $a' = fa + m$. Accordingly, we only have to prove that, for any $a \in G$, $m \in \mu$,

$$\delta\gamma(fa + m) = \delta\gamma a.$$

But $$\delta\gamma(fa + m) = \delta\gamma\left(fa\left(1 + \frac{m}{\delta a}\right)\right) = \delta\gamma f \cdot \delta\gamma a \cdot \delta\gamma\left(1 + \frac{m}{fa}\right),$$

since $\gamma f = 1$ and $m' \in \mu$, taking into account that fa is neither infinite nor infinitesimal. Thus it only remains to be shown that $\delta\gamma(1 + m') = 1$ for any $m' \in \mu$, where we observe that $1 + m' \in b$ for such m'. But, for $m' \in \mu$, the ideal $(1 + m')$ belongs to H_0 and so $\delta\gamma(1 + m') = 1$, as required.

We now define a mapping φ from C onto D as follows. For any $c \in C$ choose $a \in G$ such that $c = \beta\alpha a$ and define $d = \varphi c$ by $d = \delta\gamma a$. As we have just seen, this definition is independent of the particular choice of a. The mapping φ is onto D, since $\delta\gamma$ is onto D and since any $a \in G$ belongs to the range of $\beta\alpha$. Moreover, φ is multiplicative and so *it represents the required epimorphism from C onto D.*

6. The Kernel of φ

The kernel C_0 of φ consists of the $c \in C$ for which there exists an $a \in G$ such that $c = \beta\alpha a$ and $\delta\gamma a = 1$; that is, $\gamma a \in H_0$. Thus γa is of the form $\gamma a = (g)$, where $g > 0$ for all real valuations and $\text{ord}_{P_\nu}(g - 1)$ is positive infinite for all standard finite primes P_ν. It follows that only nonstandard prime ideals occur in the decomposition of g and so $(a) = (fg)$, where $f \neq 0$ is standard. We conclude that $a = f\epsilon g$, where ϵ is a unit in $^*F$ and $\epsilon g \in G$. Then $c = \beta\alpha(f\epsilon g) = \beta\alpha(\epsilon g)$. Conversely, if ϵ is any unit in $^*F$, $\epsilon g \in G$, $g > 0$, for all real valuations, and $\text{ord}_{P_\nu}(g - 1)$ is positive infinite for all standard finite primes P_ν, then $c = \beta\alpha(\epsilon g)$ belongs to C_0.

Now suppose that F has r_1 real Archimedean primes and r_2 complex Archimedean primes, and let $r = r_1 + r_2 - 1$. Let $\epsilon_1, \cdots, \epsilon_r$ be a fundamental system of units in F. Then every unit ϵ *in* $^*F$ may be written

$$\epsilon_0 = \epsilon_1^{k_1} \cdots \epsilon_r^{k_r},$$

where the k_j are (rational) integers which may be either finite or infinite.

Now $\beta\alpha(\epsilon_j) = 1$ for $j = 1, \cdots, r$ and so the expression

6.1. $$\beta\alpha(\epsilon_1^{k_1} \cdots \epsilon_r^{k_r} g)$$

with g as above, does not change if we multiply by one or more ϵ_j. Accordingly, we obtain all elements of C_0 by considering the expressions 6.1 for even k_j only. Or, equivalently and varying our notation, we may take $\epsilon_1, \cdots, \epsilon_r$ as the squares of the units of a fundamental system. Then $\epsilon_1, \cdots, \epsilon_r$ are still an independent system of units and, moreover, are now totally positive, that is, positive in all real valuations. For g satisfying the same conditions as before, C_0 is obtained again from 6.1 by letting the k_j range over all integers, standard and nonstandard.

To continue, we write the group of idèles I as the direct product $\bar{I}\tilde{I}$ of the group of idèles that have component 1 at every Archimedean prime, $\bar{I}$, and the group of idèles that have component 1 at every finite prime, $\tilde{I}$. Let $\bar{\psi}$ and $\tilde{\psi}$ be the canonical maps from I to $\bar{I}$ and $\tilde{I}$, respectively, and let $\bar{\alpha} = \bar{\psi}\alpha$ and $\tilde{\alpha} = \tilde{\psi}\alpha$. For our last-mentioned choice of $\epsilon_1, \cdots, \epsilon_r$ and for g satisfying the conditions specified previously, consider the expressions

6.2. $$\tilde{\alpha}(\epsilon_1^{k_1} \cdots \epsilon_r^{k_r} g).$$

An application of the approximation theorem similar to that made earlier in this section to determine a'' shows that, for any fixed $\epsilon = \epsilon_1^{k_1} \cdots \epsilon_r^{k_r}$, we may still adjust g so as to obtain any given element of $\tilde{I}$ with positive components for the real primes and with nonzero components for the complex primes. Let this subset of $\tilde{I}$ be called $\tilde{E}$, and define $\bar{E}$ as the set of all idèles of $\bar{I}$ which are of the form $\bar{\alpha}(\epsilon_1^{k_1} \cdots \epsilon_r^{k_r} g)$. Since $\text{ord}_{P_\nu}(g - 1)$ is positive infinite for all finite primes,

$$\bar{\alpha}(\epsilon_1^{k_1} \cdots \epsilon_r^{k_r} g) = \bar{\alpha}(\epsilon_1^{k_1} \cdots \epsilon_r^{k_r})$$

and so $\bar{E}$ is also given by

6.3. $$\bar{\alpha}(\epsilon_1^{k_1}) \cdots \bar{\alpha}(\epsilon_r^{k_r})$$

where the k_j range over $^*Z$, Z being the ring of standard integers, as usual. Observe that $C_0 = \beta E$, where $E = \bar{E}\tilde{E}$.

To analyze E further, consider any expression $\bar{\alpha}(\epsilon^k)$, where $k \in Z$ and ϵ is a unit in F. As customary in algebraic number theory, we denote by $\bar{Z}$ the completion of Z as an additive group for the topology for which the nonzero ideals of Z form a fundamental system of neighborhoods of 0. $\bar{Z}$ is the direct sum of the p_ν-adic integers Z_{p_ν}, where p_ν ranges over the standard rational primes. Denoting by $\bar{\mu}$ the set of numbers of $^*Z$ which are divisible by all numbers of Z, we find that $\bar{Z}$ is isomorphic to, and may be identified with, $^*Z/\bar{\mu}$ (compare [8]). Let ζ be the canonical mapping from $^*Z$ to $\bar{Z}$.

Let F^i be the ring of integers of F as before and let μ' be the set of numbers of $^*F^i$ which are divisible by the numbers of Z. Then $\bar{F} = {}^*F^i$ is isomorphic to the direct sum (or product) $\Pi F_{p_\nu}{}^i$, where the $F_{p_\nu}{}^i$ are the completions of F^i for the standard finite primes P_ν of F [8]. The elements of $\bar{I}$ whose components are entire, for example, $\bar{\alpha}(\epsilon^k)$, may be identified with elements of $\bar{F}$, and $\bar{\alpha}$ then coincides with the canonical mapping from $^*F^i$ to $\bar{F}$. For any P_ν and for any finite natural number n, the theorem of Fermat–Euler shows that there exists a finite natural m such that

$$\epsilon^m \equiv 1 \bmod P_\nu{}^n.$$

We conclude that for all $k \in \bar{\mu}$, $\epsilon^k - 1$ is divisible by all finite powers of all P_ν and so $\epsilon^k - 1 \in \mu'$. Hence $\bar{\alpha}(\epsilon^k) = 1$ for $k \in \bar{\mu}$, and so $\bar{\alpha}(\epsilon^{k_1}) = \bar{\alpha}(\epsilon^{k_2})$ for $k_1 - k_2 \in \bar{\mu}$. This shows that, for any $Z \in \bar{Z}$, we may define ϵ uniquely and consistently with the old definition for $Z \in Z$—by putting $\epsilon^2 = \bar{\alpha}(\epsilon^k)$ for any k such that $z = \zeta k$. With this definition, $\bar{E}$ consists of the totality of products

6.4. $$\epsilon_1{}^{z_1} \cdots \epsilon_r{}^{z_r}$$

as $z_1, \cdots, z_r$ range over $\bar{Z}$.

The function ϵ^z coincides with the exponential function used in [1] for the components at finite primes. Bearing in mind that exponentiation is defined there also for the Archimedean primes, so as to take care of the real components and of the moduli of the complex components, it is now not difficult to verify that $C_0 = \beta(\bar{E}\tilde{E})$ coincides with the connected component of the identity in the idèle class group C (see in particular eq. (1) of ref. [1]). We may sum up the conclusions reached in this and the preceding section in the following

Theorem 6.5. *There exists an epimorphism φ from the group $D = H/H_0$ to the idèle class group C of F such that the kernel of φ is the connected component of the identity in C.*

It is a fundamental classical result of Chevalley's [2] that C/C_0 is isomorphic to the Galois group Γ of the maximal Abelian extension of F over F. On the other hand, it has been shown ([9], sec. 6) that $D = H/H_0$ is isomorphic to Γ. Since the theory given there is independent of Chevalley's result, it actually provides a new proof of it when taken in conjunction with Theorem 6.5. However, as we have seen, the connection between D and C which is established by the analysis of the last two sections possesses some interest of its own.

References

1. E. ARTIN, Representatives of the Connected Component of the Idèle Class Group, *International Symposium on Algebraic Number Theory, Tokyo, 1955*, pp. 51–54, Collected papers, pp. 249–252.
2. C. CHEVALLEY, Généralisation de la théorie du corps de classes pour les extensiones infinis, *J. Math. Pure Appl.* (9)*15* (1936), 359–371.
3. J. FRAYNE, A. C. MOREL, AND D. S. SCOTT, Reduced Direct Products, *Fundamenta Math. 51* (1962), 195–227.
4. S. KOCHEN, Ultraproducts in the Theory of Models, *Ann. Math.* (2)*79* (1961), 221–261.
5. W. KRULL, Idealtheorie in unendlichen Zahlkörpern, II, *Math. Z. 31* (1930), 517–557.
6. A. ROBINSON, *Non-Standard Analysis* (Studies in Logic and the Foundations of Mathematics), Amsterdam: North-Holland, 1966.
7. A. ROBINSON, A New Approach to the Theory of Algebraic Numbers, *Atti. Accad. Nazl. Lincei, Rend.* (8)*40* (1966), 222–225, 770–774.
8. A. ROBINSON, Non-Standard Theory of Dedekind Rings, *Proc. Acad. Sci. Amsterdam A70* (1967), 444–452.
9. A. ROBINSON, Non-Standard Arithmetic (address before American Mathematical Society Meeting, San Jose, April 1967), *Bull. Am. Math. Soc. 73* (1967), 818–843.

A Set-Theoretical Characterization of Enlargements

by ABRAHAM ROBINSON and ELIAS ZAKON

Introduction

Nonstandard analysis was developed in [4] within a type-theoretical version of higher-order logic. In the present paper we shall describe a purely set-theoretical approach to the subject. In particular, we shall show that the basic notions of nonstandard analysis, such as the concept of enlargement and the concepts of standard and internal entities, can all be defined in terms of certain injections (or monomorphisms) of one model of set theory into another. Even the ultrapower construction, which, in other respects, is helpful in giving a concrete picture of the situation, does not reveal some of the characteristics of internal elements as they will appear from our present approach. In conformity with the spirit of axiomatic set theory, the type-theoretical restrictions of [4] will be replaced by a simpler and less restrictive condition imposed on quantifiers (see Section 3). The equivalence of our present approach with that of [4] follows directly from set-theoretical relations between a full structure, as defined in [4], and a model of set theory (called "superstructure" below).

We are pleased to acknowledge that there are certain points of contact between the theory presented here and the formulation outlined by Kreisel during this symposium.

1. Preliminaries. Terminology and Notation

An *ordered pair* and *n-tuple* are defined, as usual, by $(a, b) = \{\{a, b\}, \{b\}\}$ and $(x_1, \cdots, x_n) = ((x_1, \cdots, x_{n-1}), x_n)$, $(x_1) = x_1$. An *n-ary relation* is any set of ordered n-tuples for a fixed n. For any set R, we define its *domain* $D_R = D(R) = \{x \mid (\exists y)(x, y) \in R\}$, and its *range*

$D'_R = D'(R) = \{y \mid (\exists x)(x, y) \in R\}$. The *image* of a set X under R (briefly, "the R-image of X") is defined by $R[X] = \{y \mid (\exists x \in X)(x, y) \in R\}$. Here the quantifier "$(\exists x \in X)$" means "there is an $x \in X$ such that. . . ." Equivalently, $R[X]$ is the range D'_S of th relation $S = R \cap (X \times D'_R)$. The set X in this definition may, but need not, be a subset of D_R. The *inverse image*, $R^{-1}[X]$, is the image of X under the *inverse relation* $R^{-1} = \{(y, x) \mid (x, y) \in R\}$. A binary relation R is called a *mapping* (*function*) if, for each $x \in D_R$, $R[\{x\}]$ has only one element, denoted by $R(x)$ and called the *function value* at x. In all cases, and especially if x is a set, $R(x)$ must be distinguished from $R[x]$. The *composition* $R \circ S$ of two binary relations R and S is the relation $\{(x, y) \mid (\exists z)(x, z) \in S, (z, y) \in R\}$. For n-ary relations R, we also define the operations of *grouping* and *permuting* the arguments. If $\mathcal{P}$ is a permutation on n elements, we denote by $\mathcal{P}R$ the relation obtained from R by applying the permutation $\mathcal{P}$ to each n-tuple $(x_1, \cdots, x_n) \in R$. Instead, we can *group* the n-tuples in various manners; say, split a quintuple $(a, b, c, d, e) \in R$ into a triple (a, b, c) and a pair (d, e): $((a, b, c), (d, e))$. Then the quintary relation R becomes a binary one, with D_R a set of triples and D'_R a set of ordered pairs. If the same grouping $\mathcal{G}$ is applied to all n-tuples in R, the resulting relation is denoted by $\mathcal{G}R$.

Given a set A, we define inductively the sets $A_0 = A$ and $A_{n+1} = P(\bigcup_{k=0}^{n} A_k)$, $n = 0, 1, \cdots$, where $P(X)$ is the set of all subsets of X. This construction can be continued transfinitely, but we shall not need it. The union of all A_n, $\bigcup_{n=0}^{\infty} A_n$, is called the *superstructure* on A, denoted $\hat{A}$. Clearly, $A_n \in A_{n+1}$ for $n = 0, 1, \cdots$, and also $A_n \subset A_{n+1}$ for $n \geq 1$. Elements of $A_n - A_{n-1}$ ($n \geq 1$) are said to be *of type n* (in $\hat{A}$); those *of type 0*, that is, elements of A_0, are also called "individuals." For convenience we shall assume that "individuals" are objects other than the empty set $\varnothing$ but possessing no elements (so called "Urelements"), although the theory can also be developed under other assumptions. Thus, if $a \in A_0$, then $x \notin a$ for all $x \in \hat{A}$ (but the formula $x \notin a$ or $x \in a$ is always meaningful). No entities other than sets and individuals exist inside $\hat{A}$. Under these assumptions, A_0 is disjoint from other A_n. The latter increase with n, so that $\bigcup_{k=0}^{n} A_k = A_n \cup A_0$; hence $x \in y \in A_{n+1}$ implies $x \in A_n \cup A_0$. If $a,b \in A_n$ then $(a, b) = \{\{a, b\}, \{b\}\} \in A_{n+2}$; thus (a, b) is in $\hat{A}$ when $a, b \in \hat{A}$. Similarly for n-tuples, by induction, hence for n-ary relations of *bounded type* (that is, such that all n-tuples are in *one* A_n for some n).[1] We have $\varphi \in A_n$ for $n \geq 1$. As $A_n \in A_{n+1} \subset \hat{A}$,

[1] Equivalently, a set R is an n-ary relation in $\hat{A}$ iff $R \subseteq (A_0 \cup A_m)^n$ for some m, where X^n is the Cartesian product of n factors equal to X.

each A_n is an element of $\hat{A}$. We write C^n for $C \times C \times \cdots \times C$ (n times), and $x \in y \in c$ for $(x \in y,\ y \in c)$.

Although we use the language of naïve set theory, it is clear that, with small adjustments (such as replacing individuals by nonempty sets with no elements in $\hat{A}$), all can be formalized in any existing axiomatic theory.

2. Monomorphisms

Let A, B be two sets with superstructures $\hat{A}$, $\hat{B}$, respectively. A one-to-one map $\Phi: \hat{A} \to \hat{B}$ is called a *monomorphism* of $\hat{A}$ into $\hat{B}$ if, writing $^*x$ for $\Phi(x)$, we have:

2.1. *For any* $x \in \hat{A}$, $^*\{x\} = \{^*x\}$.

2.2. *If* $X, Y \in \hat{A}$, *then* $^*(X - Y) = {}^*X - {}^*Y$ *and* $^*(X \times Y) = {}^*X \times {}^*Y$.

2.3. *If* $\mathcal{P}$ *is a grouping or permutation on n elements, then* $^*(\mathcal{P}R) = \mathcal{P}(^*R)$ *for any n-ary relation* $R \in \hat{A}$; *hence* $^*(R^{-1}) = (^*R)^{-1}$.

2.4. *For any binary relation* $R \in \hat{A}$, $^*D_R = D(^*R)$; *hence* $^*D'_R = D'(^*R)$.

2.5. *If* $C \in \hat{A}$ *and* $R = \{(x, y) \mid x \in y \in C\}$, *then* $^*R = \{(x, y) \mid x \in y \in {}^*C\}$.

A monomorphism Φ is said to be *normal* if it also preserves all identity relations; that is, we have:

2.5′. *If* $C \in \hat{A}$ *and* $R = \{(x, x) \mid x \in C\}$, *then* $^*R = \{(x, x) \mid x \in {}^*C\}$.[2]

Postulate 2.3 could be split into a few "simpler" ones, dealing with ordered pairs and triples only. We often identify $^*x$ with x *if x is an individual* ($x \in A_0$); then we have:

2.6. *For any* $x \in A_0$, $^*x = x$; *hence* $X \subseteq {}^*X$ *for* $X \subseteq A_0{}^n$ $(n \geq 1)$.

Note that, except for individuals, Φ is a set function; so it is imperative to distinguish $^*X = \Phi(X)$ from $\Phi[X] = \{^*x \mid x \in X\}$. We now derive a few simple consequences of Postulates 2.1–2.5.

[2] Postulates 2.5 and 2.5′ may be limited to the case where C is one of the sets A_n. The general case then follows by Corollary 2.7(d), proved below.

Corollaries 2.7. *Under any monomorphism* $\Phi: \hat{A} \to \hat{B}$, *we have:*

(a) $^*\varnothing = \varnothing$.
(b) $X \subseteq Y$ *iff* $^*X \subseteq {}^*Y$.
(c) $x \in Y$ *iff* $^*x \in {}^*Y$.
(d) $^*(X \cap Y) = {}^*X \cap {}^*Y$.
(e) $^*(X \cup Y) = {}^*X \cup {}^*Y$.
(f) $^*\{x_1, \cdots, x_n\} = \{^*x_1, \cdots, {}^*x_n\}$.
(g) $^*(x_1, \cdots, x_n) = (^*x_1, \cdots, {}^*x_n)$.
(h) $(x_1, \cdots, x_n) \in R$ *iff* $(^*x_1, \cdots, {}^*x_n) \in {}^*R$.
(i) *If* $R \in \hat{A}$ *is an n-ary relation, so is* $^*R$.
(j) *For any binary relation* $R \in \hat{A}$ *and any* $X \in \hat{A}$, $^*(R[X]) = (^*R)[^*X]$.
(k) *If* $x \in Q \in {}^*A_n$, *then* $x \in {}^*A_0 \cup {}^*A_{n-1}$.
(l) *If* $(x_1, \cdots, x_m) \in Q \in {}^*A_n$, *then* $x_k \in {}^*A_0 \cup {}^*A_{n-1}$, $k = 1, \cdots, m$.

Proof. (a) By Postulate 2.2, $^*\varnothing = {}^*(X - X) = {}^*X - {}^*X = \varnothing$. As $X \subseteq Y$ is equivalent to $X - Y = \varnothing$, also (b) follows. Hence replacing $x \in Y$ by $\{x\} \subseteq Y$, we obtain (c). To obtain (d), we note that $X \cap Y = X - (X - Y)$ and apply Postulate 2.2. Similarly, $X \cup Y = Z - [Z - (X \cap Y)]$, with $Z = X \cup Y \in \hat{A}$, yields (e). This, combined with Postulate 2.1, yields (f): $^*\{x_1, \cdots, x_n\} = \bigcup_1^n \{^*x_k\} = \{^*x_1, \cdots, {}^*x_n\}$; and hence also (g) easily follows, by the definition of an ordered pair and ordered n-tuple. Combining (g) with (c), we obtain (h). For the proof of (i), we note that R is an n-ary relation in $\hat{A}$ iff $R \subseteq (A_0 \cup A_m)^n$ for some m; but then, by (b), (e), and Postulate 2.2, we have $^*R \subseteq (^*A_0 \cup {}^*A_m)^n$, so that $^*R$, too, is a set of n-tuples. To prove (j), let $S = R \cap (X \times D'_R)$. Then (see Section 1) $R[X] = D'_S$. Hence, by Postulates 2.4, 2.2, and 2.7 (d), $^*(R[X]) = {}^*D'_S = D'(^*S) = D'(^*R \cap (^*X \times D'_{*R})) = (^*R)[^*X]$, as required. Next, for (k), let $R = \{(x, y) \mid x \in y \in A_n\}$. By Postulate 2.5, $^*R = \{(x, y) \mid x \in y \in {}^*A_n\}$. As $x \in y \in A_n$ implies $x \in A_0 \cup A_{n-1}$ (see Section 1), the definition of R yields $D_R \subseteq A_0 \cup A_{n-1}$. Hence, by (b), (e), and Postulate 2.4, $D(^*R) = {}^*A_0 \cup {}^*A_{n-1}$. Now, if $x \in Q \in {}^*A_n$, we have $(x, Q) \in {}^*R$ by the formula for $^*R$ obtained above, and hence $x \in D(^*R) \subseteq {}^*A_{n-1} \cup {}^*A_0$. Thus (k) is proved. Finally, (l) easily follows from (k) by the definition of an ordered pair and n-tuple. This completes the proof.

NOTE 1. By (f), we have $^*X = \Phi(X) = \Phi[X]$ if X is *finite*.
We now extend Postulate 2.5 to $(n + 1)$-ary relations.

2.8. *If* $C \in \hat{A}$ *and* $R = \{(x_1, \cdots, x_n, y) \mid (x_1, \cdots, x_n) \in y \in C\}$, *then, under any monomorphism* Φ *on* $\hat{A}$, $*R = \{(x_1, \cdots, x_n, y) \mid (x_1, \cdots, x_n) \in y \in *C\}$.

Proof. As $C \in \hat{A}$, we have $C \in A_m$ for some m, and so the defining condition of R, $(x_1, \cdots, x_n) \in y \in C$, implies that all x_k and y in the $(n + 1)$-tuples belonging to R are in the set $D = A_0 \cup A_m$. Hence $R = \{(x, y) \mid x \in y \in C\} \cap D^{n+1}$, where $x = (x_1, \cdots, x_n)$; indeed, the added term D^{n+1} ensures that $(x, y) \in R$ is an $(n + 1)$-*tuple* also in the new formula for R, as it is in the original one; thus the two formulas coincide. Now, by Postulates 2.5, 2.2, and 2.7(d), we have $*R = \{(x, y) \mid x \in y \in *C\} \cap *D^{n+1} = \{(x_1, \cdots, x_n, y) \mid (x_1, \cdots, x_n) \in y \in *C\} \cap *D^{n+1}$. But $*D^{n+1}$ is redundant here because the condition $(x_1, \cdots, x_n) \in y \in *C$, combined with $*C \in *A_m$ and $*D = *A_m \cup *A_0$, *implies* that each x_k and y is in $*D$, as follows from Corollary 2.7(l). Thus, dropping $*D^{n+1}$, we obtain the result.

NOTE 2. If Φ is normal (but not otherwise), Postulate 2.8 holds also with *repeating* variables, such as in $R = \{(x, x, y) \mid (x, x) \in y \in C\}$. In this particular case, we put $R = T \cap (S \times C)$, where $S = \{(u, u) \mid u \in A_0 \cup A_m\}$, $T = \{(x, u, y) \mid (x, u) \in y \in C\}$, and the result follows by Postulates 2.8 and 2.5′.

In more complicated cases, we use *several* identity maps S.

The following example illustrates a typical procedure to be used later.

2.9. *Let* $E = \{(x, y, z) \mid (x, a, b, y, z, c) \in d\}$ *where* $a, b, c, d \in \hat{A}$ *are fixed. Then* $*E = \{(x, y, z) \mid (x, *a, *b, y, z, *c) \in *d\}$.[3]

Proof. Replacing a, b, c, d by variables s, t, u, v, consider the *binary* relation $R = \{((s, t, u, v), (x, y, z)) \mid (x, s, t, y, z, u) \in v \in C\}$ choosing $C \in \hat{A}$ such that $d \in C$, $*d \in *C$. By Postulates 2.8 and 2.3, we easily obtain $*R = \{((s, t, u, v), (x, y, z)) \mid (x, s, t, y, z, u) \in v \in *C\}$. Let $X = \{(a, b, c, d)\}$. Then $R[X]$ consists of all triples (x, y, z) such that the pair $((a, b, c, d), (x, y, z))$ is in R, that is, satisfies $(x, a, b, y, z, c) \in d$. In other words, $R[X] = R[\{(a, b, c, d)\}]$ is exactly the set E given above. Similarly, by definition, $*R[*X] = *R[\{(*a, *b, *c, *d)\}] = \{(x, y, z) \mid (x, *a, *b, y, z, *x) \in *d\}$. Thus, by Corollary 2.7(j), this set equals $*E$, as asserted.

Obviously, this proof does not depend on the number of the variables x, y, z and the constants a, b, c, d, and on their arrangement.[4] By Note 2, we may admit repeating variables if Φ is normal.

[3] Observe that, when passing from E to $*E$, we replace each *constant* c by $*c$, leaving the variables and the rest unchanged.

[4] Moreover, with slight modifications, the proof also works if E has the form $\{(x, y, z) \mid (x, a, b, y, d, c) \in z\} \cap C$, $C \in \hat{A}$.

3. **Internal and Standard Elements. The Metatheorem**

Given a monomorphism $\Phi: \hat{A} \to \hat{B}$ as in Section 2, we define $^*\hat{A} = \bigcup_{n=0}^{\infty} {}^*A_n = \bigcup_{n=0}^{\infty} \Phi(A_n)$, and call all elements of $^*\hat{A}$ the Φ-*internal* (briefly, *internal*) elements (of $\hat{B}$). As previously noted, $\Phi(A_n) \neq \Phi[A_n]$ in general. Thus $^*\hat{A}$ is different from $\Phi[\hat{A}] = \bigcup_{n=0}^{\infty} \Phi[A_n] = \{ ^*x \mid x \in \hat{A} \}$. Elements of the form $^*x$ $(x \in \hat{A})$, that is, those of $\Phi[\hat{A}]$, are called Φ-*standard* (briefly, *standard*) elements. As $x \in A_n$ implies $^*x \in {}^*A_n$ [Corollary 2.7(c)], *all standard elements are internal,* that is, $\Phi[\hat{A}] \subset {}^*\hat{A}$, but the converse is not true.[5] If Postulate 2.6 is assumed, all elements of A_0 (individuals of $\hat{A}$) are standard, hence internal. They belong to $^*A_0$, called the set of *individuals in* $^*\hat{A}$[6]; however, $^*A_0$ may also have other (internal) elements. From Corollary 2.7(k) we immediately obtain:

3.1. *If a set* $Q \in \hat{B}$ *is internal, so are all its elements.*

The converse fails, as follows from well-known examples.

To shorten further proofs we shall now adopt a first-order logical language L, using the connectives $\wedge$, $\vee$, $\supset$, $\cdot\equiv\cdot$, and $\neg$ for "and," "or," "implies," "iff," and "not," respectively, with other details as in [4, p. 6ff.]. For simplicity, we assume that all constants of L are in one-to-one correspondence with all elements of $\hat{A}$, and identify such constants with the corresponding elements, so that these become a part of L and denote themselves. Atomic formulas in L are those of the form $(x_1, \cdots, x_n) \in y$, where y and the x_k are constants or *distinct* variables (constants need not be distinct); if, however, the monomorphism Φ is normal, then also repeating variables are allowed in atomic formulas. Well-formed formulas (wff) and well-formed sentences (wfs) then are defined inductively, as in [4, p. 7]. We use the abbreviations "$(\forall x \in C)$" and "$(\exists x \in C)$" for "$(\forall x)[[x \in C] \supset \cdots]$" and "$(\exists x)[[x \in C] \wedge \cdots]$," respectively (read: "*for every* x *in* $C, \cdots$" and "*there is an* x *in* C *such that* $\cdots$"); here C is supposed to be an element of $\hat{A}$, that is, a *constant* of L (not a variable). We single out those wff in which all quantifiers (if any) are of that particular *form; that is, each quantifier specifies the domain of its variable in the manner described above, with* $C \in \hat{A}$. Henceforth *only such formulas will be admitted and called "wff" or "wfs."* This is tantamount to singling out a certain set of wff in L, without changing

[5] This, combined with Postulate 3.1, implies that internal elements are exactly all elements of standard elements; that is, $x \in {}^*\hat{A}$ iff $x \in {}^*X$ for some $X \in \hat{A}$. This extremely simple characterization of internal elements has been made possible by our present set-theoretical approach based on monomorphisms.

[6] Members of $^*A_1$ need not be genuine "Urelements," but they have no *internal* elements (see Postulate 3.3). Thus, *inside* $^*\hat{A}$, they behave like individuals.

the nature of L as a *first-order* language. Note that it suffices to use quantifiers with $C = A_n \cup A_0$; for if $C \in A_{n+1}$, the quantifiers $(\forall x \in C)$ and $(\exists x \in C)$ can be written $(\forall x \in A_n \cup A_0)[[x \in C] \supset \cdots]$ and $(\exists x \in A_n \cup A_0)[[x \in C] \wedge \cdots]$, respectively.

By the *Φ-transform* of such a wff α, denoted $*\alpha$, we mean the formula (not necessarily in L) obtained from α by replacing in it every constant c by $*c$, but leaving the variables and the rest of the formula unchanged (see the footnote to Postulate 2.9); $*\alpha$ has a self-evident interpretation in $*\hat{A}$ (not in $\hat{A}$). For example, the Φ-transform of the formula $(\forall x \in C)[[x \in a] \vee [y \in b]]$, where a, b, C are constants, is $(\forall x \in *C)[[x \in *a] \vee [x \in *b]]$. The symbol $\{x \in C \mid \alpha(x)\}$, where $\alpha(x)$ is a wff containing x as a free variable, means "the set of all elements x of C, satisfying $\alpha(x)$." Similarly, for sets of n-tuples, $\{(x_1, \cdots, x_n) \in C \mid \alpha(x_1, \cdots, x_n)\}$, and for sets in $*\hat{A}$ such as $\{x \in *C \mid *\alpha(x)\}$. We shall sometimes use semiformal abbreviations of wff such as $a \subseteq b$ for $(\forall x \in a)[x \in b]$, $a \notin b$, for $\neg[a \in b]$, $x \in a \cup b$, for $[x \in a] \vee [x \in b]$, and so on. (Note that these expressions are also defined if a and b are individuals; then, by our conventions, $a \cup b = a \cap b = \varnothing$.) We shall now prove jointly two metamathematical propositions:

Meta-theorem 3.2. (a) *A well-formed sentence α is true in $\hat{A}$ iff its Φ-transform $*\alpha$ is true in $*\hat{A}$.* (b) *If $\alpha(x_1, \cdots, x_m)$ is a wff, with $x_1, \cdots, x_m$ its only free variables, and if, for some $C \in \hat{A}$, $E = \{(x_1, \cdots, x_m) \in C \mid \alpha(x_1, \cdots, x_m)\}$, then $*E = \{(x_1, \cdots, x_m) \in *C \mid *\alpha(x_1, \cdots, x_m)\}$.*

Proof. If α is an atomic sentence, it has the form $(a_1, \cdots, a_n) \in b$ $(a_k, b \in \hat{A})$. By Corollaries 2.7 it is equivalent to $(*a_1, \cdots, *a_n) \in *b$. Thus (a) is true for atomic sentences. A simple induction process (over the number of brackets $[\cdots]$ that show how α is built from atomic formulas) extends (a) to all wfs *containing no quantifiers*. Next, take an atomic wff $\alpha(x_1, \cdots, x_p, a_1, \cdots, a_q)$, with $x_1, \cdots, x_p$ its variables, and a_k its constants. Then it has the form $(x_1, \cdots, x_p, a_1, \cdots, a_{q-1}) \in a_q$, or $(x_1, \cdots, x_{p-1}, a_1, \cdots, a_q) \in x_p$, with the x_k and a_i possibly permuted. In all such cases, (b) follows in the manner exemplified in Postulate 2.9, that is, by regrouping the relation $R = \{(x_1, \cdots, x_p, y_1, \cdots, y_q) \mid \alpha(x_1, \cdots, x_p, y_1, \cdots, y_k)\}$, and computing $R[\{(a_1, \cdots, a_q)\}]$. This proves (b) for atomic formulas.

Next we show that, if (b) holds for two wff α and β, then it also holds for $\alpha \wedge \beta$ and $\neg\alpha$ (this takes care of *all* logical connectives). Let $\alpha = \alpha(x_1, \cdots, x_m, y_1, \cdots, y_n)$, $\beta = \beta(x_1, \cdots, x_m, z_1, \cdots, z_p)$, $m \geq 0$, where the x_k are the *common* free variables of α and β (if any). For brevity, we put $x = (x_1, \cdots, x_m)$, $y = (y_1, \cdots, y_n)$, $z = (z_1, \cdots, z_p)$. With x, y, z so defined, let $R = \{(x, y) \in D^2 \mid \alpha\}$ and $S = \{(x, z) \in D^2 \mid \beta\}$, where $D \in \hat{A}$ is to be fixed later; we treat R and S as *binary* relations.[7] Even so, using Postulate 2.3 and our assumptions as to α and β, we easily obtain [since (b) holds for α and β]

3.2.1. $*R = \{(x, y) \in *D^2 \mid *\alpha\}, \qquad *S = \{(x, z) \in *D^2 \mid *\beta\}.$

Now let $E = \{(x, y, z) \in C \mid \alpha \wedge \beta\}$, $C \in \hat{A}$. Again, by Postulate 2.3, it does not matter whether E is treated as an $(m + n + p)$-ary or ternary relation. For, if E is $(m + n + p)$-ary ,we may safely assume that the set C, too, consists of $(m + n + p)$-tuples only (otherwise, drop the redundant other elements of C!). Then we can regroup *both* E and C as ternary relations, and by Postulate 2.3 it suffices to show that $*E = \{(x, y, z) \in *C \mid *\alpha \wedge *\beta\}$, as a set of *triples.*[8] This is what we shall prove now.

As $C \in \hat{A}$, we have $C \in A_r$ for some r. Hence, for any $(x, y, z) \in E$, the assumed condition $(x, y, z) \in C$ implies $x,y,z \in A_r \cup A_0 = D$ (*we thus fix D in Postulate* 3.2.1). Noting this, we obtain at once: $E = \{(x, y, z) \in C \mid \alpha \wedge \beta\} = \{(x, y, z) \in C \mid (x, y) \in R, (x, z) \in S\} = C \cap \{(x, y, z) \mid (x, y) \in R, z \in D\} \cap \{(x, y, z) \mid (x, z) \in S, y \in D\} = C \cap (R \times D) \cap \mathcal{P}(S \times D)$, where $\mathcal{P}$ is the permutation $(x, z, y) \rightarrow (x, y, z)$. Hence, by Postulates 2.2, 2.7(d), and 2.3, we have $*E = *C \cap (*R \times *D) \cap \mathcal{P}(*S \times *D)$ or, by Postulate 3.2.1, after simplifications, $*E = \{(x, y, z) \in *C \mid *\alpha \wedge *\beta\}$, as required. Thus, indeed, if (b) holds for α and β, it also holds for $\alpha \wedge \beta$.

Now let $E = \{(x, y, z) \in C \mid \neg\alpha(x, y, z)\} = C - \{(x, y, z) \in C \mid \alpha\}$. Then, by Postulate 2.2, $*E = *C - \{(x, y, z) \in *C \mid *\alpha\} = \{(x, y, z) \in *C \mid \neg *\alpha\}$. Thus, again using induction over the number of brackets, we can complete the proof of both (a) and (b) *for wff without quantifiers.*

If there are quantifiers, we use induction over their number. Suppose that both (a) and (b) hold for formulas with n quantifiers ($n \geq 0$), and let α be a wfs with $n + 1$ quantifiers in prenex form: $\alpha = \alpha(qx_{n+1})(qx_n) \cdots (qx_1)\ \beta(x_1, \cdots, x_{n+1})$, where the (qx_k) are quantifiers and $\beta = \beta(x_1, \cdots, x_{n+1})$ is a wff without quantifiers. We may assume that (qx_{n+1}) is an *existential* quantifier, $(\exists x_{n+1} \in Q)$, $Q \in \hat{A}$; otherwise, we achieve this by replacing α by $\neg\alpha$. Thus, writing y for x_{n+1}, we have $\alpha = (\exists y \in Q)(qx_n) \cdots (qx_1)\ \beta(x_1, \cdots, x_n, y)$. In other words, α states that the set

$$D = \{y \in Q \mid (qx_n) \cdots (qx_1)\ \beta(x_1, \cdots, x_n, y)\} \qquad (Q \in \hat{A})$$

[7] If α and β have no free variables in common ($m = 0$), then R and S become *unary* relations, and the proof is trivial. We omit this trivial case and assume henceforth that $m > 0$.

[8] Owing to Postulate 2.3, no generality is lost by assuming this particular arrangement of the variables: $(x_1, \cdots, x_m, y_1, \cdots, y_n, z_1, \cdots, z_p)$.

is not empty. As the right-hand expression contains only n quantifiers, our inductive assumption yields [by Corollary 2.7(a)]

3.2.2. $*D = \{y \in *Q \mid (*qx_n) \cdots (*qx_1) \quad *\beta(x_1, \cdots, x_n, y)\} \neq \varnothing$,

where $(*qx_k)$ stands for $(\exists x_k \in *Q_k)$ or $(\forall x_k \in *Q_k)$, $Q_k \in \hat{A}$. But Formula 3.2.2 means exactly that $*\alpha$ holds in $*\hat{A}$. Conversely, if $*\alpha$ holds, then $*D \neq \varnothing$ implies that $D \neq \varnothing$, hence that α is true. This completes the inductive process as far as (a) is concerned.

For (b), consider a set $E = \{(x_1, \cdots, x_m) \in C \mid \alpha(x_1, \cdots, x_m)\}$ and suppose that $\alpha = \alpha(x_1, \cdots, x_m) = (qy_{n+1}) \cdots (qy_1) \; \beta(x_1, \cdots, x_m, y_1, \cdots, y_{n+1})$, that is, α contains $n + 1$ quantifiers. We may again assume that (qy_{n+1}) is an existential quantifier, $(\exists y \in Q)$, $Q \in \hat{A}$. Thus, with $y = y_{n+1}$,

3.2.3. $E = \{(x_1, \cdots, x_m) \in C \mid (\exists y \in Q)(qy_n) \cdots (qy_1)\beta\}, \qquad Q \in \hat{A}$,
where $\beta = \beta(x_1, \cdots, x_m, y_1, \cdots, y_n, y)$ has no quantifiers.

Now consider also the set $R = \{((x_1, \cdots, x_m), y) \in C \times Q \mid (qy_n) \cdots (qy_1)\beta\}$, treating it as a *binary* relation. Our inductive assumption again yields $*R = \{((x_1, \cdots, x_m), y) \in *C \times *Q \mid (*qy_n) \cdots (*qy_1)*\beta\}$. The domain of $*R$ is, by definition, $D_{*R} = \{(x_1, \cdots, x_m) \in *C \mid (\exists y \in *Q)(*qy_n) \cdots (*qy)*\beta\}$. Similarly, D_R is exactly the set E in Formula 3.2.3. Hence, by Postulate 2.4, $*E = *D_R = D(*R) = \{(x_1, \cdots, x_m) \in *C \mid *\alpha\}$, and the proof is complete.

We shall denote by K_A or K the set of all sentences that are true in $\hat{A}$ and can be written as *wfs* (in the restricted sense). All such sentences will be called *K-sentences*. In a wider sense, we apply the name "K-sentence" also to any Φ-transform of a K-sentence, under any monomorphism Φ. We shall now give a few examples of applications of Meta-theorem 3.2.[9]

3.3. *No internal elements $x \in *\hat{A}$ can belong to any $y \in *A_0$.*

Proof. As A_0 consists of individuals, we have for every n the K-sentence: $(\forall y \in A_0)(\forall x \in A_n)$, $x \notin y$. By Meta-theorem 3.2, then, $(\forall y \in *A_0)(\forall x \in *A_n)$, $x \notin y$; hence $x \notin y$ for every $x \in *\hat{A}$, as asserted.

3.4. *The union, difference, and intersection of any finite number of members of $*\hat{A}$ is an internal set. So also is the union or intersection of any internal set family $U \in *\hat{A}$, even if U is infinite.*

[9] The propositions to be proved are known for "internal relations" as defined in [4, p. 42]. We prove them for "*φ-internal elements*" defined by monomorphisms, mainly to demonstrate the economy gained by the use of Theorem 3.2 and the set-theoretical approach.

Proof. For any n, we have the K-sentence $(\forall X, Y \in A_n \cup A_0)(\exists Z \in A_{n+1})(\forall x \in A_n \cup A_0)[x \in Z] \cdot\equiv\cdot [x \in X \cup Y]$. Hence, by Theorem 3.2, for any $X, Y \in {}^*A_n \cup {}^*A_0$, there is a $Z \in {}^*A_{n+1}$ (hence $Z \in {}^*\hat{A}$) containing the same elements $x \in {}^*A_n \cup {}^*A_0$ as does $X \cup Y$. But, by Corollary 2.7(k), *all* elements of Z, X, and Y are in ${}^*A_n \cup {}^*A_0$. Thus $Z = X \cup Y \in {}^*\hat{A}$; similarly for $X - Y$ and $X \cap Y$. Since every $X, Y \in {}^*\hat{A}$ are in *one* ${}^*A_n \cup {}^*A_0$ for a large n, all is proved for *two* sets, hence for finitely many sets. Next, if $U \in {}^*A_{n+2}$, we use the K-sentence $(\forall U \in A_{n+2})(\exists Z \in A_{n+1})(\forall x \in A_n \cup A_0)[x \in Z] \cdot\equiv\cdot [(\exists Y \in A_{n+1}), x \in Y \in U]$. The rest is obvious.

Similarly we obtain: If two binary relations R and S are internal, so is their composition $R \circ S$.

3.5. *For any binary relations $R,S \in \hat{A}$, we have ${}^*(R \circ S) = {}^*R \circ {}^*S$.*

Proof. Choose n such that $R,S \in A_n$ and put $D = A_0 \cup A_n$. Then $(x, y) \in R \cup S$ implies $x,y \in D$; $(x, y) \in {}^*R \cup {}^*S$ implies $x,y \in {}^*D$; and so the definitions of $R \circ S$ and ${}^*R \circ {}^*S$ can be formally written as $R \circ S = \{(x, y) \in D^2 \mid (\exists z \in D)[(x, z) \in S] \wedge [(z, y) \in R]\}$, and similarly for ${}^*R \circ {}^*S$. By applying Theorem 3.2(b) to $R \circ S$, we immediately obtain ${}^*(R \circ S) = {}^*R \circ {}^*S$, as required.

3.6. *If the monomorphism Φ is normal and if $f \in \hat{A}$ is a mapping, so also is ${}^*f$.*

Proof. By Corollary 2.7(i), ${}^*f$ is certainly a binary relation. The requirement that it be a mapping is equivalent to ${}^*f \circ {}^*f^{-1} = {}^*R$, where ${}^*R = \{(x, x) \mid x \in D'_{*f}\}$ is the identity map on D'_{*f}. This, however, follows from $f \circ f^{-1} = R = \{(x, x) \mid x \in D'_f\}$ by Theorem 3.5 and Postulates 2.4 and 2.5′ (2.5′ applies since Φ is normal).

3.7. *A monomorphism Φ on $\hat{A}$ is normal iff $x \in {}^*\hat{A}$ implies $\{x\} \in {}^*\hat{A}$; more precisely, $x \in {}^*A_n$ implies $\{x\} \in A_{n+1}$.*

Proof. The identity map I_n on A_n coincides with the relation $\{(x, y) \in A_n{}^2 \mid (\forall Z \in A_{n+1})\ x \in Z \cdot\equiv\cdot y \in Z\}$. By Theorem 3.2(b), we have ${}^*I_n = \{(x, y) \in {}^*A_n{}^2 \mid (\forall Z \in {}^*A_{n+1})\ x \in Z \cdot\equiv\cdot y \in Z\}$. If Φ is normal, ${}^*I_n$ must be the identity map on ${}^*A_n$ (by Postulate 2.5′) and so, for each $x \in {}^*A_n$, ${}^*I_n(x) = x$ and ${}^*I_n[\{x\}] = \{x\} = \{y \in {}^*A_n \mid (x, y) \in {}^*I_n\}$. We can now use the K-sentence $(\forall x \in A_n)(\exists Z \in A_{n+1})(\forall y \in A_n)[y \in Z \cdot\equiv\cdot (x, y) \in I_n]$ to obtain that for each $x \in {}^*A_n$ there is $Z \in {}^*A_{n+1}$ such that $Z = {}^*I_n[\{x\}] = \{x\}$, whence $\{x\} \in {}^*A_{n+1}$, as asserted. Conversely, if $x \in {}^*A_n$ implies $\{x\} \in {}^*A_{n+1}$, then ${}^*I_n$ must be the identity relation on ${}^*A_n$; indeed, if $x \neq y$ $(x, y \in {}^*A_n)$, then $Z = \{x\} \in {}^*A_{n+1}$, $x \in Z$, $y \notin Z$, whence $(x, y) \notin {}^*I_n$; thus ${}^*I_n$ can only contain pairs (x, y) with $x = y$.

4. Concurrent Relations. Enlargements

A binary relation $R \in \hat{A}$ is said to be *concurrent* (see [4], pp. 41–42) if, for any finite number of elements $a_1, \cdots, a_m$ of its domain D_R, there is some $b \in \hat{A}$ such that $(a_k, b) \in R$, $k = 1, \cdots, m$. A monomorphism $\Phi: \hat{A} \to \hat{B}$ is referred to as *enlarging* if, for every concurrent relation $R \in \hat{A}$, there is some $b \in {}^*\hat{A}$ such that $({}^*x, b) \in {}^*R$ for all $x \in D_R$ simultaneously; we then also say that Φ "*bounds*" all concurrent relations. In this case ${}^*\hat{A}$ is called an *enlargement* of $\hat{A}$ (a *normal enlargement* if Φ is normal). The enlargement ${}^*\hat{A}$ (and the monomorphism Φ) are said to be *comprehensive* (see [6]) if, for any sets $C, D \in \hat{A}$ and any mapping $f: C \to {}^*D$, there is an *internal* map $g: {}^*C \to {}^*D$ ($g \in {}^*\hat{A}$) such that $f(x) = g({}^*x)$ for every $x \in C$. The existence of enlargements can be proved in various ways (see [4], p. 30ff., for higher-order structures). We shall use a construction based on ultrapowers; this will also lead to a construction of a set-theoretical monomorphism as defined in Section 2.

For this purpose (although not for the later work) it is convenient to replace the language L of Section 3 by a language L' as follows. The constants of L' are again all elements of $\hat{A}$, which thus denote themselves. The logical connectives are as before. The atomic formulas in L' are those of the form $x \mathrel{\dot{\in}} y$ or $x \doteq y$, where x and y are variables or constants (possibly, constant n-tuples, treated as *single* elements of $\hat{A}$). The symbols $\dot{\in}$ and $\doteq$ are interpreted in $\hat{A}$ as the ordinary membership $\in$ and equality $=$, respectively; in other structures they may have a different meaning. Well-formed formulas (wff) and sentences (wfs) are defined as usual for first-order logic (see [4], p. 7), *without the restriction imposed on quantifiers in L*. We denote by K' the set of all wfs in L' which hold in $\hat{A}$, and call them *K'-sentences*. Two important K'-sentences are $(\forall x)(\forall y)[x \doteq y] \cdot\equiv\cdot (\forall z)[x \mathrel{\dot{\in}} z \cdot\equiv\cdot y \mathrel{\dot{\in}} z]$ and $(\forall x)(\forall y)(\forall z)(\forall u)[[x \mathrel{\dot{\in}} y] \wedge [x \doteq z] \wedge [y \doteq u]] \supset [z \mathrel{\dot{\in}} u]$. They show that $\doteq$ is an equivalence relation with substitutivity property (with respect to $\dot{\in}$) in $\hat{A}$, hence in any other model M of K'.[10] Thus, replacing each element $x \in M$ by its equivalence class $[x] = \{y \mid y \doteq x\}$ and setting $[x] \mathrel{\dot{\in}} [z]$ iff $x \mathrel{\dot{\in}} z$, we can obtain a new model of K' in which $\doteq$ is the ordinary identity relation. We shall henceforth always assume that this is the case. Another K'-sentence states that, for any X, Y, there is Z such that Z is the "union" of X and Y, that is, $(\forall x)[[x \mathrel{\dot{\in}} X \vee x \mathrel{\dot{\in}} Y] \cdot\equiv\cdot x \mathrel{\dot{\in}} Z]$; so we define $Z = X \mathbin{\dot{\cup}} Y$ in M; the "dot" is to

[10] For brevity, we denote by M both the *structure* and the *set of its objects;* "$x \in M$" means that x is one of such objects.

distinguish $X \dot{\cup} Y$ from ordinary set-theoretical unions; similarly for other set-theoretical concepts.

Moreover, if M is a model of K', there is a one-to-one correspondence $\psi: \hat{A} \to M$ between the constants of K' and some elements of M, with the property that K'-sentences containing constants hold in M when each constant c is replaced by $\psi(c)$; we call ψ the *interpretation map* for M. Setting $\psi(A_n) = {}^*A_n$, $n = 0, 1, \cdots$ (where $\hat{A} = \bigcup_{n=0}^{\infty} A_n$), we call an element $x \in M$ *internal* if $x \dot{\in} {}^*A_n$ for some n. Using suitable K'-sentences, one can easily show that Corollary 2.7(k) and Postulate 3.3 hold in M with $\in$ replaced by $\dot{\in}$[11] (we say that x is an "element" of X if $x \dot{\in} X$); also $X \dot{\in} {}^*A_1$ iff $X \dot{\subseteq} {}^*A_0$; that is, $(\forall x)[[x \dot{\in} X] \supset [x \dot{\in} {}^*A_0]]$. Thus we can replace each $X \dot{\in} {}^*A_1$ by the genuine *set* $\{x \mid x \dot{\in} X\}$; then we do the same with each $X \dot{\in} {}^*A_2$, and so on. Proceeding inductively, we can achieve that, for *internal* members of M, $\dot{\in}$ becomes the ordinary $\in$. We call M so modified a *collapsed* model of K'. In such models we use the ordinary set-theoretical notation. We now obtain:

Theorem 4.1. *For every collapsed model M of K', there is a normal monomorphism Φ on $\hat{A}$ into a superstructure $\hat{B}$ such that the Φ-internal members of $\hat{B}$ are exactly the internal members of M.*

Proof. Let $\Phi: \hat{A} \to M$ be the interpretation map of M. As noted above, members of $\Phi(A_0)$ have no elements in M; so we may treat them as a set of individuals B and replace M by the superstructure $\hat{B}$ (this does not affect the map Φ and the *internal* members of M because they have internal elements only). We shall now verify that $\Phi: \hat{A} \to \hat{B}$ satisfies the Postulates 2.1–2.5′.

1. Fix $a \in \hat{A}$ and let $P = \{a\}$. Then we have the K'-sentence $[a \in P] \wedge (\forall x)[[x \in P] \cdot\equiv\cdot [x = a]]$. Hence, writing again $^*a$ for $\Phi(a)$, $[{}^*a \in {}^*P] \wedge (\forall x)[[x \in {}^*P] \cdot\equiv\cdot [x = {}^*a]]$, by the properties of interpretation maps. This means that $\{{}^*a\} = {}^*P = {}^*\{a\}$, proving Postulate 2.1.

2. Fix $C, D \in \hat{A}$ and let $P = C - D$. Then we easily obtain ${}^*P = {}^*C - {}^*D$ from the K'-sentence: $(\forall x)[[x \in P] \cdot\equiv\cdot [x \in C \wedge \neg [x \in D]]$. A similar proof yields ${}^*C \times {}^*D = {}^*(C \times D)$ on noting that $(x, y) \in P$ can be written as a wff.[12] This proves Postulate 2.2.

3. For Postulate 2.3 it suffices to consider groupings and permutations on two and three elements. Let $\mathcal{P}: (x, y) \to (y, x)$, and fix a binary relation $R \in \hat{A}$. Then, by the already proved property 2.2, $^*R$ is a binary

[11] More precisely: If $x \dot{\in} Q$ and $Q \dot{\in} {}^*A_n$, then $x \dot{\in} {}^*A_0$ or $x \dot{\in} {}^*A_{n-1}$. If $x \dot{\in} {}^*A_0$, then $y \dot{\notin} x$ for all y.

[12] Indeed, we have $(x, y) \in$ P iff $(\exists z)(\exists u)(\exists v)[[z \in P] \wedge [z = \{u, v\}] \wedge [u = \{x, y\}] \wedge [v = \{y\}]]$, where "$z = \{u, v\}$" stands for "$[u \in z] \wedge [v \in z] \wedge (Vw)[[w \in z] \supset [w = u] \vee [w = v]]$"; similarly for "$u = \{x, y\}$ and "$v = \{y\}$."

relation, too [indeed, Corollary 2.7(i) is proved from Postulate 2.2 alone]. Now we easily obtain $*(R^{-1}) = (*R)^{-1}$ from the K'-sentence $(\forall x)(\forall y)[(x, y) \in R^{-1} \cdot\equiv\cdot (y, x) \in R]$; similarly for the other cases involved. Thus Φ satisfies Postulate 2.3.

4. Postulate 2.4 follows from the K'-sentence $(\forall x)[[(\exists y)(x, y) \in R] \cdot\equiv\cdot [x \in P]]$, where $P = D_R$.

5. If R is the identity map on $C \in \hat{A}$, the K'-sentence $(\forall x)(\forall y)[[x \in C \wedge y \in C \wedge x = y] \cdot\equiv\cdot (x, y) \in R]$ shows that $*R$ is the identity map on $*C$. This proves Postulate 2.5′. Similarly for Postulate 2.5.

Thus Φ is, indeed, a normal monomorphism. As it is also the interpretation map of M, the Φ-internal elements are, by definition, the internal elements of M. Thus all is proved.

NOTE 1. From Theorem 3.2(a) it now follows that the internal elements of $\hat{B}$ (hence of M) form a model $*\hat{A}$ of K as defined in Section 3. On the other hand, given a monomorphism $\Phi: \hat{A} \to \hat{B}$, the Φ-internal elements *do not, in general, constitute a model of* K', even if we define in $*\hat{A}$ an equality relation by setting $x \doteq y$ iff $[x \in Z \cdot\equiv\cdot y \in Z]$ for all $Z \in *\hat{A}$. This is the main reason why the language L and the set K of Section 3 are more useful in applications than L' and K'.

NOTE 2. It also follows that every monomorphism $\Phi: \hat{A} \to \hat{B}$ can be transformed into a normal one by defining the equivalence relation as in Note 1 and replacing each element of $*\hat{A}$ by its equivalence class.

NOTE 3. If the model M in Theorem 4.1 is not organized, the same proof shows that Φ still satisfies Postulates 2.1–2.5, but with ordinary set-theoretical concepts replaced by their "dotted" counterparts. Theorem 4.1, combined with known results, now yields:

Theorem 4.2. *For every superstructure* $\hat{A}$, *there is a superstructure* $\hat{B}$ *and a monomorphism* $\Phi: \hat{A} \to \hat{B}$ *which is normal, enlarging, and comprehensive. Thus* $\hat{A}$ *always has a normal comprehensive enlargement,* $*\hat{A}$, *generated by a monomorphism* Φ.

Proof. Let I be an infinite index set and let F be an ultrafilter of subsets of I. Let M be the family of all maps $f: I \to \hat{A}$. Given $f, g \in M$, we write $f \doteq g$ iff the set $\{i \in I \mid f(i) = g(i)\}$ is a member of F; similarly, we put $f \,\dot{\in}\, g$ iff $\{i \in I \mid f(i) \in g(i)\} \in F$. Then M, with $\doteq$ and $\dot{\in}$ so defined, is the F-ultrapower of $\hat{A}$ (as a model of K'). As is well known (see [1], [2], [3], or [5], p. 242), such an ultrapower must itself be a model of K'.[13] Passing to equivalence classes and collapsing, we can transform

[13] The interpretation map $\varphi: \hat{A} \to M$ is given by setting, for each $a \in \hat{A}$, $\varphi(a) = *a$, where $*a$ is the constant function $*a: I \to M$, with $*a(i) = a$ for all $i \in I$.

M into a structure to which Theorem 4.1 applies. Thus we obtain a normal monomorphism $\Phi: \hat{A} \to \hat{B}$, and the corresponding set $^*\hat{A}$. Finally, as is shown in [6], the index set I and the ultrafilter F can be so chosen that all concurrent relations $R \in \hat{A}$ are "bounded" in M, and also the comprehensiveness condition is satisfied. (Although "internal" elements are defined in [6] in terms of higher-order logic, the argument carries over, almost verbally, to our case as well.) Thus the monomorphism $\Phi: \hat{A} \to \hat{B}$ becomes comprehensive and enlarging, and the proof is complete.

References

1. T. FRAYNE, D. C. MORE, AND D. S. SCOTT, Reduced Direct Products, *Fundamenta Math. 51* (1962), 195–227.
2. H. J. KEISLER, A Survey of Ultraproducts, *Logic, Methodology and Philosophy of Science, Proc. 1964 Intern. Conf.* Amsterdam: North-Holland, 1965, pp. 112–124.
3. S. KOCHEN, *Ultraproducts in the Theory of Models, Ann. Math.* (2)*79* (1961), 221–261.
4. A. ROBINSON, *Non-Standard Analysis* (Studies in Logic and the Foundations of Mathematics). Amsterdam: North-Holland, 1966.
5. A. ROBINSON, *Introduction to Model Theory and to the Metamathematics of Algebra.* Amsterdam: North-Holland, 1965.
6. A. ROBINSON, Non-Standard Theory of Dedekind Rings, *Proc. Acad. Sci. Amsterdam A70* (1967), 444–452.

Germs

by ABRAHAM ROBINSON

1. Introduction

Let S and T be a pair of topological spaces and let p and q be any two elements of S and T, respectively. We define an equivalence relation in the set of subsets of S by putting $V_1 \sim V_2$ for any $V_1 \subset S$, $V_2 \subset S$ if there exists an open neighborhood U of p such that $V_1 \cap U = V_2 \cap U$. The equivalence classes with respect to this relation are called the *germs of sets* (or, briefly, *set germs*) *at* p. For any two set germs of p, α and β, we define an inclusion relation, $\subset$, by putting $\alpha \subset \beta$ if there exist $V_1 \subset \alpha$, $V_2 \subset \beta$ and an open neighborhood U of p such that $\alpha \cap U \subset \beta \cap U$. Then $\alpha = \beta$ if and only if $\alpha \subset \beta$ and $\beta \subset \alpha$. Similarly, we define formal operations of union and intersection for set germs α, β at p by putting $\alpha \cup \beta = \gamma$ (or $\alpha \cap \beta = \gamma$) if γ is the (uniquely defined) set germ such that there exist $V_1 \in \alpha$, $V_2 \in \beta$, $V_3 \in \gamma$ for which $V_1 \cup V_2 = V_3$ (or $V_1 \cap V_2 = V_3$).

Let F be the class of functions into T whose domain is a subset of S and includes an open neighborhood of p. For any $A \subset S$, $f \in F$, $f \mid A$ shall denote the restriction of f to A (so that the domain of definition of $f \mid A$ is the intersection of A with the domain of definition of f). We define an equivalence relation in F by putting $f_1 \sim f_2$ for any $f_1 \in F$, $f_2 \in F$ if there exists an open neighborhood U of p such that $f_1 \mid U = f_2 \mid U$. The equivalence classes with respect to this relation are called the *germs of functions* (briefly, *function germs*) at p. If ϕ is a function germ, then the set germ α is called the *local variety* or *variety germ* of ϕ at p (for the given $q \in T$) if there exist $f \in \phi$, $V \in \alpha$ and an open neighborhood U of p which belongs to the domain of definition of f such that

$$V = \{x \mid x \in U \text{ and } f(x) = q\}.$$

If T is endowed with a ring structure, this induces a ring structure also on the function germs, as follows. If ϕ and ψ are function germs, choose $f_1 \in \phi$ and $f_2 \in \psi$ and an open neighborhood U of B such that both f_1 and f_2 are defined on U, and let $\phi + \psi$ at $\phi\psi$ be the function germs which include the functions $f_1 + f_2$, f_1f_2 as defined on U by pointwise addition and multiplication, respectively.

The language of germs is in wide use at present since it provides a convenient framework for the analysis of the local behavior of functions. However, it will be observed that set germs are not actual sets of points, and function germs are not actual functions; in other words, the individual points are lost in the passage to germs. In the present paper, we shall show how to remedy the situation by the use of nonstandard analysis. Thus we shall replace set germs and function germs by actual sets and actual functions, and we shall give an effective application of this procedure.

2. Nonstandard Germs

Retaining the notation of Section 1, let $^*S$ and $^*T$ be enlargements of S and T, respectively (supposed embedded, as usual, in a common enlarged universe, $^*M$). The monad of p in $^*S$, $\mu(p)$, is defined as the intersection of the extensions to $^*S$, of the standard open neighborhoods of p in S, $\mu(p) = \bigcap_\nu {}^*U_\nu$ (see [2], Chap. 4). Let $f \in F$; then we shall, for clarity, denote the extension of f to $^*S$ by $^*f$ (and not by f, as in [2]).

An *n.s.* (nonstandard) *set germ at* p is defined as the intersection $^*V \cap \mu(p)$ for any subset V of S. Similarly, an *n.s. function germ at* p is defined as the restriction of $^*f$ to $\mu(p)$, $^*f \mid \mu(p)$, for any $f \in F$. In this connection, observe that $^*f$, and hence the corresponding n.s. function germ, is defined *on* $\mu(p)$. For any set germ α, define $\rho(\alpha)$ as $^*V \cap \mu(p)$ for arbitrary $V \in \alpha$; and for any function germ ϕ define $\sigma(\phi)$ as $^*f \mid \mu(p)$ for arbitrary $f \in \phi$. Using elementary considerations from nonstandard analysis, it is not difficult to verify that these rules establish one-to-one mappings, ρ and σ between the standard and nonstandard germs of sets and of functions, respectively. Moreover, $\rho(\alpha) \subset \rho(\beta)$ if and only if $\alpha \subset \beta$ and $\rho(\alpha \cup \beta) = \rho(\alpha) \cup \rho(\beta)$, $\rho(\alpha \cap \beta) = \rho(\alpha) \cap \rho(\beta)$. Notice that there is no need to *define* the symbols $\subset$, $\cup$, $\cap$ for n.s. set germs since these are actual sets. Similarly, if T is endowed with a ring structure, then $\sigma(f + g) = \sigma(f) + \sigma(g)$, $\sigma(f, g) = \sigma(f)\sigma(g)$.

For any finite set of function germs, $\phi_1, \cdots, \phi_k$, the variety germ of $(\phi_1, \cdots, \phi_k)$ at p is defined by

$$v(\phi_1, \cdots, \phi_k) = v(\phi_1) \cap \cdots \cap v(\phi_k),$$

where the $v(\phi_j)$ on the right side denote the variety germs of individual function germs introduced in the preceding section, for a given $q \in T$. The corresponding notion for a set of n.s. function germs, $\psi_1, \cdots, \psi_k$ is that of the n.s. variety germ of $(\psi_1, \cdots, \psi_k)$, which is defined by

$$v(\psi_1, \cdots, \psi_k) = \{x \mid x \in \mu(p) \quad \text{and} \quad \psi_j(x) = q, j = 1, \cdots, k\}.$$

This definition extends immediately to an infinite set of n.s. function germs $\{\psi_\nu\}$. For finite sets $\phi_1, \cdots, \phi_k$ only we have

$$\sigma(v(\phi_1, \cdots, \phi_k)) = v(\rho(\phi_1), \cdots, \rho(\phi_k)).$$

3. Rückert's Nullstellensatz

Suppose now that S is the n-dimensional complex space C^n, for some (standard) positive integer n where T coincides with the field of complex numbers, C, and $g = 0$, while $p = (0, \cdots, 0)$ is the origin of C^n. We consider only *analytic function germs*, that is, function germs which contain functions $f(z_1, \cdots, z_n)$ that are analytic in a neighborhood of the origin.

A central result in the theory of analytic function germs is

Theorem 3.1 (**Rückert's Nullstellensatz**). *Let $\phi_1, \cdots, \phi_k$, $k \geq 1$, and ψ, be analytic function germs (at the origin of C^n) such that*

3.2. $$v(\psi) \supset v(\phi_1, \cdots, \phi_k).$$

Then there exist analytic function germs $\psi_1, \cdots, \psi_k$ and a positive integer ρ such that

3.3. $$\psi\rho = \psi_1\phi_1 + \cdots + \psi_k\phi_k.$$

That is, 3.2 implies that a power of ψ belongs to the ideal generated by $\phi_1, \cdots, \phi_k$ in the ring of analytic function germs at the origin.

Theorem 3.1 is the analogue, for function germs, of the classical

Theorem 3.4 (**Hilbert's Nullstellensatz**). *Let $\phi_1, \cdots, \phi_k$, $k \geq 1$, and ψ be polynomials of n variables such that*

3.5. $$v(\psi) \supset v(\phi_1, \cdots, \phi_k),$$

where $v(\psi)$ and $v(\phi_1, \cdots, \phi_k)$ are the algebraic variables of ψ and $(\phi_1, \cdots, \phi_k)$. Then there exist polynomials $\psi_1, \cdots, \psi_k$ and a positive integer ρ such that

3.6. $$\psi^\rho = \psi_1\phi_1 + \cdots + \psi_k\phi_k;$$

that is, a power of ψ belongs to the ideal generated by $\phi_1, \cdots, \phi_k$.

Actually, the statement of Theorem 3.4 is ambiguous. If we suppose that the coefficients of the polynomials in question are in a field Φ and that 3.5 holds if we take the varieties in any extension of Φ, then we obtain a weak form of Hilbert's Nullstellensatz. If, on the other hand, we suppose only that 3.5 holds for the varieties in the algebraic closure $\bar{\Phi}$ of Φ, then we obtain the strong form of the theorem (for example, if both the coefficients and the varieties are taken in the field of complex numbers).

An expeditious proof of the weak form of Hilbert's Nullstellensatz, mentioned above, is as follows. Suppose that $\phi_1, \cdots, \phi_k$, and ψ belong to the ring of polynomials $\Phi[z_1, \cdots, z_n]$, where Φ is a given field and suppose that 3.6 does not hold for any ρ. That is to say, the ideal $J \subset R[z_1, \cdots, z_n]$ which is generated by $\phi_1, \cdots, \phi_k$ does not include any power of ψ. By Zorn's lemma, there exists an ideal J', $J \subset J' \subset R[z_1, \cdots, z_n]$, which is maximal with respect to the property of excluding all powers of ψ. A familiar elementary argument shows that J' is prime. Let Φ' be the field of quotients of the quotient ring $R[z_1, \cdots, z_n]/J'$ and let $\zeta_1, \cdots, \zeta_n$ be the canonical images of $z_1, \cdots, z_n$ in Φ'. Then $\zeta = (\zeta_1, \cdots, \zeta_n)$ is a generic point for J'; that is, for any $f(z_1, \cdots, z_n) \in R[z_1, \cdots, z_n]$, $f(\zeta_1, \cdots, \zeta_n) = 0$ if and only if $f(z_1, \cdots, z_n) \in J'$. Hence, in particular, $\phi_j(\zeta, \cdots, \zeta_n) = 0$, $j = 1, \cdots, k$, but $\psi(\zeta_1, \cdots, \zeta_n) \neq 0$. This shows that the variety of ψ in Φ' does not include the variety of $(\phi_1, \cdots, \phi_k)$ in Φ' and proves the weak form of Theorem 3.4.

Although we shall not make use of this fact subsequently, we may mention in passing that the model completeness of the notion of an algebraically closed field now provides an easy transition to the strong form of Theorem 3.4. For $\zeta_1, \cdots, \zeta_n$ as introduced above may be regarded also as elements of the algebraic closure $\bar{\Phi}'$ of Φ', and so the sentence "there exist $w_1, \cdots, w_n$ such that $\phi_1(w_1, \cdots, w_n) = 0, \cdots, \phi_k(w_1, \cdots, w_n) = 0, \psi(w_1, \cdots, w_n) \neq 0$" holds in Φ'. The model completeness of the notion of an algebraically closed field now shows that the sentence in quotation marks holds also in the algebraic closure $\bar{\Phi}$ of Φ, and so the variety of ψ in $\bar{\Phi}$ cannot include the variety of $(\phi_1, \cdots, \phi_k)$ in $\bar{\Phi}$. This completes the proof.

4. Proof of Rückert's Theorem

It is natural to ask whether there does not exist an analogous proof of Theorem 3.1.

The argument applied previously to polynomials still shows that if G_n is the ring of analytic function germs of n variables at the origin, and $\phi_1, \cdots, \phi_k$ and ψ are elements of G_n such that the ideal $J = (\phi_1,$

$\cdots, \phi_k) \subset G_n$ excludes all powers of ψ, then there exists a prime ideal J' such that $J \subset J' \subset G_n$, where J' excludes all powers of ψ (and is, in fact, maximal with respect to this property). The same holds true in the ring of n.s. analytic function germs of n variables at the origin, to be denoted by Γ_n, since the mapping ρ introduced in Section 2 provides an isomorphism between G_n and Γ_n. On the other hand, it does not now make sense to talk of a generic point for a prime ideal J in G_n. However, we are going to show that the corresponding fact is still true in Γ_n, more precisely.

Theorem 4.1. *Let J be a prime ideal in Γ_n, $J \neq \Gamma_n$, where n is a standard positive integer and let μ be the monad of the origin in C^n. Then there exists a point $\zeta = (\zeta_1, \cdots, \zeta_n) \in \mu$ such that for any $\phi(z_1, \cdots, z_n) \in \Gamma_n$, $\phi(\zeta_1, \cdots, \zeta_n) = 0$ if and only if $\phi \in J$.*

A point ζ as described in the theorem will be called a *generic point* for J.

In order to prove Theorem 4.1 we require the following results of Weierstrass', which form the starting point also of the standard proof of Rückert's Nullstellensatz.

Let $f(z_1, \cdots, z_n)$ be a function of n complex variables which is analytic at the origin, and let

$$f(z_1, \cdots, z_n) = \sum_{j=0}^{\infty} f_j(z_1, \cdots, z_n),$$

where $f_k(z_1, \cdots, z_n)$ is the homogeneous polynomial of degree j in the power-series expansion of $f(z_1, \cdots, z_n)$ around the origin, $j = 1, 2, 3, \cdots$. $f(z)$ is said to be *regular in z_n of order $k > 0$* if $f_j(z_1, \cdots, z_n) \equiv 0$ for $j < k$ and $z_n{}^k$ has a nonvanishing coefficient in $f_k(z_1, \cdots, z_n)$. If

$$f(z_1, \cdots, z_n) = \sum_{j=k}^{\infty} f_j(z_1, \cdots, z_n),\ f_k(z_1, \cdots, z_n) \not\equiv 0,$$

then it is easy to find a nonsingular linear transformation $z_j \to z_j'$ that transforms f into a function which is regular in z_n' of order k.

A *Weierstrass polynomial of degree $k > 0$ in z_n* is a function $h(z_1, \cdots, z_n)$ of the form

4.2. $$h(z_1, \cdots, z_{n-1}, z_n) = z_n{}^k + a_1(z_1, \cdots, z_{n-1})z_n{}^{k-1} + \cdots + a_k(z_1, \cdots, z_{n-1}),$$

where the $a_j(z_1, \cdots, z_{n-1})$ are analytic in a neighborhood of the origin and such that $a_j(0, \cdots, 0) = 0, j = 1, \cdots, k$. Notice that such a polynomial is regular in z_n of order k.

With these notions, we have the following theorems.

***Theorem* 4.3 (*Weierstrass' Preparation Theorem*).** *If $f(z_1, \cdots, z_n)$, $n \geq 1$ is analytic in the neighborhood of the origin and is regular in z_n of order $k > 0$, then in some neighborhood of the origin there exist a Weierstrass polynomial $h(z_1, \cdots, z_{n-1}, z_n)$ and an analytic function $u(z_1, \cdots, z_n)$, $u(0, \cdots, 0) \neq 0$, such that*

4.4. $$f(z_1, \cdots, z_n) = u(z_1, \cdots, z_n)h(z_1, \cdots, z_{n-1}, z_n).$$

***Theorem* 4.5 (*Weierstrass' Division Theorem*).** *Let $h(z_1, \cdots, z_{n-1}, z_n)$ be a Weierstrass polynomial of order k in z_n in some neighborhood of the origin and let $f(z_1, \cdots, z_n)$ be any function which is analytic at the origin. Then, in a suitable neighborhood V of the origin,*

4.6. $$f(z_1, \cdots, z_n) = g(z_1, \cdots, z_n)h(z_1, \cdots, z_n) + v(z_1, \cdots, z_n),$$

where $g(z_1, \cdots, z_n)$ and $v(z_1, \cdots, z_n)$ are analytic in V and $v(z_1, \cdots, z_n)$ is a polynomial of degree $< k$ with respect to z_n. That is to say,

4.7. $$v(z_1, \cdots, z_n) = b_1(z_1, \cdots, z_{n-1})z_n^{k-1} + \cdots + b_k(z_1, \cdots, z_{n-1}),$$

where the b_j are analytic at the origin.

We shall make use of these results in the specified enlargement. However, the polynomials and analytic functions that we shall consider will be assumed throughout to be (the extensions of) standard functions.

Proof of Theorem 4.1. By induction with respect to n. Suppose $n = 1$. If $J = (0)$, then any point $\zeta = (\zeta_1)$ with $0 \neq \zeta \in \mu$ will do for the ζ of the theorem. If $J \neq (0)$, let $0 \neq \phi(\zeta_1) \in J$. Then $\phi(0) = 0$, for otherwise $\phi(\zeta_1)$ would possess an inverse in Γ_1 and $J = \Gamma_1$. Hence $\phi(z_1) = z_1^k\psi(z_1)$, where $\psi(0) \neq 0$ and so $z_1^k \in J$, $z_1 \in J$. It follows that $\zeta = (0)$ will do in this case.

Since every n.s. analytic function germ is the restriction of an analytic function to μ, we may transfer to n.s. analytic function germs the notion of regularity of order k with respect to z_n. Disregarding the case $\zeta = (0)$, which is again trivial, let $0 \neq \phi\ (z_1, \cdots, z_n) \in J$, so that, as before, $\phi(0, \cdots, 0) = 0$ and, if necessary, perform a standard linear transformation $z_j \to z_j'$ so that the transform of ϕ is a regular function in z_n' of order $k \geq 1$. The transformation carries prime ideals into prime ideals and zeros of functions into zeros of the corresponding functions, and so we may as well suppose from the outset that J contains an element $\phi(z_1, \cdots, z_n)$ which is regular with respect to z_n, of order $k \geq 1$.

An n.s. Weierstrass polynomial germ of degree $k > 0$ in z_n is an n.s. function germ $h(z_1, \cdots, z_n)$ which is of the form 4.2, where the $a_j(z_1, \cdots, z_{n-1})$ are now n.s. analytic function germs such that $a_j(0, \cdots, 0) = 0$. Thus an n.s. Weierstrass polynomial germ is the restriction to μ of a Weierstrass polynomial, Theorem 4.3 shows that the function germ $\phi(z_1, \cdots, z_n) \in J$ introduced above can be written in the form 4.4, where $h(z_1, \cdots, z_n)$ is an n.s. Weierstrass polynomial germ and $u(z_1, \cdots, z_n)$ is an invertible element of Γ_n; that is, $u(0, \cdots, 0) \neq 0$. Since J is prime, but $J \neq \Gamma_n$, we may conclude that $h(z_1, \cdots, z_n)$ belongs to J.

Let Π_n be the set of elements of Γ_n which are polynomials with respect to z_n, that is, such that there is a bound on the powers of z_n which appear in their power-series expansion. Then Π_n is an integral domain which, as far as its algebraic structure is concerned, may be identified with $\Gamma_{n-1}[z_n]$. In particular, Π_n contains the n.s. Weierstrass polynomial germs as given by 4.2. Thus $h(z_1, \cdots, z_n) \in J$ belongs to Π_n and, in the sense just explained, may also be regarded as an element of $\Gamma_{n-1}[z_n]$.

Let $J' = J \cap \Gamma_{n-1}$; then J' also is a prime ideal which does not contain an element invertible in Γ_{n-1}. Thus $J' \neq \Gamma_{n-1}$, and so, by the assumption of induction, J' possesses a generic point $\zeta' = (\zeta_1, \cdots, \zeta_{n-1}) \in \mu$, that is, for any $f(z_1, \cdots, z_{n-1}) \in \Gamma_{n-1}$, $f(z_1, \cdots, z_{n-1}) \in \zeta'$ if and only if $f(\zeta_1, \cdots, \zeta_{n-1}) = 0$. Accordingly, the evaluation map of Γ_{n-1} with $^*C$ at the point ζ' is a homomorphism with kernel J'. This yields a monomorphism from Γ_{n-1}/J' into $^*C$, to be denoted by τ.

Consider the quotient ring Γ_n/J. Since $J \cap \Gamma_{n-1} = J'$ we may regard Γ_n/J as an extension of Γ_{n-1}/J'. We claim that this extension is finite algebraic. Indeed, let λ be the canonical map $\Gamma_n \to \Gamma_n/J$. Since $h \in J$, we then have, by 4.2,

4.8. $$0 = \lambda(h) = (\lambda(z_n))^k + \lambda(a_1)(\lambda(z_n))^{k+1} + \cdots + \lambda(a_k)$$

But $\lambda(a_j) \in \Gamma_{n-1}/J'$ for $j = 1, \cdots, k$. This shows that $\lambda(z_n)$ is algebraic over Γ_{n-1}/J'.

Now let $f(z_1, \cdots, z_n)$ be any other element of Γ_n. The division theorem, 4.5, shows (by restriction to μ) that f can be written in the form 4.6, where v is given by 4.7 and g and the b_j are n.s. analytic function germs, $g \in \Gamma_n$, $b_j \in \Gamma_{n-1}$. Hence

$$\begin{aligned}\lambda(f) &= \lambda(g)\lambda(h) + \lambda(v)\\ &= \lambda(b_1)(\lambda(z_n))^{k-1} + \cdots + \lambda(b_k).\end{aligned}$$

This shows that $\lambda(f)$ is a polynomial of $\lambda(z_n)$ with coefficients in Γ_{n-1}/J' and proves our assertion.

Let Δ be the field of quotients of Γ_{n-1}/J'. Then

$$p(x) = x^k + \lambda(a_1)x^{k-1} + \cdots + \lambda(a_k) \in \Delta[x]$$

and $$p(\lambda(z_n)) = 0.$$

It follows that $\lambda(z_n)$ is a root of an irreducible factor of $p(x)$ in $\Delta[x]$, to be denoted by $q(x)$. There is a canonical extension of the mapping τ from Γ_{n-1}/J' into $^*C$, to a mapping from Δ into $^*C$, which will still be denoted by τ. Applying τ to the coefficients of $p(x)$ and $q(x)$, respectively, we obtain polynomials $p_\tau(x)$ and $q_\tau(x)$ with coefficients in $^*C$ such that $q_\tau(x)$ divides $p_\tau(x)$ in $\Delta[x]$. Let ζ_n be any root of $q(x)$ in $^*C$. Then we claim that ζ_n must be infinitesimal.

Indeed, $q_\tau(\zeta_n) = 0$ entails $p_\tau(\zeta_n) = 0$. Now

$$\begin{aligned} p_\tau(x) &= x^k + \tau(\lambda(a_1))x^{k-1} + \cdots + \tau(\lambda(a_k)) \\ &= x^k + a_1(\zeta_1, \cdots, \zeta_{n-1})x^{k-1} + \cdots + a_k(\zeta_1, \cdots, \zeta_{n-1}), \end{aligned}$$

and since $a_j(0, \cdots, 0) = 0$, $j = 1, \cdots, k$, and $(\zeta_1, \cdots, \zeta_{n-1})$ is in the monad of the origin of C^{n-1}, it follows that $a_j(\zeta_1, \cdots, \zeta_{n-1})$ is infinitesimal. Suppose now that ζ_n is not infinitesimal. Then

$$\zeta_n = d_1 + d_2\zeta_n^{-1} + \cdots + d_k\zeta^{1-k},$$

where $d_j = -a_j(\zeta_1, \cdots, \zeta_{n-1})$, so that the terms on the right side are all infinitesimal. This shows that ζ_n also must be infinitesimal, proving our assertion by contradiction. It follows that $(\zeta_1, \cdots, \zeta_n)$ belongs to the monad of the origin of C^n, μ.

In particular, let $q(x)$ be a polynomial with coefficients in Γ_{n-1}/J', which is irreducible over the field of quotients of that ring and such that $q(\lambda(z_n)) = 0$. Then $q(x)$ is determined uniquely up to nonzero factors from Γ_{n-1}/J'. Let $\zeta_n \in {}^*C$ be any root of $q(\zeta_n) = 0$. We are going to show that $\zeta = (\zeta_1, \cdots, \zeta_n)$ is a generic point for J.

As we have seen already, $\zeta \in \mu$. Also

$$\begin{aligned} 0 = p_\tau(\zeta_n) &= \zeta_n^k + a_1(\zeta_1, \cdots, \zeta_{n-1})\zeta_n^{k-1} + \cdots + a_k(\zeta_1, \cdots, \zeta_{n-1}) \\ &= h(\zeta_1, \cdots, \zeta_n), \end{aligned}$$

and so, by 4.6,

$$f(\zeta_1, \cdots, \zeta_n) = r(\zeta_1, \cdots, \zeta_n)$$

for any $f(z_1, \cdots, z_n) \in \Gamma_n$, where v is given by 4.7, with $b_j \in \Gamma_{n-1}$, $j = 1, \cdots, k$. At the same time, $\lambda(f(z_1, \cdots, z_n)) = \lambda(v(z_1, \cdots, z_n))$, since $h \in J$. Accordingly, the proof of Theorem 4.1 will be complete as soon as we show that $r(\zeta_1, \cdots, \zeta_n) = 0$ if and only if $r(z_1, \cdots, z_n) \in J$ for any $r(z_1, \cdots, z_n)$ of the form 4.7, with $b_j(z_1, \cdots, z_n) \in \Gamma_{n-1}$, $j =$

1, · · · , k, that is, for any $r \in \Gamma_{n-1}[z_n]$. In other words, we have to verify that for any $r(z_1, \cdots, z_n) \in \Gamma_{n-1}[z_n]$ the evaluation map η at $\zeta_1, \cdots, \zeta_n$ is a homomorphism with kernel $J \cap \Gamma_{n-1}[z_n]$, that is, η induces a monomorphism from $\Gamma_{n-1}[z_n]/J \cap \Gamma_{n-1}[z_n]$ into $^*C$. But we know that the restriction of η to Γ_{n-1} induces a monomorphism, called previously τ, from Γ_{n-1}/J' into $^*C$. Thus we only have to show that the addition and specification $\lambda(z_n) \to \zeta_n$ extends τ to a monomorphism from $\Gamma_{n-1}[z_n]/J \cap \Gamma_{n-1}[z_n]$ into $^*C$. But this follows from the choice of ζ_n, since $\lambda(z_n)$ and ζ_n are roots of the irreducible polynomials $q(x)$ and $q_\tau(x)$, respectively, which correspond under the mapping τ. This completes the proof of Theorem 4.1.

Rückert's Nullstellensatz (Theorem 3.1) now follows immediately by the argument used in the proof of Theorem 3.4. As we compare our procedure with a standard proof of Theorem 3.1 (for example, [1]) we may notice the great advantage obtained from the introduction of the notion of a generic point. In particular, we did not require the result that Γ_n is Noetherian. In Section 5 we shall consider a situation where the rings under consideration are in fact no longer Noetherian, but where the conclusion of Theorem 3.1 is valid all the same.

5. **Function Germs of an Unbounded Number of Variables**

Let $\nu = \{I\}$ be an index set of arbitrary cardinality and let C^I be the direct product of $\bar{\bar{I}}$ copies of the complex numbers indexed in I. We consider the class Φ of functions from C^I into C which depend only on a finite number of coordinates $\nu \in I$, and we may, for simplicity, use ordinary integers 0, 1, 2, · · · to indicate the coordinates on which the function in question may depend. The elements of Φ will be called *cylindrical functions*. The generalization of the theory of the preceding sections to functions of this kind presents itself naturally in a nonstandard setting. Accordingly, we shall approach the problem in this framework.

Passing to enlargements, C is extended to $^*C$ and C^I is extended to $^*C^{*I}$, where we have to bear in mind that any *internal* points are to be regarded as elements of $^*C^{*I}$. The *monad of* $^*C^{*I}$, μ, will be defined as the set of internal points $a \in {}^*C^{*I}$ such that the νth coordinate of a, a_ν, is infinitesimal for all standard ν. μ is, in fact, the monad of the origin in the product topology of $^*C^{*I}$ [2]. By an *n.s. set germ* we mean an intersection $A \cap \mu$ where $A = {}^*B$ is the extension of any $B \subset C^I$ to $^*C^{*I}$ (that is, A is any *standard* set).

By an *n.s. function germ* (more precisely, a *nonstandard cylindrical analytic function germ at the origin*) we mean the restriction of a cylindrical function $f(z_1, \cdots, z_n)$ which is analytic for sufficiently small $|z_1|, \cdots, |z_n|$ to the monad μ. The n.s. function germs constitute an integral domain, to be denoted by Γ. If J is an ideal in Γ which excludes all (standard) powers g^0 of some $g \in \Gamma$, then there exists a prime ideal $J' \supset J$ which still excludes all powers of g.

Let $A \subset \Gamma$. The *n.s. variety germ* of A, $v(A)$, is defined as the set of all points $\zeta \in \mu$ such that $f(\zeta) = 0$ for all $f \in A$ [where we write $f(\zeta)$ briefly for $f(\cdots, \zeta_\nu, \cdots)$]. The n.s. variety germ of the ideal generated by A coincides with the n.s. variety germ of A. If A consists of a single element, $A = \{f\}$, then $v(A)$ is an n.s. set germ. For if $f = f(z_1, \cdots, z_n)$, then f is the restriction of a standard function $^*g(z_1, \cdots, z_n)$ which is analytic for $|z_j| < \rho$, $j = 1, \cdots, n$, for sufficiently small standard positive ρ, and $v(A)$ is then the restriction to μ of the set of zeros of $^*g$ which satisfy $|z_j| < \rho, j = 1, \cdots, n$. A similar argument shows, more generally, that if A is a finite set, or a finitely generated ideal, then $v(A)$ is an n.s. set germ.

Let J be an ideal in Γ, at point $\zeta \in \mu$ will be said to be *generic for* J if for any $f \in \Gamma$, $f(\zeta) = 0$ if and only if $f \in J$. We are going to prove

Theorem 5.1. *Let J be a prime ideal in Γ, $J \neq \Gamma$. Then J possesses a generic point.*

Proof. Given the prime ideal J, let G be any finite subset of Γ, put $H = J \cap G$, $K = G - H$, and suppose that H and K are not empty. Since G is finite, there exists a finite subset I' of I such that the elements of G are independent of coordinates with subscripts in $I - I'$. Let Γ' be the set of elements of Γ which depend only (that is, at most) on coordinates with subscripts in I', and let $J' = \Gamma' \cap J$. Then J' includes H and excludes K and hence is a prime ideal in Γ' which is different from Γ'. Let $I' = \{\nu_1, \cdots, \nu_m\}$; then Theorem 4.1 shows that there exist infinitesimal complex numbers $\zeta_{\nu_1}, \cdots, \zeta_{\nu_m}$ such that for any $f(z_{\nu_1}, \cdots, z_{\nu_m}) \in \Gamma'$, $f(\zeta_{\nu_1}, \cdots, \zeta_{\nu_m}) = 0$ if and only if $f(z_{\nu_1}, \cdots, z_{\nu_m}) \in J'$. It follows that, for any standard positive integer n, the following statement holds in the enlargement.

5.2. "There exist a positive $\rho < 1/n$ and complex numbers ζ_{ν_j}, $j = 1, \cdots, m$ such that $|\zeta_{\nu_j}| < \rho$ and the elements of $G = H \cup K$ are analytic for $|z_{\nu_j}| < \rho$, $j = 1, \cdots, m$ and $f(\zeta_{\nu_1}, \cdots, \zeta_{\nu_m}) = 0$ for all $f \in H$ while $f(\zeta_{\nu_1}, \cdots, \zeta_{\nu_m}) \neq 0$ for all $f \in K$."

Bearing in mind that the set G is finite and consists of standard functions only, we see that 5.2 must hold also in the standard domain, for certain standard ρ, $\zeta_{\nu_1}, \cdots, \zeta_{\nu_m}$, provided we replace the n.s. function germs $f(z_{\nu_1}, \cdots, z_{\nu_m})$ by standard analytic functions whose restrictions they are.

Let Γ_0 be obtained from Γ by replacing each element f of Γ by an appropriate standard analytic function f_0 from which f is attained by restriction, $f \leftrightarrow f_0$. Let J_0 be the image of J under this mapping. Then the argument just applied to 5.2 shows that for every finite subset H_0 of J_0 and for every finite subset K_0 of $\Gamma_0 - J_0$ there exist a standard $\rho < 0$ and standard complex $\zeta_{\nu_1}, \cdots, \zeta_{\nu_m}$ such that $|\zeta_{\nu_j}| < \rho$ and the elements of $G_0 = H_0 \cup K_0$ depend only on $z_{\nu_1}, \cdots, z_{\nu_m}$ and are analytic for $|z_{\nu_j}| < \rho$ and $f(\zeta_{\nu_1}, \cdots, \zeta_{\nu_m}) = 0$ for all $f \in H_0$ while $f(\zeta_{\nu_1}, \cdots, \zeta_{\nu_m}) \neq 0$ for all $f \in k_0$.

Consider the binary relation $R(x, y)$ which is defined as follows. x is an array $\langle n, H_0, K_0\rangle$ where n is a (standard) positive integer, H_0 is a finite subset of J_0, and K_0 is a finite subset of $\Gamma_0 - J_0$; and y is a set of complex numbers $\{\zeta_\nu\}$ indexed in I such that if $z_{\nu_1}, \cdots, z_{\nu_m}$ are the variables on which the functions of $H_0 \cup K_0$ depend, then $|\zeta_{\nu_j}| < 1/n$ for $j = 1, \cdots, m$, and the functions of $H_0 \cup K_0$ are analytic for $(\zeta_{\nu_1}, \cdots, \zeta_{\nu_m})$ and $f(\zeta_{\nu_1}, \cdots, \zeta_{\nu_m}) = 0$ for $f \in H_0$ and $f(\zeta_{\nu_1}, \cdots, \zeta_{\nu_m}) \neq 0$ for $f \in K_0$. Then the discussion following the statement of 5.2 shows that $R(x, y)$ is concurrent. It follows that there exists a set of $\zeta_\nu \in {}^*C$ indexed in ${}^*I$ such that the relations $R(\langle n, H_0, K_0\rangle, \{\zeta_\nu\})$ hold for all n, H_0, K_0 as specified above, simultaneously. Accordingly, $|\zeta_\nu| < 1/n$ for $\nu \in I$ and all standard n; that is, ζ_ν is infinitesimal for $\nu \in I$ and hence $\{\zeta_\nu\} = \zeta$ belongs to the monad μ. Also, since every $f \in \Gamma$ is the restriction of some $f_0 \in \Gamma_0$ it follows that $f(\zeta) = 0$ if and only if $f \in J$. This shows that ζ is a generic point for J and proves Theorem 5.1.

Similarly as before, the argument of Section 3 above leads from Theorem 5.1 to the following Nullstellensatz.

Theorem 5.3. *Let $A \subset I$ and let g be an element of Γ such that $g(\zeta) = 0$ for all $\zeta \in v(A)$. Then there exist a standard positive integer ρ and n.s. function germs $f_1, \cdots, f_n \in A$, $g_1, \cdots, g_n \in \Gamma$ such that*

$$g^0 = g_1 f_1 + \cdots + g_n f_n.$$

For finite Γ, Theorem 5.3 includes Theorem 3.1, but is in fact seen to be no stronger than Theorem 3.1 if account is taken of the fact that I_n is Noetherian. For it then follows that for any $A \subset I_n$, $v(A) = v(A')$ for some finite subset A' of A, so that Theorem 5.3 reduces to Theorem 3.1 in this particular case. However, for infinite I, the ring Γ is no longer Noetherian, and Theorem 5.3, which is formulated in nonstandard language, then provides a natural generalization of Theorem 3.1.

References

1. R. C. GUNNING AND H. ROSSI, *Analytic Functions of Several Complex Variables.* Englewood Cliffs, N. J.: Prentice-Hall, 1965.
2. A. ROBINSON, *Non-Standard Analysis* (Studies in Logic and the Foundations of Mathematics). Amsterdam: North-Holland, 1966.

Elementary Embeddings of Fields of Power Series*

ABRAHAM ROBINSON

Department of Mathematics, Yale University, New Haven, Connecticut 06520

Communicated by P. Roquette

Received August 5, 1969

Let K be a field of characteristic 0 and let $Q = K((t))$ be the field of formal Laurent series with coefficients in K. Let $^*K$ and $^*Q$ be obtained from K and Q by means of a nontrivial ultrapower construction. Then the elements of $^*Q$ are "Laurent series" with subscripts in the corresponding nonstandard integers and coefficients in $^*K$. We show that the field of ordinary Laurent series with coefficients in $^*K$ can be embedded in $^*Q$ elementarily, consistently with the natural embedding of Q in $^*Q$. As an application we derive an algebraic result due originally to M. J. Greenberg.

1. INTRODUCTION

Let K be a field of characteristic 0 and let $^*K$ be an ultrapower of K. Thus, $^*K$ is obtained by a canonical construction (here supposed known to the reader) which depends on an index set I and on an ultrafilter D on the set of subsets of I, $^*K = K^I/D$. For fixed I and D, we may regard this construction as a transformation Φ_{ID} such that $^*K = \Phi_{ID}K$. The same transformation Φ_{ID}, when applied to the natural numbers, N, yields a nonstandard model $^*N$ of N where we shall assume explicitly that $^*N$ is a proper extension of N. More generally, Φ_{ID} acts on any given mathematical structure, M, yielding a structure $^*M = \Phi_{ID}M$. Then $^*M$ satisfies the same sentences as M and this applies even if we include sentences in a higher order language, provided these are interpreted in $^*M$ "in Henkin's sense" ([5, 6]).

Let $P = K[[t]]$ be the ring of formal power series of a variable t with coefficients in K, and let Q be the field of quotients of P, $Q = K((t))$. Thus, Q is the field of Laurent series over K. We introduce a valuation v into P and Q, as usual, by putting $v(q) = m$ if $a_m t^m$, $m \leqslant 0$, is the first monomial with nonvanishing coefficient a_m for $q \neq 0$, together with

* Research supported in part by the National Science Foundation, Grant No. GP-8625.

$v(o) = \infty$. Then the valuation group of Q is the group of rational integers, Z. Q together with Z may be regarded, model-theoretically, as a first order structure, which is given by the relations of equality in Q and of equality and order in Z, and by the operations of addition and multiplication in Q and of addition in Z, together with the one place operation $v(x)$ which provides the valuation of the elements of Q in Z [with some fixed convention for $v(o)$]. It will be convenient to assume that Q and Z are disjoint. The properties of belonging to Q and Z, respectively, can be defined in terms of the other extralogical constants.

Applying the transformation Φ_{ID} to P and Q, respectively, we obtain their "nonstandard extensions" $^*P = \Phi_{ID}P$ and $^*Q = \Phi_{ID}Q$. The elements of $^*P$ and $^*Q$ may be regarded as power series or Laurent series with coefficients in $^*K$, $\sum a_n t^n$, where, however, the exponents n range over nonstandard as well as over standard elements of $^*Z$. On the other hand, the series which are elements of $^*P$ and $^*Q$ are limited by the condition that they must be internal ([5, 6]). At the same time $^*Q$ is the field of quotients of $^*P$ just as Q is the field of quotients of P.

There is a natural embedding of any structure M into its ultrapower $^*M = \Phi_{ID}M = M^I/D$, which is obtained by mapping any $a \in M$ on the constant function $f \equiv a$ in M^I. This embedding is *elementary* (or *arithmetical*), that is to say if X is any sentence which is formulated in the Lower Predicate Calculus in terms of the relations and operations of M *and* in terms of the individuals of M, then X holds in M if and only if it holds in $^*M$. In particular, we thus obtain an elementary embedding Λ of Q into $^*Q$ whose restriction to P embeds P elementarily in $^*P$. We write $^*q = \Lambda q$ for any $q \in Q$.

In addition, we shall consider the ring of ordinary power series over $^*K$, $P' = {}^*K[[t]]$ and its field of quotients $Q' = {}^*K((t))$. The natural embedding of K in $^*K$, together with $t \to t$, yields an embedding of Q in Q', which will be denoted by Θ. We shall show that this is an elementary embedding. For this and other purposes we require several notions and results, which are due to *Ax* and Kochen.

Let F be a field with valuation in an ordered Abelian group G. We shall always take it for granted that the valuation is onto G, $G = \text{ord}\, F$. F (or more precisely, the system $\langle F, G\rangle$) will be called a Hensel field if it satisfies the following conditions.

(1.1) G possesses a smallest positive element; and the quotient group G/nG is the cyclic group of n elements for all positive integers n.

(1.2) F satisfies *Hensel's lemma*. Explicitly, let $f(x) \in O_v[x]$ be a monic polynomial with coefficients in the valuation ring O_v of F and let $\bar{f}(x) \in \bar{F}[x]$ be the image of $f(x)$ with coefficients in the residue class field $\bar{F}$ of F.

Suppose that $f(x) = r(x)\,s(x)$ where $r(x)$ and $s(x)$ are relatively prime polynomials in $\bar{F}[x]$. Then there exist $g(x), h(x) \in O_v[x]$ whose images in $\bar{F}[x]$ are $r(x)$ and $s(x)$, respectively, such that $f(x) = g(x)\,h(x)$ and such that the degree of $g(x)$ is equal to the degree of $r(x)$.

The following results are proved by *Ax* and Kochen ([1], [2]).

(1.3) Let F and F' be two Hensel fields $F \subset F'$, with residue class fields $\bar{F}$ and $\bar{F}'$ of characteristic 0. Suppose that $\bar{F}'$ is an elementary extension of $\bar{F}$. Then F' is an elementary extension of F.

(1.4) Let F be a Hensel field with valuation in a group G, and let ρ be the canonical mapping of the valuation ring O_v of F onto its residue class field $\bar{F}$. Let C be a field which is contained in O_v such that $\rho C = \bar{F}$ and let a be an element of F such that $v(a)$ is a minimal positive element of G. Then if F_0 is a subfield of F which contains $C(a)$ and which is algebraically closed in F, F_0 (or more precisely, $\langle F_0, \operatorname{ord} F_0 \rangle$) is a Hensel field.

We observe for future reference that if Hensel's lemma 1.2 is satisfied in a valued field then this implies the validity of the statement which is obtained from (1.2) by dropping the attribute "monic" for $f(x)$ ([4]).

Returning to the consideration of the fields Q and Q', let us identify Q with its Θ-image in Q' so that Q becomes a substructure of Q'. Then the residue class fields of Q and Q' are K and $^*K$ respectively and $^*K$ is an elementary extension of K. Hence, by (1.3), Q' is an elementary extension of Q or, more precisely, Θ is an elementary embedding of Q in Q'.

We have now related Q to both $^*Q$ and Q'. It will be the chief purpose of this paper to relate Q' to $^*Q$ by means of the following result.

1.5. Main Theorem. *There exists an elementary embedding Γ of Q' into $^*Q$. The composite mapping $\Gamma \circ \Theta$ coincides with Λ.*

The proof of 1.5 is given in section 2 below. In section 3 we apply 1.5 to derive an algebraic result due to M. J. Greenberg.

2. Proof of the Main Theorem

Retaining the notation of section 1, we establish a first connection between Q' and $^*Q$ by defining a mapping Ψ from a subset B of $^*Q$ onto Q' as follows. We call an element $q \in {}^*Q$ bounded from below if $v(q)$ is not negative infinite and hence is either finite, or positive infinite, or equal to ∞. The set of elements of $^*Q$ which are bounded from below is a subring of $^*Q$, which will be denoted by B. Let $q \in B$ so that $q = \sum_{n=\nu}^{\infty} a_n t^n (n \in {}^*Z)$ where ν may be chosen negative and finite. We then define Ψq by $\sum_{n=\nu}^{\infty} a_n f^n (n \in Z)$, in other words, we "chop off" the monomials of q whose exponents are infinite. It is not difficult to see that Ψ is a homomorphism of B into Q'. We propose to show that Ψ is onto Q'.

To see this, let $q' = \sum_{n=\nu}^{\infty} a_n' t^n (n \in Z)$ be any element of Q'. Now $^*K$ and $^*Q$ are ultrapowers and, as such, are (sequentially) comprehensive ([6, 7]). Thus, there exists an *internal sequence* of elements of $^*K$, $\{a_\nu, a_{\nu+1}, \ldots, a_n, \ldots\}$, $n \in {}^*Z$, such that for all finite n, $a_n = a_n'$. Then the corresponding power series in $^*Q$, $q = \sum_{n=\nu}^{\infty} a_n t^n$ is such that $\Psi q = q'$. In particular, for every $r \in {}^*K \subset {}^*Q$, $\Psi r = r$.

For any $q \in {}^*Q$, $v(\Psi q) = v(q)$ provided $v(q)$ is finite, while $v(\Psi q) = \infty$ if $v(q)$ is either infinite or is equal to ∞.

Next, we establish a lemma.

2.1. LEMMA. *Let $r_1, \ldots, r_k \in {}^*K$, where k is a finite positive integer and let $q_1, \ldots, q_k \in Q$. Employing a familiar notation, let $^*q_j = \Lambda q_j$, $j = 1, \ldots, k$. Then $r_1 q_1 + \cdots + r_k q_k = 0$ implies $r_1 {}^*q_1 + \cdots + r_k {}^*q_k = 0$.*

Proof. Write $q_j = \sum_{n=-\infty}^{\infty} a_{jn} t^n (n \in Z)$ where the a_{jn} are equal to zero up to some integer $n = n(j)$, $j = 1, \ldots, k$. Consider the following matrix with k rows and with columns whose number is infinite in both directions

$$\begin{pmatrix} \cdots & a_{1n} & a_{1,n+1} & a_{1,n+2} & \cdots \\ \cdots & a_{2n} & a_{2,n+1} & a_{2,n+2} & \cdots \\ \cdots & \cdot & \cdot & \cdot & \cdots \\ \cdots & a_{kn} & a_{k,n+1} & a_{k,n+2} & \cdots \end{pmatrix}. \tag{2.2}$$

Since the rank of each of the matrices $A_l = (a_{jn})$, $j = 1, \ldots, k$, $n = -l, \ldots, -1, 0, 1, \ldots l$, l any natural number, is at most k, there exist integers $l_1, \ldots, l_k$ such that all columns of matrix (2.2) depend linearly over K on the columns with second subscript $l_1, \ldots, l_k$. On passing from K to $^*K$ we conclude that for infinite ω, also, the column $(a_{1\omega}, a_{2\omega}, \ldots, a_{k\omega})$ (transposed) depends linearly on the columns with second subscript $l_1, \ldots, l_k$. That is to say, there exist $s_1, \ldots, s_k \in {}^*K$ such that

$$a_{j\omega} = s_1 a_{jl} + s_2 a_{jl_2} + \cdots + s_k a_{jl_k}, \qquad j = 1, \ldots, k. \tag{2.3}$$

But then

$$\sum r_j a_{j\omega} = \sum\sum r_j s_i a_{jl_i} = \sum\sum s_i r_j a_{jl_i} = 0$$

and hence $r_1 {}^*q_1 + \cdots + r_k {}^*q_k = 0$. This proves 2.1.

Let Q_0 be the Λ-image of Q in $^*Q$ and let B_0 be the compositum of Q_0 and of $^*K$ (regarded as a subfield of $^*Q$). Then both Q_0 and $^*K$ belong to B. Let q be any element of the ring generated by Q_0 and $^*K$ in $^*Q$.

Then $q \in B$ and q is of the form

$$q = r_1{}^*q_1 + \cdots + r_k{}^*q_k\,, \quad r_1\,,\ldots, r_k \in {}^*K, \quad q_1\,,\ldots, q_k \in Q. \tag{2.4}$$

Suppose that $q \neq 0$. Then we claim that $\Psi q \neq 0$. Suppose on the contrary that $\Psi q = 0$. Then

$$0 = \Psi q = \Psi(r_1{}^*q_1 + \cdots + r_k{}^*q_k) = r_1q_1 + \cdots + r_kq_k$$

and so, by 2.1., $0 = r_1{}^*q_1 + \cdots + r_k{}^*q_k = q$, contrary to assumption. This proves that $\Psi q \neq 0$, $v(q)$ is finite. Since every element of B_0 is a quotient of expressions of the form (2.4), we conclude that it is bounded from below and so $B_0 \subset B$.

Let S be the set of subfields of ${}^*Q$ which are included in B and which include B_0. S is not empty since it includes B_0 and is partially ordered by the relation of inclusion, $B_1 \prec B_2$ if $B_1 \subset B_2$. Any totally ordered subset of S possesses an upper bound, i.e., its union (direct limit). Accordingly Zorn's lemma applies and S includes maximal elements. We choose one of them, calling it $\bar{B}$.

We wish to show that $\bar{B}$ is algebraically closed in ${}^*Q$.

Suppose on the contrary that there exist an element $q \in {}^*Q - \bar{B}$ and an irreducible polynomial $p(x) \in \bar{B}[x]$ as given by

$$\begin{gathered} p(x) = a_0 + a_1x + \cdots + a_mx^m, \qquad m > 0, \qquad a_m \neq 0, \\ a_j \in \bar{B}, \qquad j = 0,\ldots, m \end{gathered} \tag{2.5}$$

such that $p(q) = 0$. Then $m \geqslant 2$. Also, q is bounded from below. For suppose $v(q)$ is negative infinite. Since

$$a_m = -\frac{a_0}{q^m} - \frac{a_1}{q^{m-1}} - \cdots$$

we have

$$v(a_m) \geqslant \min_{0 \leqslant j \leqslant m-1} (v(a_j) - (m-j)\, v(q)). \tag{2.6}$$

But then the right hand side of this inequality is positve infinite while $v(a_m)$ is finite (since both a_m and a_m^{-1} belong to B). This shows that q is bounded from below.

Consider now the field $\bar{B}(q)$. Its elements are given by expressions

$$b = b_0 + b_1q + \cdots + b_lq^l, \quad l < m, \quad b_j \in \bar{B}, \quad j = 0,\ldots, l. \tag{2.7}$$

Then $v(b) \geqslant \min_{0 \leqslant j < l} (v(b_j) + jv(q))$ and so b is bounded from below. Hence $\bar{B}(q) \subset B$. But this contradicts the maximality of $\bar{B}$ and proves that $\bar{B}$ is algebraically closed in ${}^*Q$.

Now $\bar{B}$ is a field with a valuation inherited from $^*Q$. $\bar{B}$ contains the field $^*K$ which is isomorphic to the residue class field of $^*Q$ and hence also of $\bar{B}$. $\bar{B}$ also contains the element $t \in {}^*Q$ whose valuation is 1, i.e., the positive minimal element of the valuation group $^*Z$ of $^*Q$. In addition, as we have just seen, $\bar{B}$ is algebraically closed in $^*Q$. Hence, by (1.4), $\bar{B}$ is a Hensel field and, by (1.3), $^*Q$ is an elementary extension of $\bar{B}$.

Now let $\bar{\Psi}$ be the restriction of Ψ to $\bar{B}$. Since $\bar{B}$ is a field, $\bar{\Psi}\bar{B} = \bar{Q}$ also is a field. $\bar{Q}$ is thus a subfield of Q'. We propose to prove that $\bar{Q} = Q'$.

We show first that Q' is algebraic over $\bar{Q}$. Supposing the contrary, let $q \in Q' - \bar{Q}$ where q is transcendental over $\bar{Q}$. Choose a $b \in B$ such that $\Psi b = q$. Then $b \in B - \bar{B}$, and b is transcendental over $\bar{B}$ since $\bar{B}$ is algebraically closed in $^*Q$.

Consider the ring $\bar{B}[b] \subset B$. The elements of $\bar{B}(b)$ outside $\bar{B}$ are of the form $p(b)$ where $p(x)$ is given by (2.5). Suppose that $\Psi p(b) = 0$ for some such $p(b)$, i.e.,

$$\Psi(a_0) + \Psi(a_1)q + \cdots + \Psi(a_m)q^m = 0. \tag{2.8}$$

But $\Psi(a_j) = \bar{\Psi}(a_j)$, $j = 0,\ldots, m$, and $a_m \neq 0$ implies $\bar{\Psi}(a_m) \neq 0$. Thus, q would be algebraic over $\bar{Q}$, which is contrary to assumption. We conclude that $v(p(b))$ is not positive infinite and the same therefore applies to all nonzero elements of $\bar{B}[b]$. It follows that the field of quotients of $\bar{B}[b]$ is a subset of B and this contradicts the maximality of $\bar{B}$. Accordingly, Q' is algebraic over $\bar{Q}$.

We shall now show—essentially as our last step in the proof of 1.5—that $\bar{Q}$ is algebraically closed in Q'. This will prove that $\bar{Q} = Q'$ since we have already established that Q' is algebraic over $\bar{Q}$. Our method is an adaptation of known procedures in valuation theory.

Suppose that $\bar{Q}$ is not algebraically closed in Q'. Then there exists an element $q \in Q' - \bar{Q}$ which satisfies an irreducible polynomial $p(x) \in \bar{Q}[x]$,

$$\begin{aligned} (x) = a_0 + a_1x + a_2x^2 + \cdots + a_mx^m, \qquad m \geqslant 2, \\ a_m \neq 0, \qquad a_j \in \bar{Q}, \qquad j = 0,\ldots, m. \end{aligned} \tag{2.9}$$

We may suppose that $p(x)$ is *primitive*, i.e., that $v(a_j) \geqslant 0$ for $j = 0,\ldots, m$ and that at least one of the $v(a_j)$ is positive. For if this is not the case from the outset we may achieve it by division by an a_j for which $v(a_j)$ is minimal.

Let

$$q = k_\nu t^\nu + k_{\nu+1}t^{\nu+1} + \cdots + k_n t^n + \cdots, \qquad k_j \in {}^*K. \tag{2.10}$$

Then this power series must be effectively infinite for if it were finite, then q would belong to the compositum B_0 of Q_0 and of $^*K$ and hence, would be included in $\bar{Q}$. Accordingly, there exists a sequence of partial sums of q, $\{q_n\}$, where

$$q_n = k_\nu t^\nu + k_{\nu+1} t^{\nu+1} + \cdots + k_{m_n} t^{m_n}, \qquad n = 0, 1, 2,...$$

such that $\lim q_n = q$ in the metric of the valuation of Q' and such that if $\pi_n = v(q - q_n)$, $n = 0, 1, 2,...,$ then $\pi_0 < \pi_1 < \pi_2 < \cdots < \pi_n < \cdots$. Since $p(x)$ is continuous in the metric just mentioned and since $p(q) = 0$ we conclude that $\lim v(p(q_n)) = \infty$. On the other hand, let

$$p_j(x) = \frac{1}{j!}\frac{d^j p}{dx^j}, \qquad j = 1,..., m,$$

then $p_j(q) \neq 0$ since $p(x)$ is irreducible. Hence, if $\nu_j = v(p_j(q))$, $j = 1,..., m$, then ν_j is a finite natural number and $\nu_j = v(p_j(q_n))$ for sufficiently high n.

We now claim that, for sufficiently high n,

$$\nu_1 + \pi_n < \nu_j + j\pi_n, \qquad j = 2,..., m. \tag{2.11}$$

To see this, we observe to begin with that the difference of any two terms on the right hand side of (2.11) remains ultimately (i.e., for sufficiently high n) either positive or negative. Indeed, consider such a difference

$$\nu_j + j\pi_n - (\nu_i + i\pi_n) = \nu_j - \nu_i + (j - i)\pi_n \qquad i \neq j \tag{2.12}$$

where we may suppose without loss of generality that $j > i$. Since π_n is strictly increasing the expression on the right hand side of (2.12) is strictly increasing and hence either stays ultimately positive or stays ultimately negative, as asserted. It follows that for sufficiently high n just one $\nu_j + j\pi_n$ is, and stays, strictly smaller than all others.

Consider now the Taylor expansion of $p'(x)$ about q_n taken at q. This is

$$p'(q) = p'(q_n) + \sum_{j=2}^{m} jp_j(q_n)(q - q_n)^{j-1}.$$

Hence, for sufficiently high n,

$$v(p'(q) - p'(q_n)) = \min_{j=2,3,...,m} (\nu_j + (j-1)\pi_n).$$

On the other hand, for sufficiently high n,

$$v(p'(q) - p'(q_n)) \geqslant \min(v(p'(q)), v(p'(q_n))) = \nu_1 .$$

Hence $\nu_1 \leqslant \nu_j + (j-1)\pi_n$, $j = 2,..., m$ or, since π_n is strictly increasing, $\nu_1 < \nu_j + (j-1)\pi_n$. This proves (2.11).

Let $\lambda_n = v(p(q_n))$. We claim that, for sufficiently large n,

$$\lambda_n = \nu_1 + \pi_n \tag{2.13}$$

To see this, consider the expression of $p(q)$ as a Taylor series about q_n. This is

$$p(q) = p(q_n) + \sum_{j=1}^{m} p_j(q_n)(q - q_n)^j.$$

Hence, taking into account that $p(q) = 0$,

$$-p(q_n) = \sum_{j=1}^{m} p_j(q_n)(q - q_n)^j.$$

Hence, for sufficiently high n, bearing in mind (2.11),

$$\lambda_n = v(p(q_n)) = \min_{j=1,2,\ldots,m} (\nu_j + j\pi_n) = \nu_1 + \pi_1 ,$$

as asserted.

For any given n, consider the polynomial

$$\phi(x) = p(q_n + t^{\lambda_n - \nu_1}x)/p(q_n).$$

Expanding the right hand side, we obtain

$$\phi(x) = 1 + \sum_{j=1}^{m} s_j x^j$$

where $s_j = (p_j(q_n)/p(q_n))t^{j(\lambda_n - \nu_1)}$, $j = 1,\ldots, m$. Then for sufficiently high n

$$v(s_j) = v(p_j(q_n)) - v(p(q_n)) + j(\lambda_n - \nu_1) = \nu_j - j\nu_1 + (j - 1)\lambda_n .$$

Hence, $v(s_1) = 0$ while, for $j \geqslant 2$, taking into account (2.11) and (2.13),

$$v(s_j) = \nu_j - j\nu_1 + (j - 1)(\nu_1 + \pi_n) = \nu_j + j\pi_n - (\nu_1 + \pi_n) \geqslant 1.$$

Now let $\bar{\phi}(x)$ be the image of $\phi(x)$ under the canonical mapping $\rho(x)$ of $\bar{Q}$ into its residue class field which is (isomorphic to) $^*K$. Then, for sufficiently high n,

$$\bar{\phi}(x) = 1 + \bar{s}_1 x, \qquad \text{where} \quad \bar{s}_1 = \rho(s_1), \qquad \bar{s}_1 \neq 0$$

since $v(s_1) = 0$, $v(s_j) > 0$ for $j = 2,\ldots, m$. Thus $\bar{\phi}(x) = (1 + \bar{s}_1 x) \cdot 1$ is a representation of $\bar{\phi}(x)$ as a product of relatively prime polynomials in $^*K[x]$. Also, since $\bar{Q}$ is isomorphic to $\bar{B}$ it is a Hensel field and hence, satisfies (1.2) even when $f(x)$ (here $\phi(x)$) is not monic. We conclude that

$\phi(x)$ has a linear factor in $\bar{B}[x]$, contrary to assumption. This shows that $\bar{Q}$ is algebraically closed in Q' and hence, coincides with it. Accordingly, $\Gamma = \bar{\Psi}^{-1}$ maps Q' onto $\bar{B}$. Since $^*Q$ is an elementary extension of $\bar{B}$ it follows that Γ is an elementary embedding of Q' into $^*Q$. Also, if $q \in Q$, $^*q = \Lambda q$, then $^*q$ is in the domain of Ψ_0 and, hence, of $\bar{\Psi}$ and $\bar{\Psi}{}^*q = \Theta q$. It follows that $\Gamma \circ \Theta q = \bar{\Psi}^{-1} \circ \bar{\Psi}{}^*q = \Lambda q$. This completes the proof of 1.5.

3. Application

We shall prove (compare [3])

3.1. Theorem. Let

$$f_j(x_1, ..., x_n), \quad j = 1, ..., m; \quad g_j(x_1, ..., x_n), \quad j = 1, ..., m_1;$$
$$h_j(x_1, ..., x_n), \quad j = 1, ..., m_2; \quad k_j(x_1, ..., x_n), \quad j = 1, ..., m_3,$$

be polynomials with coefficients in $K((t))$, $n \geqslant 1$, where K is a field of characteristic zero and where one or more of the classes of g_j, h_j, or k_j may be empty. Let

$$\alpha_j, j = 1, ..., m_1; \qquad \beta_j, j = 1, ..., m_2; \qquad \gamma_j, j = 1, ..., m_3$$

be given (standard rational) integers. Suppose that for all integers λ, there exist elements $q_1, ..., q_n$ of $K[[t]]$ such that

$$v(f_j(q_1, ..., q_n)) \geqslant \lambda, \qquad j = 1, ..., m \tag{3.2}$$

$$\begin{cases} v(g_j(q_1, ..., q_n)) = \alpha_j, & j = 1, ..., m_1; \\ v(h_j(q_1, ..., q_n)) < \beta_j, & j = 1, ..., m_2; \\ v(k_j(q_1, ..., q_n)) > \gamma_j, & j = 1, ..., m_3. \end{cases} \tag{3.3}$$

Then there exist elements $p_1, ..., p_n$ of $K[[t]]$ such that

$$f_j(p_1, ..., p_n) = 0, \qquad j = 1, ..., m \tag{3.4}$$

$$\begin{cases} v(g_j(p_1, ..., p_n)) = \alpha_j, & j = 1, ..., m_1; \\ v(h_j(p_1, ..., p_n)) < \beta_j, & j = 1, ..., m_2; \\ v(k_j(p_1, ..., p_n)) > \gamma_j, & j = 1, ..., m_3. \end{cases} \tag{3.5}$$

Proof. Supposing that the conditions of the theorem are satisfied, we pass from $Q = K((t))$ to the nonstandard extension $^*Q$. Then f_j, g_j, h_j, k_j are replaced by $^*f_j$, $^*g_j$, $^*h_j$, $^*k_j$, respectively. Making λ positive infinite, let $q_1, ..., q_n \in {}^*Q$ satisfy (3.2) and (3.3); stars having been appended to the f, g, h, k. Now apply the transformation Ψ, i.e., "chop off" the monomials with nonstandard exponents in the equations and inequalities of (3.2) and (3.3) (see section 2 above). This transform the $*f$, $*g$, $*h$, $*k$ back into the corresponding f, g, h, k while mapping $q_1, \ldots, q_n$ on elements $q_1', \ldots, q_n'$ of Q'. At the same time, λ in (3.2) has to be replaced by ∞. Thus, we now have

$$f_j(q_1', \ldots, q_n') = 0, \qquad j = 1, \ldots, m \tag{3.6}$$

$$\begin{cases} v(g_j(q_1', \ldots, q_n')) = \alpha_j, & j = 1, \ldots, m_1; \\ v(h_j(q_1', \ldots, q_n')) < \beta_j, & j = 1, \ldots, m_2; \\ v(k_j(q_1', \ldots, q_n')) > \gamma_j, & j = 1, \ldots, m_3. \end{cases} \tag{3.7}$$

Next, apply the mapping Γ of Theorem 1.5. This again sends the f, g, h, k into the corresponding $*f$, $*g$, $*h$, $*k$. At the same time $q_1', \ldots, q_n'$ are mapped on certain elements $\bar{q}_1, \ldots, \bar{q}_n$ of $*Q$. Thus, in $*Q$,

$$*f_j(\bar{q}_1, \ldots, \bar{q}_n) = 0, \qquad j = 1, \ldots, m \tag{3.8}$$

$$\begin{cases} v(*g_j(\bar{q}_1, \ldots, \bar{q}_n)) = \alpha_j, & j = 1, \ldots, m_1; \\ v(*h_j(\bar{q}_1, \ldots, \bar{q}_n)) < \beta_j, & j = 1, \ldots, m_2; \\ v(*k_j(\bar{q}_1, \ldots, \bar{q}_n)) > \gamma_j, & j = 1, \ldots, m_3. \end{cases} \tag{3.9}$$

Now consider the following statement, which can be formulated in the Lower Predicate Calculus in the vocabulary of a field with valuation, with individual constants for the elements of Q.

"There exist entire elements $x_1, \ldots, x_n$ such that $f_j(x_1, \ldots, x_n) = 0$, $j = 1, \ldots, m$; while $v(g_j(x_1, \ldots, x_n)) = \alpha_j$, $j = 1, \ldots, m_1$; $v(h_j(x_1, \ldots, x_n)) < \beta_j$, $j = 1, \ldots, m_2$; and $v(k_j(x_1, \ldots, x_n)) > \gamma_j$, $j = 1, \ldots, m_3$."

We have just seen that this statement holds in $*Q$, with the appropriate interpretation $*f_j$ for f_j, $*g_j$ for g_j, $*h_j$ for h_j. Accordingly, the statement holds also in $Q = K((t))$. This completes the proof of 3.1.

Acknowledgments

I am indebted to Simon Kochen for a valuable conversation on the subject of the present paper; to Michael Artin for drawing my attention to Greenberg's work, and to Paul Eklof for suggesting substantial improvements in the presentation.

References

1. J. Ax and S. Kochen, Diophantine problems over local fields, I, II, *Amer. J. Math.* **87** (1965), 605–630, 631–648.

2. J. Ax and S. Kochen, Diophantine problems over local fields, III, decidable fields, *Ann. Math.* **83** (1966), 437–456.
3. M. J. Greenberg, Rational points in Henselian discrete valuation rings, *Publ. Math. Inst. Hautes Etud. Sci.* **31** (1966), 59–64 (563–568).
4. P. Ribenboim, "Théorie des Valuations," Les Presses de l'Université de Montréal, Montreal 101, P.Q., 1964.
5. A. Robinson, Non-standard Analysis, Amsterdam 1966.
6. A. Robinson, Non-standard theory of Dedekind rings, *Proc. Roy. Acad. Sci. Neth., ser. A* **70**, 444–452.
7. A. Robinson, Compactification of groups and rings and Non-standard analysis, *J. Symbolic Logic*, **34** (1969), 576–588.

COMPACTIFICATION OF GROUPS AND RINGS AND NONSTANDARD ANALYSIS[1]

ABRAHAM ROBINSON

§1. Introduction. Let G be a separated (Hausdorff) topological group and let $^*G$ be an enlargement of G (see [8]). Thus, $^*G$ (i) possesses the same formal properties as G in the sense explained in [8], and (ii) every set of subsets $\{A_\nu\}$ of G with the finite intersection property—i.e. such that every nonempty finite subset of $\{A_\nu\}$ has a nonempty intersection—satisfies $\bigcap {}^*A_\nu \neq \varnothing$, where the $^*A_\nu$ are the extensions of the A_ν in $^*G$, respectively. In terms of the definition of an enlargement, (ii) follows from the fact that the relation $R(x, y) = \{\langle x, y\rangle \mid y \in x, x \in \{A_\nu\}\}$ is concurrent.

By a *compactification* of G we mean a triple $\langle \Delta, \Gamma, \phi\rangle$ where Δ is a compact and separated topological group, Γ is a subgroup of Δ which is dense in Δ and ϕ is a continuous homomorphic mapping from G onto Γ. We shall show in the present paper that, for any such compactification *there exists a homomorphic mapping* λ *from* $^*G$ *onto* Δ *which maps* $G \subset {}^*G$ *on* Γ. The reader may check that only properties (i) and (ii) above will be used in the proof of this assertion, although elsewhere we shall rely also on other properties of enlargements and on special kinds of enlargements.

A similar result applies to topological rings.

Several homomorphisms constructed elsewhere [9], [10] in order to obtain the P-adic and (restricted) adelic completions of a ring of algebraic integers can be subsumed under the above result. The same is true of the representations of Bohr compactifications obtained by L. Kugler [6]. We shall see also that our method enables us to discuss, and sometimes to identify, compactifications which solve the universal mapping problems for particular groups and rings.

Thanks are due to Feit, Jacobson, Kakutani, Luxemburg and Massey for relevant discussions.

§2. Enlargements and compactifications. For a given group G as described in §1, let βG be the Čech-Stone compactification of G as a topological space (see e.g. [3] for the relevant notions used here). The elements p of βG are in one-to-one correspondence with the z-ultrafilters on G. A base $B(\beta G)$ for the closed sets of βG is obtained by collecting all subsets $\bar{A}$ of βG for which there is a z-set $A \subset G$ such that $\bar{A} = \{p \mid A \in F(p)\}$, where $F(p)$ is the z-ultrafilter which corresponds to p. G may be regarded as a subspace of βG. The points of G correspond to the fixed z-ultrafilters.

We consider at the same time a space αG whose points are in one-to-one corre-

Received June 18, 1968.

[1] Research supported in part by the National Science Foundation, Grant No. GP-8625. The author is indebted to a referee for pointing out a number of errors in the typescript and for suggesting several changes in the presentation.

spondence with the prime z-filters on G. We may suppose that $\beta G \subset \alpha G$ and, consistently with our previous notation, we may denote by $F(p)$ the prime z-filter which corresponds to a point $p \in \alpha G$.

It is known that every prime z-filter $F(p)$ on G is contained in (i.e., can be extended to) a unique z-ultrafilter $F(q)$. This induces a mapping $\psi: p \to q$ from αG onto βG, which reduces to the identical mapping on $\beta G \subset \alpha G$. We define the topology of αG as the weakest topology which is compatible with the continuity of ψ. Thus, a base $B(\alpha G)$ for the closed sets of αG is given by

$$B(\alpha G) = \{X \mid X = \psi^{-1} Y \text{ for some } Y \in B(\beta G)\}.$$

αG is compact for if $\{X_\nu\}$ is a subset of $B(\alpha G)$ with the finite intersection property, $X_\nu = \psi^{-1} Y_\nu$, $Y_\nu \in B(\beta G)$ then $\{Y_\nu\}$ has the finite intersection property; hence $\bigcap Y_\nu \neq \varnothing$; hence $\bigcap X_\nu \neq \varnothing$.

Now let $^*G$ be an enlargement of G. There is a natural mapping σ from $^*G$ onto αG, as follows. For any $a \in {}^*G$ consider the set of subsets of G which is given by

$$F = \{X \mid a \in {}^*X, X \text{ is a } z\text{-set}\}.$$

It is not difficult to check that F is a prime z-filter. For example, if X_1 and X_2 are two z-sets such that $X_1 \cup X_2 \in F$ then $a \in {}^*(X_1 \cup X_2) = {}^*X_1 \cup {}^*X_2$; hence a belongs to at least one of the two sets $^*X_1$, $^*X_2$, e.g., $a \in {}^*X_1$, $X_1 \in F$. Hence $F = F(p)$ for some $p \in \alpha G$, and we define $\sigma a = p$.

The mapping σ is onto. For let $F(p)$ be any prime z-filter on G, $F(p) = \{A_\nu\}$, and let $\{B_\nu\}$ be the set of z-sets in G which do not belong to $F(p)$. Let $C_\nu = G - B_\nu$ and consider the set $\{A_\nu\} \cup \{C_\nu\}$. We claim that this set has the finite intersection property. For let $A_1, \cdots, A_k \in \{A_\nu\}$, $C_1, \cdots, C_m \in \{C_\nu\}$. Then $A_1 \cap \cdots \cap A_k \in \{A_\nu\}$ since $F(p) = \{A_\nu\}$ is a z-filter and $C_1 \cap \cdots \cap C_m \in \{C_\nu\}$ since $F(p)$ is a prime z-filter. Thus, we only have to show that for any $A \in F(p)$ and any other z-set $B \notin F(p)$, there exists an element $b \in A$ which does not belong to B. But this is obvious, for otherwise $A \subset B$, $B \in F(p)$.

It follows, by property (ii) of $^*G$ (see §1, above) that there exists an $a \in {}^*G$ such that $a \in {}^*A_\nu$ for all $A_\nu \in F(p)$ and $a \notin {}^*B_\nu$ for the remaining z-sets. We conclude that $F(p) = \{X \mid a \in {}^*X, X \text{ is a } z\text{-set}\}$ and, hence, that $p = \sigma a$.

We define the topology of $^*G$ as the weakest topology which is compatible with the continuity of σ. Then $^*G$ is compact. In general, $^*G$ is *not* a topological group. Put $\rho = \psi\sigma$.

Let A be a z-set in G. Then $\overline{A} = \{x \mid A \in F(x)\}$ while $\psi^{-1}\overline{A}$ consists of all $y \in \alpha G$ such that $F(y)$ can be extended to some $F(x) \in \overline{A}$, i.e. to some z-ultrafilter which includes A. But this is the case precisely if A meets all elements of $F(y)$ and so $\psi^{-1}\overline{A}$ consists of the elements $y \in \alpha G$ such that for all $B \in F(y)$, $A \cap B \neq \varnothing$. Moreover, $\rho^{-1}\overline{A} = \sigma^{-1}\psi^{-1}\overline{A}$ is the intersection of all $^*B_\nu$ such that $B_\nu \subset G$ is a cozero set which contains A. Indeed, if $a \notin \bigcap {}^*B_\nu$ then there exists a z-set C such that $A \cap C = \varnothing$ while $a \in C$, $C \in F(\sigma a)$ and so $\sigma a \notin \psi^{-1}\overline{A}$, $a \notin \sigma^{-1}\psi^{-1}\overline{A}$. Conversely, if $a \notin \sigma^{-1}\psi^{-1}\overline{A}$ then $\sigma a \notin \psi^{-1}\overline{A}$. Hence, there exists a z-set $C \in F(\sigma a)$ such that $A \cap C = \varnothing$. Thus, $a \in C$ where C is complementary to a set which includes A, and so $a \notin \bigcap {}^*B_\nu$.

In particular, if $A = \{q\}$ for some $q \in G$ then $\rho^{-1}\overline{A} = \bigcap {}^*B_\nu$ where B_ν ranges over the cozero sets which contain q. Now let Q be any other neighborhood of q in G.

Then there exists a cozero set B_ν which includes q but is included in Q and so $\rho^{-1}\bar{A}$ is the intersection of all extensions to $*G$ of neighborhoods of q in G, $\rho^{-1}\{q\} = \mu(q)$ where $\mu(q)$ is the *monad* of q (see [8]).

Let A be any cozero set in G. We put $\tilde{A} = \beta G - \overline{G - A}$ so that $\rho^{-1}\tilde{A} = *G - \rho^{-1}(\overline{G - A})$. Now $\rho^{-1}(\overline{G - A})$ is the intersection of all extensions to $*G$ of cozero sets that contain $G - A$ and so $\rho^{-1}\tilde{A}$ is the union of the complement of these sets in $*G$, i.e., it is the union of all extensions to $*G$ of z-sets $C_\nu \subset A$, $\rho^{-1}\tilde{A} = \bigcup *C_\nu$. The totality of these $\rho^{-1}\tilde{A}$ constitutes a base E for the open sets of $*G$.

Now let $\langle \Delta, \Gamma, \phi \rangle$ be any compactification of G. Since ϕ maps G continuously into Δ, it may be extended uniquely to a continuous mapping Φ from βG into Δ. Then $\Phi(\beta G)$ is compact and contains Γ, while Γ is dense in Δ. Hence $\Phi(\beta G) = \Delta$, the mapping Φ is onto. It follows that $\lambda = \Phi\rho$ is a continuous mapping from $*G$ onto Δ.

2.1. THEOREM. *The mapping λ is a homomorphism.*

PROOF. Let $a, b \in *G$. Then we have to show that

$$\lambda a \lambda b = \lambda(ab).$$

At any rate, the restriction of λ to $G \subset *G$ is a homomorphism. In order to see this, we observe that σ reduces to the identity on G. Indeed, for $a \in G$, $F(\sigma a)$ is just the set of all z-sets which contain a, i.e., it is the fixed z-ultrafilter which corresponds to a as a point of αG. But $\lambda = \Phi\psi\sigma$ and so λ reduces to ϕ on G, λ is a homomorphism on G.

Suppose now that there exist $a, b \in *G$ such that $\lambda a \lambda b = c$, $\lambda(ab) = d$ where $c \neq d$. Since Δ is separated, by assumption, we may choose open neighborhoods U_c and U_d of c and d respectively such that $U_c \cap U_d = \varnothing$. Since λ is continuous there exists an open neighborhood V_d of ab such that $\lambda V_d \subset U_d$ and since multiplication is continuous in Δ there exist open neighborhoods U_a and U_b of λa and λb respectively such that $U_a U_b \subset U_c$ and, further, there exist open neighborhoods V_a and V_b of a and b respectively such that $\lambda V_a \subset U_a$ and $\lambda V_b \subset U_b$.

Now G is dense in $*G$ as can be seen either from the fact that G is dense in αG or by observing that the nonempty elements of E are unions of sets $*A$ which are extensions of subsets A of G. For such A, $*A \cap G = A$ and so $*A \cap G = \varnothing$ would imply $A = \varnothing$ and, further, $*A = \varnothing$ by transfer from G to $*G$. G being dense in $*G$ there exist elements $a' \in V_a \cap G$ and $b' \in V_b \cap G$. For such a', b' we have

$$\lambda(a'b') = \lambda a' \lambda b' \in U_a U_b \subset U_c.$$

Suppose now that $\lambda(a'b') \in U_d$ or, which is the same, that $a'b' \in V_d$. Then $\lambda(a'b') \in U_c \cap U_d$, contrary to the assumption that $U_c \cap U_d$ is empty. We conclude that $a'b' \notin V_d$. Accordingly, in order to prove our theorem we only have to show that if V_a, V_b, and V_d are open neighborhoods of a, b, $ab \in *G$, respectively, then there exist $a', b' \in G$ such that $a' \in V_a$, $b' \in V_b$, and $a'b' \in V_d$. And since E is a base for the open sets of $*G$ it is actually sufficient to prove our assertion for the case that V_a, V_b, V_d belong to E. Then $V_a = \bigcup *C_\nu^{(a)}$ where the $C_\nu^{(a)}$ are z-sets in G, and similarly, $V_b = \bigcup *C_\nu^{(b)}$, $V_d = \bigcup *C_\nu^{(d)}$ for certain z-sets $C_\nu^{(b)}$, $C_\nu^{(d)}$. Hence, there exist z-sets $C^{(a)}$, $C^{(b)}$, $C^{(d)}$ such that $a \in *C^{(a)} \subset V_a$, $b \in *C^{(b)} \subset V_b$, $ab \in *C^{(d)}$. Suppose now that $(C^{(a)}C^{(b)}) \cap C^{(d)} = \varnothing$. The corresponding assertion for $*G$, which must then be true in $^*G$, is that $(^*C^{(c)*}C^{(b)}) \cap {}^*C^{(d)} = \varnothing$. But this is contradicted by the fact that $ab \in (^*C^{(a)*}C^{(b)}) \cap {}^*C^{(d)}$. We conclude that $(C^{(a)}C^{(b)}) \cap C^{(d)} \neq \varnothing$, there exist elements $a', b' \in G$ such that $a' \in C^{(a)} \subset V_a$, $b' \in C^{(b)} \subset V_b$, $a'b' \in C^{(d)} \subset V_d$. This completes the proof of Theorem 2.1.

§3. **Universal compactifications.** Using the notation of the preceding section, let K be the kernel of λ. Then Δ is isomorphic to the quotient group $^*G/K$ and Γ is isomorphic to $G/(K \cap G)$. For a given G, consider the totality of compactifications $\langle \Delta_\nu, \Gamma_\nu, \phi_\nu \rangle$ of G, supposed indexed in a set $I = \{\nu\}$, and let λ_ν be the mapping from $^*G$ onto Δ_ν introduced above, with kernel K_ν. Let $K_0 = \bigcap K_\nu$. We propose to show that K_0 gives rise to a compactification of G which possesses the universal mapping property for the class of structures under consideration.

Let $\Delta_0 = {}^*G/K_0$ and let λ_0 be the natural mapping from $^*G$ onto Δ_0. For any $\langle \Delta_\nu, \Gamma_\nu, \phi_\nu \rangle$, $\nu \in I$, let $H_\nu = \lambda_0 K_\nu$. Then $\Delta_\nu \simeq \Delta_0/H_\nu$ and this induces a natural homomorphism η_ν from Δ_0 onto Δ_ν such that $\lambda_\nu = \eta_\nu \lambda_0$. We topologize Δ_0 by means of the weakest topology which makes all the mappings η_ν continuous. Thus, a subbase S for the open sets of Δ_0 is given by the sets $\eta_\nu^{-1}Q$ as ν ranges over I and Q ranges over the open sets of Δ_ν. We propose to show that Δ_0 is a topological group. This is nontrivial because, as observed earlier, $^*G$ is not a topological group, generally speaking.

Let $B(\Delta_0)$ be the base for the open sets of Δ_0 which is obtained by taking finite intersections of elements of S. Let $ab = c$ for $a, b, c \in \Delta_0$. In order to establish the continuity of multiplication in Δ_0 it is sufficient to show that for any $U \in B(\Delta_0)$ which contains c there exist $U_a, U_b \in B(\Delta_0)$, $a \in U_a$, $b \in U_b$ such that $U_a U_b \subset U$. Let $\{1, 2, \cdots, k\}$ be any finite subset of I, let

3.1. $$U = \eta_1^{-1}U_1 \cap \eta_2^{-1}U_2 \cap \cdots \cap \eta_k^{-1}U_k$$

where $U_1, U_2, \cdots, U_k$ are open sets in $\Delta_1, \Delta_2, \cdots, \Delta_k$ respectively, and let $c_j = \eta_j c$, $j = 1, \cdots, k$. Then $c_j \in U_j$ and there exist open neighborhoods $U_j(a)$, $U_j(b)$ of $a_j = \eta_j a$ and $b_j = \eta_j b$, respectively, such that $U_j(a)U_j(b) \subset U_j$. Let

$$U_a = \eta_1^{-1}U_1(a) \cap \cdots \cap \eta_k^{-1}U_k(a),$$
$$U_b = \eta_1^{-1}U_1(b) \cap \cdots \cap \eta_k^{-1}U_k(b).$$

Then U_a and U_b are elements of $B(\Delta_0)$ which are neighborhoods of a and b, respectively. Also, if $a' \in U_a$, $b' \in U_b$, then

$$\eta_j(a'b') = \eta_j a' \eta_j b' \in U_j(a)U_j(b) \subset U_j.$$

Hence, $a'b' \in \eta_j^{-1}U_j$, $j = 1, \cdots, k$, $a'b' \in U$. A similar argument shows that the operation of inversion in Δ_0 also is continuous. Accordingly, Δ_0 is a topological group.

λ_0 *is continuous*. For let $a \in \Delta_0$ and let U be an open neighborhood of a where we may suppose again that U belongs to $B(\Delta_0)$, and is given by 3.1. Let $a = \lambda_0 b$, then we have to find a neighborhood V of b such that $\lambda_0 V \subset U$. Now let $a_j = \lambda_j b = \eta_j \lambda_0 b = \eta_j a$, $j = 1, \cdots, k$. Choose open neighborhoods V_j of a_j such that $\lambda_j V_j \subset U_j$ and let $V = V_1 \cap \cdots \cap V_k$. Then $\eta_j \lambda_0 V_j = \lambda_j V_j \subset U_j$ and so $\lambda_0 V_j \subset \eta_j^{-1}U_j$. Hence

$$\lambda_0 V = \lambda_0(V_1 \cap \cdots \cap V_k) \subset \lambda_0 V_1 \cap \cdots \cap \lambda_0 V_k \subset U,$$

as required. The continuity of λ_0 in turn implies that Δ_0 is compact.

Let e_0 be the identity in Δ_0. In order to prove that Δ_0 is separated, we have to show that $\{e_0\}$ is closed in Δ_0. Let $a \in \Delta_0$, $a \neq e_0$, and let $A = \lambda_0^{-1}\{a\}$. Then $A \cap K_0 = \varnothing$. Choose $a' \in A$, then there exists a K_ν such that $a' \notin K_\nu$. Hence $\lambda_\nu a' \neq e_\nu$ where e_ν is the identity in Δ_ν. Since Δ_ν is separated there exists an open neighborhood U_ν of $\lambda_\nu a'$ such that $e_\nu \notin U_\nu$. But then $K_\nu \cap \lambda_\nu^{-1}U_\nu = \varnothing$ and so $K_0 \cap \lambda_\nu^{-1}U_\nu = \varnothing$, $K_0 \cap \lambda_0^{-1}\eta_\nu^{-1}U_\nu = \varnothing$. Let $U = \eta_\nu^{-1}U_\nu$, then $a \in U$ since $\eta_\nu a = \eta_\nu\lambda_0 a' = \lambda_\nu a' \in U_\nu$. Also, $K_0 \cap \lambda_0^{-1}U = \varnothing$ and so $\{e_0\} \cap U = \varnothing$. This shows that $\{e_0\}$ is closed, so that Δ_0 is separated.

Let $a, b \in {}^*G$ and such that $\rho a = \rho b$. Then $\lambda_\nu a = \lambda_\nu b$ for all ν in the index set I. Suppose now that $\lambda_0 a \neq \lambda_0 b$. Then $ab^{-1} \notin K_0$ and so, for some $\nu \in I$, $ab^{-1} \notin K_\nu$, $\lambda_\nu a \neq \lambda_\nu b$, $\rho a \neq \rho b$. This contradiction shows that $\rho a = \rho b$ implies $\lambda_0 a = \lambda_0 b$ and so there is a mapping Φ_0 from βG onto Δ_0 such that $\lambda_0 = \Phi_0\rho$. Moreover, for any $p \in \beta G$, $\eta_\nu\Phi_0 p = \Phi_\nu p$ where Φ_ν is the continuous extension of ϕ_ν to βG. For if $p = \rho a$ then the assertion is that $\eta_\nu\lambda_0 a = \lambda_\nu a$ and this follows immediately from the definition of η_ν.

The mapping ρ is open for if X is an open subset of ${}^*G$ and $a \in X$ then there exists a set $V \in E$ which contains A and is contained in X. Then $V = \rho^{-1}\tilde{A}$ where A is a cozero set in G and $\tilde{A}$ is open in βG. Hence $\rho V = \tilde{A}$ is open and contains ρa and is contained in ρX. Since ρa ranges over ρX as a ranges over X this shows that ρ is open. Thus, if U is an open subset of Δ_0 then $\Phi_0^{-1}U = \rho(\rho^{-1}\Phi_0^{-1}U) = \rho\lambda_0^{-1}U$ also must be open, and so Φ_0 is continuous.

Now let ϕ_0 be the restriction of Φ_0 to G as a subset of βG. Then, for any $a \in G \subset \beta G$, $\phi_0 a = \phi_0\rho a = \lambda_0 a$ and so ϕ_0 maps G continuously and homomorphically on a subgroup Γ_0 of Δ_0. Γ_0 is dense in Δ_0.

We have proved

3.2. THEOREM. $\langle\Delta_0, \Gamma_0, \phi_0\rangle$ *is a compactification of G. For any other compactification of G,* $\langle\Delta, \Gamma, \phi\rangle$ *there exists a continuous homomorphism* η *from* Δ_0 *onto* Δ *which maps* Γ_0 *on* Γ *such that* $\phi = \eta\phi_0$.

Any compactification $\langle\Delta_0', \Gamma_0', \phi_0'\rangle$ of G which possesses the properties ascribed to $\langle\Delta_0, \Gamma_0, \phi_0\rangle$ by Theorem 3.2 will be said to be a universal compactification of G. The universal compactification of a separated topological group is essentially unique. For suppose that $\langle\Delta_0, \Gamma_0, \phi_0\rangle$ possesses the properties in question. Then there exist continuous homomorphisms η' and η from Δ_0 onto Δ_0' and from Δ_0' onto Δ_0, respectively such that $\phi_0' = \eta'\phi_0$, $\phi_0 = \eta\phi_0'$. Hence $\phi_0 = \eta\eta'\phi_0$ and this shows that $\eta\eta'$ is the identity on Γ_0 and, hence, on Δ_0. It follows that η' is an isomorphism from Δ_0 onto Δ_0'. But η' is also a homeomorphism since both η' and its inverse, η, are continuous. This proves our assertion.

An exactly parallel theory can be developed for general (commutative or non-commutative) rings. In that case, G, Γ, Δ are rings while K, K_ν, K_0 are ideals.

It is well known that the existence of universal compactifications for groups or rings can be subsumed also under more general results. However, our present method has the advantage of being, in its own way, rather concrete. We shall see that this will enable us, in certain cases, to identify the algebraic structure of the universal compactification of a given ring.

§4. The structure of K. Let K be the normal subgroup of $^*G$ which was introduced at the beginning of §3 above as the kernel of λ. We shall determine a specific normal subgroup of $^*G$, to be denoted by $\tilde{K}$ such that $\tilde{K} \subset K$ and hence, in particular, $\tilde{K} \subset K_0$.

Let $q \in \beta G$, then we define the monad of q, $\mu(q)$, as $\rho^{-1}\{q\}$. For $q \in G \subset \beta G$ this coincides with the definition of a monad given elsewhere (see [8], compare §2 above). Quite generally, $\mu(q) = \bigcap {^*B_\nu}$ where $\{B_\nu\}$ is the set of cozero sets such that $G - B_\nu \notin F(q)$. For let $p \in \mu(q)$ but suppose that $p \notin {^*B_\nu}$ for some cozero set B_ν. Then $p \in {^*(G - B_\nu)}$ and so $G - B_\nu$ belongs to the prime z-filter $F(\sigma p)$ and to the z-ultrafilter $F(\rho p)$. But $\rho p = q$ and so $G - B_\nu \in F(q)$, contrary to assumption. Thus, $\mu(q) \subset \bigcap {^*B_\nu}$.

Conversely, suppose that $p \notin \mu(q)$. Then $F(\sigma p)$ is not contained in $F(q)$ and so $p \in {^*A}$ for some z-set A such that $A \notin F(q)$. Putting $B = G - A$ we then have $p \notin {^*B}$ although $G - B \notin F(q)$. Hence, $p \notin \bigcap {^*B_\nu}$, $\bigcap {^*B_\nu} \subset \mu(q)$, $\bigcap {^*B_\nu} = \mu(q)$.

The monads $\rho^{-1}\{q\}$ are disjoint and, accordingly, define an equivalence relation in $^*G$, to be denoted by $\simeq$. We now define $\tilde{K}$ as follows.

4.1. Definition. $\tilde{K}$ is the set of all $a \in {^*G}$ for which there exist $a_1, \cdots, a_n$, $b_1, \cdots b_n \in {^*G}, n \in N, n \geq 1$, such that $a_j \simeq b_j, j = 1, \cdots, n$, and $a_1 a_2 \cdots a_n = a$, $b_1 b_2 \cdots b_n = e$, where e is the identity in G and in $^*G$.

4.2. Theorem. *$\tilde{K}$ is a normal subgroup of $^*G$.*

Proof. $\tilde{K}$ is closed under multiplication. For suppose $a, a' \in \tilde{K}$. Then there exist elements of $^*G$, $a_1, \cdots, a_n, a_{n+1}, \cdots, a_m, b_1, \cdots, b_n, b_{n+1}, \cdots, b_m$, $m > n$, such that $a_j \simeq b_j$, $j = 1, \cdots, m$, $a_1 \cdots a_n = a$, $a_{n+1} \cdots a_m = b$, $b_1 \cdots b_n = b_{n+1} \cdots b_m = e$. Then $a_1 \cdots b_m = aa'$, $b_1 \cdots b_m = e^2 = e$, $aa' \in \tilde{K}$.

$\tilde{K}$ is also closed under inversion. For the mapping $x \to x^{-1}$ is a homeomorphism in G and, hence, maps cozero sets on cozero sets in G and monads on monads in $^*G$. It follows that $a \simeq b$ if and only if $a^{-1} \simeq b^{-1}$.

Suppose now that $a \in \tilde{K}$ so that a satisfies the condition of Definition 4.1. Then $a_j^{-1} \simeq b_j^{-1}$, $j = 1, \cdots, n$, and $a_n^{-1} \cdots a_2^{-1} a_1^{-1} = a^{-1}$ while $b_n^{-1} \cdots b_2^{-1} b_1^{-1} = e$. This shows that a^{-1} belongs to $\tilde{K}$, $\tilde{K}$ is a subgroup of $^*G$.

To see that $\tilde{K}$ is normal, suppose that a satisfies the condition of 4.1 and let $g \in {^*G}$. Then $g \simeq g, g^{-1} \simeq g^{-1}, ga_1, \cdots a_n g^{-1} = gag^{-1}, gb_1, \cdots b_n g^{-1} = geg^{-1} = e$ and so $gag^{-1} \in K$. This completes the proof of Theorem 4.2.

Observe that if $a \simeq a'$ then a belongs to the same coset with respect to $\tilde{K}$ as a'. For if $a \simeq a^{-1}$ then $a^{-1} \simeq a^{-1}$, and since $a^{-1}a = e$ we conclude that $a^{-1}a' \in \tilde{K}$.

Now let $\langle \Delta, \Gamma, \phi \rangle$ be any compactification of G and let K be the kernel of the mapping λ as in §§2 and 3 above. Suppose $a \in \tilde{K}$ so that a satisfies the condition of 4.1. Then $\rho a_j = \rho b_j$, $j = 1, \cdots, n$, and so $\lambda a_j = \Phi \rho a_j = \Phi \rho b_j = \lambda b_j$ and $\lambda a = \lambda a_1 \cdots \lambda a_n = \lambda b_1 \cdots \lambda b_n = \lambda e$, $a \in K$. Thus,

4.3. Theorem. *$\tilde{K}$ is a subgroup of K.*

A similar construction can be carried out for rings. Supposing, for the remainder of this section, that G is a ring; we vary the strict analogy with the case of a group by defining $\tilde{K}$ as follows.

4.4. Definition. $\tilde{K}$ is the set of all $a \in {^*G}$ for which there exist a polynomial $p(x_1, \cdots, x_n)$ with integer coefficients, $n \geq 1$, and elements $a_1, \cdots, a_n, b_1, \cdots, b_n \in {^*G}$ such that $a_j \simeq b_j$, $j = 1, \cdots, n$, and $p(a_1, \cdots, a_n) = a$, $p(b_1, \cdots, b_n) = 0$. If G

is noncommutative, the variables $x_1, \cdots, x_n$ are to be assumed noncommuting.

As before, it is not difficult to see that $\tilde{K}$ is an ideal and that, for any compactification $\langle\Delta, \Gamma, \phi\rangle$ of G, $\tilde{K}$ is contained in the kernel K. In particular, $\tilde{K}$ is contained in K_0 (see §3 above). Suppose now that we can find a compactification $\langle\tilde{\Delta}, \tilde{\Gamma}, \tilde{\phi}\rangle$ for which the corresponding kernel is actually $\tilde{K}$. Then $\tilde{K}$ contains K_0 and hence $\tilde{K} = K_0$. And since $\tilde{\Delta} = {}^*G/\tilde{K}$ and $\Delta_0 = {}^*G/K_0$ it follows that Δ_0 is isomorphic to $\tilde{\Delta}$. We shall see later that this situation actually arises in connection with the ring of integers in an algebraic number field.

§5. **Discrete groups.** Suppose now that G is a group which is endowed with the discrete topology. This assumption leads to a considerable simplification of our theory. Every subset of G is now a z-set and every prime z-filter is an ultrafilter. Thus, αG and βG coincide and the mapping ψ reduces to the identity. The topology introduced previously for ${}^*G$ reduces to its S-topology (compare [8]), i.e. we obtain a base for the open sets (and also for the closed sets) of ${}^*G$ by taking the extensions ${}^*A$ of all $A \subset G$. More specifically, the definition of ρ now tells us that $a \in \rho^{-1}\tilde{A}$ for any $A \subset G$ if and only if A belongs to the ultrafilter determined by a, i.e. if and only if $a \in {}^*A$. Accordingly, we have $\rho^{-1}\tilde{A} = {}^*A$ and the totality of these sets constitutes a base for the closed as well as for the open sets of ${}^*G$. For any ultrafilter $F(q)$, $\rho^{-1}\{q\} = \mu(q)$ now belongs to a class of objects studied extensively by Luxemburg [7] under the name of *discrete monads*.

Suppose in particular that $G = Z$, the discrete additive group of integers. Then it is not difficult to see that inversion, $x \to -x$, is continuous in ${}^*G = {}^*Z$ (and, indeed, quite generally). On the other hand, multiplication is not continuous in ${}^*Z$. For if it were then ${}^*Z$ would be a topological group. And since the set $\{0\} \subset Z$ coincides with its extension to ${}^*Z$, $\{0\}$ is open in ${}^*Z$. But then, for any infinite positive $\omega \in {}^*Z$, the set $\{0 + \omega\} = \{\omega\}$ would have to be open in ${}^*Z$ and hence, would be equal to a union $\bigcup {}^*A_\nu$ for standard sets $A_\nu \subset Z$. Thus, $\{\omega\} = {}^*A$ for some $A \subset Z$, which is impossible since ω is not standard.

There is no way to define addition in βZ so as to obtain a compactification $\langle\beta Z, Z, \phi\rangle$ of Z. For if there were such a compactification, then ϕ would be the identical map on Z and so Φ would be the identical map on βZ. Hence, $\Phi^{-1}\{0\} = \{0\}$ and this, together with $\rho^{-1}\{0\} = \{0\}$, entails $\lambda^{-1}\{0\} = K = \{0\}$, implying that λ is one-to-one. On the other hand, it is not difficult to see that in any enlargement ${}^*Z$ of Z the monad $\mu(q)$ of a free ultrafilter $F(q)$ is actually infinite. Thus, if $a \in \beta Z - Z$, then $\lambda^{-1}\{a\}$ is infinite, a contradiction.

There are several other ways for showing that there is no continuous extension of addition from Z to βZ. One of them, which was pointed out to me by Luxemburg some time ago, is an immediate consequence of duality theory. I recall that in 1957 when the positive interest in nonstandard models of arithmetic was in its early stages, there were suggestions that βZ (or βN) might be made to serve as a nonstandard model of arithmetic, but the idea was dropped for the reason mentioned above, i.e. it turned out to be impossible to define a suitable operation of addition.

We shall now discuss several compactifications of Z. Let T be the additive group of real numbers modulo 1 and let τ be the natural homomorphism from R onto T.

Since T is compact, every element $a \in {}^*T$ is infinitely close, in the metric of ${}^*T$, to an element ${}^0a$ of T.

Let θ be a fixed standard positive irrational number. We define a mapping λ from ${}^*Z$ into T by $\lambda a = {}^0(\tau(a\theta))$. Then λ is a homomorphism. The kernel of λ, K consists of all $a \in {}^*Z$ such that $a\theta$ is infinitely close to an integer. Kronecker's theorem in one dimension states that we can approximate any real number r, $0 \leq r < 1$, as closely as we please by some $a\theta$ modulo 1 and this implies that λ is onto. K may be defined also as the intersection of a sequence of sets ${}^*H_n$, $n \in N$, where

$$H_n = \{x \in Z \mid \|x\theta\| < 1/n\},$$

$\|y\|$ being the distance of y from the nearest integer.

We are going to show that λ is continuous. Let $b = \lambda a$ and let U be an open neighborhood of b. We may suppose that U is of the form

5.1. $$U_n(b) = \{x \mid d(x, b) < 1/n\}$$

where $d(x, y)$ is the distance function on T, since the sets 5.1 constitute a fundamental system of neighborhoods of b. To find a neighborhood V of a such that $\lambda V \subset U_n(b)$ we choose $a_0 \in Z$ such that $\|(a_0 - a)\theta\| < 1/3n$ and define

$$W = \{x \in Z \mid \|(x - a_0)\theta\| < 1/2n\}.$$

For any $a' \in {}^*W$ we then have

$$\|a'\theta - a\theta\| \leq \|(a' - a_0)\theta\| + \|(a_0 - a)\theta\| < 5/6n.$$

It follows that $d(\lambda a', \lambda a) = d(\lambda a', b) < 1/n$, so that the choice $V = {}^*W$ establishes the continuity of λ.

Let ρ be the mapping from ${}^*Z$ onto βZ introduced previously (§2) for groups in general. We claim that $\rho a = \rho b$ implies $\lambda a = \lambda b$ for all $a, b \in {}^*Z$.

Indeed, $\rho a = \rho b$ signifies that a and b determine the same ultrafilter F on Z. Thus, if $\lambda a = t$ then $\|a\theta - r\| < 1/n$, $n = 1, 2, 3, \cdots$, for some standard real r such that $\tau r = t$. But then the set $S = \{x \mid \|x\theta - r\| < 1/n\}$ belongs to F and, hence, $b \in {}^*S$. Thus, $\|b\theta - r\| < 1/n$, $n = 1, 2, 3, \cdots$, $\lambda b = t$, as asserted.

We may now define a mapping Φ from βZ onto T by setting $\Phi a = \lambda a'$ for any $a \in \beta Z$ and for any $a' \in {}^*Z$ such that $a = \rho a'$. Then $\lambda = \Phi\rho$. Φ is continuous for if U is open in T then $\lambda^{-1}U = \bigcup {}^*A_\nu$ where $\{A_\nu\}$ is a set of subsets of Z. Hence

$$\Phi^{-1}U = \rho(\lambda^{-1}U) = \bigcup \rho^*A_\nu = \bigcup \rho(\rho^{-1}\bar{A}_\nu) = \bigcup \bar{A}_\nu$$

so $\Phi^{-1}U$ is open.

Let ϕ be the restriction of Φ to Z, then ϕ is a monomorphism and ϕZ is dense in T. Thus, $\langle T, \phi Z, \phi\rangle$ is a compactification of Z with kernel K (i.e. K is the kernel of the mapping λ).

Our next construction applies to an arbitrary infinite discrete group G. Let $S = \{H_\nu\}$ be a set of normal subgroups of G which satisfies the following conditions:

5.1. (i) The elements of S are of finite index in G.

(ii) If $H_1, H_2 \in S$ then $H_1 \cap H_2 \in S$.

(iii) $\bigcap H_\nu = \{e\}$ where e is the identity in G.

Note that $\{e\} \notin S$ since G is supposed to be infinite.

Let $\mu = \bigcap {}^*H_\nu$ where H_ν ranges over the elements of S. The ${}^*H_\nu$ are normal subgroups of ${}^*G$ and so, therefore, is μ. Let $\Delta = {}^*G/\mu$ and let λ be the natural mapping from $^*G$ onto Δ. By (iii), $\mu \cap G = \{e\}$ and so the restriction of λ to G is a monomorphism, $\Gamma = \lambda G$ is an isomorphic image of G.

Let $B(\Delta) = \{\lambda(a^*H_\nu)\}$ where a ranges over $G \subset {}^*G$ and H_ν ranges over S. We claim that $B(\Delta)$ may serve as a base for the open sets of a topology in Δ. For let $b \in \lambda(a_1{}^*H_1) \cap \lambda(a_2{}^*H_2)$, i.e. $b = \lambda(a_1h_1) = \lambda(a_2h_2)$ where $h_1 \in {}^*H_1$, $h_2 \in {}^*H_2$. Then a_1h_1 and a_2h_2 are in the same coset with respect to μ and hence with respect to $^*H_1$ and also with respect to $^*H_2$. It follows that they are in the same coset with respect to $^*H_3 = {}^*H_1 \cap {}^*H_2$. But $H_3 \in S$ by (ii) and the index of H_3 is finite, by (i). Accordingly, the coset in question can be written as $a_3{}^*H_3$ for some *standard* a_3. Then $\lambda(a_3{}^*H_3) \in B(\Delta)$ and $b \in \lambda(a_3{}^*H_3)$, $B(\Delta)$ satisfies the conditions for a base. It is not difficult to verify that, with the topology defined by $B(\Delta)$, Δ becomes a topological group. From now on, this topology will be assumed implicitly.

The mapping λ is continuous. Given any $\lambda(a^*H_\nu) \in B(\Delta)$, we have to show that $\lambda^{-1}\lambda(a^*H_\nu)$ is open. But $\mu \subset {}^*H_\nu$ and a is standard (i.e. $a \in G$) and so $\lambda^{-1}\lambda(a^*H_\nu) = a^*H_\nu = {}^*(aH_\nu)$, which is an open set in $^*G$.

Since $^*G$ is compact, we may now conclude that Δ is compact. Also, as in the previous example, $\rho a = \rho b$ implies $\lambda a = \lambda b$ since $\rho a = \rho b$ implies $ab^{-1} \in \mu$. Hence, there is a unique mapping Φ from βG onto Δ such that $\lambda = \Phi\rho$. Φ is continuous. Let ϕ be the restriction of Φ to $G \subset \beta G$, then $\Gamma = \phi G$. In order to show that $\langle \Delta, \Gamma, \phi \rangle$ is a compactification of G it remains to verify only that Δ is separated and that Γ is dense in Δ. But Δ is separated since $\mu = \lambda^{-1}\{e'\}$ is closed in the topology of $^*G$ where e' is the identity in Δ. And Γ is dense in Δ since every $\lambda(a^*H_\nu) \in B(\Delta)$ contains the element $\lambda a \in \Gamma$.

For example, let $G = Z$ and let $S = \{H_\nu\}$, $\nu = 1, 2, \cdots, \nu \in N$, where H_ν is the ideal (p^ν) for a given rational prime p. The corresponding Δ is the additive group of p-adic numbers.

The groups Δ which occur in the compactifications of the additive group Z are known as *monothetic groups*. The classical theory of duality for locally compact abelian groups provides a comprehensive analysis of monothetic groups ([1], [5]) and it may well be possible to get a more direct analysis of this kind from the theory presented here. In particular it would be interesting to obtain in this way the *universal monothetic group*.

The reader may consult [10] for a connection between our theory and the theory of profinite groups.

§6. **Discrete rings.** Reinterpreting the symbols used in the preceding section, let G be a discrete infinite ring and let $S = \{H_\nu\}$ be a set of ideals in G which satisfies the following conditions:

6.1. (i) For any $H_\nu \in S$ the number of residue classes of G modulo H_ν is finite.

(ii) If $H_1, H_2 \in S$ then $H_1 \cap H_2 \in S$.

(iii) $\bigcap H_\nu = \{0\}$.

Let $\mu = \bigcap {}^*H_\nu$ and put $\Delta = {}^*G/\mu$. The topology of Δ shall be defined by the basic open sets $\lambda(a + {}^*H_\nu)$ where a ranges over $G \subset {}^*G$, H_ν ranges over S, and λ is the natural mapping from $^*G$ onto Δ. Then Δ is a compact and separated topological ring and λ is continuous and induces a continuous mapping Φ from βG onto Δ such that $\lambda = \Phi\rho$. Finally, if ϕ is the restriction of Φ to $G \subset \beta G$ then ϕ is an isomorphism and $\Gamma = \phi G$ is dense in Δ. Thus, $\langle \Delta, \Gamma, \phi \rangle$ is a compactification of G.

In particular, let G be the ring of integers in an algebraic number field (a finite extension of the field of rational numbers) and let $S = \{P^\nu\}$, $\nu = 1, 2, \cdots, \nu \in N$, where P is any given prime ideal in G. Then 6.1 is satisfied and we may construct the corresponding compactification $\langle \Delta_P, \Gamma_P, \phi_P \rangle$ of G such that $\Delta_P = {}^*G/\mu_P$, $\mu_P = \bigcap {}^*P^\nu$. As shown in [9] Δ_P is the ring of P-adic integers over G.

Suppose next that H_ν ranges over all ideals in G, where G is the ring of integers in an algebraic number field, as before. Put $\bigcap {}^*H_\nu = \bar{\mu}$. Then $\bar{\mu}$ can be characterized also as the set of $a \in {}^*G$ which are divisible by all rational integers other than zero. Let $\langle \bar{\Delta}, \bar{\Gamma}, \bar{\phi} \rangle$ be the corresponding compactification of G. It is shown in [9] that $\bar{\Delta}$ can be injected algebraically into the direct product of the P-adic integers over ${}^*G$, $\prod \Delta_P$. ($\prod \Delta_P$ is the ring of restricted entire adèles, or adèles without archimedean components.) Moreover, the reader may check that in that injection, the topology of $\bar{\Delta}$, as introduced above, coincides with the product (adèle) topology of $\prod \Delta_P$ and that the injection $\bar{\phi}: G \to \bar{\Gamma}$ coincides with the diagonal embedding of G into $\prod \Delta_P$. And since G, embedded in this way, is dense in $\prod \Delta_P$ and $\bar{\Delta}$ is compact we conclude that the injection of $\bar{\Delta}$ must be *onto* $\prod \Delta_P$. We recall that in [9] the conclusion that ${}^*G/\bar{\mu}$ is mapped onto $\prod \Delta_P$ was reached only on the assumption that the enlargement ${}^*G$ is comprehensive, e.g. that it is an ultrapower.

G being the ring of integers in an algebraic number field as before, consider the following condition.

6.2. There exists an ultrafilter on G, to be denoted by $F(d)$—where d is the corresponding element in βG—such that $\rho^{-1}\{d\}$ contains for every infinite $k \in {}^*N$ an infinite $h \in {}^*N$ which is smaller than k. Thus, briefly, $\rho^{-1}\{d\} \cap ({}^*N - N)$ is to be coinitial with ${}^*N - N$.

Defining $\bar{\mu}$ as above, defining $\tilde{K}$ as at the end of §4 and assuming that 6.2 is satisfied, we shall prove

6.3. THEOREM. $\bar{\mu} = \tilde{K}$.

PROOF. Let $p_1 = 2, p_2 = 3, \cdots, p_n, \cdots$ be the sequence of rational primes, and let $c \in \bar{\mu}$. Then $\operatorname{ord}_{p_n} c = m$ (the greatest integer m such that c is divisible by p_n^m) is infinite for all finite n. Hence, by a basic lemma of Nonstandard Analysis, $\operatorname{ord}_{p_n} c$ must be infinite for all n up to and including some infinite $n = \omega$. Moreover, the integer $\zeta = \min_{n \le \omega} \operatorname{ord}_{p_n} c$ must be infinite since, by the choice of ω, $\operatorname{ord}_{p_n} c$ is not finite for any $n \le \omega$. Accordingly, c is divisible by $\gamma = p_1^\zeta p_2^\zeta \cdots p_\omega^\zeta$. I shall now show that, for an element $d \in \beta G$ as specified in 6.2, $\rho^{-1}\{d\}$ contains elements b_1 and b_2 whose difference divides γ.

Let $k = \min (p_1^\zeta, p_2^\zeta, \cdots, p_\omega^\zeta, p_\omega)$ so that k is infinite. Choose $b_1 \in \rho^{-1}\{d\}$ such that b_1 is infinite and smaller than k and then choose $b_2 \in \rho^{-1}\{d\}$ such that b_2 is infinite and $b_2 < b_1$. (Even if we did not require specifically that b_2 be infinite it would still have to be so.) Since $b_1 - b_2 < m \le p_\omega$ all the prime divisors of $b_1 - b_2$ are among $p_1, p_2, \cdots, p_\omega$. And since $b_1 \le p_j^\zeta$, it follows that $\operatorname{ord}_{p_j}(b_1 - b_2) < \zeta$ for $j = 1, 2, \cdots, \omega$. Accordingly, γ is divisible by $b_1 - b_2$. But $b_1 \simeq b_2$ and so $b_1 - b_2 \in \tilde{K}$. Since $\tilde{K}$ is an ideal we conclude that $\gamma \in \tilde{K}$, $\bar{\mu} \subset \tilde{K}$.

On the other hand, independently of 6.2, $\tilde{K} \subset \bar{\mu}$ since $\bar{\mu}$ is the kernel of a compactification of G. Hence $\tilde{K} = \bar{\mu}$, as asserted.

I do not know how common is the property 6.2 among ultrafilters. For our present purposes it will be sufficient to verify that it is satisfied in at least one instance. More precisely we shall show

6.4. THEOREM. *There exists an enlargement* $*G$ *such that for every* $d \in \beta G$, $\rho^{-1}\{d\} \cap (*N - N)$ *is coinitial with* $*N - N$ *unless it is empty.*

To find such a $*G$, consider a chain of enlargements of G

$$G = G_0 \subset G_1 \subset G_2 \subset \cdots \subset G_n \subset \cdots$$

where each G_n, $n \geq 1$, is an enlargement of G_{n-1} and may be constructed, for example, as an ultrapower. Then the union $\bigcup G_n = *G$ is an enlargement of G_0. In establishing this fact, the only point that requires some attention is the introduction of type restrictions [8] (which may, or may not, be considered desirable).

Now let $F(d)$ be an ultrafilter on G, $d \in \beta G$, such that $\rho^{-1}\{d\}$ contains at least one infinite natural number, m. Let k be any infinite natural number. Then k is contained in some $G_{n(k)}$. Consider the relation $R(x, y)$ which is defined as follows:

"x is an ordered pair $\langle z, w\rangle$ where z is a finite natural number, w is an element of $F(d)$ and y is a natural number which is contained in w and is smaller than k."

The relation $R(x, y)$ is formulated in the language appropriate to $G_{n(k)}$, though not to G_0. We claim that it is concurrent. For let $x_1 = \langle z_1, w_1\rangle, \cdots, x_j = \langle z_j, w_j\rangle$ be ordered pairs of the kind just described, and let $z = \max z_j$, $w = \bigcap w_j$. Then w cannot be empty interpreted as an element of G_0 since its interpretation in $*G$ contains m. Thus, there exists a *finite* natural number y which is greater than z and belongs to w. Being finite, y must be smaller than k, showing that $R(x, y)$ is concurrent. We conclude that $G_{n(k)+1}$ contains a natural number b which is greater than all standard natural numbers but smaller than k and which belongs to the extensions to $G_{n(k)+1}$ of all elements of $F(d)$. Then b belongs, a fortiori, to the extensions to $*G$ of all elements of $F(d)$, $b \in \rho^{-1}\{d\}$. This proves 6.4.

In any $*G$ such as exists according to 6.4, $\bar{\mu} = \tilde{K}$ and hence, $\bar{\Delta} = \tilde{\Delta}$. It follows that $\tilde{\Delta}$ is isomorphic to $\prod \Delta_P$, the direct product of P-adic integers. But the universal compactification of a ring is canonical (see §3 above) and so we have the following "standard" result, which is independent of any particular enlargement.

6.5. THEOREM. *Let G be the ring of integers in an algebraic number field and let* $\langle \tilde{\Delta}, \Gamma, \phi\rangle$ *be the universal compactification of G. Then* $\tilde{\Delta}$ *is isomorphic to the direct product of all P-adic integers over G.*

The following variant of our proof of 6.5 is of some interest. As mentioned, the equation $\bar{\Delta} = \prod \Delta_P$ was proved explicitly in [9] only on the assumption that $*G$ is *comprehensive.*[2] However, it is easy to check that in actual fact the proof given there relies only on the following condition.

6.6. Let $a_0, a_1, a_2, \cdots$ be a sequence of elements of $*G$ which is indexed in N. Then there exists an internal sequence $\{s_n\}$ in $*G$ (which is indexed in $*N$) such that $s_n = a_n$ for all finite n.

Thus we shall have proved Theorem 6.5 as soon as we can establish the existence of an enlargement $*G$ which satisfies 6.1 as well as 6.6. Now it is shown in [9] that 6.6 is satisfied by all ultrapower enlargements. Accordingly we choose $*G$ as the

[2] We take this opportunity to indicate the following corrections in [9]: p. 446, line 8. Insert before "which"—"from $*A$ into $*B$"; p. 446, line 26. For "$g_a(x)$" read "$g_y(x)$".

union of a chain of structures $\{G_\nu\}$ which is indexed in ω_1 (the initial ordinal of $\aleph_1$) and which is constructed as follows: $G_0 = G$; $G_\nu = \bigcup_{\xi<\nu} G_\xi$ for any limit ordinal ν; $G_{\nu+1}$ is an ultrapower enlargement of G_ν for any successor ordinal $\nu + 1$.

We see, as in the proof of 6.4, that $^*G$ has the property described in that theorem and, hence, satisfies 6.1. At the same time, since ω_1 is regular, every sequence $a_0, a_1, a_2, \cdots$, of elements of $^*G$ which is indexed in N must be contained already in some G_ν. It follows that $G_{\nu+1}$ contains an internal sequence $\{s_n\}$ such that $s_n = a_n$ for finite n, and the extension of that sequence to $^*G$ satisfies 6.6. This completes the argument.

In [4] the authors characterize the ring of all adèles over an algebraic number field (which is locally compact but not compact) by a universal mapping property. It might be possible to establish a connection between that result and our Theorem 6.5.

§7. **Conclusion.** Although our examples were concerned only with the compactification of discrete groups and rings, we recall that our general theory covered all separated groups and rings. It appears that the investigation of compactifications of nondiscrete groups or rings by our method may be of considerable interest. We may regard this kind of analysis as a branch of topological model theory, i.e., of the theory of relational structures which are imposed on underlying topological spaces.

In conclusion, we wish to point out that Theorem 2.1 above has a simple metamathematical corollary. Let X be a positive sentence on groups or rings, i.e. a sentence formulated in the Lower Predicate Calculus with equality in terms of the group or ring operations, by means of conjunction, disjunction and quantification. Then it is well known that whenever X is true in a given group (ring) G or $^*G$, X is true also in all homomorphic images of G or $^*G$. But if X is true in G then it is true also in $^*G$. Hence

7.1. THEOREM. *Let X be a positive sentence which holds in a group (ring) G and let $\langle \Delta, \Gamma, \phi \rangle$ be a compactification of G. Then X holds also in Δ.*

BIBLIOGRAPHY

[1] H. ANZAI and S. KAKUTANI, *Bohr compactifications of a locally compact abelian group.* I, II, ***Proceedings of the Imperial Academy of Tokyo***, vol. 19 (1943), pp. 476–480, 533–539.

[2] J. E. FENSTAD, *A note on "standard" versus "non-standard" topology*, ***Proceedings of the Royal Academy of Science***, North-Holland, Amsterdam, vol. 70 (1967), pp. 378–380.

[3] L. GILLMAN and M. JERISON, ***Rings of continuous functions***, Van Nostrand, New York, 1960.

[4] O. GOLDMAN and C. H. SAH, *On a special class of locally compact rings*, ***Journal of algebra***, vol. 4 (1966), pp. 71–95.

[5] E. HEWITT and K. A. ROSS, ***Abstract harmonic analysis***. Vol. I: ***Structure of topological groups. Integration theory, group representations***, Die Grundlehren der mathematischen Wissenschaften, Band 115, Academic Press, New York, and Springer-Verlag, Berlin, 1963.

[6] L. KUGLER, *Non-standard analysis of almost periodic functions*, Dissertation, University of California, Los Angeles, Calif., 1966.

[7] W. A. J. LUXEMBURG, *A new approach to the theory of monads*, Mimeographed notes, California Institute of Technology, Pasadena, Calif., 1967.

[8] A. ROBINSON, ***Non-standard analysis***, Studies in logic and foundations of mathematics, North-Holland, Amsterdam, 1966.

[9] A. Robinson, *Non-standard theory of Dedekind rings*, ***Proceedings of the Royal Academy of Science***, North-Holland, Amsterdam, vol. 70 (1967), pp. 444–452.

[10] A. Robinson, *Non-standard arithmetic*, ***Bulletin of the American Mathematical Society***, vol. 73 (1967), pp. 818–843.

Added in Proof. [11] A. L. Stone, *Nonstandard analysis in topological algebra*, ***Applications of model theory to algebra, analysis, and probability***, Holt, Rinehart and Winston, New York, 1969, pp. 285–307.

YALE UNIVERSITY

ALGEBRAIC FUNCTION FIELDS AND NON-STANDARD ARITHMETIC[1])

Abraham ROBINSON
Yale University

1. Introduction. Let $^*Q$ be a non-standard model of the field of rational numbers Q, for a higher order language (see Robinson [1966]). In particular, $^*Q$ may be an ultrapower of Q. Let α be an element of $^*Q$ which is not contained in Q. Then it is easy to verify that α is transcendental over Q. Thus, the field $A = Q(\alpha) \subset {}^*Q$ is the field of rational functions with rational coefficients.

Another method for finding an algebraic function field (of one variable, always) which is a subfield of $^*Q$ is as follows. Let Γ be a plane curve which is given by an equation $F(x, y) = 0$ where $F(x, y) \in Q[x, y]$. Let (α, β) be a non-standard rational point on $^*\Gamma$, i.e. such that at least one of α, β is non-standard. Then $A = Q(\alpha, \beta)$ is an algebraic function field which is also a subfield of $^*Q$. For example, let $F(x, y) = x^2 + y^2 - 1$, so that Γ is the unit circle and let $\alpha^2 + \beta^2 = 1$ where α and β are rational but non-standard. In this case we really obtain nothing new since $Q(\alpha, \beta) = Q(\omega)$ where ω is related to α and β by

$$\omega = \frac{\alpha}{1 - \beta}, \quad \alpha = \frac{2\omega}{\omega^2 + 1}, \quad \beta = \frac{\omega^2 - 1}{\omega^2 + 1}.$$

Thus, $Q(\alpha, \beta)$ is again the field of rational functions with rational coefficients. However, it is well known that no such reduction is possible if the curve Γ is of positive genus.

An internal valuation of $^*Q$ is given either (i), by a non-standard prime number in $^*Q$ or (ii), by a standard prime number in Q (and $^*Q$) or (iii), by the archimedean valuation of $^*Q$. We shall refer to these valuations also as valuations of the first, second, and third kind respectively. We avoid the use of the term "infinite prime" because it has incompatible interpretations in non-standard analysis and in valuation theory.

[1]) Research supported in part by the National Science Foundation Grant No. GP-18728.

Let A be an algebraic function field over Q which is also a subfield of $^*Q$ (see above). We shall show that, in certain circumstances, *an internal valuation of* $^*Q$ *induces a valuation of* A *and that, moreover, all valuations of* A *can be obtained in this way.*

In order to avoid repetitions, we shall prove the corresponding assertion immediately for the case where Q is replaced by any algebraic number field (finite algebraic extension of Q) K. In this case also, $^*K$ may, for example, be chosen as an ultrapower of K. An internal valuation of K now is given either (i), by a non-standard prime ideal in $^*K$ or (ii) by the canonical extension to $^*K$ of a standard prime ideal in K or (iii), by the canonical extension to $^*K$ of an archimedean valuation of K. We shall refer to these valuations again as valuations of the first, second, or third kind, respectively.

2. Valuations induced in A. Let K be an algebraic number field, $^*K$ a non-standard model of K, and A an algebraic function field over K which is contained in $^*K$. We consider valuations of the first kind in $^*K$. Thus, let P be a non-standard prime ideal in $^*K$. For any $x \in {}^*K$ we denote by $v_P x$ the order (exponential value) of x at P. Then $v_P x \in {}^*Z$, where $^*Z$ is the non-standard model of the rational integers which is included in $^*K$.

Suppose now that $v_P x$ *is not identically zero on* $A - \{0\}$. Choose $\alpha \in A$ such that $v_P\alpha > 0$ ($\alpha \neq 0$). Let $\beta \neq 0$ be another element of A. Then we claim that $v_P\beta/v_P\alpha$ is a standard rational number.

We observe first that $\alpha \notin K$ since the non-standard ideal P cannot divide any standard number in $^*K$ (i.e. $v_P x = 0$ on $K - \{0\}$). Accordingly, β is algebraic over $K(\alpha)$, there exists a non-zero polynomial $f(x, y) \in K[x, y]$ such that $f(\alpha, \beta) = 0$. Let

2.1. $f(x, y) = \sum c_{ij}x^i y^j, \quad c_{ij} \in K.$

Since $f(\alpha, \beta) = 0$, there exist distinct non-zero terms $c_{ij}x^i y^j$ and $c_{kl}x^k y^l$ such that

2.2. $v_P(c_{ij}\alpha^i\beta^j) = v_P(c_{kl}\alpha^k\beta^l).$

But c_{ij} and c_{kl} belong to K, so $v_P c_{ij} = v_P c_{kl} = 0$ and $(i - k)v_P\alpha = (l - j)v_P\beta$. Now $l - j \neq 0$ for if $l - j = 0$ then $i - k \neq 0$ and $v_P\alpha = 0$, contrary to our choice of α. Hence

2.3. $v_P\beta = \dfrac{i - k}{l - j}\, v_P\alpha.$

Thus, $v_P\beta$ is a standard rational multiple of $v_P\alpha$. Putting

2.4. $w_P x = v_P x / v_P\alpha$

we see that 2.4 defines a valuation of A in the additive group of rational numbers, with $w_P x = 0$ for $x \in K - \{0\}$. Thus w_P is a valuation of the algebraic function field A over K. It follows that w_P is discrete, and this can also be seen directly from 2.3. We shall say that w_P (or any equivalent valuation) is *induced by* v_P *in* A. The valuation is, up to equivalence, independent of our particular choice of α (subject to the stated conditions).

Next, we consider valuations of the second kind in $^*K$. Let then P be a standard prime ideal in $^*K$, i.e. the canonical extension of a prime ideal of K to $^*K$. Using the same notation as before suppose that $v_P x$ *is not finite* (in the sense of Non-standard Analysis) for some $x \in A - \{0\}$.

Choose $\alpha \in A$ such that $v_P\alpha$ is positive infinite ($\alpha \neq 0$). Then $\alpha \notin K$ since $v_P x$ is finite on $K - \{0\}$. Let $\beta \neq 0$ be an element of A. As before, there exists a polynomial $f(x, y) \in K[x, y]$ as given by 2.1 such that $f(\alpha, \beta) = 0$. Hence, again, 2.2 is satisfied for two distinct non-zero terms of $f(x, y)$. From 2.2,

2.5. $$(i - k)v_P\alpha = (l - j)v_P\beta + (v_P c_{kl} - v_P c_{ij}).$$

Now $l - j = 0$ would imply that $v_P\alpha$ is finite, which is contrary to our choice of α. Hence $l - j \neq 0$ and

2.6. $$\frac{v_P\beta}{v_P\alpha} = \frac{i - k}{l - j} + \frac{v_P c_{ij} - v_P c_{kl}}{(l - j)v_P\alpha}.$$

The numerator in the second term on the right-hand side of 2.6 is finite while the denominator is infinite. It follows that the term is infinitesimal. Hence, taking standard parts on both sides of 2.6, we obtain, introducing $w_P\beta$ by definition

2.7. $$w_P\beta = {}^\circ\!\left(\frac{v_P\beta}{v_P\alpha}\right) = \frac{i - k}{l - j}.$$

Notice that 2.6 also contains some additional information on the proximity of $^\circ(v_P\beta/v_P\alpha)$ to $(i - k)/(l - j)$ since it shows that the difference between these quantities cannot exceed some finite multiple of $(v_P\alpha)^{-1}$ (or, equivalently if $v_P\beta \neq 0$, some finite multiple of $(v_P\beta)^{-1}$).

We claim that $w_P x$, as given by 2.7, is a valuation of A over K in the additive group of standard rational numbers. Indeed, for $\beta \neq 0$, $w_P\beta$ is standard rational, by definition. Also, for $\beta = K - \{0\}$, $w_P\beta = 0$ since $v_P\beta$ is then finite. Also, for $\beta, \gamma \in A$, $\beta \neq 0$, $\gamma \neq 0$,

$$w_P(\beta\gamma) = {}^\circ\left(\frac{v_P(\beta\gamma)}{v_P\alpha}\right) = {}^\circ\left(\frac{v_P\beta + v_P\gamma}{v_P\alpha}\right) = {}^\circ\left(\frac{v_P\beta}{v_P\alpha}\right) + {}^\circ\left(\frac{v_P\gamma}{v_P\alpha}\right)$$
$$= w_P\beta + w_P\gamma,$$

which shows that the rule for the value of a product is satisfied. To consider the value of a sum, we may suppose that $v_P\beta \leq v_P\gamma$ and $\beta + \gamma \neq 0$. We then have $w_P\beta \leq w_P\gamma$ and

$$w_P(\beta + \gamma) = {}^\circ\left(\frac{v_P(\beta + \gamma)}{v_P\alpha}\right) \geq {}^\circ\left(\frac{v_P\beta}{v_P\alpha}\right) = w_P\beta.$$

Thus, w_P is a valuation of A over K. We call it the valuation which is *induced by v_P in A*. The valuation is, up to equivalence, again independent of our particular choice of α.

Finally, we have to consider valuations of the third kind in $^*K$. Such a valuation is given by the absolute values $|x|$ provided by an embedding of $^*K$ in the corresponding non-standard model of the complex numbers, $^*C$. The value of $|x|$ depends on the embedding but we shall not indicate this in the notation.

Suppose that $|x|$ *does not remain finite* everywhere on A. That is to say, $|x|$ is infinite for some $x \in A$ and hence, considering inverses, that it is infinitesimal but not zero somewhere on A. We choose a fixed $\alpha \neq 0$ in A such that $|\alpha|$ is infinitesimal. Then $\alpha \notin K$.

Let $\beta \neq 0$ be an element of A. Again there exists a polynomial $f(x, y) \in K[x, y]$ as given by 2.1 such that $f(\alpha, \beta) = 0$. Pick a term $c_{ij}x^iy^j$ such that $|c_{ij}\alpha^i\beta^j|$ is as large as possible. Then $|c_{ij}\alpha^i\beta^j| \neq 0$, and for any other term $c_{kl}x^ky^l$ the ratio

$$\text{2.8.} \qquad \frac{|c_{kl}\alpha^k\beta^l|}{|c_{ij}\alpha^i\beta^j|}$$

cannot exceed 1. At the same time, this ratio cannot be infinitesimal for all such terms for if it were then $|f(\alpha, \beta) - c_{ij}\alpha^i\beta^j|/|c_{ij}\alpha^i\beta^j|$ also would be infinitesimal. And this is impossible, for this ratio is actually equal to 1 since $f(\alpha, \beta) = 0$. Accordingly, there exists a monomial $c_{kl}x^ky^l$ distinct from $c_{ij}x^iy^j$ such that the ratio 2.8 is not infinitesimal. It follows that the natural logarithm of 2.8 is a finite real number, $-\mu$, where $\mu \geq 0$,

$$\ln|c_{kl}| - \ln|c_{ij}| + (k - i)\ln|\alpha| + (l - j)\ln|\beta| = -\mu.$$

Now $\ln|c_{kl}|$ and $\ln|c_{ij}|$ are finite since c_{kl} and c_{ij} are standard while $\ln|\alpha|$ is negative infinite, since $|\alpha|$ is infinitesimal. Hence

$$(k - i)\ln|\alpha| + (l - j)\ln|\beta| = v$$

where v is finite, and $l - j \neq 0$, otherwise $\ln|\alpha|$ also would have to be finite. Then

$$2.9. \quad \frac{\ln|\beta|}{\ln|\alpha|} = \frac{i - k}{l - j} + \frac{v}{(l - j)\ln|\alpha|}$$

where the last term on the right-hand side is infinitesimal. We put

$$2.10. \quad w\beta = {}^{\circ}\left(\frac{\ln|\beta|}{\ln|\alpha|}\right) = \frac{i - k}{l - j}$$

so that $w\beta$ is standard rational for all $\beta \in A$, $\beta \neq 0$, and we assert that this defines a valuation of A over K. Here again, 2.9 gives a measure for the difference between $w\beta$ and $\ln|\beta|/\ln|\alpha|$ by showing that this difference cannot exceed a finite multiple of $(\ln|\alpha|)^{-1}$ (or, for $\ln|\beta| \neq 0$, if $(\ln|\beta|)^{-1}$).

Let $\beta, \gamma \in A$, $\beta \neq 0$, $\gamma \neq 0$. Then

$$w(\beta\gamma) = {}^{\circ}\left(\frac{\ln|\beta\gamma|}{\ln|\alpha|}\right) = {}^{\circ}\left(\frac{\ln|\beta|}{\ln|\alpha|} + \frac{\ln|\gamma|}{\ln|\alpha|}\right) = {}^{\circ}\left(\frac{\ln|\beta|}{\ln|\alpha|}\right) + {}^{\circ}\left(\frac{\ln|\gamma|}{\ln|\alpha|}\right)$$

$$= w\beta + w\gamma.$$

Also, assuming $-\ln|\beta| \leq -\ln|\gamma|$ and $\beta + \gamma \neq 0$, we have $w\beta \leq w\gamma$ and

$$2.11. \quad w(\beta + \gamma) = {}^{\circ}\left(\frac{\ln|\beta + \gamma|}{\ln|\alpha|}\right) = {}^{\circ}\left(\frac{\ln|\beta|}{\ln|\alpha|} + \frac{\ln|1 + \gamma/\beta|}{\ln|\alpha|}\right).$$

Since $\beta + \gamma \neq 0$ and $1 + \gamma/\beta \in A$, $\ln|1 + \gamma/\beta|/\ln|\alpha|$ is finite. We claim that

$$2.12. \quad {}^{\circ}\left(\frac{\ln|1 + \gamma/\beta|}{\ln|\alpha|}\right) \geq 0.$$

Since $\ln|\alpha|$ is negative infinite, 2.12 will be established, if we can show that $\ln|1 + \gamma/\beta|$ cannot be positive infinite, i.e. that it is either finite or negative infinite. By assumption, $\ln|\beta| - \ln|\gamma| \geq 0$, i.e. $|\gamma/\beta| \leq 1$. This shows that $\ln|1 + \gamma/\beta| \leq \ln 2$ and proves 2.12. Hence, from 2.11

$$w(\beta + \gamma) \geq {}^{\circ}\left(\frac{\ln|\beta|}{\ln|\alpha|}\right) = w\beta.$$

Finally, since $\ln|\beta|$ is finite for all $\beta \in K - \{0\}$, $w\beta = 0$ for all such β. This shows that wx is a valuation of A over K. We shall say that *wx is induced by the given archimedean valuation.* Once again, wx is, up to equivalence, independent of the particular choice of α.

Examples which show that all three kinds of induced valuations may actually occur can be obtained already for $K = Q$, $A = Q(\omega)$, where $\omega \in {}^*Q - Q$. Choosing ω as a non-standard prime number in $^*Q$, we see that $P = (\omega)$ yields a valuation of the first kind which induces a valuation of A. Choosing $\omega = p^{\nu}$ where p is a standard rational prime and ν is a non-standard positive integer, we obtain a valuation of the second kind which induces a valuation of A, from $P = (\omega) = (p^{\nu})$. Finally, for $\omega = \nu^{-1}$, where ν is an arbitrary non-standard positive integer, the archimedean valuation of $^*Q$ also induces a valuation of A.

We shall show in the next two sections that, actually, *all* valuations of an algebraic function field $A \subset {}^*K$ are induced. However, it is entirely possible that different valuations of $^*K$, and even valuations of different kinds, may induce the same valuation of A.

3. All valuations of $A = K(\omega)$ over K are induced. For K an algebraic number field, as before, suppose first that $A = K(\omega) \subset {}^*K$ where $\omega \in {}^*K - K$ so that $K(\omega)$ is the field of rational functions of one variable with coefficients in K. We shall require the following lemma.

3.1. LEMMA. *Let α be any algebraic integer which belongs to $^*K$. If $|\alpha|$ is finite in all archimedean valuations of $^*K$ then α must be standard, $\alpha \in K$.*

Proof. Choose a fixed embedding of $^*K$ in the corresponding non-standard model of the complex numbers, $^*C$. Let $\alpha = \alpha^{(1)}, \alpha^{(2)}, \ldots, \alpha^{(n)}$ be the conjugates of α in $^*C$. Suppose that $|\alpha^{(1)}|, |\alpha^{(2)}|, \ldots, |\alpha^{(n)}|$ are all finite. Then we have to show that α is standard.

Let b be a finite upper bound for the $|\alpha^{(j)}|$, $j = 1, \ldots, n$, and let $s_1, \ldots, s_n$ be the fundamental symmetrical functions of $\alpha^{(1)}, \ldots, \alpha^{(n)}$. Then $|s_k| \leq \binom{n}{k} b$, $k = 1, \ldots, n$. Thus, the s_k are finite rational integers in $^*Q$. They are therefore standard, $s_k \in Q$, $k = 1, \ldots, n$. Consider the polynomial

$$f(x) = x^n - s_1 x^{n-1} + \ldots + (-1)^n s_n.$$

This is a standard polynomial, whose roots are therefore also standard. But α is one of these roots. Accordingly α is standard. This proves the assertion.

A valuation of $K(\omega)$ is given either by a prime polynomial $p(t) \in K[t]$ or it is the valuation "at infinity". Consider the former case. The corresponding valuation of $K(t)$ will be denoted by V_p. It is obtained as follows. For any $q(t) \in K[t]$, write

3.2. $q(t) = (p(t))^m s(t), \quad m \geq 0$

where $p(t)$ is prime to $s(t)$. Then there exist elements $g(t)$ and $h(t)$ of $K[t]$ such that

3.3. $g(t)p(t) + h(t)s(t) = 1.$

The valuation of $q(t)$ is now defined by $V_p q = m$ and by $V_p f = V_p q - V_p r$ for any $f(t) \in K(t)$, $f(t) = q(t)/r(t)$. The corresponding valuation of $K(\omega)$, which will again be denoted by V_p is given by the canonical mapping of $K(t)$ on $K(\omega)$, i.e. by $V_p(f(\omega)) = V_p(f(t))$ for any $f(t) \in K(t)$.

For any $q(t) \in K[t]$ consider $q(\omega) \in K[\omega] \subset {}^*K$. Since $q(\omega)$ is an algebraic number in ${}^*K$, the ideal $(q(\omega))$ can be written as a product of prime powers $\prod P_j^{\nu_j}$, with starfinite index set (i.e. the index set is "finite" in the non-standard sense).

Suppose first that there is a non-standard prime ideal P_j in the numerator of the prime power representation of $p(\omega)$, i.e. $\nu_j > 0$. We put $P_j = P$, $\nu_j = \nu$, and we consider the valuation v_P in ${}^*K$, which is of the first kind. By assumption, $v_P(p(\omega)) = \nu > 0$.

We observe that $v_P\omega \geq 0$. For suppose $v_P\omega < 0$ and let

3.4. $p(t) = a_0 + a_1 t + \ldots + a_\lambda t^\lambda, \quad a_j \in K, a_\lambda \neq 0.$

Since $v_P a_j = 0$ for $a_j \neq 0$ we then have $v_P(p(\omega)) = \lambda v_P \omega < 0$, which is contrary to assumption. It follows that $v_P\omega > 0$ and, hence, that $v_P(k(\omega)) \geq 0$ for all $k(t) \in K[t]$.

In particular, for any $q(t) \in K[t]$, we then have $v_P(g(\omega)) \geq 0$, $v_P(h(\omega)) \geq 0$, $v_P(s(\omega)) \geq 0$ where g, h, and s are given by 3.2 and 3.3. Now $v_P(s(\omega)) > 0$ would imply

$$v_P 1 = v_P(g(\omega)p(\omega) + h(\omega)s(\omega)) > 0$$

although $v_P 1 = 0$. Hence $v_P(s(\omega)) = 0$ and, from 3.2,

3.5. $v_P(q(\omega)) = m v_P(p(\omega)).$

Now let $\alpha = p(\omega)$ for the construction of the valuation w_P induced by v_P in A. We claim that w_P coincides with V_p on A. Indeed, for $x = q(\omega) \in K[\omega]$, we have, from 2.4,

$$w_P(q(\omega)) = \frac{v_P(q(\omega))}{v_P(p(\omega))} = \frac{m v_P(p(\omega))}{v_P(p(\omega))} = m = V_p(q(\omega)).$$

Hence for $f(t) \in K(t)$, $f(t) = q(t)/r(t)$, $q(t)$, $r(t) \in K[t]$

$$w_P(f(\omega)) = w_P(q(\omega)) - w_P(r(\omega)) = V_p(q(\omega)) - V_p(r(\omega)) = V_p(f(\omega)),$$

which proves our assertion.

Suppose next, that for some P_j in the numerator of the prime power representation of $p(\omega)$, P_j is a standard prime ideal and ν_j is infinite, hence positive infinite. Putting $P_j = P$ and $\nu_j = \nu$, we observe that, this time, $v_P\omega$ while not necessarily non-negative is at any rate not negative infinite. To see this, let $p(t)$ be given by 3.4. The numbers $v_P a_j$, $a_j \neq 0$, being standard, cannot be negative infinite. If $v_P\omega$ were negative infinite we should again have $v_P(p(\omega)) = \lambda v_P\omega$, so that $v_P(p(\omega))$ also would be negative infinite, contrary to assumption. It follows that no monomial $a\omega^\mu$, μ finite, $a \in K$, can be negative infinite, and hence that $v_P(k(\omega))$ cannot be negative infinite for any $k(t) \in K[t]$.

Let $q(t) \in K[t]$ and let g, h, and s be given by 3.2 and 3.3. Then we claim that $v_P(s(\omega))$ cannot be positive infinite. For if this were the case then $v_P(g(\omega)p(\omega))$ and $v_P(h(\omega)s(\omega))$ would both be positive infinite. Since

$$v_P 1 = v_P(g(\omega)p(\omega) + h(\omega)s(\omega)) \geq \min(v_P(g(\omega)p(\omega)), v_P(h(\omega)s(\omega)))$$

it would then follow that $v_P 1$ is positive infinite, which is not true. We conclude that $v_P(s(\omega))$ is finite and, hence, standard, $v_P(s(\omega)) = \mu$, say. By 3.2, then,

$$v_P(q(\omega)) = m v_P(p(\omega)) + \mu.$$

Let $\alpha = p(\omega)$ for the construction of the valuation w_P which is induced by v_P, a valuation of the second kind in A. We claim that $w_P = V_p$ on A and we see that again it will be sufficient to show this for $x = q(\omega) \in K[\omega]$. From 2.7,

$$w_P(q(\omega)) = {}^\circ\left(\frac{v_P(q(\omega))}{v_P(p(\omega))}\right) = {}^\circ\left(\frac{m v_P(p(\omega))}{v_P(p(\omega))} + \frac{\mu}{\nu}\right) = m = V_p(q(\omega)),$$

since μ/ν is infinitesimal. This proves our assertion.

It only remains for us to dispose of the case that the numerator of $(p(\omega)) = \prod P_i^{\nu_i}$ contains neither a non-standard prime ideal, nor a standard prime ideal with infinite exponent. We claim, first of all that in this case the number of prime ideals in the numerator of $p(\omega)$ must be finite in the absolute sense. Let the set of these ideals be S. At any rate, S is internal. Range all prime ideals of K in a sequence, $\{P^{(j)}\} = \sigma$ without repetitions. If S is not finite then, for every standard natural number μ, there exists a $j > \mu$ such that $P^{(j)} \in S$. On the other hand, S is starfinite, so there must exist a first natural number, μ_0, such that $P^{(j)} \notin S$ for all $j > \mu_0$. Since we have just seen that μ_0 cannot be finite, it must be infinite. But then $P^{(\mu_0)}$ is a non-standard prime ideal which belongs to S, which is contrary to assumption. Thus,

S is finite. Since, in the case under consideration, the prime ideals P_i which belong to S all appear in the prime power representation of $(p(\omega))$ with finite exponents, the numerator of $(p(\omega))$ must be standard, so

$$(p(\omega)) = J_1/J_2$$

where J_1 and J_2 are entire ideals in $^*K$ and J_1 is standard. Let γ be the absolute norm of J_1 then γ is standard and is divisible by J_1, $(\gamma) = J_1J_3$ say, where J_3 is standard and entire. Hence

$$(p(\omega)) = \frac{J_1J_3}{J_2J_3} = \frac{(\gamma)}{J_2J_3} = \frac{(\gamma)}{(\delta)}$$

where δ is some integer in $^*K$. It follows that $p(\omega) = \gamma/\varepsilon\delta$ where ε is a unit in $^*K$, $\varepsilon\delta = \gamma/p(\omega)$. Now $\varepsilon\delta$ is an algebraic integer, hence, by 3.1 there exists an archimedean valuation of $^*K$ for which the absolute value of $\varepsilon\delta$ (to be denoted simply by $|\varepsilon\delta|$) is infinite. We conclude that $|p(\omega)| = \gamma/|\varepsilon\delta|$ is infinitesimal.

We put $\alpha = |p(\omega)|$ and we construct a valuation wx on A according to the procedure given in section 3 above for valuations induced in A by valuations of $^*K$ which are of the third kind. We are going to show that w coincides with V_p on A. Again it is sufficient to prove $wx = V_px$ for any $x = q(\omega)$ where $q(t) \in K[t]$, and where g, h, and s are given by 3.2 and 3.3. Then $V_p(q(\omega)) = m$.

Let $p(t)$ be given by 3.4. If $|\omega|$ were infinite then

$$|p(\omega)| = |\omega|^{\lambda}\left|\frac{a_0}{\omega^{\lambda}} + \frac{a_1}{\omega^{\lambda-1}} + \ldots + a_{\lambda}\right|$$

would be infinite also. But $|p(\omega)|$ is infinitesimal, so $|\omega|$ must be finite. It follows that $|g(\omega)|$, $|h(\omega)|$, $|s(\omega)|$ also are finite. But then $|s(\omega)|$ cannot be infinitesimal, since, by 3.3,

$$1 = |1| = |g(\omega)p(\omega) + h(\omega)s(\omega)| \leq |g(\omega)||p(\omega)| + |h(\omega)||s(\omega)|$$

so that $|1|$ would be infinitesimal as well. Hence, $|s(\omega)|$ is neither infinite nor infinitesimal, implying that $\mu = \ln|s(\omega)|$ is finite. Using the definition of w by 2.10 we therefore obtain

$$w(q(\omega)) = {}^{\circ}\left(\frac{\ln|q(\omega)|}{\ln|p(\omega)|}\right) = {}^{\circ}\left(\frac{m\ln|p(\omega)|}{\ln|p(\omega)|} + \frac{\mu}{\ln|p(\omega)|}\right) = m = V_p(q(\omega)).$$

This completes our argument. We have shown that V_p is always induced by a valuation of $^*K$. We still have to consider the valuation of $K(\omega)$ "at infinity", but this is reduced to the previous case by the substitution $\omega' = \omega^{-1}$. We conclude that *all* valuations of $A = K(\omega)$ over K are induced by valuations of $^*K$ of the first, second, or third kind.

4. All valuations of A over K are induced. We now tackle the case of a general algebraic function field A over K, $A \subset {}^*K$. Our argument depends heavily on the following "standard" lemma.

4.1. LEMMA. *Let A be an algebraic function field over a field K, where K is algebraically closed in A. Let S be a set of valuations of A over K with the following property. For every element $\omega \in A - K$ and for every valuation V of $K(\omega)$ over K there exists a $V' \in S$ which reduces to V on $K(\omega)$. Then S is the set of all valuations of A over K.*

Leaving aside the proof of 4.1 for the moment, we show first that it implies the statement which titles this section. In our case, K is certainly algebraically closed in A since it is even algebraically closed in $^*K$. We now identify S with the set of valuations which are induced in A by (internal) valuations of $^*K$. If $\omega \in A - K$ and if V is a valuation of $K(\omega)$ over K then, as shown in section 3 there exists a valuation v of $^*K$ which induces the valuation V on $K(\omega)$ over K. But the same v induces a valuation V' on A over K which reduces to V on $K(\omega)$. We conclude that S consists of all valuations of A over K, i.e. every valuation of A over K is induced by some valuation of $^*K$. This confirms our assertion.

In order to prove 4.1, we shall find it convenient to identify the elements of S with the corresponding places (prime divisors). Using a terminology which is suggested by the particular application made above, we call the elements of S *induced* places or *induced* prime divisors. Products of powers of induced prime divisors will be called induced divisors. The set of induced divisors will be denoted by D. Then $S \subset S_0$, $D \subset D_0$ where S_0 and D_0 are the sets of all prime divisors, and the sets of all divisors, in A over K, respectively. Every induced divisor, in particular every induced prime divisor has a *degree* in the usual sense.

Let α be any element of A, (α) the corresponding divisor. Then (α) can be written in the form R/Q where Q and R are elements of D_0 without common factor. If now we exclude from Q and R those prime factors which do not belong to S, we obtain elements of D which will be denoted by Q_i and R_i respectively. If $Q \neq Q_i$ we then have $d(Q) > d(Q_i)$ where $d(X)$ denotes the degree of a divisor X. Suppose now that $D_0 \neq D$ so that $D_0 - D$

contains a divisor C. Then C appears in the denominator of (α) for some $\alpha \in A$. For this α, $Q \neq Q_i$, $d(Q) > d(Q_i)$. Now $d(Q) = n$ where n is the degree of A over $K(\alpha)$. Accordingly, in order to show that $D_0 - D$ is empty, i.e. in order to prove 4.1, we only have to establish that $d(Q_i) = n$ also, for all $\alpha \in A$.

By an induced zero (pole) of an element $\alpha \in A$ we mean an induced prime divisor which belongs to the numerator (denominator) of α.

4.2. LEMMA. *On the assumptions of* 4.1, *an element* $\alpha \in A$ *without induced poles is a constant.*

Proof of 4.2. Suppose $\alpha \in A$ is not a constant. Then there exists a valuation V of A such that $V\alpha < 0$. Let V' be the restriction of V to $K(\alpha)$, and let $V'' \in S$ be an extension of V' to all of A such as exists on the assumptions of 4.1 and 4.2. Then $V''\alpha < 0$, α has an induced pole.

Let Q be an induced entire divisor. An element $\alpha \in A$ will be called an induced multiple of Q if for every prime divisor P which appears in Q with non-zero exponent v, the order (value) of α at P, $V_p\alpha$ is not smaller than v, and if for all other *induced* prime divisors the order of Q is non-negative. Notice that, a priori, α may be a multiple of the induced divisor Q without being an induced multiple of Q. For a given induced Q, the induced multiples of Q^{-1} constitute a K-module $M_i(Q)$. We claim that $M_i(Q)$ has a finite rank, $l_i(Q)$, over K which satisfies the inequality

4.3. $$l_i(Q) \leq d(Q) + 1.$$

The proof of 4.3 as well as the subsequent steps given before for the proof of 4.1 are entirely analogous to the corresponding proofs for ordinary multiples. Compare Van der Waerden [1967].

Since we have supposed that Q is entire, $Q = \prod P_j^{v_j}$ where the P_j are induced and the v_i are non-negative. Let $\delta_j = d(P_j)$ and let π_j be a corresponding prime element of A–i.e. $V_{P_j^{\pi_k}} = 1$ where V_{P_j} is normed with smallest positive element 1–for any one of the P_j. δ_j is, as usual, the degree of the residue class field W_j/J_j over K where $W_j \subset A$ and $J_j \subset W_j$ are the valuation ring and the valuation ideal for P_j, respectively. Let $\delta = d(Q) = \sum \delta_j v_j$ and let $\alpha_1, \ldots, \alpha_\lambda$ be elements of $M_i(Q)$. Then the expansion of any of the a_σ at P_j begins with the v_jth negative power of π_j (at most). Hence, at most δ linear conditions are required on $k_1, \ldots, k_\lambda \in K$ in order to ensure that for the linear combination $\alpha = k_1\alpha_1 + \ldots + k_\lambda\alpha_\lambda$ no negative powers of π_j occur at any of the P_j. But then α has no induced pole and so $\alpha \in K$, by 4.2. Suppose now that $\alpha_1, \ldots, \alpha_\lambda$ are linearly independent over K, then the dimension of the $\alpha = k_1\alpha_1 + \ldots + k_\lambda\alpha_\lambda$ for which no negative power of π_j occur at any of the P_j cannot be less than $\lambda - \delta$. But since $\alpha \in K$, this dimension is at most 1, $\lambda - \delta \leq 1$, $\lambda \leq \delta + 1$. This proves 4.3.

Let $\beta_1, \ldots, \beta_n$ be a basis of A over $K(\alpha)$ where $\alpha \in A - K$. Suppose that, in the first place, β_1 has a pole for an induced divisor P for which α does not have a pole, $V_P\, \alpha \geq 0$, $V_P\, \beta < 0$. Let V' be the restriction of V_P to $K(\alpha)$. Since $V_P\, \alpha \geq 0$, V' is not the valuation of $K(\alpha)$ at "infinity", V' is defined by a prime polynomial $p(t) \in K[t]$. Then $V'(p(\alpha)) = V_P(p(\alpha)) > 0$ so that we obtain a β'_1 with $V_P\beta'_1 > 0$ if we multiply β by a sufficiently great power of $p(\alpha)$. We apply the same procedure for all other induced divisors P for which β_1 has, and α does not have a pole, and we then do the same also for $\beta_2, \ldots, \beta_n$ in turn. In this way, we finally obtain a basis of A over $K(\alpha)$ none of whose elements has a pole for an induced divisor for which α does not have a pole. We denote the elements of this basis, without fear of confusion, again by $\beta_1, \ldots, \beta_n$. Since all the induced poles of $\beta_1, \ldots, \beta_n$ are now among the induced poles of α, $\beta_1, \ldots, \beta_n$ are induced multiples of some positive power Q_i^μ of Q_i where Q_i is, as before, obtained from the denominator Q of $(\alpha) = R/Q$ by excluding the prime divisors which do not belong to S. Now let m be an integer greater than μ and consider the set of elements $\beta_j\alpha^k$, $j = 1, \ldots, n$, $k = 0, \ldots, m - \mu - 1$. Their number is $n(m - \mu)$, they are linearly independent over K, and they are all induced multiples of Q_i^m. Hence, by 4.3

$$n(m - \mu) \leq l_i(Q_i^m) \leq d(Q_i^m) + 1 = d(Q_i) + 1$$

and so

4.4. $m(n - d(Q_i)) \leq n\mu + 1.$

If $n - d(Q_i)$ were positive then, by choosing m sufficiently large we could make the left-hand side of 4.4 greater than its right-hand side. Hence, $n \leq d(Q_i)$. But we also have $d(Q_i) \leq d(Q)$ and since $d(Q) = n$, we conclude that $d(Q_i) = n$. This completes the proof of 4.1.

5. Conclusion. We conclude with some further discussion of the circumstances under which the theory of the present paper applies. Let K be an algebraic number field and let $^*K$ be a non-standard model of K, as before. Let $f(x, y)$ be an absolutely irreducible polynomial with coefficients in K and let Γ be the irreducible curve which is given by $f(x, y) = 0$. By a K-point we mean a point (α, β) with coefficients in K, with a similar definition for $^*K$. A point of $*\Gamma$ is non-standard if not both of its coordinates are standard.

5.1. THEOREM. *$*\Gamma$ contains a non-standard $*K$-point if and only if the number of K-points on Γ is infinite.*

Proof. Suppose $*\Gamma$ contains a non-standard $*K$-point, (α, β). The assertion "Γ contains a K-point (x, y)" is then true in the non-standard model, where "Γ" denotes $*\Gamma$ and "K" denotes $*K$. It therefore holds also in the standard model, for some $x = \alpha_1$, $y = \beta_1$, $\alpha_1, \beta_1 \in K$. Next, the assertion "Γ contains a K-point (x, y) which is different from (α_1, β_1)" holds in the non-standard model and therefore also in the standard model, for some $x = \alpha_2$, $y = \beta_2$, $\alpha_2, \beta_2 \in K$, $(\alpha_1, \beta_1) \neq (\alpha_2, \beta_2)$. Next, the assertion "$\Gamma$ contains a K-point (x, y) which is different from (α_1, β_1) and from (α_2, β_2)" holds in the non-standard and, hence also in the standard model, for some (α_3, β_3) which is distinct from (α_1, β_1) and from (α_2, β_2). Continuing in this way, we see that Γ contains an infinite sequence (α_1, β_1), (α_2, β_2), (α_3, β_3), ... of distinct K-points. This shows that the condition is sufficient.

To show that this condition is necessary, range all elements of K in a sequence $\{\eta_n\}$, without repetitions. Then it is true in K that "for every positive integer n, there exists a point (α, β) on Γ such that for some $m > n$ either $\alpha = \eta_m$ or $\beta = \eta_m$". But the statement in quotes must still be true in $*K$. Choosing n infinite, we obtain a non-standard point (α, β) on $*\Gamma$.

The question whether there can be an infinite number of rational points on a curve of genus greater than 1 is the subject of a famous conjecture of Mordell's.

Suppose now that the number of K-points of Γ–which is given by $f(x, y) = 0$, $f(x, y) \in K[x, y]$ and absolutely irreducible–is infinite. Then there exists a non-standard $*K$-point (α, β) on $*\Gamma$. Let ξ be an indeterminate over K and let η be algebraic over $K(\xi)$ such that $f(\xi, \eta) = 0$. Then $K(\xi, \eta)$ is the algebraic function field which belongs to Γ.

We wish to show that $K(\alpha, \beta)$ is isomorphic to $K(\xi, \eta)$ under a map $\alpha \to \xi$. However, this is true only subject to the trivial restriction that *Γ is not a straight line parallel to one of the coordinate axes.*

Let $f(x, y) = \sum f_j(x)y^j$ where $f_j(x) \in K[x]$. At least one of the $f_j(x)$ cannot be a constant, otherwise Γ (being irreducible) would reduce to a straight line parallel to the x-axis. Suppose α is algebraic over K. Then $f_j(\alpha) = 0$ for all j, otherwise β would be algebraic over K also, contrary to the assumption that (α, β) is a non-standard point. But then Γ reduces to straight lines, and hence to a single straight line, parallel to the y-axis, which is again contrary to assumption. Thus, α is transcendental over K and we have a natural isomorphism ψ: $K(\xi) \to K(\alpha)$ which reduces to the identity on K, such that $\xi \overset{\psi}{\to} \alpha$. Since $f(\xi, \eta) = 0$ and $f(\alpha, \beta) = 0$, ψ can be extended to an isomorphism $K(\xi, \eta) \to K(\alpha, \beta)$ for which $\eta \to \beta$. This shows that $K(\alpha, \beta)$ can be identified with the algebraic function field which belongs to Γ.

References

Chevalley, C., 1951, Introduction to the Theory of Algebraic Functions of one variable, *Mathematical Surveys VI.*

Robinson, A., 1966, *Non-standard Analysis*, Studies in Logic and the Foundations of Mathematics (North-Holland, Amsterdam, second printing 1969).

Van der Waerden, B. L., 1967, *Algebra* (5th ed. of Moderne Algebra), 2nd part (Berlin-Heidelberg-New York).

Received 22 March 1971

The nonstandard $\lambda{:}\phi_2^4(x)$: model. I. The technique of nonstandard analysis in theoretical physics*

Peter J. Kelemen

Courant Institute of Mathematical Sciences and Department of Mathematics, University College, New York University, New York

Abraham Robinson

Department of Mathematics, Yale University, New Haven, Connecticut

(Received 22 May 1972; revised manuscript received 21 July 1972)

The methods of nonstandard analysis are demonstrated as a preliminary step for the construction of the nonstandard $\lambda{:}\phi_2^4$: model. Elementary quantum mechanical problems are solved and the renormalization of the scalar field (Yukawa interaction) is investigated.

1. INTRODUCTION

In the applications of analysis, one often speaks of infinitesimal increments and of infinitesimal volume elements. Since Weierstrass, the above phrases were understood to be shorthand formulations of more complicated expressions involving limits. However, some time ago one of the authors[1,2] showed that expressions involving infinitesimals can be taken literally if one refers them to a suitable number system that contains infinitely large and infinitely small elements. The subject that arose out of this realization has come to be known as nonstandard analysis (n.s.a.). It is described informally in Sec. 2 below.

We feel that n.s.a. can be used advantageously in physics. Thus, many calculations can be simplified by its use through the avoidance of passages to the limit at certain stages. Also, using infinitely large numbers one can give a rigorous meaning to self-energies and renormalization. Finally, one may treat certain nonseparable Hilbert spaces with the same ease as separable ones.

Using n.s.a., we may retain results calculated by standard techniques whenever desirable, and, moreover, we may reinterpret them in the nonstandard system. On the other hand, the method is conservative; that is to say, any final result that has been obtained by nonstandard techniques but is itself formulated in standard terms might have been obtained by standard methods, though perhaps at the cost of a considerable effort.

Here we shall consider cases in which the basic assumptions of a problem were formulated originally in standard language and are then translated into the language of n.s.a. and solved by its methods. Several examples of this technique are given in Secs. 3 and 4 below. If we were to include an assumption that can be formulated only in nonstandard terms, then the result might not be amenable to a standard formulation and would have to be interpreted directly. We intend to develop this approach in a future paper.

2. NONSTANDARD ANALYSIS

Let R be the system of real numbers. An *ordered field* is an ordered number system which shares with R all the usual properties involving the operations of addition, subtraction, multiplication, and division. It has been known for a long time that there exist ordered fields which are extensions of R. Such fields are *non-archimedean*, i.e., they contain positive numbers which are greater than any natural number (in R), while their reciprocals are smaller than any positive element of R. However, generally speaking, one cannot extend most of the familiar functions of analysis [e.g., e^x, $\sin x$, $\ln x$, $J_n(x)$] to these new systems so as to preserve their usual properties. Nonstandard analysis is based on the existence of *particular* ordered fields which include R as a subfield and for which the extended functions in question are, in fact, available and, moreover, are provided automatically by certain model theoretic procedures. One of these is the so-called *ultrapower construction*[3] which, for the case of a countable index set, runs as follows. Let N be the set of natural numbers and let F be a free ultrafilter[4] on N. Then the field in question, $^*R$, is defined as the system of all sequences of real numbers R^N, where two sequences are regarded as equal if they coincide on a set of natural numbers which belongs to F. Functions on $^*R$ are defined termwise on representative sequences selected from the equivalence classes just defined, and a relation is said to hold between sequences if it holds termwise for an index which belongs to the ultrafilter. R can be embedded in $^*R$ by identifying any real $r \in R$ with the (equivalence class of the) sequence $(r, r, r, \cdots)$ in $^*R$.

While the ultrapower construction sketched above provides a relatively concrete realization of the type of structure required for nonstandard analysis, the class of these structures is quite large and contains fields which cannot be obtained in this way. They can be characterized in the following way.

A *nonstandard model of analysis* is a proper extension $^*R$ of the system of real numbers R, such that

Transfer Theorem: Any true assertion X about R is still valid in $^*R$, provided we reinterpret X in $^*R$ as follows:

For every class of objects in R (e.g., functions of one numerical variable, relations between numbers, relations between functions, functionals, i.e., mappings from functions to numbers), there exists a subclass, said to be the class of the corresponding *internal* objects. In particular, the class of internal entities of $^*R$, contains an element corresponding to each object in R. For example, the class $^*\Phi$ of internal functions of one numerical variable in $^*R$ contains, in particular, extensions of all functions on R; but these functions do not exhaust $^*\Phi$. However, all individuals (numbers) of $^*R$ are regarded as internal. Then the assertion X, supposed true in R, is still valid in $^*R$, provided we reinterpret each quantifier in X (e.g., "there exists a function," "for all relations") as referring only to the corresponding *internal* objects in $^*R$ ("there exists an *internal* function," "for all *internal* relations").

See Ref. 2. for a more rigorous formulation of these notions. An object which is not internal is said to be *external*. The following examples are instructive.

(1) Let N be the set of natural numbers. Then $N \subset R$. The corresponding *internal* subset $^*N$ of $^*R$ contains N. The elements of $^*N$-N are called the *infinite* natural numbers. Any $a \in {}^*N$-N is greater than all elements of N. If $^*R$ has been obtained by the ultrapower construction sketched above, then an example of an infinite natural number is given by the sequence $(0,1,2,3,\cdots)$. The axiom of induction is satisfied in $^*N$ if *set* (property) is interpreted as *internal set* (property). Accordingly, every nonempty internal subset of N possesses a smallest element. The set $^*N$-N does not have a smallest element and, accordingly, is *external*.

(2) If $a \in {}^*R$ is numerically smaller than any positive $r \in {}^*R$, then a is said to be infinitely small or *infinitesimal*. The set of infinitely small numbers in $^*R$ is said to be the *monad of zero*. More generally, for any $r \in R$ the set of all numbers $a \in {}^*R$ which differ from r only by an infinitesimal amount is said to be the *monad* of r, $\mu(r)$. If $a \in \mu(r)$, then we write $r = {}^0a$ and we call r the *standard part* of a. All monads are external. The numbers of R ("standard numbers") are isolated points in the interval topology of $^*R$.

(3) If a and b are any numbers in $^*R$, finite or infinite, then the interval $a < x < a + b$ is internal. The interval of all $a \in {}^*R$, a positive infinite, is external.

(4) The extensions[5] of x^2 and e^x to $^*R$, $^*(x^2)$ and $^*(e^x)$, are internal and even standard. (We may omit the star on the extended functions, by convention.) The function

$$f(x) = \begin{cases} 0 & \text{for } x \text{ infinite} \\ 1 & \text{for } x \text{ finite} \end{cases}$$

is external. Among the functions which are internal but not standard are various representations of the Dirac delta function, e.g., $\delta(x) = (\pi\eta x)^{-1} \sin\eta^2 x$, where η is an infinite natural number

or

$$\delta(x) = \begin{cases} \eta & \text{for } -\frac{1}{2}\eta^{-1} \le x \le \frac{1}{2}\eta^{-1} \\ 0 & \text{otherwise} \end{cases}.$$

The first of these representations is analytic in $^*R$; the second is not. In either case we have[6]

$${}^0\left[\int_{^*R} \delta(x) f(x) dx\right] = f(0)$$

for any $f(x)$, which extends a bounded function which is defined and continuous in the neighborhood of 0 in R. The validity of this equation is, in fact, a condition which has to be imposed on any reasonable interpretation of the delta function. On the other hand, $\int_{^*R} \delta^2(x)dx$ depends on our particular choice of the representations.

(5) Let $f(x)$ be a real function in R, defined for an interval $a < x < b$, and let x_0 be a point in that interval. $f(x)$ possesses an automatic extension to $^*R$, as stated. It can then be shown that $f(x_0 + \eta)$ is infinitely close to $f(x_0)$, in symbols $f(x_0 + \eta) \simeq f(x_0)$, for all infinitesimal η if and only if $f(x)$ is continuous at x_0 in the classical (Weierstrass) sense, i.e., if and only if the following condition is satisfied:

(C) For every positive ϵ (in R), there exists a positive δ (in R) such that $|f(x_0 + h) - f(x_0)| < \epsilon$ provided $h < \delta$.

Now let $f(x)$ be an *internal* function in $^*R$ Then condition (C) when interpreted in $^*R$ according to the rule adopted earlier, refers to positive ϵ and δ in $^*R$. If $f(x)$ satisfies the condition in this form, then we say that $f(x)$ is *Q continuous* (at x_0). On the other hand, if $f(x)$ satisfies (C) with ϵ and δ still assumed *in* R, then we say that $f(x)$ is *S continuous*. It turns out that the condition that $f(x_0 + \eta) \simeq f(x_0)$ for infinitesimal η is equivalent to S continuity. Thus, if $f(x)$ is standard then all these definitions coincide. For example, the two δ functions defined above are both Q continuous at the origin, but not S continuous. By contrast, the function

$$f(x) = \begin{cases} 0, & x \le 0 \\ \eta, & x > 0 \end{cases},$$

where η is infinitesimal, is S continuous at the origin, but not Q continuous.

In the sequel to this paper (Ref. 7) we shall require the notion of an *enlargement* which generalizes the kind of nonstandard model discussed here (see Ref. 2). However, the above indications should be sufficient for the study of the present paper.

3. EXAMPLES FROM QUANTUM MECHANICS

In this section we demonstrate the use of n.s.a. by finding the bound state solution of the one-dimensional Schrödinger equation

$$\left(\frac{d^2}{dx^2} + E - V(x)\right) \phi(x) = 0, \tag{1}$$

for the square well, infinite square well, δ-function, and singular square well potentials. The infinite square well and the δ-function potentials are limiting cases of "physical" potentials, but themselves are outside the framework of quantum mechanics. This only means that we have to prescribe conditions other than the continuity of the logarithmic derivative of the wavefunction or to treat these problems by taking limits.

In the n.s. translation we have

$$\left(\frac{d^2}{dx^2} + E - V(x)\right) \phi(x) = 0$$

where

$$V(x) = \begin{cases} V_0, & x < 0 \\ V_1, & 0 < x < L \\ V_0, & L < x \end{cases}, \tag{2}$$

and the conditions (i) $V_0 > E > V_1$, (ii) $\phi(x)$ is infinitesimal for $x \in {}^*R$-R_0, where R_0 is the set of all finite reals; and (iii) the logarithmic derivative of $\phi(x)$ is Q continuous.

In our formulation all four potentials are "physical," since all four are "finite" square well potentials in the nonstandard universe. They differ only in the values of V_0, V_1, and L. For a finite square well potential the solution of (1) is well known. Since (2) contains only internal objects, by the transfer theorem its solution is given by

$$c_n\phi_n(x)=\begin{cases}\exp(k_nx), & x<0\\ \dfrac{p_n-ik_n}{2p_n}\exp(ip_nx)+\dfrac{p_n+ik_n}{2p_n}\exp(-ip_nx), & 0<x<L,\\ \dfrac{k_n^2+p_n^2}{2k_np_n}\sin p_n\exp(k_nL-k_nx), & L<x\end{cases}$$

where c_n is determined by normalization, p_n and k_n satisfy the equations $2k_np_n(k_n^2+p_n^2)^{-1}=\tan p_nL$, and $E_n=p_n^2+V_1=V_0-k_n^2$. Observe that for $n\in N$:

(α) For the finite square well V_0, V_1 and $L\in R$, thus $k_n\in R$ and ${}^0[\exp(k_nx)]=\exp(k_n{}^0x)$, ${}^0\phi_n(x)$ is the well-known solution of (1).

(β) For the infinite square well $V_1=0$, $V_0=+\eta$ (η positive infinite), and $L\in R$. Thus $p_n=(E_n)^{1/2}$ is finite, and ${}^0[2k_np_n(k_n^2+p_n^2)^{-1}]={}^0[2(E_n)^{1/2}(\eta+E_n)^{1/2}\eta^{-1}]=0={}^0\tan p_nL$, i.e., ${}^0p_nL=n\pi$ or ${}^0p_n=n\pi L^{-1}$. Again, ${}^0\phi_n(x)$ is the known solution of (1).

(γ) For $V_0=0$, $V_1=-AL^{-1}$, $0<A\in R$, and $0<L$ infinitesimal, i.e., the δ-function potential of strength $-A$, we have $2(-E_n)^{1/2}(AL^{-1}+E_n)^{1/2}A^{-1}=\tan[L(AL^{-1}+E_n)^{1/2}]$. Thus, ${}^0(-E_n)^{1/2}={}^0\{1/2AL^{-1}(AL^{-1}+E_n)^{-1/2}\tan[L(AL^{-1}+E_n)^{1/2}]\}$ or ${}^0(-E_n)^{1/2}={}^{1/2}A$, again leading to the known solution of (1).

The transfer theorem ensures that all representations of the δ function lead to the same ${}^0\phi(x)$, i.e., to the unique solution of (1).

(δ) For $V_0=0$, $V_1=-AL^{-2}$, $0<A\in R$, and $0<L$ infinitesimal, i.e., the singular square well, we have

$$2(-E_n)^{1/2}(AL^{-2}+E_n)^{1/2}A^{-1}L^2=\tan[(A+L^2E_n)^{1/2}]. \tag{3}$$

Since $E_n<0$ and $A+L^2E_n>0$, there are only a finite number of bound state solutions. They have infinite energies, since if E_n were finite in (3) the right-hand side would not be infinitesimal while the left-hand side would be. Evaluating c_n from normalization we find that $c_n=b_nL^{-1/2}$, where b_n is finite. Therefore, $\phi_n(x)$ is infinite in the monad of zero and one cannot take the standard part of $\phi(0)$.

A more interesting case occurs when the potential in (1) is given by

$$V_a(x)=\begin{cases}\infty, & x<a, a>0\\ -Ax^{-2}, & x>a, A>\frac{1}{4}\end{cases}.$$

As $a\to 0$, $V_a(x)$ becomes a singular potential.

In the nonstandard formulation we have

$$\left(\frac{d}{dx^2}+E-V_a(x)\right)\phi(x)=0,$$

where

$$V_a=\begin{cases}\infty, & x<a,\ a>0\\ -Ax^{-2}, & x>a,\ A>\frac{1}{4},\end{cases} \tag{4}$$

and the conditions: (i) $E<0$; (ii) $\phi(x)$ is infinitesimal for $x\in{}^*R\text{-}R_0$; (iii) $\phi(x)$ is Q continuous at $x=a$.

Note that $V_a(x)$ is not an n.s. physical potential.

Let $E=-k^2<0$, $z=kx$, and $\phi(x)=\psi(z)$. Then by the transfer theorem the two linearly independent solutions of (4) are

$$\psi_1(z)=\begin{cases}0, & z<ak\\ I_{(\frac{1}{4}-A)^{1/2}}(z) & z>ak\end{cases},$$

$$\psi_2(z)=\begin{cases}0, & z<ak\\ K_{(\frac{1}{4}-A)^{1/2}}(z), & z<ak\end{cases}.$$

Because of condition (ii) only $\psi_2(z)$ is acceptable. Denote the zeros of $K_{(\frac{1}{4}-A)^{1/2}}(z)$ by $z_0, z_1, \ldots, z_m$. By the transfer theorem m is finite. Condition (iii) requires that $k_i=z_ia^{-1}$, i.e., $E_i=-z_i^2a^{-2}$ for $i=0,1,\ldots,m$. Thus, we have a finite set of discrete eigenvalues. When a is infinitesimal; the E_i are infinite as expected on physical grounds.

Next we determine the normalization constants. Using the transfer theorem we find that[8]

$$\phi_i(x,a)=\begin{cases}0, & x<a\\ b_i^{1/2}a^{-1}x^{1/2}K_{(\frac{1}{4}-A)^{1/2}}(a^{-1}z_ix) & x\geq a\end{cases},$$

where

$$b_i^{-1}=2\mathrm{Re}\,[K_{(\frac{1}{4}-A)^{1/2}-1}(z_i)K_{(\frac{1}{4}-A)^{1/2}+1}(z_i)].$$

For standard $a\neq 0$, ${}^0\phi_i(x,a)$ is the well-known solution of (1). For infinitesimal a, $\phi_i(x,a)$ is infinite for some points in the monad of zero, and infinitesimal for x positive and not infinitesimal. We also see that ${}^0\langle\phi_i(x,a_1)\,|\,\phi_j(x,a_2)\rangle=0$ when a_1 is finite and a_2 is infinitesimal. That is $\phi_i(x,a)\to 0$ in the weak topology as $a\to 0$, or $\phi_i(x,a)$ rotates out of $H\subset{}^*H$ into ${}^*H\text{-}H$. The renormalized operator $\theta=a^2[(d^2/dx^2)+(A/x^2)]$ also rotates in such a manner that $\langle\phi_i(x,a)\,|\,\theta\phi_i(x,a)\rangle$ remains finite and independent of a.

4. THE SCALAR FIELD

In this section we give the n.s. version of the scalar field interacting with a (nonrecoiling) nucleon.[9] The form factor $f(\mathbf{k}^2)$ is taken to be the characteristic function

$$x_\eta(\mathbf{k}^2)\equiv\begin{cases}1 & \text{for } \mathbf{k}^2\leq\eta^2\\ 0 & \text{otherwise}\end{cases},$$

where η is some infinite integer.

The equivalent potential, under these conditions, differs infinitesimally from the Yukawa potential. We renormalize the resulting theory.

Following Ref. 9, we introduce the definitions:

(i) Let $\mathcal{H}=F=\sum_{n\in{}^*N}F_n$ be the n.s. Fock space. F will have ${}^*N$ mutually orthogonal axis, and the vectors in F have infinite, finite, or infinitesimal norms.

(ii) The Hamiltonian $H(\eta)=H_0+H_I(\eta)$, where

$$H_0=m_0\int d\mathbf{p}\,\psi^*(\mathbf{p})\psi(\mathbf{p})d\mathbf{p}+\int d\mathbf{k}\omega(\mathbf{k})a^*(\mathbf{k})a(\mathbf{k}),$$

$$H_I(\eta)=\lambda(2\pi)^{3/2}\int d\mathbf{p}\int d\mathbf{k}\times x_\eta(\mathbf{k}^2)(2\omega(\mathbf{k}))^{1/2}\psi^*(\mathbf{p}+\mathbf{k})\psi(\mathbf{p})[a(\mathbf{k})+a^*(-\mathbf{k})],$$

where $\psi(\mathbf{p})$, $\psi^*(\mathbf{p})$ and $a(\mathbf{k})$, $a^*(\mathbf{k})$ are the destruction and creation operators for the nucleons and mesons, respectively, $\omega(\mathbf{k}) = (\mathbf{k}^2 + \mu^2)^{1/2}$, μ is the mass of the meson, and m_0 is the bare mass of the nucleon, and λ is a coupling constant.

(iii) The following commutation rules are satisfied:

$$[\psi(\mathbf{p}), \psi(\mathbf{p}')]_+ = [\psi^*(\mathbf{p}), \psi^*(\mathbf{p}')]_+ = 0,$$

$$[a(\mathbf{k}), a(\mathbf{k}')] = [a^*(\mathbf{k}), a^*(\mathbf{k}')] = 0,$$

$$[\psi(\mathbf{p}), a(\mathbf{k})] = [\psi(\mathbf{p}), a^*(\mathbf{k})] = 0,$$

$$[\psi^*(\mathbf{p}), a(\mathbf{k})] = [\psi^*(\mathbf{p}), a^*(\mathbf{k})] = 0,$$

$$[\psi(\mathbf{p}), \psi^*(\mathbf{p}')]_+ = \delta^3(\mathbf{p} - \mathbf{p}') \text{ and } [a(\mathbf{k}), a^*(\mathbf{k}')] = \delta^3(\mathbf{k} - \mathbf{k}'),$$

where

$$\delta(x) \equiv \begin{cases} 0 & x \neq 0 \\ \text{undefined for } x = 0 \end{cases}$$

and

$$\int \delta(x) g(x)\, dx = g(0) \quad \forall g \in {}^*L_2.$$

(iv) The vector $|\eta, 0\rangle$ satisfying $\psi(\mathbf{p})|\eta, 0\rangle = a(\mathbf{k})|\eta, 0\rangle = 0$ $\forall \mathbf{p}$ and $\mathbf{k}$ is the physical vacuum.

(v) The physical nucleon state $|\eta, 1, \mathbf{p}\rangle$ is defined by the equation $H(\eta)|\eta, 1, \mathbf{p}\rangle = m(\eta)|\eta, 1, \mathbf{p}\rangle$ and $\| |\eta, 1, p\rangle \| = 1$, where $m(\eta)$ is the physical mass.

With the aid of the transfer theorem using the results of pp. 341–44 of Ref. 9, we get

(i) $H(\eta)|\eta, 0\rangle = 0$;

(ii) $H(\eta)a^*(\mathbf{k})|\eta, 0\rangle = \omega(\mathbf{k})|\eta, 0\rangle$;

(iii) $\psi(\mathbf{p})^*|\eta, 0\rangle$ is not an eigenstate of $H(\eta)$;

(iv) $|\eta, 1, p\rangle = \sum_{n \subset {}^*N} \int d\mathbf{q} d\mathbf{k}_1 \cdots d\mathbf{k}_n$
$\times\, c_p^{(n)}(\eta, \mathbf{q}; \mathbf{k}_1, \ldots, \mathbf{k}_n)(1/n!)a^*(k_1) \cdots a^*(k_n)|\eta, 0\rangle$,

where

$$c_p^{(n)}(\eta, \mathbf{q}; \mathbf{k}_1, \ldots, \mathbf{k}_n) = z^{1/2}\delta^3\left(\mathbf{q} + \sum_{i=1}^n \mathbf{k}_i - \mathbf{p}\right)\frac{(-\lambda)^n}{n!} \times \prod_{i=1}^n \frac{x_\eta(\mathbf{k}_i^2)}{[2(2\pi)^3\omega^3(\mathbf{k}_i)]^{1/2}},$$

where

$$z = \frac{\lambda^2}{4\pi^2}\exp\left(\ln\mu - \ln[\eta + (\eta^2 + \mu^2)^{1/2}] + \frac{\eta}{\sqrt{\eta^2 + \mu^2}}\right)$$

{i.e., infinitesimal for finite λ and behaves as $\eta^{-\lambda^2/4\pi^2}$ if $m(\eta) = m_0 - (\lambda^2/4\pi^2)[\eta - \mu \tan(\eta/\mu)]$; ${}^0[\tan(\eta/\mu)] = \frac{1}{2}\pi$}.

For finite cutoff d the one-particle state $|d, 1, p\rangle$ is in $F \subset {}^*F$. As d becomes infinite the one-particle state rotates to ${}^*F - F$. As $|\mathbf{k}_i|$ increases the factor $[2(2\pi)^3\omega^3(\mathbf{k}_i)]^{-1/2}$ behaves as $|\mathbf{k}_i|^{-3/2}$. The volume grows as $|\mathbf{k}_i|^2$. Therefore, one is more likely to find mesons with large momentum than with small momentum. Hence as d increases the n meson state rotates out of $F_n \subset {}^*F_n$ into ${}^*F_n - F_n$. To $|d, 1, \mathbf{p}\rangle$ the contribution ratio from the $n + 1$ and n meson states is proportional to $[\lambda d/(n + 1)^{1/2}]$. Thus when $d = \eta$ (infinite), the main part of $\langle\eta, 1, \mathbf{p}\rangle$ will come from $\cup_{k \in I_k} {}^*F_{K+k}$ for some infinite integer K and $I_k = \{1, 2, \ldots, k\}$ for some finite k.

In accord with pp. 347–48 of Ref. 9 we define

$$S_\eta = i\lambda(2\pi)^{-3/2} \int d\mathbf{p} \int d\mathbf{k}\, \chi_\eta(\mathbf{k}^2)[2\omega^3(\mathbf{k})]^{-1/2} \times \psi(\mathbf{p} + \mathbf{k})[a^*(\mathbf{k}) - a\psi^*(\mathbf{p})]$$

and

$$\psi_r^*(\mathbf{p}) = e^{iS_\eta}\psi^*(\mathbf{p})e^{-iS_\eta} \text{ and } a_r^*(\mathbf{k}) = e^{iS_\eta}a^*(\mathbf{k})e^{-iS_\eta}.$$

Then $H_r(\eta) = H_{0_r} + H_{I_r}(\eta)$, where

$$H_{0_r} = m_0 \int d\mathbf{p}\psi_r(\mathbf{p})\psi_r(\mathbf{p}) + \int d\mathbf{k}\omega(\mathbf{k})a_r^*(\mathbf{k})a_r(\mathbf{k}),$$

$$H_{I_r}(\eta) = \lambda^2(2\pi)^{-3} \int d\mathbf{q} \int d\mathbf{p} \times \chi_\eta(\mathbf{k}^2)[2\omega^2(\mathbf{k})]^{-1}\psi_r^*(\mathbf{p} + \mathbf{k})\psi_r^*(\mathbf{q})\psi_r(\mathbf{p})\psi_r(\mathbf{q} + \mathbf{k}),$$

i.e., with this rotation of the n.s. Fock space the Hamiltonian no longer contains a self-interaction term, but contains interaction between "dressed" nucleons (nucleons with mesons clouds that contain most probably an infinite number of mesons with infinite momentum).

The equivalent static potential is

$$V(\mathbf{x} - \mathbf{x}') = \lambda^2(2\pi)^{-3} \int d\mathbf{k}[2\omega^2(\mathbf{k})]^{-1} \exp[i\mathbf{k} \cdot (\mathbf{x} - \mathbf{x}')]$$
$$= \frac{\lambda^2}{8\pi}\frac{\exp\{-\mu|\mathbf{x} - \mathbf{x}'|\}}{|\mathbf{x} - \mathbf{x}'|} + \frac{i\lambda^2}{8\pi^2|\mathbf{x} - \mathbf{x}'|} \int_0^\pi \frac{\eta^2 \exp[i2\theta + i\eta(\cos\theta + i\sin\theta)]}{\eta^2 \exp(i^2\theta) + \mu^2}.$$

The second term is infinitesimal for infinite η and ${}^0|\mathbf{x} - \mathbf{x}'| \neq 0$. For θ in the monads of 0 and π the integrand is finite and, otherwise, the integrand is infinitesimal. Hence, if ${}^0|\mathbf{x} - \mathbf{x}'| \neq 0$, ${}^0V(\mathbf{x} - \mathbf{x}')$ is the Yukawa potential. Moreover, the second term is infinitesimal compared to the first term even when ${}^0|\mathbf{x} - \mathbf{x}'| = 0$.

5. CONCLUSION

We demonstrated the techniques of n.s.a. by working simple examples. In Sec. 3 we translated well-known one-dimensional bound state problems into n.s. language. We showed that potentials that are commonly called "idealized" or "limiting," are "physical" in the n.s. formulation. We showed how one recovers the standard results.

In Sec. 4 we showed the use of n.s.a. in investigating the properties of the wavefunction as a given parameter tends to some limit. We saw how the wavefunction "rotated" out of ordinary Hilbert space.

In Sec. 5 we found a field theory in which the equivalent potential differs infinitesimally from the Yukawa potential. We saw how the vacuum vector "rotated" out of Fock space into the n.s. Fock space as the form factor became one on an infinite set. One may renormalize the infinite cutoff Yukawa theory, by defining the mass renormalized Hamiltonian H_{ren} on a separable Hilbert space $\mathcal{H}_{\text{ren}}$, extracted from ${}^*F$. This construction is carried out for the $\lambda{:}\phi_2^4{:}$ model in the sequel to this paper.[7]

*Research supported in part by the National Science Foundation, Grants Nos. GP-29218 and GP-32996X.

[1] A. Robinson, Proc. Roy. Acad. 64, 432 (1961).

[2] For a detailed treatment of nonstandard analysis see A. Robinson, *Non-Standard*

Analysis (North-Holland, Amsterdam, 1966).

[3] W. A. J. Luxemburg, *Non-Standard Analysis* (California Institute of Technology, Pasadena, Calif., 1962).

[4] F is a free-ultrafilter on N if: (i) F is a nonempty family of subsets of N; (ii) the empty set $\phi \notin F$; (iii) $N_1 \in F$ and $N_2 \in F$, then $N_1 \cap N_2 \in F$; (iv) $N_1 \in F$ and $N_1 \supset N_2$, then $N_2 \in F$; (v) $N_1 \subset N$, then either $N_1 \in F$ or $N - N_1 \in F$; (vi) the intersection of an infinite number of distinct elements of F is empty. Note that (i)–(iv) establish the equivalence classes of sequences, (v) ensures that every sequence will belong to an equivalence class, and (vi) ensures that the equivalence classes are not based on a single element of the sequence. For further detail see Ref. 3 or A. Vörös, *Introduction to Non-Standard Analysis* (to be published).

[5] The * in the upper left corner carries an object from the standard universe into the corresponding object of the nonstandard universe.

[6] $\int_{*R}$ is the extension of the linear functional $\int_R$.

[7] P. J. Kelemen and A. Robinson, J. Math. Phys. (N.Y) **13**, 1875 (1972).

[8] G. N. Watson, *Theory of Bessel Functions* (Cambridge, London, 1966), p.134.

[9] S. S. Schweber, *An Introduction of Relativistic Quantum Field Theory* (Harper and Row, New York, 1961), pp. 339–348.

The nonstandard $\lambda{:}\phi_2^4(x)$: model. II. The standard model from a nonstandard point of view*

Peter J. Kelemen
Courant Institute of Mathematical Sciences and Department of Mathematics, University College, New York University, New York

Abraham Robinson
Department of Mathematics, Yale University, New Haven, Connecticut
(Received 22 May 1972)

As a second step in the construction of the nonstandard $\lambda{:}\phi_2^4$: model we analyze Glimm and Jaffe's work from the nonstandard point of view.

1. INTRODUCTION

In this paper we analyze the $\lambda:\phi_2^4(x)$: model of quantum field theory using the tools of nonstandard analysis, n.s.a. The model was selected because it shows both the conceptual and the technical difficulties that one encounters in building a nontrivial model of quantum field theory. These problems will become more transparent through explicit constructions within the framework of n.s.a.

The $\lambda:\phi_2^4(x)$: model was investigated by Glimm and Jaffe[1] in three papers to which we will refer as I, II, and III. For comparison between the nonstandard and standard treatment of this model, we will restrict ourself to the subject matter covered and to the assumptions made in these papers. This permits us to concentrate on those aspects of this model where the nonstandard approach is advantageous.

In I and II it is found that with a space cutoff $g(x)$ imposed the theory is meaningful. There exists a self-adjoint operator on a Fock space, the Hamiltonian, that generates the time translations, provided the time interval is sufficiently short. The Hamiltonian possesses an isolated lowest eigenvalue, E_g of multiplicity one. The corresponding eigenvector Ω_g, the vacuum vector is an element of Fock space. However, as the support of $g(x)$ grows, the vacuum vector seems to move out of the Fock space, $E_g \to -\infty$ and the Hamiltonian ceases to be an operator.

In consequence, one is forced to change Hilbert spaces and to redefine the operators of the model. This renormalization is carried out in III using the GNS construction.

In the n.s. treatment both for finite and for infinite cutoffs the Hamiltonian is an n.s. self-adjoint operator on an n.s. Fock space with a unique vacuum vector. For finite cutoff the vacuum vector is an element of the standard Fock space which is imbedded in the n.s. Fock space. To renormalize the theory we map a certain subspace of the n.s. Fock space onto a standard Hilbert space and redefine the operators.

To carry out this program, we begin by describing n.s. objects such as $^*L_2(^*R)$, n.s. operators, etc. (Sec. 2). In Sec. 3 we outline the $\lambda:\phi_2^4(x)$: model. In Sec. 4 we build the n.s. model, and in Sec. 5 we renormalize it. In Sec. 6 we summarize our results and sketch lines of development that we intend to follow up in the future.

2. NONSTANDARD PRELIMINARIES

The method of n.s. extension that we are using in this paper is provided by model theory.[2] We consider R (reals), C (complex numbers), arithmetic, analysis, $L_2(R^n)$, D and D' (the spaces of test functions and distributions), F (Fock space), operators, and linear functionals on F as given within the framework of some structure M. A model of some nontrivial enlargement $^*M$ of M serves as our n.s. extension.[3]

For what follows, it is unnecessary to construct an explicit model for some specific enlargement $^*M$ of M.

However, it is convenient to picture the n.s. objects. The model we select is such that $^*R$, $^*C$, $^*S$, $^*F$ may be visualized through an ultrafilter construction.[4] This means that a vector of $^*F$ can be pictured as an infinite sequence of vectors of F (reduced with respect to a certain equivalence relation.)

For convenience we restate the main theorem of n.s.a.

Transfer theorem: All true assertions about analysis remain true in the nonstandard model provided we reinterpret them as referring to *internal objects* only.

See Ref. 2 for the notion of internal and external objects and Ref. 4 for an informal discussion of these concepts. We recall that any set, function, operator, operator algebra, etc. is either internal or external but not both. Among the internal objects are the nonstandard extensions of all standard objects. Thus, every function, set, operator, etc., in standard analysis possesses a canonical extension to the nonstandard model. The ultrapower method provides a relatively concrete construction of internal objects.

3. THE $\lambda:\phi_2^4$: MODEL

The model developed in I and II is a spin-zero boron field ϕ, with a nonlinear, $\lambda:\phi_2^4$:, self-interaction with a space cutoff in two dimensional space–time. The field $\phi(x,t)$ is a bilinear-form-valued solution of

$$\left(\frac{\partial^2}{\partial t^2} - \frac{\partial^2}{\partial x^2} + m_o^2\right)\phi(x,t) = -4\lambda g(x)\phi^3(x,t), \tag{1}$$

where $g(x)$ is a smooth positive function that equals one on some bounded interval and vanishes outside some larger bounded interval containing the smaller one. In III the interval on which $g(x)=1$ is increased indefinitely in some prescribed manner, i.e., through a divergent sequence of intervals. An infinite sequence of standard intervals is replaced by a unique n.s. interval. Hence, "removing the cutoff" is equivalent to selecting a cutoff in the n.s. model.

The Hamiltonian corresponding to (1) is given by $H(g) = H_0 + \lambda\int_R :\phi^4(x):g(x)dx$, where H_0 is the free particle Hamiltonian. The corresponding vacuum vector is Ω_g. In I and II it is shown that for all finite cutoffs, Ω_g belongs to F. Hence, $\Omega_g \in {}^*F$ even when supp$g = (-\eta, \omega)$ where η and ω are infinite positive numbers. It is believed that as the length of the cutoff increases, Ω_g converges weakly to zero.[5] From the transfer theorem it would then follow that as the length of the n.s. cutoff increases, Ω_g would converge weakly to zero in $^*F$, i.e., there is no unique n.s. vacuum.

On the other hand in II it was shown that if $x_0 \in \text{supp} g_1(x)$ and $\text{supp} g_1(x)$ is large enough, then $\phi_{g_1}(x_0, t_0) = \phi_{g_2}(x_0, t_0)$ when $\text{supp} g_1(x) \subset \text{supp} g_2(x)$. Thus for finite (x_0, t_0), $\phi_g(x_0, t_0) = \phi(x_0, t_0)$ provided that $(-\eta, \eta) \subset \text{supp} g(x)$ for some infinite positive η. Note that $\phi(x, t)$ is an internal operator for each finite (x, t), but the collection of $\phi(x, t)$ for all finite (x, t) is an external set.

After the above intuitive remarks we could proceed by translating into n.s. language all the theorems and lemmas of I, II, and III. In some cases the standard proofs plus the Transfer theorem would provide the n.s. proofs. In others, especially where limits are taken to remove the momentum cutoffs, the n.s. proofs would be shorter.[6] But a different approach is more useful here. We apply the Transfer theorem only to the results of I and II in building an unrenormalized n.s. theory. (This exemplifies the fact that in building an n.s. theory one may incorporate as many of the standard results as desired.) We then extract from $^*F$ a standard Hilbert space $\mathcal{H}_{\text{ren}}$ which is identical with F_{ren}. In the sequel to this paper we will add external assumptions to the unrenormalized theory and find new interpretations.

4. THE n.s. MODEL

In this section we compile those n.s. definitions and theorems which define the n.s. model. The numbers in brackets after our numbering refer to the page on which the standard counterparts are found.

Definition 1 [II-364]: The Fock space $^*F$ is the Hilbert space completion of the symmetric tensor algebra over $L \equiv {}^*L_2({}^*R)$,

$$^*F = G(L) \equiv \oplus_{n \in {}^*N} {}^*F_n,$$

where $^*F_n \equiv L \otimes_s L \otimes_s \cdots \otimes_s L$ (n factors) is the space of n noninteracting particles. For $\psi \in {}^*F$, $\psi = \{\psi_0, \psi_1, \cdots\}$ we have $\|\psi\|^2 = \sum_{n \in {}^*N} \|\psi_n\|^2$.

Definition 2 [II-364]: The no particle space $^*F_0 = {}^*C$ (the complex numbers and $\Omega_0 = \{1, 0, 0, \cdots\} \in {}^*F$ is the bare vacuum or bare no-particle state vector.

Definition 3 [II-366]: The Hamiltonian $H(g)$ acts on $^*F$ and can be written as $H(g) = H_0 + \lambda \int_{*R} : \phi^4(x) : g(x) dx = H_0 + H_{I,g}$, where H_0 is the free particle Hamiltonian; $H_{I,g}$ is the interaction cutoff Hamiltonian; and $g(x)$ is an internal smooth positive function that equals one on an internal set B that contains in the interval $(-\eta, \eta)$, $\eta \in {}^*N$, and vanishes off an internal set that contains B.

Definition 4 [I-1946]: The domain $D_0 \equiv \bigcap_{n \in {}^*N} D(H_0^n)$.

Theorem 1 [I-1949]: (a) $H(g)$ is self-adjoint with domain $D(H(g)) = D(H_0 \cap D(H_{I,g})$; (b) $H(g)$ is essentially self-adjoint on D_0.

Definition 5 [II-364]: E_g is the lowest eigenvalue of the equation

$$H(g)\Omega_g = E_g \Omega_g, \qquad \|\Omega_g\| = 1.$$

Theorem 2 [II-368]: There exists a vacuum vector Ω_g for $H(g)$.

Theorem 3 [II-372]: The lower bound of $H(g)$ is a simple eigenvalue. (Note that we got the above result with the transfer theorem. Therefore, the gap between E_g and the rest of the spectrum may be infinitesimal but not zero.

Definition 6 [II-382]: Let $f(x, t)$ be the extension of a $C\infty$ function that vanishes off the rectangle $-n \le x, t \le n$ for some $n \in N \subset {}^*N$. Then $A_g(f) \equiv \int_{*R} \phi_g(x, t) f(x, t) dx$ and $\phi_g(f)$ is the closure of the operator defined by

$$(\psi, \phi_g(f)\psi) = \int_{*R} (\psi, A_g(f)\psi) dt, \qquad \psi \in D[(H(g) + b)^{1/2}],$$

where b is a suitably large constant.

Theorem 4 [II-388]: $A_g(f)$ and $\phi_g(f)$ are self-adjoint operators.

Theorem 5 [II-388]: $\phi_g(f) = \phi(f)$ provided $(-\eta, \eta) \subset \text{supp} g$ for some $\eta \in {}^*N - N$.

Theorem 6 [II-385]:

$$\pi(f) = \phi\left(-\frac{d}{dt} f\right) = i[H(g), \phi(f)].$$

Remark: $\phi(f)$ is an internal operator when $f(x, t)$ is the extension of a $C\infty$ function with support in the rectangle $-n < x, t < n$ for some $n \in N$; but the collection of all such operators is an external set. The importance of this fact cannot be over emphasized. Since $\phi(f)$ is internal its properties are determined by the standard operator that extends to $\phi(f)$. But there is no standard theorem that determines the properties of this external set of operators. This is why a renormalized nontrivial theory may exist.

5. THE RENORMALIZED MODEL

We reproduce the renormalized model of III. From $^*F$ we extract a standard Hilbert space $\mathcal{H}_{\text{ren}}$. Our method is equivalent to the Gel'fand–Naimark–Segal GNS construction. We redefine the operator of $^*F$ on $\mathcal{H}_{\text{ren}}$. Our construction illuminates the one employed in III.

To make the connection between the GNS construction and our extraction of $\mathcal{H}_{\text{ren}}$ more transparent first we discuss a case in which the linear functional used in the GNS construction is simpler than the one used in III. Let Ω_g be the vacuum vector for the cutoff $g(x)$, let $\hat{C}\infty \equiv \{f(x) | f(x) = {}^*h(x), h(x) \in C\infty$ and has compact support$\}$. Define $S \equiv \{z \in {}^*F | z = e^{i\phi(f_1)} \cdots e^{i\phi(f_k)} \Omega_g$, $f_j \in \hat{C}\infty j = 1, 2, \ldots, k \in N\}$, and let $\mathcal{H}$ be the subspace of $^*F$ spanned by S. Note that each element of $\hat{C}\infty$ and of $\mathcal{H}$ is internal, but that both $\hat{C}\infty$ and $\mathcal{H}$ are external objects. We extract $\mathcal{H}_0$ from $\mathcal{H}$ by discarding all vectors which have infinite norms. To get $\mathcal{H}_{\text{ren}}$ from $\mathcal{H}_0$ we collapse into a single vector those vectors which differ from each other by a vector of infinitesimal norm; and redefine the innerproduct by passing from (z_1, z_2) to $^0(z_1, z_2)$. Equivalently, map $\mathcal{H}_0$ into $\mathcal{H}_{\text{ren}}$ a subspace of some standard Hilbert space by the rule that if $z_1, z_2 \in \mathcal{H}_0$ and $z_1 \to b_1$, $z_2 \to b_2$, then $^0(z_1, z_2)_{*F} = (b_1, b_2)_{\mathcal{H}_{\text{ren}}}$.

The elements of the C^*-algebra $\mathfrak{A}$, generated by $\{e^{i\phi(f)} | \cdot f \in \hat{C}\infty\}$, are operators on $^*F$. From the Riesz representation theorem we infer that the linear functionals on $^*F$ are innerproducts. In particular, the positive linear functional $\phi(A)$ of the GNS construction in this concrete case is $^0(\Omega_g, A\Omega_g)$. Constructing the quotient space with the left ideal I, $I = \{\tau \in \mathfrak{A} \mid \phi(\tau^*\tau) = 0\}$ amounts to our "collapsing" of the vectors

The linear functional of the GNS construction employed in III is more complicated, because it uses an averaging process. As we will see the averaging serves two purposes. It ensures, first, that the energy per unit volume is finite space translation invariant.

To give the n.s. version of Sec. 2 of III we also take the cutoff function $g(x) \in \hat{C}^\infty$ to be nonnegative and equal to 1 on $[-3, 3]$, and define $g_n(x) \equiv g(x/n)$, $n \in {}^*N$. The corresponding vacuum vector is denoted by Ω_n. As in III we fix an $h(x) \in \hat{C}^\infty$ with support in $[-1,]$ that has the property $\int h(x)dx = 1$. We use the notation $E_j(\alpha) = e^{i\psi_\alpha(f_j)}$ where $\psi_\alpha(f_j) = \int_{*R}\psi(x+\alpha, t)f_j(x)dx$ and ψ stands for either ϕ or π. We define $S(\alpha, n) \equiv \{z(\alpha) \in {}^*F \mid z(\alpha) = E_1(\alpha)\cdots E_k(\alpha)\Omega_n,\ f_j \in \hat{C}^\infty,\ j = 1, 2, \cdots, k \in N\}$, and $\mathcal{H}_0(\alpha, n) \subset {}^*F$ the subspace that contains only finite normed vectors and spanned by $S(\alpha, n)$. To get $\mathcal{H}_{\text{ren}}(n)$ we average the $\mathcal{H}_0(\alpha, n) - s$. The vectors in $\mathcal{H}_{\text{ren}}(n)$ satisfy the conditions that if $z_1(0)$ and $z_2(0)$ map into b_1 and b_2, respectively, then

$$(b_1, b_2)_{\mathcal{H}_{\text{ren}}(n)} = {}^0[(1/n)\int h(\alpha/n)(z_1(\alpha), z_2(\alpha))_{*F}\, d\alpha].$$

For the proper choice of n, say η, $\mathcal{H}_{\text{ren}}(\eta)$ is identical to F_{ren} of III. The only difference between the construction of the two spaces is that in III $\mathfrak{A}$ is an abstract algebra while in the n.s. model it is an operator algebra. This is the case, because the space $\mathcal{H}_0(0, n)$ on which $\mathfrak{A}$ is defined and leaves invariant is nonstandard. The innerproduct of the GNS construction in III is defined by the mapping of $\mathcal{H}_0(0, \eta)$ onto $\mathcal{H}_{\text{ren}}(\eta)$, since for any standard bounded operator A, $\omega_n(A) = {}^0(\Omega_n, {}^*A\Omega_n)_{*F}$, and since a convergent subset of ω_n means selecting the corresponding Ω_η. Note that η is infinite, i.e., $\eta \in {}^*N - N$. Therefore $g_\eta(x)$ is equal to 1 on an infinite interval that contains $[-3\eta, 3\eta]$. But, by definition, the support of $g_n(x)$ is contained in $[-k, k]$ for some $K \in {}^*N$. Hence we have an infinite cutoff n.s. model, which means that the renormalized standard model without cutoff may not be unique. From the construction of $\mathcal{H}_{\text{ren}}(\eta)$ we see the effects of the averaging by $h(\alpha)$. The mapping of $\mathcal{H}_0(0, \eta)$ onto $\mathcal{H}_{\text{ren}}(\eta)$ leaves the norms invariant. $E_j^*(\alpha)E_j(\alpha) = I$, so that

$$\|b_j\|_{\mathcal{H}_{\text{ren}}(\eta)} = {}^0([1/\eta)\int h(\alpha/\eta)(E_1(\alpha)\cdots E_k(\alpha)\Omega_\eta,$$
$$E_1(\alpha)\cdots E_k(\alpha)\Omega_\eta)_{*F}d\alpha] = {}^0[(1/\eta)\int h(\alpha/n)(\Omega_\eta, \Omega_\eta)_{*F}d\alpha]$$
$$= {}^0[(\Omega_\eta, \Omega_\eta)_{*F}(1/\eta)h(\alpha/\eta)d\alpha] = (\Omega_\eta, \Omega_\eta)_{*F}$$
$$= (E_1(0)\cdots E_k(0)\Omega_\eta, E_1(0)\cdots E_k(0)\Omega_\eta)_{*F}.$$

But the map changes the angles between some of the vectors. They become larger, i.e., their innerproduct smaller.

It is easier to demonstrate some of the properties of F_{ren} on $\mathcal{H}_{\text{ren}}(\eta)$. For examples,

(i) H_{ren} is defined through the action of $H(g_\eta)$ on $\mathcal{H}_0(0, \eta)$. Thus, the spectrum of H_{ren} is nonnegative because the spectrum of $H(g_\eta)$ is nonnegative.

(ii) Finite time translation invariance follows from the finite propagation speed and from

$$e^{i\tau H(g_\eta)}\Omega_\eta = \Omega_\eta.$$

(iii) To see that the model is finite translation invariant it is sufficient to observe that

$$0 = {}^0[(1/\eta)\int_I h(\alpha/\eta)(\Omega_\eta, E_1(\alpha)\cdots E_k(\alpha)\Omega_\eta)_{*F}d\alpha]$$

for all finite intervals I; which is evident, since η is infinite and $|(\Omega_\eta, E_1(\alpha)\cdots E_k(\alpha)\Omega_\eta)_{*F}| \le 1$ so that

$$|(1/\eta)\int_I h(\alpha/\eta)(\Omega_\eta, E_1(\alpha)\cdots E_k(\alpha))_{*F}d\alpha| \le (1/\eta)$$
$$\times \text{length}(I) = \text{infinitesimal}.$$

But there is yet another way to see this.

$$(1/\eta)\int_{*R}h(\alpha/\eta)(\Omega_\alpha, E_1(\alpha)\cdots E_k(\alpha)\Omega_\eta)_{*F}d\alpha$$
$$= (1/\eta)\int_{*R}h(\alpha/\eta)(\Omega_{\eta(\alpha)}E_1(0)\cdots E_k(0)\Omega_{\eta,\alpha})_{*F}d\alpha),$$

where $\Omega_{\eta,\alpha}$ is the vacuum vector of the $g[(x-\alpha)/\eta]$ cutoff. The last formula is interpreted the following way. For each $\alpha \in [-\eta, \eta]$ we construct a renormalized Hilbert space corresponding to the vacuum $\Omega_{\eta,\alpha}$ by the procedure given in the beginning of this section, and then we average over the renormalized Hilbert spaces. Clearly adding or deleting Hilbert space corresponding to a finite interval cannot affect the average.

6. CONCLUSIONS

In Sec. 4 we constructed an n.s. $\lambda:\phi_2^4$: model with an infinite cutoff. In Sec. 5 we recovered the renormalized Fock space.

Nonstandard analysis allowed us to work with operators on a Hilbert space, instead of an abstract operator algebra, and to employ intuitive ideas which are not available in the standard approach. It illuminated several interesting features of the renormalization. In III a renormalization by averaging is employed in addition to the energy renormalization by the subtraction of an infinite constant. This averaging "opens" the Hilbert space, i.e., it diminishes the inner product of two vectors. Vectors that are close together in $\mathcal{H}_{\text{ren}}(\eta)$ came from vectors that were even closer in $^*F$, and, hence, the finite space translation follows. This "opening" also decreases the energy density. On the other hand, this averaging procedure does not decrease the cardinality of the set of the basis vectors of $\mathcal{H}_{\text{ren}}(\eta)$. If the renormalized model constructed without averaging is only locally Fock, then the renormalized model constructed with averaging can be locally Fock only.

As it was pointed out, $\mathcal{H}_{\text{ren}}$ constructed without an averaging is already an external subspace of $^*F$. Hence, no standard theorem about F is transferable to F_{ren} directly by the use of the Transfer theorem. In particular, there is some hope that no analog of the Haag theorem will apply to F_{ren}. This statement demonstrates that it may be advantageous to investigate standard models with n.s. methods, bringing into play the distinction between external and internal objects.

Other approaches to the problem of finding a renormalized $\lambda:\phi_2^4$: model suggest themselves: (i) Retaining an infinite momentum cutoff may remove some of the difficulties. (ii) Using periodic boundary conditions, of infinite period, both in momentum and in position space would allow one to use n.s. Fourier series. (iii) Quantizing position space in an infinite box with rigid wall has its obvious advantages.

Field theory is probably best formulated on a nonseparable Hilbert space. The logical candidate is $^*F$. Having an established n.s. model (Sec. 4) one should check whether or not it satisfies a modified n.s. version of the Wightman axioms. We found in this paper that modification by external assumption is necessary. One can only require invariance for finite translations. Thus in the n.s. version of the Wightman axioms one should use the phrase "finite Lorentz transformation." What modifications, if any, are needed to assure that we do not need to average by $h(x)$ is not clear. Hopefully, one of the three approaches mentioned in the preceding paragraph will provide the answer.

ACKNOWLEDGMENTS

We would like to thank H. Georgi and J. Glimm for many helpful suggestions.

---

* Research supported in part by the National Science Foundation, Grant No. GP-29218, and in part by the Air Force Office of Scientific Research under Grant No. AFOSR-71-2013.

[1] J. Glimm and A. Jaffe Phys. Rev. **176**, 1945 (1968); Ann. Math. **91**, (1970); Acta Math. **125**, (1970).

[2] A. Robinson, *Non-Standard Analysis* (North-Holland, Amsterdam, 1966).

[3] The * in the upper left corner carries an object from the standard universe into the corresponding object of the nonstandard universe.

[4] P. J. Kelemen and A. Robinson, J. Math. Phys. (N.Y.) **13**, 1870 (1972).

[5] The vector moved out of F but remains in $^*F$. The occurrence of this phenomenon for the sharp cutoff case follows from F. Guerra, Phys. Rev. Letters **28**, 1213 (1972).

[6] For this type of proof see p. 63 of Ref. 2.

A Limit Theorem on the Cores of Large Standard Exchange Economies

(nonstandard analysis/competitive equilibria/mathematical model)

DONALD J. BROWN* and ABRAHAM ROBINSON†

* Cowles Foundation for Research in Economics at Yale University and the † Mathematics Department, Yale University, New Haven, Connecticut 06520

Communicated by Tjalling C. Koopmans, March 10, 1972

ABSTRACT **This note introduces a new mathematical tool, nonstandard analysis, for the analysis of an important class of problems in mathematical economics—the relation between bargaining and the competitive price system.**

I. INTRODUCTION

An exchange economy consists of a set of traders, each of whom is characterized by an initial endowment and a preference relation. In addition, one usually assumes that the set of traders is finite. Edgeworth's conjecture [6] that, as the number of traders in an exchange economy increases the core approaches the set of competitive equilibria, has been formalized in two disparate ways by mathematical economists.

One approach has been to talk about a sequence of economies growing without bound, and to look at the relationship between the core and the set of competitive equilibria for very large economies. This was the method of Debreu-Scarf [5].

The other approach has been to consider an exchange economy having an infinite number of traders, to define the notions of core and competitive equilibrium in this economy, and to show the equivalence between these two concepts. Aumann's work on continuous economies [2] has been of this nature.

Here, we report the results obtained by a new method for the resolution of Edgeworth's conjecture, based on nonstandard analysis, which synthesizes the asymptotic method of Debreu-Scarf and the infinite method of Aumann. We have shown that within nonstandard analysis the concept of the core and competitive equilibrium are the same. As a consequence of this theorem, we have derived a number of asymptotic results concerned with unbounded families of standard exchange economies.

In Section II, we give a brief introduction to the essentials of nonstandard analysis; in Section III, we describe the economic model and state our theorem concerning Edgeworth's conjecture; in Section IV, we give an asymptotic result or limit theorem for the cores of large standard economies.

II. NONSTANDARD ANALYSIS

Let R be the system of real numbers. Any statement about R (involving individuals, subsets of R, functions on R, relations on R, sets of relations on R, etc.) can be expressed in a formal language, which includes: names for all these entities; connectives: $\neg$, $\wedge$, $\vee$, $\Longrightarrow$; variables for entities of different types; and quantifiers $\exists$ and $\forall$ over different types of variables.

Let K be the set of all statements in this formalized language that are true for R. Then, it can be shown that there exist: a proper extension, $^*R$, of R; a designated subset of the set of all subsets of $^*R$; a designated subset of the set of all functions from $^*R$ to $^*R$; a designated subset of the set of all relations on $^*R$; etc., such that the following holds. Every statement that is true in R remains true in $^*R$ provided we reinterpret the existential quantifiers that occur in such a statement as follows: "there exists a set" shall mean "there exists an *internal* set", "there exists a family of relations," shall mean "there exists an *internal* family of relations," similarly for all other types of entities. Here *internal means an element of the appropriate designated sets of entities.* (Note that this qualification does not apply to individuals.)

We shall call: the elements of $^*R$, *real numbers;* the elements of R, *standard real numbers;* and the elements of $^*R$ that do not belong to R, *nonstandard real numbers.* As a proper extension of R, $^*R$ must be a nonarchimedean field, that is, it contains numbers whose absolute values are greater than all standard real numbers, which will be called infinite numbers. All other numbers will be called *finite.* The reciprocals of *infinite numbers* together with zero are said to be *infinitesimals* or *infinitely small numbers.* Every finite number, x, is infinitely close to a unique standard real number, ^0x, which is called the *standard part* of x. The *monad* of a number, $\mu(x)$, is the set of all numbers that are infinitely close to x.

$^*R$ also contains a set of numbers, denoted as $^*N$, that has the same properties as N (the set of natural numbers) in the sense that any statement true about N is true about $^*N$ when reinterpreted in terms of internal entities. $^*N$ is a proper extension of N. We shall call: the elements of $^*N$, *natural numbers;* the elements of N, *finite natural numbers;* and the elements of N^* that do not belong to N, *infinite natural numbers.* An internal set that has ω elements, where $\omega \in {}^*N$, is said to be *starfinite.*

A structure $^*R$ of the required kind may be constructed as an ultrapower of R over the natural numbers. That is to say, $^*R$ consists of all sequences of real numbers; equality between such sequences as well as any other kind of relation is defined with respect to a free ultrafilter in the Boolean algebra $\mathcal{P}(N)$, the power set of the natural numbers. Similarly internal sets, internal relations, etc., can be identified with sequences of sets, relations, etc. again reduced with respect to the ultrafilter. Compare refs. 7–9.

We shall not use the construction just mentioned, but we assume that the following particular property of the $^*R$ just constructed is satisfied. Let f be a function with domain

N, whose range values are numbers (internal sets of numbers) of $^*R$. Then, there exists an internal sequence of numbers (internal sets of numbers) of $^*R$, such that the values of the sequence agree with the values of f on all finite j. That is, if $f: N \to {}^*R$, then there exists an internal $g: {}^*N \to {}^*R$ such that for all $j \in N, g(j) = f(j)$.

$^*R_n$ is the n-fold cartesian product of $^*R$ and $^*\Omega_n$ is the positive orthant of $^*R_n$. Let $\bar{x}$, $\bar{y}$ be vectors in $^*R_n$. The *monad of* $\bar{x}$, $\mu(\bar{x})$ is the set of points whose distance from $\bar{x}$ is an infinitesimal. If $\bar{y} \in \mu(\bar{x})$, we shall write $\bar{x} \simeq \bar{y}$; $\bar{x} \geq \bar{y}$ means $x_i \geq y_i$ for all i; $\bar{x} > \bar{y}$ means $x_i \geq y_i$ for all i and $x_i > y_i$ for some i; $\bar{x} \gg \bar{y}$ means $x_i > y_i$ for all i. $\bar{x} \gtrsim \bar{y}$ means that $\bar{x} \geq \bar{y}$ or $\bar{x} \simeq \bar{y}$. $\bar{x} \gnsim \bar{y}$ means that x_i is greater than y_i by a noninfinitesimal amount for all i.

III. ECONOMIC MODEL AND EDGEWORTH'S CONJECTURE

Let T be an initial segment of $^*N$, where $|T|$, the number of elements in T, is some infinite natural number ω. That is, $T = \{1, 2, \ldots, \omega\}$ and $\omega \in {}^*N - N$. T is to be interpreted as the set of traders in the economy. If S is any internal subset of T, then $|S|$ will denote the number of elements in S.

A *nonstandard exchange economy*, $\mathcal{E}$, consists of a pair of functions I and P, where $I: T \to {}^*\Omega_n$ and $P: T \to {}^*\Omega_n \times {}^*\Omega_n$. Denoting the functions I and P, respectively, as $\{\bar{x}_t\}_{t=1}^{\omega}$ and $\{>_t\}_{t=1}^{\omega}$, $\mathcal{E} = \langle \{\bar{x}_t\}_{t=1}^{\omega}, \{>_t\}_{t=1}^{\omega} \rangle \cdot \bar{x}_t$ is the initial endowment of the tth trader and $>_t$ is his preference relation over $^*\Omega_n$, the space of commodities. The nonstandard exchange economies that we will consider are assumed to have the following properties:

(*i*) The function indexing the initial endowments, $I(t)$, is internal.

(*ii*) $I(t)$ is standardly bounded, i.e. there exists a standard vector $\bar{r}_0$ such that for all $t, I(t) \leq \bar{r}_0$.

(*iii*) $\frac{1}{\omega} \sum_{1}^{\omega} I(t) \gnsim \bar{0}$.

(*iv*) The relation, Q, where $Q = \{\langle t, >_t \rangle | t \in T, >_t \subseteq {}^*\Omega_n \times {}^*\Omega_n\}$, is internal. For all t,

(α) $>_t$ is irreflexive, i.e. if $\bar{x} >_t \bar{y}$ then $\bar{x} \neq \bar{y}$

(β) If $\bar{x} > \bar{y}$ then $\bar{x} >_t \bar{y}$

(γ) If $\bar{x} \not\simeq \bar{y}$ and $\bar{x} >_t \bar{y}$ then there exists a standard $\delta > 0$ such that if $\bar{z} \in S(\bar{x}, \delta)$, the open ball with center $\bar{x}$ and radius δ, then $\bar{z} >_t \bar{y}$.

Equivalently, if $\bar{x} \not\simeq \bar{y}$ and $\bar{x} >_t \bar{y}$ and $\bar{z} \in \mu(\bar{x})$, then $\bar{z} >_t \bar{y}$. It is shown in [3] that these assumptions are consistent.

An *assignment* Y is an internal function $Y(t)$ from T, the set of traders, into $^*\Omega_n$.

An *allocation or final allocation* is a standardly bounded assignment $Y(t)$ from the set of traders, T, into $^*\Omega_n$ such that $\frac{1}{\omega} \sum_{t=1}^{\omega} Y(t) \simeq \frac{1}{\omega} \sum_{t=1}^{\omega} I(t)$.

(*iv*) implies for all internal X, Y, $\in {}^*\Omega_n^T$, where T is the set of traders, that

(*v*) $\{t | X(t) >_t Y(t)\}$ is an internal set of traders.

A *coalition*, S, is defined as an internal set of traders. It is said to be negligible if $|S|/\omega \simeq 0$. Note that if S is negligible, then for all allocations $X(t)$, $\frac{1}{\omega} \sum_{t \in S} X(t) \simeq \bar{0}$. $|s|$ is the cardinality of s.

A coalition, S, is *feasible* with respect to an allocation Y if $\frac{1}{\omega} \sum_{t \in S} Y(t) \simeq \frac{1}{\omega} \sum_{t \in S} I(t)$.

An allocation Y *dominates* an allocation X via a coalition S if S is feasible with respect to Y and if for all $t \in S$, $X(t) \not\simeq Y(t)$ and $Y(t) >_t X(t)$.

The *core* is defined as the set of all allocations X that are not dominated by any allocation Y via any non-negligible coalition.

A *price vector*, $\bar{p}$, is a finite nonstandard vector in $^*\Omega_n$ such that $\bar{p} \gnsim \bar{0}$.

The tth trader *budget set*, $B_{\bar{p}}(t)$, is $\{\bar{x} \in {}^*\Omega_n | \bar{p} \cdot \bar{x} \lesssim \bar{p} \cdot I(t)\}$.

$\bar{y}$ is said to be *maximal in* $B_{\bar{p}}(t)$ if $\bar{y} \in B_{\bar{p}}(t)$ and there does not exist an $\bar{x} \in B_{\bar{p}}(t)$, such that $\bar{x} \not\simeq \bar{y}$ and $\bar{x} >_t \bar{y}$.

A *competitive equilibrium* is defined as a pair $\langle \bar{p}, X \rangle$, where $\bar{p}$ is a price vector and X an allocation such that $X(t)$ is maximal in $B_{\bar{p}}(t)$ for almost all the traders. That is, $\exists\, K = \{t | X(t) \text{ is maximal in } B_{\bar{p}}(t)\}$ and $|K|/\omega \simeq 1$.

THEOREM 1. *If $\mathcal{E}$ is a nonstandard exchange economy satisfying the above assumptions, then an allocation X is in the core of $\mathcal{E}$ if and only if there exists a price vector, $\bar{p}$, such that $\langle \bar{p}, X \rangle$ is a competitive equilibrium of $\mathcal{E}$.*

The proof of this theorem is given in [3].

IV. LIMIT THEOREM

A standard exchange economy $\mathcal{E}$ of size m consists of m traders, where m is a standard natural number, whose initial endowments and preferences are restricted to the standard commodity space $\Omega_n \cdot \Omega_n$ is the positive orthant of R_n, the n-fold cartesian product of R. Let $\mathcal{E} = \langle I(t), >_t \rangle$ where for all t, $I(t) \in \Omega_n$ and $>_t \subseteq \Omega_n \times \Omega_n$, $t \in \mathcal{E}$ will refer to the tth trader's endowment and preference relation.

A *competitive equilibrium* for $\mathcal{E}$ is a pair $\langle \bar{p}, X \rangle$ such that: $\bar{p} \in \Omega_n$; $\sum_{t=1}^{m} X(t) = \sum_{t=1}^{m} I(t)$; for each t, $X(t) \in \Omega_n$ and $\bar{p} \cdot X(t) \leq \bar{p} \cdot I(t)$; there does not exist a $\bar{y} \in \Omega_n$ where $\bar{p} \cdot \bar{y} \leq \bar{p} \cdot I(t)$ and $\bar{y} >_t X(t)$. $\bar{p}$ is called a *price vector* and $X(t)$ a *competitive allocation.*

$X(t)$ is an *allocation* if for each t, $X(t) \in \Omega_n$ and $\sum_{t=1}^{m} X(t) = \sum_{t=1}^{m} I(t)$. An allocation $X(t)$ is *blocked* by an allocation $Z(t)$ if there exists a coalition of traders, S, such that $\sum_{t \in S} Z(t) = \sum_{t \in S} I(t)$ and for all $t \in S$, $Z(t) >_t X(t)$. The *core* of $\mathcal{E}$ is the set of unblocked allocations.

Let $\mathcal{G} = \{\mathcal{E}_i\}_{i \in \mathcal{I}}$ be an unbounded family of standard exchange economies, i.e., for each $k \in N$, there exists $i \in \mathcal{I}$ such that $|\mathcal{E}_i|$, the number of traders in the ith economy, is greater than k. Suppose $\mathcal{G}$ satisfies the following conditions:

(*a*) The initial endowments of the traders in $\mathcal{G}$ are uniformly bounded from above.

(*b*) The initial endowments of the traders in $\mathcal{G}$ are uniformly bounded away from zero in each commodity.

(*c*) Each trader's preference relation is "irreflexive," "continuous," and "strongly monotonic." $>_t$ is *irreflexive* if for all $\bar{x} \in \Omega_n$, $\bar{x} \not>_t \bar{x}$. $>_t$ is *continuous* if for all $\bar{x}, \bar{y} \in \Omega_n$ the sets $\{\bar{z} \in \Omega_n | \bar{z} >_t \bar{x}\}$ and $\{\bar{z} \in \Omega_n | \bar{y} >_t \bar{y}\}$ are open sets in R^m. $>_t$ is *strongly monotonic* if $\bar{x} > \bar{y}$ implies that $\bar{x} >_t \bar{y}$.

(*d*) The family of all trader's preference relations in $\mathcal{G}$ is equicontinuous on Ω_n.

A family, $\mathfrak{F}$, of preference relations is said to be *equicontinuous* on Ω_n if $(\forall \in > 0)\ (\exists \delta > 0)\ (\forall t \in \mathfrak{F})\ (\forall \bar{x}_t, \bar{y}_t \in \Omega_n)\ [|\bar{x}_t - \bar{y}_t| \geq \in \wedge\ \bar{x}_t >_t \bar{y}_t \Longrightarrow (\forall \bar{w} \in S(\bar{x}_t; \delta))\ \bar{w} >_t \bar{y}_t]$. An example of such a family is the set of preference relations $>_f$ defined by a family of equicontinuous utility functions, where $\bar{x} >_f \bar{y}$ if and only if $f(\bar{x}) > f(\bar{y})$.

Given a standard exchange economy $\mathcal{E} = \langle I(t), >_t \rangle$, an allocation $X(t)$ for $\mathcal{E}$, and a price vector $\bar{p}$ in Ω_n, we define the following sets for each positive real number δ:

$$E_\delta^{\bar{p}}(X) = \{t \in \mathcal{E} \mid \bar{p}\cdot X(t) - \bar{p}\cdot I(t) \geq \delta\}$$

$$F_\delta^{\bar{p}}(X) = \{t \in \mathcal{E} \mid (\exists \bar{y} \in \Omega_n)\bar{p}\cdot\bar{y} \leq \bar{p}\cdot I(t) \wedge \bar{y} >_t X(t) \wedge |\bar{y} - X(t)| \geq \delta\}$$

$$G_\delta^{\bar{p}}(X) = E_\delta^{\bar{p}}(X) \cup F_\delta^{\bar{p}}(X)$$

THEOREM 2. *Let $\mathcal{G}$ be an unbounded family of standard exchange economies satisfying the assumption stated above. For every $\delta > 0$, there exists an $m \in N$ where for all economies, $\mathcal{E}$, in $\mathcal{G}$ and for all allocations $X(t)$; if $|\mathcal{E}| > m$ and $X(t)$ is in the core of $\mathcal{E}$, then there exists a price vector $\bar{p}$, such that $|G_\delta^{\bar{p}}(X)|/|\mathcal{E}| < \delta$.*

For a given economy $\mathcal{E}$ and allocation $X(t)$, $E_\delta^{\bar{p}}(X)$ is the set of traders in $\mathcal{E}$ who if they purchase $X(t)$ at prices $\bar{p}$ violate their budget constraint by δ. $F_\delta^{\bar{p}}(X)$ is the set of traders in $\mathcal{E}$ who can buy a commodity bundle $\bar{y}_t$ at prices $\bar{p}$, without violating their budget constraint, which they prefer to the commodity bundle assigned to them by the allocation X; and $\bar{y}_t$ is at least a distance δ from $X(t)$. G_δ^{p} is just the union of these two exceptional sets.

Theorem 2 simply says that for core allocations in large economies there exists prices such that the average number of exceptional traders is small.

The proof of this theorem is given in [4].

We would like to note that although Theorem 2 is a consequence of Theorem 1, it is not "equivalent" to Theorem 1. A less intuitive result in terms of sequences of economies that is "equivalent" to Theorem 1 is given in [3].

An excellent discussion of the published literature pertaining to Edgeworth's conjecture is given in [1]. Edgeworth's original analysis of this problem may be found in [6].

We are happy to acknowledge several very helpful discussions with Herbert E. Scarf. This research was supported in part by the National Science Foundation (GP 29218) and the Office of Naval Research.

1. Arrow, K. J. & Hahn, F. H. (1971) in *General Competitive Analysis* (Holden-Day Inc., San Francisco, pp. 205–206.
2. Aumann, R. J. (1964) "Markets with a Continuum of Traders," *Econometrica* **32**, 39–50.
3. Brown, D. J. & Robinson, A., "Nonstandard Exchange Economics," *Econometrica*, in press.
4. Brown, D. J. & Robinson, A., "The Cores of Large Standard Exchange Economies," *J. Econ. Theor.* (submitted).
5. Debreu, G. & Scarf, H., "A Limit Theorem on the Core of an Economy," *Int. Econ. Rev.* **4**, 235–246.
6. Edgeworth, F. Y. (1881) *Mathematical Psychics* (C. Kegan Paul, London), pp. 20–56.
7. Kochen, S. (1961) "Ultraproducts in the Theory of Models," *Ann. of Math.* **79**, 221–261.
8. Luxemburg, W. A. J. (1964) *Non-Standard Analysis* (lectures on A. Robinson's theory of infinitesimals and infinitely large numbers), Pasadena, California, Institute of Technology.
9. Robinson, A. (1971) "Non-Standard Theory of Dedekind Rings," *Proc. Acad. Sci.*, Amsterdam, A70, pp. 444–452.

Correction. In the article "A Limit Theorem on the Cores of Large Standard Exchange Economies," by Brown, D. J. & Robinson, A., which appeared in the May 1972 issue of *Proc. Nat. Acad. Sci. USA* **69**, 1258–1260, the following corrections should be made.

Page 1259, column 2, line 3 from the bottom to page 1260, column 1, line 2:

The definition of equicontinuity of preference relations should be weakened so as to read:

"$(\forall \bar{x} \geq 0)\ (\forall \bar{y} > \bar{0})\ (\exists\, \delta > 0)\ (\forall n \in N)\ (\forall t \in \mathcal{E}_n)\ [\bar{z} \in S(\bar{x} + \bar{y}, \delta) \wedge \bar{w} \in S(\bar{x}, \delta) \Longrightarrow \bar{z} >_t \bar{w}]$. An example of such a family is the set of preference relations $>_f$ defined by a finite family of continuous and strongly monotonic utility functions, where $\bar{x} >_f \bar{y}$ if and only if $f(\bar{x}) > f(\bar{y})$."

This weakening of the definition of equicontinuity entails the following additional corrections:

Page 1259, column 1, lines 16 and 19 (from bottom): for "$\bar{x} >_t \bar{y}$", *read* "$\bar{x} > \bar{y}$". Page 1259, column 2, line 3 from top: For "$Y(t) >_t X(t)$", *read* "and if $\bar{z} \simeq Y(t)$, then $\bar{z} >_t X(t)$".

Page 1259, column 2, line 11 from top: For "$\bar{x} >_t \bar{y}$", *read* "if $\bar{z} \simeq \bar{x}$, then $\bar{z} >_t \bar{y}$".

Page 1260, column 1, third line of equation (line 8) should read:

$$\text{"} \wedge\ (\forall \bar{z} \in S(\bar{y}, \delta))\bar{z} >_t X(t)\text{"}$$

Page 1260, column 1, line 7 from bottom: For "which they prefer," *read* "such that they prefer all $\bar{z}$ which are at most a distance δ from $\bar{y}_t$."

Page 1260, column 1, line 5 from bottom: Delete "and $\bar{y}_t$ is at least a distance δ from $X(t)$."

FUNCTION THEORY ON SOME NONARCHIMEDEAN FIELDS

ABRAHAM ROBINSON, Yale University

1. Introduction. Archimedes' axiom states that for any two positive numbers a and b, a smaller than b, the continued addition of a to itself ultimately yields numbers which are greater than b. More formally, if F is an ordered abelian group or, more particularly, an ordered field, then Archimedes' axiom is as follows.

1.1. *If $0 < a < b$, where a and b are elements of F then there exists a natural number n such that*

$$\underbrace{a + a + \cdots + a}_{n \text{ times}} > b.$$

Throughout the history of mathematics, Archimedes' axiom has been associated with the foundations of the Differential and Integral Calculus. Already in Greek science the method which, much later, was dubbed the method of exhaustion and which, to a large extent, anticipated the ε, δ method in the calculation of areas and volumes, depended on the validity of Archimedes' axiom, which was formulated explicitly for this purpose. On the other hand, when a method of infinitely small and infinitely large numbers is used, as in Nonstandard Analysis, then it is just the nonarchimedean nature of the system which is essential for its success or, more precisely, the superposition of a nonarchimedean field on the archimedean field of real numbers.

Although Nonstandard Analysis (see [**4**] or [**6**]) may perhaps be regarded as the most successful effort in this direction, many other systems have been introduced for the same purpose. Thus, not long ago, D. Laugwitz [**2**] considered a theory of functions on the field L of generalized power series with real coefficients and **real exponents**. The same field was investigated many years earlier by T. Levi-Civita [3], also because of its nonarchimedean character, and by A. Ostrowski [**5**], in connection with the theory of valuations.

Laugwitz raised the question whether the functions considered by him satisfy the intermediate value theorem and the mean value theorem of the Differential Calculus. We shall show in the present paper that although these theorems are not valid here in full generality, they are true under rather wide conditions. In order to obtain these results, we shall embed L in the residue class field ${}^{\rho}R$ of a certain subring of a nonstandard model of Analysis, ${}^{*}R$. It appears that ${}^{\rho}R$ has many interesting properties which make it a suitable subject for investigation quite apart from the particular problem just mentioned. In particular, the behavior of a function on ${}^{\rho}R$ is closely connected with the theory of asymptotic expansions, although we shall not pursue this topic in the present paper.

2. Ordered fields and fields with valuation. An **ordered field** F is a commutative field in which an ordering relation $x < y$ (or, equivalently, $y > x$) is defined and satisfies the following conditions.

2.1. *The ordering is transitive,* $x < y$ *and* $y < z$ *implies* $x < z$, *and irreflexive,* $x < y$ *implies* $x \neq y$.

2.2. *The ordering is total, if* $x \neq y$ *then either* $x < y$ *or* $y < x$ *(but not both, by* 2.1).

2.3. *The ordering is related to addition by the requirement that* $x < y$ *implies* $x + z < y + z$; *and to multiplication by the requirement that* $x < y$ *and* $0 < z$ *implies* $xz < yz$.

An ordered field can be characterized also by means of the set of its **positive elements** $P = \{x \mid x > 0\}$. Thus, suppose that a subset P of a field F possesses the following properties.

2.4. $0 \notin P$; for all $x \neq 0$, $x \in P$ or $-x \in P$.

2.5. *If* $x, y \in P$, *then* $x + y \in P$ *and* $xy \in P$.

Then the relation *defined* by

$$x < y \text{ if and only if } y - x \in P$$

satisfies the conditions 2.1–2.3 and P is just the set of positive elements of the field according to this relation.

We shall suppose that the reader is familiar with the elementary properties of ordered fields, e.g., that an ordered field is of characteristic 0 and that $x^2 > 0$ for all $x \neq 0$. As usual, we write $x \leqq y$ or $y \geqq x$ if either $x < y$ or $x = y$.

The rational numbers form an ordered field Q whose positive elements are the fractions (ratios) of natural numbers different from zero, and the real numbers form an ordered field R whose positive elements are just the squares other than zero. In both cases the ordering is unique. Moreover, both Q and R are archimedean, i.e., they satisfy Archimedes' Axiom 1.1.

Perhaps the simplest example of a non-archimedean field is as follows. Let $R(t)$ be a simple transcendental extension of the field of real numbers R. Thus $R(t)$ may be identified with the field of rational functions of the indeterminate t with coefficients in R, each element of $R(t)$ may be written in the form

$$f = \frac{p(t)}{q(t)} = \frac{a_0 + a_1 t + \cdots + a_n t^n}{b_0 + b_1 t + \cdots + b_m t^m}, \tag{2.6}$$

where $q(t) \neq 0$, at least one of the b_j is different from 0. We may then suppose the first $b_j \neq 0$ is actually equal to 1, for if this is not the case from the outset, we may achieve it by multiplying the numerator and denominator on the right hand side of (2.6) by b_j^{-1}. Thus, if $f \neq 0$, we may write

$$f = \frac{a_k t^k + \cdots + a_n t^n}{t^j + b_{j+1} t^{j+1} + \cdots + b_m t^m}, \qquad a_k \neq 0, \tag{2.7}$$

$$0 \leqq k \leqq n,\ 0 \leqq j \leqq m.$$

We now determine an ordering in $R(t)$ by defining that $f \neq 0$ is **positive** if and only if $a_k > 0$. To make sure that this is a good definition one first has to check that it is independent of the particular representation (2.7) chosen for the given f. Next one verifies that the set of positive elements of $R(t)$ defined in this way satisfies the conditions of 2.4 and 2.5. We suppose that these rather simple tasks have been carried out so that $R(t)$ becomes indeed an ordered field with the above definition. Moreover, this ordered field is nonarchimedean. For, by our definitions, $0 < t$, $t < 1$ (since $1 - t$ is positive) and, for any positive integer n,

$$\underbrace{t + t + \cdots + t}_{n \text{ times}} < 1$$

(since $1 - nt$ is positive). This shows that 1.1 is not satisfied.

In any ordered field, the absolute value of a number a is defined to be $|a| = a$ if $a \geqq 0$, otherwise $|a| = -a$. Then $|ab| = |a|\ |b|$ and $|a + b| \leqq |a| + |b|$ (triangle inequality).

Let F be a nonarchimedean ordered field. Then F is of characteristic 0 and, hence, contains the field of rational numbers Q. An element $a \in F$ is said to be **infinite** if $|a| > q$ for all $q \in Q$. Also, $a \in F$ is said to be **infinitely small** or **infinitesimal** if $|a| < q$ for all positive $q \in Q$. $a \in F$ is **finite** if it is not infinite. This will be the case if and only if $|a| < q$ for some $q \in Q$.

The finite elements of F constitute a subring F_0 of F. The infinitesimal elements of F constitute a proper ideal F_1 within F_0. F_1 is maximal in F_0 as can be seen by the following argument. Suppose that $F_1 \subset J \subset F_0$ where J is an ideal in F_0, such that $J - F_1 \neq \varnothing$. Let $a \in J - F_1$ then a is not infinitesimal. We conclude without difficulty that a^{-1} is finite, so $a^{-1} \in F_0$, $aa^{-1} = 1 \in J$. But then $J = F_0$, F_1 is maximal in F_0.

It follows that $F' = F_0/F_1$ is a field. F' is called the **residue class field of the ordering**. The canonical mapping $F_0 \xrightarrow{\psi} F'$ induces an ordering in F' according to the rule that, for any $a \in F'$, $a \neq 0$, a is to be positive in F' if and only if one (and hence, all) of the elements of $\psi^{-1}a$ is (are) positive. It is not difficult to show that F' is archimedean according to this ordering and (hence) that it is isomorphic and order-isomorphic to a subfield of R.

The cosets of F_1 as an additive subgroup of F are called **monads**. If a is any element of F then we denote the monad containing it by $\mu(a)$. In particular, $\mu(0) = F_1$. The monads which are subsets of F_0 may be identified with the elements of F'.

As a tool in our investigation of nonarchimedean fields we shall require also the notion of a **field with valuation**, more particularly, the notion of a **field with non-**

archimedean valuation in the real numbers. This concept is given by a field F together with a mapping $v(x)$ from $F - \{0\}$ into the real numbers R such that the following conditions are satisfied:

2.8. *For all* $x \neq 0$, $y \neq 0$ *in* F, $v(xy) = v(x) + v(y)$.
2.9. *For all* x, y *in* F *such that* $x \neq 0$, $y \neq 0$, $x + y \neq 0$,

$$v(x + y) \geqq \min(v(x), v(y)).$$

If we add to R an **element** ∞ (usually called "a symbol") with the rules $x + \infty = \infty + x = \infty + \infty = \infty$ and the stipulation that $\infty > x$ for all real x, then the auxiliary definition $v(0) = \infty$ ensures that the equations of 2.8 and 2.9 are satisfied without any restriction on x and y.

The set $O_F = \{x \in F \mid v(x) \geqq 0\}$ is a subring of F, the **valuation ring**, and the set $J_F = \{x \in F \mid v(x) > 0\}$ constitutes a maximal ideal in O_F, the **valuation ideal.** The field $\bar{F} = O_F / J_F$ is called the residue class field of the given valuation.

Let c be an arbitrary but fixed constant greater than 1. Then the definition of distance

$$d(x, y) = c^{-v(x-y)},$$

where $c^{-\infty}$ is interpreted as 0, turns F into a metric space. If every Cauchy sequence in that space has a limit then F is said to be **complete** for the given valuation.

See [**1**], [**7**] or [**8**] for basic facts in valuation theory. From now on such facts will be taken for granted.

3. The field L. The field $R(t)$ is inadequate for the development of the calculus because we cannot extend to it even some of the most common functions defined in the field of real numbers, e.g., the function $y = \sqrt{x}$. Passing to the field of formal Laurent series $\sum_{k=-n}^{\infty} a_k t^k$, $a_k \in R$, does not remedy the situation. Following Laugwitz, we therefore consider the field of generalized power series L, which is defined as follows:

The elements of L are the formal expressions

$$\sum_{k=0}^{\infty} a_k t^{\nu_k} \qquad a_k, \nu_k \in R, \qquad \nu_k \uparrow \infty, \tag{3.1}$$

(where the last symbol implies $\nu_0 < \nu_1 < \nu_2 < \cdots$). Two expressions (3.1) are, by definition regarded as equal if for any term at a_ν which occurs in one but not in the other, $a = 0$. We shall also write $a_0 t^{\nu_0} + a_1 t^{\nu_1} + \cdots + a_k t^{\nu_k}$ for an expression for which $a_{k+1} = a_{k+2} = \cdots = 0$.

The **sum** of two expressions $\sum a_k t^{\nu_k}$ and $\sum b_k t^{\mu_k}$ as in (3.1) is the expression $\sum c_k t^{\lambda_k}$ which is defined as follows. The sequence $\{\lambda_k\}$ is the set theoretical union of the sequences $\{\nu_k\}$ and $\{\mu_k\}$ arranged in increasing order. If a particular λ_m occurs both in $\{\nu_k\}$ and in $\{\mu_k\}$, e.g., $\lambda_m = \nu_p = \mu_q$ then $c_m = a_p + b_q$; if $\lambda_m = \nu_p$ but λ_m does not occur in $\{\mu_k\}$ then $c_m = a_p$; and if $\lambda_m = \mu_q$ but λ_m does not occur in $\{\nu_k\}$ then

$c_m = b_q$. Thus, briefly, the sum $\Sigma\, c_k t^{\lambda_k}$ is obtained by the formal addition of the terms of $\Sigma\, a_k t^{\nu_k}$ and $\Sigma\, b_k t^{\mu_k}$. Similarly, the **product** $\Sigma\, c_k t^{\lambda_k}$ of $\Sigma a_k t^{\nu_k}$ and $\Sigma\, b_k t^{\mu_k}$ as in 3.1 is obtained by formal multiplication. Thus, the sequence $\{\lambda_k\}$ consists of the sums $\nu_p + \mu_q$ arranged in increasing order and $c_k = \Sigma\, a_p b_q$ where p and q range over the natural numbers such that $\nu_p + \mu_q = \lambda_k$. It is not difficult to see that all these sums are finite and that the resulting expression satisfies the conditions of (3.1). Moreover, our definitions of sums and products are compatible with the relation of equality introduced earlier, and they turn L into a ring whose zero and unit elements may be written as $0t^0 + 0t^1 + 0t^2 + \cdots$, or 0, and as $1\,t^0 + 0t^1 + 0t^2 + \cdots$, or 1.

Now let $\alpha = 1 + \Sigma_{k=1}^{\infty}\, a_k t^{\nu_k}$, $0 < \nu_1 < \nu_2 < \cdots \to \infty$, i.e., α is an element of L as in 3.1 with $\nu_0 = 0$, $a_0 = 1$. We wish to show that α possesses a multiplicative inverse in L. For this purpose we define β as the formal expansion in powers of t of the expression

$$1 - \left(\sum_{k=1}^{\infty} a_k t^{\nu_k}\right) + \left(\sum_{k=1}^{\infty} a_k t^{\nu_k}\right)^2 - \left(\sum_{k=1}^{\infty} a_k t^{\nu_k}\right)^3 + \cdots.$$

Again it is not difficult to see that this expansion can be worked out and that it is of the form $\beta = 1 + \Sigma_{k=1}^{\infty}\, b_k t^{\mu_k}$ where $0 < \mu_1 < \mu_2 < \cdots \to \infty$, so that β belongs to L.

We now claim that $\alpha\beta = 1$. To see this, consider the identity

$$(1 + \gamma)\,(1 - \gamma + \gamma^2 - \gamma^3 + \cdots + \gamma^{2m}) = 1 + \gamma^{2m+1} \tag{3.2}$$

which holds in L for arbitrary natural m. We may substitute $\Sigma_{k=1}^{\infty}\, a_k t^{\nu_k}$ for γ and expand on both sides of (3.2). This yields an equation

$$\alpha\beta' = \gamma\ , \tag{3.3}$$

where β' is the expansion of

$$1 - \left(\sum_{k=1}^{\infty} a_k t^{\nu_k}\right) + \left(\sum_{k=1}^{\infty} a_k t^{\nu_k}\right)^2 - \cdots + \left(\sum_{k=1}^{\infty} a_k t^{\nu_k}\right)^{2m}$$

and γ' is the expansion of $1 + (\Sigma_{k=1}^{\infty}\, a_k t^{\nu_k})^{2m+1}$. But then β' differs from β only in powers of t whose exponent is at least $(2m + 1)\nu_1$ and γ differs from 1 only in powers of t whose exponent also is at least $(2m + 1)\nu_1$. Since m is an arbitrary natural number, we conclude that $\alpha\beta = 1$, $\beta = \alpha^{-1}$.

Now let $\alpha \in L$ be different from zero, otherwise arbitrary. Then $\alpha = \Sigma_{k=0}^{\infty}\, a_k t^{\nu_k}$, where we may assume that $a_0 \neq 0$. Putting $\alpha = a_0 t^{\nu_0}\, \alpha'$ where

$$\alpha' = 1 + \sum_{k=1}^{\infty} (a_k/a_0) t^{\nu_k - \nu_0},$$

we then obtain $a_0^{-1}\, t^{-\nu_0}\, \alpha'^{-1}$ as the multiplicative inverse of α.

Thus, L is a field. We introduce an ordering of L by defining that an element

$\alpha \in L$, $\alpha \neq 0$ is positive if and only if the nonvanishing coefficient a_k with lowest subscript m in the expression $\alpha = \sum_{k=0}^{\infty} a_k t^{\nu_k}$ is positive. Also, L obtains a valuation by defining $v(\alpha) = \nu_m$ (so that $a_m \neq 0$, $a_k = 0$ for $k < m$), for $\alpha \neq 0$, together with $v(0) = \infty$ in accordance with our general convention.

In this valuation, the valuation ring O_L consists of all elements of L which can be written as $\sum a_k t^{\nu_k}$ with $\nu_0 \geqq 0$, and this is also the ring of **finite** elements of L in the ordering of L; and the valuation ideal J_L consists of all $\sum a_k t^{\nu_k}$ with $\nu_0 > 0$ and coincides with the set of **infinitesimal** elements of L. Thus, the residue class field of L with respect to its valuation coincides with the residue class field of L with respect to its ordering and is, in fact, the field of real numbers R. Also, since $J_L \neq \{0\}$, L is nonarchimedean.

There is a natural (and obvious) embedding (injection) of R into L: $a \to a = at^0 + 0t^1 + 0t^2 + \cdots$ and this extends, equally obviously, to an embedding of $R[t]$ into L:

$$a_0 + a_1 t + \cdots + a_n t^n \to a_0 t^0 + a_1 t^1 + a_2 t^2 + \cdots + a_n t^n + 0t^{n+1} + \cdots$$

and hence, to an embedding of $R(t)$ into L. The embedding is order preserving for the ordering of $R(t)$ defined in section 2 above.

It is shown in [**5**] that L is complete. It is also shown there that the field L' which is obtained by taking complex coefficients in place of the real coefficients in L, is algebraically closed. Since $L' = L(\sqrt{-1})$ it follows (compare [**7**]) that L is real-closed, i.e., that every positive element of L possesses a square root in L and that every polynomial of odd degree in $L[x]$ possesses a root in L. It follows in particular that a positive element of L possesses roots of all orders $n = 2, 3, 4, \cdots$. The same result is established by elementary means in [**2**] and will be used later in this paper.

Now let $f(x)$ be a real-valued infinitely differentiable function of a real variable which is defined in an interval $a < x < b$, $a, b \in R$. On passing from R to L, we find that the interval $a < x < b$ in L consists of points $x = \xi + \sum_{k=1}^{\infty} a_k t^{\nu_k}$, $0 < \nu_1 < \to \infty$, of three kinds,

(i) $$a < \xi < b,$$

(ii) $$\xi = a, \quad \sum_{k=1}^{\infty} a_k t^{\nu_k} > 0, \text{ and}$$

(iii) $$\xi = b, \quad \sum_{k=1}^{\infty} a_k t^{\nu_k} < 0.$$

In all these cases ξ is the unique real number which is infinitely close to x, i.e., such that $x - \xi$ is infinitely small and (by analogy with the terminology in Nonstandard Analysis) we call ξ the **standard part** of x, $\xi = {}^0x$.

Laugwitz extends the function $f(x)$ to values of x in L with standard part ξ, $a < \xi < b$ by using the formal Taylor expansion of $f(x)$,

$$f(x+h) = \sum_{n=0}^{\infty} \frac{1}{n!} f^{(n)}(x)h^n.$$

Thus, he defines for $x = \xi + \sum_{k=1}^{\infty} a_k t^{\nu_k}$,

$$(3.4) \qquad {}^L f(x) = \sum_{n=0}^{\infty} \frac{1}{n!} f^{(n)}(\xi) \left(\sum_{k=1}^{\infty} a_k t^{\nu_k} \right)^n ,$$

where it is understood that ${}^L f(x)$ is the element of L which is obtained by expanding the right hand side of (3.4) and rearranging it in powers of t. Once again, the condition $\nu_0 > 0$ shows that this can be done.

We shall show in the following sections that the definition proposed by Laugwitz is obtained in a natural way by relating L to a nonstandard model of Analysis.

4. The field ${}^\rho R$. Let $^*R$ be a nonstandard model of Analysis (cf. [4] and [6]). We shall suppose that $^*R$ is **sequentially comprehensive**. That is to say, if $a_0, a_1, a_2, \cdots, a_n, \cdots, n \in N$, is a sequence of entities of $^*R$ (of the same type, if type restrictions are adopted), e.g., a sequence of numbers of $^*R$, then there exists an internal sequence $\{s_n\}$ in $^*R$ (where n now ranges over $^*N$) such that $s_n = a_n$ for all finite n.

There exist sequentially comprehensive $^*R$. More particularly, all $^*R$ which are ultrapowers are sequentially comprehensive. Thus, suppose $^*R = R^I/D$ where D is a free ultrafilter on the index set I. Every internal entity of $^*R$ is represented by (is an equivalence class of) functions $f(\nu)$ on I. Let $f_n(\nu)$ represent a_n, $n = 0, 1, 2, \cdots$, and for each $\nu \in I$, consider $s(\nu) = \{f_n(\nu)\}$. Then $s(\nu)$, ν ranging over I represents an internal sequence $\{s_n\}$ in $^*R$. We claim that for each finite k, the value of that sequence is just a_k. Now, in order to obtain the value of $\{s_n\}$ for $n = k$, we have to substitute the function $f(\nu) \equiv k$ for each n in $f_n(\nu)$. This yields precisely $f_k(\nu)$, i.e., a_k.

Supposing, from now on, that $^*R$ is sequentially comprehensive, we wish to show that the set of infinite natural numbers, $^*N - N$, cannot be coinitial with ω^*. In other words:

4.1. THEOREM. *Let $a_0 > a_1 > a_2 > \cdots > a_n > \cdots, n \in N$ be a strictly decreasing sequence of infinite natural numbers, internal or external. Then there exists an infinite natural number a, such that $a_n > a$ for all $n \in N$.*

Proof. Since $^*R$ is sequentially comprehensive, we may suppose that, for all $n \in N$, $a_n = s_n$ where $\{s_n\}$ is an internal sequence of numbers of $^*R$. Consider the internal sequence

$$t_n = \frac{n}{\min(s_0, s_1, \cdots, s_n)}, \qquad n \in {}^*N.$$

Then $0 \leqq t_n < 1$ for all finite n but $t_n > 1$ for large infinite n. Hence there exists a smallest m, which must then be infinite, such that $0 \leqq t_m < 1$ does not hold.

Thus, for $k = m - 1$,

$$0 \leqq \frac{k}{\min(s_0, s_1, \cdots, s_k)} < 1.$$

This shows that $k < a_0, k < a_1, \cdots, k < a_n, \cdots$ for all finite n and proves the theorem.

Now let ρ be an arbitrary but fixed positive infinitesimal number in $^*R$. We define subsets M_0 and M_1 of $^*R$ by

$$M_0 = \{x \in {}^*R \mid |x| \leqq \rho^{-n} \text{ for some finite positive integer } n\},$$

$$M_1 = \{x \in {}^*R \mid |x| \leqq \rho^{n} \text{ for all finite positive integers } n\}.$$

Evidently, $M_1 \subset M_0$ and $M_0 \supset R$. Both M_0 and M_1 are rings under the operations of $^*R$. For if $|x| \leqq \rho^{-n}$, $|y| \leqq \rho^{-m}$, with $n \leqq m$ say, then

$$|x + y| \leqq |x| + |y| \leqq 2\rho^{-m} \leqq \rho^{-(m+1)}$$

and $|xy| \leqq \rho^{-(n+m)}$, so M_0 is a ring. And if $|x| \leqq \rho^n$, $|y| \leqq \rho^n$ then $|x \pm y| \leqq 2\rho^n \leqq \rho^{n-1}$, $|xy| \leqq \rho^{2n}$. Since, in the definition of M_1, n is arbitrary, this shows that M_1 also is a ring.

Moreover, M_1 is an ideal in M_0, for if $x \in M_1$ and $y \in M_0$ then $|y| \leqq \rho^{-n}$ for some natural number n, and since $|x| \leqq \rho^{m+n}$ for all natural n, it follows that $|xy| \leqq \rho^m$ for all natural m, $xy \in M_1$. M_1 is a proper ideal since it does not contain 1. Finally, M_1 is a *maximal* ideal in M_0. For let $J \supset M_1$ be another ideal in M_0 such that $J - M_1$ is not empty, and let $x \in J - M_1$. Then $|x| > \rho^m$ for some finite natural number m and so $|x^{-1}| < \rho^{-m}$, $x^{-1} \in M_0$. Hence $1 = xx^{-1} \in J$, $J = M_0$, showing that M_1 is maximal.

We conclude that the quotient ring $^\rho R = M_0/M_1$ is a field. Moreover, the canonical map

$$\psi: M_0 \to {}^\rho R \tag{4.2}$$

induces an ordering in $^\rho R$. For let $x \in M_0 - M_1$, $x > 0$, and let $x + y$, $y \in M_1$ be any other element of the coset of x with respect to M_1. Then $|x| > \rho^m$ for some finite natural number m and $|y| \leqq \rho^n$ for all finite natural numbers n. Hence $|y| < |x|$, and so $x + y \geqq x - |y| = |x| - |y| > 0$, all elements of the coset of x are positive. Accordingly, we may define an ordering in $^\rho R$ by defining that an element $\alpha \in {}^\rho R$, $\alpha \neq 0$, is positive if and only if the elements of $\psi^{-1}\alpha$ are positive. Then the sum and product of positive elements of $^\rho R$ are positive but $0 \in {}^\rho R$ is not positive. This shows that our definition turns $^\rho R$ into an ordered field. We also observe that for any $\alpha \in {}^\rho R$, $\psi^{-1}\alpha$ is an interval in M_0 and $^*R$. Finally, since M_1 contains only the single standard number 0, ψR provides an embedding of R (as a subfield of $^*R$) in $^\rho R$.

Next, we define a valuation in $^\rho R$, as follows. For any $\alpha \in {}^\rho R$, $\alpha \neq 0$, let x and $x + y$ be elements of $\psi^{-1}\alpha$, $y \in M_1$ and consider $\log_\rho|x|$ and $\log_\rho|x + y|$. Since $|x|$

and $|x+y|$ are greater than some positive, and smaller than some negative power of ρ, $\log_\rho|x|$ and $\log_\rho|x+y|$ are finite and possess standard parts. We claim that

$$^0(\log_\rho|x|) = {}^0(\log_\rho|x+y|),$$

i.e., that

$$\log_\rho|x+y| - \log_\rho|x| = \log_\rho|1+y/x|$$

is infinitesimal. But $\log_\rho|1+(y/x)| = \ln|1+(y/x)|/\ln\rho$. Since y/x is infinitesimal and $\ln|w|$ is a standard function which is continuous at $w=1$, $\ln|1+(y/x)|$ is infinitesimal. Hence $\log_\rho|1+(y/x)|$ also is infinitesimal, as asserted.

Accordingly, we obtain a unique definition of a function $v(\alpha)$ for $\alpha \in {}^\rho R$, $\alpha \neq 0$, by putting $v(\alpha) = {}^0(\log_\rho|x|)$ for any $x \in \psi^{-1}\alpha$. We claim that this defines a valuation of the field ${}^\rho R$.

Let $\alpha, \beta \in {}^\rho R$, $\alpha \neq 0$, $\beta \neq 0$ and let $x \in \psi^{-1}\alpha$, $y \in \psi^{-1}\beta$. Then

$$^0(\log_\rho|xy|) = {}^0(\log_\rho|x|) + {}^0(\log_\rho|y|)$$

and so $v(\alpha\beta) = v(\alpha) + v(\beta)$, as required. Next, suppose $\alpha+\beta \neq 0$, then we have to show that $v(\alpha+\beta) \geqq \min(v(\alpha), v(\beta))$ or, equivalently, that

$$^0(\log_\rho|x+y|) \geqq \min({}^0(\log_\rho|x|), {}^0(\log_\rho|y|)). \tag{4.3}$$

We may suppose without essential loss of generality that $\log_\rho|x| \geqq \log_\rho|y|$. Then (4.3) will hold precisely if there is an infinitesimal η such that

$$\log_\rho|x+y| \geqq \log_\rho|y| - \eta,$$

i.e., such that

$$\log_\rho\left|1+\frac{x}{y}\right| \geqq -\eta.$$

Putting $x/y = w$, we have to show $\log_\rho|1+w| \geqq -\eta$ for $\log_\rho|w| \geqq 0$, (where we may rule out $w=-1$ because of $\alpha+\beta \neq 0$). Put $\sigma = \log_\rho|w|$, $|w| = \rho^\sigma$, where $\sigma \geqq 0$, then

$$|1+w| \leqq 1+|w| = 1+\rho^\sigma \leqq 2\rho^\sigma = \rho^{\sigma+\log_\rho 2}$$

$$\log_\rho|1+w| \geqq \sigma + \log_\rho 2 \geqq \log_\rho 2.$$

But $\ln_\rho 2$ is (negative) infinitesimal, and so (4.3) is proved. We supplement the definition of $v(x)$ as usual by putting $v(0) = \infty$.

The valuation ring of the valuation just defined will be denoted by O_ρ. It is not difficult to see that O_ρ includes the ψ-images of all finite elements of ${}^*R$. However, O_ρ includes other elements as well. For example, let $\lambda = \psi\ln\rho$. Then $v(\lambda) = {}^0(\log_\rho|\ln\rho|) = {}^0(\ln|\ln_\rho|/\ln\rho)$. But the expression in the parentheses on the right hand side is infinitesimal, for $\ln \rho$ is (negative) infinite and

$$\lim_{x \to \infty} \frac{\ln x}{x} = 0.$$

Hence $v(\lambda) = 0$.

We shall now show that the field ${}^{\rho}R$ is *complete* for the valuation defined above. Defining the distance between two elements of ${}^{\rho}R$, α and β, by $d(\alpha, \beta) = c^{-v(\alpha-\beta)}$ (see the end of section 2 above) let $\{\alpha_n\}$ be a Cauchy sequence in this metric.

$$\lim_{\substack{n \to \infty \\ m \to \infty}} d(\alpha_n, \alpha_m) = 0. \tag{4.4}$$

Then we have to show that $\{\alpha_n\}$ converges to a limit α in ${}^{\rho}R$.

Choose elements $x_n \in \psi^{-1}\alpha_n$, $n = 0, 1, 2, \cdots, n \in N$. Since ${}^*R$ is sequentially comprehensive there exists an internal sequence $\{s_n\}$ of elements of ${}^*R$ such that $s_n = x_n$ for all finite n. We shall write x_n in place of s_n also for infinite n. By (4.4)

$$\lim_{\substack{n \to \infty \\ m \to \infty}} v(\alpha_n - \alpha_m) = \infty.$$

Equivalently, given any finite natural number k, there exists a finite natural $j = j_k$ such that

$$\log_{\rho} | x_n - x_m | > k \text{ for } n, m > j_k, \ n, m \in N. \tag{4.5}$$

Now since 4.5 holds for all finite n and m greater than j, it holds for all $n > j$, $m > j$, $n + m$ finite, $j = j_k$. A standard argument of Nonstandard Analysis, which was exemplified in the proof of 4.1, now shows that there exists an *infinite* natural $\omega = \omega_k$ such that (4.5) holds for all $n > j$, $m > j$, and $n + m < 2\omega_k$ and hence, in particular, for all $n > j$, $m > j$ and $n < \omega_k$, $m < \omega_k$. Moreover, by determining $\omega_0, \omega_1, \omega_2, \cdots$ one after the other, we may evidently assume that $\omega_0 > \omega_1 > \omega_2 > \cdots$. Appealing to 4.1, we may then choose an infinite natural number Ω which is smaller than ω_0, ω_1, ω_2 and—obviously, being infinite, larger than $j_0, j_1, j_2, \cdots$. Then,

$$\log_{\rho} | x_n - x_{\Omega} | > k \text{ for } n > j_k, \qquad n \in N, \quad k \in N. \tag{4.6}$$

(4.6) shows in the first place, that $x_{\Omega} \in M_0$. To see this, choose $n > j_0$ then $\log_{\rho} | x_n - x_{\Omega} | > 0$, so $| x_n - x_{\Omega} |$ is finite. Also, $x_n \in M_0$, so $| x_n | \leq \rho^{-m}$ for some positive integer m and $| x_{\Omega} | \leq | x_{\Omega} - x_n | + | x_n | \leq 2\rho^{-m} < \rho^{-(m+1)}$, $x_{\Omega} \in M_0$.

Now let $\alpha = \psi x_{\Omega}$, then we wish to show that $\lim_{n \to \infty} \alpha_n = \alpha$ or, which is equivalent, that

$$\lim_{n \to \infty} v(\alpha_n - \alpha) = \infty. \tag{4.7}$$

But this is an immediate consequence of (4.6), since (4.6) implies

$$^{0}(\log_{\rho}|x_n - x_\Omega|) > k-1 \text{ for } n > j_k, \quad n \in N, \quad k \in N$$

and this is the same as

$$v(\alpha_n - \alpha) > k - 1 \text{ for } n > j_k, \quad n \in N, \quad k \in N$$

which is just an explicit expression for the validity of (4.7). Thus, we have shown that ${}^{\rho}R$ is complete.

Let $\bar{\rho} = \psi\rho$ and consider any infinite series in ${}^{\rho}R$ of the form

$$a_0\bar{\rho}^{v_0} + a_1\bar{\rho}^{v_1} + a_2\bar{\rho}^{v_2} + \cdots, \qquad a_n \in R \subset {}^{\rho}R, \tag{4.8}$$

$$v_0 < v_1 < v_2 < \cdots \to \infty,$$

where the v_j are standard real. The partial sums of (4.8) are

$$\sigma_k = a_0\bar{\rho}^{v_0} + a_1\bar{\rho}^{v_1} + \cdots + a_k\bar{\rho}^{v_k}, \qquad k = 0, 1, 2, \cdots.$$

The value of any monomial in (4.8) is, for $a_j \neq 0$, $v(a_j\bar{\rho}^{v_j}) = v(a_j) + v(\bar{\rho}^{v_j}) = 0 + v_j = v_j$, with $v(a_j\bar{\rho}^{v_j}) = \infty$ for $a_j = 0$. Hence $v(\sigma_k) = v_j$ where j is the smallest subscript $\leqq k$ for which $a_j \neq 0$, if any, otherwise $v(\sigma_k) = \infty$. Also, for $0 \leqq k < l$

$$\sigma_l - \sigma_k = a_{k+1}\bar{\rho}^{v_{k+1}} + \cdots + a_l\bar{\rho}^{v_l}$$

and so $v(\sigma_l - \sigma_k) \geqq v_{k+1}$. This shows that $\{\sigma_k\}$ is a Cauchy sequence, and the limit of that sequence, σ is just the sum of (4.8). Also, $v(\sigma) = v_j$ where j is the lowest subscript for which $a_j \neq 0$ or, if there is no such j, i.e., if all a_j vanish, $v(\sigma) = \infty$ and this is the case if and only if $\sigma = 0$. As usual in the theory of infinite series, we identify (4.8) with its sum in ${}^{\rho}R$. It is then not difficult to verify that the sum of two numbers of ${}^{\rho}R$, σ and τ, given by (4.8) and

$$b_0\bar{\rho}^{\mu_0} + b_1,\bar{\rho}^{\mu_1} + b_2\bar{\rho}^{\mu_2} + \cdots, b_n \in R \subset {}^{\rho}R, \tag{4.9}$$

$$\mu_0 < \mu_1 < \mu_2 < \cdots \to \infty,$$

is represented by an expression

$$c_0\bar{\rho}^{\lambda_0} + c_1\bar{\rho}^{\lambda_1} + c_2\bar{\rho}^{\lambda_2} + \cdots$$

which is obtained from (4.8) and (4.9) just as the sum $\Sigma\, c_k t^{\lambda_k}$ was obtained from $\Sigma\, a_k t^{v_k}$ and $b_k t^{\mu_k}$ as elements of L in section 3 above. The product of (4.8) and (4.9) also is obtained by the procedure described in section 3, with $\bar{\rho}$ for t. It follows that the mapping

$$\Phi\colon a_0 t^{v_0} + a_1 t^{v_1} + a_2 t^{v_2} + \cdots \to a_0\bar{\rho}^{v_0} + a_1\bar{\rho}^{v_1} + a_2\bar{\rho}^{v_2} + \cdots, \tag{4.10}$$

$$a_j \in R, \; v_0 < v_1 < v_2 < \cdots \to \infty,$$

where the v_j are standard real, is a homomorphism from L into ${}^{\rho}R$. This homomorphism is an injection since $\Phi\alpha = 0$ implies $a_0 = a_1 = a_2 = \cdots = 0$ (see above) and, hence, $\alpha = 0$. It follows that ΦL is a field which is isomorphic to L and we write $\Phi L = {}^{\rho}L$. Evidently, Φ is analytic (i.e., value preserving, $v(\Phi(\alpha)) = v(\alpha)$). But Φ is also order preserving, as can be shown by verifying that, for any $\alpha \in L$, $\Phi\alpha > 0$ if and only if $\alpha > 0$. Now for $\alpha \neq 0$, $\alpha > 0$ if and only if the first nonvanishing a_j is positive, so we only have to show that an expression as in (4.8), $\bar{\sigma} = a_0\bar{\rho}^{\nu_0} + a_1\bar{\rho}^{\nu_1} + a_2\bar{\rho}^{\nu_2} + \cdots$ is positive provided (without loss of generality) $a_0 > 0$. Now, we may write $\bar{\sigma} = \psi\sigma$, where $\sigma = a_0\rho^{\nu_0} + \tau$, ${}^0(\log_\rho|\tau|) \geqq \nu_1$. It follows that if ν is an arbitrary standard real number between ν_0 and $\nu_1, \nu_0 < \nu < \nu_1$ then $\log_\rho|\tau| > \nu$, $|\tau| < \rho^\nu$, $a_0\rho^{\nu_0} > |\tau|$ and so

$$\sigma = a_0\rho^{\nu_0} + \tau \geqq a_0\rho^{\nu_0} - |\tau| > 0$$

and hence, $\bar{\sigma} > 0$. Thus Φ is order preserving, as asserted.

5. Functions in ${}^{\rho}R$. Let $f(x)$ be any real-valued function defined for $a < x < b$, $a, b \in R$. On passing to ${}^*R$, $f(x)$ is extended automatically to a function ${}^*f(x)$ which is defined for $a < x < b$ in ${}^*R$. We wish to find a natural extension of the function $f(x)$ as we pass from R to ${}^{\rho}R$.

Such an extension can be obtained, under certain conditions, as follows. Let ξ be any element of ${}^{\rho}R$ between a and b, $a < \xi < b$. Let ψ be the canonical homomorphism from M_0 to ${}^{\rho}R$ as before (see (4.2) above). Then we define

$$ {}^{\rho}f(\xi) = \psi({}^*f(x)) \text{ for } x \in \psi^{-1}\xi, \qquad a < x < b \tag{5.1}$$

provided the expression on the right hand side of (5.1) is independent of the particular choice of x subject to the stated conditions ($a < x < b$, $\psi x = \xi$).

Suppose in particular that $f(x)$ satisfies a Lipschitz condition in any closed subinterval of $a < x < b$. Thus, for any $a < a' < b' < b$ there exists a $k = k(a', b')$ such that for any $a' \leqq x_1 < x_2 \leqq b'$,

$$|f(x_2) - f(x_1)| \leqq k|x_2 - x_1|. \tag{5.2}$$

Passing from R to ${}^*R$, we see that (5.2) still holds, for standard a', b' and for arbitrary x_1, x_2 in the interval $\langle a', b'\rangle$, if we affix a star to $f(x_2)$ and $f(x_1)$. In particular, it therefore holds for two points x_1, x_2 of ${}^*R$ which are infinitely close to some standard x_0, $a < x_0 < b$ (where the constant k may depend on x_0).

Now let $\xi \in {}^{\rho}R$ be infinitely close to $x_0 \in R$. Then if x_1, x_2 belong to $\psi^{-1}\xi$, both x_1 and x_2 are infinitely close to x_0 in ${}^*R$, and (5.2) applies for an appropriate standard k. But then $x_2 - x_1 \in M_1$ and so, by (5.2), $f(x_2) - f(x_1) \in M_1$, $\psi f(x_2) = \psi f(x_1)$. This shows that in this case, (5.1) provides a unique definition for ${}^{\rho}f(\xi)$.

In particular, the Lipschitz condition is satisfied if $f(x)$ has a continuous derivative for $a < x < b$ or, more particularly, if $f(x)$ is infinitely differentiable in that interval. Suppose that this is the case and consider the restriction of ${}^{\rho}f(x)$ to points

$$\xi = a_0 + a_1\bar{\rho}^{\nu_1} + a_2\bar{\rho}^{\nu_2} + \cdots, \; 0 < \nu_1 < \nu_2 < \cdots, \qquad a < a_0 < b.$$

We may compare ${}^{\rho}f(x)$ for such a point with the function which is obtained by transferring Laugwitz' definition from L to ${}^{\rho}L$, i.e., with the function

$$F(x) = \Phi({}^{L}f(\Phi^{-1}x)).$$

We propose to show that ${}^{\rho}f(x)$ actually *coincides* with $F(x)$ for such argument values,

$$(5.3) \qquad {}^{\rho}f(\xi) = \Phi({}^{L}f(\Phi^{-1}\xi)).$$

In order to verify this identity, we observe that, except for rearrangements (which can be justified without difficulty within ${}^{\rho}R$), the right hand side of (5.3) is simply the formal Taylor expansion in ${}^{\rho}R$ of $f(x)$ about the point a_0. Thus, our claim is that

$$(5.4) \qquad \begin{aligned} {}^{\rho}f(\xi) = f(a_0) &+ \frac{f'(a_0)}{1!}(a_1\bar{\rho}^{\nu_1} + a_2\bar{\rho}^{\nu_2} + \cdots) + \frac{f''(a_0)}{2!}(a_1\bar{\rho}^{\nu_1} + a_2\bar{\rho}^{\nu_2} + \cdots)^2 \\ &+ \cdots + \frac{f^{(n)}(a_0)}{n!}(a_1\bar{\rho}^{\nu_1} + a_2\bar{\rho}^{\nu_2} + \cdots)^n + \cdots, \end{aligned}$$

in other words, that the Taylor series of ${}^{\rho}f$ about a_0 converges at ξ. Put $\eta = a_1\bar{\rho}^{\nu_1} + a_2\bar{\rho}^{\nu_2} + \cdots$ and choose $h \in \psi^{-1}\eta$, so that $a_0 + h \in \psi^{-1}\xi$. By Taylor's formula with Lagrange's remainder term

$$\begin{aligned} {}^{*}f(a_0 + h) &= {}^{*}f(a_0) + \frac{{}^{*}f'(a_0)}{1!}h + \frac{{}^{*}f''(a_0)}{2!}h^2 + \cdots \\ &+ \frac{{}^{*}f^{(n)}(a_0)}{n!}h^n + \frac{{}^{*}f^{(n+1)}(a_0 + \theta h)}{(n+1)!}h^{n+1}, \end{aligned}$$

where $0 \leqq \theta \leqq 1$. Now on the right hand side of this identity ${}^{*}f^{(k)}(a_0) = f^{(k)}(a_0)$ since a_0 is standard. Also, since $f(x)$ is infinitely differentiable, $f^{(n+1)}(x)$, and hence ${}^{*}f^{(n+1)}(x)$, is bounded by a standard real number in any standard closed subinterval of $\langle a, b\rangle$ and hence, is bounded by a standard number B in the monad of a_0. Hence

$$(5.5) \qquad \left| \frac{{}^{*}f^{(n+1)}(a_0 + \theta h)}{(n+1)!} h^{n+1} \right| \leqq B \left| h \right|^{n+1}.$$

Let ν be any standard positive number less than ν_1. Then (5.5) together with the fact that $v(\eta) = {}^{0}(\log_{\rho}|h|) \geqq \nu_1$ shows that

$$\left| \frac{{}^{*}f^{(n+1)}(a_0 + \theta h)}{(n+1)!} h^{n+1} \right| < \rho^{(n+1)\nu}.$$

Hence

$$\left| {}^{*}f(a_0 + h) - \sum_{k=0}^{n} \frac{f^{(k)}(a_0)}{k!} h^k \right| < \rho^{(n+1)\nu},$$

and so

$$v\left({}^{\rho}f(\xi) - \sum_{k=0}^{n} \frac{f^{(k)}(a_0)}{k!}\eta^k\right) \geqq (n+1)v.$$

This shows that

$${}^{\rho}f(\xi) = \lim_{n\to\infty} \sum_{k=0}^{n} \frac{f^{(k)}(a_0)}{k!}\eta^k,$$

proving (5.4).

The identity (5.3) is of interest in itself since it provides a natural justification of Laugwitz' definition within a more comprehensive framework. Beyond that, by relating Laugwitz' theory to that wider framework, we are able to make use of the resources of Nonstandard Analysis in order to provide satisfactory answers to several problems which were left open by Laugwitz. We shall turn to this task in our next section.

6. The intermediate value theorem in L. In view of (5.3), the function ${}^{\rho}f(x)$, with values restricted to ${}^{\rho}L$, behaves in exactly the same way as the function ${}^{L}f(x)$ on a corresponding interval in L. Consider the real valued function $f(x)$ which is defined in the interval $-1 < x < 1$ by

$$f(x) = \begin{cases} e^{-1/|x|} & \text{for } x \neq 0, \\ 0 & \text{for } x = 0. \end{cases} \tag{6.1}$$

Then $f(x)$ is infinitely differentiable in the entire interval of definition, including $x = 0$. At that point $f^{(n)}(x) = 0$ for $n = 0, 1, 2, \cdots$.

Let $x_1 = 0$, $x_2 = \frac{1}{2}$. Then ${}^{\rho}f(x_1) = f(x_1) = 0$, ${}^{\rho}f(x_2) = f(x_2) = 1/\sqrt{e^2}$. If ${}^{\rho}f(x)$ satisfied the intermediate value theorem, there would exist a $\xi \in {}^{\rho}L$, $0 < \xi < \frac{1}{2}$, such that ${}^{\rho}f(\xi) = \bar{\rho}$. We shall show that there is no such ξ.

Suppose first that ξ is infinitely close to 0,

$$\xi = a_0\bar{\rho}^{\nu_0} + a_1\bar{\rho}^{\nu_1} + a_2\bar{\rho}^{\nu_2} + \cdots,\ 0 < \nu_0 < \nu_1 < \cdots \to \infty,\ a_1 > 0.$$

Then, by (5.4)

$${}^{\rho}f(\xi) = f(0) + \frac{f'(0)}{1!}\xi + \frac{f''(0)}{2!}\xi^2 + \cdots = 0$$

so ${}^{\rho}f(\xi)$ cannot be equal to $\bar{\rho}$.

Suppose next that ξ is not infinitely close to 0. Then $\xi = a_0 + \eta$, where $0 < a_0 \leqq \frac{1}{2}$, $v(\eta) > 0$ and so, by (5.4), ${}^{\rho}f(\xi) = f(a_0) + \xi$, where $v(\xi) > 0$. This shows that ${}^{\rho}f(\xi)$ is infinitely close to $f(a_0)$, which is a standard real number different from 0, and so again ${}^{\rho}f(\xi)$ cannot be equal to $\bar{\rho}$, which is infinitesimal.

By contrast, if $f(x)$ is continuous in an interval $a < x < b$ and if the definition

(5.1) is effective in an interval $x_1 \leqq x \leqq x_2$ where $a < x_1 < x_2 < b$, $x_1, x_2 \in {}^\rho R$, then the intermediate value theorem does apply in ${}^\rho R$. That is to say, under these conditions:

6.2. THEOREM. *If* ${}^\rho f(x_1) < \eta < {}^\rho f(x_2)$ for $\eta \in {}^\rho R$, *then there exists a* $\xi \in {}^\rho R$, $x_1 < \xi < x_2$ *such that* ${}^\rho f(\xi) = \eta$.

To see this, we only have to choose elements of $^*R$, x_1', x_2', η' such that $\psi x_1' = x_1$, $\psi x_2' = x_2$, $\psi \eta' = \eta$. Then $^*f(x_1') < \eta' < {}^*f(x_2')$ and so, by the intermediate value theorem for $^*f(x)$ there exists a $\xi \in {}^*R$, $x_1' < \xi' < x_2'$ such that $^*f(\xi') = \eta'$. Putting $\xi = \psi\xi'$ we then have ${}^\rho f(\xi) = \psi({}^*f(\xi')) = \psi\eta' = \eta$. This shows that the intermediate value theorem is satisfied in this case.

For the remainder of this section, it will be our main purpose to show that the intermediate value theorem holds also in ${}^\rho L$ for functions ${}^\rho f(x)$ which are obtained from infinitely differentiable functions $f(x)$ in R—and hence, holds also in L for the corresponding functions ${}^L f(x)$—subject to rather mild restrictions, as follows.

6.3. THEOREM. *Let* $f(x)$ *be a real-valued function which is defined and infinitely differentiable for* $a < x < b$, $a, b \in R$ *and let* ${}^\rho f(x)$ *be defined by* 5.1. *Suppose that for every* $x \in R$, $a < x < b$, *there is a positive integer* n *such that* $f^{(n)}(x) \neq 0$. *For any* $x_1, x_2 \in {}^\rho L$, $a < x_1 < x_2 < b$, *let* a_0 *and* b_0 *be the uniquely defined elements of* R *which are infinitely close to* x_1 *and* x_2 *respectively, i.e.,*

$$x_1 = a_0 + a_1\bar{\rho}^{\nu_1} + a_2\bar{\rho}^{\nu_2} + \cdots, \quad 0 < \nu_1 < \nu_2 < \cdots \to \infty,$$

$$x_2 = b_0 + b_1\bar{\rho}^{\mu_1} + b_2\bar{\rho}^{\mu_2} + \cdots, \quad 0 < \mu_1 < \mu_2 < \cdots \to \infty,$$

and suppose that $a < a_0 \leqq b_0 < b$. *Let* η *be an element of* ${}^\rho L$ *such that* ${}^\rho f(x_1) < \eta < {}^\rho f(x_2)$.

Then there exists a $\xi \in {}^\rho L$, $x_1 < \xi < x_2$, *such that* ${}^\rho f(\xi) = \eta$.

Proof. Comparing 6.3 with 6.2 (which applies to the situation described in 6.3) we see that we only have to show that the $\xi \in {}^\rho R$ mentioned in 6.2 belongs more particularly to ${}^\rho L$. Choosing x_1', x_2', η' as in the proof of 6.2 such that $\psi x_1' = x_1$, $\psi x_2' = x_2$, $\psi\eta = \eta$ we have, for some $\xi' \in {}^*R$, $x_1' < \xi' < x_2'$, $^*f(\xi') = \eta'$ and hence ${}^\rho f(\xi) = \eta$ where $\xi = \psi\xi'$. Now $x_1' < \xi' < x_2'$ implies that ξ' is finite and has a standard part, ${}^0\xi' = d_0$, where $a < a_0 \leqq d_0 \leqq b_0 < b$. At the same time, η must be of the form $e_0 + e_1\bar{\rho}^{\lambda_1} + e_2\bar{\rho}^{\lambda_2} + \cdots, 0 < \lambda_1 < \lambda_2 < \cdots \to \infty$ since it is in ${}^\rho L$ and finite. Hence, ${}^0\eta' = e_0$ and $f(d_0) = e_0$.

Suppose now that $f'(d_0) \neq 0$. Then the inversion theorem is applicable. It follows that there exist $h_1 > 0$, $h_2 > 0$, $k_1 > 0$, $k_2 > 0$, such that $f(x)$ is a one-to-one mapping of the interval D defined by $d_0 - h_1 < x < d_0 + h_2$ on the interval E defined by $e_0 - k_1 < y < e_0 + k_2$ such that the inverse function $g(y) = f^{-1}(y)$ is infinitely differentiable on E. Passing to $^*R$, we find that the infinitely differentiable function $^*f(x)$ maps $^*D$ in one-to-one correspondence on $^*E$ such that $^*g(y)$ is the inverse of this mapping and is infinitely differentiable as well (in the sense of $^*R$). Hence, $^*f(\xi') = \eta'$ entails $^*g(\eta') = \xi'$ and so

$$\xi = \psi\xi' = \psi(^*g(\eta')) = {}^\rho g(\eta) \in {}^\rho L,$$

proving our assertion in this case.

Dropping the restriction that $f'(d_0) \neq 0$ (but not excluding this case) we put $F(x) = f(x) - f(d_0)$ and define $H(x)$ for $a < x < b$ by

$$H(x) = \begin{cases} \dfrac{F(x)}{x - d_0} & \text{for } x \neq d_0 \\ F'(d_0) & \text{for } x = d_0. \end{cases}$$

Also, on the assumption of our theorem, there is an $n \geqq 0$ such that

$$F(d_0) = F'(d_0) = \cdots = F^{(n)}(d_0) = 0, \qquad F^{(n+1)}(d_0) \neq 0.$$

Then $F(x) = H(x)\,(x - d_0)$, and so

$$F'(x) = H(x) + H'(x)\,(x - d_0) \qquad \text{for} \quad x \neq d_0 \tag{6.4}$$

and, more generally,

$$F^{(k)}(x) = kH^{(k-1)}(x) + H^{(k)}(x)\,(x - d_0) \tag{6.5}$$

for $k = 1, 2, \cdots$, $x \neq d_0$, $a < x < b$, by induction.

We now wish to show that, for $x \neq d_0$,

$$H^{(\lambda)}(x) = \frac{F^{(\lambda+1)}(d_0)}{\lambda + 1} + \frac{F^{(\lambda+2)}(d_0)}{1!(\lambda + 2)}(x - d_0) + \cdots + \frac{F^{(\lambda+m)}(d_0)}{(m-1)!\,(\lambda + m)}(x - d_0)^{m-1} + G_{\lambda,m}(x - d_0)^m \tag{6.6}$$

provided $\lambda \geqq 1$, $m \geqq 1$, where $G_{\lambda,m}$ is a linear combination with rational coefficient of values of $F^{(\lambda+m+1)}(x)$ taken at points x' in the interval $\langle d_0, x \rangle$.

For $\lambda = 1$, we have the Taylor expansion for $F'(x)$

$$F'(x) = F'(d_0) + \frac{F''(d_0)}{1!}(x - d_0) + \cdots + \frac{F^{(1+m)}(d_0)}{m!}(x - d_0)^m + \frac{F^{(2+m)}(d_0 + \theta_1(x - d_0))}{(m+1)!}(x - d_0)^{m+1}$$

where $0 \leqq \theta_1 \leqq 1$, while the Taylor expansion for $F(x)$ yields

$$H(x) = F'(d_0) + \frac{F''(d_0)}{2!}(x - d_0) + \cdots + \frac{F^{(1+m)}(d_0)}{(m+1)!}(x - d_0)^m + \frac{F^{(2+m)}(d_0 + \theta_0(x - d_0))}{(m+2)!}(x - d_0)^{m+1}, \tag{6.7}$$

where $0 \leqq \theta_0 \leqq 1$. Hence, from (6.4),

$$\begin{aligned} H'(x) &= \frac{F'(x) - H(x)}{(x - d_0)} \\ &= \frac{F''(d_0)}{2} + \cdots + \frac{F^{(1+m)}(d_0)}{(m-1)!\,(1+m)}(x - d_0)^{m-1} + G_{1,m}(x - d_0)^m, \end{aligned}$$

where

$$G_{1,m} = \frac{F^{(2+m)}(d_0 + \theta_1(x - d_0))}{(m+1)!} - \frac{F^{(2+m)}(d_0 + \theta_0(x - d_0))}{(m+2)!},$$

as required.

Suppose now that the assertion has been proved up to some $\lambda \geqq 1$, for all $m \geqq 1$. In order to prove the corresponding formula for $\lambda + 1$, we write down the appropriate Taylor expansion for $F^{(\lambda+1)}(x)$, so

$$\begin{aligned} F^{(\lambda+1)}(x) = F^{(\lambda+1)}(d_0) &+ \frac{F^{(\lambda+2)}(d_0)}{1!}(x - d_0) + \cdots + \frac{F^{(\lambda+m+1)}(d_0)}{m!}(x - d_0)^m \\ &+ \frac{F^{(\lambda+m+2)}(d_0 + \theta_{\lambda+1}(x - d_0))}{(m+1)!}(x - d_0)^{m+1}, \end{aligned}$$

where $0 \leqq \theta_{\lambda+1} \leqq 1$. Then, by (6.5) and (6.6) (with $m + 1$ for m)

$$\begin{aligned} H^{(\lambda+1)}(x) &= \frac{F^{(\lambda+1)}(x) - (\lambda+1)H^{(\lambda)}(x)}{x - d_0} \\ &= \frac{F^{(\lambda+2)}(d_0)}{\lambda+2} + \cdots + \frac{F^{(\lambda+m+1)}(d_0)}{(m-1)!\,(\lambda+1+m)}(x - d_0)^{m-1} + G_{\lambda+1,m}(x - d_0)^m, \end{aligned}$$

where

$$G_{\lambda+1,m} = \frac{F^{(\lambda+m+2)}(d_0 + \theta_{\lambda+1}(x - d_0))}{(m+1)!} - (\lambda+1)G_{\lambda,m+1}.$$

This proves (6.6). We now obtain immediately, for $\lambda \geqq 1$

$$\lim_{x \to d_0} H^{(\lambda)}(x) = \frac{F^{(\lambda+1)}(d_0)}{\lambda+1} \tag{6.8}$$

and this is true also for $\lambda = 0$, by (6.7). On the other hand, we may calculate the derivatives of $H(x)$ *at* d_0. We have, from (6.7), which is valid also for $m = 0$,

$$\begin{aligned} H'(d_0) = \lim_{x \to d_0} \frac{H(x) - H(d_0)}{x - d_0} = \lim_{x \to d_0} \frac{H(x) - F'(d_0)}{x - d_0} &= \lim_{x \to d_0} \frac{F''(d_0 + \theta_0(x - d_0))}{2} \\ &= \frac{F''(d_0)}{2}, \end{aligned}$$

where θ_0 may depend on x. Thus, $H(x)$ has a continuous derivative everywhere in its interval of definition.

Suppose now that we have proved that $H(x)$ has continuous derivatives up to order $\lambda \geqq 1$ in the entire interval of definition $a < x < b$ such that $H^{(\lambda)}(d_0) = F^{(\lambda+1)}(d_0)/(\lambda+1)$. We then make use of (6.6) for $m = 2$, where we observe that $G_{\lambda,2}$ (as a linear combination with fixed rational coefficients of values of $F^{(\lambda+3)}$ for arguments x' in the interval $\langle d_0, x \rangle$) remains bounded in the neighborhood of x_0. Hence, for such x,

$$H^{(\lambda)}(x) = \frac{F^{(\lambda+1)}(d_0)}{\lambda+1} + \frac{F^{(\lambda+2)}(d_0)}{\lambda+2}(x-d_0) + O(x-d_0)^2$$

and so

$$\lim_{x \to d_0} \frac{H^{(\lambda)}(x) - H^{(\lambda)}(d_0)}{x - d_0} = \frac{F^{(\lambda+2)}(d_0)}{\lambda+2} = \lim_{x \to d_0} H^{(\lambda+1)}(x).$$

This shows that $H(x)$ possesses continuous derivatives of all orders in its interval of definition. In particular

$$H^{(\lambda)}(d_0) = F^{(\lambda+1)}(d_0)/(\lambda+1), \qquad \lambda = 0, 1, \cdots \tag{6.9}$$

and so

$$H(d_0) = H'(d_0) = \cdots = H^{(n-1)}(d_0) = 0, \qquad H^{(n)}(d_0) = \frac{F^{(n+1)}(d_0)}{n+1} \neq 0.$$

If $n > 0$, we may repeat our procedure, obtaining from $H(x)$ a function $H_1(x)$ in the same way in which we obtained $H(x)$ from $F(x)$. Thus, putting

$$H_1(x) = \begin{cases} H(x)/(x-d_0) = F(x)/(x-d_0)^2 & \text{for } x \neq d_0 \\ H'(d_0) & \text{for } x = d_0, \end{cases}$$

we find that $H_1(x)$ is infinitely differentiable for $a < x < b$ and

$$H_1(d_0) = H_1'(d_0) = \cdots = H_1^{(n-2)}(d_0) = 0, \ H_1^{(n-1)}(d_0) = \frac{F^{(n+1)}(d_0)}{n(n+1)} \neq 0.$$

Continuing in this way, we obtain after $n-1$ more steps the function

$$H_n(x) = \begin{cases} H_{n-1}(x)/(x-d_0) = F(x)/(x-d_0)^{n+1} = \dfrac{f(x) - f(d_0)}{(x-d_0)^{n+1}}, & \text{for } x \neq d_0 \\ \dfrac{f^{(n+1)}(d_0)}{(n+1)!} \neq 0, \ \text{for } x = d_0 \end{cases}$$

which is infinitely differentiable for $a < x < b$.

Suppose first that n is even, $n+1$ is odd. Then the function $w^{1/(n+1)}$, with the determination that $(H_n(d_0))^{1/(n+1)}$ be real, is infinitely differentiable in the neighborhood of $H_n(d_0)$ and so the function $P(x) = (H_n(x))^{1/(n+1)}$ is infinitely differentiable in some neighborhood of $x = d_0$, for $d_0 - h < x < d_0 + h$, say. The function

$$Q(x) = P(x)(x - d_0) = (f(x) - f(d_0))^{1/(n+1)}$$

therefore is also infinitely differentiable in the same interval, and

$$Q'(x) = P(x) + P'(x)(x - d_0), \quad Q'(d_0) = P(d_0) \neq 0.$$

Passing to $*R$ we see that, for $x = \xi'$,

$$*Q(\xi') = (f(\xi') - f(d_0))^{1/(n+1)} = (\eta' - e_0)^{1/(n+1)}.$$

Hence

$${}^{\rho}Q(\xi) = \psi((\eta' - e_0)^{1/(n+1)}) = (\eta - e_0)^{1/(n+1)} \in {}^{\rho}L$$

since L, and hence ${}^{\rho}L$, is real closed (see section 3 above). Hence, applying the inversion theorem to $Q(x)$ at $x = d_0$ (exactly as we applied it earlier to $f(x)$ on the assumption that $f'(d_0) \neq 0$), and letting $S(y)$ be the inverse function to $Q(x)$ at $x = d_0$, $y = 0$, we obtain

$$\xi = \psi\xi' = \psi(*S((\eta' - e_0)^{1/(n+1)})) = {}^{\rho}S(\eta - e_0)^{1/(n+1)} \in {}^{\rho}L.$$

This disposes of the case that n is even.

Suppose finally that n is odd, $n + 1$ is even. We may assume without loss of generality that $H_n(d_0) = f^{(n+1)}(d_0)/(n + 1)!$ is positive, otherwise we consider $-f(x)$ in place of $f(x)$. Then $f(x) - f(d_0) = H_n(x)(x - d_0)^{n+1}$ must be positive, for $x \neq d_0$ in a sufficiently small neighborhood of d_0. Introducing $P(x) = (H_n(x))^{1/(n+1)}$ with the postiive determination for $(H_n(d_0))^{1/(n+1)}$, and $Q(x) = P(x)(x - d_0)$ we then have again that $P(x)$ and $Q(x)$ are infinitely differentiable in a neighborhood of $x = d_0$, and that $Q'(d_0) = P(d_0) \neq 0$. Also,

$$*Q(\xi') = \pm (f(\xi') - f(d_0))^{1/(n+1)} = \pm (\eta' - e_0)^{1/(n+1)}$$

leading to ${}^{\rho}Q(\xi) = \pm (\eta - e_0)^{1/(n+1)}$, which is again an element of ${}^{\rho}L$. Finally, introducing the inverse function $S(y)$ of $Q(x)$ with $S(0) = d_0$, as before, we have

$$\xi = \psi\xi' = \psi(S(\pm (\eta' - e_0)^{1/(n+1)})) = {}^{\rho}S(\pm (\eta - e_0)^{1/(n+1)}) \in {}^{\rho}L.$$

The proof of Theorem 6.3 is now complete.

Although the counterexample given at the beginning of the section shows that some restriction on the behavior of the derivatives of $f(x)$ is required, the particular set of conditions given in 6.3, is not strictly necessary. Thus, if $f(x) = \text{const.}$, then the conditions of the theorem are not satisfied but its conclusion is, trivially. Nevertheless, 6.3 includes a large number of interesting cases, e.g., all non-constant real analytic functions $f(x)$.

7. The mean value theorem. Suppose the function $f(x)$ is continuously differentiable for $a < x < b$. Let D be the set of points $\xi \in {}^\rho R$ such that ξ is infinitely close to a point a_0 *in the interior* of that interval, $a < a_0 < b$. Then $f'(x)$ is bounded in any closed subinterval $a' \leqq x \leqq b'$ of $a < x < b$ and so $f(x)$ satisfies a Lipschitz condition in that interval. Taking $a' < a_0 < b'$ we see, therefore, that the definition 5.1 is effective. We claim, moreover, that the resulting function ${}^\rho f(x)$ is continuous, in the sense of the metric of ${}^\rho R$, at all points $\xi \in D$.

To see this, let $\{\xi_n\}$ be a sequence of elements of D such that $\lim_{n\to\infty} \xi_n = \xi$ and choose a number ξ' and a sequence $\{\xi_n'\}$ in ${}^*R$ such that $\psi\xi' = \xi$, $\psi\xi_n' = \xi_n$, $n=0, 1, 2, \cdots$. Since $\lim \xi_n = \xi$, there exist $a', b' \in R$ such that $a' < \xi < b'$, $a' < \xi' < b'$, $a' < \xi_n < b'$, $a' < \xi_n' < b'$, $n = 0, 1, 2, \cdots$. Let m be a bound for $f'(x)$ in the closed interval $a' \leqq x \leqq b'$ within R and, hence within ${}^*R$. Then

$$^*f(\xi_n') - {}^*f(\xi') = {}^*f'(\xi' + \theta(\xi_n' - \xi'))(\xi_n' - \xi')$$

for some $0 \leqq \theta \leqq 1$ and, hence

$$|{}^\rho f(\xi_n) - {}^\rho f(\xi)| \leqq m |\xi_n - \xi|.$$

This, together with $\lim \xi_n = \xi$ implies $\lim {}^\rho f(\xi_n) = {}^\rho f(\xi)$, proving our assertion.

Suppose next that $f(x)$ is twice continuously differentiable for $a < x < b$. In this case, we propose to show that ${}^\rho f(x)$ is differentiable in D (in the sense of the metric of ${}^\rho R$) and that on D,

$$\frac{d}{dx}\,{}^\rho f(x) = {}^\rho(f'(x)). \tag{7.1}$$

For ξ in D and $\eta \neq 0$ such that $\xi + \eta$ also belongs to D, choose ξ' and η' for which $\psi\xi' = \xi$, $\psi\eta' = \eta$. Then there exists a $\theta' \in {}^*R$, $0 \leqq \theta \leqq 1$, such that

$$\frac{f(\xi' + \eta') - f(\xi')}{\eta'} = f'(\xi' + \theta'\eta'). \tag{7.2}$$

Applying the mapping ψ to (7.2), we obtain

$$\frac{{}^\rho f(\xi + \eta) - {}^\rho f(\xi)}{\eta} = ({}^\rho f'(x))_{x=\xi+\theta\eta}, \tag{7.3}$$

where $\theta = \psi\theta'$. Now let η tend to zero. Then the right hand side of (7.3) tends to $({}^\rho f'(x))_{x=\xi}$ since ${}^\rho f'(x)$ is continuous on D. This proves (7.1).

In particular, if $f(x)$ is infinitely differentiable, then ${}^\rho f(\xi)$ and $({}^\rho f'(x))_{x=\xi}$ belong to ${}^\rho L$ for $\xi \in {}^\rho L$. It follows that in that case ${}^\rho f(x)$ is defined and infinitely differentiable in $D \cap {}^\rho L$. Accordingly, the same is true of the function

$${}^L f(x) \text{ for } x = a_0 + a_1 t^{\nu_1} + a_2 t^{\nu_2} + \cdots,\ 0 < \nu_1 < \nu_2 < \cdots \to \infty, \qquad a < a_0 < b.$$

(7.3), in combination with (7.1) shows also that the mean value theorem holds in

${}^{\rho}R$ under the stated conditions, more particularly for infinitely differentiable $f(x)$. However, here again we may show that the mean value theorem breaks down, for certain infinitely differentiable functions, both in ${}^{\rho}L$ and in L. The function (6.1) which provided an example for the breakdown of the intermediate value theorem, will do also for the present issue as can be seen by considering the ratio of increments $(f(\xi_2)-f(\xi_1))/(\xi_2-\xi_1)$ for $\xi_2 = 1/(2+\bar{\rho})$, $\xi_1 = -\frac{1}{2}$. There is no $\xi_3 \in {}^{\rho}L$ in the closed interval from ξ_1 to ξ_2 such that $({}^{\rho}f(x))'$ is equal to that ratio for $x = \xi_3$.

We shall prove, as our principal positive result in this area:

7.4. THEOREM. *Let $f(x)$ be a real valued function which is defined and infinitely differentiable for $a < x < b$*; $a, b \in R$ and let *${}^{\rho}f(x)$ be defined by* 5.1. *Suppose that for every x, $a < x < b$, there is an integer $n \geqq 2$ such that $f^{(n)}(x) \neq 0$. For any $x_1, x_2 \in {}^{\rho}L$, $a < x_1 < x_2 < b$, let a_0 and b_0 be the uniquely defined elements of R which are infinitely close to x_1 and x_2 respectively, i.e.*,

$$x_1 = a_0 + a_1\bar{\rho}^{\nu_1} + a_2\bar{\rho}^{\nu_2} + \cdots,\quad 0 < \nu_1 < \nu_2 < \cdots \to \infty,$$

$$x_2 = b_0 + b_1\bar{\rho}^{\mu_1} + b_2\bar{\rho}^{\mu_2} + \cdots,\quad 0 < \mu_1 < \mu_2 < \cdots \to \infty,$$

and suppose that $a < a_0 \leqq b_0 < b$.

Then there exists a $\xi \in {}^{\rho}L$, $x_1 \leqq \xi \leqq x_2$ such that

$$\frac{{}^{\rho}f(x_2) - {}^{\rho}f(x_1)}{x_2 - x_1} = \left(\frac{d}{dx}\,{}^{\rho}f(x)\right)_{x=\xi}$$

Here again there is an exactly corresponding theorem for the function ${}^{L}f(x)$ in L. The conditions of the theorem are not necessary since they exclude all functions of constant slope, for which the conclusion of the theorem is obviously correct. However, the theorem is nevertheless of a rather general character, including, for example, all other real analytic functions.

For the proof, we require the following auxiliary consideration.

Assume that the conditions of (7.4) are satisfied and choose $x_1' \in \psi^{-1}x_1$, $x_2' \in \psi^{-1}x_2$. Then we claim that ${}^*f'(x)$ attains its maximum in the interval $x_1' \leqq x \leqq x_2'$ either at x_1' or at x_2' or at some *standard* point x_0, $x_1' \leqq x_0 \leqq x_2'$ (but, possibly, also elsewhere).

Suppose that ${}^*f(x)$ attains its maximum neither at x_1' nor at x_2' but at a point $\bar{x}, x_1' < \bar{x} < x_2'$. Let x_0 be the standard part of $\bar{x}$, $x_0 = {}^0\bar{x}$. Suppose that $x_0 < x_1$ (so that $x_0 = a_0$). Depending on whether the first non-vanishing derivative of $f'(x)$ at x_0 is either positive or negative, ${}^*f'(x)$ will be either strictly increasing or strictly decreasing in some interval $x_0 \leqq x \leqq x_0 + h$, where h is standard and positive. Since $\bar{x}$ and x_1' belong to that interval, the latter case would involve ${}^*f'(x_1') > {}^*f'(\bar{x})$, contrary to our choice of $\bar{x}$. Accordingly, we have to assume that ${}^*f'(x)$ increases strictly for $x_0 \leqq x \leqq x_0 + h$. Now x_2' cannot belong to that interval for then ${}^*f'(x_2') > {}^*f'(\bar{x})$, which is again impossible. It follows that $\bar{x} < x_0 + h < x_2'$ and

$^*f'(x_0 + h) > f(\bar{x})$ which is also impossible. We therefore conclude that $x_0 \geqq x_1$ and, by similar reasoning, $x_0 \leqq x_2$. The discussion of the variation of $^*f'(x)$ in the neighborhood of x_0 shows that we must exclude both $x_0 < \bar{x}$ and $x_0 > \bar{x}$ and so we conclude that $\bar{x} = x_0$.

Thus we have shown than $^*f'(x)$ attains its maximum at x_1' or at x_2' or at some standard point $x_1' < x_0 < x_2'$ (although several of these cases may occur simultaneously). Accordingly $^*f'(x)$ attains its maximum in the interval $x_1' \leqq x \leqq x_2'$ in all cases at a point ζ_1' such that $\psi\zeta_2' = \zeta_2 \in {}^\rho L$. By a similar argument (or, by applying the conclusion to $-f(x)$) we find that $^*f'(x)$ attains its minimum in the same interval at a point ζ_2 such that $\psi\zeta_1' = \zeta_1 \in {}^\rho L$. Passing from $^*R$ to $^\rho R$, we then conclude that $^\rho(f'(x))$ attains its maximum and minimum in $x_1 \leqq x \leqq x_2$ at points ζ_1 and ζ_2 which belong to $^\rho L$.

By a well-known formula of the Integral Calculus, which can be transferred from R to $^*R$, we have

$$^*f'(\zeta_2')(x_2' - x_1') \leqq \int_{x_1'}^{x_2'} {}^*f'(x)dx \leqq {}^*f'(\zeta_1')(x_2' - x_1'),$$

i.e.,

$$^*f'(\zeta_2')(x_2' - x_1') \leqq {}^*f(x_2') - {}^*f(x_1') \leqq {}^*f'(\zeta_1')(x_2' - x_1').$$

We apply the mapping ψ to this chain of inequalities and obtain

$$^\rho(f'(\zeta_2))(x_2 - x_1) \leqq {}^\rho f(x_2) - {}^\rho f(x_1) \leqq {}^\rho(f'(\zeta_1))(x_2 - x_1)$$

or, equivalently

$$^\rho(f'(\zeta_2)) \leqq \frac{^\rho f(x_2) - {}^\rho f(x_1)}{x_2 - x_1} \leqq {}^\rho(f'(\zeta_1)).$$

But this shows that $({}^\rho f(x_2) - {}^\rho f(x_1))/(x_2 - x_1)$ is intermediate between $^\rho(f'(\zeta_2))$ and $^\rho(f'(\zeta_1))$. Hence, by the intermediate value Theorem 6.3, there exists a $\xi \in {}^\rho L$ which belongs to the closed interval with endpoint ζ_1 and ζ_2 and, hence, belongs to $x_1 \leqq x \leqq x_2$, such that

$$\frac{^\rho f(x_2) - {}^\rho f(x_1)}{x_2 - x_1} = {}^\rho(f'(\xi))$$

and this is the same as

$$\frac{^\rho f(x_2) - {}^\rho f(x_1)}{x_2 - x_1} = \left(\frac{d}{dx}\,{}^\rho f(x)\right)_{x=\xi},$$

by (7.1). The proof of 7.4 is now complete.

8. Conclusion. As Laugwitz points out, his method for extending a function $f(x)$ from R to L applies only in the infinitesimal neighborhood of a point at which $f(x)$ is infinitely differentiable and hence, possesses at least a formal Taylor series. However, if we consider points in the infinitesimal neighborhood of the endpoints of the interval of definition $a < x < b$ of $f(x)$, e.g., $x = a + a_1 t^{\nu_1} + a_2 t^{\nu_2} + \cdots$, $0 < \nu_1 < \nu_2 < \cdots$, $a_1 > 0$, then we can still define ${}^L f$ at x, provided f possesses an asymptotic expansion at $x = a$ as x tends to a from the right. Similarly, if $f(x)$ is defined in a semi-infinite interval, for $x > a$ say, we can define ${}^L f(x)$ for positive infinite x provided $f(x)$ possesses an asymptotic expansion as $x \to +\infty$. In all of these cases, ${}^L f(x)$ can again be obtained "automatically" as $\Phi^{-1}({}^\rho f(\Phi x))$ (see (5.3) above). However, ${}^\rho f(x)$ exists also in many cases where no asymptotic expansion as a generalized power series is available, e.g., ${}^\rho \log x$ exists for positive infinitesimal and infinite x. Conversely, we may investigate the asymptotic expansion of a function $f(x)$ at a singular point (even when it contains logarithmic terms, as happens frequently in the theory of ordinary differential equations) by means of the function ${}^\rho f(x)$. Going further in the direction of concrete applications, ${}^\rho R$ also provides us with a convenient framework for the discussion of matched asymptotic expansions for the solution of singular perturbation problems.

This research was supported in part by the National Science Foundation Grant No. GP-18728.

References

1. N. Jacobson, Lectures in Abstract Algebra, vol. III, Princeton-Toronto-New-York-London, 1964.

2. D. Laugwitz, Eine nichtarchimedische Erweiterung angeordneter Körper, Math. Nachr., 37 (1968) 225–236.

3. T. Levi-Civita, Sugli infiniti ed infinitesimi attuali quali elementi analitici (1892–1893), Opere matematiche, vol. 1, Bologna 1954, pp. 1–39.

4. W. A. J. Luxemburg, What is Nonstandard Analysis, California Institute of Technology, 1968, to be published.

5. A. Ostrowski, Untersuchungen zur arithmetischen Theorie der Körper, Math. Z., 39 (1935) 269–404.

6. A. Robinson, Non-standard Analysis, Studies in Logic and the Foundations of Mathematics, Amsterdam, 1966.

7. B. L. v. der Waerden, Algebra, 5th edition, Berlin-Heidelberg-New York, 1966/1967.

8. O. Zariski and P. Samuel, Commutative Algebra, vol. 2, Princeton-Toronto-New-York-London, 1960.

Nonstandard Points on Algebraic Curves*

Abraham Robinson

Department of Mathematics, Yale University, New Haven, Connecticut 06520

Communicated by A. E. Ross

Received June 30, 1971

Let Γ be an algebraic curve which is given by an equation $f(x, y) = 0$, $f(x, y) \in k[x, y]$ where k is an algebraic number field and $f(x, y)$ is irreducible. Suppose that there exists an $*\Gamma$ a nonstandard point $(\xi, \eta) \in *k \times *k$. Then $k(\xi, \eta)$ is (isomorphic to) the algebraic function field of Γ and, at the same time, is a subfield of $*k$. Correlating the divisors of the function field $k(\xi, \eta)$ and of the number field $*k$, we develop an analogue of the Artin–Whaples theory of the product formula. This leads to one of Siegel's basic inequalities for rational points on algebraic curves.

1. Introduction

We shall be concerned with the arithmetic of algebraic curves from the point of view of nonstandard analysis. Let k be an algebraic number field and let Γ be an irreducible algebraic curve defined by $f(x, y) = 0$ where $f(x, y) \in k[x, y]$. Our starting point is the observation that if $A = (\xi, \eta)$ is a nonstandard zero of $f(x, y)$, i.e., if ξ and η are not both standard then A is by necessity a generic point of Γ. Thus, $k(\xi, \eta)$ may be regarded as the function field of Γ. If, in addition, ξ and η belong to the field $*k$, which is the nonstandard extension of k in the given framework, then $k(\xi, \eta) \subset *k$. We may then investigate the connection between the *internal* valuations of $*k$ as a (nonstandard) number field and the valuations (divisors) of $k(\xi, \eta)$ as a function field.

The reader will perceive that the work of the present paper is related to the analysis which led A. Weil to his decomposition theorem and C. L. Segel to the result that the number of integers on a curve of positive genus is finite. We shall not attempt a detailed correlation but shall show

* Research supported in part by the National Science Foundation Grant No. GP-29218.

how one of the decisive inequalities of Siegel's theory follows naturally from our results.

The fact that a valuation of $^*k$ may induce a valuation of $k(\xi, \eta)/k$ was pointed out in Ref. [4]. In that paper it was also shown that all valuations of $k(\xi, \eta)/k$ may be obtained in this way from valuations of $^*k$. The same conclusion will be obtained here as a corollary of more powerful results. A general introduction to nonstandard arithmetic is to be found in Ref. [3].

2. Nonstandard Extensions of Algebraic Curves

Let k be an algebraic number field which is embedded in some definite way in the field of complex numbers, C. Let $^*C$ be a higher order nonstandard extension of C. For the purposes of this paper, $^*C$ need not be an enlargement, and it would be sufficient to take it as a countable ultrapower of C with respect to an arbitrary free ultrafilter.

Let $f(x, y) = 0$, $f(x, y) \in k[x, y]$, be the equation of an irreducible algebraic curve Γ, and let $A = (\xi, \eta)$ be a nonstandard point on $^*\Gamma$. Thus, $f(\xi, \eta) = 0$ where ξ and η are not both standard. We claim that A is a generic point of $^*\Gamma$ (or, if we follow the usage customary in algebraic geometry, of Γ). Indeed, suppose that there is a $g(x, y) \in k[x, y]$ which does not vanish identically on Γ such that $g(\xi, \eta) = 0$. Then ξ and η are algebraic and hence, belong to C. This is contrary to our assumption.

Suppose now that Γ has an infinite number of points with coordinates in k. Let us range them, without repetition, in a sequence $\{A_n\}$. For any infinite natural number ω, A_ω has coordinates ξ, η which belong to $^*k$. Then ξ and η cannot both be standard, for if they were then $A_n = (\xi, \eta)$ for some finite n. Then $A_n = A_\omega$. This is impossible since $A_n \neq A_m$ for all standard $n \neq m$ and, hence, also for all $n \neq m$ in $^*N$ (where $^*N$ is the extension of the natural numbers, N). Accordingly, $A_\omega = (\xi, \eta)$ is a generic point for Γ over k.

Conversely, suppose that Γ has a generic point A with coordinates ξ, η in $^*k$. Then ξ and η cannot be both algebraic (and one of them can be algebraic only if Γ is a straight line $x = \text{const}$ or $y = \text{const}$). Then we claim that Γ has an infinite number of points with coordinates in k. For suppose that we have shown that Γ has the points $A_0, \ldots, A_n$ with coordinates in k (including the possibility that the set $\{A_0, \ldots, A_n\}$ may be empty). Then the assertion that "there is a point A on Γ which is different from $A_0, A_1, \ldots, A_n$" must be true for k since it is true for $^*k$ (for $A = A_\omega$). This proves our assertion.

3. Internal Valuations

Let v be any *internal* valuation of $^*k$. Then v may be given by a nonstandard prime ideal, or by a standard prime ideal, or it may be an "Archimedean" valuation of $^*k$, i.e., the extension to $^*k$ of an Archimedean valuation of k. There are no nonstandard internal valuations of $^*k$ which are Archimedean since the set of these valuations on k is finite.

Suppose first that $v = v_P$ is given by a nonstandard prime ideal P. Let $\alpha \in k^\times$ where $k^\times$ is, as usual, the multiplicative group of k. Since the number of prime divisors of (α) (i.e., prime ideals which have nonzero exponent in the prime power representation of (α)) is finite it does not increase or passing from k to $^*k$. Hence, $v_P(\alpha) = 0$, v_P is trivial on k.

Suppose next that $v = v_P$ is given by a standard prime ideal P. In this case, v_P is not trivial on k. We derive another valuation v_P' which is trivial on k in the following way.

Let o_P' be the set of all $\alpha \in {}^*k$ such that $v_P(\alpha)$ is not negative infinite, together with 0. The inclusion of 0 is implicit if we regard $v_P(o) = \infty$ as positive infinite, and we shall do so from now on. Then o_P' is a valuation ring which includes o_P, the valuation ring of v_P and which also includes k. Let v_P' be the corresponding valuation. Then v_P' is trivial on k. Its valuation ideal, m_P', consists of all $\alpha \in {}^*k$ for which $v_P(\alpha)$ is positive infinite, including 0. Finally, while the value group for v_P is $^*Z$, the nonstandard extension of the rational integers Z, the value group for v_P' is $^*Z/Z$, which will be denoted also by Z_∞. More precisely, if ψ is the canonical mapping $^*Z \to Z_\infty$ then we may write $v_P'(\alpha) = \psi v_P(\alpha)$ for all $\alpha \in {}^*k^\times$.

Now let $|\, x \,|_P$ be one of the Archimedean valuations of $^*k$, where P is merely a symbol for the particular valuation under consideration. For any $\alpha \in {}^*k$, $|\, \alpha \,|_P \in {}^*R$, where $^*R$ is the nonstandard extension of the field of real numbers, R. Let $R_\infty = {}^*R/R_0$ be the additive quotient group of $^*R$ with respect to R_0 where R_0 is the group of finite numbers in $^*R$, and let ψ be the canonical mapping from $^*R$ to R_∞.

Then ψ induces an order in R_∞ since R_0 is isolated in $^*R$. We shall regard R_∞ as an ordered group with this particular order.

We propose to show that the definition

$$v_P'(\alpha) = -\psi \ln |\, \alpha \,|_P \qquad \text{for} \quad \alpha \in {}^*k^\times \tag{3.1}$$

defines a *non-Archimedean* valuation of $^*k$ with value group R_∞.

Indeed, for any $\alpha, \beta \in {}^*k^\times$ we have

$$\begin{aligned} v_P'(\alpha\beta) &= -\psi \ln |\, \alpha\beta \,|_P = -\psi(\ln |\, \alpha \,|_P + \ln |\, \beta \,|_P) \\ &= -\psi \ln |\, \alpha \,|_P - \psi \ln |\, \beta \,|_P = v_P'(\alpha) + v_P'(\beta). \end{aligned}$$

Accordingly, it only remains for us to show that $v_P'(\alpha + \beta) \geqslant \min(v_P'(\alpha), v_P'(\beta))$, provided $\alpha + \beta \neq 0$.

We may assume without loss of generality that $v_P'(\alpha) \leqslant v_P'(\beta)$, i.e., that $\ln | \alpha |_P \geqslant \ln | \beta |_P$. Then $v_P'(\alpha + \beta) = v_P'(\alpha) + v_P'(1 + \beta/\alpha) = -\psi \ln | \alpha |_P - \psi \ln | 1 + \beta/\alpha |_P$. Accordingly, we only have to verify that $-\psi \ln | 1 + \beta/\alpha |_P$ is not strictly negative in R_∞, i.e., that $\ln | 1 + \beta/\alpha |_P$ is not positive infinite. By assumption, $\ln | \alpha |_P \geqslant \ln | \beta |_P$, i.e., $| \beta/\alpha |_P \leqslant 1$. This shows that $\ln | 1 + \beta/\alpha |_P \leqslant \ln 2$ and proves our assertion.

In order to see that all elements of R_∞ are values of elements of $^*k$ it suffices to consider the positive integers α (standard or nonstandard). For such α, $| \alpha |_P = \alpha$ and $\ln | \alpha + 1 |_P - \ln | \alpha |_P = \ln | 1 + 1/\alpha |_P$ is always finite. It follows that each coset of R_0 in $^*R$ contains an infinite number of $v_P'(\alpha)$, provided it contains *negative* numbers. In order to find values $v_p'(\beta)$ in any coset R_0 of $^*R$ which contains only positive numbers, we then have to put only $\beta = \alpha^{-1}$, where α is a positive integer.

The valuation ring of v_P' is given by the condition

$$o_P' = \{\alpha \in {}^*k \mid v_P'(\alpha) \geqslant 0\} = \{\alpha \in {}^*k \mid \psi \ln | \alpha |_P \leqslant 0\}$$

which applies also to $\alpha = 0$, if we put $\ln 0 = -\infty$. Thus, in order that $\alpha \in o_P'$, $\ln | \alpha |_P$ must be either finite or negative infinite, i.e., $| \alpha |_P$ must be finite. Similarly, the valuation ideal of v_P', m_P', consists of all $\alpha \in {}^*k$ such that $| \alpha |_P$ is infinitesimal, i.e., m_P' is the monad of zero in that valuation. Since $k^\times \subset o_P' - m_P'$ it follows that v_P' is trivial on k.

4. Induced Valuations

Now let K be a finitely generated field of transcendence degree 1 (i.e., a function field of one variable) over k such that K is a subfield of $^*k$. For example, K may have been obtained as the function field $k(\xi, \eta)$ of a curve Γ, as in Section 2. We note that K is by necessity an *external* subfield of $^*k$. Notice that k is algebraically closed in K.

Let v be any non-Archimedean valuation of $^*k/k$ where v may be, to begin with, internal or external. Let o_v and m_v be the valuation ring and the valuation ideal of v, respectively. Since v is trivial on k, it induces a valuation of K/k, which will be denoted by V. The valuation ring and ideal of V are given by $O_V = o_v \cap K$ and $M_V = m_v \cap K$, respectively. V is nontrivial on K if and only if $K - o_v \neq \varnothing$, i.e., if and only if $v(\alpha)$ is positive for some $\alpha \in K^\times$. In that case, V must be a discrete valuation of rank 1 in K, i.e., a prime divisor of K.

Let us now apply these considerations to the valuations introduced in the preceding section. If $v = v_P$ where P is a nonstandard prime ideal in $^*k$, then v_P induces a nontrivial valuation of K provided $v_P(\alpha) > 0$ for some $\alpha \in K^\times$. In particular, if α_0 is such that $v_P(\alpha_0)$ is positive and as small as possible then the valuation induced by v_P in K is normalized with value group Z by putting

$$V_P(\alpha) = v_P(\alpha)/v_P(\alpha_0) \qquad \text{for} \quad \alpha \in K^\times. \tag{4.1}$$

Suppose next, that $v = v_P$ is given by a standard prime ideal. In order that v_P' be nontrivial on K it is necessary and sufficient that $v_P(\alpha)$ be infinite for at least one element $\alpha \in K^\times$. If this is the case then the group of values $v_P'(\alpha)$, $\alpha \in K^\times$, which is a subgroup of Z_∞ is isomorphic to Z. Choosing $\alpha_0 \in K^\times$ so that $v_P'(\alpha_0)$ is positive and as small as possible, we normalize the valuation induced in K by putting

$$V_P(\alpha) = v_P'(\alpha)/v_P'(\alpha_0) \qquad \text{for} \quad \alpha \in K^\times. \tag{4.2}$$

In this formula, the ratio on the right side is to be interpreted as the uniquely determined standard integer n such that $v_P'(\alpha) = n \cdot v_P'(\alpha_0)$. Since $Z_\infty = {}^*Z/Z$ we may write also $v_P(\alpha) = n \cdot v_P(\alpha_0) + r$, where r is a finite integer. Hence

$$V_P(\alpha) = {}^0(v_P(\alpha)/v_P(\alpha_0)) \qquad \text{for} \quad \alpha \in K^\times. \tag{4.3}$$

Suppose finally that $v = v_P$ where P stands for an Archimedean valuation. Here, the valuation induced in K is nontrivial if and only if $\ln |\alpha|_P$ is infinite for some $\alpha \in K^\times$, and this will be the case if and only if $|\alpha|_P$ is infinite for some $\alpha \in K^\times$. On this assumption, the group of values $v_P'(\alpha)$, $\alpha \in K^\times$, which is now a subgroup of R_∞ is again isomorphic to Z. We choose $\alpha_0 \in K^\times$ so that $v_P'(\alpha_0)$ is positive and as small as possible. Then the induced valuation is normalized by defining $V_P(\alpha)$ by (4.2). Here again, the right side should be interpreted as the uniquely defined standard integer n such that $v_P'(\alpha) = nv_P'(\alpha_0)$. By (3.1), this is equivalent to

$$\psi(\ln |\alpha|_P - n \ln |\alpha_0|_P) = 0,$$

i.e., $\ln |\alpha|_P - n \ln |\alpha_0|_P = r$, where r is finite. Hence, $r/\ln |\alpha_0|_P$ is infinitesimal and so, this time,

$$V_P(\alpha) = {}^0(\ln |\alpha|_P/\ln |\alpha_0|_P) \qquad \text{for} \quad \alpha \in K^\times. \tag{4.4}$$

However, (4.2) is more precise than either (4.3) or (4.4) and we note for future reference that

$$V_P(\alpha) = v_P(\alpha)/v_P(\alpha_0) - r/v_P(\alpha_0), \tag{4.5}$$

where r is a finite integer if P is a standard prime ideal, and

$$V_P(\alpha) = \ln | \alpha |_P / \ln | \alpha_0 |_P - r/\ln | \alpha_0 |_P , \qquad r \text{ finite}, \qquad (4.6)$$

if P stands for an Archimedean valuation.

It is quite possible that two different valuations v_P in $^*k$ induce the same nontrivial valuation in K.

5. The Case of a Rational Function Field

Suppose now that K is (isomorphic to) the field of rational functions of one variable over k. Such a K is obtained by adjoining to k any element of $^*k - k$. We show that in this case all valuations of K/k are induced by internal valuations of $^*k$.

Let $K = k(\omega)$ where $\omega \in {}^*k - k$. Let χ be the canonical mapping from $k(x)$ onto $k(\omega)$ where x is an indeterminate. χ maps x on ω and reduces to the identity on k. Let $p(x) \in k[x]$ be of positive degree and irreducible in $k[x]$. $p(x)$ defines a valuation w_p of $k(x)$ according to the following rule. For any $q(x) \in k[x]$, $q(x) \not\equiv 0$, let $q(x) = (p(x))^m\, s(x)$ where $s(x)$ is prime to $p(x)$. Then $w_p(q(x)) = m$. w_p is extended to $k(x)$ in the natural way. There is a corresponding valuation of $k(\omega)$ which is given by $W_p = w_p\chi^{-1}$, or $W_p(q(\omega)) = W_p(q(x)) = m$. In particular, $W_p(p(\omega)) = 1$.

Put $\alpha = p(\omega)$ and let $(\alpha) = \prod P_j^{\nu_j}$ be the prime ideal factorization of the ideal (α) in $^*k$, where the number of P_j which are effectively present may be finite, or infinite but starfinite. Suppose, in the first place that (α) has a nonstandard prime ideal $P = P_j$ in its numerator (i.e., with $\nu_j > 0$). We claim that V_p coincides with W_p.

Let

$$p(x) = a_0 + a_1 x + \cdots + a_n x^n, \qquad a_n \neq 0, \quad n \geqslant 1. \qquad (5.1)$$

We show that $v_P(\beta) \geqslant 0$ for all $\beta \in k[\omega]$. Indeed, taking first $\beta = \omega$ suppose that $v_P(\omega) < 0$. But $v_P(a_j) = 0$, $j = 0, 1, \ldots, n$ for $a_j \neq 0$. Hence, $v_P(p(\omega)) = \min_j v_P(a_j\omega^j) < 0$, contrary to assumption. Hence, $v_P(\beta) \geqslant 0$ for $\beta = \omega$ and the same is then true for any

$$q(\omega) = b_0 + b_1\omega + \cdots + b_l\omega^l \in k[\omega].$$

We claim next that $v_P(s(\omega)) = 0$, where $q(x) = (p(x))^m\, s(x)$ as above. For we then have an identity $p(x)\, q(x) + s(x)\, h(x) = 1$ with $g(x)$, $h(x)$ in $k[x]$.

Hence,

$$0 = v_P(1) = v_P(p(\omega)\, g(\omega) + s(\omega)\, h(\omega))$$
$$\geqslant \min(v_P(p(\omega)\, g(\omega)), v_P(s(\omega)\, h(\omega))).$$

But $v_P(p(\omega)\, g(\omega)) = v_P(p(\omega)) + v_P(g(\omega)) \geqslant 1$, so $v_P(s(\omega)) = v_P(h(\omega)) = 0$. Hence, $v_P(q(\omega)) = mv_P(p(\omega)) = mv_P(\alpha)$. This shows that $v_P(y)$ is positive and as small as possible for $y = \alpha$. Accordingly, we may choose $\alpha = p(\omega)$ as the α_0 of (4.1). This yields, for $\beta = q(\omega)$,

$$V_P(\beta) = v_P(\beta)/v_P(\alpha_0) = m = W_p(\beta).$$

Thus, V_P and W_p coincide on $k[\omega]$ and, hence, on $k(\omega)$.

Now suppose that some standard ideal $P = P_j$ occurs in the prime power representation of (α) with positive infinite exponent $\nu = \nu_j$, $\alpha = p(\omega)$, where $p(x)$ is given by (5.1). Then $v_P(\omega)$ cannot be negative infinite because $v_P(a_j)$ is finite for $a_j \neq 0$, and so $v_P(\alpha) = \min(v_P(a_j\omega^j))$ would be negative infinite. Hence, $v_P(q(\omega))$ is finite, or positive infinite, for all $q(x) \in k[x]$. Also, since

$$0 = v_P(1) = v_P(p(\omega)\, g(\omega) + s(\omega)\, h(\omega))$$
$$\geqslant \min(\nu + v_P(g(\omega)), v_P(s(\omega)) + v_P(h(\omega))),$$

we conclude that $v_P(s(\omega))$ is finite. Thus, if $\beta = q(\omega) = (p(\omega))^m\, s(\omega)$ then $v_P(\beta) = m\nu + r$ where r is a finite integer and $\nu = v_P(\alpha)$. Hence, using the canonical map ${}^*Z \xrightarrow{\psi} Z_\infty$, we obtain $v_P'(\beta) = mv_P'(\alpha)$. This shows that $v_P'(\alpha)$ is positive and as small as possible, so that α may be chosen as the α_0 of (4.2). Then

$$V_P(\beta) = v_P'(\beta)/v_P'(\alpha_0) = m = W_p(\beta). \tag{5.2}$$

This shows that, in this case also, V_P and W_P coincide on $k[\omega]$ and, hence, on $k(\omega)$.

Finally, suppose that $\alpha = p(\omega)$—where $p(x)$ is given by (5.1)—is infinitesimal for the Archimedean valuation P of ${}^*k$. In that case, $|\, \omega\, |_P$ must be finite, for if $|\, \omega\, |_P$ were infinite then

$$|\, a_n\, |_P = |\, p(\omega)/\omega^n - a_0/\omega^n - \cdots - a_{n-1}/\omega\, |_P$$

would be infinitesimal. And that is impossible since a_n is standard and different from 0. Hence, $|\, \omega\, |_P$ is finite and the same applies to all $|\, q(\omega)|_P$ where $q(x) \in k[x]$. Referring again to the identity $p(x)\, g(x) + s(x)\, h(x) = 1$ we then have $|\, p(\omega)\, g(\omega)|_P \simeq 0$ and so $|\, s(\omega)\, h(\omega)|_P \simeq 1$. Thus, both

$| s(\omega)|_P$ and $| h(\omega)|_P$ must be finite but not infinitesimal. Hence, for $\beta = q(\omega) \neq 0$,

$$\ln | \beta |_P = \ln |(p(\omega)^m s(\omega)|_P = m \ln | \alpha |_P + r,$$

where $r = | s(\omega)|_P$ is finite. The canonical map $^*R \xrightarrow{\psi} R_\infty$ now yields (compare (3.1))

$$v_P'(\beta) = mv_P'(\alpha)$$

so that $v_P'(\alpha)$, which if positive is, subject to this condition, as small as possible. Putting $\alpha = \alpha_0$ in (4.2), we obtain again Eq. (5.2).

We shall now show that, for a given prime $p(x) \in k[x]$ at least one of the three cases considered above must occur. That is to say, either (α) has a nonstandard prime divisor in its numerator, or it has a standard prime divisor with infinite exponent in its numerator, or $| \alpha |_P$ is infinitesimal for some Archimedean P, or several of these possibilities occur simultaneously.

Let μ be a sequence which enumerates the prime ideals of k without repetitions. Then $^*\mu$ enumerates the (internal) prime ideals of $^*k$ without repetitions and if $^*\mu = \{P_0, P_1, ..., P_n, ...\}$ then P_n is standard for finite n and nonstandard for infinite n.

Now suppose that $(\alpha) = (p(\omega))$ does not contain any nonstandard prime ideals in its numerator. In that case, the set S_α of prime ideals in the numerator of (α) must be finite (in the absolute sense). For S_α is, at any rate, an internal set which is starfinite. If S_α is empty, there is nothing more to be proved. If not, there exists a first natural number n such that $P_j \notin S_\alpha$ for all $j \geqslant n$. Then $n \geqslant 1$ and $P_{n-1} \in S_\alpha$. It follows that n must be finite, so S_α must be finite. We conclude that if (α) has neither a nonstandard prime ideal in its numerator nor any infinite power of a standard prime ideal then the numerator of (α) is finite and we may write $(\alpha) = J_1/J_2$ where J_1 is just the product $\prod P_j^{\nu_j}$ taken over the P_j for which $\nu_j > 0$ and is, therefore, standard and J_2 is some entire ideal in $^*k$. Since the class number of k is finite, it is preserved on passing to $^*k$. Thus, there exists a standard ideal J_3 such that J_1J_3 is principal, and, hence, J_2J_3 is principal as well. Moreover, $J_1J_3 = (\gamma)$ where γ is a standard integer, $\gamma \in k$, while $J_2J_3 = (\delta)$ where δ may be a standard or nonstandard integer. Then$\alpha = \gamma/\epsilon\delta$ where ϵ is a unit in $^*k$, or $\alpha = \gamma/\zeta$, $\zeta = \gamma/\alpha$.

We now claim that $| \alpha |_P$ must be infinitesimal for at least one of the Archimedean valuations of $^*k$. For if not then $| \zeta |_P = | \gamma/\alpha |_P$ is finite for all the Archimedean valuations of $^*k$. In other words, if $\zeta = \zeta^{(1)}, ..., \zeta^{(l)}$ are the conjugates of ζ over the rationals, then $| \zeta^{(1)} |, ..., | \zeta^{(l)} |$ are all finite, where the absolute values are taken for the specified embedding of $*k$ in $*C$. But this implies that the fundamental symmetric functions of $\zeta^{(1)},\dots,\zeta^{(l)}$, to be denoted by $s_1,\dots,s_l$, are all finite integers, and since they are rational they must, therefore, be standard rational. Also, ζ is a root of the polynomial $z^l - s_1 z^{l-1} + \cdots + (-1)^l s_l$, and since the coefficients of this polynomial are all standard, ζ must itself be standard, $\zeta \in k$. It then follows that $\alpha = p(\omega)$ is standard, which is contrary to the assumption that the degree of $p(x)$ is positive. Accordingly, we have shown that indeed at least one of the three cases considered earlier must occur and hence that W_p is induced by at least one internal valuation of $*k$.

However, with one exception, all valuations of $k(\omega)/k = K/k$ are given by prime polynomials $p(x) \in k[x]$, the exception being the valuation "at infinity." But this valuation is given by the polynomial $p(x') = x'$ if we put $\omega' = 1/\omega$ and consider K as the field $k(\omega')$. Accordingly, we have shown that *all* valuations of K/k are induced by internal valuations of $*k$.

6. The Size of a Divisor

We now return to the general case of a function field $K \subset *k$ of transcendence degree 1 over k.

Let R_+ be the multiplicative group of positive real numbers, so that $*R_+$ is its nonstandard extension. Within $*R_+$ let R_l be the set of numbers which are finite but not infinitesimal. We may call R_l the set of *logarithmically finite* numbers, since they are just the positive real numbers whose natural logarithm is finite. R_l is a multiplicative subgroup of $*R_+$. Let $R_{+\infty}$ be the quotient group $*R_+/R_l$, and let θ be the canonical mapping from $*R_+$ onto $R_{+\infty}$.

Let λ be the mapping $y = \ln x$ from $*R_+$ onto $*R$. This is an isomorphism from the multiplicative group $*R_+$ onto the additive group $*R$. It induces an isomorphism λ_0 from $R_{+\infty}$ onto R_∞, and this isomorphism also sets up an order in $R_{+\infty}$.

Let D be the set of divisors of K as a function field. For any $\alpha \in K^\times$ we denote the divisor of α as a function by $[\alpha]$. We shall define a mapping σ from D into $R_{+\infty}$.

Let A be any entire divisor in D, $A = \prod A_j^{\omega_j}$ where the A_j are prime divisors and the ω_j are positive. Using the approximation theorem, it is not difficult to produce two elements α and β of $K^\times$ such that each A_j appears in the numerators of $[\alpha]$ and $[\beta]$ exactly in the ω_jth power and such that $[\alpha]$ and $[\beta]$ have no other factors in common in their numerators. It follows that A is the greatest common divisor (g.c.d.) of the numerators of $[\alpha]$ and $[\beta]$. Now let (α) and (β) be the (entire or fractional) ideals generated by α and β, respectively, in $^*k$, in agreement with our earlier notation. Then the g.c.d. of the numerators of (α) and (β) is an entire internal ideal in $^*k$, which will be denoted by $J_{\alpha,\beta}$. Let $J_{\alpha,\beta} = \prod P_j^{\nu_j}$ be the prime power factorization of $J_{\alpha,\beta}$. For each P_j, let NP_j be the norm of P_j relative to the (extended) field of rational numbers, $^*Q$. Put $| J_{\alpha,\beta} | = \prod (NP_j)^{-\nu_j}$. Multiply $| J_{\alpha,\beta} |$ by the product of $\max(| \alpha |_p , | \beta |_P)$ for all the Archimedean valuations P of $^*k$ for which both $| \alpha |_P$ and $| \beta |_P$ are infinitesimal. Call the result $s(\alpha, \beta)$. Thus,

$$s(\alpha, \beta) = | J_{\alpha,\beta} | \prod \max(| \alpha |_P , | \beta |_P), \tag{6.1}$$

where the product on the right hand side is taken over the indicated primes. We define $\sigma(A)$ by

$$\sigma(A) = \theta(s(\alpha, \beta)) = \theta(| J_{\alpha,\beta} |) \prod \theta(\max(| \alpha |_P , | \beta |_P)), \tag{6.2}$$

where θ is the canonical mapping $^*R_+ \to R_{+\infty}$ as before, and we call $\sigma(A)$ the *size* of A.

We have to show that $\sigma(A)$ is independent of our particular choice of α and β. In other words, we have to show that if α' and β' are two other elements of $K^\times$ which satisfy the same condition as α and β relative to A (see above) then $\theta(s(\alpha', \beta')) = \theta(s(\alpha, \beta))$, or, equivalently, the ratio $s(\alpha, \beta)/s(\alpha', \beta')$ is logarithmically finite. Bearing in mind the symmetry between (α, β) and (α', β'), we may break down our problem as follows:

(i) Let P be an Archimedean valuation for which both $| \alpha |_P$ and $| \beta |_P$ are infinitesimal. Then $| \alpha' |_P$ and $| \beta' |_P$ are infinitesimal and the ratio $\max(| \alpha' |_P , | \beta' |_P)/\max(| \alpha |_P , | \beta |_P)$ is logarithmically finite.

(ii) The ideal $J_{\alpha',\beta'}/J_{\alpha,\beta}$ is standard.

Indeed, if (i) is satisfied, as well as the assertion obtained from interchanging (α, β) and (α', β') then the product on the right side of (6.2), $\prod \theta(\max(| \alpha |_P , | \beta |_P)$, is equal to the corresponding product for (α', β'). Also, if $J_{\alpha',\beta'}/J_{\alpha,\beta}$ is standard then $| J_{\alpha',\beta'} |/| J_{\alpha,\beta} |$ is logarithmically finite and so $\theta(| J_{\alpha',\beta'} |) = \theta(| J_{\alpha,\beta} |)$. This implies $\theta(s(\alpha, \beta)) = \theta(s(\alpha', \beta'))$.

Let

$$[\alpha] = \prod Q_j^{\varkappa_j}, \qquad [\beta] = \prod R_j^{\lambda_j}, \qquad [\alpha'] = \prod S_j^{\mu_j}, \qquad [\beta'] = \prod T_j^{\rho_j}, \tag{6.3}$$

where we include only the divisors whose exponents are different from 0. Suppose that P is an Archimedean valuation for which both $| \alpha |_P$ and $| \beta |_P$ are infinitesimal. Then P induces in K/k a valuation Q for which both α and β have positive order. In other words Q (regarded as a divisor) coincides with one of the A_j and also with one of the Q_j and one of the R_j. We may suppose without loss of generality that $j = 0$ for all these cases. Then $\omega_0 = \chi_0 = \lambda_0 > 0$. But A divides also $[\alpha']$ and $[\beta']$, more precisely $\omega_0 = \mu_0 = \rho_0$ if we assume that Q coincides with S_0 and T_0. Choosing α_0 as in Section 4, we then have (see (4.6)).

$$\begin{aligned} \omega_0 \ln | \alpha_0 |_P &= \ln | \alpha |_P - r = \ln | \beta |_P - r_2 \\ &= \ln | \alpha' |_P - r_3 = \ln | \beta' |_P - r_4 , \end{aligned} \tag{6.4}$$

where $\ln | \alpha_0 |_P$, $\ln | \alpha |_P$, $\ln | \beta |_P$, $\ln | \alpha' |_P$, $\ln | \beta' |_P$ are negative infinite and r_1, r_2, r_3, r_4 are finite. Hence,

$$\begin{aligned} | \alpha |_P &= e^{r_1}(| \alpha_0 |_P)^{\omega_0}, \qquad & | \beta |_P &= e^{r_2}(| \alpha_0 |_P)^{\omega_0}, \\ | \alpha' |_P &= e^{r_3}(| \alpha_0 |_P)^{\omega_0}, \qquad & | \beta' |_P &= e^{r_4}(| \alpha_0 |_P)^{\omega_0}. \end{aligned} \tag{6.5}$$

Since $| \alpha_0 |_P$ is infinitesimal, we conclude that

$$\max(| \alpha |_P , | \beta |_P) = c_1(| \alpha_0 |_P)^{\omega_0}, \qquad \max(| \alpha' |_P , | \beta' |_P) = c_2(| \alpha_0 |_P)^{\omega_0},$$

where c_1 and c_2 are logarithmically finite. This proves (i).

In order to confirm (ii), we consider the ideals (α), (β), (α'), (β') in $^*k$. Suppose that the nonstandard prime ideal P appears in the factorizations of (α) and (β) with positive exponents ν_1 and ν_2. Then P induces a valuation Q in K which, as a divisor, must coincide with one of the Q_j (see (6.3)), and also with one of the R_j, $Q = Q_0 = R_0$, say. But then Q coincides also with one of the A_j, e.g. $Q = A_0$. Choosing α_0 as in Section 4 (for the case of a nonstandard prime ideal), we obtain from (4.1),

$$\omega_0 = \chi_0 = \lambda_0 = \nu_1 / v_P(\alpha_0) = \nu_2 / v_P(\alpha_0), \tag{6.6}$$

where $v_P(\alpha_0)$ is a standard or nonstandard positive integer. Hence $\nu_1 = \nu_2$. At the same time, it follows that Q appears also among the prime divisors of the numerators of $[\alpha']$ and $[\beta']$, e.g. $Q = S_0 = T_0$, where $\mu_0 = \rho_0 = \omega_0$. Then $\omega_0 v_P(\alpha_2) = \nu_3 = \nu_4$, where $\nu_3 = \nu_4$ is the exponent of P in the factorizations of (α') and (β'). Hence, $\nu_1 = \nu_2 = \nu_3 = \nu_4$.

We have shown that every nonstandard prime ideal which divides $J_{\alpha,\beta}$, divides also $J_{\alpha',\beta'}$, and in exactly the same power. Let H be the set of prime ideals which divide $J_{\alpha,\beta}$ in a higher power than $J_{\alpha',\beta'}$. Then H is internal and, as we have just seen, it can contain only standard prime ideals. H may be empty. If not, we claim that, at any rate, H must be finite.

To see this, let $^*\mu = \{P_0, P_1, P_2, \ldots\}$ be the sequence of all internal prime ideals of $^*k$ introduced in Section 5. Let n be the smallest integer such that $P_j \notin H$ for all $j \geqslant n$. If n were infinite then P_{n-1} would belong to H. This is impossible since P_{n-1} is nonstandard. Hence, n is finite and H must be finite also.

Now let $P \in H$. Then P divides the numerators of both (α) and (β), in the ν_1th and ν_2th powers, say, $\nu_1 > 0$, $\nu_2 > 0$. If ν_1 and ν_2 are finite then P occurs only in a finite power in the numerator of $J_{\alpha,\beta}/J_{\alpha',\beta'}$. Suppose then that at least one of the ν_1, ν_2 is infinite. Then P induces a valuation Q in K which, as a divisor, appears in the numerators of both $[\alpha]$ and $[\beta]$. Accordingly, we may assume that $Q = A_0 = Q_0 = R_0$ (see (6.3)), where $\omega_0 = \chi_0 = \lambda_0$. It follows that, e.g. $Q = S_0 = T_0$ where $\omega_0 = \mu_0 = \rho_0$. Choose α_0 as in Section 4, so that (4.5) applies and

$$\omega_0 v_P(\alpha_0) = \nu_1 - r_1 = \nu_2 - r_2 = \nu_3 - r_3 = \nu_4 - r_4,$$

where r_1, r_2, r_3, r_4 are finite and where P divides (α') and (β') exactly in the ν_3th and ν_4th powers, respectively. It follows that

$$\min(\nu_1, \nu_2) - \min(\nu_3, \nu_4) = r$$

where r is finite, i.e., the difference between the exponents of P in $J_{\alpha,\beta}$ and in $J_{\alpha',\beta'}$ is finite. We conclude that the numerator of $J_{\alpha,\beta}/J_{\alpha',\beta'}$ is a standard ideal and, by symmetry, that its denominator is standard also. This proves (ii).

Suppose in particular that $\alpha = \beta$, so that A is just the numerator of $[\alpha]$. By (6.2),

$$\sigma(A) = \theta(s(\alpha, \alpha)) = \theta\left(\prod \min(|\alpha|_P, 1)\right), \tag{6.7}$$

where P now ranges over all internal valuations, Archimedean and non-Archimedean and where, for non-Archimedean P, $|\alpha|_P$ is defined by $(NP)^{-\nu}$ if ν is the exponent of P in the prime ideal factorization of P. On the other hand, the *height* of α as a number in $^*k$ is, by definition (compare Ref. [2])

$$H(\alpha) = \prod \max(|\alpha|_P, 1). \tag{6.8}$$

Since the product of all $|\alpha|_P$ is 1, by the product formula, we, therefore, have $s(\alpha, \alpha) = (H(\alpha))^{-1}$ and

$$\sigma(A) = \theta((H(\alpha))^{-1}) = (\theta(H(\alpha)))^{-1}. \tag{6.9}$$

We show next that the mapping σ is multiplicative on the semigroup of entire divisors (which is the set on which σ has been defined so far). Let A and B be two entire divisors. We first choose α and β, as before, and then (again using the approximation theorem) we choose α' and α' so as to satisfy the corresponding conditions in relation to B and also such that in the notation of (6.3), no prime divisor Q_j or R_j occurs either as an S_j or as a T_j unless it divides both A and B. As a result, every prime divisor of AB appears in the numerators of $[\alpha\alpha']$ and of $[\beta\beta']$ exactly in the power in which it appears in AB. Accordingly, we have, in addition to (6.2),

$$\sigma(B) = \theta(s(\alpha'\beta')) = \theta(|\, J_{\alpha',\beta'} \,|) \prod \theta(\max(|\, \alpha' \,|_P \,, |\, \beta' \,|_P)), \tag{6.10}$$

$$\sigma(AB) = \theta(s(\alpha\alpha', \beta\beta')) = \theta(|\, J_{\alpha\alpha',\beta\beta} \,|) \prod \theta(\max(|\, \alpha\alpha' \,|_P \,, |\, \beta\beta' \,|_P)), \tag{6.11}$$

where the products on the right side are taken, each time, over the appropriate Archimedean valuations.

In order to confirm that $\sigma(AB) = \sigma(A)\, \sigma(B)$ it is, therefore, sufficient to prove the following assertions.

(iii) Let P be any Archimedean valuation. Then $\max(|\, \alpha\alpha' \,|_P \,, |\, \beta\beta' \,|_P)$ is infinitesimal if and only if at least one of $\max(|\, \alpha \,|_P \,, |\, \beta \,|_P)$ and $\max(|\, \alpha' \,|_P \,, |\, \beta' \,|_P)$ is infinitesimal. And if this is the case then the ratio $\max(|\, \alpha \,|_P \,, |\, \beta \,|_P) \max(|\, \alpha' \,|_P \,, |\, \beta' \,|_P)/\max(|\, \alpha\alpha' \,|_P \,, |\, \beta\beta' \,|_P)$ is logarithmically finite.

(iv) The ideal $J_{\alpha,\beta} J_{\alpha',\beta'}/J_{\alpha\alpha',\beta\beta'}$ is standard.

The proof is parallel to that of the assertions labeled (i) and (ii) earlier in this section. We first show that if $\max(|\, \alpha \,|_P \,|, \, \beta \,|_P)$ is infinitesimal or $\max(|\, \alpha' \,|_P \,, |\, \beta' \,|_P)$ is infinitesimal then it induces a valuation (prime divisor) Q of K/k which divides A or B, respectively. But then Q divides AB and this makes $\max(|\, \alpha\alpha' \,|_P \,, |\, \beta\beta' \,|_P)$ infinitesimal. Conversely, if $\max(|\, \alpha\alpha' \,|_P \,, |\, \beta\beta' \,|_P)$ is infinitesimal, i.e., if both $|\, \alpha\alpha' \,|_P$ and $|\, \beta\beta' \,|_P$ are infinitesimal then P induces a Q which divides AB and, hence, divides at least one of the factors A or B. It follows that at least one of $\max(|\, \alpha \,|_P \,, |\, \beta \,|_P)$ and $\max(|\, \alpha' \,|_P \,, |\, \beta' \,|_P)$ is infinitesimal. Supposing that this is the situation, let $\omega_0 \geqslant 0$ and $\zeta_0 \geqslant 0$ be the powers in which Q divides A and B, respectively (so that at least one of ω_0, ζ_0 is positive). Then Q divides AB in the power $\omega_0 + \zeta_0$. Hence, by (4.6), for an appropriate α_0,

$$\omega_0 \ln |\, \alpha_0 \,|_P = \ln |\, \alpha \,|_P - r_1 = \ln |\, \beta \,|_P - r_2 \,, \tag{6.12}$$

where r_1, r_2 are finite, provided $\omega_0 > 0$. But this is still true if $\omega_0 = 0$, for in that case Q cannot, at any rate, appear in the denominators of $[\alpha]$

or of $[\beta]$ so that $\ln|\alpha|_P$ and $\ln|\beta|_P$ must be finite. Similarly,

$$\zeta_0 \ln|\alpha_0|_P = \ln|\alpha'|_P - r_3 = \ln|\beta'|_P - r_4, \tag{6.13}$$

where r_3 and r_4 are finite, and from the consideration of AB,

$$(\omega_0 + \zeta_0) \ln|\alpha_0|_P = \ln|\alpha\alpha'|_P - r_5 = \ln|\beta\beta'|_P - r_6, \tag{6.14}$$

where r_5, r_6 are finite.

Hence,

$$\max(|\alpha|_P, |\beta|_P) = c_1(|\alpha_0|_P)^{\omega_0}, \qquad \max(|\alpha'|_P, |\beta'|_P) = c_2(|\alpha_0|_P)^{\zeta_0}$$

and

$$\begin{aligned}\max(|\alpha\alpha'|_P, |\beta\beta'|_P) &= c_3(|\alpha_0|_P)^{\omega_0+\zeta_0} = c_3(|\alpha_0|_P)^{\omega_0}(|\alpha_0|_P)^{\zeta_0} \\ &= c_4 \max(|\alpha|_P, |\beta|_P) \max(|\alpha'|_P, |\beta'|_P)\end{aligned}$$

where c_1, c_2, c_3, c_4 are logarithmically finite. This proves (iii).

In order to establish (iv), we again consider first a nonstandard prime ideal P. If P divides $J_{\alpha,\beta}$ then it divides the numerators of (α) and of (β). Accordingly, P induces a valuation Q in K/k which divides $[\alpha]$ and $[\beta]$. It follows that Q cannot divide the denominators of $[\alpha']$ and $[\beta']$ and, hence, divides the numerators of $[\alpha\alpha']$ and $[\beta\beta']$. This in turn implies that P divides $(\alpha\alpha')$ and $(\beta\beta')$ and, in consequence divides $J_{\alpha\alpha',\beta\beta'}$. For the same reason, if P divides $J_{\alpha',\beta'}$ then it divides also $J_{\alpha\alpha',\beta\beta'}$.

Now suppose that P divides $J_{\alpha\alpha',\beta\beta'}$. Then P induces a valuation Q of K/k which divides the numerators of $[\alpha\alpha']$ and $[\beta\beta']$. This implies that Q divides AB and, hence, divides at least one of A, B. Assuming without loss of generality that Q divides A, we conclude that Q divides the numerators of $[\alpha]$ and of $[\beta]$ and, hence, that P divides the numerators of (α) and of (β), Q divides $J_{\alpha,\beta}$. Thus, in any case, P divides $J_{\alpha\alpha',\beta\beta'}$ if and only if P divides $J_{\alpha,\beta}J_{\alpha',\beta'}$, and, in that case, Q divides AB.

On this assumption, let $\omega_0 \geqslant 0$ and $\zeta_0 \geqslant 0$ be the powers in which Q divides A and B, respectively. Then Q divides AB in the power $\omega_0 + \zeta_0 > 0$. Hence, by (4.1), for an appropriate α_0,

$$\omega_0 v_P(\alpha_0) = \nu_1, \qquad \zeta_0 v_P(\alpha_0) = \nu_2, \qquad (\omega_0 + \zeta_0)\, v_P(\alpha_0) = \nu_1 + \nu_2, \tag{6.15}$$

where ν_1 and ν_2 are the powers in which P divides the numerators of (α) and (β) on one hand and of (α') and (β') on the other hand. It follows that P divides $J_{\alpha,\beta}$ in the ν_1th power, $J_{\alpha',\beta'}$ in the ν_2th power, and $J_{\alpha\alpha',\beta\beta'}$ in the $(\nu_1 + \nu_2)$th power. This shows that the ideal $J_0 = J_{\alpha,\beta}J_{\alpha',\beta'}/J_{\alpha\alpha',\beta\beta'}$ contains only standard prime divisors in both numerator and denominator.

Making use of the sequence $^*\mu$, as before, we see that the number of these ideals must be finite.

Let P be one of them. If P divides $J_{\alpha,\beta}$ and $J_{\alpha',\beta'}$ only in finite powers then it can only occur in a finite power in the numerator of J_0. If P divides $J_{\alpha\alpha',\beta\beta'}$ only in a finite power then it occurs only in a finite power in the denominator of J_0. If one or the other of these cases does not apply then P induces a valuation Q of K/k, which is a prime divisor. Then Q divides A or B (or both), and, in any case, divides AB. Suppose that Q divides A, then we may assume that $Q = A_0 = Q_0 = R_0$ (so that $\omega_0 = \chi_0 = \lambda_0$ is positive). Then by (4.5), for an appropriate α_0,

$$\omega_0 v_P(\alpha_0) = v_P(\alpha) - r_1 = v_P(\beta) - r_2\,,$$

where r_1 and r_2 are finite. At the same time,

$$\zeta_0 v_P(\alpha_0) = v_P(\alpha') - r_3 - v_P(\beta') - r_4\,,$$

$$(\zeta_0 + \omega_0)\, v_P(\alpha_0) = v_P(\alpha\alpha') - r_5 = v_P(\beta\beta') - r_6\,,$$

where Q divides B in the ζ_0th power, $\zeta_0 \geqslant 0$. It follows that P divides $J_{\alpha,\beta}$, $J_{\alpha',\beta'}$, $J_{\alpha\alpha',\beta\beta'}$, respectively, in powers with exponents $\omega_0 v_P(\alpha_0)$, $\zeta_0 v_P(\alpha_0)$, and $(\omega_0 + \zeta_0)\, v_P(\alpha_0)$, up to finite quantities. Thus, in this case also, P can appear in either the numerator or the denominator of J_0 only in a finite power. This complete the proof that J_0 is standard, as asserted by (iv).

It follows that if $A = \prod A_j^{\omega_j}$ is the representation of A as a product of powers of prime divisors, then

$$\sigma(A) = \prod (\sigma(A_j))^{\omega_j}. \tag{6.16}$$

We extend the definition of $\sigma(A)$ to fractional divisors, by defining that $\sigma(A)$ is given by (6.16) in this more general case also. Then σ is still multiplicative. Also, for any $\alpha \in K^\times$, $\sigma([\alpha]) = 1$ where 1 is the identity in $R_{+\infty}$. This can be seen most easily from (6.8) and (6.9). For if $[\alpha] = A/B$, where A and B have no divisors in common then $\sigma([\alpha]) = \sigma(A)/\sigma(B)$, and $\sigma(A) = H(\alpha) = H(1/\alpha) = \sigma(B)$.

When defining the size of an entire divisor A at the beginning of this section we assumed for convenience that the prime factors of A were contained in $[\alpha]$ and $[\beta]$ in exactly the same powers as in A. However, by examining the arguments leading to the proof of (i) and (ii) it is not difficult to see that this assumption can be relaxed. Thus, (6.2) remains valid, with $s(\alpha, \beta)$ defined by (6.1), as long as A is the g.c.d. of the numerators of $[\alpha]$ and $[\beta]$. Taking into account the definition of $|J_{\alpha,\beta}|$, we may then write

$$s(\alpha, \beta) = \prod \max(\min(|\alpha|_P, 1) \min(|\beta|_P, 1)), \tag{6.17}$$

where P ranges over all internal valuations of $^*k$, Archimedean or non-Archimedean.

Now let α, β be any two elements of K^x, and let A and B be the denominators of α and β, respectively. Let E be the lowest common multiple of A and B. Thus, $E = AB/(A, B)$, where (A, B) is the g.c.d. of A and B. Then (see (6.7))

$$\sigma(A) = \theta\left(\prod \min(|\alpha^{-1}|_P, 1)\right) \qquad \sigma(B) = \theta\left(\prod \min(|\beta^{-1}|_P, 1)\right)$$

and

$$\sigma((A, B)) = s(\alpha^{-1}, \beta^{-1}) = \prod \max(\min(|\alpha^{-1}|_P, 1), \min(|\beta^{-1}|_P, 1))$$

Hence,

$$\begin{aligned}\sigma(E) = \frac{\sigma(A)\,\sigma(B)}{\sigma((A, B))} &= \theta\left(\left(\prod \min(\min(|\alpha^{-1}|_P, 1), \min(|\beta^{-1}|_P, 1))\right)^{-1}\right)\\ &= \theta\left(\left(\prod \min(|\alpha^{-1}|_P, |\beta^{-1}|_P, 1)\right)^{-1}\right),\\ &= \theta\left(\prod \max(|\alpha|_P, |\beta|_P, 1)\right).\end{aligned}$$

But $\prod \max(|\alpha|_P, |\beta|_P, 1)$ is, by definition (see Ref. [2]), just the height of the triple $(\alpha, \beta, 1)$, $H(\alpha, \beta, 1)$. Hence,

$$\sigma(E) = \theta(H(\alpha, \beta, 1)). \tag{6.18}$$

We shall make use of (6.18) in Section 8.

A reader who is not familiar with the intricacies of Nonstandard Analysis may have wondered why we did not define $\sigma(A)$ for a prime divisor A simply by reference to the set of valuations P_j which induce A. However, this set may actually be external, so that the corresponding product $\prod (NP_j)^{-\nu_j}$ need not exist.

7. The Size of a Divisor and its Absolute Value

Let q be a fixed standard real number greater than 1. Let A be any divisor in K (as a function field) and let $A = \prod A_j^{\nu_j}$ be the representation of A as a product of prime divisors. Then the absolute value of A, to be denoted by $|A|$, is defined classically by

$$|A| = q^{-\Sigma \nu_j f_j}, \tag{7.1}$$

where f_j is the *degree* of A_j, i.e., the degree of the residue class field of A_j over k.

For any $\alpha \in K^{\times}$, $|[\alpha]| = 1$. Since, as we have seen $\sigma([\alpha]) = 1$ and since $\sigma([\alpha])$ is itself a kind of product of absolute values of α, we may hope that there is a direct relationship between $|A|$ and $\sigma(A)$. On the other hand, $\sigma(A)$ is not a real number (not even a nonstandard real number) and so we cannot apply the result of Artin and Whaples on the uniqueness of the product formula directly. However, we shall still make use of the ideas of their paper (Ref. [1]) in order to prove that the expected connection exists.

THEOREM 7.2. *Suppose that K is the field of rational functions of one variable over k. Thus, $K = k(\omega)$ for some $\omega \in {}^*k - k$. Then there exists a positive infinite real number λ such that for any divisor A in K,*

$$\sigma(A) = \theta(|A|^{\lambda}). \tag{7.3}$$

Proof. $\tau = \omega^{-1}$ determines a prime divisor of degree 1 in K, the prime divisor "at infinity." We denote it by A_∞. Then $|A_\infty| = q^{-1}$. Choose a positive real $a \in {}^*R_+$ such that $\sigma(A_\infty) = \theta(a)$ where θ is, as before, the canonical mapping ${}^*R_+ \to R_{+\infty}$, and put $\lambda = -\log^q a$. Since A_∞ is induced by some valuation of K, a must be negative infinite, so λ is positive infinite. Also, $a = q^{-\lambda}$ and so $\sigma(A_\infty) = \theta(q^{-\lambda}) = \theta(|A_\infty|^{\lambda})$. Thus, (7.3) is satisfied for $A = A_\infty$. We claim that (7.3) holds also for all other divisors A in K. Since σ, θ, and the absolute value, $|\ |$, are all multiplicative, it is sufficient to prove our assertion for prime A. So let A be a prime divisor in K other than A_∞. Then A is given by a prime polynomial $p(x) \in k[x]$ (compare Section 5). The absolute value of A is $|A| = q^{-n}$ where $n \geqslant 1$ is the degree of $p(x)$. Accordingly, we have to show only that $\sigma(A) = \theta(q^{-n\lambda}) = (\theta(q^{-\lambda}))^n = (\sigma(A_\infty))^n = \sigma(A_\infty{}^n)$, in other words, that $\sigma(A/A_\infty{}^n) = 1$. But this is obvious since $[p(\omega)] = A/A_\infty{}^n$ and, as we have seen, $\sigma([p(\omega)]) = 1$.

We now return to the general case of a function field K of transcendence degree 1 over k, $K \subset {}^*k$. For any $\alpha \in K^{\times}$ and for any prime divisor Q in K, we define $\sigma_Q(\alpha) \in R_{+\infty}$ as $(\sigma(Q))^{\nu}$ where ν is the order (exponential value) of α in the valuation Q, $\nu = W_Q(\alpha)$. Then σ_Q is multiplicative, $\sigma_Q(\alpha\beta) = \sigma_Q(\alpha)\,\sigma_Q(\beta)$. Also, for prime Q—or, more generally, for entire

$Q-\sigma(Q) \leqslant 1$. Hence, for $\alpha, \beta, \alpha+\beta \in K^{\times}$, $W_Q(\alpha+\beta) \geqslant \min(W_Q(\alpha), W_Q(\beta))$ implies that $\sigma_Q(\alpha+\beta) \leqslant \max(\sigma_Q(\alpha), \sigma_Q(\beta))$.

For any $\omega \in K - k$ let $K' = k(\omega)$ so that Theorem 7.2 applies to $k(\omega)$. We again write A_∞ for the divisor determined by $\tau = \omega^{-1}$. We also choose, for each divisor A in K, a $\lambda(A) \in {}^*R$ so that $\sigma(A) = \theta(|A|^{\lambda(A)})$. Observe that $\lambda(A)$ is determinate only up to the addition of a finite real number. Observe also that it is entirely possible that $\lambda(A) = 0$ since we have not yet shown that every prime divisor in K is induced by some valuation of ${}^*k$. We may exclude $\lambda(A) < 0$.

Let $\alpha_1, \alpha_2, ..., \alpha_f$ be a basis of K over $k(\omega)$. We may choose the α_j in such a way that they are entire for all divisors Q in K which do not divide A_∞ (regarded as a divisor in K). For such Q, $\sigma_Q(\alpha_j) \leqslant 1$, $j = 1,..., f$.

For a specified finite positive integer d, let S_d be the set of linear combinations $\sum_{j=1}^{f} p_j(\omega)\, \alpha_j$ where the $p_j(\omega) \in k[\omega]$ range over the polynomials of degrees not exceeding d. Then S_d is a vector space of rank $(d+1)f$ over k. For any $\xi \in S_d$, $\xi \neq 0$ and for any prime divisor Q in K, we have

$$\sigma_Q(\xi) = \sigma_Q\left(\sum_{j=1}^{f} p_j(\omega)\, \alpha_j\right) \leqslant \max(\sigma_Q(p_j(\omega)\, \sigma_Q(\alpha_j))).$$

But $\sigma_Q(p_j(\omega)) \leqslant 1$, since Q does not divide A_∞, and $\sigma_Q(\alpha_j) \leqslant 1$. Hence

$$\sigma_Q(\xi) \leqslant 1. \tag{7.4}$$

Let

$$A_\infty = \prod Q_{i\infty}^{e_i}, \tag{7.5}$$

where A_∞ is regarded as a divisor in K and (7.5) is its decomposition into prime divisors. Let Q be one of the $Q_{i\infty}$. Then we still have

$$\sigma_Q(\xi) \leqslant \max_j (\sigma_Q(p_j(\omega)\, \sigma_Q(\alpha_j))).$$

Also $W_Q(p_j(\omega)) \leqslant W_Q(\omega^d)$ and so $\sigma_Q(p_j(\omega)) = \sigma_Q(\omega^d)$. Also, since each $Q_{i\infty}$ divides A_∞, we have $W_Q(\alpha_j) \geqslant W_Q(\omega^t)$, $j = 1,..., f$, for a sufficiently large integer t, and, so $\sigma(\alpha_j) \leqslant \sigma(\omega^t)$. Hence,

$$\sigma_Q(\xi) \leqslant \sigma_Q(\omega^{t+d}), \tag{7.6}$$

where t depends on the initial choice of ω but not of d.

Now let Q be any prime divisor in K and let d_Q be its degree. Then there exists a positive integer r such that

$$rd_Q < (d+1)f \leqslant (r+1)\, d_Q\,.$$

Choose an element $\zeta \in S_d$ such that $W_Q(\zeta) \leqslant W_Q(\xi)$ for all $\xi \in S_d$. Such a ζ exists since Q is discrete and, for any $\xi \in S_d$ as above $W_Q(\xi) \geqslant \min(\omega^m \alpha_j)$, where $m = 0, 1,..., d$ and $j = 1,..., f$. Accordingly, for any $\xi \in S_d$, $W_Q(\xi/\zeta) \geqslant 0$, i.e., ξ/ζ belongs to the valuation ring O_Q of Q. Since no more than rd_Q elements of O_Q can be independent modulo Q^r there exists a $\xi_0 \in S_d$ such that ξ_0/ζ is congruent to 0 modulo Q^r, although $\xi_0 \neq 0$. Thus, $W_Q(\xi_0/\zeta) \geqslant r$. We define $| \alpha |_Q = q^{-W_Q(\alpha) d_Q}$, as usual, and we then have

$$| \xi_0/\zeta |_Q \leqslant q^{-r d_Q} \leqslant \frac{q^{d_Q}}{(r+1)\, d_Q} \leqslant q^{d_Q - (d+1) f}.$$

Hence,

$$| \xi_0 |_Q \leqslant q^{d_Q - (d+1) f} \, | \zeta |_Q .$$

Raising this inequality to the power $\lambda(Q)$ and applying θ, we obtain

$$\sigma_Q(\xi_0) \leqslant \theta(q^{(d_Q - (d+1) f) \lambda(Q)}) \, \sigma_Q(\zeta). \tag{7.7}$$

We now multiply (7.7) by (7.4) or (7.6), whichever is applicable, for $\xi = \xi_0$ and for all divisors $Q' \neq Q$ in K. This yields

$$\sigma_Q(\xi_0) \prod \sigma_{Q'}(\xi_0) \leqslant \theta(q^{(d_Q - (\alpha+1) f) \lambda(Q)}) \, \sigma_Q(\zeta) \prod \sigma_{Q'}(\omega^{t+d}). \tag{7.8}$$

Now $\sigma_Q(\zeta) \leqslant 1$ or $\sigma_Q(\zeta) \leqslant \sigma_Q(\omega^{t+d})$ accordingly as Q does not, or does, divide A_∞. In any case, we obtain from (7.8),

$$\sigma_Q(\xi_0) \prod \sigma_{Q'}(\xi_0) \leqslant \theta(q^{(d_Q - (d+1) f) \lambda(Q)}) \prod \sigma_{Q'}(\omega^{t+d}) \tag{7.9}$$

where, in the last product, Q' ranges over all prime divisors of A_∞. At the same time, the product on the left side of (7.9) is just $\sigma([\xi_0])$ and this is equal to 1, as pointed out at the end of Section 7.

For any $Q' = Q_{i\infty}$ we now introduce $\mu(Q')$ by the condition

$$\sigma_{Q'}(\omega) = \theta(q^{\mu(Q')}).$$

Here again, $\mu(Q')$ is not defined uniquely but only up to positive quantities, and we make it determinate by choice. At any rate, we may assume that $\mu(Q') \geqslant 0$. Substituting in (7.9), we then obtain

$$1 \leqslant \theta(q^{(d_Q - (\alpha+1) f) \lambda(Q) + (t+d) \Sigma \mu(Q')}). \tag{7.10}$$

Thus, the exponent on the right side of this inequality cannot be negative infinite. In other words, there exists a finite real number b such that

$$(f - d_Q)\,\lambda(Q) - t \sum \mu(Q') + d(f\lambda(Q) - \sum \mu(Q') < b. \quad (7.11)$$

We are now going to show that $\sum \mu(Q')$ must be positive infinite. Indeed, let P be any valuation of $^*k$ which induces the valuation A_∞ in $k(\omega)$. We know from Section 5 that such valuations exist. Then P induces a valuation Q' in K which extends A_∞. In other words, Q' divides A_∞ as a divisor in K. Since the contribution of any P to $\sigma(Q')$ is infinitesimal, it then follows that $\sigma_{Q'}(\omega)$ is infinite and, hence, that $\mu(Q')$ is positive infinite. It follows that $\sum \mu(Q')$ is positive infinite.

Returning to (7.11), we observe that this inequality holds for arbitrary finite positive integers d, with b depending on d. We are going to show that Either

$$f\lambda(Q) - \sum \mu(Q') \leqslant 0 \quad (7.12)$$

or, if this is not the case, then $f - 1/\lambda(Q) \sum \mu(Q')$ is positive infinitesimal and $\lambda(Q)$ is positive infinite.

Suppose, contrary to (7.12), that

$$f\lambda(Q) - \sum \mu(Q') > 0.$$

Then $\lambda(Q)$ is positive infinite. Suppose further that

$$f - (1/\lambda(Q)) \sum \mu(Q') > \epsilon,$$

where ϵ is a standard positive number. Then $\sum \mu(Q') < (f - \epsilon)\,\lambda(Q)$. It follows that the left side of (7.11) is greater than $[(f - d_Q) - t(f - \epsilon) + d\epsilon]\,\lambda(Q)$. But by making d sufficiently large, we can make the expression in the square brackets greater than 1. For such d, the left side of (7.11) is then greater than $\lambda(Q)$, i.e., it is infinite, and this is impossible.

Now let $\alpha \in k(\omega)$, $\alpha \neq 0$, and let B be an arbitrary prime divisor in $k(\omega)$ and $B = \prod B_j^{d_j}$ its decomposition into prime divisors in K. Then if f_j is the degree of inertia of B_j, and we put $|\,\alpha\,|_B$ for the absolute value of α in the valuation B in $k(\omega)$, we have $|\,\alpha\,|_{B_j} = |\,\alpha\,|_B^{d_j f_j}$ for each j. It then follows that $\sigma_{B_j}(\alpha) \equiv \theta(|\,\alpha\,|^{d_j f_j \lambda(B_j)})$. Comparing this with (7.1) for the case $Q' = Q_j = B_j$, $A_\infty = B$, $\alpha = \omega$, we see that there we have $\sigma_{Q_j}(\alpha) = \theta(q^{\mu(Q_j)}) = \theta(|\,\alpha\,|_B^{\mu(Q_j)})$. It follows that the difference $d_j f_j \lambda(Q_j) - \mu(Q_j)$ is finite and so therefore is $\sum d_j f_j \lambda(Q_j) - \sum \mu(Q_j)$. Observe that $\sum \mu(Q_j)$ is the quantity denoted earlier by $\sum \mu(Q')$. We put $\sum \mu(Q_j) = \mu_\infty$. μ_∞ is positive infinite.

Now, for general $\alpha \in k(\omega)$, $\alpha \neq 0$, we have $\sigma([\alpha]) = 1$ and, also,

$$\sigma([\alpha]) = \prod \sigma_{B_j}(\alpha) = \theta\left(\prod_B \prod_{B_j|B} | \alpha |_B^{d_j f_j \lambda(B_j)}\right),$$

where, in the first product, we have adopted the convention that B_j ranges over all divisors in K. Hence,

$$\theta\left(\prod | \alpha |_B^{\Sigma_j d_j f_j \lambda(B_j)}\right) = 1. \tag{7.13}$$

Suppose in particular that $\alpha = p(\omega)$ is an irreducible polynomial which determines the prime divisor B. Then $[\alpha] = BA_\infty^{-n}$ where n is the degree of $p(\omega)$. Hence, using the notation of (7.5) and writing g_j for the degree of inertia of Q_j, we have $| \alpha |_B = q^{-n}$, $| \alpha |_{A_\infty} = q^n$ and so, from (7.13)

$$\theta(q^{-n(\Sigma d_j f_j \lambda(B_j) - \Sigma e_j g_j \lambda(Q_j))}) = 1.$$

This shows that

$$\sum d_j f_j \lambda(B_j) - \sum e_j g_j \lambda(Q_j) \tag{7.14}$$

is finite. Moreover, the second sum in (7.14) can be replaced by μ_∞ which (see above) differs from it only be a finite amount. Thus,

$$\sum d_j f_j \lambda(B_j) - \mu_\infty \tag{7.15}$$

is finite. At the same time, the alternatives described by (7.12) apply for each $B_i = Q$, where $\sum \mu(Q') = \mu_\infty$. Suppose first that $f\lambda(B_i) - \mu_\infty \leqslant 0$ for all B_i. Then

$$\sum d_j f_j \lambda(B_j) - f\lambda(B_i) \tag{7.16}$$

is not negative infinite.

Choose l such that $\lambda(B_i)$ is a maximum for $i = l$. Using the classical relation $\sum d_j f_j = f$ we obtain from (7.16).

$$\sum_j d_j f_j \lambda(B_j) - f\lambda(B_l) = \sum_j d_j f_j (\lambda(B_j) - \lambda(B_l)) \tag{7.17}$$

is not negative infinite. Since none of the terms on the right side can be positive, it follows that they are all finite. Writing $\lambda(B_j) = \lambda(B_l) + b_j$ where b_j is finite and substituting in (7.15), we find that $f\lambda(B_l) - \mu_\infty$ is finite. Hence, $\lambda(B_l)$ is positive infinite and $f - \mu_\infty/\lambda(B_l)$ is infinitesimal. But the same conclusion applies if $f\lambda(B_i) - \mu_\infty > 0$ for some i, for then this is true also for $i = l$ and the second alternative of (7.12) applies.

In any case, dividing (7.15) by $\lambda(B_l)$,

$$\sum d_j f_j(\lambda(B_j)/\lambda(B_l)) - \mu_\infty/\lambda(B_l) \simeq 0$$

and so

$$\sum d_j f_j(\lambda(B_j)/\lambda(B_l)) - f \simeq 0$$

and

$$\sum d_j f_j(\lambda(B_j)/\lambda(B_l) - 1) \simeq 0. \tag{7.18}$$

But $\lambda(B_j) \leqslant \lambda(B_l)$ for all j, and so we may conclude that $\lambda(B_j)/\lambda(B_l) - 1$ is infinitesimal in all cases, i.e., $\lambda(B_j)/\lambda(B_l) \simeq 1$. Dividing by $\mu_\infty/\lambda(B_l) \simeq f$, we obtain finally

$$\lambda(B_j)/\mu_\infty \simeq 1/f. \tag{7.19}$$

Now every prime divisor $Q = B_j$ in K divides a prime divisor B in $k(\omega)$ where B, as a valuation, is obtained by restricting B_j to $k(\omega)$. Hence, applying (7.19) to any two prime divisors Q, Q' in K and dividing, we obtain

$$\lambda(Q')/\lambda(Q) \simeq 1 \tag{7.20}$$

We have proved the following theorem.

THEOREM 7.21. *Let Q, Q' be two prime divisors in K, and let $\lambda(Q)$, $\lambda(Q')$ be elements of $^*R$ such that*

$$\sigma(Q) = \theta(| Q |^{\lambda(Q)}) \qquad \textit{and} \qquad \sigma(Q') = \theta(| Q' |^{\lambda(Q')})$$

Then $\lambda(Q)$ and $\lambda(Q')$ are positive infinite and the ratio $\lambda(Q')/\lambda(Q)$ is infinitely close to 1.

COROLLARY 7.22. *Theorem* 7.21 *holds also for arbitrary entire divisors Q, Q' in K.*

Proof. It is clearly sufficient to prove Corollary 7.22 on the assumption that Q is prime, while Q' is an arbitrary entire divisor. Let $Q' = \prod Q_j^{\nu_j}$, and let d' be the degree of Q' and d_j the degree of Q_j. We choose appropriate λ_j so that $\sigma(Q_j) = \theta(| Q_j |^{\lambda_j})$. Then $\lambda_j = \lambda(Q)(1 + \epsilon_j)$ where ϵ_j is infinitesimal. Hence,

$$\sigma(Q') = \prod (\sigma(Q_j))^{\nu_j} = \theta\left(\prod | Q_j |^{\lambda_j \nu_j}\right) = \theta(| Q' |^{\lambda(Q)})\, \theta\left(\prod | Q_j |^{\lambda(Q)\nu_j \epsilon_j}\right).$$

Also $\sigma(Q') = \theta(|\, Q' \,|^{\lambda(Q')})$, by assumption, and so

$$\theta\left(|\, Q' \,|^{\lambda(Q)-\lambda(Q')} \prod |\, Q_j \,|^{\lambda(Q)\nu_j\epsilon_j}\right) = 1.$$

Remembering that $|\, Q' \,| = q^{-d'}$, $|\, Q_j \,| = q^{-d_j}$, we deduce that

$$d'(\lambda(Q) - \lambda(Q')) + \sum_j \lambda(Q)\, d_j\nu_j\epsilon, \tag{7.23}$$

is finite. Dividing by $\lambda(Q)$ we then obtain

$$d'(1 - \lambda(Q')/\lambda(Q)) + \sum_j d_j\nu_j\epsilon_j \simeq 0.$$

But $\epsilon_j \simeq 0$ and so $1 - \lambda(Q')/\lambda(Q) \simeq 0$, as asserted.

COROLLARY 7.24. *Every prime divisor Q in K is induced by at least one valuation P in $^*k$.*

Indeed since $\lambda(Q)$ is infinite, the product which defines $\sigma(Q)$ cannot be empty.

As mentioned in the introduction, Corollary 7.24 was obtained already in Ref. [4] by an entirely different method.

By interpreting the symbols used in (7.1) in terms of their definitions in Sections 6 and 7 (compare also the end of Section 6) and by putting $q = e$, the base of natural logarithms, we arrive at the following reformulation of Corollary 7.22.

THEOREM 7.25. *Let A be the g.c.d. of the numerators of two function divisors, $[\alpha]$ and $[\beta]$ in K. Suppose that A is different from the identity and let d be the degree of A. Let A' be the g.c.d. of the numerators of $[\alpha']$ and $[\beta']$ in K, where A' is again different from the identity, with degree d'. Then* (*see* (6.1)).

$$\ln s(\alpha', \beta')/\ln s(\alpha, \beta) \simeq d'/d. \tag{7.25}$$

8. TRANSLATION

In this section we shall show that our theory leads almost immediately to some of the basic inequalities of the theory of rational points on algebraic curves (compare Refs. [5–7]).

To begin with, we consider the situation in the standard field of real numbers, R. Let $\{s_n\}$ and $\{s_n'\}$ two positive sequences such that, as $n \to \infty$, $\{s_n\}$ and $\{s_n'\}$ either both tend to 0 or both tend to ∞. Then $\{s_n\}$ and $\{s_n'\}$

are said to be *quasiequivalent*—and we write $\{s_n\} \approx \{s_n'\}$—(compare Ref. [2]) if, for any $\epsilon > 0$ there exist positive constants c_1, c_2 such that

$$c_1 s_n^{1-\epsilon} \leqslant s_n' \leqslant c_2 s_n^{1+\epsilon}. \tag{8.1}$$

Evidently, (8.1) is equivalent to

$$\frac{\ln c_1}{\ln s_n'} + (1-\epsilon)\frac{\ln s_n}{\ln s_n'} \leqslant 1 \leqslant \frac{\ln c_2}{\ln s_n'} + (1+\epsilon)\frac{\ln s_n}{\ln s_n'}. \tag{8.2}$$

Moreover—subject to the assumption that $\{s_n\}$ and $\{s_n'\}$ tend to 0 or ∞, simultaneously—it is not difficult to see that (8.2), in turn, is equivalent to

$$\lim_{n\to\infty} \ln s_n/\ln s_n' = 1. \tag{8.3}$$

For if (8.3) is satisfied then $\ln s_n/\ln s_n'$ is close to 1 for sufficiently large n. Accordingly, (8.2) is certainly satisfied, and we only have to adjust c_1 and c_2 so as to satisfy (8.2) also for the remaining n, which are finite in number. The converse can be verified with equal ease.

Passing to $^*R$—while the sequences $\{s_n\}$ and $\{s_n'\}$ are still supposed to be standard—we see that one of our preliminary conditions amounts to the assumption that s_ω and s_ω' are either all infinite or all infinitesimal for infinite ω, while (8.3) is replaced by

$$\ln s_\omega/\ln s_\omega' \simeq 1. \tag{8.4}$$

Now (compare Section 2) let the irreducible curve Γ be given by $f(x, y) = 0$, $f(x, y) \in k[x, y]$, where k is an algebraic number field, as before, and suppose that Γ possesses an infinite sequence of distinct points (ξ_n, η_n), $n = 0, 1, 2,\ldots$ with coordinates in k. As we have shown, $(\xi_\omega, \eta_\omega)$ is then a generic point on Γ, for any infinite ω. Accordingly, $K = k(\xi_\omega, \eta_\omega) \subset {}^*k$ is a function field of the kind considered in the preceding sections.

Now let $g(x, y) \in k(x, y)$, so that $g(\xi_\omega, \eta_\omega)$ is finite (i.e., the denominator of $g(x, y)$ is not divisible by f) and not a constant (i.e., not an element of k). Let d_g be the degree of $g(\xi_\omega, \eta_\omega)$ in K. Then the numerator of the function divisor $[g]$ also is of degree d_g. As pointed out in Section 6 (see (6.7)–(6.9)) the height of $g(\xi_\omega, \eta_\omega)$ is given by

$$H(g(\xi_\omega, \eta_\omega)) = (s(g(\xi_\omega, \eta_\omega), g(\xi_\omega, \eta_\omega)))^{-1}. \tag{8.5}$$

Let $h(x, y) \in k(x, y)$ be another function subject to the same conditions, of degree d_h, so that

$$H(h(\xi_\omega, \eta_\omega)) = (s(h(\xi_\omega, \eta_\omega), h(\xi_\omega, \eta_\omega)))^{-1}. \tag{8.6}$$

Since $d_g > 0$, $d_h > 0$, the left sides of (8.5) and (8.6) are the reciprocals of infinitesimal numbers and must, therefore, be infinite. Substituting them for their right sides in (7.25), for $a = \beta = g(\xi_\omega, \eta_\omega)$, $\alpha' = \beta' = h(\xi_\omega, \eta_\omega)$, $d = d_g$, $d' = d_h$, we obtain

$$\frac{\ln(H(h(\xi_\omega, \eta_\omega)))^{1/d_g}}{\ln(H(g(\xi_\omega, \eta_\omega)))^{1/d}.} \simeq 1. \tag{8.7}$$

Accordingly, the argument developed at the beginning of this section applies to the sequences

$$s_n = (H(g(\xi_\omega, \eta_\omega)))^{1/d_g} \quad \text{and} \quad s_n' = (H(h(\xi_\omega, \eta_\omega)))^{1/d_h}$$

(provided we omit the finite number of points at which g or h vanish or become infinite). This proves the following theorem.

THEOREM 8.8. *Let Γ be an irreducible algebraic curve which is given by an equation $f(x, y) = 0$, $f(x, y) \in k[x, y]$, where k is an algebraic number field. Let $g(x, y)$ and $h(x, y)$ be elements of $k(x, y)$ whose images in the function field K/k of Γ are finite, nonconstant, and of degrees d_g and d_h, respectively. Let (ξ_n, η_n) be a sequence of points of Γ with coordinates in k such that $g(\xi_n, \eta_n)$ and $h(\xi_n, \eta_n)$ all are neither infinite nor equal to zero. Then the sequences $(H(g(\xi_n, \eta_n)))^{1/d_g}$ and $(H(h(\xi_n, \eta_n)))^{1/d_h}$ are quasiequivalent.*

Notice that Theorem 8.8 is a standard theorem. The words "infinite" or "finite" used in it do not belong to nonstandard analysis but indicate the presence or absence of a factor in the denominators of $g(x, y)$ or $h(x, y)$ which becomes zero after substitution.

Now let E be the l.c.m. (least common multiple) of the denominators of the function divisors $[\xi_\omega]$ and $[\eta_\omega]$. Standard theory (consider $\xi_\omega + r\eta_\omega$ for suitable $r \in k$)! shows that the degree of E, d, is equal to the order of the curve Γ or, which is the same, the degree of K over $k(\xi_\omega)$. Hence, from (6.18) and (7.21),

$$\ln H(g(\xi_\omega, \eta_\omega))/\ln(H(\xi_\omega, \eta_\omega, 1))^{d_g/d} \simeq 1. \tag{8.9}$$

We have proved the following theorem.

THEOREM 8.10. *With the assumptions of Theorem 8.8, let d be the order of the curve Γ. Then the sequence $H(g(\xi_n, \eta_n))$ is quasiequivalent to $(H(\xi_n, \eta_n, 1)^{d_g/d}$.*

Suppose now that $\xi_n = x_n/z_n$, $\eta_n = y_n/z_n$ where the x_n, y_n, z_n are algebraic integers in k and where the elements in each of the triples (x_n, y_n, z_n) have no common factor. Then

$$H(\xi_n, \eta_n, 1) = H(x_n, y_n, z_n) \geqslant \prod{}' \tfrac{1}{3}(| x_n |_P + | y_n |_P + | z_n |_P), \quad (8.11)$$

where $\prod'$ shall indicate that the product is to be taken over the Archimedean valuations only. Suppose also that $g(x, y) = 1/F(x, y)$ where it is assumed that $F(\xi_n, \eta_n)$ is an integer for all n. Suppose, furthermore, that there is a $\delta > 0$ (standard, of course) such that $| F(\xi_n, \eta_n)|_P > \delta$ for all n and for all Archimedean valuations P of k. Then

$$\begin{aligned} H(g(\xi_n, \eta_n)) &= \prod \max(| g(\xi_n, \eta_n)|_P, 1) = \prod{}'' (| 1/F(\xi_n, \eta_n)|_P) \\ &= \left(\prod{}' (| 1/F(\xi_n, \eta_n)|_P)\right)^{-1}, \end{aligned} \quad (8.12)$$

where $\prod''$ indicates that the product is to be taken over the non-Archimedean valuations while $\prod'$ refers to the Archimedean valuations, as before. Theorem 8.10 now shows that for any given $\epsilon > 0$ there exists a $c > 0$ such that, for all n,

$$\left(\prod{}' (| 1/F(\xi_n, \eta_n)|_P)\right)^{-1} \geqslant c \left(\prod{}' (| x_n |_P + | y_n |_P + | z_n |_P)\right)^{(d_g/d)-\epsilon}.$$

Hence,

$$\prod{}' \left| \frac{1}{F(x_n/z_n, y_n/z_n)} \right|_P \leqslant (1/c) \left(\prod{}' (| x_n |_P + | y_n |_P + | z_n |_P)\right)^{(-d_g/d)+\epsilon}, \quad (8.13)$$

where we may, if we wish, replace $\leqslant$ by $<$. This is Siegel's basic inequality [5, p. 51].

Acknowledgment

I am indebted to Iacopo Barsotti for several stimulating conversations on the subject of this paper.

References

1. E. Artin and G. Whaples, Axiomatic characterization of fields by the product formula for valuations, *Bull. Amer. Math. Soc.* **51** (1945), 469–492.
2. S. Lang, "Diophantine Geometry," Wiley, New York/London, 1962.
3. A. Robinson, Nonstandard arithmetic, *Bull. Amer. Math. Soc.* **73** (1967), 818–843.
4. A. Robinson, "Algebraic function fields and nonstandard arithmetic," Contributions to Non-Standard Analysis (ed. W. A. J. Luxemburg and A. Robinson), North Holland Publishing Company, Amsterdam/London, 1972, pp. 1–14.

5. C. L. SIEGEL, Über einige Anwendungen diophantischer Approximationen, Abhandlungen der Preussischen Akademie der Wissenschaften, Physikalisch-Mathematische Klasse, Berlin, 1929, pp. 209–266.
6. A. WEIL, Number theory and algebraic geometry, "Proceedings of the International Congress of Mathematicians, 1950," Vol. 2, pp. 90–100, Cambridge, MA. Harvard Univ. Press, 1952.
7. A. WEIL, Arithmetic on algebraic varieties, *Ann. of Math.* **53** (1957), 412–444.

ENLARGED SHEAVES

Abraham Robinson

Yale University, New Haven

1. The present note is related to [4], but we shall not assume familiarity with that paper. The reader may consult [1], [2] for standard results from the theory of functions of several complex variables used here and to [3] for basic concepts in nonstandard analysis.

Let $\mathcal{S} = (S,D,\pi)$ be a sheaf of rings with base space D and map $\pi : S \to D$. Let ${}^*\mathcal{S} = ({}^*S, {}^*d, \pi)$ be an enlargement of $\mathcal{S}$ where, as usual, we have not appended the star to the extended π. For any section f in $\mathcal{S}$ with domain $U \subset D$ (where U is always assumed open) and for any point $z \in U$, we denote by $\gamma_z(f)$ the germ generated by f at z. Thus, the mapping $(f,z) \to \gamma_z(f)$ is defined provided z is in the domain of f. The mapping extends to ${}^*\mathcal{S}$ where it is applicable also to nonstandard internal f and to nonstandard z in the domain of f.

Again, let z be a standard point of the base space D and let $\mu(z)$ be its monad. For any standard section f which includes z and hence also $\mu(z)$ in its domain, we denote by $\beta_z(f)$ the restriction of f to $\mu(z)$. Thus, $\beta_z(f)$ is, generally speaking, external. We may extend this definition also to nonstandard internal f, provided z is included in the domain of f. Let

$$G_z = \{y \mid y = \gamma_z(f) \text{ for some section } f\}$$

and let

$$M_z = \{y \mid y = \beta_z(f) \text{ for some section } f\}$$

where M_z is defined only for standard z and where, in both cases, the f that are to be taken into account are standard or nonstandard internal. Also, for

---

Research supported in part by the National Science Foundation, Grant No. GP - 34088.

standard z, let

$$^{o}G_z = \{y \in G_z | y = \gamma_z(f), \quad f \quad \text{standard}\}$$

and let

$$^{o}M_z = \{y \in M_z | y = \beta_z(f), \quad f \quad \text{standard}\} .$$

Then $^{o}G_z$ is simply the stalk of γ at z and $G_z = {}^{*}({}^{o}G_z)$ is the stalk of ${}^{*}\gamma$ at z. Both $^{o}G_z$ and G_z are endowed with a ring structure such that $^{o}G_z$ is (or may be regarded as) a subring of G_z. Moreover, M_z has a ring structure which is given by pointwise addition, subtraction, and multiplication on $\mu(z)$ and includes $^{o}M_z$ as a subring.

For standard z, there is a natural map $\lambda : M_z \to G_z$ as follows. Let U be any (internal) open neighborhood of z such that $U \subset \mu(z)$. For any $g \in M_z$, $g|U$ is a section which determines an element f of the stalk of the sheaf at z, i.e. of G_z, where the particular choice of U is irrelevant. We then put $f = \lambda(g)$. λ is a ring homomorphism and the restriction $^{o}\lambda$ of λ to $^{o}M_z$ is an isomorphism onto $^{o}G_z$ since each element of the stalk of γ at z determines standard sections which coincide on $\mu(z)$. To this extent, $^{o}M_z$ reflects the properties of the stalk of germs $^{o}G_z$ and this fact has been used in [4] for a discussion of the Rückert Nullstellensatz. At the same time, since they are genuine functions, the elements of M_z also reflect the properties of (standard) sections. In the present paper we offer some remarks concerning $^{o}M_z$ and M_z from the latter point of view, and in this context we prefer to call the elements of M_z monadic sections rather than nonstandard germs as in [4].

As usual in this area, we shall make use of the Weierstrass preparation and division theorems. As for the latter, a glance at the proof given on page 70 of [1], shows that if $h(z_1, \ldots, z_n)$ is a Weierstrass polynomial in z_n of degree $k > 0$ which is holomorphic for $|z_j| < \delta$, $j = 1, \ldots, n-1$, $\delta > 0$, then there exists a δ', $0 < \delta' > \delta$ such that for every function which is holomorphic for $|z_j| < \delta'$, $j = 1, \ldots, n$, we have a representation

$$f(z_1, \ldots, z_n) = g(z_1, \ldots, z_n)h(z_1, \ldots, z_n) + r(z_1, \ldots, z_n)$$

where $g(z_1, \ldots, z_n)$ is holomorphic for $|z_j| < \delta'$, and $r(z_1, \ldots, z_n)$ is a polynomial in z_n of degree $< k$ with coefficients which are holomorphic functions of $z_1, \ldots, z_{n-1}$ for $|z_j| < \delta'$.

2. From now on, we shall assume that $\mathcal{S} = (S,D,\pi)$ is the sheaf of germs of holomorphic functions on an open set D in n-dimensional complex space. Let $g_1, \ldots, g_k$ be elements of ${}^{O}M_z$ for some $z = (z_1, \ldots, z_n) \in D$ and let f be an element of ${}^{O}M_z$. If $\lambda(f)$ is included in the ideal generated in ${}^{O}G_z$ by $\lambda(g_1), \ldots, \lambda(g_k)$, i.e. if

$$\lambda(f) = \sum_{j=1}^{k} p_j \lambda(g_j) \quad \text{with} \quad p_j \in {}^{O}G_z$$

then the isomorphism provided by λ ensures that a corresponding fact is true within ${}^{O}M_z$, that is to say, putting $p_j = \lambda(q_j)$ we have

$$f = \sum_{j=1}^{k} q_j g_j \ .$$

Thus, f belongs to the ideal generated by $g_1, \ldots, g_k$. The following theorem states that this conclusion can be extended to nonstandard internal functions f which are holomorphic on $\mu(z)$. More precisely - and more generally - we shall prove

<u>2.1. Theorem</u>. Let

$$G_j = (g_{j1}, g_{j2}, \ldots g_{jm}),\ m \geq 1,\ j = 1, \ldots, k,\ k \geq 1$$

where $g_{ji} \in {}^{O}M_z$ and let $F = (f_1, \ldots, f_m)$ where $f_j \in M_z$ (so that the $f_j(z_1, \ldots, z_n)$ are holomorphic but not necessarily standard on $\mu(z)$). Suppose that $\lambda(F)$ belongs to the module generated by $\lambda(G_1), \ldots, \lambda(G_k)$ over G_z

where

$$\lambda(F) = (\lambda(f_1), \ldots, \lambda(f_m)), \lambda(G_j) = (\lambda(g_{j1}), \ldots, \lambda(g_{im})) .$$

Then F belongs to the module generated by $G_1, \ldots, G_k$ over M_z, i.e.

$$F = \sum_{j=1}^{k} p_j G_j$$

for certain monadic sections p_j in M_z .

For the proof, we first reformulate 2.1, reinterpreting the g_{ji} and f_j as holomorphic functions, standard or internal respectively which have domains <u>including</u> $\mu(z)$. On this assumption we have to establish -

2.1'. Suppose that there exist internal functions $p_j(z_1, \ldots, z_n)$, $j = 1, 2, \ldots, k$, which are holomorphic in some internal (but not necessarily standard) open neighborhood U of z such that

$$F = \sum_{j=1}^{k} p_j G_j$$

on U. Then there exist internal functions $q_j(z_1, \ldots, z_n)$ which are holomorphic on sets including $\mu(z)$ such that

$$F = \sum_{j=1}^{k} q_j G_j \quad .$$

<u>Proof of 2.1'</u>. We may suppose that $z = \underline{0} = (0, \ldots, 0)$. There is nothing to prove for $n = 0$ (for in that case the g_{ji}, f_j and p_j are constants, by convention). Suppose that the assertion has been proved for $n - 1$, $n > 1$, and for all positive integers m and k, and let n be the number of variables and $m = 1$. Replacing the notation g_{j1}, f_1, p_1 by g_j, f, and p, respectively we then have

$$f = \sum_{j=1}^{k} p_j g_j$$

in some internal open neighborhood U of $\underline{0}$.

Suppose first that $k = 1$. We may assume, if necessary after first carrying out a suitable standard nonsingular linear transformation of the independent variables, that g_1 is regular of order o in z_n. Then $g_1 = eh$ where e and h are both standard with domains including $\mu(\underline{0})$ and e is invertible on $\mu(\underline{0})$ while h is a Weierstrass polynomial,

$$h(z_1, \ldots, z_n) = z_n^{\rho} + a_1 z_n^{\rho-1} + \ldots + a_{\rho} ,$$

where the a_j are standard functions of $z_1, \ldots, z_{n-1}$ with $a_j(0, \ldots, 0) = 0$ and holomorphic in a standard neighborhood of the origin in $(z_1, \ldots, z_{n-1})$ - space. Since $f = p_1 g_1$ we have $f = p_1 eh$. At the same time, by the Weierstrass division theorem, there is a representation $f = sh + r$ where, according to a remark in section 1 above, s and h are holomorphic in domains which include $\mu(\underline{0})$. Thus, $p_1 eh = sh + r$ in some internal neighborhood of $\underline{0}$. But then, by the uniqueness of the representation $f = sh + r$, we must have $p_1 e = s$ and $r = 0$, and so $p_1 = e^{-1}s$ is holomorphic in a domain which includes $\mu(\underline{0})$. This remains true if we first had to introduce the linear transformation which made g_1 regular in z_n, for a standard nonsingular transformation maps monads on monads.

Still assuming $m = 1$, suppose that we have proved the assertion of the theorem for positive integers less than some $k \geq 2$. We may assume that the functions $g_1, \ldots, g_k$ are regular in z_n for if this is not the case from the outset we may achieve it for all g_j simultaneously by means of a linear transformation of the independent variables, as before. Furthermore, we may assume that none of the g_j vanish identically (otherwise we omit the g_j in question, unless they all vanish, in which case f also vanishes identically). We may then write

$$g_j = e_j h_j, \ j = 1, \ldots, k,$$

as before, where we may suppose that h_k is the Weierstrass polynomial of highest degree, ρ. The e_j are all holomorphic and invertible in some standard neighborhood of $\underline{0}$, while

$$h_j = z_n^{\rho_j} + a_1^j z_n^{\rho_j - 1} + \ldots + a_{\rho_j}^j, \ \rho_j \le \rho, \ j = 1, \ldots, k, \ \rho_k = \rho,$$

where the a_i^j are standard and holomorphic in the neighborhood of the origin of $(z_1, \ldots, z_{n-1})$ - space and vanish at the origin. By the division theorem, $f = sh_k + r$ where s and r are holomorphic in $\mu(\underline{0})$. Accordingly, it only remains to be shown that r can be written as a linear combination of the h_j, thus, $r = \Sigma s_j h_j$ where the s_j are internal and holomorphic on $\mu(\underline{0})$.

Let

$$r = b_1 z_n^{\rho-1} + b_2 z_n^{\rho-2} + \ldots + b_\rho .$$

Also, let A be the ideal in ${}^{\circ}G_{\underline{o}}$ which is generated by $\gamma_{\underline{0}}(h_1), \ldots, \gamma_{\underline{0}}(h_j)$, and let B be the module consisting of $(\rho + 1)$ - tuples of germs of standard holomorphic functions in $(z_1, \ldots, z_{n-1})$-space, $(c_o, \ldots, c_\rho)$ such that $c_o z_n^{\rho} + c_1 z_n^{\rho-1} + \ldots + c_\rho \in A$. By the finite basis theorem for modules of germs, B has a basis (set of generators) $B_1, \ldots, B_\ell$, $B_j = (\gamma'(b_0^j), \gamma'(b_1^j), \ldots, \gamma'(b_\rho^j))$, where the b_i^j are standard holomorphic functions in the neighborhood of the origin of $(z_1, \ldots, z_{n-1})$ - space and the γ' indicates that the germs are taken at that point. Let

$$d^j = b_0^j z_n^{\rho} + b_1^j z_n^{\rho-1} + \ldots + b_\rho^j ;$$

then $\gamma_{\underline{0}}(d_j)$ belongs to A; $j = 1, \ldots, \ell$, and so

$$d^j = \sum_j s_i^j h_i$$

where the s_i^j are standard and holomorphic on the monad $\mu(\underline{0})$.

So far, everything we have said about A and B involves standard functions only. However, by the definition of B we also have that $(0,b_1,b_2, \ldots, b_\rho)$ belongs to ${}^*B$ and so, by the assumption of our induction,

$$(0,b_1,b_2, \ldots, b\rho) = \sum_j k_j(b_0^j,b_1^j, \ldots, b_\rho^j) \quad ,$$

where the k_j are internal and holomorphic on the monad of $(z_1, \ldots, z_{k-1})$-space. It follows that

$$r = \sum_j k_j(b_0^j z_n^\rho + b_1^j z_n^{\rho-1} + \ldots + b_\rho^j) = \sum_{j,i} k_j s_i^j h_i \quad .$$

But the $k_j s_i^j$ are internal and holomorphic on $\mu(\underline{0})$ and so our assertion is proved for this case. This is still true if we first have to carry out a linear transformation on the independent variables in order to make $g_1, \ldots, g_k$ regular in z_n.

Suppose finally that the assertion has been proved up to and including some $m \geq 1$ and that

$$F = \sum_{j=1}^{k} p_j G_j$$

where

$$F = (f_1, \ldots, f_m, f_{m+1}), \; G_j = (g_{j1}, \ldots, g_{jm}, g_{j,m+1})$$

where the f_j are internal and the g_{ji} are standard, and both the f_j and the g_{ji} are holomorphic on $\mu(\underline{0})$ while the p_j are internal but holomorphic only on some internal neighborhood of the origin, to begin with. In particular,

$$f_{m+1} = \sum p_j g_{j,m+1} \; .$$

By what has been proved already we may replace the p_j in this equation by functions p_j^{m+1} which are internal and holomorphic on $\mu(\underline{0})$, thus

$$f_{m+1} = \sum_j p_j^{m+1} g_{j,m+1} \ .$$

Then the last component in

$$F' = F - \sum_j p_j^{m+1} G_j = \sum_j (p_j - p_j^{m+1}) g_j$$

vanishes and so, by the inductive assumption on m, there exist functions q_j which are internal and holomorphic on the monad of the origin such that

$$(f_1', \ldots, f_m', 0) = \sum_j q_j (g_{j1}, \ldots, g_{jm}, 0)$$

where

$$F' = (f_1', \ldots, f_m', 0),\ f_i' = f_i - \sum_j p_j^{m+1} g_{j,m+1} \ .$$

But

$$\sum_j (p_j - p_j^{m+1}) g_{j,m+1} = 0$$

in some internal neighborhood of $\underline{0}$ and so

$$\sum q_j g_{j,m+1} = 0$$

in the same neighborhood and, hence on the entire monad $\mu(\underline{0})$. Thus, on $\mu(\underline{0})$,

$$(f_1', \ldots, f_m', 0) = \sum_j q_j (g_{j1}, \ldots, g_{jm}, g_{j,m+1})$$

and further

$$(f_1, \ldots, f_m, f_{m+1}) = \sum_j (q_j + p_j^{m+1})(g_{j1}, \ldots, g_{jm}, g_{j,m+1}) \ .$$

This completes the proof of 2.1' and hence, of 2.1.

Remarks. In the above proof, we adopted the convention that, for $n = 0$, sections and germs are just complex numbers. Alternatively, we might have considered the case $n = 1$ separately. This would have led to a module B of $(\rho+1)$-tuples of constants in the last part of the proof.

2.2. Corollary. Suppose that the monadic sections $f_1, \ldots, f_m$ in

Theorem 2.1. are finite on $\mu(z)$. Then we may choose the functions $p_1, \ldots, p_k$ so as to be finite on $\mu(z)$.

A standard argument of nonstandard analysis ("Robinson's lemma") shows that the assumption that an internal function is finite on $\mu(z)$ is equivalent to the condition that the function be standardly bounded on $\mu(z)$. Accordingly we may replace "finite," by standardly bounded" both in the hypothesis and in the conclusion of 2.2.

In order to prove 2.2, we only have to check through the proof of 2.1 (or, rather, 2.1'). All the standard holomorphic functions introduced in the course of the proof must be finite (or, equivalently, standardly bounded) on $\mu(\underline{0})$. In addition, we introduce internal functions s and r by means of Weierstrass' division theorem, $f = sh + r$ where h is a standard Weierstrass polynomial and f is standardly bounded on $\mu(\underline{0})$ and, hence, on some polydisc including $\mu(\underline{0})$. Another glance at the integral formula for s on page 70 of [1] (where our s is called g) shows that s and hence r, also are standardly bounded on $\mu(\underline{0})$. Moreover, an appeal to Cauchy's integral formula shows that the coefficients of r as a polynomial of z_n must then be standardly bounded on the monad of the origin of $(z_1, \ldots, z_{n-1})$-space. By means of these facts it is now easy to verify that the p_j also may be chosen so as to be standardly bounded on $\mu(\underline{0})$.

2.1 and 2.2 together provide an easy proof of the following classical result.

2.3 Closure of modules theorem. Suppose that $F = (f_1, \ldots, f_m)$ has holomorphic components on an open neighborhood U of a point z and suppose that F can be uniformly approximated on compact subsets of U by holomorphic functions $(g_1, \ldots, g_m)$ such that $(\gamma_z(g_1), \ldots, \gamma_z(g_m))$ belongs to a submodule A of the m-module of germs at z. Then $(\gamma_z(f_1), \ldots, \gamma_z(f_m))$ also belongs to A.

Proof. Suppose without loss of generality that $z = \underline{0}$. Let $G_1, \ldots, G_\ell$ be a basis of A. Choose a closed polydisc $P\colon |z_j| \leq \delta$ which is a subset of U such that the functions of $G_1, \ldots, G_\ell$ are holomorphic on P.

By transferring the hypothesis of the theorem we see that there exist internal functions $h_1, \ldots, h_m$ which are holomorphic on P such that

$$(\gamma(h_1), \ldots, \gamma(h_m)) \in {}^*A$$

and such that

$$\sup|f_j - h_j| \simeq 0, \; j = 1, \ldots, m \quad \text{on} \quad P.$$

Then the h_j are standardly bounded on P. Hence, by 2.1 and 2.2 there exists a standard δ', $0 < \delta' < \delta$ such that $(h_1, \ldots, h_m) = \sum_j p_j G_j$ for $|z_j| \leq \delta'$, where the p_j are holomorphic and standardly bounded for $|z_j| \leq \delta'$. Writing $G_j = (g_{j1}, \ldots, g_{jm})$ we then have

2.4. $$f_i - \sum_j p_j g_{ji} \simeq 0 \quad \text{for} \quad |z_j| \leq \delta', \quad i = 1, \ldots, m$$

But since the p_j are standardly bounded they have (as for functions of a singlecomplex variable) standard parts

$$q_j(t_1, \ldots, z_n) = {}^\circ p_j(z_1, \ldots, z_n) .$$

Then, by 2.4,

$$f_i - \sum_j q_j g_{ji} \simeq 0$$

for standard points such that $|z_j| < \delta'$. But $f_i - \sum_j q_j g_{ji}$ is standard and so

$f_i - \sum_j q_j g_{ji} = 0$ for $|z_j| < \delta'$. This proves that $(\gamma_{\underline{0}}(f_1), \ldots, \gamma_{\underline{0}}(f_m))$ belongs to A.

The reader may find it instructive to compare our argument with the standard proof given in [1], chapter 2, section D.

3. As might be expected by now, the question of the coherence of a given sheaf is determined entirely by the behavior of its monadic sections. In order to illustrate this fact we shall conclude this paper by stating the non-standard version of the assertion that every locally finitely generated subsheaf of the sheaf of germs of holomorphic functions on a given domain is coherent (Oka's theorem). If f is a monadic section on $\mu(z)$ and $\zeta \in \mu(z)$ we write $\gamma_\zeta(f)$ for the uniquely determined germ at ζ of a holomorphic function g such that $\beta_z(g) = f$. In particular $\lambda(f) = \gamma_z(f)$. We also write $\gamma_z(F)$ for $(\gamma_z(f_1), \ldots, \gamma_z(f_m))$ if $F = (f_1, \ldots, f_m)$.

<u>3.1. Theorem</u>. Let $G_1, \ldots, G_k$, $k \geq 1$, be m-tuples of standard monadic sections of holomorphic functions at a standard point z,

$$G_j = (g_{j1}, \ldots, g_{jm}), \quad m \geq 1.$$

Then there exist k-tuples of standard monadic sections at z,

$$F_1, \ldots, F_\ell, \; F_j = (f_{j1}, \ldots, f_{jk})$$

such that

$$\sum_{i=1}^{k} f_{ji} G_i = 0 \quad \text{on} \quad \mu(z), \; j = 1, \ldots, h$$

and such that the following condition is satisfied. Let $\zeta \in \mu(z)$ and let $f = (f_1, \ldots, f_k)$ consist of internal functions which are holomorphic in an internal neighborhood of ζ such that $\Sigma\ \gamma_\zeta\ (f_j)\ \gamma_\zeta\ (G_j) = 0$. Then $\gamma_\zeta(f)$ belongs to the module generated by $\gamma_\zeta(F_1)^j$, $\gamma_\zeta(F_\iota)$ over G_ζ.

List of References

1. Gunning, R. C. and Rossi, H. Analytic Functions of Several Complex Variables, Englewood Cliffs, N. J. 1965.
2. Hörmander, L. An Introduction to Complex Analysis in Several Variables, Toronto-New York-London 1966.
3. Robinson, A. Nonstandard Analysis, Studies in Logic and the Foundations of Mathematics, Amsterdam 1966.
4. ___________. Germs, Applications of Model Theory to Algebra, Analysis, and Probability (ed. W. A. J. Luxemburg) New York, etc. 1969 pp. 138-149.

The Cores of Large Standard Exchange Economies*

Donald J. Brown†

Department of Economics, University of California at Berkeley, Berkeley, California

and

Abraham Robinson

Mathematics Department, Yale University, New Haven, Connecticut 06520

Received January 21, 1972

I. Introduction

An exchange economy consists of a set of traders, each of whom is characterized by an initial endowment and a preference relation. In addition, one usually assumes that the set of traders is finite. Edgeworth's conjecture that, as the number of traders in an exchange economy increases, the core approaches the set of competitive equilibria has been formalized by Debreu-Scarf [4].

Their approach was to consider a sequence of replicated economies and to look at the relationship between the core and the set of competitive equilibria for very large replications.

Here we extend the results of Debreu-Scarf to sequences of economies more general than those obtained by replication. We have shown in [3] that within nonstandard analysis the concept of the core and competitive equilibrium are the same. As a consequence of this theorem we shall derive an asymptotic result concerning the cores of a sequence of standard (finite) exchange economies.

In Section II, we give a brief introduction to the essentials of nonstandard analysis; in Section III, we describe the economic model and state our theorem concerning Edgeworth's conjecture; in Section IV,

* The research described in this paper was carried out in part by grants from the National Science Foundation (GP-29218 and GS-40786X) and the Office of Naval Research. The first author would also like to acknowledge the support of a Guggenheim Fellowship for the year 1973–1974. We wish to thank Herbert E. Scarf and the referees of this journal for several useful suggestions.

† On leave from the Cowles Foundation for Research in Economics at Yale University.

we give an asymptotic result or limit theorem for the cores of large standard economies; in Section V, we consider the relationship between our work and that of Debreu-Scarf.

II. Nonstandard Analysis

Let R be the system of real numbers. Any statement abour R (involving individuals, subsets of R, functions on R, relations on R, sets of relations on R, etc.) can be expressed in a formal language which includes names for all these entities, connectives ($\neg$, $\wedge$, $\vee$, $\Rightarrow$), variables for entities of different types, and quantifiers $\exists$ and $\forall$ over different types of variables.

Let K be the set of all statements in this formalized language which are true for R. Then it can be shown that there exists a proper extension, $^*R$, of R, a designated subset of the set of all subsets of $^*R$, a designated subset of the set of all functions from $^*R$ to $^*R$, a designated subset of the set of all relations on $^*R$, etc., such that the following holds. Every statement which is true in R remains true in R provided we reinterpret the existential quantifiers which occur in that statement as follows: "There exists" a set shall mean "there exists" an *internal* set; "there exists" a family of relations, shall mean "there exists" an *internal* family of relations; similarly for all other types of entities. Here *internal means an element of the appropriate designated sets of entities.* (Note that this qualification does not apply to individuals.) Such a statement is said to be true in $^*R$ by *transfer*.

We shall call the elements of $^*R$ *real numbers*, the elements of R *standard real numbers*, and the elements of $^*R$ which do not belong to R *nonstandard real numbers*. As a proper extension of R, $^*R$ must be a nonarchimedean field; that is, it contains numbers whose absolute values are greater than all standard real numbers, which will be called infinite numbers. All other numbers will be called *finite*. The reciprocals of *infinite numbers* together with zero are said to be *infinitesimals* or *infinitely small numbers*. Every finite number, x, is infinitely close to a unique standard real number, $^\circ x$, which is called the *standard part* of x. The *monad* of a number, $\mu(x)$, is the set of all numbers which are infinitely close to x.

$^*R$ also contains a set of numbers, denoted as $^*N$, which has the same properties as N (the set of natural numbers) in the sense that any statement true about N is true about $^*N$ when reinterpreted in terms of internal entities. $^*N$ is a proper extension of N. We shall call the elements of $^*N$ *natural numbers*, the elements of N *finite natural numbers*, and the elements of $^*N$ which do not belong to N *infinite natural numbers*. An internal set which has ω elements, where $\omega \in {}^*N$, is said to be *starfinite*.

A structure $^*R$ of the required kind may be constructed as an ultrapower of R over the natural numbers. That is to say, $^*R$ consists of all sequences of real numbers; equality between such sequences as well as any other kind of relation is defined with respect to a free ultrafilter in the boolean algebra, $B(N)$, the power of set of the natural numbers. Similarly, internal sets, internal relations, etc., can be identified with sequences of sets, relations, etc., again reduced with respect to the ultrafilter. (Compare Ref. [6–8]).

We shall not use the construction just mentioned, but we assume that the following particular property of the $^*R$ just constructed is satisfied. Let f be a function with domain N whose range values are numbers (internal sets of numbers) of $^*R$. Then there exists an internal sequence of numbers (internal sets of numbers) of $^*R$, such that the values of the sequence agree with the values of f on all finite j. That is, if $f\colon N \to {}^*R$, then there exists an internal $g\colon {}^*N \to {}^*R$ such that, for all $j \in N$, $g(j) = f(j)$.

$^*R_n$ is the n-fold cartesian product of $^*R$ and $^*\Omega_n$ is the positive orthant of $^*R_n$. $^*R_n$ has the same "formal" properties as R_n; in particular, it is an euclidean vector space. Let $\bar{x}$, $\bar{y}$ be vectors in $^*R_n$. The monad of $\bar{x}$, $\mu(\bar{x})$, is the set of points whose distance from $\bar{x}$ is an infinitesimal. If $\bar{y} \in \mu(\bar{x})$, we shall write $\bar{x} \simeq \bar{y}$, $\bar{x} \geqslant \bar{y}$ means $x_i \geqslant y_i$ for all i; $\bar{x} > \bar{y}$ means $x_i \geqslant y_i$ for all i and $x_i > y_i$ for some i; $\bar{x} \gg \bar{y}$ means $x_i > y_i$ for all i. $\bar{x} \gtrsim \bar{y}$ means that $x_i \geqslant y_i$ or $x_i \simeq y_i$ for all i. $\bar{x} \underset{\not\sim}{>} \bar{y}$ means that x_i is greater than y_i by a finite amount for some i and that $x_i \simeq y_i$ for all i. $\bar{x} \underset{\not\sim}{\gg} \bar{y}$ means that x_i is greater than y_i by a finite amount for all i.

III. Economic Model and Edgeworth's Conjecture

Let T be an initial segment of $^*N$, where $|T|$, the number of elements in T, is ω some infinite natural number. That is, $T = \{1, 2, \ldots, \omega\}$ and $\omega \in {}^*N - N$. T is to be interpreted as the set of traders in the economy. If S is any internal subset of T, then $|S|$ will denote the number of elements in S.

A *nonstandard exchange economy*, E, consists of a pair of functions I and P, where $I\colon T \to {}^*\Omega_n$ and $P\colon T \to B({}^*\Omega_n \times {}^*\Omega_n)$.[1] Denoting the function P as $>_t$, $\langle I(t), >_t \rangle = E$. $I(t)$ is the initial endowment of the tth trader, and $>_t$ is his preference relation over $^*\Omega_n$, the space of commodities. The nonstandard exchange economies which we will consider are assumed to have the following properties.

[1] If A is a set, then $B(A)$ will be the power set of A.

(i) The function indexing the initial endowments, $I(t)$, is internal

(ii) $I(t)$ is standardly bounded, i.e., there exists a standard vector $\bar{r}_0$ such that, for all t, $I(t) \leqslant \bar{r}_0$.

(iii) $(1/\omega) \sum_1^\omega I(t) \gtrapprox \bar{0}$.

(iv) The relation, Q, where $Q = \{\langle t, >_t \rangle \mid t \in T, >_t \subseteq {}^*\Omega_n \times {}^*\Omega_n\}$ is internal. For all t,

(α) $>_t$ is irreflexive, i.e., if $\bar{x} >_t \bar{y}$, then $\bar{x} \neq \bar{y}$;
(β) if $\bar{x} > \bar{y}$, then $\bar{x} >_t \bar{y}$;
(γ) if $\bar{x} \not\simeq \bar{y}$ and $\bar{x} > \bar{y}$, then there exists a standard $\delta > 0$ such that, if $\bar{z} \in S(\bar{x}, \delta)$, open ball with center $\bar{x}$ and radius δ, then $\bar{z} >_t \bar{y}$.

Equivalently, if $\bar{x} \not\simeq \bar{y}$ and $\bar{x} > \bar{y}$ and $\bar{z} \in \mu(\bar{x})$, then $\bar{z} >_t \bar{y}$. It is shown in [3] that these assumptions are consistent.

An *assignment* Y is an internal function $Y(t)$ from T, the set of traders, into ${}^*\Omega_n$.

An *allocation or final allocation* is a standardly bounded assignment $Y(t)$ from the set of traders, T, into ${}^*\Omega_n$ such that $(1/\omega) \sum_{t=1}^\omega Y(t) \simeq (1/\omega) \sum_{t=1}^\omega I(t)$.

(iv) implies that, for all internal $X, Y \in {}^*\Omega_n{}^T$, where T is the set of traders, that

(v) $\{t \mid X(t) >_t Y(t)\}$ is an internal set of traders.

A *coalition*, S, is defined as an internal set of traders. It is said to be negligible if $|S|/\omega \simeq 0$. Note that if S is negligible then, for all allocations $X(t)$, $(1/\omega) \sum_{t \in S} X(t) \simeq \bar{0}$.

A coalition, S, is feasible with respect to an allocation Y if $(1/\omega) \sum_{t \in S} Y(t) \simeq (1/\omega) \sum_{t \in S} I(t)$.

$$\bar{x} \gg_t \bar{y} \quad \text{iff} \quad (\forall \bar{w} \in \mu(\bar{x}))\ \bar{w} >_t \bar{y}$$

An allocation Y *dominates* an allocation X via a coalition S if S is feasible with respect to Y and if, for all $t \in S$, $Y(t) \gg_t X(t)$.

The *core* is defined as the set of all allocations X which are not dominated by any allocation Y via any nonnegligible coalition.

A *price vector*, $\bar{p}$, is finite vector in ${}^*\Omega_n$ such that $\bar{p} \gtrapprox \bar{0}$. The tth trader *budget set*, $B_{\bar{p}}(t)$, is $\{\bar{x} \in {}^*\Omega_n \mid \bar{p} \cdot \bar{x} \lesssim \bar{p} \cdot I(t)\}$.

$\bar{y}$ is said to be *maximal in* $B_{\bar{p}}(t)$ if $\bar{y} \in B_{\bar{p}}(t)$ and there does not exist an $\bar{x} \in B_{\bar{p}}(t)$ such that $\bar{x} \gg_t \bar{y}$.

A *competitive equilibrium* is defined as a pair $\langle \bar{p}, X \rangle$, where $\bar{p}$ is a price vector, X is an allocation, and there exists an internal set of traders, K, where $|K|/\omega \simeq 1$, such that $X(t)$ is maximal in $B_{\bar{p}}(t)$ for all $t \in K$.

THEOREM 1. *If E is a nonstandard exchange economy satisfying the above assumptions, then an allocation X is in the core of E if and only if there exists a price vector,* $\bar{p}$, *such that* $\langle \bar{p}, X \rangle$ *is a competitive equilibrium of E.*

The proof of this theorem is given in [3].

IV. LIMIT THEOREM

A standard exchange economy $\tilde{E}$ of size m consists of m traders, where m is a standard natural number, whose initial endowments and preferences are restricted to the standard commodity space Ω_n. Ω_n is the positive orthant of R_n, the n-fold cartesian product of R. Let $\tilde{E} = \langle I(t), >_t \rangle$ where, for all t, $I(t) \in \Omega_n$ and $>_t \in B(\Omega_n \times \Omega_n)$, $t \in \tilde{E}$ will refer to the tth traders endowment and preference relation.

A competitive equilibrium for $\tilde{E}$ is a pair $\langle \bar{p}, X \rangle$ such that $\bar{p} \in \Omega_n$; $\sum_{t=1}^{m} X(t) = \sum_{t=1}^{m} I(t)$; for each t, $X(t) \in \Omega_n$ and $\bar{p} \cdot X(t) \leqslant \bar{p} \cdot I(t)$; there does not exist a $\bar{y} \in \Omega_n$ where $\bar{p} \cdot \bar{y} \leqslant \bar{p} \cdot I(t)$ and $\bar{y} >_t X(t)$. $\bar{p}$ is called a *price vector* and $X(t)$ a *competitive allocation.*

$X(t)$ is an *allocation* if for each t, $X(t) \in \Omega_n$ and $\sum_{t=1}^{m} X(t) = \sum_{t=1}^{m} I(t)$. An allocation $X(t)$ is *blocked* by an allocation $Z(t)$ if there exists a coalition of traders, S, such that $\sum_{t \in S} Z(t) = \sum_{t \in S} I(t)$ and, for all $t \in S$, $Z(t) >_t Y(t)$. The *core* of $\tilde{E}$ is the set of unblocked allocations.

Let $H = \{\tilde{E}_i\}_{i \in I}$ be an unbounded family of standard exchange economies; i.e., for each $k \in N$, there exists $i \in I$ such that $|\tilde{E}_i|$, the number of traders in the ith economy, is greater than k. Suppose H satisfies the following conditions:

(1) The initial endowments of the traders in H are uniformly bounded from above.

(2) The initial endowments of the traders in H are uniformly bounded away from zero in each commodity.

(3) Each trader's preference relations are "irreflexive," "con- "continuous," and "strongly monotonic." $>_t$ is *irreflexive* if, for all $\bar{x} \in \Omega_n$, $\bar{x} \not>_t \bar{x}$. $>_t$ is *continuous* if for all $\bar{x}, \bar{y} \in \Omega_n$ the sets $\{\bar{z} \in \Omega_n \mid \bar{z} >_t \bar{x}\}$ and $\{\bar{z} \in \Omega_n \mid \bar{y} >_t \bar{z}\}$ are open sets in R_n. $>_t$ is *strongly monotonic* if $\bar{x} > \bar{y}$ implies that $\bar{x} >_t \bar{y}$.

(4) The family of all trader's preference relations in H is equimonotonic on Ω_n.

A family of preference relations is said to be *equimonotonic* on Ω_n if

$$(\forall \bar{x} \geqslant \bar{0})(\forall \bar{y} > \bar{0})(\exists \delta > 0)(\forall_i \in I)(\forall t \in \tilde{E}_i)$$
$$[\bar{z} \in S(\bar{x} + \bar{y}, \delta) \wedge \bar{w} \in S(\bar{x}, \delta) \Rightarrow \bar{z} >_t \bar{w}].$$

An example of such a family is the set of preference relations $>_f$ defined by a finite set of continuous and strongly monotonic utility functions, where $\bar{x} >_f \bar{y}$ if and only if $f(\bar{x}) > f(\bar{y})$.

Given an exchange economy $E' = \langle I(t), >_t \rangle$ (standard or nonstandard), an assignment $X(t)$ for E', and a price vector $\bar{p}$ in $^*\Omega_n$, we define the following sets for each positive real number δ:

$$D_\delta^{\bar{p}}(X) = \{t \in E' \mid \bar{p} \cdot X(t) - \bar{p} \cdot I(t) \geqslant \delta\},$$
$$F_\delta^{\bar{p}}(X) = \{t \in E' \mid (\exists \bar{y} \in \Omega_n)\, \bar{p} \cdot \bar{y} \leqslant \bar{p} \cdot I(t) \wedge (\forall \bar{w} \in S(\bar{y}, \delta))\, \bar{w} >_t X(t)\},$$
$$G_\delta^{\bar{p}}(X) = D_\delta^{\bar{p}}(X) \cup F_\delta^{\bar{p}}(X).$$

$D_\delta^{\bar{p}}(X)$ is the set of traders in E' who, if they purchase $X(t)$ at prices $\bar{p}$, violate their budget constraint by δ. $F_\delta^{\bar{p}}(X)$ is the set of traders in E' who can buy a commodity bundle $\bar{y}_t$ at prices $\bar{p}$, without violating their budget constraint, such that they prefer all $\bar{z}$ which are at most a distance δ from $\bar{y}_t$ to the commodity bundle assigned to them by $X(t)$.

THEOREM 2. *Let* $K = \langle \tilde{E}_i, X_i \rangle_{i \in I}$ *and* $H = \{\tilde{E}_i\}_{i \in I}$. *Suppose* H *is an unbounded family of exchange economies satisfying the above stated assumptions. Let* X_i *be a family of allocations, where* X_i *is in the core of* $\tilde{E}_i$, *and the* X_i *are uniformly bounded from above. Then, for every* $\delta > 0$, *there exists an* $m \in N$ *such that, for all economies* $\tilde{E}_i$ *in* H, *if* $| \tilde{E}_i | > m$, *then there exists a price vector* $\bar{p}$, *such that* $| G_\delta^{\bar{p}}(X_i) | / | \tilde{E}_i | < \delta$.

The notions of the core which we have defined for standard and nonstandard exchange economies differ in their definition of allocation and blocking. But the concept of the core as it is defined in standard exchange economies is also meaningful in nonstandard exchange economies. We shall call this the Q-core (quasistandard core). Suppose $E = \langle I(t), >_t \rangle$ is a nonstandard exchange economy satisfying all the assumptions of Section III. Let $\phi = \{Y \in {}^*\Omega_n{}^T \mid Y$ be an assignment and $\sum_{t \in T} Y(t) \leqslant \sum_{t \in T} I(t)\}$. If $Z, Y \in \phi$, then Z *Q-blocks* Y via a coalition $S \subseteq T$ iff $\sum_{t \in S} Z(t) = \sum_{t \in S} I(t)$ and for all $t \in S$, $Z(t) >_t T(t)$. If $Y \in \phi$ and Y is not Q-blocked by any $Z \in \phi$, then Y is said to be in the *Q-core*.

We shall prove Theorem 2 by first proving a lemma that the standardly bounded assignments in the Q-core of E are contained in the core of E. We then take the nonstandard extension of K, $^*K$. The economies in $^*H - H$ are nonstandard exchange economies. By assuming that

Theorem 2 is false, there will exist a $\langle E, X\rangle$ in $^*K$ where X is a standardly bounded assignment in the Q-core of the nonstandard exchange economy E, which is not a competitive allocation. Therefore, by Theorem 1, it is not in the core of E. This contradicts our lemma and proves Theorem 2.

LEMMA. *If* $(\exists\delta \gtrsim\!\!\!\!/\; 0)(\forall t \in T)(\forall j)\ I^j(t) \geqslant \delta$. *Then every standardly bounded assignment in the Q-core of E is contained in the core of E.*

Proof. Suppose X is a standardly bounded assignment in the Q-core of E and not in the core of E. Then there exists a coalition S and an allocation Y such that for all $t \in S$, $Y(t) \gg_t X(t)$ also $(1/\omega)\sum_{t\in S} Y(t) \simeq (1/\omega)\sum_{t\in S} I(t)$, $|S|/\omega \not\simeq 0$. If $(1/\omega)\sum_{t\in S} Y(t) \leqslant (1/\omega)\sum_{t\in S} I(t)$, then $\sum_{t\in S} Y(t) \leqslant \sum_{t\in S} I(t)$. Hence the assignment

$$Z(t) = \begin{cases} Y(t), & t \in S \\ I(t), & \text{otherwise} \end{cases}$$

is in ϕ and Z Q-blocks X, which contradicts the assumption that X is in the Q-core of E. So suppose for some j that $(1/\omega)\sum_{t\in S} Y^j(t) > (1/\omega)\sum_{t\in S} I^j(t)$. By assumption there exists a $\delta \gtrsim\!\!\!\!/\; 0$ such that $(1/\omega)\sum_{t\in S} I^j(t) \geqslant (|S|/\omega)\,\delta$. Hence $(1/\omega)\sum_{t\in S} Y^j(t) \gtrsim\!\!\!\!/\; 0$. Let $B_n = \{t \in S \mid Y^j(t) \geqslant 1/n\}$ for all $n \in N$; then B_n is internal for all $n \in N$ and $B_n \subseteq B_{n+1}$. Suppose for all $n \in N$, $|B_n|/\omega \simeq 0$. Then $\nu \in {}^*N - N$ such that $|B_\nu|/\omega \simeq 0$. This implies that

$$\begin{aligned}(1/\omega)\sum_{t\in S} Y^j(t) &= (1/\omega)\sum_{t\in B_\nu} Y^j(t) + (1/\omega)\sum_{t\in S/B_\nu} Y^j(t)\\ &\simeq (1/\omega)\sum_{t\in S/B_\nu} Y^j(t) \leqslant (1/\nu)(|S|/\omega) \simeq 0.\end{aligned}$$

This is true since allocations are standardly bounded. Therefore we have the contradiction that $0 \simeq (1/\omega)\sum_{t\in S} Y^j(t) \gtrsim\!\!\!\!/\; 0$. Consequently, there exists $n \in N$ such that $|B_n|/\omega \gtrsim\!\!\!\!/\; 0$. Let $\alpha = |B_n|$,

$$\gamma = (1/\omega)\sum_{t\in S} Y^j(t) - (1/\omega)\sum_{t\in S} I^j(t),$$

and for all $t \in B_n$, $\epsilon_t = \omega\gamma/\alpha$. Note that γ is a positive infinitesimal. We now define

$$Z^j(t) = \begin{cases} Y^j(t) - \epsilon_t & \text{for } t \in B_n, \\ Y^j(t) & \text{for } t \in S/B_n, \end{cases} \qquad i \neq j.$$

Let $Z^i(t) = Y^i(t)$ for all $t \in S$. Finally, let $Z(t) = I(t)$ for $t \in T/S$. Now $|B_n|/\omega \gtrsim\!\!\!\!/\; 0$ implies that ω/α is finite, hence $\epsilon_t = \omega\gamma/\alpha$ is infinitesimal. Therefore for all $t \in S$, $Z(t) \simeq Y(t)$. Hence, $Z(t) >_t X(t)$ for all $t \in S$

since $Z(t) \gg_t X(t)$. But $Z \in \phi$, since $(1/\omega) \sum_{t \in S} Z^j(t) = (1/\omega) \sum_{t \in B_n} Z^j(t) + (1/\omega) \sum_{t \in S/B_n} Z^j(t) = (1/\omega) \sum_{t \in S} Y^j(t) - \gamma = (1/\omega) \sum_{t \in S} I^j(t)$, i.e.,

$$\sum_{t \in S} Z^j(t) = \sum_{t \in S} I^j(t).$$

If there is more than one j for which $(1/\omega) \sum_{t \in S} Y^j(t) > (1/\omega) \sum_{t \in S} I^j(t)$, repeat the same construction.

Therefore we have demonstrated the existence of a $Z \in \phi$ which Q-blocks X, contradicting the assumption that $X \in Q$-core of E. Hence, if X is in the Q-core of E and standardly bounded, then X is the core of E.

The proof of Theorem 2 follows immediately. Suppose the theorem is false; then for some $\delta > 0$ and for every $m \in N$, there exists $\langle \tilde{E}_i, X_i \rangle \in K$ such that $| \tilde{E}_i | > m$ and X_i have the property that for all price vectors $\bar{p}$, $| G_\delta^{\bar{p}}(X_i)|/| \tilde{E}_i | \geqslant \delta$. If we now transfer this statement to $*K$, the nonstandard extension of K, then for some positive real number δ and for all $m \in *N$, there exists $\langle E', X \rangle \in *K$ such that $| E' | > m$, X is in the core of E' if E' is a standard economy, and X is a standardly bounded assignment in the Q-core of E' if E' is a nonstandard economy. In any case, X has the property that, for all price vectors $\bar{p}$, $| G_\delta^{\bar{p}}(X)|/| E' | \geqslant \delta$. Choosing m to be some nonstandard integer $\omega \in *N - N$, there exists $\langle E', X \rangle \in *K - K$ satisfying the above. But E is a nonstandard exchange economy and X is not a competitive allocation. Hence X is not in the core of E by Theorem 1. Since X is a standardly bounded assignment in the Q-core of E, this contradicts our Lemma. Note that the assumptions on H, the unbounded family of standard exchange economies, are sufficient to guarantee that all of the nonstandard exchange economies in $*H$ satisfy the conditions of Theorem 1.

We would like to point out that, although Theorem 2 is a consequence of Theorem 1, it is not "equivalent" to Theorem 1. A less intuitive result in terms of sequences of economies which is "equivalent" to Theorem 1 is given in [3].

V. Relationship with the Debreu-Scarf Theorem

Let E be a standard exchange economy having m traders. The r-fold replica of E, denoted $E^{(r)}$, consists of r copies of E. That is, $E^{(r)}$ has mr traders consisting of m types of traders where each type has r traders and all members of a type have the same initial endowment and preference relation. Moreover each type has the same preference and initial endowment as some trader in the original economy E. Every allocation in E defines an allocation in $E^{(r)}$ by replication, i.e., each member of a type is given the same commodity bundle as his counterpart in E. If X is an allocation in E, then $X^{(r)}$ will denote the replicated allocation in $E^{(r)}$. The version of the Debreu-Scarf Theorem which best compares with our work is as follows.

THEOREM 3 (*Debreu-Scarf*). *If each trader's utility function has the properties of strong monotonicity, strong convexity, and continuity, and each trader has strictly postive initial endowments, then an allocation X is a competitive allocation for E iff $X^{(r)}$ is in the core of $E^{(r)}$ for all $r \geqslant 1$. (Note that the original Debreu-Scarf theorem assumed the weaker condition of insatiability in lieu of strong monotonicity.)*

Theorem 1 and our Lemma allow us to prove Theorem 3 under much weaker assumptions on E.

THEOREM 4. *If each trader's preference relation is irreflexive, strongly monotonic, and continuous, and each trader has strictly positive initial endowments, then an allocation X is a competitive allocation for E iff $X^{(r)}$ is in the core of $E^{(r)}$ for all $r \geqslant 1$.*

Proof. Suppose $X^{(r)}$ is in the core of $E^{(r)}$ for all $r \geqslant 1$. Then by transfer $X^{(\rho)}$ is a standardly bounded assignment in the Q-core of $E^{(\rho)}$ for all $\rho \in {}^*N - N$. Let $\omega \in {}^*N - N$; then $E^{(\omega)}$ satisfies all the assumptions of Theorem 1 and the Lemma. Therefore $X^{(\omega)}$ is in the core of $E^{(\omega)}$ by the Lemma, and there exists a price vector $\bar{p}$ such that $\langle \bar{p}, X^{(\omega)} \rangle$ is a competitive equilibrium for $E^{(\omega)}$. In [3] it is shown that $\bar{p}$ is a standard price vector in the interior of Ωn. We claim that $\langle \bar{p}, X^{(1)} \rangle$ is a competitive equilibrium for the original economy $E^{(1)}$. If not, at least one trader, t, has a $\bar{y}$ in his budget set which he prefers to $X^{(1)}(t)$. In the ωth fold-replica $E^{(\omega)}$, there will then be at least $m\omega$ traders who prefer $\bar{y}$ and everything in a standard ball about $\bar{y}$ to their assignments by $X^{(\omega)}$, and $\bar{y}$ is in their budget set. This contradicts the assertion that $\langle \bar{p}, X^{(\omega)} \rangle$ is a competitive equilibrium for $E^{(\omega)}$.

We would like to point out that Theorems 3 and 4 are about a single economy E and in fact are characterizations of those core allocations of E which are competitive. Also, the principal important difference in the two theorems is the dropping of the convexity assumption. Scarf [9] has shown that a sufficient condition for an exchange economy to have a nonempty core is that as a cooperative game it be balanced. One might hope to find an interesting class of nonconvex balanced economies and show the existence of a competitive equilibrium by applying Theorem 4.

An excellent discussion of the published literature pertaining to

Edgeworth's conjecture is given in [1]. Edgeworth's original analysis of this problem may be found in [5].

Hildenbrand has pointed out to us that he has independently proved a theorem similar to our Theorem 2 and that he has also proved Theorem 4. His theorems appear in his forthcoming book, "Core and Equilibrium in a Large Economy."

Preliminary versions of Theorems 1 and 2 were first announced by Brown-Robinson in "A limit theorem on the cores of large standard exchange economies" *Proc. Nat. Acad. Sci. U.S.A.* **69**, 1258–1260, 1972. A correction to this article appeared in *Proc. Nat. Acad. Sci. U.S.A.* **69**, 3068, 1972.

References

1. K. J. Arrow and F. H. Hahn, "General Competitive Analysis," Holden-Day, San Francisco, CA, 1971.
2. R. J. Aumann, Markets with a continuum of traders, *Econometrica* **32** (1964), 39–50.
3. D. J. Brown and A. Robinson, Nonstandard exchange economies, *Econometrica*, to be published.
4. G. Debreu and H. Scarf, A limit theorem on the core of an economy, *Int. Econ. Rev.* (1963), 235–246.
5. F. Y. Edgeworth, "Mathematical Psychics," Kegan Paul, London, 1881.
6. S. Kochen, Ultraproducts in the theory of models, *Ann. Math.* **79** (1961), 221–261.
7. W. A. J. Luxemburg, A general theory of monads, presented of the Symposium on Applications of Model Theory to Analysis and Algebra, Pasadena, May 1967.
8. A. Robinson, Non-standard theory of Dedekind rings, *Proc. Acad. Sci. Amsterdam A* **70** (1967), 444–452.
9. H. Scarf, The core of an *N*-person game, *Econometrica* **35** (1967), 50–69.

NONSTANDARD EXCHANGE ECONOMIES

BY DONALD J. BROWN AND ABRAHAM ROBINSON[1]

Edgeworth's conjecture that as the number of traders in an exchange economy increases the core approaches the set of competitive equilibria has been formalized both as a theorem about a sequence of finite economies, and as a theorem about an economy having an infinite number of agents.

This paper, using nonstandard analysis, provides a synthesis of these two approaches. It is shown that the core and the set of competitive equilibria are equivalent within a nonstandard exchange economy. This theorem implies an asymptotic theorem concerning the core and competitive equilibria of sequences of finite economies.

1. INTRODUCTION

AN EXCHANGE ECONOMY consists of a set of traders, each of whom is characterized by an initial endowment and a preference relation. In addition, one usually assumes that the set of traders is finite. However, in order to state theorems precisely concerning the asymptotic or limiting properties of the core (such theorems will be called limit theorems), economies have been studied which have an infinite number of traders. For instance, there are the denumerable economies of Debreu-Scarf [**3**] and the nondenumerable economies of Aumann [**1**].

The concepts of interest, here the core and competitive equilibrium, can be defined even in infinite economies. Hence, one has the option of taking an imprecise statement about the core, such as Edgeworth's conjecture that "the core approaches the set of competitive allocations as the number of traders increases" and translating it into a theorem, T', about the core of an infinite economy or expressing it as a limit theorem, T, about the cores of large but finite economies.

Of course, these two approaches need not be incompatible in that one might be able to define a limiting process such that a statement T about the asymptotic behavior of a family of large but finite economies is reflected in a statement T' about a certain infinite economy, in the sense that T is true if and only if T' is true. In this context Edgeworth's conjecture gives rise to two theorems, T and T'.

Although we are primarily interested in the truth of T, we shall establish T by proving T' and by showing that T is a consequence of a general metamathematical argument. Because of this argument, it is not surprising that an area of mathematical logic, model theory, provided an appropriate framework for defining the limiting process described above.

[1] This research was supported in part by the National Science Foundation (GP18728) and the Office of Naval Research. We are happy to acknowledge several stimulating conversations with H. Scarf. Also we would like to acknowledge Aumann's seminal paper [**1**] in that the idea of the proof of Theorem 1 is based on his paper.

The fundamental result that we will need from model theory is the existence of a particular kind of extension of the real numbers, R, called the nonstandard numbers and denoted $^*R$. $^*R$ has all the formal properties of R in that any mathematical property of R which can be described in some given language, which may be the first order predicate calculus or some higher language, can be translated into a mathematical property of $^*R$. Moreover, the sentence, φ, expressing the fact that this property holds for R is true if and only if its interpretation in $^*R$, $^*\varphi$, is a true sentence about $^*R$. In addition, $^*R$ contains nonzero infinitesimals, i.e., nonzero numbers whose absolute value is less than any positive real number, and infinite integers, i.e., integers which are greater than every real integer.

It is the existence of infinitesimals and infinite integers that permits us to formulate precisely the notion of perfect competition, the economic concept which underlies the Edgeworth conjecture.

The idealized notion of perfect competition is that the action of any finite set of traders in terms of their willingness to buy or sell at the competitive prices should have a negligible effect on these prices. Clearly this cannot be the case for any finite economy, but perfect competition is a meaningful concept for an economy having ω traders, where ω is an infinite integer, and where the endowment for each trader relative to that of the whole economy is infinitesimal.

After defining a nonstandard exchange economy and the notions of the core and competitive equilibrium in such an economy, we will prove that Edgeworth's conjecture is true in the following sense:

THEOREM 1: *If $\mathscr{E}$ is a nonstandard exchange economy, then the allocation X is in the core of $\mathscr{E}$ if and only if there exists a price vector, $\bar{p}$, such that $\langle \bar{p}, X \rangle$ is a competitive equilibrium.*

We then define competitive equilibrium and core allocations for a family of large but finite exchange economies, $\mathscr{G}$. Allocations, price vectors, and preference relations for $\mathscr{G}$ consist of sequences of allocations, price vectors, and preference relations belonging to the economies which make up $\mathscr{G}$.

We show that:

THEOREM 2: *The allocation X is in the core of $\mathscr{G}$ if and only if there exists a price vector, $\bar{p}$, such that $\langle \bar{p}, X \rangle$ is a competitive equilibrium for $\mathscr{G}$.*

Theorems 1 and 2 correspond to T' and T discussed earlier. Theorem 2 is another formulation of Edgeworth's conjecture, but in terms of the relationship between core allocations and equilibria of large but finite economies.

2. DEFINITIONS AND ASSUMPTIONS

The basic mathematical notions that we will need will come from an area of model theory which is termed nonstandard analysis. An informal introduction to these ideas follows. The interested reader may consult [**6**, **7**, and **8**] for details.

Let R be the field of real numbers. We consider the properties of R in a higher-order language, L, which includes symbols for all individuals (that is, numbers) of R, all subsets of R, all relations of two, three, . . . , variables between individuals of R, all functions from individuals to individuals of R, and, more generally, all relations and functions of a finite type (for example, functions from sets of numbers into relations between numbers) that can be defined beginning with the numbers of R. Let K be the set of all sentences formulated in L which hold (are true) in R. Then there exists a structure $*R$ with the following properties:

2.1: $*R$ is a model of K "in Henkin's sense." That is, all sentences of K hold in $*R$, provided, however, that we interpret all quantifiers other than those referring to individuals in a nonstandard fashion, as follows. Within the class of all entities of any given type other than 0 (the type of individuals) there is distinguished a certain subclass of entities called *internal*. And, for such a type, the quantifiers "for all x" and "there exists an x" are to be interpreted as "for all internal x," "there exists an internal x" (of the given type). Such sentences are said to be true in $*R$ by transfer. R can be injected into $*R$, and so $*R$ may, and will, be regarded as an extension of R.

2.2: Every *concurrent* binary relation $S(x, y)$ in R possesses *a bound* in $*R$. The relation $S(x, y)$, of any type is called *concurrent* if for any $a_1, \ldots, a_n$, $n \geqslant 1$, for which there exists $b_1, \ldots, b_n$ such that $S(a_1, b_1), \ldots, S(a_n, b_n)$ hold in R, there also exists a b such that $S(a_1, b), \ldots, S(a_n, b)$ hold in R. The entity b_s in $*R$ is called *a bound* for $S(x, y)$ if $S(a, b_s)$ holds in $*R$ for every a for which there exists a b such that $S(a, b)$ holds in R.

2.3: For any mapping $f(y)$ from a set A (of any type) in R into the extension $*B$ of a set B in R, there exists an internal mapping $\varphi(y)$ from $*A$, the extensions of A, into $*B$ which coincides with $f(y)$ on A.

Any structure $*R$ which satisfies 2.1, 2.2, and 2.3 is called a *comprehensive enlargement of* R. In particular $*R$ can be constructed as an ultrapower of R. (See [**4**] and [**5**] for the notions of an ultrapower.) It is shown in [7] that $*R$ has the following properties:

(i) The set of individuals of $*R$ is an ordered non-archimedean field. In this paper we shall call all individuals of $*R$ *non-standard numbers*.

(ii) R, the real numbers, is a proper ordered subfield of the non-standard numbers. Elements of R will be called *standard numbers*. Observe that according to our present usage every standard number is a nonstandard number.

(iii) There exist nonstandard numbers, in particular nonstandard integers, which are greater than every standard number. These numbers are called *infinite nonstandard numbers* and *infinite integers* respectively.

(iv) There exist numbers in $*R$ which are in absolute value less than any positive real number, in fact they, except zero, are just the reciprocals of the infinite non-standard numbers. These numbers are called *infinitesimals*.

(v) The system of internal entities in $^*R$ has the following property: If S is an internal set of relations, then all elements of S are internal. More generally, if S is an internal n-ary relation, $n \geqslant 1$ and the n-tuple $(S_1, \ldots, S_n)$ satisfies (belongs to) S, then $S_1, \ldots, S_n$ are internal.

A nonstandard number is said to be finite if in absolute value it is less than some standard number. If r is a finite nonstandard number, then there exists a unique standard number called the *standard part of* r, denoted by ${}^0r$, such that $r - {}^0r$ is an infinitesimal. Hence there are three kinds of nonstandard numbers: (i) the infinitesimals which in absolute value are less than every standard number, (ii) the finite nonstandard numbers which in absolute value are less than some standard number, and (iii) the infinite nonstandard numbers which are greater than every standard number. Note that according to this taxonomy every infinitesimal is a finite nonstandard number.

We will denote the n-dimensional vector space over $^*R$ by $^*R_n$. A commodity bundle $\bar{x}$ is a point in the nonnegative orthant $^*\Omega_n$ of $^*R_n$. Let $^*\rho$ be the nonstandard extension of the Euclidean metric ρ. Then a set of points $S(\bar{x}, r)$ in $^*R_n$ will be called an *S-ball* if there exists $\bar{x} \in {}^*R_n$ and a positive standard number r such that $S(\bar{x}, r) = \{\bar{y} | {}^*\rho(\bar{x}, \bar{y}) < r\}$. It is shown in [7] that the S-balls may serve as a basis for a topology in $^*R_n$. This topology will be called the *S-topology*.

The *monad of* $\bar{x}$, $\mu(\bar{x})$, is the set of points whose distance from $\bar{x}$ is an infinitesimal. If $\bar{y} \in \mu(\bar{x})$, we shall write $\bar{x} \simeq \bar{y}$. $\bar{x} > \bar{y}$, $\bar{x} \geqslant \bar{y}$, and $\bar{x} \geqslant \bar{y}$ will have their conventional meanings: $\bar{x} \gtrsim \bar{y}$ means that $\bar{x}$ is greater than $\bar{y}$ or $\bar{y}$ exceeds $\bar{x}$ by at most an infinitesimal amount in each coordinate; and $\bar{x} \gtrapprox \bar{y}$ means that $\bar{x}$ is greater than $\bar{y}$ by a noninfinitesimal amount in each coordinate.

The *S-convex hull* of a set B is defined as the set of all finite convex combinations of elements of B, i.e., vectors of the form $\bar{y} = \alpha_1\bar{x}_1 + \ldots + \alpha_n\bar{x}_n$ where $\bar{x}_i \in B$, $\alpha_i \geqslant 0$, $\Sigma_{i=1}^n \alpha_i = 1$, the α_i are nonstandard numbers, and n is a standard integer. B is defined as *S-convex* if B contains its S-convex hull.

A *nonstandard exchange economy* is defined as a pair of indexed sets, $\{\bar{x}_t\}_{t=1}^{\omega}$ and $\{\succ_t\}_1^{\omega}$ where for all t, $\bar{x}_t \in {}^*\Omega_n$, $\succ_t \subseteq {}^*\Omega_n \times {}^*\Omega_n$, and ω is an infinite (nonstandard) integer. Interpret $\bar{x}_t$ as the initial endowment of the tth trader and interpret $\succ_t$ as his preference relation. The nonstandard exchange economies which we will consider are assumed to have the following properties:

(i) The function indexing the initial endowments, $I(t)$, is internal.

(ii) $I(t)$ is standardly bounded, i.e., there exists a standard vector $\bar{r}_0$ such that for all t, $I(t) \leqslant \bar{r}_0$.

(iii) $(1/\omega)\,\Sigma_1^{\omega}\, I(t) \gtrapprox \bar{0}$.

(iv) The relation, Q, where $Q = \{\langle t, \succ_t\rangle | t \in T, \succ_t \subseteq {}^*\Omega_n x {}^*\Omega_n\}$ is internal.

(α) $\succ_t$ is irreflexive.

(β) If $\bar{x} \geqslant \bar{y}$, then $\bar{x} \succ_t \bar{y}$, i.e., $\succ_t$ is monotonic.

(γ) For all finite $\bar{x}, \bar{y} \in {}^*\Omega_n$, if $\bar{x} \gtrapprox \bar{0}$, then for every $\bar{w} \simeq \bar{x} + \bar{y}$, $\bar{z} \simeq \bar{y}$, $\bar{w} \succ \bar{z}$, where $\bar{x} \gtrapprox \bar{0}$ means that $\bar{x}$ is greater than $\bar{0}$ by a noninfinitesimal amount in at least one component.

An *assignment* is an internal function from T, the set of traders, into $^*\Omega_n$.

An *allocation or final allocation* is a standardly bounded assignment $Y(t)$ from the set of traders, $\{1, 2, \ldots, \omega\}$, into $*\Omega_n$ such that

$$\frac{1}{\omega}\sum_{t=1}^{\omega} Y(t) \simeq \frac{1}{\omega}\sum_{t=1}^{\omega} I(t).$$

Condition (iv) implies for all internal $X(t), Y(t) \in *\Omega_n^T$, where T is the set of traders, that:

(v) $\{t | X(t) \succ_t Y(t)\}$ is an internal set of traders.

A *coalition*, S, is defined as an internal set of traders. It is said to be negligible if $|S|/\omega \simeq 0$. Note that if S is negligible then for all allocations

$$X(t), \frac{1}{\omega}\sum_{t\in S} X(t) \simeq \bar{0}.$$

A coalition, S, is effective for an allocation Y if $(1/\omega)\,\Sigma_{t\in S}\,Y(t) \simeq (1/\omega)\,\Sigma_{t\in S}\,I(t)$.

$$\bar{x} \gg_t \bar{y} \text{ iff } (\forall\ \bar{w} \in \mu(\bar{x}))\bar{w} \succ_t \bar{y}.$$

An allocation Y *dominates* an allocation X via a coalition S if for all $t \in S$, $Y(t) \gg_t X(t)$, and S is effective for Y.

The *core* is defined as the set of all allocations which are not dominated via any non-negligible coalition.

A *price vector*, $\bar{p}$, is a finite nonstandard vector such that $\bar{p} \gtrsim \bar{0}$.

The tth traders budget set, $B_{\bar{p}}(t)$, is $\{\bar{x} \in *\Omega_n | \bar{p}\cdot\bar{x} \lesssim \bar{p}\cdot I(t)\}$.

We say that $\bar{y}$ is *maximal in* $B_{\bar{p}}(t)$ if $\bar{y} \in B_{\bar{p}}(t)$ and there does not exist an $\bar{x} \in B_{\bar{p}}(t)$ such that $\bar{x} \gg_t \bar{y}$.

A *competitive equilibrium* is defined as a pair $\langle \bar{p}, X\rangle$, where $\bar{p}$ is a price vector and X an allocation such that there exists an internal set of traders K for which $|K|/\omega \simeq 1$ and $X(t)$ is maximal in $B_{\bar{p}}(t)$ for all $t \in K$.

3. THEOREMS

We shall first demonstrate that there exist nonstandard economies satisfying the assumptions we wish to make concerning the initial endowments and preferences of the traders.

Let f be a monotonic continuous function of Ω_n into R and define the relation $\succ_t$ on Ω_n as $\bar{x} \succ_f \bar{y}$ if $f(\bar{x}) > f(\bar{y})$ for all $\bar{x}, \bar{y} \in \Omega_n$. Then it is shown in [7] that $*f: D \to *R$ is S-continuous where $*f$ is the nonstandard extension of f and D a standardly bounded subset of $*\Omega_n$. Now $*f$ induces a preference relation $\succ_{*f}$, on $*\Omega_n$ where for all $\bar{x}, \bar{y} \in *\Omega_n, \bar{x} \succ_{*f} \bar{y}$ if $*f(\bar{x}) > *f(\bar{y})$. Note that $\succ_{*f} = *(\succ_{*f})$.

It is obvious that $\succ_{*f}$ is irreflexive and monotonic. Suppose $\bar{x}, \bar{y}$ are finite vectors in $*\Omega_n$ and $\bar{x} \gtrsim \bar{0}$. Then $*f({}^0\bar{x} + {}^0\bar{y}) = f({}^0\bar{x} + {}^0\bar{y}) > f({}^0\bar{y}) = *f({}^0\bar{y})$. Since $f({}^0\bar{x} + {}^0\bar{y})$ and $f({}^0\bar{y})$ are standard numbers and $*f({}^0\bar{y}) \simeq *f(\bar{y}), *f({}^0\bar{x} + {}^0\bar{y}) \simeq *f(\bar{x} + \bar{y})$; the continuity of f implies that assumption (iv) (γ) is satisfied.

An example of a function having all of the above properties is $f(\bar{x}) = x_1 + \ldots + x_n$ for all $x \in \Omega_n$. Hence $*f(\bar{x}) = x_1 + \ldots + x_n$ for all $x \in *\Omega_n$.

CONSISTENCY LEMMA: *Let $\mathscr{G}$ be a countably infinite standard economy which has the following properties for all traders, t, in $\mathscr{G}$:*

(i) *There exists a standard vector $\bar{r}_0$ such that $I(t) \leqslant \bar{r}_0$.*

(ii) $I(t) > \bar{0}$.

(iii) $\succ_t = \succ_f$, *where f is a monotonic continuous function of Ω_n into R and for all $\bar{x}, \bar{y} \in \Omega_n$ we use the same f for every t. If $\bar{x} \geqslant \bar{y}$, then $f(\bar{x}) > f(\bar{y})$.*

Then there exists a nonstandard exchange economy $\mathscr{G}$ satisfying the assumptions of Section (2).

PROOF: Let N be the nonnegative standard integers; then $I: N \to R_n$. Thus its nonstandard extension $^*I$ maps $^*N$ into $^*R_n$. $^*I$ restricted to $\{1, 2, \ldots, \omega\}$ for any infinite integer ω is a function which satisfies assumptions (i), (ii), and (iii) in Section (2). We then assign the same preference relation $\succ_{^*f}$ to each of the traders $1, 2, \ldots, \omega$. This provides the desired model.

In our analysis of a nonstandard exchange economy the existence of an equilibrium price vector will be established by invoking a separating hyperplane theorem. Since our notions of convexity and topology are not internal notions, we cannot prove separating hyperplane theorems in $^*R_n$ by simple "transfer" from R_n.

A set D in $^*R_n$ is said to be *nearstandard* if every vector in D is finite. We shall establish a separation theorem for nearstandard S-convex sets. The S-interior of a subset, A, of $^*R_n$ is the union of all the subsets of A which are open in the S-topology. The S-closure of A is the complement of the S-interior of the complement of A.

LEMMA 1: *If A is a subset of $^*R_n$ and $\bar{0} \notin$ S-interior of A, S-int(A), then for all $\bar{x} \in$ S-int(A), $\bar{\mu}(0) \cap \bar{\mu}(x) = \varnothing$.*

PROOF: Suppose there exists an $\bar{x} \in S$-int (A) such that $\mu(\bar{0}) \cap \mu(\bar{x}) \neq \varnothing$; then $^*\rho(\bar{0}, \bar{x}) \simeq 0$. But $\mu(\bar{x}) \subset S(\bar{x}, r) \subset A$ for some standard r. Hence $S(\bar{0}, r) \subset A$, a contradiction.

LEMMA 2: *If A a subset of $^*R_n$ and $\bar{0} \notin$ S-int (A), then $\bar{0} \notin$ standard part of S-int (A), ^{0}(S-int(A)). ^{0}(S-int (A)) $= \{{}^0\bar{x} | \bar{x} \in$ S-int (A), $\bar{x}$ finite$\}$.*

PROOF: If $\bar{0} \in {}^0(S$-int $(A))$, then there exists an $\bar{x} \in S$-int (A) such that $\bar{0} \in \mu(\bar{x})$, i.e., $\mu(\bar{x}) = \mu(\bar{0})$. But this contradicts Lemma 1.

To show that the S-interior of an S-convex set in $^*R_n$ is S-convex, we will adapt a well-known proof that the interior of a convex set in R_n is convex [**9**].

Let $\bar{S}(\bar{x}, \delta)$ denote the closed S-ball centered on $\bar{x}$ with radius δ and $B = \bar{S}(\bar{0}, 1)$; then $\bar{S}(\bar{x}, \delta) = \bar{x} + \delta B$. The S-closure, S-cl (A), and S-int (B) can then be expressed as S-cl $(A) = \cap\{A + \delta B | \delta \gtrapprox 0\}$ and S-int $(A) = \{\bar{x} |$ there exists a $\delta \gtrapprox 0, x + \delta B \subset A\}$.

LEMMA 3: *Let A be a S-convex set in* $^*R_n$. *Let* $\bar{x} \in S\text{-}int(A)$ *and* $\bar{y} \in S\text{-}cl(A)$. *Then* $(1-\lambda)\bar{x} + \lambda\bar{y}$ *belongs to S-int (A) for* λ *such that* $0 \leqslant \lambda \lessapprox 1$.

PROOF: It is sufficient to prove the theorem for standard λ. Suppose λ standard and $\in [0, 1)$; then we must show that $(1-\lambda)\bar{x} + \lambda\bar{y} + \delta B \subseteq A$ for some standard $\delta > 0$. We have $\bar{y} \in A + \delta\beta$ for all standard $\delta > 0$, because $\bar{y} \in S$-cl (A). Thus, for every standard $\delta > 0$, $(1-\lambda)\bar{x} + \lambda\bar{y} + \delta B \subset (1-\lambda)\bar{x} + \lambda(A + \delta B) + \delta B = (1-\lambda)[\bar{x} + \delta(1+\lambda)(1-\lambda)^{-1}B] + \lambda A$. But $\bar{x} + \delta(1+\lambda)(1-\lambda)^{-1}B \subset A$ for sufficiently small δ. Hence, $(1-\lambda)\bar{x} + \lambda\bar{y} + \delta B \subset (1-\lambda)A + \lambda A = A$ for some standard $\delta > 0$.

LEMMA 4: *If A is a nearstandard S-convex set in* $^*R_n$, *then S-int (A) is S-convex.*

PROOF: Take $\bar{y}$ to be in *S-int* (A) in Lemma 3, for $0 \leqslant \lambda \lessapprox 1$. For $1 \simeq \lambda$, $(1-\lambda)\bar{x} + \lambda\bar{y} \simeq \bar{y}$. But $y \in S$-int (A); hence we are done.

LEMMA 5: *If A is an S-convex set in* $^*R_n$, *then* ^0A *is a convex set in* R_n.

PROOF: Suppose $\bar{x}, \bar{y} \in {}^0A$, and α a standard number such that $0 \leqslant \alpha \leqslant 1$. Then there exist infinitesimals $\bar{\varepsilon}_1$, $\bar{\varepsilon}_2$ such that $\bar{x} + \bar{\varepsilon}_1$, $\bar{y} + \bar{\varepsilon}_2 \in A$. But A is S-convex which implies that $\alpha(\bar{x} + \bar{\varepsilon}_1) + (1-\alpha)(\bar{y} + \bar{\varepsilon}_2) \in A$. We can express $\alpha(\bar{x} + \bar{\varepsilon}_1) + (1-\alpha)(\bar{y} + \bar{\varepsilon}_2)$ as $\alpha\bar{x} + (1-\alpha)\bar{y} + \bar{\varepsilon}_3$, where $\bar{\varepsilon}_3 = \alpha\bar{\varepsilon}_1 + (1-\alpha)\bar{\varepsilon}_2 \simeq \bar{0}$. Thus $\alpha\bar{x} + (1-\alpha)\bar{y} \in {}^0A$.

S-SEPARATION LEMMA: *If A is a nearstandard S-convex set in* $^*R_n$ *and* $\bar{0} \notin S\text{-}int(A)$, *then there exists a standard* $\bar{p} \neq \bar{0}$ *such that for all* $\bar{y} \in S\text{-}int(A)$, $\bar{p} \cdot \bar{y} \gtrsim 0$.

PROOF: By Lemma 4 S-int (A) is S-convex. By Lemma 5 $^0(S\text{-int}(A))$ is a convex set in R_n. By Lemma 2 $\bar{0} \notin {}^0(S\text{-int}(A))$. Hence, there exists a $\bar{p} \in R_n$ such that $\bar{p} \neq \bar{0}$ and for all $\bar{x} \notin {}^0(S\text{-int}(A))$, $\bar{p} \cdot \bar{x} \geqslant 0$ (see [2] for a proof of this result). Every $\bar{y} \in S$-int (A) can be expressed as $\bar{y} = \bar{x} + \bar{\varepsilon}$ where $\bar{x} \in {}^0(S\text{-int}(A))$ and $\bar{\varepsilon} \simeq \bar{0}$. Thus, for all $y \in S$-int (A), we have that $\bar{p} \cdot \bar{y} = \bar{p} \cdot \bar{x} + \bar{p} \cdot \bar{\varepsilon} \geqslant 0 + \bar{p} \cdot \bar{\varepsilon} \simeq 0$, i.e., $p \cdot y \gtrsim 0$.

Let $\mathscr{E}$ be the nonstandard exchange economy where the set of traders is an internal set T and $|T| = \omega$, an infinite integer. A set of traders U is said to be full if U is internal and $|T - U|/\omega \simeq 0$. Let $X(t)$ be a fixed allocation in the core of $\mathscr{E}$; then $F_n^{(t)} = \{\bar{x} \in {}^*\Omega_n | \bar{x}$ finite, $(\forall \bar{w} \in S_{1/n}(\bar{x} + I(t))\bar{w} \succ_t X(t)\}$, $G_n(t) = F_n(t) - I(t)$, $G(t) = \bigcup_{n \in N} G_n(t)$, and $F(t) = \bigcup_{n \in N} F_n(t)$. Let $\Delta(U)$ denote the S-convex hull of the union $\bigcup_{t \in T} G(t)$. Note that $F(t)$ and $G(t)$ are S-open.

PRINCIPAL LEMMA: *There exists a full set of traders U such that* $\bar{0} \notin S\text{-}int(\Delta(U))$.

PROOF: For each finite $\bar{x} \in {}^*R_n$, let $G^{-1}(\bar{x})$ be the set of all traders t for which $G(t)$ contains $\bar{x}$. Then $G^{-1}(\bar{x}) = \bigcup_{n \in N} G_n^{-1}(\bar{x})$, where $G_n^{-1}(x) = \{t \in T | (\forall \bar{w} \in S_{1/n}(\bar{x} + I(t))\bar{w} \succ_t X(t)$ and $\bar{x} + I(t) \in {}^*\Omega_n\}$.

It follows from assumption (v) that for each n and every $\bar{x}$, $G_n^{-1}(\bar{x})$ is internal. Let M be the set of all those standard rational points $\bar{r} \in {}^*R_n$ (i.e., points with standard rational coordinates) such that for all $n \in N$, $G_n^{-1}(\bar{r})$ is negligible. Since the standard rational points can be put into a one-to-one correspondence with the standard integers, we can express M as $\{\bar{r}_i\}_{i\in N}$. For each $\bar{r}_i$ there exists a $v_i \in {}^*N$ such that $G^{-1}(\bar{r}_i) \subseteq G_{v_i}^{-1}(\bar{r}_i)$, where $G_{v_i}^{-1}(\bar{r}_i)$ is internal and negligible.

This follows from the assumption that we are working in a comprehensive enlargement and from the following theorem proved in [7]: Let $\{S_n\}_{n\in{}^*N}$ be an internal sequence such that S_n is infinitesimal for all finite n. Then there exists an infinite integer v such that S_n is infinitesimal for all $n < v$.

The same argument can be used to show that if

$$B_n = \bigcup_{i=0}^{n} G_{v_i}^{-1}(\bar{r}_i), \quad n \in N,$$

then there exists an infinite integer ρ such that B_ρ is negligible and $B_n \subseteq B_\rho$ for all $n \in N$. Define $U = T - B_\rho$; then U is full.

Suppose $\bar{0} \in S\text{-int}\,(\Delta(U))$. Then there exists an $\bar{x} \gtrapprox \bar{0}$ such that $-\bar{x} \in \Delta(U)$. By the definition of $\Delta(U)$, $-\bar{x}$ is a S-convex combination of k points in $\bigcup_{t\in U} G(t_i)$, where k is finite. That is, there are traders $t_1, \ldots, t_k \in U$ (not necessarily distinct), points $\bar{x}_i \in G(t_i)$, and positive standard numbers $\beta_1, \ldots, \beta_k$ such that $\Sigma_{i=1}^k \beta_i\bar{x}_i = -\bar{x} \lessapprox \bar{0}$. We may assume that for all i, $\beta_i\bar{x}_i \gtrapprox \bar{0}$. Then $\bar{x}_i \in G(t_i)$ implies that for some $n \in N$, $\{\forall\, \bar{w} \in S_{1/n}(\bar{x}_i + I(t_i))\bar{w} \succ_{t_i} X(t_i)\}$. Hence, in all cases there exist standard rational points $\bar{r}_i \in G(t_i)$ sufficiently close to the $\bar{x}_i$, and positive standard rational numbers γ_i sufficiently close to the $\bar{r}_i$ so that we still have

$$\sum_{i=1}^{k} \gamma_i\bar{r}_i \lessapprox \bar{0} \quad \text{and} \quad \sum_{i=1}^{k} \gamma_i = 1.$$

Without loss of generality, we can assume that the r_i are noninfinitesimal.

Let $-\bar{r} = \Sigma_{i=1}^k \gamma_i\bar{r}_i$ and pick an arbitrary trader t_0 in U. Since $\bar{r} \gtrapprox \bar{0}$, we have $\alpha\bar{r} + I(t_0) \gtrapprox X(t_0)$ for sufficiently large positive finite rational α. Hence, by the monotonic assumption $\alpha\bar{r} + I(t_0) \succ\!\!\succ_t X(t_0)$, i.e., $\alpha\bar{r} \in G(t_0)$. Now set $\bar{r}_0 = \alpha\bar{r}$, $\alpha_0 = 1/(\alpha + 1)$, $\alpha_i = \alpha\gamma_i/(\alpha + 1)$ for $i \doteq 1, \ldots, k$. Then $\alpha_i > 0$ for all i and $\Sigma_0^k \alpha_i = 1$; furthermore $\Sigma_{i=0}^k \alpha_i\bar{r}_i = (\alpha/(\alpha + 1))\bar{r} + (\alpha/(\alpha + 1))\,\Sigma_{i=1}^k \gamma_i\bar{r}_i = (\alpha/(\alpha + 1))(r - \bar{r}) = \bar{0}$, and for all i, $r_i \in G(t_i)$. Then $t_i \in G^{-1}(\bar{r}_i)$, and since $t_i \in U$, it follows that $\bar{r}_i \notin M$. Hence for all i, $\exists n_i \in N$ such that $|G_{n_i}^{-1}(\bar{r}_i)|/\omega \not\simeq 0$.

We shall show that for a sufficiently small positive standard number δ, we can find disjoint subsets A_i' of $G_{n_i}^{-1}(\bar{r}_i)$ such that $|A_i'|/\omega \simeq \delta\alpha_i$. Let

$$\theta = \min_{1\leqslant i\leqslant k} \left\{\frac{|G_{n_i}^{-1}(\bar{r}_i)|}{\omega\alpha_i}\right\}$$

and $\delta = \theta/k$. Clearly, this δ will do. Let $A' = \bigcup_{i=0}^k A_i'$ and define the following internal function:

$$Y(t) = \begin{cases} \bar{r}_i + I(t) & \text{for} \quad t \in A_i' \\ I(t) & \text{for} \quad t \notin A' \end{cases}.$$

It is obvious that $Y(t) \in {}^*\Omega_n$ for $t \notin A'$, and for $t \in A_i'$ it follows from $\bar{r}_i \in G(t)$, i.e., $A_i' \subset G_{n_i}^{-1}(\bar{r}_i)$; hence $\bar{r}_i + I(t) \in F_{n_i}(t) \subset {}^*\Omega_n$. Next,

$$\frac{1}{\omega}\sum_{t\in A'} Y(t) = \frac{1}{\omega}\sum_{i=0}^{k}\sum_{t\in A_i'} Y(t) = \frac{1}{\omega}\sum_{i=0}^{k}\sum_{t\in A_i'}\{\bar{r}_i + I(t)\} = \frac{1}{\omega}\sum_{i=0}^{k}\sum_{t\in A_i'}\bar{r}_i$$

$$+ \frac{1}{\omega}\sum_{i=0}^{k}\sum_{t\in A_i'} I(t) = \frac{1}{\omega}\sum_{i=0}^{k}|A_i|\bar{r}_i + \frac{1}{\omega}\sum_{t\in A'} I(t) \simeq \delta\sum_{i=0}^{k}\alpha_i\bar{r}_i$$

$$+ \frac{1}{\omega}\sum_{t\in A'} I(t) \simeq \bar{0} + \frac{1}{\omega}\sum_{t\in A'} I(t) = \frac{1}{\omega}\sum_{t\in A'} I(t)$$

and therefore A' is effective for Y. Since $Y(t) = I(t)$ for $t \notin A'$, it follows that Y is an allocation. Finally, from $A_i' \subset G_{n_i}^{-1}(\bar{r}_i)$ it follows that $\bar{r}_i + I(t) \succ_t X(t)$ for $t \in A_i'$, i.e., $Y(t) \succ_t X(t)$ for $t \in A'$. Since A' is not negligible, we have shown that X is not in the core, contrary to assumption.

THEOREM 1: *If $\mathscr{E}$ is a nonstandard exchange economy satisfying the assumptions of Section 2, then an allocation X is in the core of $\mathscr{E}$ if and only if there exists a price vector, $\bar{p}$, such that $\langle \bar{p}, X\rangle$ is a competitive equilibrium of $\mathscr{E}$.*

PROOF: Suppose $\langle \bar{p}, X\rangle$ is a competitive equilibrium for $\mathscr{E}$, i.e., there exists an internal set K such that $|K|/\omega \simeq 1$ and $X(t)$ are maximal in $B_{\bar{p}}(t)$ for all $t \in K$. If $X(t)$ is not in the core of $\mathscr{E}$, then there exists an allocation $Y(t)$ and a nonnegligible coalition S such that $(1/\omega)\,\Sigma_{t\in S}\, Y(t) \simeq (1/\omega)\,\Sigma_{t\in S}\, I(t)$ and for all $t \in S$, $Y(t) \succ_t X(t)$. Let $R = S \cap K$; then R is internal and we see that: (i) $|R|/\omega \simeq |S|/\omega$; (ii) for all $t \in R$, $Y(t) \succ_t X(t)$; and (iii) for every $t \in R$, $\bar{p}\cdot Y(t) \gtrapprox \bar{p}\cdot I(t)$. The third result implies that for every $t \in R$, $\bar{p}\cdot(Y(t) - I(t)) \gtrapprox 0$. Since R is an internal set, there exists $t_0 \in R$ such that for all $t \in R$, $\bar{p}\cdot(Y(t) - I(t)) \geqslant \bar{p}\cdot(Y(t_0) - I(t_0))$. But $\bar{p}\cdot(Y(t_0) - I(t_0)) \gtrapprox 0$; hence there exists $\varepsilon \gtrapprox 0$ such that $\bar{p}\cdot(Y(t_0) - I(t_0)) \gtrapprox \varepsilon \gtrapprox 0$. Consequently

$$\frac{1}{\omega}\sum_{t\in R}\bar{p}\cdot(Y(t) - I(t)) \geqslant \frac{1}{\omega}\sum_{t\in R}\bar{p}\cdot(Y(t_0) - I(t_0)) \geqslant \frac{|R|}{\omega}\varepsilon \gtrapprox 0.$$

Therefore,

$$\bar{p}\cdot\sum_{t\in R} Y(t)/\omega = \frac{1}{\omega}\sum_{t\in R}\bar{p}\cdot Y(t) \gtrapprox \bar{p}\cdot\sum_{t\in R} I(t)/\omega$$

which contradicts the effectiveness of S. Since

$$\frac{1}{\omega}\sum_{t\in S} Y(t) = \frac{1}{\omega}\sum_{t\in R} Y(t) + \frac{1}{\omega}\sum_{t\in S-R} Y(t) \simeq \frac{1}{\omega}\sum_{t\in R} I(t) + \frac{1}{\omega}\sum_{t\in S-R} I(t)$$

$$= \frac{1}{\omega}\sum_{t\in S} I(t),$$

S is effective. But

$$\sum_{t\in S-R} Y(t)/\omega \simeq \sum_{t\in S-R} I(t)/\omega \simeq 0;$$

hence,

$$\bar{p}\cdot\sum_{t\in R} Y(t)/\omega \simeq \bar{p}\cdot\sum_{t\in R} I(t)/\omega.$$

Suppose X is in the core of $\mathscr{E}$ and let U be the full set of traders in the principal lemma. Then by the principal lemma $0 \notin S\text{-int}(\Delta U)$; hence by the S-separation lemma there exists a standard $\bar{p} \neq \bar{0}$ such that for all $\bar{x} \in S\text{-int}(\Delta U)$, $\bar{p}\cdot\bar{x} \gtrsim 0$. Therefore for all $\bar{y} \in G(t)$, $\bar{p}\cdot\bar{y} \gtrsim 0$ since $S\text{-int}(\Delta U) \supseteq S\text{-int}(G(t)) = G(t)$. This is equivalent to saying that $\bar{p}\cdot\bar{x} \gtrsim \bar{p}\cdot I(t)$ for all $\bar{x} \in F(t)$. Suppose $\bar{z}$ is a standard vector such that $\bar{z} \geqslant \bar{0}$. Then $\bar{z} + X(t) \in F(t)$ for all $t \in T$, by assumption (iv) (γ). Therefore $\bar{p}\cdot(\bar{z} + X(t)) \gtrsim \bar{p}\cdot I(t)$. But if for some i, $p_i = 0$, we can choose a $\bar{z} \in \Omega_n$ such that $\bar{p}\cdot(\bar{z} + X(t)) \lesssim \bar{p}\cdot I(t)$. Thus $\bar{p} \geqslant \bar{0}$.

We will now show that for all $t \in T$, $\bar{p}\cdot X(t) \gtrsim \bar{p}\cdot I(t)$. Suppose for some t that $\bar{p}\cdot X(t) \lesssim \bar{p}\cdot I(t)$. Then there exists $\bar{z} \in \Omega_n$ such that $\bar{z} \geqslant \bar{0}$, $X(t) + \bar{z} \succ\!\succ_t X(t)$, and $\bar{p}\cdot(X(t) + \bar{z}) \lesssim \bar{p}\cdot I(t)$. Since $X(t) + \bar{z} \in F(t)$, this is a contradiction.

We can now show that except for at most a negligible set of t, $\bar{p}\cdot X(t) \simeq \bar{p}\cdot I(t)$. If for some non-negligible internal set, S, we have $\bar{p}\cdot X(t) \gtrapprox \bar{p}\cdot I(t)$, then $(1/\omega)\,\Sigma_{t\in S}\,\bar{p}\cdot X(t) \gtrapprox (1/\omega)\,\Sigma_{t\in S}\,\bar{p}\cdot I(t)$, which contradicts the assumption that X is in the core. Thus $\bar{p}\cdot X(t) \simeq \bar{p}\cdot I(t)$ except for at most a negligible internal set of traders.

To complete the proof we must show that $X(t)$ is maximal in t's budget set, i.e., that $\bar{p}\cdot\bar{x} \gtrapprox \bar{p}\cdot I(t)$ for $\bar{x} \in F(t)$. We first show that $\bar{p} \gtrapprox \bar{0}$. Suppose not; let $p^1 \simeq 0$, say. Since $\bar{p}$ is standard and $\bar{p} \neq \bar{0}$, some coordinate of $\bar{p}$ is not infinitesimal; say $p^2 \gtrapprox 0$. But $(1/\omega)\,\Sigma_{t\in T}\,I^2(t) \gtrapprox 0$. Since X is an allocation it follows that $(1/\omega)\,\Sigma_{t\in T}\,X^2(t) \gtrapprox 0$, so there must be a nonnegligible internal set of traders, S, for whom $X^2(t) \gtrapprox 0$. Now for any trader t, it follows from assumption (iv) (γ) that $X(t) + (1, 0, \ldots, 0) \succ\!\succ_t X(t)$. Choosing $t \in S$, we see that for some sufficiently small $\varepsilon \gtrapprox 0$, $X(t) + (1, -\varepsilon, 0, \ldots, 0) \succ\!\succ_t X(t)$. Hence, $X(t) + (1, -\varepsilon, 0, \ldots, 0) \in F(t)$. Therefore, $\bar{p}\cdot I(t) \lesssim \bar{p}\cdot[X(t) + (1, -\varepsilon, 0, \ldots, 0)] = \bar{p}\cdot X(t) + p^1 - \varepsilon p^2 \lessapprox \bar{p}\cdot X(t)$, i.e., $\bar{p}\cdot I(t) \lessapprox \bar{p}\cdot X(t)$ for all $t \in S$. Since $|S|/\omega \not\simeq 0$, this contradicts $\bar{p}\cdot X(t) \simeq \bar{p}\cdot I(t)$ except for at most a negligible internal set of t. Therefore, $\bar{p} \gtrapprox \bar{0}$.

Now suppose $\bar{x} \in F(t)$ and that $I(t) \gtrapprox \bar{0}$; then $\bar{p}\cdot I(t) \gtrapprox 0$ because $\bar{p} \gtrapprox \bar{0}$. Since $\bar{p}\cdot\bar{x} \gtrsim \bar{p}\cdot I(t)$, it follows that $\bar{p}\cdot\bar{x} \gtrapprox 0$; hence there is a j such that $x^j \gtrapprox 0$; let $j = 1$, say. It then follows that $\bar{x} - (\varepsilon, 0, \ldots, 0) \in F(t)$ for sufficiently small $\varepsilon \gtrapprox 0$. Then $\bar{p}\cdot I(t) \lesssim \bar{p}\cdot[\bar{x} - (\varepsilon, 0, \ldots, 0)] = \bar{p}\cdot\bar{x} - \varepsilon p^1 \lesssim \bar{p}\cdot\bar{x}$, i.e., $\bar{p}\cdot I(t) \lessapprox \bar{p}\cdot\bar{x}$. If $I(t) \simeq \bar{0}$ and $\bar{x} \gtrapprox 0$, then clearly $\bar{p}\cdot\bar{x} \gtrapprox 0 \simeq 0\,\bar{p}\cdot I(t)$. Finally, suppose $I(t) \simeq \bar{0}$ and $\bar{x} \simeq \bar{0}$. Since $\bar{x} \in F(t)$, this means that $I(t) \in F(t)$. If the set of traders t for whom $I(t) \in F(t)$, S, is negligible, then it can be ignored; if, on the other hand, it is non-negligible, then $I(t)$ dominates X via S, contradicting the membership of X in the core. This completes the proof of the theorem.

Theorem 1 poses Edgeworth's conjecture as an equivalence theorem for core allocations and competitive allocations in an infinite (more particularly, nonstandard) economy. It is well-known that the conjecture is not generally true for finite economies. In view of the unfamiliar nature of nonstandard exchange economies, it seems worthwhile to restate the main theorem as an equivalent result on a family of finite economies. We represent each trader by a sequence of traders in the given finite economies in keeping with the idea that the ultimate behavior of the sequence of finite economies should reflect the behavior of traders selected from each of the economies. The details are as follows.

Let $\mathscr{G}$ be a countable family of large but finite economies, i.e., $\mathscr{G} = \{\mathscr{E}_n\}_{n=1}^{\infty}$, where for any n, $|\mathscr{E}_n|$, the number of traders in $\mathscr{E}_n$, $< |\mathscr{E}_{n+1}|$; $|\mathscr{E}_n| < \infty$; and $|\mathscr{E}_1| > 0$. $\mathscr{E}_n$ is completely specified by the initial endowments and preferences of the traders in $\mathscr{E}_n$. Hence let

$$\mathscr{E}_n = \{I_n(t), P_n(t)\}$$

where for all n, $I_n(t) \in \Omega_m^{|\mathscr{E}_n|}$ and $P_n(t) \in \mathscr{P}(\Omega_m \times \Omega_m)^{|\mathscr{E}_n|}$. We shall assume that $\mathscr{E}$ has the following properties.

1. There exists $\bar{r}_0 \in \Omega_m$ such that for all n and every $t \in \mathscr{E}_n$, $I_n(t) \leqslant \bar{r}_0$.
2. $\liminf_{n\to\infty} (1/|\mathscr{E}_n|) \Sigma_{t\in\mathscr{E}_n} I_n(t) > \bar{0}$.
3. For all n and every $t \in \mathscr{E}_n$, we shall assume that: (α) $P_n(t)$ is irreflexive. (β) if $\bar{x} \geqslant \bar{y}$, then $\bar{x}P_n(t)\bar{y}$. All sequences $\{\bar{x}_n\}_1^{\infty}$, $\{\bar{y}_n\}_1^{\infty}$, are convergent and take their values in Ω_m. (γ) for all $\{\bar{x}_n\}_1^{\infty}$, $\{\bar{y}_n\}_1^{\infty}$, $\{t_n\}_1^{\infty}$, if $\liminf_{n\to\infty} \bar{y}_n > \bar{0}$, then there exists $\delta > 0, n_0 \in N$, such that $n \geqslant n_0$, $\bar{z}_n \in S(\bar{x}_n + \bar{y}_n, \delta)$, and $\bar{w}_n \in S(\bar{x}_n, \delta)$ imply $\bar{z}_n P_n(t_n)\bar{w}_n$.

Let $\mathscr{H}$, $\{\mathscr{E}_{S_n}\}_1^{\infty}$, be any subsequence of economies of $\mathscr{G}$, $\{\mathscr{E}_n\}_1^{\infty}$; then the following notions are defined with respect to $\mathscr{H}$.

$\{Y_n(t)\}_{n=1}^{\infty}$ is an allocation for $\mathscr{H}$, if for all n, $Y_n(t) \in \Omega_m^{|\mathscr{E}_{S_n}|}$, and there exists an $\bar{r}_1 \in \Omega_m$ such that for all n and every $t \in \mathscr{E}_{S_n}$, $Y_n(t) \leqslant \bar{r}_1$. Also

$$\lim_{n\to\infty} \left(\frac{1}{|\mathscr{E}_{S_n}|} \sum_{t\in\mathscr{E}_{S_n}} I_{S_n}(t) - \frac{1}{|\mathscr{E}_{S_n}|} \sum_{t\in\mathscr{E}_{S_n}} Y_n(t) \right) = \bar{0}.$$

$\{\mathscr{S}_n\}_1^{\infty}$ is a *coalition* of $\mathscr{H}$, if for all n, $\mathscr{S}_n \subseteq \mathscr{E}_{S_n}$. We say that $\{\mathscr{S}_n\}_1^{\infty}$ is *negligible* if $\lim_{n\to\infty} (|\mathscr{S}_n|/|\mathscr{E}_{S_n}|) = 0$. Note that if $\{\mathscr{S}_n\}_1^{\infty}$ is negligible, then for all allocations $\{Y_n(t)\}_1^{\infty}$,

$$\lim_{n\to\infty} \frac{1}{|\mathscr{E}_{S_n}|} \sum_{t\in\mathscr{S}_n} Y_n(t) = 0.$$

A coalition, $\{\mathscr{S}_n\}_1^{\infty}$, is *effective* for an allocation $\{Y_n(t)\}_1^{\infty}$ if

$$\lim_{n\to\infty} \left(\frac{1}{|\mathscr{E}_{S_n}|} \sum_{t\in\mathscr{S}_n} I_{S_n}(t) - \frac{1}{|\mathscr{E}_{S_n}|} \sum_{t\in\mathscr{S}_n} Y_n(t) \right) = \bar{0}.$$

An allocation $\{Y_n(t)\}_1^{\infty}$, dominates an allocation $\{X_n(t)\}_1^{\infty}$ via a coalition $\{\mathscr{S}_n\}_1^{\infty}$, if for all $\{t_n\}_1^{\infty}$ such that $t_n \in \mathscr{S}_n$, $\liminf_{n\to\infty} |Y_n(t_n) - X_n(t_n)| > 0$, and for some $\delta > 0$ and $n_0 \in N$ if $n \geqslant n_0$, $\bar{w} \in S(Y_n(t_n), \delta)$, then $\bar{w}P_{S_n}(t_n)X_n(t_n)$, where $\{\mathscr{S}_n\}_1^{\infty}$ is effective for $\{Y_n(t)\}_1^{\infty}$.

The *core of* $\mathscr{H}$ is the set of all allocations which are not dominated via any nonnegligible coalition.

An allocation for $\mathscr{G}$ is said to be in the *core of* $\mathscr{G}$ if for every subsequence of economies, $\mathscr{H}$, the allocation for $\mathscr{G}$ restricted to $\mathscr{H}$ is in the core of $\mathscr{H}$.

$\{\bar{p}_n\}_1^\infty$ is a price vector if for all n, $\bar{p}_n \in \Omega_m$, and $\lim_{n\to\infty} \bar{p}_n > 0$.

$\langle\{X_n(t)\}_1^\infty, \{\bar{p}_n\}_1^\infty\rangle$ is a *competitive equilibrium for* $\mathscr{H}$ if the following conditions hold: (i) $\{X_n(t)\}_1^\infty$ is an allocation; (ii) $\{\bar{p}_n\}_1^\infty$ is a price vector; and (iii) there does not exist a coalition $\{\mathscr{S}_n\}_1^\infty$ such that $\liminf_{n\to\infty} (|\mathscr{S}_n|/|\mathscr{E}_{S_n}|) > 0$, and for all $\{t_n\}_1^\infty$, where $t_n \in \mathscr{S}_n$, there exists $\{\bar{y}_n\}_1^\infty$ for which $\liminf_{n\to\infty} (\bar{p}_n \cdot I_{S_n}(t_n) - \bar{p}_n \cdot \bar{y}_n) > 0$, $\liminf_{n\to\infty} |\bar{y}_n - X_n(t_n)| > 0$, and for some $\delta > 0$, $n_0 \in N$, if $n \geqslant n_0$, $\bar{w} \in S(\bar{y}_n, \delta)$, then $\bar{w} P_{S_n}(t_n) X_n(t_n)$; (iv) there does not exist a coalition $\{\mathscr{S}_n\}_1^\infty$ such that $\liminf_{n\to\infty} (|\mathscr{S}_n|/|\mathscr{E}_{S_n}|) > 0$, and for all $\{t_n\}_1^\infty$, where $t_n \in \mathscr{S}_n$, for which $\liminf_{n\to\infty} (\bar{p}_n \cdot X_{S_n}(t_n) - \bar{p}_n \cdot I_{S_n}(t_n)) > 0$.

$\langle\{X_n(t)\}_1^\infty, \{\bar{p}_n\}_1^\infty\rangle$ is a *competitive equilibrium for* $\mathscr{G}$ if $\langle\{X_n(t)\}_1^\infty, \{\bar{p}_n\}_1^\infty\rangle$ restricted to $\mathscr{H}$, where $\mathscr{H}$ is any subsequence of economies of $\mathscr{G}$, is a competitive equilibrium for $\mathscr{H}$.

Since $\mathscr{G}$ is a pair of sequences of functions, $\{I_n(t)\}_1^\infty$ and $\{P_n(t)\}_{n=1}^\infty$, we may define $^*\mathscr{G}$ as the nonstandard extensions of these two sequences. Hence $^*\mathscr{G} = \langle\{^*I_n(t)\}_1^\infty, \{^*P_n(t)\}_1^\infty\rangle$, where n now ranges over the nonstandard integers, $^*N$. For each infinite integer $\omega \in {}^*N - N$ we define the nonstandard exchange economy $\mathscr{E}_\omega = \{^*I_\omega(t), {}^*P_\omega(t)\}$.

LEMMA 6: *$\{X_n(t)\}_1^\infty$ is in the core of $\mathscr{G}$, $\mathscr{C}(\mathscr{G})$, if and only if for all infinite integers ω, $X_\omega(t)$ is in the core of $\mathscr{E}_\omega$, $\mathscr{C}(\mathscr{E}_\omega)$.*

PROOF: Suppose there exists an infinite integer ω such that $X_\omega(t) \notin \mathscr{C}(\mathscr{E}_\omega)$. Then there exists an allocation $Y_\omega(t)$, a coalition $\mathscr{S}_\omega \subseteq \mathscr{E}_\omega$, and $\varepsilon_1, \varepsilon_2 \gtrapprox 0$ such that $(1/\omega)\sum_{t\in\mathscr{E}_\omega} Y_\omega(t) \simeq (1/\omega)\sum_{t\in\mathscr{E}_\omega} I_\omega(t)$, $(1/\omega)\sum_{t\in\mathscr{S}_\omega} Y_\omega \simeq (1/\omega)\sum_{t\in\mathscr{S}_\omega} I_\omega(t)$, and $|\mathscr{S}_\omega|/|\mathscr{E}_\omega| \geqslant \varepsilon_2$ and for all $t \in \mathscr{S}_\omega$, for all $\bar{w} \in S(Y_\omega(t), \varepsilon_1)$, $\bar{w} P_\omega(t) X_\omega(t)$. Therefore for all $n \in N$ and every positive $\delta \in R$, the following sentence is true in our nonstandard universe:

$$^*U:(\exists \nu \in {}^*N)(\exists Y\nu \in {}^*\Omega_m^{|\mathscr{E}_\nu|})(\exists \mathscr{S}_\nu \subseteq \mathscr{E}_\nu)$$

$$\Bigg[\nu > n \wedge \left|\frac{1}{\nu}\sum_{t\in\mathscr{E}_\nu} Y_\nu(t) - \frac{1}{\nu}\sum_{t\in\mathscr{E}_\nu} I_\nu(t)\right|$$

$$< \delta \wedge \left|\frac{1}{\nu}\sum_{t\in\mathscr{S}_\nu} Y_\nu(t) - \frac{1}{\nu}\sum_{t\in\mathscr{S}_\nu} I_\nu(t)\right|$$

$$< \delta \wedge (\forall t \in \mathscr{S}_\nu)(\forall \bar{w} \in S(Y_\nu(t), \varepsilon_1))\bar{w} P_\nu(t) X_\nu(t) \wedge |\mathscr{S}_\nu|/|\mathscr{E}_\nu| \geqslant \varepsilon_2\Bigg].$$

Hence this sentence is true when translated in U, our standard universe. Therefore, for every $n \in N$ and $\delta \in R, \delta > 0$, there exists $m \in N$, $Y_m \in \Omega_m^{|\mathscr{E}_m|}$, and $\mathscr{S}_m \subseteq |\mathscr{E}_m|$ such that $m > n$,

$$\left|\frac{1}{m}\sum_{t\in\mathscr{E}_m} Y_m(t) - \frac{1}{m}\sum_{t\in\mathscr{E}_m} I_m(t)\right| < \delta \wedge \left|\frac{1}{m}\sum_{t\in\mathscr{S}_m} Y_m(t) - \frac{1}{m}\sum_{t\in\mathscr{S}_m} I_m(t)\right| < \delta$$

$$\wedge (\forall t \in \mathscr{S}_m)(\forall \bar{w} \in S(Y_m(t), \varepsilon_1))\bar{w} P_m(t) X_m(t) \wedge |\mathscr{S}_m|/|\mathscr{E}_m| \geqslant \varepsilon_2.$$

Consequently there exists a subsequence of economies, $\mathscr{H}$, which has the allocation $\{Y_m(t)\}_{m=1}^{\infty}$ and a nonnegligible coalition $\{\mathscr{S}_m\}_1^{\infty}$ such that $\{Y_m(t)\}_{m=1}^{\infty}$ dominates $\{X_n(t)\}_1^{\infty}$ via $\{\mathscr{S}_m\}_1^{\infty}$; hence $\{X_n(t)\}_1^{\infty} \notin \mathscr{C}(\mathscr{G})$.

Suppose $\{X_m(t)\}_1^{\infty} \notin \mathscr{C}(\mathscr{G})$; then there exists a subsequence of economies $\{\mathscr{E}_{S_m}\}_1^{\infty}$, $\mathscr{H}$, a coalition $\{\mathscr{S}_m\}_1^{\infty}$, and an allocation $\{Y_m\}_1^{\infty}$ such that

$$\lim_{m\to\infty} \frac{1}{|\mathscr{E}_{S_m}|}\left(\sum_{t\in\mathscr{E}_{S_m}} Y_m(t) - \sum_{t\in\mathscr{E}_{S_m}} I_{S_m}(t)\right) = 0,$$

$$\lim_{m\to\infty} \left(\frac{1}{|\mathscr{E}_{S_m}|}\sum_{t\in\mathscr{S}_m} Y_m(t) - \frac{1}{|\mathscr{E}_{S_m}|}\sum_{t\in\mathscr{S}_m} I_{S_m}(t)\right) = 0,$$

$$\liminf_{m\to\infty} |\mathscr{S}_m|/|\mathscr{E}_{S_m}| > 0.$$

For all $\{t_m\}_1^{\infty}$ such that for every m, $t_m \in \mathscr{S}_m$,

$$\liminf_{m\to\infty} |Y_m(t_m) - X_m(t_m)| > 0,$$

and for some $\delta > 0$ and $m_0 \in N$ if $m \geqslant m_0$, $\bar{w} \in S(Y_m(t_m), \delta)$, then $\bar{w}P_{S_m}(t_m)X_m(t_m)$. Hence there exists an infinite integer ω for which the following statements hold about the nonstandard extensions of $\{X_m(t)\}_1^{\infty}$, $\{Y_m(t)\}_1^{\infty}$, $\{P_{S_m}(t)\}_1^{\infty}$, $\{\mathscr{S}_m\}_1^{\infty}$, and $\{I_{S_m}(t)\}_1^{\infty}$:

$$\frac{1}{\omega}\sum_{t\in\mathscr{E}_\omega} X_\omega(t) \simeq \frac{1}{\omega}\sum_{t\in\mathscr{E}_\omega} I_\omega(t),$$

$$\frac{1}{\omega}\sum_{t\in\mathscr{E}_\omega} Y_\omega(t) \simeq \frac{1}{\omega}\sum_{t\in\mathscr{E}_\omega} I_\omega(t),$$

$$\frac{1}{\omega}\sum_{t\in\mathscr{S}_\omega} Y_\omega(t) \simeq \frac{1}{\omega}\sum_{t\in\mathscr{S}_\omega} I_\omega(t),$$

$|\mathscr{S}_\omega|/|\mathscr{E}_\omega| \not\simeq 0$, for all $t \in \mathscr{S}_\omega$, $Y_\omega(t) \gg X_\omega(t)$. Hence $X_\omega(t) \notin \mathscr{C}(\mathscr{E}_\omega)$.

LEMMA 7: *$\langle\{X_n(t)\}_1^{\infty}, \{\bar{p}_n\}_1^{\infty}\rangle$ is a competitive equilibrium for $\mathscr{G}$ if and only if for all infinite integers ω, $\langle X_\omega(t), \bar{p}_\omega\rangle$ is a competitive equilibrium for $\mathscr{E}_\omega$.*

PROOF: Suppose there exists an infinite integer $\omega \in N^* - N$ such that $\langle X_\omega(t), \bar{p}_\omega\rangle$ is not a competitive equilibrium for $\mathscr{E}_\omega$. Then there exists an internal set of traders, $\mathscr{S}_\omega$, and $\varepsilon_1, \varepsilon_2 \gtrsim 0$ such that $|\mathscr{S}_\omega|/|\mathscr{E}_\omega| \geqslant \varepsilon_2$, and for all $t \in \mathscr{S}_\omega$, there exists a $\bar{y}_t \in B_{\bar{p}}(t)$ such that $\bar{w}P_\omega(t)X_\omega(t)$ for all $\bar{w} \in S(\bar{y}_t, \varepsilon_1)$. Hence for all $n \in N$ and for all $\delta \in R$, $\delta > 0$, the following sentence is true in $*U$, our nonstandard universe:

$$(\exists v \in {}^*N)(\exists \bar{p}_v \in {}^*\Omega_m)(\exists \mathscr{S}_v \subseteq \mathscr{E}_v)[v > n \wedge (\forall t \in \mathscr{S}_v)(\exists \bar{y}_t \in {}^*\Omega_m)\{(|\bar{p}_v \cdot \bar{y}_t - \bar{p}_v \cdot I_v(t)| < \delta \vee \bar{p}_v \cdot \bar{y}_t \leqslant \bar{p}_v \cdot I_v(t))$$
$$\text{if } \bar{w} \in S(\bar{y}_t, \varepsilon_1) \text{ then } \bar{w}P_v(t)X_v(t)\} \wedge |\mathscr{S}_v|/|\mathscr{E}_v| \geqslant \varepsilon_2].$$

Therefore, this sentence is true when translated in U, our standard universe.

Hence, for every $n \in N$ and $\delta \in R$, $\delta > 0$, there exists $m > n$, $\bar{p}_m \in \Omega_m$, $\mathscr{S}_m \subseteq \mathscr{E}_m$ such that $m > n$

$$(\forall t \in \mathscr{S}_m)(\exists \bar{y}_t \in \Omega_m)\{(|\bar{p}_m \cdot \bar{y}_t - \bar{p}_m \cdot I_m(t)| < \delta \vee \bar{p}_m \cdot \bar{y}_t \leqslant \bar{p}_m \cdot I_m(t))$$
$$\text{if } \bar{w} \in S(\bar{y}_t, \varepsilon_1) \text{ then } \bar{w}P_m(t)X_m(t)\} \wedge |\mathscr{S}_m|/|\mathscr{E}_m| \geqslant \varepsilon_2.$$

Consequently there exists a subsequence of economies, $\mathscr{H}$, such that $\langle\{X_n(t)\}_1^\infty, \{\bar{p}_n\}_1^\infty\rangle$ restricted to $\mathscr{H}$ is not a competitive equilibrium for $\mathscr{H}$; hence $\langle\{H_n(t)\}_1^\infty, \{\bar{p}_n\}_1^\infty\rangle$ is not a competitive equilibrium for $\mathscr{G}$.

Suppose $\langle\{X_n(t)\}_1^\infty, \{\bar{p}_n\}_1^\infty\rangle$ is not a competitive equilibrium for $\mathscr{G}$; then there exists subsequences $\{X_m(t)\}_1^\infty$, $\{\bar{p}_m\}_1^\infty$, a sequence $\{\mathscr{S}_m\}_1^\infty$ and $\varepsilon_1, \varepsilon_2 > 0$ such that $\liminf_{m\to\infty} (|\mathscr{S}_m|/|\mathscr{E}_m|) \geqslant \varepsilon_2$ and for all $\{t_m\}_1^\infty$, where $\liminf_{n\to\infty} |\bar{y}_m - X_m(t_m)| > 0$, $\liminf_{n\to\infty} (\bar{p}_m \cdot I_m(t) - \bar{p}_m \cdot \bar{y}_m) > 0$, and for some $m_0, \bar{w}P_m(t_m)X_m(t_m)$ for all $m \geqslant m_0$ and $w \in S(\bar{y}_m, \varepsilon_1)$. Hence, there exists an infinite integer ω for which the following statements hold about the nonstandard extensions of $\{X_n(t)\}_1^\infty$, $\{I_n(t)\}_1^\infty$, $\{\mathscr{S}_m\}_1^\infty$, and $\{P_n(t)\}_1^\infty$:

$$\frac{1}{\omega}\sum_{t\in\mathscr{S}_\omega} I_\omega(t) \simeq \frac{1}{\omega}\sum_{t\in\mathscr{S}_\omega} X_\omega(t), \qquad |\mathscr{S}_\omega|/|\mathscr{E}_\omega| \not\simeq 0,$$

for all $t \in \mathscr{S}_\omega$, there exists $\bar{y}_t \in B_{\bar{p}_\omega}(t)$ such that $\bar{y}_t \gg X_\omega(t)$. Hence, $\langle X_\omega(t), \bar{p}_\omega\rangle$ is not a competitive equilibrium for $\mathscr{E}_\omega$.

THEOREM 2: *$\{X_n(t)\}_1^\infty$ is in the core of $\mathscr{G}$ if and only if there exists $\{\bar{p}_n\}_1^\infty$ such that $\langle\{X_n(t)\}_1^\infty, \{\bar{p}_n\}_1^\infty\rangle$ is a competitive equilibrium for $\mathscr{G}$.*

PROOF: We need only show that every $\mathscr{E}_\omega$ is a nonstandard exchange economy satisfying the assumptions of Theorem 1. Then the proof is an immediate consequence of Lemmas 6 and 7. All the assumptions are obviously met except possibly (iv) (γ), which we shall show holds also. Suppose (iv) (γ) is false in some $\mathscr{E}_w$; then there exists a trader t, and standard $\bar{x}, \bar{y} \in {}^*\Omega_m$, where $\bar{x} \geqslant \bar{0}$, such that for all $\alpha \gtrsim 0$ there exists $\bar{x}' \in S(\bar{x} + \bar{y}, \alpha)$ where $\bar{x}' \not\succ_t \bar{y}$. Therefore, for all $n \in N$ and every positive α in R, the following sentence is true in our nonstandard universe ${}^*U$:

$$(\exists v \in {}^*N)(\exists t \in \mathscr{E}_v)(\exists \bar{x}' \in {}^*\Omega_m)[v > n \wedge \bar{x}' \in S(\bar{x} + \bar{y}, \alpha) \wedge \bar{x}' \not\succ_t \bar{y}].$$

Hence this sentence is true when translated in U, our standard universe. Therefore, for every $n \in N$; $\alpha \in R$; $\alpha > 0$, there exists $l > n$, t_l in $\mathscr{E}_l$, $\bar{x}'_l \in \Omega_m$ such that $\bar{x}'_l \in S(x + y, \alpha)$ and $\bar{x}'_l$ not preferred to $\bar{y}$. But this contradicts our assumption (3) (γ) about the preference of $\mathscr{G}$. To complete the proof, suppose $\langle\{X_n(t)\}_1^\infty, \{\bar{p}_n\}_1^\infty\rangle$ is a competitive equilibrium for $\mathscr{G}$; then by Lemma 7 $\langle X_\omega(t), \bar{p}_\omega\rangle$ is a competitive equilibrium for $\mathscr{E}_\omega$ for every infinite integer ω. Hence by Theorem 1 for all infinite ω, X_ω is in the core of $\mathscr{E}_\omega$ and consequently by Lemma 6 $\{X_n(t)\}_1^\infty$ is in the core of $\mathscr{G}$. Suppose $\{X_n(t)\}_1^\infty$ is in the core of $\mathscr{G}$; then by Lemma 6 $X_\omega(t)$ is in the core of $\mathscr{E}_\omega$ for all infinite integers ω. Hence, by Theorem 1, there exists $\bar{p}_\omega$ such that $\langle X_\omega(t), \bar{p}_\omega\rangle$ is a competitive equilibrium for $\mathscr{E}_\omega$ for every infinite integer ω. Consequently for every positive $\delta_1, \delta_2 \in R$ the following sentence is true in $^*U$:

$$(\exists \nu \in {}^*N)(\forall \theta \in {}^*N)[\theta \geqslant \nu \Rightarrow (\exists \bar{p}_\theta \in {}^*\Omega_m)[(\forall \mathscr{S}_\theta \subseteq \mathscr{E}_\theta)[(\forall t \in \mathscr{S}_\theta)(\exists \bar{y}_t \in {}^*\Omega_m)$$
$$\times \{[\bar{p}_\theta \cdot \bar{y}_t \leqslant \bar{p}_\theta \cdot I_\theta(t) \vee |\bar{p}_\theta \cdot \bar{y}_t - \bar{p}_\theta \cdot I_\theta(t)| < \delta_2] \wedge \bar{y}_t P_\theta(t) X_\theta(t)\}$$
$$\wedge |\bar{y}_t - X_\theta(t)| \geqslant \delta_1] \Rightarrow |\mathscr{S}_\theta|/|\mathscr{E}_\theta| < \delta_2]].$$

Translating this sentence in U for $\delta_1 = \delta_2 = 1/n$, $n = 1, 2, \ldots$, we generate a sequence of prices $\{\bar{p}_j\}_1^\infty$ and integers $\{n_j\}_{j=1}^\infty$ where n_j is the first integer such that the sentence is true for $\delta_1 = \delta_2 = 1/j$, $\bar{p}_j$ is a corresponding price, and $n_j < n_{j+1}$. Let $n_0 = 0$ and assign economies $n_j + 1$ through n_{j+1} prices $\bar{p}_j$ where $j = 0, 1, 2, \ldots$. We claim that $\langle \{X_n(t)\}_1^\infty, \{\bar{p}_n\}_1^\infty \rangle$ is a competitive equilibrium for $\mathscr{G}$. Suppose not; then there exists $\{\mathscr{S}_n\}_1^\infty, \{t_n\}_1^\infty, \{\bar{y}_n\}_1^\infty, \varepsilon_1, \varepsilon_2$ such that $\liminf_{n\to\infty} (|\mathscr{S}_n|/|\mathscr{E}_n|) > \varepsilon_2$, all n, $\{t_n\}_1^\infty \in \mathscr{S}_n$ and $\bar{y}_n \in \Omega_m$, and $\liminf_{n\to\infty} (\bar{p}_n \cdot I_n(t_n) - \bar{p}_n \cdot \bar{y}_n) > 0$, $\liminf_{n\to\infty} |\bar{y}_n - X_n(t_n)| > 0$, for some n_0, $\bar{w} P_n(t_n) X_n(t_n)$ for all $n \geqslant n_0$ and $\bar{w} \in S(\bar{y}_m, \varepsilon)$. But for all j such that $(1/j) < \min \{\liminf_{n\to\infty} (|\mathscr{S}_n|/|\mathscr{E}_n|), \liminf_{n\to\infty} (\bar{p}_n \cdot I_n(t_n) - \bar{p}_n \cdot \bar{y}_n), \liminf_{n\to\infty} |\bar{y}_n - X_n(t_n)|\}$ and $j > n_0$, this contradicts the properties of $\langle X_{n_j}(t), \bar{p}_{n_j} \rangle$ for economy $\mathscr{E}_{n_j}$.

Yale University

Manuscript received June, 1971.

REFERENCES

[1] AUMANN, R. J.: "Markets with a Continuum of Traders," *Econometrica*, 32 (1964), 39–50.
[2] BERGE, C., AND A. GHOUILA-HOURI: *Programming, Games and Transportation Networks*. New York: John Wiley, 1962.
[3] DEBREU, G., AND H. SCARF: "A Limit Theorem on the Core of an Economy," *International Economic Review*, 4 (1963), 235–246.
[4] FRAYNE, T., A. C. MOREL, AND D. S. SCOTT: "Reduced Direct Products," *Fundementa Mathematicae*, 51 (1962), 195–227.
[5] KOCHEN, S.: "Ultraproducts in Theory of Models," *Annals of Mathematics*, 79 (1961), 221–261.
[6] LUXEMBURG, W. A. J., ED.: *Symposium on Applications of Model Theory to Analysis and Algebra*. New York: Holt, Rinehart, and Winston, 1967.
[7] ROBINSON, A.: *Non-Standard Analysis, Studies in Logic and the Foundations of Mathematics*. Amsterdam: North-Holland, 1966.
[8] ———: "Non-Standard Theory of Dedekind Rings," in *Proceedings of the Academy of Science*, Amsterdam, A70 (1967), 444–452.
[9] ROCKAFELLAR, T.: *Convex Analysis*. Princeton, New Jersey: Princeton University Press, 1970.

On the Finiteness Theorem of Siegel and Mahler Concerning Diophantine Equations

A. ROBINSON*

Department of Mathematics, Yale University, New Haven, Connecticut 06520

AND

P. ROQUETTE

Mathematisches Institut, Universität Heidelberg, 6900 Heidelberg, Germany

Received November 26, 1974

DEDICATED TO PROFESSOR K. MAHLER ON THE OCCASION OF HIS 70TH BIRTHDAY

In this paper we present a new proof, involving so-called nonstandard arguments, of Siegel's classical theorem on diophantine equations: Any irreducible algebraic equation $f(x, y) = 0$ of genus $g > 0$ admits only finitely many integral solutions. We also include Mahler's generalization of this theorem, namely the following: Instead of solutions in integers, we are considering solutions in rationals, but with the provision that their denominators should be divisible only by such primes which belong to a given finite set. Then again, the above equation admits only finitely many such solutions. From general nonstandard theory, we need the definition and the existence of enlargements of an algebraic number field. The idea of proof is to compare the natural arithmetic in such an enlargement, with the functional arithmetic in the function field defined by the above equation.

1. INTRODUCTION

We work over a given algebraic number field K of finite degree. We consider a plane algebraic curve Γ, defined by an irreducible algebraic equation

$$f(x, y) = 0,$$

whose coefficients are contained in K. In his classical paper [22], Siegel has proved the following theorem.

* Deceased April 11, 1974.

If Γ has genus $g > 0$, then there are only finitely many points (x, y) on Γ, whose coordinates are algebraic integers in K.

Mahler [11] has extended this result by considering also those points (x, y) in Γ whose coordinates admit finitely many prime divisors in their denominators. More precisely, let

$$\mathfrak{S} = \{\mathfrak{p}_1, \ldots, \mathfrak{p}_s\}$$

be a finite set of prime divisors of the field K, and let us consider those algebraic numbers in K whose denominators are divisible by primes from $\mathfrak{S}$ only. For brevity, these numbers will be called the *quasi-integers* in K, with respect to $\mathfrak{S}$ as the set of admissible denominatorial prime divisors. It is clear that every integer in the ordinary sense is also a quasi-integer. Now, the above theorem remains true if, in this theorem, the notion of integer is replaced by the notion of quasiinteger, with respect to a given finite set $\mathfrak{S}$ of prime divisors of K.

We shall refer to this statement as the *Siegel–Mahler Theorem*. Mahler [11] gave a proof for curves of genus $g = 1$, and he conjectured that the theorem is still true for curves of higher genus. This conjecture was verified by Lang [7] and, again, by LeVeque [9].

In the present paper we shall give a nonstandard proof of the Siegel–Mahler Theorem, i.e., a proof which uses the methods of nonstandard analysis or, as we should rather say in our case, of *nonstandard arithmetic*. Our main purpose is to exhibit the usefulness of these nonstandard methods in dealing with problems on diophantine equations.

From general nonstandard theory we shall use the *existence* of enlargements of our field K, for a higher order language [14]. For the convenience of the reader we shall explain the definition and the basic properties of such enlargements in Section 2. We shall try to keep these explanations self-contained, so that this paper can be followed also by those readers who are not yet acquainted with nonstandard arguments. Such a reader might perhaps prefer to look through Section 2 before proceeding further in this section, where we are now going to state a nonstandard version of the Siegel–Mahler Theorem.

$^*K$ denotes an enlargement of the field K, fixed throughout the following discussion. $^*K$ is a certain field extension of K whose basic properties are explained in Sections 2 and 3; in particular, let us note that K is algebraically closed in $^*K$. The elements of K are called standard, while the elements of $^*K$ not in K are called nonstandard. A point $(x, y) \in {}^*K \times {}^*K$ is nonstandard if not both its coordinates x and y are standard.

Now let us assume that, contrary to the assertion of the Siegel–Mahler Theorem, there are infinitely many points on Γ whose coordinates are quasi-integers in K, with respect to a given finite set $\mathfrak{S}$ of prime divisors of K. Then, according to the enlargement principle (see Section 2), there exists a nonstandard point (x, y) on Γ with the same property; i.e., the coordinates x and y are quasi-integers in $^*K$ with respect to the same finite set $\mathfrak{S}$ of prime divisors. By construction, $\mathfrak{S}$ consists solely of standard prime divisors; hence, we conclude that *the denominators of x and y are not divisible by any nonstandard prime divisor of* $^*K$.

As we have noted above, the field K is algebraically closed in $^*K$. Hence, since the point (x, y) is nonstandard, at least one of its coordinates is transcendental over K. We conclude that (x, y) is a *generic* point over K of the curve Γ. Therefore, the field $F = K(x, y)$ is K-isomorphic to the field of K-rational functions on Γ. We may identify F with this function field; thus, the inclusion $F \subset {}^*K$ might be interpreted *as a representation of the functions in F by means of (nonstandard) algebraic numbers.* The genus of F in the sense of algebraic function fields equals the genus g of the curve Γ.

By construction, F is generated by the two functions x and y which, when regarded as algebraic numbers in $^*K$, do not admit any nonstandard prime divisor in their denominators. *We assert that this is impossible if* $g > 0$; if this assertion is proved, then we obtain a contradiction to our assumption about infinitely many quasi-integral points on Γ, and the theorem of Siegel–Mahler follows.

Thus we are faced with proving the following theorem, referring to a given algebraic number field K and its enlargement $^*K$.

THEOREM 1.1. *Let F be an algebraic function field of one variable over K, and assume that F is embedded into $^*K$, so that $K \subset F \subset {}^*K$.*

If F has genus $g > 0$, then every nonconstant function $x \in F$ admits at least one nonstandard prime divisor of $^*K$ in its denominator.

We have shown above that the Siegel–Mahler Theorem is an immediate consequence of Theorem 1.1. It would be equally easy to show that, conversely, our Theorem 1.1 follows from the Siegel–Mahler Theorem. Thus 1.1 can be regarded as the nonstandard equivalent of the Siegel–Mahler Theorem.

There is another version of Theorem 1.1 which refers to the prime divisors of the function field $F \mid K$, instead of to the nonconstant functions. Here, the notion of prime divisor of $F \mid K$ is to be understood in the usual sense; these prime divisors correspond to the nontrivial valuations of F over K. We shall refer to these prime divisors of $F \mid K$ as the *functional prime divisors*, in contrast to the *arithmetical prime divisors* of $*K$ which belong to the nontrivial internal valuations of $*K$.

Now every *nonstandard* arithmetical prime divisor $\mathfrak{p}$ of $*K$ is trivial on K (see Section 3). Hence, $\mathfrak{p}$ induces in F a valuation which, if it is not entirely trivial on F, belongs to some functional prime divisor P of $F \mid K$. We say that P is *induced* by $\mathfrak{p}$, and we write $\mathfrak{p} \mid P$. There arises the question whether every functional prime divisor of F is induced by some nonstandard arithmetical prime divisor of $*K$.

If this is the case, then the assertion of Theorem 1.1 follows immediately. Namely, according to the general theory of algebraic function fields, every nonconstant function $x \in F$ has at least one pole, say P, which is a functional prime divisor in the above sense. If we know that P is induced by a nonstandard arithmetical prime $\mathfrak{p}$ of $*K$, then we have found a $\mathfrak{p}$ which induces a pole of x and hence divides the denominator of x.

Conversely, it follows from Theorem 1.1 that every functional prime divisor P of $F \mid K$ is induced by some nonstandard arithmetical prime divisor of $*K$. This is because, by the Riemann-Roch Theorem, there exists a function $x \in F$ admitting P as its *only* pole. Applying Theorem 1.1 to this function, we find a nonstandard prime divisor $\mathfrak{p}$ of $*K$ which divides the denominator of x and hence induces in F a pole of x. Since P is the only pole of x, we see that $\mathfrak{p} \mid P$.

The foregoing arguments show that the following theorem is an equivalent version of Theorem 1.1.

THEOREM 1.2. *In the same situation as in Theorem* 1.1, *we assume again that* $g > 0$. *Then every functional prime divisor* P *of* $F \mid K$ *is induced by some nonstandard, arithmetical prime divisor* $\mathfrak{p}$ *of* $*K$.

If the hypothesis $g > 0$ is not satisfied, then Theorems 1.1 and 1.2 need not be true. For instance, consider the rational function field $F = K(x)$, where x is some nonstandard integer in $*K$. Then $g = 0$. The denominator of x in $*K$ does not contain any (nonarchimedean) prime divisor at all, standard or nonstandard; hence, Theorem 1.1 is not true for this function $x \in F$. Theorem 1.2 is not true for the pole of x in F.

However, our methods of proof will also yield some information in the case $g = 0$. We shall show that in this case there are *at most two* functional prime divisors of $F \mid K$ which are exceptional in the sense of Theorem 1.2, i.e., which are not induced by any nonstandard arithmetical prime. This will yield a parametrization of those functions $x \in F$ which do not have any nonstandard arithmetical prime in their denominators. These results can be regarded as the nonstandard version of Siegel's parametrization of curves of genus 0 which admit infinitely many integral points [22]. For details, we refer to Section 8 below.

2. General Remarks on Enlargements

Nonstandard methods are based on the fact that every mathematical structure M admits what is called an *enlargement*. Such an enlargement $^*M$ is an extension of M such that the following properties hold, which we shall state as "principles."

The first of these principles expresses the fact that $^*M$ is a *model* of M.

2.1. Principle of Permanence. *Every mathematical statement about M has an interpretation in $^*M$, and this interpretation is true in $^*M$ if and only if the original statement is true in M.*

Let us explain this in more detail. Mathematical statements about M are envisaged as being expressed in a formal language of higher order predicate calculus over M. This language contains names for all individuals[1] in M, as well as for all entities of higher type in M (e.g., sets of individuals, relations between individuals, relations between sets, etc.). Starting from these and from a sufficient supply of variables, every sentence in this language is built up in finitely many steps, according to the well-known rules of predicate calculus, with the help of the logical connectives and quantifiers. Quantification is permitted not only with respect to individuals but also with respect to entities of any given type.

In most cases, of course, mathematical statements about M are *not* explicitly expressed in this formal language. Instead, one usually prefers to use the more informal common language to describe mathematical statements, provided it is clear that a translation into the formal language exists. For instance, the Siegel-Mahler Theorem in Section 1 can be regarded as a mathematical statement about K in the above sense.

The basic property of nonstandard models, as given above, refers to the concept of "interpretation" of mathematical statements about M, in the enlargement $^*M$. This interpretation is meant in a special way, namely according to the following provisions.

(i) The interpretation of the logical connectives ("and," "or," "not," "implies") is the usual one.

(ii) The name of any individual in M is the same in $^*M$, and quantification with respect to individuals ("*there exists* a number," "*for all* numbers") has its usual meaning in $^*M$.

(iii) The names of other entities in M (sets, relations, relations between sets, etc.) also denote corresponding entities in $^*M$ which are then called

[1] The individuals of a mathematical structure are the elements of its underlying set. If M is an algebraic number field, then the individuals of M are the algebraic numbers of that field.

standard entities. However, quantification with respect to these entities ("*there exists* a relation," "*for all* sets," etc.) does not refer to the class of all entities of that type in $^*M$ but to a certain subclass of them whose members are called *internal* entities (sets, relations, etc.). Among these are the standard entities.[2]

As an example, let us consider the case $M = \mathbf{N}$, the natural numbers, and Peano's principle of mathematical induction which implies that every nonempty, bounded subset of $\mathbf{N}$ has a maximal element. This statement contains a quantifier with respect to sets. According to (iii), its interpretation in $^*\mathbf{N}$ refers to *internal* sets only. Hence, every nonempty bounded internal subset of $^*\mathbf{N}$ contains a maximal element; this is a true statement in $^*\mathbf{N}$ since it is the interpretation of a true statement in $\mathbf{N}$. The reader should note that there are nonempty, bounded subsets of $^*\mathbf{N}$, necessarily external, which do *not* contain a maximal element; for instance, $\mathbf{N}$ is bounded in $^*\mathbf{N}$, every *nonstandard* number $x \in {}^*\mathbf{N}$ being a bound of $\mathbf{N}$. (This last argument uses the fact that $^*\mathbf{N}$ is a proper extension of $\mathbf{N}$, i.e., that there are nonstandard numbers in $^*\mathbf{N}$ (see 2.4 below).)

It should be observed that the notions of "standard" and "internal" entities belong to the definition of enlargement (in much the same way as the notion of "open" sets belong to the definition of topological space). More precisely, an enlargement $^*M$ of M is defined to be a higher order structure, extending M, in which certain entities are distinguished as being "standard" or "internal" respectively, and such that the basic principles 2.1–2.5 hold.

The next principle is concerned with *relations* in the structure M. Let R denote such a relation, say an n-ary relation between individuals. According to (iii), R also denotes a certain standard n-ary relation in the enlargement $^*M$. Let $a_1, \ldots, a_n$ be n individuals of M, and consider the statement that $R(a_1, \ldots, a_n)$ holds. According to the principle of permanence, this statement is true in $^*M$ if and only if it is true in M. In other words, the new relation R in $^*M$ is an extension of the original relation R in M. Thus we obtain the following principle which, as we have seen, is actually a consequence of 2.1 but which we prefer to state separately as a convenient reference.

2.2. EXTENSION PRINCIPLE. *Any relation in M extends naturally and uniquely to a standard relation of the same type in $^*M$. The extended relation in $^*M$ is usually denoted by the same symbol as the original relation*

[2] The terminology of "internal" and "standard" is also applied to individuals. Namely, every individual in $^*M$ is internal; the individuals in M and only those are standard. With this terminology, (ii) can be regarded as a special case of (iii).

*in M. Every property of the original relation which is expressible in the language of M does also hold for the extended standard relation, provided it is interpreted in *M as explained above.*

This principle holds not only for relations between individuals but also for relations between entities of higher type. We have used it already in Section 1 in the case $M = K$, an algebraic number field, where we have said that **K is a field extension of K*. In order to see this, consider the two ternary relations which represent addition ($a + b = c$) and multiplication ($a \cdot b = c$) in the field K. These relations extend to certain standard ternary relations in $^*K$, also denoted in the same way as addition and multiplication. Now, the original relations in K satisfy all the axioms which make K into a field. Therefore, these field axioms are also satisfied by their standard extensions. This shows that, indeed, $^*K$ is a field extension of K. As to the nature of this extension, we have already noted in Section 1 that *K is algebraically closed in *K*. This is an immediate consequence of the principle of permanence. Namely, let $g(X) \in K[X]$ be any polynomial; we have to show that $g(X)$ has a root in $^*K$ if and only if it has a root in K. In fact, the statement that $g(X)$ has a root belongs to the language of K and hence it is true in K if and only if it is true in $^*K$.

We observe that principles 2.1 and 2.2 do not imply that $^*M$ is a proper extension of M; they are trivially valid in M instead of $^*M$. In contrast, the next principle asserts that $^*M$ is sufficiently large such as to contain certain nonstandard entities; this explains its name of "enlargement."

We consider binary relations in M. Let R denote such a binary relation, say a relation between individuals. Let a be an individual in M. If there exists b in M such that $R(a, b)$ holds, then a is said to be in the *left domain* of R. The relation R is said to be *concurrent* in M if the following holds: Given finitely many elements $a_1, \ldots, a_n$ in the left domain of R, then there exists b in M such that the relations $R(a_i, b)$ hold simultaneously for $1 \leqslant i \leqslant n$.

2.3 ENLARGEMENT PRINCIPLE FOR BINARY RELATIONS. *If the relation R is concurrent in M, then there exists x in *M such that R(a, x) holds simultaneously for all a in M which are contained in the left domain of R.*

In other words, the (possibly infinite) system of conditions $R(a, x)$ can be solved in the enlargement $^*M$, provided any finite subsystem can be solved in M already. Of course, in writing $R(a, x)$ we have to interpret R as a binary relation in $^*M$ according to 2.2.

If M is *infinite*, then the enlargement principle guarantees the existence of *nonstandard* individuals in M. For, consider the relation of inequality $a \neq b$ between individuals of M, the left and right domain of this relation being M itself. Since M is infinite, this relation is concurrent and we conclude from 2.3 the existence of $x \in {}^*M$ which is different from every $a \in M$; i.e., x is nonstandard.

More generally, let S be any set in M, say a set of individuals. According to (iii), S determines a certain standard set $^*S$ in $^*M$ whose characteristic properties are identical with the characteristic properties of S in M, if these are interpreted in $^*M$. From this it is clear that $^*S$ is an *extension* of S; more precisely, S consists exactly of the standard elements which are contained in $^*S$. There arises the question whether $^*S$ is a *proper* extension of S, i.e., whether $^*S$ contains nonstandard elements. If S is finite, say with n elements, then $^*S$ coincides with S. This is because the statement that S consists of n elements belongs to the language of M and hence remains true in its interpretation in $^*M$, which says that $^*S$ consists of n elements. On the other hand, if S is infinite, then we may consider the relation of inequality $a \neq b$, *restricted to the set* S; similarly as above, we deduce from 2.3 that there exists a nonstandard element in $^*S$. Thus we have the following.

2.4. ENLARGEMENT PRINCIPLE FOR SETS. *Every set S in M determines naturally and uniquely a certain standard set $^*S$ in $^*M$; if S is described by a sentence in the language of M, then this same sentence, if interpreted in $^*M$, yields a description of $^*S$. The original set S consists precisely of the standard elements which belong to $^*S$. If S is infinite, and only in this case, does $^*S$ contain nonstandard elements*; *i.e., $^*S$ is then a proper extension of S.*

We have used 2.4 already in Section 1 in order to deduce the existence of certain nonstandard points on our curve Γ.

Enlargement principles 2.3 and 2.4 are valid not only for relations and sets of individuals but also for relations and sets of entities of higher type. Perhaps a further comment is necessary to explain the situation of S with respect to $^*S$, if S is a set of entities of higher type (e.g., a set of relations or functions, etc.) In this case, the elements $a \in S$ are entities of higher type in M, and *a priori* they are *not* contained in the structure $^*M$. However, every entity extends naturally and uniquely to a standard entity of $^*S$; if we assign to every $a \in S$ its corresponding standard entity, then we obtain an *injection* from S into $^*S$. It is with respect to this injection that $^*S$ is to be regarded as an extension of S, in the context of 2.4. Often it will be convenient, during the investigation of a given set S, to identify the elements of S with their corresponding standard entities, so that S becomes a *subset* of $^*S$, viz., the subset of standard entities in $^*S$. For instance, we shall not distinguish between the prime divisors of the number field K and their corresponding standard extensions to $^*K$.

It should be noted that every infinite set S in M, if regarded as a subset of $^*S$ in the way as explained above, is necessarily *external*. To see this, consider a denumerable infinite subset of S, which may be identified with the set $\mathbf{N}$ of natural numbers. There exists a *surjective* map $\varphi: S \to \mathbf{N}$. In view of extension principle 2.2, φ extends uniquely to a standard map $\varphi : {}^*S \to {}^*\mathbf{N}$. As a standard map, $\varphi : {}^*S \to {}^*\mathbf{N}$ is *a fortiori* internal and hence maps internal subsets of $^*S$ onto internal subsets of $^*\mathbf{N}$. Thus, if S were internal, then $\varphi(S) = \mathbf{N}$ would be internal too. But we have already remarked above that $\mathbf{N}$ is an external subset of $^*\mathbf{N}$, as a consequence of Peano's principle of mathematical induction. Thus we have the following.

2.5. PRINCIPLE OF EXTERNITY. *Every infinite set S which consists of only standard elements is necessarily external in $^*M$. In other words, every infinite internal set in $^*M$ contains a nonstandard element.*

The enlargements $^*M$ of a given structure M are not unique. There are many ways of realizing enlargements, e.g., by ultrapower methods. Such explicit methods might perhaps yield a more graphic illustration of the notions of "standard" and "internal" entities, but it does not make any difference in our arguments. This is why we prefer to define enlargements by their relevant properties 2.1–2.5 only, without reference to their possible modes of construction.

Assume that, for a given higher order structure M, we have chosen a definite enlargement $^*M$. Let S be a set in M, either a set of individuals or a set of entities of higher type. We know that S determines naturally and uniquely a certain standard set $^*S$ in $^*M$, consisting of internal entities of the same type as the entities in S. Now, let us regard S not only as a set, but also as a higher order substructure of M and, similarly, $^*S$ as a substructure of $^*M$ (with the notions of "internal" and "standard" as those which are induced by $^*M$). Then it is easy to see that $^*S$ is an enlargement of S. Thus, *our fixed enlargement $^*M$ of M contains naturally an enlargement $^*S$ for every substructure S of M.* In other words, we have a well-defined functor

$$S \mapsto {}^*S$$

from substructures of M to substructures of $^*M$ such that $^*S$ is an enlargement of S. By the basic property 2.1 of $^*M$, this functor is faithful, not only with respect to the inclusion relation $S \subset S'$ between substructures, but also with respect to all those relations between substructures which can be expressed in the language of M.

In view of this it will be convenient, during the discussions of this paper, to adopt the following viewpoint. We work in a fixed universe M, in the sense that M as a higher order structure contains all mathematical structures which are of interest in number theory, at least in the usual treatments and in any case in this paper. More precisely, M should contain all algebraic number fields and their completions with respect to their various valuations. We choose a fixed enlargement $^*M$ of M. This being settled, we regard every structure S occurring in our arguments as a substructure of M and, therefore, using the remark above, its enlargement $^*S$ is *uniquely defined* as a substructure of $^*M$. In this way, we have eliminated all ambiguity concerning the choice of enlargements, and the star symbol will have a well-defined meaning throughout our discussion.

As to the possible choice of the universe M, we may take $M = \mathbf{N}$; more precisely, M is to be the full higher order structure based on the set $\mathbf{N}$ of natural numbers. The substructures of this universe are precisely those which can be described in the language of the natural numbers. Thus, M contains the integers $\mathbf{Z}$ (which can be described as pairs of natural numbers—more precisely, as certain equivalence classes of such pairs). It follows that M contains the rational numbers $\mathbf{Q}$ (pairs of integers) and the real numbers $\mathbf{R}$ as well as the p-adic numbers $\mathbf{Q}_p$ (sequences of rationals). If K is our algebraic number field of finite degree and if $u_1, \ldots, u_n$ is a basis of K, then we can describe the elements of K by means of their coordinates with respect to this basis, i.e., as n-tuples of rational numbers. Thus K is also contained in M.

The above explanations should give the reader a general idea of the underlying theory and the nature of our arguments pertaining to nonstandard arithmetic. He may consult [14] for a more systematic treatment of this theory, including an existence proof for enlargements (based on the compactness theorem of algebraic logic). We also refer the reader to Section 3, where he will find an opportunity for exercises in applying the general principles of this section, which will perhaps help him understand nonstandard methods.

3. Prime Divisors and Divisors in the Enlargement of an Algebraic Number Field

As before, K denotes an algebraic number field of finite degree. $^*K$ is its enlargement as explained in Section 2. In this section we are going to discuss the arithmetic properties of $^*K$, as far as they are relevant for our purpose. Also, we want to fix our notations which we are going to use in this paper.

The arithmetic structure of the field K can be described by means of its *prime divisors*, which are defined in terms of valuations. More precisely, a prime divisor $\mathfrak{p}$ of K is defined to be a class of nontrivial valuations of K, with respect to the ordinary equivalence relation for valuations. There are two types of prime divisors of K: namely, the archimedean primes, which correspond to the archimedean valuations, and the non-archimedean primes, which correspond to the nonarchimedean valuations.

Let $\mathfrak{p}$ be a nonarchimedean prime divisor of K. Among the valuations belonging to $\mathfrak{p}$ there is exactly one which is normalized such that its value group is $\mathbf{Z}$, the additive group of integers. This valuation is denoted by $v_{\mathfrak{p}}$ and is called the $\mathfrak{p}$-*adic ordinal function* of K. In addition to this ordinal function, we also consider the normalized *absolute value*, defined by the formula

$$\| x \|_{\mathfrak{p}} = N\mathfrak{p}^{-v_{\mathfrak{p}}(x)}; \tag{3.1}$$

this is a multiplicative valuation belonging to $\mathfrak{p}$. As usual, $N\mathfrak{p}$ denotes the *norm* of $\mathfrak{p}$, i.e., the number of elements in the $\mathfrak{p}$-adic residue field.

Now let $\mathfrak{p}$ be an archimedean prime divisor of K. Among all the valuations belonging to $\mathfrak{p}$ there is exactly one which induces in $\mathbf{Q}$ the ordinary absolute value; this valuation is denoted by $| x |_{\mathfrak{p}}$. We also consider the normalized absolute value, defined by the formula

$$\| x \|_{\mathfrak{p}} = \begin{cases} | x |_{\mathfrak{p}} & \text{if } \mathfrak{p} \text{ is real,} \\ | x |_{\mathfrak{p}}^{2} & \text{if } \mathfrak{p} \text{ is complex.} \end{cases} \tag{3.2}$$

As usual, $\mathfrak{p}$ is called real or complex according to whether the $\mathfrak{p}$-adic completion of K is isomorphic to the field $\mathbf{R}$ of real numbers or to the field $\mathbf{C}$ of complex numbers.

The absolute values of K are normalized in such a way that the following product formula holds for every nonzero element $x \in K$:

$$\prod_{\mathfrak{p}} \| x \|_{\mathfrak{p}} = 1. \tag{3.3}$$

Here, $\mathfrak{p}$ ranges over all prime divisors of K, archimedean or nonarchimedean. The product in (3.3) is essentially a finite product. For, given $0 \neq x \in K$, there are only finitely many prime divisors $\mathfrak{p}$ for which $\| x \|_{\mathfrak{p}} \neq 1$.

Sometimes in the literature, the term of prime divisor is restricted to denote nonarchimedean primes only. We shall not follow this terminology since we have to take the archimedean primes also into consideration, and it seems more natural to put them on equal footing with the nonarchimedean ones. This is a well-established procedure in number theory. In order to obtain unified formulas, we extend the above notations to include the case when $\mathfrak{p}$ is archimedean:

$$v_{\mathfrak{p}}(x) = -\log \| x \|_{\mathfrak{p}} ,$$
$$N\mathfrak{p} = e \text{ (base of the logarithm).}$$

With these notations, (3.1) holds also for archimedean primes.

Now, let V be the set of all prime divisors of K. According to the general principles explained in Section 2, its enlargement $^*V$ is to be interpreted as the set of all internal prime divisors of $^*K$. Here, the notion of internal prime divisor is to be defined as an equivalence class of nontrivial internal valuations of $^*K$. As we know, every true statement in K concerning prime divisors yields a true statement in $^*K$ which concerns internal prime divisors by means of its interpretation. Therefore, the following statements hold.

There are two types of internal prime divisors of $^*K$: archimedean and nonarchimedean. Let $\mathfrak{p}$ be a nonarchimedean internal prime divisor. Among all the (internal) valuations of $^*K$ which belong to $\mathfrak{p}$ there is exactly one which is normalized such that its value group is $^*\mathbf{Z}$, the additive group of standard or nonstandard integers. This valuation is denoted by $v_{\mathfrak{p}}$ and is called the $\mathfrak{p}$-*adic ordinal function* of $^*K$. Thus, for $0 \neq x \in {}^*K$, its $\mathfrak{p}$-adic ordinal $v_{\mathfrak{p}}(x)$ is a standard or nonstandard integer. $v_{\mathfrak{p}}$ is an additive valuation of $^*K$ in the sense of Krull, with the extra condition that $v_{\mathfrak{p}}$ is *internal*, a notion which is inherent in the structure of enlargement. The $\mathfrak{p}$-adic residue field of $^*K$ is not necessarily finite; however, it is *starfinite* in the following sense. There is an element $N \in {}^*\mathbf{N}$ and an internal bijection from the residue field onto the initial interval $1 \leqslant \nu \leqslant N$ in $^*\mathbf{N}$. It is clear that this notion of "starfinite" is the interpretation of the ordinary notion of "finite." The above number $N \in {}^*\mathbf{N}$ is uniquely determined and is called the *norm* of $\mathfrak{p}$: notation $N\mathfrak{p}$. In addition to the ordinal function, we also consider the normalized $\mathfrak{p}$-*adic absolute value*, defined by formula (3.1). This formula is now interpreted in $^*K$. That is, x is an element in $^*K$ and $\| x \|_{\mathfrak{p}}$ is a nonnegative element in $^*\mathbf{Q}$.

Now let $\mathfrak{p}$ be an archimedean internal prime divisor of $^*K$. Among all the valuations belonging to $\mathfrak{p}$ there is exactly one which induces in $^*\mathbf{Q}$ the ordinary standard absolute value. This valuation is denoted by $| x |_{\mathfrak{p}}$. We also consider the normalized absolute value $\| x \|_{\mathfrak{p}}$, defined by formula (3.2), which is now to be interpreted in $^*K$. The prime $\mathfrak{p}$ is called real or complex according to whether the $\mathfrak{p}$-*adic* completion of $^*K$ is internally isomorphic to $^*\mathbf{R}$ or to $^*\mathbf{C}$.

Now, *product formula* (3.3) holds in $^*K$, since it is the interpretation of a true formula in K. In this interpretation, x denotes any nonzero element in $*K$ and $\mathfrak{p}$ ranges over all internal prime divisors of $*K$. The product in (3.3) is essentially starfinite, for, given $0 \neq x \in *K$, the set of $\mathfrak{p} \in *V$ with $\| x \|_{\mathfrak{p}} \neq 1$ is starfinite.

Perhaps it is useful to insert a few general remarks about starfinite products. Let A be any internal abelian group, written multiplicatively, and let a_i be an internal sequence in A with starfinite support. That is, the index i ranges over an internal index set I, the map $i \to a_i$ is an internal map from I to A, and the set of those i for which $a_i \neq 1$ is starfinite. Under these conditions, the product $\prod_{i \in I} a_i$ is well defined as an element in A. This definition is the interpretation of the obvious definition of finite products in the usual sense. Starfinite products satisfy all the rules which are satisfied by finite products, as long as these rules can be expressed in the language of M. To be sure, the number of "factors" in a starfinite product need not be finite, and thus it is not a product in the sense of ordinary algebra. Nevertheless, the starfinite product is a certain operator which is inherent in the structure of enlargement. The notation and the name "product" are justified since it satisfies the usual rules for finite products. (The situation is much the same as in ordinary analysis, where one considers infinite "products", these being not products in the algebraic sense but defined by limit operations.) It goes without saying that similar remarks apply to starfinite sums, etc., in abelian groups which are additively written.

Let us continue with the discussion of prime divisors. According to the general extension principle 2.2, every prime divisor $\mathfrak{p}$ of K extends naturally and uniquely to a standard prime divisor of $*K$. This standard extension is denoted with the same symbol $\mathfrak{p}$, and it enjoys the same properties as the original prime divisor of K, as long as these properties are expressed in the language of K. For example, both norms $N\mathfrak{p}$ coincide, regardless of whether we look at $\mathfrak{p}$ as a prime divisor of K or as a standard prime divisor of $*K$. Also, if $a \in K$, then its absolute value $\| a \|_{\mathfrak{p}}$ is same whether $\mathfrak{p}$ is regarded as a prime of K or as a standard prime in $*K$.[3]

Due to this notational convention, every prime $\mathfrak{p} \in V$ now appears also as a standard prime in $*V$. That is, the set V appears as a subset of $*V$, viz., the set of standard primes. (Note that V is an *external* subset of $*V$.) Since there are infinitely many prime divisors of K, we know that V is an infinite set and, hence, due to enlargement principle 2.4, we conclude that $*V$ is a proper extension of V. In other words, *there are nonstandard prime divisors of* $*K$. The following lemma gives some of their fundamental properties.

[3] If $\mathfrak{p}$ is a prime divisor of K, then it can be shown that its standard extension is a $\mathfrak{p}$-extension in the sense of [19].

LEMMA 3.1. *Every nonstandard prime divisor* $\mathfrak{p}$ *of* $^*K$ *is trivial on* K. *In particular,* $\mathfrak{p}$ *is nonarchimedean. The norm* $N\mathfrak{p}$ *is infinitely large. If* $x \in {}^*K$ *is such that* $\| x \|_\mathfrak{p} > 1$, *then* $\| x \|_\mathfrak{p}$ *is infinitely large.*

An element in $^*\mathbf{R}$ is called infinitely large if it is greater than any standard real number.

Proof. Let a be a nonzero element in K. The set S of those $\mathfrak{q} \in V$ for which $\| a \|_\mathfrak{q} \neq 1$ is finite. Hence, this set is not enlarged in $^*V$, due to enlargement principle 2.4. That is, if $\mathfrak{q}$ is any internal prime divisor in $^*V$ for which $\| a \|_\mathfrak{q} \neq 1$, then $\mathfrak{q} \in S$. In particular, since $S \subset V$, we see that every such $\mathfrak{q}$ is standard.

Now, since $\mathfrak{p}$ is assumed to be nonstandard, it follows that $\| a \|_\mathfrak{p} = 1$. This being true for any $0 \neq a \in K$, we see that $\mathfrak{p}$ is trivial on K. As archimedean valuations are nontrivial on $\mathbf{Q}$, hence nontrivial on K, it follows that $\mathfrak{p}$ is nonarchimedean.

Since $\mathfrak{p}$ is trivial on K, we see that K is isomorphically contained in the $\mathfrak{p}$-adic residue field of $^*K$. In particular, the $\mathfrak{p}$-adic residue field of $^*K$ is infinite. On the other hand, we know from the above that this residue field is starfinite, and that there exists an internal bijection from it to the initial intervall $1 \leqslant \nu \leqslant N\mathfrak{p}$ in $^*\mathbf{N}$. We conclude that $N\mathfrak{p}$ is nonstandard and, in fact, infinitely large.

If $\| x \|_\mathfrak{p} > 1$, then $v_\mathfrak{p}(x) < 0$. Now, $v_\mathfrak{p}(x)$ is an element in $^*\mathbf{Z}$. Therefore, from $v_\mathfrak{p}(x) < 0$ we conclude $v_\mathfrak{p}(x) \leqslant -1$. (This conclusion is valid in $^*\mathbf{Z}$ since it is valid in $\mathbf{Z}$.) In view of (3.1) we obtain $\| x \|_\mathfrak{p} \geqslant N\mathfrak{p}$; hence $\| x \|_\mathfrak{p}$ is infinitely large too. Q.E.D.

Now let $c \geqslant 1$ be a standard real number. Consider those elements $x \in K$ which are contained in the "parallelotope," given by the conditions

$$\| x \|_\mathfrak{p} \leqslant c \tag{3.4}$$

for all $\mathfrak{p} \in V$. A well-known theorem of algebraic number theory says that the set of these $x \in K$ is *finite*. Hence, in view of 2.4, this set is not enlarged in $^*K$. Therefore, if an element $x \in {}^*K$ satisfies the conditions (3.4) for all $\mathfrak{p} \in {}^*V$, then x is contained in K already; i.e., x is standard. In other words, if the element $x \in {}^*K$ is *nonstandard*, then the conditions (3.4) are not all satisfied; i.e., there exists at least one prime $\mathfrak{p} \in {}^*V$ such that

$$\| x \|_\mathfrak{p} > c.$$

This prime $\mathfrak{p} \in {}^*V$ might depend on the choice of $c \in \mathbf{R}$. However, we claim there is a prime $\mathfrak{p} \in {}^*V$ such that $\| x \|_\mathfrak{p} > c$ holds simultaneously for all standard $c \in \mathbf{R}$. To show this, we distinguish two cases: Let $\mathfrak{S}_x$ denote the set of those $\mathfrak{p} \in {}^*V$ for which $\| x \|_\mathfrak{p} > 1$.

Case 1. $\mathfrak{S}_x$ *is finite.* In this case, we put successively $c = 1, 2, 3,...$ and, by the above remark, find a sequence $\mathfrak{p}_1, \mathfrak{p}_2, \mathfrak{p}_3, ..., \in {}^*V$ such that $\| x \|_{\mathfrak{p}_n} > n$ for $n \in \mathbf{N}$. Every member $\mathfrak{p}_n$ of this sequence is contained in the finite set $\mathfrak{S}_x$. Hence, there is an infinite subsequence which is constant; i.e., there is $\mathfrak{p} \in \mathfrak{S}_x$ such that $\mathfrak{p} = \mathfrak{p}_n$ for infinitely many $n \in \mathbf{N}$. By construction, this prime $\mathfrak{p}$ satisfies $\| x \|_\mathfrak{p} > n$ for infinitely many (and hence all) standard natural numbers n. In other words, $\| x \|_\mathfrak{p}$ is infinitely large.

Case 2. $\mathfrak{S}_x$ *is infinite.* We remark that, according to its definition, $\mathfrak{S}_x$ is an *internal* set. We know from 2.5 that every infinite internal set contains a nonstandard element. Thus, there is a nonstandard prime $\mathfrak{p}$ which lies in $\mathfrak{S}_x$, i.e., for which $\| x \|_\mathfrak{p} > 1$. Lemma 3.1 now shows that $\| x \|_\mathfrak{p}$ is infinitely large.

We have proved the following.

LEMMA 3.2. *Let* $x \in {}^*K$ *be nonstandard. Then there exists a prime divisor* $\mathfrak{p} \in {}^*V$ *such that* $\| x \|_\mathfrak{p}$ *is infinitely large.*

We shall often prefer to work with the *logarithmic values*

$$w_\mathfrak{p}(x) = -\log \| x \|_\mathfrak{p} = v_\mathfrak{p}(x) \log(N\mathfrak{p}). \tag{3.5}$$

To say that $\| x \|_\mathfrak{p}$ is infinitely large is equivalent to saying that $w_\mathfrak{p}(x)$ is infinitely small, i.e., less than every (positive or negative) standard real number.

If $\mathfrak{p}$ is nonarchimedean, then $w_\mathfrak{p}$ is an additive valuation of the field ${}^*K$ in the sense of Krull; it differs from the normalized ordinal function $v_\mathfrak{p}$ by the factor $\log(N\mathfrak{p})$ only. Hence, in the nonarchimedean case, we have the following rules which express the properties of an additive valuation:

$$w_\mathfrak{p}(xy) = w_\mathfrak{p}(x) + w_\mathfrak{p}(y),$$
$$w_\mathfrak{p}(x + y) \geqslant \min(w_\mathfrak{p}(x), w_\mathfrak{p}(y)).$$

If $\mathfrak{p}$ is archimedean, then we still have the first of these rules, which expresses the fact that $w_\mathfrak{p} : {}^*K \to {}^*\mathbf{R}$ is a homomorphism of the multiplicative group of ${}^*K$ into the additive group of ${}^*\mathbf{R}$. The second rule has to be modified in the archimedean case, namely as follows.

Recall that $\| x \|_\mathfrak{p} = | x |_\mathfrak{p}$ if $\mathfrak{p}$ is real, and $\| x \|_\mathfrak{p} = | x |_\mathfrak{p}^2$ if $\mathfrak{p}$ is complex. Now, the ordinary absolute value $| x |_\mathfrak{p}$ satisfies

$$| x + y |_\mathfrak{p} \leqslant | x |_\mathfrak{p} + | y |_\mathfrak{p} \leqslant 2 \max(| x |_\mathfrak{p}, | y |_\mathfrak{p}).$$

If we square this relation, then the factor 2 is replaced by 4. Hence in any case, real or complex, we have

$$\| x + y \|_\mathfrak{p} \leqslant 4 \max(\| x \|_\mathfrak{p}, \| y \|_\mathfrak{p})$$

and therefore

$$w_{\mathfrak{p}}(x + y) \geqslant -\log(4) + \min(w_{\mathfrak{p}}(x), w_{\mathfrak{p}}(y)).$$

Compared with the corresponding rule in the nonarchimedean case, the additional term $-\log(4)$ appears here. This term, although not negligible, is nevertheless standard, hence finite, and therefore it vanishes if we consider the infinitary orders of magnitude only. Let us explain this in more detail.

A real number $a \in {}^*\mathbf{R}$ is called *finite* if there exists a positive standard number $c \in \mathbf{R}$ such that

$$-c \leqslant a \leqslant c.$$

In particular, every standard real number is finite. If the above inequalities hold for every standard $c > 0$, then a is *infinitesimal*. Every finite number a is infinitely close to a standard number ${}^\circ a$, which is to say that $a = {}^\circ a + h$ with infinitesimal h. Namely, ${}^\circ a$ is the standard real number which represents the Dedekind cut in $\mathbf{R}$ determinded by a.

The finite numbers form an additive subgroup of ${}^*\mathbf{R}$, which we denote by $\mathbf{R}_{\text{fin}}$.[4] If two numbers $a, b \in {}^*\mathbf{R}$ differ by a finite number only, then a, b are said to be of the same order of magnitude; notation: $a \doteq b$.[5] This means that a and b determine the same residue class in the factor group

$$\dot{\mathbf{R}} = {}^*\mathbf{R}/\mathbf{R}_{\text{fin}}.$$

Note that $\dot{\mathbf{R}}$ carries naturally an order relation, which it inherits from ${}^*\mathbf{R}$ such that the natural projection ${}^*\mathbf{R} \to \dot{\mathbf{R}}$ is order preserving. If $a, b \in {}^*\mathbf{R}$, then we write

$$a \mathrel{\dot{\leqslant}} b$$

in order to indicate that the order of magnitude of a is less or equal to the order of magnitude of b. Explicitly, this means that there is a finite number $c \in \mathbf{R}_{\text{fin}}$ such that $b - a \geqslant c$. It is easily verified that this indeed defines an order relation in the factor group $\dot{\mathbf{R}}$; this is due to the fact that, by definition, $\mathbf{R}_{\text{fin}}$ is an *isolated* subgroup of ${}^*\mathbf{R}$. (That is, if $c, d \in \mathbf{R}_{\text{fin}}$, then $\mathbf{R}_{\text{fin}}$ contains every $u \in {}^*\mathbf{R}$ which lies between c and d.)

[4] On other occasions [17], the group of finite numbers has been denoted by $\mathbf{R}_0$. However, we wish to reserve the index "0" for another purpose, namely for the group of divisors of size or degree 0 (see below). This is why we use the index "fin" to denote the group of finite elements, not only with respect to $\mathbf{R}$, but also with respect to other groups in due course.

[5] This is the notation which has been proposed by Hasse [5] in this context.

This being said, let us return to our above formula involving $-\log(4)$. Since this is a standard number, its order of magnitude vanishes. We conclude that for every prime $\mathfrak{p} \in {}^*V$ the following formulas hold:

$$w_\mathfrak{p}(xy) \doteq w_\mathfrak{p}(x) + w_\mathfrak{p}(y),$$
$$w_\mathfrak{p}(x + y) \mathrel{\dot{\geqslant}} \min(w_\mathfrak{p}(x), w_\mathfrak{p}(y)).$$

These formulas say that the map

$$\dot{w}_\mathfrak{p} : {}^*K \to \dot{\mathbf{R}},$$

which is obtained from $w_\mathfrak{p} : {}^*K \to {}^*\mathbf{R}$ by applying the projection ${}^*\mathbf{R} \to \dot{\mathbf{R}}$, is a *valuation in the sense of Krull*. This valuation is trivial on K. Namely, if $0 \neq x \in K$, then $w_\mathfrak{p}(x) = -\log \| x \|_\mathfrak{p}$ is standard and hence finite; therefore, we have $w_\mathfrak{p}(x) \doteq 0$. (This holds also if $\mathfrak{p}$ is nonstandard, since then $w_\mathfrak{p}(x) = 0$ in view of Lemma 3.1.) Thus we have seen the following.

Every prime divisor $\mathfrak{p} \in {}^*V$ defines naturally a Krull valuation $\dot{w}_\mathfrak{p}$ of ${}^*K$, which is trivial on K and whose values are contained in the group $\dot{\mathbf{R}}$. By definition, $\dot{w}_\mathfrak{p}(x)$ is the order of magnitude of the logarithmic value $w_\mathfrak{p}(x) = -\log \| x \|_\mathfrak{p}$.

If $\mathfrak{p}$ is standard (archimedean or not), then it can be shown that the value group of $\dot{w}_\mathfrak{p}$ is the full group $\dot{\mathbf{R}}$; its residue field is isomorphic to the $\mathfrak{p}$-adic completion of K. If $\mathfrak{p}$ is nonstandard, then the value group of $\dot{w}_\mathfrak{p}$ is a proper subgroup of $\dot{\mathbf{R}}$; the valuation $\dot{w}_\mathfrak{p}$ is in fact equivalent to the original valuation $w_\mathfrak{p}$, and both have isomorphic value groups and residue fields. Since we shall not make use of these facts explicitly, we leave the proofs to the reader.

If $x \in {}^*K$ is nonstandard, then by Lemma 3.2 there exists at least one prime $\mathfrak{p} \in {}^*V$ such that $w_\mathfrak{p}(x) \mathrel{\dot{<}} 0$; this implies in particular that $\dot{w}_\mathfrak{p}$ does not vanish on x. Thus we see that K is the exact field of constants with respect to the set of valuations $\dot{w}_\mathfrak{p}$. Therefore, these valuations may be used to build a divisor theory which describes ${}^*K$ relative to K as its ground field. The situation is much the same as in the corresponding case with function fields, where there is also a field of constants. In the rest of this section, we shall develop this divisor theory of ${}^*K$.

First, let us discuss the ordinary notion of divisor in the algebraic number field K. This notion is defined as usual, with the provision, however, that the archimedean prime divisors should also be included. This leads to the following definition: The divisor group $\mathfrak{D}$ of K is the direct sum

$$\mathfrak{D} = \mathfrak{D}' \oplus \mathfrak{D}'',$$

where $\mathfrak{D}'$ is the free **R**-module generated by the archimedean primes, and $\mathfrak{D}''$ the free **Z**-module generated by the non-archimedean primes. This definition implies that every divisor $\mathfrak{a} \in \mathfrak{D}$ has a unique representation in the form

$$\mathfrak{a} = \sum_{\mathfrak{p}} \alpha_{\mathfrak{p}} \cdot \mathfrak{p}, \tag{3.6}$$

where $\mathfrak{p}$ ranges over the primes in V, and the coefficients $\alpha_{\mathfrak{p}}$ satisfy the following conditions:

(i) $\alpha_{\mathfrak{p}} \in \mathbf{R}$ if $\mathfrak{p}$ is archimedean;
(ii) $\alpha_{\mathfrak{p}} \in \mathbf{Z}$ if $\mathfrak{p}$ is nonarchimedean;
(iii) $\alpha_{\mathfrak{p}} \neq 0$ for finitely many $\mathfrak{p} \in V$ only.

Thus, the divisor group $\mathfrak{D}$ can be represented as the group of all functions $\alpha : V \to \mathbf{R}$ satisfying these conditions (i)–(iii).

Every nonzero element $x \in K$ determines a divisor $(x) \in \mathfrak{D}$, namely its *principal* divisor, which is defined by

$$(x) = \sum_{\mathfrak{p}} v_{\mathfrak{p}}(x) \cdot \mathfrak{p}. \tag{3.7}$$

The map $x \mapsto (x)$ yields a homomorphism $K \to \mathfrak{D}$ from the multiplicative group of K to the additive group $\mathfrak{D}$. Its kernel consists precisely of the roots of unity which are contained in K; in particular, this kernel is finite. Its cokernel $\mathfrak{C} = \mathfrak{D}/K$ is called the *divisor class group* of K, whose structure we shall discuss later.

Now let us interpret the above notions in the enlargement. $^*\mathfrak{D}$ is the group of all internal divisors. The internal prime divisors $\mathfrak{p} \in {}^*V$ are contained in $^*\mathfrak{D}$, and every $\mathfrak{a} \in {}^*\mathfrak{D}$ admits a unique representation in form (3.6), $\mathfrak{p}$ ranging over the primes in $^*V$. The coefficients $\alpha_{\mathfrak{p}}$ in (3.6) satisfy the conditions (i)–(iii) which now have to be interpreted in the enlargement ; more precisely,

*(i) $\alpha_{\mathfrak{p}} \in {}^*\mathbf{R}$ if $\mathfrak{p}$ is archimedean;
*(ii) $\alpha_{\mathfrak{p}} \in {}^*\mathbf{Z}$ if $\mathfrak{p}$ is nonarchimedean;
*(iii) the set of those $\mathfrak{p} \in {}^*V$ for which $\alpha_{\mathfrak{p}} \neq 0$ is starfinite.

Moreover, $\alpha_{\mathfrak{p}}$ depends internally on $\mathfrak{p}$, which is to say that the function $\mathfrak{p} \mapsto \alpha_{\mathfrak{p}}$ from $^*V$ to $^*\mathbf{R}$ is internal. In other words, the group $^*\mathfrak{D}$ can be represented as the group of all internal functions $\alpha : {}^*V \to {}^*\mathbf{R}$ satisfying *(i)–*(iii).

Again we have a direct sum decomposition

$$^*\mathfrak{D} = {}^*\mathfrak{D}' \oplus {}^*\mathfrak{D}'',$$

where $*\mathfrak{D}'$ is the archimedean component, and $*\mathfrak{D}''$ the nonarchimedean component of $*\mathfrak{D}$.

The principal divisor map $K \to \mathfrak{D}$ has a standard extension

$$*K \to *\mathfrak{D},$$

which is described by formula (3.7). Since the kernel of $K \to \mathfrak{D}$ is finite, it is not enlarged in $*K$. That is, the kernel of $*K \to *\mathfrak{D}$ is finite, and it consists of the roots of unity in K. The corresponding cokernel

$$*\mathfrak{C} = *\mathfrak{D}/*K$$

is the group of internal divisor classes.

As a matter of notation, the coefficients $\alpha_{\mathfrak{p}}$ in (3.6) will be denoted by $v_{\mathfrak{p}}(\mathfrak{a})$; they are called the $\mathfrak{p}$-adic ordinals of the divisor $\mathfrak{a}$. According to (3.7), this notation is coherent with the corresponding notation $v_{\mathfrak{p}}(x)$ for $x \in *K$.

We also define the $\mathfrak{p}$-adic absolute value of a divisor $\mathfrak{a} \in *\mathfrak{D}$ by the formula

$$\| \mathfrak{a} \|_{\mathfrak{p}} = N\mathfrak{p}^{-v_{\mathfrak{p}}(\mathfrak{a})},$$

in analogy to (3.1). Most often we shall work with its logarithm

$$w_{\mathfrak{p}}(\mathfrak{a}) = -\log \| \mathfrak{a} \|_{\mathfrak{p}} = v_{\mathfrak{p}}(\mathfrak{a}) \log(N\mathfrak{p}).$$

in analogy to (3.5). Recall that for archimedean $\mathfrak{p}$ we have defined $N\mathfrak{p}$ such that $\log(N\mathfrak{p}) = 1$, so that $w_{\mathfrak{p}}(\mathfrak{a}) = v_{\mathfrak{p}}(\mathfrak{a})$ in this case.

Every divisor $\mathfrak{a} \in *\mathfrak{D}$ is uniquely determined by its logarithmic absolute values $w_{\mathfrak{p}}(\mathfrak{a})$, and the divisorial relations are reflected in corresponding relations between those values. For instance, the addition of divisors $\mathfrak{a} + \mathfrak{b}$ is given by

$$w_{\mathfrak{p}}(\mathfrak{a} + \mathfrak{b}) = w_{\mathfrak{p}}(\mathfrak{a}) + w_{\mathfrak{p}}(\mathfrak{b})$$

for all $\mathfrak{p} \in *V$. The order relation $\mathfrak{a} \leqslant \mathfrak{b}$ is given by

$$w_{\mathfrak{p}}(\mathfrak{a}) \leqslant w_{\mathfrak{p}}(\mathfrak{b})$$

for all $\mathfrak{p}$. If $\mathfrak{a} \leqslant \mathfrak{b}$, then $\mathfrak{a}$ is said to *divide* $\mathfrak{b}$; this terminology takes its motivation from the arithmetic background. Accordingly, $\min(\mathfrak{a}, \mathfrak{b})$ is the greatest common divisor, and $\max(\mathfrak{a}, \mathfrak{b})$ the least common multiple of $\mathfrak{a}$ and $\mathfrak{b}$.

The number

$$\sigma(\mathfrak{a}) = \sum_{\mathfrak{p}} w_{\mathfrak{p}}(\mathfrak{a}) = \sum_{\mathfrak{p}} v_{\mathfrak{p}}(\mathfrak{a}) \log(N\mathfrak{p}) \tag{3.8}$$

is called the (additive) *size* of the divisor $\mathfrak{a}$. This formula is either to be read in $\mathfrak{D}$, in which case the size yields a homomorphism $\sigma : \mathfrak{D} \to \mathbf{R}$, or else we have to interpret the formula in $^*\mathfrak{D}$ (the sum being starfinite), in which case we obtain a homomorphism $\sigma: {}^*\mathfrak{D} \to {}^*\mathbf{R}$ which is the standard extension of the former. In any case, σ is surjective. The kernel of σ is denoted by $\mathfrak{D}_0$ resp. $^*\mathfrak{D}_0$. In view of the product formula (3.3), which can also be read as a sum formula

$$\sum_{\mathfrak{p}} w_{\mathfrak{p}}(x) = 0,$$

we see that principal divisors are contained in $\mathfrak{D}_0$ resp. $^*\mathfrak{D}_0$. Let $\mathfrak{C}_0 = \mathfrak{D}_0/K$ resp. $^*\mathfrak{C}_0 = {}^*\mathfrak{D}_0/{}^*K$ denote the corresponding divisor class groups.

We consider $\mathfrak{D}$ as a subgroup of $^*\mathfrak{D}$, viz., the subgroup of all *standard* divisors. A divisor $\mathfrak{a} \in {}^*\mathfrak{D}$ is called *finite* if there is a standard divisor $\mathfrak{c} > 0$ such that $-\mathfrak{c} \leqslant \mathfrak{a} \leqslant \mathfrak{c}$. In particular, standard divisors are finite. If the above inequalities hold for every standard $\mathfrak{c} > 0$, then $\mathfrak{a}$ is said to be *infinitesimal*; this implies that $w_{\mathfrak{p}}(\mathfrak{a}) = 0$ for nonarchimedean $\mathfrak{p}$, while $w_{\mathfrak{p}}(\mathfrak{a})$ is an infinitesimal real number for archimedean $\mathfrak{p}$. Every finite divisor $\mathfrak{a}$ is infinitely close to a standard divisor $^\circ\mathfrak{a}$, which is to say that $\mathfrak{a} = {}^\circ\mathfrak{a} + \mathfrak{y}$ with $\mathfrak{y}$ infinitesimal.

The finite divisors in $^*\mathfrak{D}$ form an isolated subgroup $\mathfrak{D}_{\mathrm{fin}}$ and the factor group

$$\dot{\mathfrak{D}} = {}^*\mathfrak{D}/\mathfrak{D}_{\mathrm{fin}}$$

is the *group of divisorial orders of magnitude*. As in the case with real numbers, we write $\mathfrak{a} \doteq \mathfrak{b}$ in order to indicate that $\mathfrak{a}$ and $\mathfrak{b}$ are of the same order of magnitude; this means that $\mathfrak{a}$ and $\mathfrak{b}$ determine the same residue class in $\dot{\mathfrak{D}}$. Also, $\mathfrak{a} \mathrel{\dot{\leqslant}} \mathfrak{b}$ means that there exists $\mathfrak{c} \in \mathfrak{D}_{\mathrm{fin}}$ such that $\mathfrak{b} - \mathfrak{a} \geqslant \mathfrak{c}$; this defines an order relation in $\dot{\mathfrak{D}}$ such that the natural projection $^*\mathfrak{D} \to \dot{\mathfrak{D}}$ is order preserving. Moreover, the operations max and min are preserved by this projection.

The group $\dot{\mathfrak{D}}$ will play a central role in our considerations. One may regard $\dot{\mathfrak{D}}$ as consisting of the same elements as $^*\mathfrak{D}$, namely internal divisors, but with the equality sign $=$ being replaced by the sign $\doteq$, which indicates the same order of magnitude. In this sense the following lemma is, in fact, a statement concerning $\dot{\mathfrak{D}}$.

LEMMA 3.3. *Let* $\mathfrak{a}, \mathfrak{b}$ *denote internal divisors in* $^*\mathfrak{D}$. *If* $\mathfrak{a} \mathrel{\dot{\leqslant}} \mathfrak{b}$, *then*

$$w_{\mathfrak{p}}(\mathfrak{a}) \mathrel{\dot{\leqslant}} w_{\mathfrak{p}}(\mathfrak{b}) \qquad \textit{for each } \mathfrak{p} \in {}^*V,$$

and conversely. In particular, it follows that the relation $\mathfrak{a} \doteq \mathfrak{b}$ *is equivalent to*

$$w_{\mathfrak{p}}(\mathfrak{a}) \doteq w_{\mathfrak{p}}(\mathfrak{b}) \qquad \textit{for each } \mathfrak{p} \in {}^*V.$$

Proof. If $\mathfrak{a} \dot{\leqslant} b$, then there exists a standard divisor $\mathfrak{c}$ such that

$$\mathfrak{b} - \mathfrak{a} \geqslant \mathfrak{c}.$$

It follows $w_\mathfrak{p}(\mathfrak{b} - \mathfrak{a}) = w_\mathfrak{p}(\mathfrak{b}) - w_\mathfrak{p}(\mathfrak{a}) \geqslant w_\mathfrak{p}(\mathfrak{c})$. Since $\mathfrak{c}$ is standard, its $\mathfrak{p}$-adic logarithmic value $w_\mathfrak{p}(\mathfrak{c})$ is a standard real number. Therefore, $w_\mathfrak{p}(\mathfrak{a}) \dot{\leqslant} w_\mathfrak{p}(\mathfrak{b})$.

Conversely, assume that $w_\mathfrak{p}(\mathfrak{a}) \dot{\leqslant} w_\mathfrak{p}(\mathfrak{b})$ for every $\mathfrak{p} \in {}^*V$. This means there is a standard real number $\gamma_\mathfrak{p}$ such that $w_\mathfrak{p}(\mathfrak{b} - \mathfrak{a}) \geqslant \gamma_\mathfrak{p}$. By definition, we have $w_\mathfrak{p}(\mathfrak{b} - \mathfrak{a}) = v_\mathfrak{p}(\mathfrak{b} - \mathfrak{a}) \log(N\mathfrak{p})$. If $\mathfrak{p}$ is nonstandard, then $N\mathfrak{p}$ is infinitely large (Lemma 3.1) and so is $\log(N\mathfrak{p})$. Therefore, if the ordinal $v_\mathfrak{p}(\mathfrak{b} - \mathfrak{a}) \in {}^*\mathbf{Z}$ would be < 0, then $v_\mathfrak{p}(\mathfrak{b} - \mathfrak{a}) \log(N\mathfrak{p})$ would be infinitely small and hence $< \gamma_\mathfrak{p}$, contradicting our assumption. We conclude that $v_\mathfrak{p}(\mathfrak{b} - \mathfrak{a}) \geqslant 0$ and hence $w_\mathfrak{p}(b - \mathfrak{a}) \geqslant 0$; this holds for every nonstandard prime $\mathfrak{p}$.

Now let S denote the set of those $\mathfrak{p} \in {}^*V$ for which $w_\mathfrak{p}(\mathfrak{b} - \mathfrak{a}) < 0$. Then S is an *internal* set (since $\mathfrak{a}$ and $\mathfrak{b}$ are internal divisors). We have just seen that S does not contain any nonstandard divisor. Therefore, we infer from 2.5 that S consists of finitely many prime divisors only. Every $\mathfrak{p} \in S$ is standard, and hence $\log(N\mathfrak{p})$ is a standard positive number. It follows that the numbers $\gamma_\mathfrak{p}/\log(N\mathfrak{p})$ for $\mathfrak{p} \in S$ are standard. Let c be a standard lower bound for these finitely many numbers; we may assume $c \in \mathbf{Z}$. Then the divisor

$$\mathfrak{c} = c \sum_{\mathfrak{p} \in S} \mathfrak{p}$$

is standard, and it satisfies $\mathfrak{c} \leqslant \mathfrak{b} - \mathfrak{a}$.

Namely, for $\mathfrak{p} \in S$ we have, by construction,

$$w_\mathfrak{p}(\mathfrak{c}) = c \log(N\mathfrak{p}) \leqslant \gamma_\mathfrak{p} \leqslant w_\mathfrak{p}(\mathfrak{b} - \mathfrak{a}).$$

For $\mathfrak{p} \notin S$ we have

$$w_\mathfrak{p}(\mathfrak{c}) = 0 \leqslant w_\mathfrak{p}(\mathfrak{b} - \mathfrak{a}),$$

by definition of S.

We have thus found a standard divisor $\mathfrak{c}$ such that $\mathfrak{b} - \mathfrak{a} \geqslant \mathfrak{c}$; this shows that $\mathfrak{a} \dot{\leqslant} \mathfrak{b}$. Q.E.D.

If $\mathfrak{a} \in {}^*\mathfrak{D}$, let us denote by $\dot{w}_\mathfrak{p}(\mathfrak{a})$ the order of magnitude of $w_\mathfrak{p}(\mathfrak{a})$. Thus $\dot{w}_\mathfrak{p}(\mathfrak{a}) \in \dot{\mathbf{R}}$. According to Lemma 3.3, $\dot{w}_\mathfrak{p}(\mathfrak{a})$ depends on the order of magnitude of $\mathfrak{a}$ only. In other words, if we regard $\mathfrak{a}$ as an element in $\dot{\mathfrak{D}}$, then $\dot{w}_\mathfrak{p}(\mathfrak{a})$ is still well defined as an element in $\dot{\mathbf{R}}$. If $\mathfrak{p}$ ranges over the primes in ${}^*V$, then we obtain a function $\mathfrak{p} \mapsto \dot{w}_\mathfrak{p}(\mathfrak{a})$ from ${}^*V$ to $\dot{\mathbf{R}}$; Lemma 3.3 shows that $\mathfrak{a} \in \dot{\mathfrak{D}}$ is uniquely determined by this function. In this way, we see that the group $\dot{\mathfrak{D}}$ can be represented faithfully as a certain group of functions from $*V$ to $\dot{\mathbf{R}}$.[6]

Now let x be a nonzero element in $*K$. Consider the principal divisor $(x) \in *\mathfrak{D}$, and its image in $\dot{\mathfrak{D}}$. The function representing this image is $\mathfrak{p} \mapsto \dot{w}_\mathfrak{p}(x)$, where $\dot{w}_\mathfrak{p}$ means the Krull valuation of $*K$ over K as introduced above; i.e., $\dot{w}_\mathfrak{p}(x)$ is the order of magnitude of $w_\mathfrak{p}(x) = -\log \| x \|_\mathfrak{p}$. Hence, *if we regard* (x) *as an element in* $\dot{\mathfrak{D}}$, *then this element comprises the information about the values of* x *at the Krull valuation* $\dot{w}_\mathfrak{p}$ *simultaneously for all* $\mathfrak{p} \in *V$. In view of this situation, the element $(x) \in \dot{\mathfrak{D}}$ is to be regarded as the "principal divisor" of x with respect to the set of Krull valuations $\dot{w}_\mathfrak{p}$. If we assign to every $x \in *K$ its principal divisor (x) in $\dot{\mathfrak{D}}$, then we obtain the "principal divisor map" $*K \to \dot{\mathfrak{D}}$ belonging to the valuations $\dot{w}_\mathfrak{p}$. (By definition, this map consists of first applying the internal principal divisor map $*K \to *\mathfrak{D}$, and then projecting $*\mathfrak{D}$ onto $\dot{\mathfrak{D}}$.) In this connection, the elements in $\dot{\mathfrak{D}}$ will be called "divisors," and $\dot{\mathfrak{D}}$ the corresponding "divisor group."

If x is standard, then (x) is standard too and hence $(x) \doteq 0$. On the other hand, if x is nonstandard, then $(x) \not\doteq 0$ in view of Lemma 3.2. Hence, Lemma 3.2 can be regarded as describing the kernel of the principal divisor map $*K \to \dot{\mathfrak{D}}$, namely, this kernel is the multiplicative group of K. That is, the sequence $1 \to K \to *K \to \dot{\mathfrak{D}}$ is exact. We are now going to describe the *image* of the principal divisor map $*K \to \dot{\mathfrak{D}}$. Let us consider the size $\sigma : *\mathfrak{D} \to *\mathbf{R}$ as defined in (3.8). If $\mathfrak{a} \doteq \mathfrak{b}$, then $\sigma(\mathfrak{a}) \doteq \sigma(\mathfrak{b})$. Thus σ defines a map $\sigma : \dot{\mathfrak{D}} \to \dot{\mathbf{R}}$ which is surjective since the original σ is surjective. Let $\dot{\mathfrak{D}}_0$ denote its kernel; it consists of those internal divisors $\mathfrak{a}$ whose size $\sigma(\mathfrak{a})$ is finite. (More precisely, $\dot{\mathfrak{D}}_0$ consists of the orders of magnitude of those divisors.) For any such $\mathfrak{a}$, we can find a divisor $\mathfrak{a}_0$ such that $\mathfrak{a}_0 \doteq \mathfrak{a}$ and $\sigma(\mathfrak{a}_0) = 0$. To see this, let $\sigma(\mathfrak{a}) = c$; this is a certain finite real number. Let $\mathfrak{p}$ be an archimedean prime; then $c\mathfrak{p}$ is a finite divisor and $\sigma(c\mathfrak{p}) = c$. Therefore, the divisor $\mathfrak{a}_0 = \mathfrak{a} - c\mathfrak{p}$ solves our problem.

We have thus shown that every element in $\dot{\mathfrak{D}}_0$ can be represented by a divisor of size 0, i.e., by a divisor in $*\mathfrak{D}_0$. In other words $\dot{\mathfrak{D}}_0$ *can be described as being the image of* $*\mathfrak{D}_0$ *in* $\dot{\mathfrak{D}}$, *consisting of the orders of magnitude of divisors with vanishing size.*

As a consequence of product formula (3.3), we have seen above that the size function σ vanishes on principal divisors. It follows that the image of the principal divisor map $*K \to \dot{\mathfrak{D}}$ is contained in $\dot{\mathfrak{D}}_0$.

[6] There arises the problem of characterizing those functions from $*V$ to $\dot{\mathbf{R}}$ which represent elements in $\dot{\mathfrak{D}}$. The exact condition for this is obtained by saying that this function must originate from an internal divisor as described above, and then by representing this internal divisor as an internal function from $*V$ to $\mathbf{R}$ satisfying *(i)–*(iii). We leave the explicit formulation of this condition to the reader.

THEOREM 3.4. *Every divisor in $\dot{\mathfrak{D}}_0$ is principal; i.e., $\dot{\mathfrak{D}}_0$ is the image of the principal divisor map $^*K \to \dot{\mathfrak{D}}$. Consequently, the following sequence is exact, exhibiting kernel and cokernel of the principal divisor map*:

$$1 \to K \to {}^*K \to \dot{\mathfrak{D}} \xrightarrow{\sigma} \dot{\mathbf{R}} \to 0.$$

Proof. Let $\mathfrak{a} \in {}^*\mathfrak{D}_0$; we have to show that its image in $\dot{\mathfrak{D}}_0$ is principal. This means that there is an element $x \in {}^*K$ such that $\mathfrak{a} \doteq (x)$, i.e., $\mathfrak{a} = (x) + \mathfrak{b}$ with some finite divisor $\mathfrak{b}$. As usual, we use the symbol $\sim$ to denote the divisor equivalence with respect to principal divisors; thus our contention is that $\mathfrak{a} \sim \mathfrak{b}$ for some finite divisor $\mathfrak{b}$. We shall exhibit a certain *standard* divisor $\mathfrak{c} \geqslant 0$ such that the following statement holds.

Every divisor $\mathfrak{a} \in {}^*\mathfrak{D}_0$ *is equivalent to some divisor* $\mathfrak{b} \in {}^*\mathfrak{D}_0$ *which satisfies* $-\mathfrak{c} \leqslant \mathfrak{b} \leqslant \mathfrak{c}$.

Since $\mathfrak{c}$ is standard, these inequalities indeed show that $\mathfrak{b}$ is finite.

Now, the above statement is the interpretation in $^*\mathfrak{D}$ of a statement in $\mathfrak{D}$, and hence it is true in $^*\mathfrak{D}$ if and only if the original statement is true in $\mathfrak{D}$. Thus it suffices to prove the original statement; this reads as follows.

Every divisor $\mathfrak{a} \in \mathfrak{D}_0$ *is equivalent to some divisor* $\mathfrak{b} \in \mathfrak{D}_0$ *which satisfies* $-\mathfrak{c} \leqslant \mathfrak{b} \leqslant \mathfrak{c}$.

Of course, this statement makes sense only if we have specified the divisor $\mathfrak{c}$. Rather than do this here, we leave this specification until the end of proof.

According to the definition of $\mathfrak{D}$, we have a direct sum decomposition $\mathfrak{D} = \mathfrak{D}' \oplus \mathfrak{D}''$, where $\mathfrak{D}'$ is the archimedean part and $\mathfrak{D}''$ the nonarchimedean part of the divisor group. If we consider only the nonarchimedean primes and their divisors, disregarding the archimedeans, then this means applying the projection $\mathfrak{D} \to \mathfrak{D}''$ which has kernel $\mathfrak{D}'$. The resulting principal divisor map $K \to \mathfrak{D} \to \mathfrak{D}''$ leads to a factor group $\mathfrak{C}'' = \mathfrak{D}''/K$; this is the nonarchimedean part of the divisor class group $\mathfrak{C}$. It is well known that $\mathfrak{C}''$ is finite, its order being the class number h of K.

If the projection $\mathfrak{D} \to \mathfrak{D}''$ is restricted to $\mathfrak{D}_0$, then it is still surjective. To see this, let $\mathfrak{a}'' \in \mathfrak{D}''$ and put $\sigma(\mathfrak{a}'') = c$. Let $\mathfrak{p}$ be an archimedean prime. Then the divisor $\mathfrak{a}'' - c\mathfrak{p}$ has size 0, hence is in $\mathfrak{D}_0$, and it is projected onto $\mathfrak{a}''$ in $\mathfrak{D}''$.

Because of the surjectivity of $\mathfrak{D}_0 \to \mathfrak{D}''$, we can find divisors $\mathfrak{c}_1, \ldots, \mathfrak{c}_h \in \mathfrak{D}_0$ whose images in $\mathfrak{D}''$ represent the different classes in $\mathfrak{C}''$. Let $\mathfrak{c}_1'', \ldots, \mathfrak{c}_h''$ denote these images. Now let $\mathfrak{a} \in \mathfrak{D}_0$. Its image in $\mathfrak{D}''$ is equivalent (modulo principal divisors in $\mathfrak{D}''$) to one $\mathfrak{c}_j''$ with $1 \leqslant j \leqslant h$. In $\mathfrak{D}$ this means an equivalence $\mathfrak{a} \sim \mathfrak{c}_j + \mathfrak{a}'$ where $\mathfrak{a}'$ has the image 0 in $\mathfrak{D}''$; i.e., $\mathfrak{a}' \in \mathfrak{D}' \cap \mathfrak{D}_0$.

It remains to discuss the divisor $\mathfrak{a}'$. Let us put $\mathfrak{D}_0' = \mathfrak{D}' \cap \mathfrak{D}_0$. This group is the kernel in $\mathfrak{D}'$ of the size map $\sigma : \mathfrak{D}' \to \mathbf{R}$. By definition, $\mathfrak{D}'$ is the free $\mathbf{R}$-module generated by the archimedean primes; i.e., $\mathfrak{D}'$ is an r-dimensional real vector space, r being the number of archimedean primes of K. Since the size $\sigma : \mathfrak{D}' \to \mathbf{R}$ is $\mathbf{R}$-linear, it follows that $\mathfrak{D}_0'$ is an $(r-1)$-dimensional hyperplane in $\mathfrak{D}'$. This hyperplane contains those principal divisors which are contained in $\mathfrak{D}'$, i.e., which have vanishing components at the nonarchimedean primes. These are precisely the principal divisors of the *units* $u \in K$. By the Dirichlet unit theorem, the group of units is finitely generated of rank $r-1$. Moreover, if $u_1, \ldots, u_{r-1}$ are $\mathbf{Z}$-independent units, then their principal divisors $(u_1), \ldots, (u_{r-1})$ form an $\mathbf{R}$-basis of $\mathfrak{D}_0'$. In fact, this last statement is equivalent to the nonvanishing of the regulator of the field K; note that the principal divisor (u) has components $v_\mathfrak{p}(u) = -\log \| u \|_\mathfrak{p}$ at the archimedean primes $\mathfrak{p}$.

It follows from the above that every divisor $\mathfrak{a}' \in \mathfrak{D}_0'$ has a unique representation of the form

$$\mathfrak{a}' = \sum_{1 \leqslant i \leqslant r-1} \lambda_i(u_i)$$

with real coefficients λ_i. Let n_i denote the largest integer in $\mathbf{Z}$ which is $\leqslant \lambda_i$, so that

$$\lambda_i = n_i + \rho_i \quad \text{with } 0 \leqslant \rho_i < 1.$$

Then

$$\begin{aligned} \mathfrak{a}' &= \sum_{1 \leqslant i \leqslant r-1} n_i(u_i) + \sum_{1 \leqslant i \leqslant r-1} \rho_i(u_i) \\ &= (u) + \mathfrak{b}' \sim \mathfrak{b}', \end{aligned}$$

where $u = u_1^{n_1} \cdots u_{r-1}^{n_{r-1}}$ is a unit, and where we have put $\mathfrak{b}' = \sum \rho_i(u_i)$. Since the coefficients ρ_i are bounded, the divisor $\mathfrak{b}'$ is contained in a bounded region. That is, we can find a divisor $\mathfrak{c}' \geqslant 0$, independent from the ρ_i, such that $-\mathfrak{c}' \leqslant \mathfrak{b}' \leqslant \mathfrak{c}'$. Explicitly, we may take

$$\mathfrak{c}' = \sum_{1 \leqslant i \leqslant r-1} \max(0, (u_i), -(u_i)).$$

We have shown that every $\mathfrak{a}' \in \mathfrak{D}_0'$ is equivalent to some $\mathfrak{b}'$ such that $-\mathfrak{c}' \leqslant \mathfrak{b}' \leqslant \mathfrak{c}'$. Hence, by what we have seen above, every $\mathfrak{a} \in \mathfrak{D}_0$ is equivalent to some $\mathfrak{c}_j + \mathfrak{b}'$ where $1 \leqslant j \leqslant h$ and $-\mathfrak{c}' \leqslant \mathfrak{b}' \leqslant \mathfrak{c}'$. Now we put

$$\mathfrak{c} = \max(0, \pm\mathfrak{c}_1, \ldots, \pm\mathfrak{c}_h) + \mathfrak{c}'.$$

Then the divisor $\mathfrak{b} = \mathfrak{c}_j + \mathfrak{b}'$ is equivalent to $\mathfrak{a}$ and satisfies $-\mathfrak{c} \leqslant \mathfrak{b} \leqslant \mathfrak{c}$.

Q.E.D.

Remarks. We could have shortened our above proof by observing that the divisor class group $\mathfrak{C}_0$ is compact, and then using the nonstandard characterization of compactness: Every class in $^*\mathfrak{C}_0$ is near-standard, hence finite (see, e.g., [14]). We have preferred the proof as given above since this exhibits explicitly its sources, namely the Dirichlet unit theorem together with the theorem about the finiteness of the class number h. In fact, it is easily seen that these two theorems are equivalent to our Theorem 3.4.

We would like to point out the similarity between Theorem 3.4 and the similar statement for rational function fields over K. In the latter case too, every divisor of degree 0 is principal. In this respect, the field extension $^*K \mid K$ behaves like the rational function field; this indicates that $^*K \mid K$ should be regarded in some sense as a function field of a *simply connected* space. In fact, our discussion in this paper will show that $^*K$, in relation to a function field which it contains, looks very much like the field of the *universal covering space*. It would be desirable to investigate this situation and to explain this similarity which, as for now, appears to be purely formal only.

4. Function Fields Embedded into the Enlargement of an Algebraic Number Field

According to Section 1, we consider the following situation: $F \mid K$ is a function field of one variable which is embedded in $^*K$; i.e., we have $K \subset F \subset {}^*K$. Our aim is to investigate the divisor theory of $F \mid K$ in relation to that of $^*K \mid K$.

As mentioned in Section 1 already, the prime divisors of $F \mid K$ will be called "functional" in order to indicate that they belong to F as a function field over K. By contrast, the internal prime divisors of $^*K$ as introduced in Section 3 will be called "arithmetical" in order to indicate that they are connected with the arithmetic properties of the field $^*K$. This terminology—functional versus arithmetical—will also be used for divisors instead of prime divisors, etc.

By definition, a functional prime divisor P is an equivalence class of nontrivial valuations of F which are trivial on K. There are no archimedean functional prime divisors. Among all valuations of F which belong to the functional prime P, there is exactly one which is normalized such that its value group is $\mathbf{Z}$. Again, this is called the P-adic *ordinal function* of F; notation v_P. The P-adic residue field of F is an extension of K of finite degree: notation $\deg(P)$. If we put

$$w_P(x) = v_P(x) \deg(P)$$

then we have the *sum formula*

$$\sum_P w_P(x) = 0, \tag{4.1}$$

which expresses the fact that the element $0 \neq x \in F$ has as many poles as it has zeros. In this formula, P ranges over all prime divisors of $F \mid K$; the sum is essentially a *finite* sum, since $w_P(x) \neq 0$ for finitely many P only.

The group D of *functional divisors* of F is defined to be the free $\mathbf{Z}$-module generated by the functional primes P. Thus every functional divisor $A \in D$ admits a unique representation

$$A = \sum_P \alpha_P \cdot P,$$

where the α_P are integers, only finitely many of them being $\neq 0$. We define

$$v_P(A) = \alpha_P$$

and

$$w_P(A) = v_P(A) \deg(P).$$

The number

$$\deg(A) = \sum_P w_P(A)$$

is called the degree of A. This defines a homomorphism deg: $D \to \mathbf{Z}$ whose kernel is denoted by D_0, the functional divisor group of degree 0.

Every $0 \neq x \in F$ determines a *principal divisor*

$$[x] = \sum_P v_P(x) P,$$

where we use brackets in order to distinguish this functional principal divisor in D from the "arithmetical" principal divisor $(x) \in {}^*\mathfrak{D}$ as introduced in Section 3.

The kernel of the principal divisor map $F \to D$ is the multiplicative group of the field of constants K. That is, the sequence $1 \to K \to F \to D$ is exact. In view of the sum formula above, the image of $F \to D$ is contained in D_0. The groups $C = D/F$ and $C_0 = D_0/F$ are the functional divisor class group and the functional divisor class group of degree 0, respectively.

This being said, we now start to investigate the connection between functional and arithmetical primes. Let $\mathfrak{p} \in {}^*V$ be an arithmetical prime. In (3.5) we have defined its logarithmic absolute value $w_\mathfrak{p}(x) = -\log \| x \|_\mathfrak{p}$. In general, this function is neither a valuation of ${}^*K$ (namely if $\mathfrak{p}$ is archimedean) nor is it trivial on K (if $\mathfrak{p}$ is standard). However, we have seen in Section 3 that the modified function $\dot{w}_{\mathfrak{p}}(x)$, which measures the order of magnitude of $w_{\mathfrak{p}}(x)$, is a valuation which is trivial on K. Hence, the restriction of $\dot{w}_{\mathfrak{p}}$ to the subfield F, if it is not entirely trivial on F, yields a valuation of $F \mid K$ which belongs to one of its functional prime divisors, say P. If this is the case, then $\mathfrak{p}$ is said to be *effective* on F, or on P; we also say that P is *induced* by $\mathfrak{p}$, notation: $\mathfrak{p} \mid P$. In using this symbol one has always to keep in mind that not the logarithmic value $w_{\mathfrak{p}}$ itself, but only its modified valuation $\dot{w}_{\mathfrak{p}}$ induces in F a valuation which is equivalent to w_P. Explicitly, the condition that $\dot{w}_{\mathfrak{p}}$ and w_P are equivalent on F means that there is a real number $\rho_{\mathfrak{p}} \dot{>} 0$ such that

$$w_{\mathfrak{p}}(x) \doteq \rho_{\mathfrak{p}} w_P(x) \tag{4.2}$$

for all $x \in F$. The order of magnitude of $\rho_{\mathfrak{p}}$ is uniquely determined by this relation. Let $\pi \in F$ be a uniformizing variable at P; i.e., $v_P(\pi) = 1$. If we put

$$e_{\mathfrak{p}} = v_{\mathfrak{p}}(\pi), \qquad f_{\mathfrak{p}} = \log(N\mathfrak{p})/\deg(P)$$

then we obtain from (4.2)

$$\rho_{\mathfrak{p}} \doteq e_{\mathfrak{p}} f_{\mathfrak{p}} \, .$$

It is clear that $e_{\mathfrak{p}}$ may be regarded as the $\mathfrak{p}$-adic ramification index, and $f_{\mathfrak{p}}$ as the $\mathfrak{p}$-adic residue degree of $^*K$ relative to F. However, we shall not consider these but work mainly with the invariant $\rho_{\mathfrak{p}}$ itself, which is to be regarded as kind of $\mathfrak{p}$-adic relative degree of $^*K$ over F.

Our first result is the following.

LEMMA 4.1. *Every functional prime divisor P of F is induced by some arithmetical prime divisor $\mathfrak{p}$ of $^*K$.*

Proof. By the theorem of Riemann-Roch there exists $x \in F$ which admits P as its only pole. That is, we have $w_P(x) < 0$, and P is the only functional prime of F with this property. Since $x \notin K$, we infer from Lemma 3.2 that there is an arithmetical prime $\mathfrak{p}$ such that $w_{\mathfrak{p}}(x) \dot{<} 0$. This inequality shows, first, that $\mathfrak{p}$ is effective on F and, secondly, that $\mathfrak{p}$ induces in F a functional prime which is a pole of x. Since there is only one pole of x, namely P, we conclude $\mathfrak{p} \mid P$. Q.E.D.

Lemma 4.1 can be found in [17] already. It shows that, in a sense, the functional divisor theory of F is induced by the arithmetical divisor theory of $^*K$. This is not literally true, however, since we have pointed out above already that the valuations $w_{\mathfrak{p}}$ have to be modified first before they can be said to induce a valuation in F. Accordingly, in formula (4.2) and in the next formulas to come, we see the sign $\doteq$ appear where in the ordinary theory one would expect the equality sign $=$. Apart from this difference, we shall develop our results in such a way that it is in complete analogy to the ordinary theory of extensions of valued fields.

Our first task is to construct what is ordinarily called the *conorm*, which in our case is an injection $i : D \to \dot{\mathfrak{D}}$ from the functional divisor group D of F into the arithmetical divisor group $\dot{\mathfrak{D}}$ of $*K$. (The appearance of $\dot{\mathfrak{D}}$ instead of $*\mathfrak{D}$ signifies that we have to use $\doteq$ instead of $=$, as has just been explained.) If $A \in D$, then we would like its image $iA \in \dot{\mathfrak{D}}$ to have the following $\mathfrak{p}$-adic values:

$$w_{\mathfrak{p}}(iA) \doteq \begin{cases} \rho_{\mathfrak{p}} w_P(A) & \text{if } \mathfrak{p} \mid P, \\ 0 & \text{if } \mathfrak{p} \text{ is not effective on } F. \end{cases} \tag{4.3}$$

It is not yet clear that such a divisor $iA \in \dot{\mathfrak{D}}$ exists; if it does exist, however, then iA is uniquely determined in view of Lemma 3.3. The following lemma solves the existence problem and more.

LEMMA 4.2. *For every functional divisor $A \in D$ there exists an arithmetical divisor $iA \in *\mathfrak{D}$ satisfying* (4.3). *This divisor is uniquely determined in its order of magnitude; the resulting map*

$$i : D \to \dot{\mathfrak{D}}$$

is an injective homomorphism which has the following properties:

strong order preservation,

$$A \leqslant B \Leftrightarrow iA \dot{\leqslant} iB;$$

maximum preservation,

$$i \max(A, B) \doteq \max(iA, iB);$$

minimum preservation,

$$i \min(A, B) \doteq \min(iA, iB);$$

principal divisor preservation,

$$i[x] \doteq (x).$$

Proof. First, we prove the existence of $iA \in *\mathfrak{D}$ satisfying (4.3). If $A = [x]$ is the principal divisor of some $x \in F$, then we can put $iA = (x)$; Eqs. (4.2) guarantee the validity of (4.3) in this case. In general, we shall try to represent A in some way by means of principal divisors; then we shall use the same representation in $*\mathfrak{D}$ to define iA.

To start with, we may assume without loss that $A \geqslant 0$. Namely, if this is not the case, then we write $A = B - C$, where $B = \max(0, A) \geqslant 0$ and $C = \max(0, -A) \geqslant 0$. If we know the existence of iB and iC, then we can put $iA = iB - iC$; note that conditions (4.3) are additive in character.

So let us assume $A \geqslant 0$; i.e., $w_P(A) \geqslant 0$ for every functional prime divisor P. There are only finitely many P with $w_P(A) > 0$. Using the approximation theorem for valuations, we can find a nonzero element $x \in F$ such that

$$w_P(x) = w_P(A) \quad \text{if } w_P(A) > 0.$$

Again, there are only finitely many P with $w_P(x) > 0$; thus we find $0 \neq y \in F$ such that

$$\begin{aligned} w_P(y) &= w_P(A) && \text{if } \ w_P(A) > 0, \\ w_P(y) &= 0 && \text{if } \ w_P(A) = 0 \ \text{ and } w_P(x) > 0. \end{aligned}$$

Now we have

$$\min(w_P(x), w_P(y)) \begin{cases} = w_P(A) & \text{if } w_P(A) > 0, \\ = 0 & \text{if } w_P(A) = 0 \text{ and } w_P(x) > 0, \\ \leqslant 0 & \text{if } w_P(x) \leqslant 0. \end{cases}$$

It follows that

$$\max(0, \min(w_P(x), w_P(y))) = w_P(A)$$

for *every* functional prime divisor P, which is to say that

$$\max(0, \min([x], [y])) = A.$$

This is the representation of A by means of principal divisors, as announced above. Now we put

$$iA = \max(0, \min((x), (y))),$$

which is the same expression in $^*\mathfrak{D}$ as A in $\mathfrak{D}$. For every arithmetical prime $\mathfrak{p}$, we have

$$w_\mathfrak{p}(iA) = \max(0, \min(w_\mathfrak{p}(x), w_\mathfrak{p}(y))).$$

Now, if $\mathfrak{p}$ is effective on P, then using (4.2) for x and y we obtain

$$w_\mathfrak{p}(iA) \doteq \rho_\mathfrak{p} \max(0, \min(w_P(x), w_P(y))) = \rho_\mathfrak{p}\, w_P(A).$$

On the other hand, if $\mathfrak{p}$ is not effective on F, then $w_\mathfrak{p}(x) \doteq 0 \doteq w_\mathfrak{p}(y)$ and it follows that $w_\mathfrak{p}(iA) \doteq 0$. We have proved that (4.3) holds.

As already said above, we have shown in Lemma 3.3 that any divisor in $\dot{\mathfrak{D}}$ is uniquely determined by its $\mathfrak{p}$-adic values in $\dot{\mathbf{R}}$. Hence, iA is uniquely determined in $\dot{\mathfrak{D}}$ by conditions (4.3), irrespective of the choice of the elements $x, y \in F$ used above to construct iA. Thus we obtain a map $i : D \to \dot{\mathfrak{D}}$. It is clear that this map is a homomorphism, since the divisor addition in D resp. $\dot{\mathfrak{D}}$ is faithfully reflected in the addition of the local P-adic resp. $\mathfrak{p}$-adic values. (We have used this remark above already by saying that conditions (4.3) are additive in character.) For similar reasons, it is clear that $i : D \to \dot{\mathfrak{D}}$ preserves the order relation of divisors, as well as the operations max and min. Formula (4.2) shows that if $A = [x]$, then $iA \doteq (x)$. It remains to prove the following strong order preservation property which at the same time yields the injectivity:

$$iA \mathrel{\dot{\leqslant}} iB \Rightarrow A \leqslant B.$$

Let P be a functional prime divisor. By Lemma 4.1 there exists an arithmetical prime divisor $\mathfrak{p}$ which is effective on P. Since $iA \mathrel{\dot{\leqslant}} iB$, we have $w_{\mathfrak{p}}(iA) \mathrel{\dot{\leqslant}} w_{\mathfrak{p}}(iB)$; in view of (4.3) this implies

$$\rho_{\mathfrak{p}} w_P(A) \mathrel{\dot{\leqslant}} \rho_{\mathfrak{p}} w_P(B),$$

which is to say that

$$\rho_{\mathfrak{p}} w_P(A) \leqslant \rho_{\mathfrak{p}} w_P(B) + c,$$

where c is a finite number. It follows that

$$w_P(A) \leqslant w_P(B) + h,$$

where $h = c/\rho_{\mathfrak{p}}$.

We know from (4.2) that $\rho_{\mathfrak{p}} \mathrel{\dot{>}} 0$; i.e., $\rho_{\mathfrak{p}}$ is infinitely large. Since c is finite, it follows that h is infinitesimal; in particular, we have $h < 1$ and therefore $w_P(A) < w_P(B) + 1$. Since $w_P(A)$ and $w_P(B)$ are (standard) integers, we have $w_P(A) \leqslant w_P(B)$.

Here P is an arbitrary functional prime divisor. Hence $A \leqslant B$. Q.E.D.

Remark. The injection $i : D \to \dot{\mathfrak{D}}$ is the nonstandard counterpart of Weil's representation of functional divisors as so-called distributions [23–25]. In this sense, our Lemma 4.2 may be compared to what is usually called the "theorem of decomposition" for Weil distributions.

Lemma 4.2 says that the map $i : D \to \dot{\mathfrak{D}}$ is injective and faithful with respect to all the relevant relations between divisors. We have already said above that this map should be regarded as an analog to the conorm mapping, in the ordinary theory of extensions of valued fields. Now, in that theory it is quite common to *identify* the divisor group of the subfield with its image under the conorm. We shall follow this procedure also in our situation.

Henceforth we identify D with its image $iD \subset \dot{\mathfrak{D}}$ *whenever this is convenient and no misunderstanding seems possible*. Accordingly, we shall regard functional divisors $A \in D$ as arithmetical internal divisors, with the provision, however, that the equality sign is replaced by $\doteq$, which measures the order of magnitude only.

Due to this identification convention, our formulas will become more lucid, leaving out the redundant symbol i. For instance, the $\mathfrak{p}$-adic logarithmic value of iA is now to be regarded as the $\mathfrak{p}$-adic logarithmic value of A itself, and to be denoted by $w_{\mathfrak{p}}(A)$. Formula (4.3) now reads

$$w_{\mathfrak{p}}(A) \doteq \begin{cases} \rho_{\mathfrak{p}} w_P(A) & \text{if } \mathfrak{p} \mid P \\ 0 & \text{if } \mathfrak{p} \text{ is not effective on } F. \end{cases} \tag{4.4}$$

This is in complete analogy to (4.2). The principal divisor property of Lemma 4.2 reads

$$[x] \doteq (x) \qquad (x \in F), \tag{4.5}$$

showing that the two principal divisors of x coincide in their order of magnitude.

Every functional divisor $A \in D$, if regarded as an element in $\dot{\mathfrak{D}}$, has a *size* $\sigma(A)$; according to Section 3 this is an element in $\dot{\mathbf{R}}$, i.e., a real number whose order of magnitude is uniquely determined by A. We thus obtain the size map $\sigma : D \to \dot{\mathbf{R}}$ which is induced by the ordinary size map $\sigma : {}^*\mathfrak{D} \to {}^*\mathbf{R}$ in the manner as described, via the embedding $D \subset \dot{\mathfrak{D}}$. On the other hand, we have the functional degree map deg: $D \to \mathbf{Z}$. There arises the question as to the connection between these two invariants σ and deg on D.

Again, we use the ordinary theory of extensions of valued fields as a guide line and motivation. In that theory, the degrees of divisors of the ground field are multiplied by a fixed number ρ, if these divisors are regarded in the extension field by means of the conorm embedding. In fact, ρ equals the degree of the field extension. Now, in our situation the field extension ${}^*K$ over F is not finite; nevertheless, we can ask whether there is such a number ρ as above. Of course, the role of the degree in the extension field ${}^*K$ is taken by the size σ. Hence we ask, more precisely, is there a number $\rho \dot{>} 0$ such that, for every functional divisor $A \in D$, we have

$$\sigma(A) \doteq \rho \deg(A)\,?$$

Now this is not true in general, the reason being the nonarchimedean type of the ordering of $\dot{\mathbf{R}}$. Nevertheless, we shall show that quite a similar statement does hold, namely:

$$\frac{\sigma(A)}{\rho} \simeq \deg(A), \tag{4.6}$$

where the symbol $\simeq$ means "infinitely close." Let us first explain this relation; afterwards we shall state our main theorem.

Let $x, y \in {}^*\mathbf{R}$. We say that x is *infinitely close* to y if $x = y + h$ with infinitesimal $h \in {}^*\mathbf{R}$. Notation: $x \simeq y$. It is clear that every infinitesimal number is finite; hence, $x \simeq y$ implies $x \doteq y$. As in the case of finite numbers, the infinitesimal numbers form an *isolated* subgroup $\mathbf{R}_{\text{inf}}$ of ${}^*\mathbf{R}$; this implies that the factor group ${}^*\mathbf{R}/\mathbf{R}_{\text{inf}}$ inherits naturally its order relation from ${}^*\mathbf{R}$. Accordingly, we shall write $x \precsim y$ in order to say that $x \leqslant y + h$ with infinitesimal h.

By definition, both relations $x \dot{\leqslant} y$ and $x \precsim y$ are of additive character; they are in general not coherent with respect to multiplication. Their behavior with respect to multiplication can be described by saying that the finite numbers $\mathbf{R}_{\text{fin}}$ form (not only an additive group but) a *valuation ring of* ${}^*\mathbf{R}$, and that $\mathbf{R}_{\text{inf}}$ is the *maximal ideal* of that valuation ring. For later references, we shall state the following lemma.

LEMMA 4.3. *Let $\rho \in {}^*\mathbf{R}$ be infinitely large. If $x \dot{\leqslant} y$ then $x/\rho \precsim y/\rho$. In particular, if $x \doteq y$ then $x/\rho \simeq y/\rho$. In other words, the map $x \mapsto x/\rho$ is an order-preserving homomorphism from the additive group $\dot{\mathbf{R}} = {}^*\mathbf{R}/\mathbf{R}_{\text{fin}}$ onto ${}^*\mathbf{R}/\mathbf{R}_{\text{inf}}$, the real numbers modulo infinitesimals.*

Proof. If $x \dot{\leqslant} y$, then $x \leqslant y + c$, where c is a finite number. It follows that $x/\rho \leqslant y/\rho + h$, where $h = c/\rho$. Since by assumption ρ is infinitely large and c is finite, it follows that h is infinitesimal. Hence, $x/\rho \precsim y/\rho$.
Q.E.D.

If $A \in D$, then we know that $\sigma(A)$ is determined modulo finite numbers. From Lemma 4.3 we infer that the ratio $\sigma(A)/\rho$ is determined modulo infinitesimals. Thus we see that statement (4.6) is at least meaningful, and it is the best what we can expect. The following theorem says that it is true.

THEOREM 4.4. *There exists an infinitely large number $\rho \in {}^*\mathbf{R}$ such that*

$$\frac{\sigma(A)}{\rho} \simeq \deg(A)$$

for all functional divisors $A \in D$. That is, the size on D is proportional to the degree, up to infinitesimals.

The number ρ is uniquely determined up to infinitesimals, in the following multiplicative sense: If $\lambda \in {}^*\mathbf{R}$ is another such number, then $\rho/\lambda \simeq 1$.

Proof. Let us start by stating the formal properties of the size which are responsible for the validity of Theorem 4.4.

(i) *The size $\sigma : D \to \dot{\mathbf{R}}$ is an order-preserving homomorphism which vanishes on principal divisors of D.*

Namely, these properties are inherited from the original size function $\sigma : {}^*\mathfrak{D} \to {}^*\mathbf{R}$, via the inclusion $D \subset \dot{\mathfrak{D}}$. As to the vanishing on principal divisors, we mean of course *functional* principal divisors $[x]$ where $x \in F$. However, due to formula (4.5), these may be identified with their arithmetical principal divisors (x); we know that the original size function $\sigma : {}^*\mathfrak{D} \to {}^*\mathbf{R}$ vanishes on those (x), as a consequence of product formula (3.3).

(ii) *The size $\sigma : D \to \dot{\mathbf{R}}$ does not vanish identically. In fact, for every $A > 0$ we have $\sigma(A) \dot{>} 0$.*

For, if $A > 0$ in D, then $A \dot{>} 0$ in $\dot{\mathfrak{D}}$. Thus we have to show, for every internal divisor, $\mathfrak{a} \dot{>} 0 \Rightarrow \sigma(\mathfrak{a}) \dot{>} 0$. By Lemma 3.3, if $\mathfrak{a} \dot{>} 0$, then there exists $\mathfrak{p} \in {}^*V$ such that $w_\mathfrak{p}(\mathfrak{a}) \dot{>} 0$. Therefore, it suffices to prove that $\mathfrak{a} \dot{\geqslant} 0 \Rightarrow \sigma(\mathfrak{a}) \dot{\geqslant} w_\mathfrak{p}(\mathfrak{a})$ for any internal divisor $\mathfrak{a}$ and any internal prime $\mathfrak{p}$. Now this rule in $\dot{\mathfrak{D}}$ is clearly inherited from the corresponding rule in ${}^*\mathfrak{D}$, namely $\mathfrak{a} \geqslant 0 \Rightarrow \sigma(\mathfrak{a}) \geqslant w_\mathfrak{p}(\mathfrak{a})$. But this is trivially true in view of definition (3.8) of the size.

This being said, we now start with the proof of Theorem 4.4. Let g denote the genus of the function field $F \mid K$. If $\deg(A) \geqslant g$, then the theorem of Riemann–Roch shows that there is a positive divisor $A' \geqslant 0$ which is equivalent to A, in the sense that $A - A' = [x]$ is principal. From (i) it follows that $\sigma(A) \doteq \sigma(A') \dot{\geqslant} 0$.

If $\deg(A) > 0$, then there exists a natural number $n \in \mathbf{N}$ such that $\deg(nA) \geqslant g$; we conclude that $\sigma(nA) = n\sigma(A) \dot{\geqslant} 0$ and hence $\sigma(A) \dot{\geqslant} 0$. Applying this to $A - B$ instead of A, we obtain

$$\deg(A) > \deg(B) \Rightarrow \sigma(A) \dot{\geqslant} \sigma(B).$$

We replace A, B by nA resp. mB, where $n, m \in \mathbf{Z}$. Thus,

$$n \deg(A) > m \deg(B) \Rightarrow n\sigma(A) \dot{\geqslant} m\sigma(B) \qquad (*)$$

This statement remains true if m, n denote arbitrary rational numbers in $\mathbf{Q}$. For, this may be reduced to the case of integers by multiplication with the least common denominator.

Now we choose a fixed positive divisor $B > 0$. Then $\deg(B) > 0$. We know from property (ii) above that $\sigma(B) \dot{>} 0$. Let $\rho \in {}^*\mathbf{R}$ be such that $\rho \doteq \sigma(B)/\deg(B)$; then ρ is infinitely large. In (*) we take $n = 1$ and $m = r/\deg(B)$, where $r \in \mathbf{Q}$; we obtain

$$\deg(A) > r \Rightarrow \sigma(A) \dot{>} r\rho \Rightarrow \sigma(A)/\rho \gtrsim r,$$

the last conclusion in view of Lemma 4.3. This statement is true for *every* rational number $r < \deg(A)$. If r tends to $\deg(A)$, we see that

$$\frac{\sigma(A)}{\rho} \gtrsim \deg(A).$$

On the other hand, formula (*) also holds if A and B are interchanged. We get, similarly,

$$\frac{\sigma(A)}{\rho} \lesssim \deg(A);$$

hence,

$$\frac{\sigma(A)}{\rho} \simeq \deg(A).$$

It remains to show the uniqueness property of ρ. Quite generally, if x and y are real numbers and x is not infinitesimal, then $x \simeq y \Rightarrow y/x \simeq 1$. For, we have $y = x + h$ with infinitesimal h, and hence $y/x = 1 + k$ where $k = h/x$ is again infinitesimal, due to the fact that x itself is not infinitesimal.

Now, if λ is as in Theorem 4.4, choose any divisor A of positive degree; then the relations

$$\frac{\sigma(A)}{\rho} \simeq \deg(A) \simeq \frac{\sigma(A)}{\lambda}$$

show that neither of the two quotients in infinitesimal; hence division leads to $\rho/\lambda \simeq 1$ since $\sigma(A)$ cancels. Q.E.D.

Remark 4.5. It is clear from our proof that Theorem 4.4 holds, not only for the size, but for an arbitrary function satisfying (i) and (ii). In other words, we have the following statement of Artin–Whaples type [1].

Let $\varphi : D \to \dot{\mathbf{R}}$ *be any nontrivial order-preserving homomorphism which vanishes on principal divisors. Then* φ *is proportional to the degree, up to infinitesimals. That is, there exists an infinitely large* ρ *such that* $\varphi(A)/\rho \simeq \deg(A)$ *for every* $A \in D$.

COROLLARY 4.6. *Let $A, B \in D$. If* $\deg(A) > 0$, *then* $\sigma(A) \gtrsim 0$ *and*

$$\frac{\sigma(B)}{\sigma(A)} \simeq \frac{\deg(B)}{\deg(A)}.$$

That is, the size quotient is infinitely close to the degree quotient.

This follows immediately from Theorem 4.4 since the factor ρ cancels out. Note that the size quotient $\sigma(B)/\sigma(A)$ is well defined up to infinitesimals.

Now let us consider a nonconstant element $x \in F$, and let us take for A the divisor of poles of x in D. That is,

$$A = -\min(0, [x]) = \max(0, -[x]).$$

It is well known that $\deg(A) = [F : K(x)]$, the right-hand side denoting the field degree of F over the rational function field generated by x. On the other hand, if we consider A as an internal divisor, we have

$$A \doteq \max(0, -(x)),$$

which is the *denominator* of x in the arithmetic sense (including the archimedean primes). To compute the size of this denominator, we notice that

$$w_{\mathfrak{p}} \max(0, -(x)) = \max(0, -w_{\mathfrak{p}}(x)) = \log \max(1, \| x \|_{\mathfrak{p}}).$$

If we put

$$H(x) = \prod_{\mathfrak{p}} \max(1, \| x \|_{\mathfrak{p}}),$$

then H is the *height function* as introduced by Hasse [4], and we have $\sigma(A) \doteq \log H(x)$. We conclude the following.

COROLLARY 4.7. *Let $x \in F$ be nonconstant. For every divisor $B \in D$ we have*

$$\frac{\sigma(B)}{\log H(x)} \simeq \frac{\deg(B)}{[F : K(x)]}.$$

In particular, taking for B the pole divisor of another nonconstant $y \in F$, we obtain the next corollary.

COROLLARY 4.8. *For nonconstant $x, y \in F$ we have*

$$\frac{\log H(y)}{\log H(x)} \simeq \frac{[F : K(y)]}{[F : K(x)]}.$$

That is, the logarithmic height quotient is infinitely close to the degree quotient.

This last formula has been obtained in [17] already. It can be regarded as the nonstandard equivalent of what is called the first basic inequality of Siegel [22].

5. Exceptional Divisors

Let P be a functional prime divisor of F. We know from Lemma 4.1 that there is at least one arithmetical prime $\mathfrak{p}$ of $^*K$ which is effective on P, i.e., $\mathfrak{p} \mid P$. The contention of Theorem 1.2 is that among these arithmetical primes $\mathfrak{p}$ there exists a *nonstandard* one provided the genus g of F is positive. Thus we have to study those functional primes P which do not admit a nonstandard $\mathfrak{p}$ with $\mathfrak{p} \mid P$. These primes P are called *exceptional*; the contention is that exceptional primes exist in the case $g = 0$ only.

It will be convenient to extend the notion of "exceptional" to divisors instead of prime divisors: namely, a functional divisor $A \in D$ is called *exceptional* if

$$A = P_1 + P_2 + \cdots + P_r,$$

where the P_i are distinct exceptional prime divisors. This definition implies that exceptional divisors are positive and without multiple components.

Our first aim is to obtain an estimate for the degree of an exceptional divisor A; this will at the same time give an upper bound for the number r of exceptional prime divisors of F (if there are any). As we know from Theorem 4.4, the degree is intimately connected with the size; this leads us to study the size of an exceptional divisor A.

An arithmetical prime $\mathfrak{p}$ is said to be effective on A if $\mathfrak{p}$ is effective on some component of A. Notation: $\mathfrak{p} \mid A$.

Lemma 5.1. *Assume $A \in D$ is an exceptional divisor of F. There are only finitely many arithmetical primes $\mathfrak{p}$ of $^*K$ which are effective on A. These primes $\mathfrak{p} \mid A$ are characterized by the condition $w_\mathfrak{p}(A) \gtrdot 0$, and we have*

$$\sigma(A) \doteq \sum_{\mathfrak{p} \mid A} w_\mathfrak{p}(A).$$

Notice that the sum on the right-hand side has finitely many terms only.

Proof. If $\mathfrak{p} \mid A$, then there is some component P of A such that $\mathfrak{p} \mid P$. Using (4.4), we conclude that

$$w_\mathfrak{p}(A) \doteq \rho_\mathfrak{p} w_P(A) \mathrel{\dot{\geqslant}} \rho_\mathfrak{p} \gtrdot 0.$$

Conversely, assume that $w_\mathfrak{p}(A) \gtrdot 0$. Then (4.4) shows, first of all, that $\mathfrak{p}$ is effective on F; i.e., there is some functional prime P such that $\mathfrak{p} \mid P$.

Moreover, (4.4) shows that $w_P(A) > 0$ for this prime P; i.e. P is a component of A and hence $\mathfrak{p} \mid A$. We have shown that $\mathfrak{p} \mid A$ if and only if $w_\mathfrak{p}(A) \dot{>} 0$, which is one of the contentions of Lemma 5.1.

In the above arguments, we had to regard A as an element of $\dot{\mathfrak{D}}$, according to the embedding $D \subset \dot{\mathfrak{D}}$ as explained in Section 4. Now let $\mathfrak{a} \in {}^*\mathfrak{D}$ be any internal divisor which represents A, i.e., such that $\mathfrak{a} \doteq A$. We then have $w_\mathfrak{p}(\mathfrak{a}) \doteq w_\mathfrak{p}(A)$ for every $\mathfrak{p}$, which means that the (standard or nonstandard) real number $w_\mathfrak{p}(\mathfrak{a})$ represents $w_\mathfrak{p}(A) \in \dot{\mathbf{R}}$. Let S denote the set of those arithmetical primes $\mathfrak{p}$ for which $w_\mathfrak{p}(\mathfrak{a}) > 0$; then S is internal and, by what we have shown above, S contains every $\mathfrak{p}$ which is effective on A. Moreover, if $\mathfrak{p} \in S$ is not effective on A, then $w_\mathfrak{p}(\mathfrak{a}) \doteq 0$.

We claim that S does not contain any nonstandard prime. For, assume $\mathfrak{p} \in S$ would be nonstandard. Then $\mathfrak{p} \nmid A$ (since A is exceptional) and hence $w_\mathfrak{p}(\mathfrak{a}) \doteq 0$. That is, the real number $w_\mathfrak{p}(\mathfrak{a}) = v_\mathfrak{p}(\mathfrak{a}) \log(N\mathfrak{p})$ would be positive and finite. But this contradicts the fact that $N\mathfrak{p}$ is infinitely large (Lemma 3.1).

Now, since S is internal and does not contain any nonstandard prime, it follows that S is *finite* (see Section 2). In particular, there are only finitely many $\mathfrak{p}$ which are effective on A.

Let us put

$$\mathfrak{a}' = \sum_{\mathfrak{p} \mid A} v_\mathfrak{p}(\mathfrak{a})\mathfrak{p}.$$

This sum contains finitely many terms only, and thus $\mathfrak{a}'$ is a well-defined internal divisor. By construction, $\mathfrak{a}'$ coincides with $\mathfrak{a}$ at the primes $\mathfrak{p} \mid A$, and $\mathfrak{a}'$ vanishes at the other primes; hence,

$$\begin{aligned} w_\mathfrak{p}(\mathfrak{a}') &\doteq w_\mathfrak{p}(A) \dot{>} 0 \quad &\text{if} \quad \mathfrak{p} \mid A \\ w_\mathfrak{p}(\mathfrak{a}') &= 0 \doteq w_\mathfrak{p}(A) \quad &\text{if} \quad \mathfrak{p} \nmid A. \end{aligned}$$

We conclude that $\mathfrak{a}' \doteq A$; i.e., $\mathfrak{a}'$ is also a representative of A in ${}^*\mathfrak{D}$. Hence,

$$\sigma(A) \doteq \sigma(\mathfrak{a}') = \sum_{\mathfrak{p} \mid A} w_\mathfrak{p}(\mathfrak{a}) \doteq \sum_{\mathfrak{p} \mid A} w_\mathfrak{p}(A).$$

Q.E.D.

COROLLARY 5.2. *Let A be exceptional as in Lemma 5.1; in addition, we assume that every component of A is of degree 1. Given any nonconstant element $x \in F$ which is A-integral, there are elements $a_\mathfrak{p} \in K$ (for $\mathfrak{p} \mid A$) such that*

$$\sigma(A) \dot{\leqslant} \sum_{\mathfrak{p} \mid A} w_\mathfrak{p}(x - a_\mathfrak{p}).$$

An element $x \in F$ is called A-integral if none of the poles of x is a component of A, i.e., if $w_P(x) \geqslant 0$ for every component P of A.

Proof. If $\mathfrak{p} \mid A$, let P denote the component of A on which $\mathfrak{p}$ is effective. Let $a_\mathfrak{p}$ denote the P-adic residue of x; since $\deg(P) = 1$, we know that $a_\mathfrak{p} \in K$. By construction, $x - a_\mathfrak{p}$ has a zero at P, i.e., $v_P(x - a_\mathfrak{p}) \geqslant 1$. On the other hand, P is a *simple* component of A (see the above definition of exceptional divisors). This implies $v_P(A) = 1 \leqslant v_P(x - a_\mathfrak{p})$; hence, $w_P(A) \leqslant w_P(x - a_\mathfrak{p})$ and therefore, in view of (4.4),

$$w_\mathfrak{p}(A) \dot{\leqslant} w_\mathfrak{p}(x - a_\mathfrak{p}).$$

Now apply Lemma 5.1. Q.E.D.

Our problem of estimating the sizes of exceptional divisors is now reduced to estimating finite sums, such as appear on the right-hand side of the formula of Corollary 5.2. To this end, we use the well-known theorem of Roth. Let us briefly recall its content.

The theorem of Roth concerns the following situation in the number field K: Let S be a finite set of prime divisors of K, let $a_\mathfrak{p}$ be elements in K belonging to the primes $\mathfrak{p} \in S$, and let $\kappa > 2$ be a real number in $\mathbf{R}$. Referring to these data we have the theorem below.

THEOREM OF ROTH. *There are only finitely many elements $x \in K$ which satisfy the approximation conditions*

$$\prod_{\mathfrak{p} \in S} \| x - a_\mathfrak{p} \|_\mathfrak{p} \leqslant \frac{1}{H(x)^\kappa} \cdot \tag{5.1}$$

Actually, the theorem of Roth in its usual formulation is somewhat more general, since the $a_\mathfrak{p}$ may be arbitrary algebraic numbers, not necessarily contained in K. Roth [21] considered the case $K = \mathbf{Q}$, and S consisting of one prime only, namely the archimedean prime of $\mathbf{Q}$. The case of several primes was settled by Ridout [13], in the case $K = \mathbf{Q}$. A proof in the general case, for an arbitrary number field K, can be found in the book of Lang [7].

By the theorem of Roth, the set of elements $x \in K$ satisfying (5.1) is finite; hence it is not enlarged in $^*K$. That is, if $x \in {^*K}$ satisfies (5.1), then x is already contained in K; i.e., x is standard. In other words, if $x \in {^*K}$ is nonstandard, then x does not satisfy (5.1). Thus we obtain the following statement in $^*K$ which, as we have seen, is the nonstandard version of Roth's theorem. Let S be a finite set of standard prime divisors, let $a_\mathfrak{p}$ be standard elements in K, belonging to the primes $\mathfrak{p} \in S$, and let $\kappa > 2$ be a standard real number.

In this situation we have the following proposition.

PROPOSITION 5.3. *For every nonstandard* $x \in {}^*K$,

$$\prod_{\mathfrak{p} \in S} \| x - a_{\mathfrak{p}} \|_{\mathfrak{p}} > \frac{1}{H(x)^{\kappa}} .$$

Taking the logarithm of both sides, this is equivalent to the additive inequality

$$\sum_{\mathfrak{p} \in S} w_{\mathfrak{p}}(x - a_{\mathfrak{p}}) < \kappa \log H(x). \tag{5.2}$$

Now, if we consider the situation of Corollary 5.2 and use (5.2) for the finite set of those $\mathfrak{p}$ which are effective on the exceptional divisor A, then we find

$$\sigma(A) \dot{\leqslant} \kappa \log H(x)$$

for every standard number $\kappa > 2$. Since $\log H(x)$ is infinitely large (Lemma 3.1), we conclude, in view of Lemma 4.3, that

$$\frac{\sigma(A)}{\log H(x)} \precsim \kappa.$$

Since $\kappa > 2$ is arbitrary standard, it follows that

$$\frac{\sigma(A)}{\log H(x)} \precsim 2.$$

On the other hand, we know from Corollary 4.7 that

$$\frac{\sigma(A)}{\log H(x)} \simeq \frac{\deg(A)}{[F : K(x)]} .$$

Therefore,

$$\frac{\deg(A)}{[F : K(x)]} \precsim 2$$

and hence

$$\deg(A) \leqslant 2\,[F : K(x)], \tag{5.3}$$

since both sides are standard integers.

We have proved formula (5.3) for every exceptional divisor A and every nonconstant $x \in F$, under the additional assumptions of Corollary 5.2: namely, (i) x is A-integral; (ii) every component of A is of degree 1. But we claim that these additional assumptions are unnecessary for the validity of (5.3). That is, (5.3) holds for every exceptional divisor A and every nonconstant $x \in F$, without further conditions.

In order to eliminate (i), we observe that formula (5.3) depends on the rational function field $K(x)$ only and not on the choice of its generator x. Hence, if x should not be A-integral, then we choose another generator y of the same field $K(x)$, such that y is A-integral; the validity of (5.3) for y then implies its validity for x, since $K(x) = K(y)$. For instance, we may choose $y = 1/(x - c)$, where $c \in K$ is selected such that c is different from the finitely many P-adic residues of x, for every component P of A which is not a pole of x.

In order to eliminate condition (ii), we use the method of constant field extension. If K' is a finite algebraic field extension of K, then let $F' = FK'$ denote the corresponding constant field extension of F. The divisor group D of F is naturally embedded into the divisor group D' of F'; it is well known that this embedding is degree preserving. That is, if we consider A as a divisor of D', then its degree (over the new constant field K') equals the degree of A when considered in D. Moreover, for every nonconstant element $x \in F$ we have $[F : K(x)] = [F' : K'(x)]$. Hence, the validity of (5.3) in $F' \mid K'$ implies its validity in $F \mid K$. Now, it is well known that K' can be chosen in such a way that every component of A splits in D' into primes of degree 1; such a field K' is called a "splitting field" of A. Thus we know that (5.3) holds for A in $F' \mid K'$ and therefore also in $F \mid K$, provided we can show that A, being exceptional in F, remains exceptional in F'. This can be shown as follows.

The notion of "exceptional" refers to the embedding $F \subset {}^*K$ of F into the enlargement ${}^*K$ of K. The constant field extension $F' = FK'$ is imbedded into the field compositum ${}^*KK'$. Now, it is easily verified that this field compositum equals the enlargement ${}^*K'$ of the field K'. Namely, let $u_1, \ldots, u_n$ be a basis of K' over K. The statement that these elements form a basis of K' over K remains true in the enlargement, which shows that the $u_1, \ldots, u_n$ form a basis of ${}^*K'$ over ${}^*K$. Hence, ${}^*K' = {}^*KK'$, and this field compositum is linearly disjoint over K. In particular, it follows that the constant field extension $F' = FK'$ is naturally embedded into ${}^*K'$; it is with respect to this embedding $F' \subset {}^*K'$ that we claim A to be exceptional. In fact, let $\mathfrak{p}'$ be any arithmetical prime of ${}^*K'$ which is effective on A; we have to show that $\mathfrak{p}'$ is standard. Let P' denote the functional prime of F' which is induced by $\mathfrak{p}'$; then P' is a component of A. Let P be the prime divisor of F which is induced by P'; then P is a component of A in $F \mid K$. Moreover, let $\mathfrak{p}$ denote the arithmetical prime induced by $\mathfrak{p}'$ in ${}^*K$. We then have the situation indicated in Fig. 1, a diagram of fields and corresponding primes. It is clear by construction that $\mathfrak{p}$ is effective on P; since P is a component of A and A is exceptional in F, it follows that $\mathfrak{p}$ is standard. In particular, $\mathfrak{p}$ is nontrivial on K and hence $\mathfrak{p}'$, which is an extension of $\mathfrak{p}$, is nontrivial on K'. Consequently, $\mathfrak{p}'$ is a standard prime

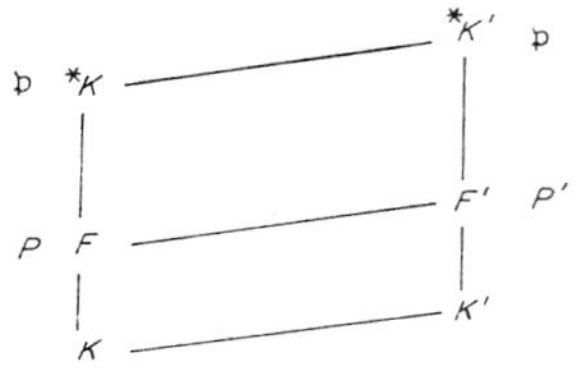

FIGURE 1

of $*K'$, since nonstandard primes are trivial on the ground field K' (Lemma 3.1).

We also observe that the constant extension $F' = FK'$ of F is *unramified*, as is well known from the general theory of function fields. Hence, since A has no multiple components as a divisor of $F \mid K$, the same is true for A in $F' \mid K'$. (Recall that the condition for A to be exceptional included the fact that A is without multiple components.)

The above discussion shows that every exceptional divisor A of $F \mid K$ remains exceptional in $F' \mid K'$. Hence, if we choose K' as a splitting field of A, we infer from the above that (5.3) is true over K', hence also over K.

We have now proved (5.3) in the general case, without the additional assumptions (i) and (ii).

Let us denote by d the *minimal degree* of F over a rational subfield; i.e.,

$$d = \min_{\substack{x \in F \\ x \notin K}} [F : K(x)]. \tag{5.4}$$

This is a certain invariant of the function field $F \mid K$. From (5.3) we infer that $\deg(A) \leqslant 2d$ for every exceptional divisor A of F. In particular, we see that the number of components of A is bounded by $2d$. It follows that there are only finitely many exceptional prime divisors and their number is $\leqslant 2d$.

Let us gather our results into the following.

THEOREM 5.4. *There are only finitely many functional prime divisors $P_1, \ldots, P_r$ of F which are exceptional in $*K$. If we put*

$$A = P_1 + \cdots + P_r,$$

then

$$\deg(A) \leqslant 2d$$

where the invariant d of $F \mid K$ is given by (5.4).

Theorem 5.4 can be regarded as the nonstandard equivalent of the so-called second fundamental inequality of Siegel [22].

COROLLARY 5.5. $\deg(A) \leqslant 2g + 2$, *where g is the genus of* $F \mid K$.

Proof. We have to verify that $d \leqslant g + 1$. In fact, if B is any functional divisor of F of degree $g + 1$ then, by the theorem of Riemann-Roch, we have $\dim(B) \geqslant 2$ and hence there are at least two different positive divisors $B', B'' \geqslant 0$ which are equivalent to B (modulo principal divisors). We have $B' - B'' = [x]$, where $x \in F$ is nonconstant, and $[F : K(x)] \leqslant \deg(B'') = \deg(B) = g + 1$. Q.E.D.

Remark 5.6. In the argument just given, we have used the fact that there is a functional divisor of degree $g + 1$. In fact, *every* standard integer is the degree of some functional divisor of F. To show this, it suffices to exhibit at least one functional divisor of degree 1. Now, *every algebraic function field* $F \mid K$ *imbedded into* $^*K$ *has infinitely many prime divisors* P *of degree* 1. This fact belongs to the fundamentals of nonstandard arithmetic and has been proved in [16] already. The argument is as follows.

Let $x_1 \in F$ be nonconstant, and let R denote the integral closure of $K[x_1]$ in F. Then R is a finitely generated K-algebra, say $R = K[x_1, ..., x_m]$. Let

$$f_j(x_1, ..., x_m) = 0 \qquad (1 \leqslant j \leqslant r)$$

be a system of defining relations of the $x_1, ..., x_m$ over K. Since these equations have a nonstandard solution—namely, $x_1, ..., x_m$—it follows from 2.4 that there are infinitely many solutions in K. That is, there are infinitely many m-tuples $a_1, ..., a_m \in K \times \cdots \times K$ such that

$$f_j(a_1, ..., a_m) = 0 \qquad (1 \leqslant j \leqslant r).$$

Every such solution defines a K-homomorphism $R \to K$ which maps x_j onto a_j $(1 \leqslant j \leqslant r)$; the kernel of this is a maximal ideal of R. *Thus,* R *has infinitely many maximal ideals* M *such that* $R/M = K$. On the other hand, we know from the general theory of function fields that R is a Dedekind ring; its maximal ideals M are in one-to-one correspondence to those prime divisors P of F whose valuation ring contains R (this means that P should not be among the poles of x_1). In this correspondence, the valuation ring of P equals the quotient ring of R with respect to M, and the residue field of P is isomorphic to R/M. We conclude that there are infinitely many functional prime divisors P of F whose residue field is K, i.e., which have degree 1.

The converse is also true: If an abstract function field $F \mid K$ of one variable has infinitely many prime divisors of degree 1, then there is a K-isomorphism from F into $^*K$. Again, this follows from 2.4 by reversing the above arguments; note that a nonstandard solution of the above equations is necessarily *generic over* K. See also Section 1.

6. Unramified Extensions: Elliptic and Hyperelliptic Function Fields

We conserve the notations and assumptions of the foregoing section. Let A be an exceptional divisor of F; i.e.,

$$A = P_1 + \cdots + P_r,$$

where the P_i are exceptional prime divisors, mutually distinct. The inequalities of Theorems 5.4 and 5.5 show that

$$\deg(A) \leqslant 2d \leqslant 2g + 2.$$

If $g = 0$, then it follows that $\deg(A) \leqslant 2$; we shall see in Section 8 that every possibility within these limits can indeed be realized by a suitable function field of genus 0. We now assume $g > 0$ and proceed to prove that in fact $A = 0$; i.e., there are no exceptional primes in F. We start from the above estimate and try to improve it such that finally $\deg(A) < 1$; then it will follow that $A = 0$. In order to obtain the desired improved estimate, we shall study *unramified extension fields of F in *K*; the application of Theorem 5.4 to such extension fields will lead to the desired result.

Let E be an extension field of F which is contained in $^*K$, so that $K \subset F \subset E \subset {}^*K$. We assume that the field degree $[E : F]$ is finite, which implies that E is an algebraic function field of one variable with K as its field of constants.

Lemma 6.1. *A functional prime P of F is exceptional if and only if every one of its extensions to E is exceptional.*

Proof. If P is exceptional, then by definition P is induced by *standard* primes $\mathfrak{p} \in {}^*V$ only. Now, if Q is an extension of P to E, then every $\mathfrak{p} \in {}^*V$ which induces Q on E will induce P on F. Hence, $\mathfrak{p}$ is standard, showing that Q is exceptional.

Conversely, assume that every extension of P to E is exceptional. Every prime $\mathfrak{p} \in {}^*V$ inducing P on F will induce on E some prime Q which is an extension of P. Hence $\mathfrak{p}$ is standard, showing that P is exceptional. Q.E.D.

Now let $Q_1, \ldots, Q_s$ be those functional primes of E which appear as extensions of one of the exceptional primes $P_1, \ldots, P_r$ appearing in A. Lemma 6.1 shows that every Q_j is an exceptional prime of E. If A is considered as a divisor of E, then it has the form $A = e_1 Q_1 + \cdots + e_s Q_s$, where e_j denotes the ramification index of Q_j over F.

Now assume that E is unramified over F. Then $A = Q_1 + \cdots + Q_s$, and we conclude that A is an exceptional divisor of E. (Recall that the definition of exceptional divisors implies that every component should be simple.) In other words, the divisor A remains exceptional in E. Therefore, Theorem 5.4 applied to E yields $\deg_E(A) \leqslant 2d_E$, where d_E denotes the field invariant appearing in Theorem 5; i.e., d_E is the minimum of the numbers $[E : K(y)]$ with $y \in E$. Moreover, $\deg_E(A)$ denotes the degree of A if considered as a divisor of E; it is well known that

$$\deg_E(A) = [E : F] \cdot \deg_F(A).$$

That is, if a divisor of F is regarded as a divisor of E, then its degree is multiplied by $[E : F]$. Combining these observations, we obtain the following corollary.

COROLLARY 6.2. *Assume E to be unramified over F. Then every exceptional divisor A of F remains exceptional in E. The degree of A in F is estimated as*

$$\deg(A) \leqslant \frac{2d_E}{[E : F]}.$$

In particular, if E is constructed such that

$$d_E < \tfrac{1}{2}[E : F], \tag{6.1}$$

then we conclude that $\deg(A) = 0$; i.e., $A = 0$.

In view of this we are now going to construct unramified extensions E of F in $*K$, such that $[E : F]$ is sufficiently large. This construction will be quite explicit and elementary if $d_F = 2$, i.e., if F is a quadratic extension of a rational function field. This case will be discussed in this section, whereas the general case will be found in Section 7.

Let us add one more preliminary remark, concerning constant field extensions. Let K' be a finite algebraic extension field of K, and let $F' = FK'$ denote the corresponding constant field extension. This is a subfield of the field compositum $*KK'$ which, as we have seen in Section 5, coincides with the enlargement $*K'$ of K'. Thus we have the situation $K' \subset F' \subset *K'$. The exceptional divisor $A = P_1 + \cdots + P_r$ of F remains exceptional in F', as has been shown in Section 5. Since F' is a constant field extension of F, the degree of A in F' coincides with the degree of A in F. Also, F' has the same genus as F. These remarks show that, in order to prove that F has no exceptional prime divisors, it is permissible to replace F by a constant field extension F' and to prove that F' has no exceptional divisors.

This simple remark will allow us to apply a suitable constant field extension before starting with our constructions; this will simplify our discussion considerably. (However, it should be remarked that these constant field extensions are not really necessary; it is possible to use constructions which are rational over K.)

This being said, we now proceed with the discussion of quadratic function fields. Thus, we assume that F is given as a quadratic extension of a rational subfield $K(x)$, i.e., $[F : K(x)] = 2$. There is a generator y of F over $K(x)$ such that $y^2 = f(x)$, where $f(x) \in K[x]$ is a polynomial without multiple roots. It is well known that the genus g of F is computed by means of the degree $m = \deg f(x)$ in the form

$$g = \begin{cases} \frac{1}{2}(m-1) & \text{if } m \text{ is odd,} \\ \frac{1}{2}(m-2) & \text{if } m \text{ is even.} \end{cases}$$

Our assumption $g > 0$ thus means that $m \geqslant 3$. If $m = 3$ or 4, then $g = 1$; i.e., F is elliptic. If $m \geqslant 5$, then F is hyperelliptic.

After applying a suitable constant field extension we may assume that the polynomial $f(x)$ has at least two roots $a, b \in K$, so that

$$y^2 = (x-a)(y-b)\,g(x),$$

where $g(x) \in K[x]$ is of degree $m-2$. Let P_a be a functional prime at which $x-a$ has a zero; then it follows from this equation that P_a is a *double zero* of $x-a$, i.e., $v_{P_a}(x-a) = 2$, and that there is no other zero of $x-a$ in F. (In other words, P_a is ramified over $K(x)$.) This shows that the principal divisor of $x-a$ has the form $[x-a] = 2P_a - X_\infty$, where X_∞ is the pole divisor of x. Similarly, $[x-b] = 2P_b - X_\infty$, where $P_b \neq P_a$ since $a \neq b$. Hence, the element

$$z = \frac{x-a}{x-b} \tag{6.2}$$

has the principal divisor

$$[z] = 2P_a - 2P_b, \tag{6.3}$$

which is seen to be divisible by 2; i.e., $[z]$ is twice some functional divisor of F. As a consequence of this property of $[z]$, we now claim the following.

There is a constant $0 \neq c \in K$ *such that* $\sqrt{cz} \in {}^*K$.

Proof. According to Section 4, we may regard P_a and P_b as divisors in $\dot{\mathfrak{D}}$, and then we have $(z) \doteq 2P_a - 2P_b$. Since the size σ vanishes on principal divisors, we conclude $0 \doteq \sigma(2P_a - 2P_b) \doteq 2\sigma(P_a - P_b)$. Let us recall that the sign $\doteq$ stands for the relation of equality in the group $\dot{\mathbf{R}}$; since this group is totally ordered, it does not admit any torsion and therefore

$$\sigma(P_a - P_b) \doteq 0.$$

Thus the divisor $P_a - P_b$ of $\dot{\mathfrak{D}}$ has vanishing size; we conclude from Theorem 3.4 that this divisor is principal in $\dot{\mathfrak{D}}$. That is, there is an element $t \in {}^*K$ such that $(t) \doteq P_a - P_b$. We have $(t^2) = 2(t) \doteq 2P_a - 2P_b \doteq (z)$ and therefore, in view of Theorem 3.4, $t^2 = cz$ with some $c \in K$. Q.E.D.

It follows from (6.2) that z and, hence, cz generate the same field as x; i.e., $K(x) = K(z) = K(cz)$. Therefore, since $t = \sqrt{cz}$, we obtain

$$K(x) \subset K(t) \text{ and } [K(t) : K(x)] = 2.$$

On the other hand, we know that $[F : K(x)] = 2$. If t would be contained in F, then $F = K(t)$, contrary to our assumption that F has genus $g > 0$. Hence, t is a quadratic irrationality over F; if we put $E = F(t)$, then $[E : F] = 2$. Thus we have constructed a certain quadratic extension field E of F inside ${}^*K$. We claim that E is *unramified* over F. In fact, E is generated over F by the quadratic radical $t = \sqrt{cz}$; hence, every prime of F which is ramified in E appears in the principal divisor $[cz] = [z]$ with an odd multiplicity. But there is no such prime; we have seen in (6.3) that every prime appearing in $[z]$ has multiplicity 2. Therefore, there is no prime ramified in E.

In our construction of this unramified extension field E, we had assumed that F should be quadratic, i.e., quadratic over a rational subfield. Now, this property is inherited by the field E. Namely, we have $E = F(t) = K(x, y, t) = K(y, t)$ and $y^2 \in K(x) \subset K(t)$. Therefore, $[E : K(t)] = 2$; i.e., E is quadratic too. Hence, this construction may be repeated and applied to E instead of F, and so on. After n steps we have the following situation.

After applying a suitable constant extension, we can construct an unramified extension $E^{(n)}$ of F inside ${}^*K$ such that

$$[E^{(n)} : F] = 2^n \quad \text{and} \quad d_{E^{(n)}} = 2.$$

If $n = 3$, we see that condition (6.1) is satisfied; hence, F does not have any exceptional prime. Theorem 1.2 is proved for quadratic function fields of genus > 0, i.e., for elliptic and hyperelliptic fields.

7. Unramified Extensions in the General Case

Now we drop our assumption that F is elliptic or hyperelliptic, and we consider the case of an arbitrary function field of genus $g > 0$. Again, our aim is to construct unramified extensions E of F within ${}^*K$, which have large degree $[E : F]$. More precisely, the degree should be large compared with d_E, namely such that the inequality (6.1) holds:

$$d_E < \tfrac{1}{2}[E : F].$$

Let n denote a standard natural number, fixed through the following discussion. Our constructions will be based on the following.

LEMMA 7.1. *Let T be a functional divisor of F such that nT is principal in F. Then there exists $t \in {}^*K$ such that $T \doteq (t)$. This element t is an nth radical over F; i.e., $t^n \in F$. The extension $F(t)$ is unramified over F.*

Proof. Because of the hypothesis, there exists $u \in F$ such that $nT = [u]$, the functional principal divisor of u. According to Section 4, we may regard T as a divisor in $\dot{\mathfrak{D}}$, and then we have $nT \doteq (u)$. We apply the size map $\sigma : \dot{\mathfrak{D}} \to \dot{\mathbf{R}}$ and recall that σ vanishes on principal divisors; it follows that $\sigma(nT) \doteq n\sigma(T) \doteq 0$. Therefore, $\sigma(T) \doteq 0$, because the group $\dot{\mathbf{R}}$ is totally ordered and hence does not admit any torsion. Now we apply Theorem 3.4, which shows that every divisor in $\dot{\mathfrak{D}}$ with vanishing size is principal. We conclude that $T \doteq (t)$ with some $t \in {}^*K$. It follows that $(u) \doteq nT \doteq (t^n)$ and therefore $t^n = cu$ with some $c \in K$. Hence t is an nth radical over F. If a functional prime P of F does not appear in the principal divisor $[u] = [cu]$, then P is unramified in $E = F(t)$. This follows immediately from the fact that the polynomial $X^n - cu$, which admits t as a root, has the discriminant $\pm n(cu)^{n-1}$, whose divisor does not contain P. On the other hand, if P does appear in $[u]$, then $v_P(u) = nv_P(T)$ is divisible by n; if we choose $x \in F$ such that $v_P(x) = v_P(T)$, then P does not appear in the principal divisor of $u' = ux^{-n}$. If we replace t by $t' = tx^{-1}$, which also generates E over F, then $t'^n = cu'$; i.e., t' is an nth radical of cu'. The above argument can be applied to u' and t' instead of u and t, showing again that P is not ramified in E. Thus, E is unramified over F. Q.E.D.

In the following, C_n denotes the so-called nth *division group* of C, consisting of those divisor classes of C which are annihilated by n. The hypothesis of Lemma 7.1 states that T should represent some class in C_n. Now let us consider *all* nth radicals $t \in {}^*K$ such that $(t) \doteq T$ with some functional divisor $T \in D$ representing a class in C_n. It is clear these $t \in {}^*K$ form a multiplicative group W_n containing all elements $\neq 0$ of F. If we assign to each $t \in W_n$ the class of its corresponding divisor $T \in D$, then we obtain a homomorphism $W_n \to C_n$ which, in view of Lemma 7.1, is surjective. Clearly, the kernel of this homomorphism is the multiplicative group of F; hence, we have an *isomorphism* $W_n/F = C_n$. The field $F(W_n)$ is unramified over F, since it is generated by the unramified extensions $F(t)$ for $t \in W_n$. If K contains the nth roots of unity, then Kummer theory shows that $F(W_n)$, being generated by nth radicals, is abelian of exponent n over F. Moreover, it follows from Kummer theory that $[F(W_n) : F]$ equals the order of the radical factor group W_n/F; hence (since C_n is finite), $[F(W_n) : F] = | C_n |$, where the right-hand side denotes the group order of C_n.

Still assuming K to contain the nth roots of unity, we claim that $F(W_n)$ is the *maximal* extension of F within $*K$, which is unramified and abelian of exponent n. By Kummer theory, any such extension is generated by nth radicals; thus, we have to show that each of these radicals is contained in W_n. So let $t \in *K$ be an nth radical over F, and assume that $F(t)$ is unramified over F. Let us put $t^n = u \in F$. Since $F(t)$ is unramified over F, it follows $v_P(u) \equiv 0 \bmod n$ for every functional prime P of F. Hence the principal divisor $[u]$ is divisible by n in the functional divisor group D; i.e., $[u] = nT$ with some $T \in D$. This shows first that the class of T is annihilated by n; i.e., this class is in C_n. Secondly, we have $nT \doteq (u) \doteq (t^n)$ and therefore $T \doteq (t)$. Thus $t \in W_n$.

We have shown the following corollary.

COROLLARY 7.2. *Assume that the nth roots of unity are contained in K. Then all the nth radicals t of Lemma* 7.1 *generate the maximal extension of F within *K, which is unramified and abelian of exponent n. The degree of this maximal extension equals* $| C_n |$, *the order of the nth division class group of F over K.*

If K' is an algebraic extension of K, then we denote by $F' = FK'$ the corresponding constant field extension of F. Also, C' is the divisor class group of F' and C_n', its nth division group. The inclusion $F \subset F'$ defines a natural map $C \to C'$ which is injective (since F' is a constant field extension of F). Hence we may regard C as a subgroup of C', and then we have $C_n = C \cap C_n'$. If we take for K' the algebraic closure of K, then it is known from the general theory of algebraic function fields that $| C_n' | = n^{2g}$. It follows that $| C_n | \leqslant n^{2g}$. If $| C_n | = n^{2g}$, then we say that all the nth division classes of F are rational over K (more precisely, the nth division classes of all constant field extensions of F are rational over K). It is well known that this implies the nth roots of unity to lie in K. Hence, we obtain the following.

COROLLARY 7.3. *Assume that all the nth division classes of F are rational over K. Let* E_n *denote the maximal extension of F within *K, which is unramified and abelian of exponent n. Then* $[E_n : F] = n^{2g}$.

This field E_n is called the *nth division field* of F within $*K$. So far, it is only defined if the nth division classes of F are rational over K. If K' is a finite algebraic extension of K then, clearly, $E_n' = E_nK'$ is the nth division field of F' within $*K'$.

In order to define E_n in the general case, with no assumptions about the rationality of division classes, we introduce the notion of semiabelian extensions. Let E be a finite extension of F within $*K$. Then E is called *semiabelian of exponent n* over F if there exists a finite extension $K' \supset K$ such that E' is abelian of exponent n over F'. (Here, $E' = EK'$ and $F' = FK'$.) Assume this to be the case and, in addition, that E is unramified over F. Then E' is unramified over F'. In view of Corollary 7.2 (applied to F') we conclude

$$[E : F] = [E' : F'] \leqslant | C_n' | \leqslant n^{2g}.$$

This relation holds for every extension E of F with the following properties:

(i) E is contained in $*K$;

(ii) E over F is unramified;

(iii) E over F is semiabelian of exponent n.

Each of these properties is preserved under field composita. Thus there exists a *maximal* extension of F with these properties. Let E_n denote this maximal extension. Then again,

$$[E_n : F] \leqslant n^{2g}. \tag{7.1}$$

If the nth division classes of F are rational over K, then we conclude that this field E_n coincides with the field E_n of Corollary 7.3; in particular, it follows that equality holds in (7.1). We claim that this is true in any case.

THEOREM 7.4. *Let E_n denote the maximal extension of F within $*K$, which is unramified and semiabelian of exponent n. Then $[E_n : F] = n^{2g}$. We shall call E_n the nth division field of F within $*K$.*

Proof. Let K' denote a finite Galois extension of K such that all the nth division classes of F are rational over K'. Consider the constant extension $F' = FK'$, which is imbedded into $*K'$, and let E_n' be the nth division field of F' within $*K'$. We know from the above that $[E_n' : F'] = n^{2g}$, and that E_n' can be characterized as being the maximal extension of F' in $*K'$ which is unramified and abelian of exponent n. This characterization shows, in particular, that every automorphism of $*K'$ which maps F' onto itself does also map E_n' onto itself.

Now let G be the Galois group of K' over K. Every automorphism of G has a standard extension to $*K'$, and hence G appears now as the Galois group of $*K'$ over $*K$. The field $F' = FK'$ is mapped under G onto itself (since this is true for each component F and K'). Therefore, by what we have said above, G maps E_n' onto itself. Hence, G induces in E_n' a certain

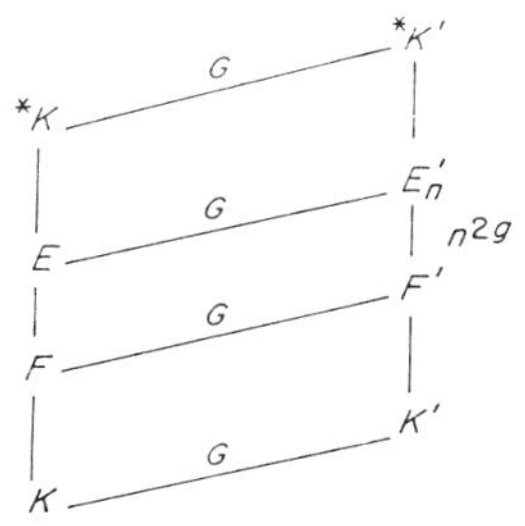

FIGURE 2

group of automorphisms. Let E denote the field of fixed elements in E_n'. This field E has the following properties (shown in Fig. 2):

(i) *E is contained in* $*K$. For, since $*K$ is the field of fixed elements in $*K'$, we have $E = *K \cap E_n'$.

(ii) *E is unramified over F.* For, we know that $E \subset E_n'$. But E_n' is unramified over F' (by definition of E_n') and F' is unramified over F (as a constant extension). Hence E_n' and therefore E too is unramified over F.

(iii) $EK' = E_n'$. This follows from Galois theory since every nontrivial automorphism of G moves the elements of K' and hence those of EK'. Now, since E_n' is abelian and of exponent n over F', it follows that *E' over F' is semiabelian and of exponent n.*

From (i)–(iii) it follows that $E \subset E_n$. Hence, from (7.1),

$$[E : F] \leqslant [E_n : F] \leqslant n^{2g}.$$

On the other hand, we infer from (iii) that

$$[E : F] = [EK' : FK'] = [E_n' : F'] = n^{2g}.$$

Thus we conclude that $E = E_n$ and $[E_n : F] = n^{2g}$. Q.E.D.

Theorem 7.4 shows the existence of unramified extensions E_n of F within $*K$, of arbitrarily large degree. Now we are interested in the degree invariants

$$d_n = d_{E_n} = \min_{\substack{y \in E_n \\ y \notin K}} [E_n : K(y)]$$

of these fields. As above, we write $d = d_F$.

LEMMA 7.5. *There exists a constant γ, depending on the genus of F only, such that $d_n \leqslant \gamma d n^{2g-2}$.*

A proof of Lemma 7.2 has been given by Siegel [22], with $\gamma = g^3$. Siegel used the analytic theory of theta functions in his proof. There is also an algebraic proof available, using Deuring's theory of correspondences of algebraic function fields and the inequality of Castelnuovo-Severi [18]. Since this proof has nothing to do with nonstandard methods, we have preferred to exclude it from the present paper and to publish it separately [20].

Putting Theorem 7.4 and Lemma 7.5 together, we obtain

$$d_n \leqslant \frac{\gamma d}{n^2} [E_n : F].$$

If n is sufficiently large, we conclude that inequality (6.1) holds:

$$d_n \leqslant \tfrac{1}{2}[E_n : F].$$

As was explained in Section 6, this shows that there are no exceptional primes of F. Theorem 1.2 is proved.

8. THE CASE $g = 0$

Although our main interest in this paper is concerned with function fields of higher genus, let us briefly review our results obtained in the case of genus zero. Thus, in this section let us assume that $g = 0$; in view of Remark 5.6 this implies that $F = K(t)$ is a rational function field.

Let $P_1, \ldots, P_r$ denote the exceptional prime divisors of F. Our aim is to describe these prime divisors, as well as the ring R of exceptional elements of F. Here an element $x \in F$ is said to be exceptional if its functional pole divisor is composed of exceptional primes only; this is equivalent to saying that, if x is regarded as an element of $^*K$, then its denominator is not divisible by any nonstandard prime. (That is, Theorem 1.1 fails to hold for x.) It is clear that the exceptional elements form a subring R of F, containing K.

According to Theorem 5.4, the divisor $A = P_1 + P_2 + \cdots + P_r$ has degree $\leqslant 2$. Hence, there are only the following four cases possible.

Case 0: $\deg(A) = 0$. In this case, there are no exceptional prime divisors, and $R = K$.

Case 1: $\deg(A) = 1$. We have $r = 1$. The only exceptional prime divisor P of F has degree 1. We can choose the generator t of $F \mid K$ such that P is the only pole of t, and of order 1. If an element $x \in F = K(t)$ has no pole except P, then x is a polynomial in $K[t]$, and conversely. Hence, $R = K[t]$.

Case 2: $\deg(A) = 2$ and $r = 2$. There are two exceptional prime divisors P_1 and P_2, each of degree 1. We can choose the generator t of $F \mid K$ such that P_1 is the pole and P_2 the zero of t, both of order 1. If an element $x \in F = K(t)$ has no pole except P_1 and P_2, then $x \in K[t, t^{-1}]$, and conversely. Hence $R = K[t, t^{-1}]$.

Case 2a: $\deg(A) = 2$ and $r = 1$. There is only one exceptional prime divisor P; it has degree 2. If t is any generator of $F \mid K$, then P is the zero of some quadratic irreducible polynomial $\varphi(t) \in K[t]$. After a suitable linear transformation of t, we may assume $\varphi(t)$ of the form $\varphi(t) = t^2 - a$, where $a \in K$, but $\sqrt{a} \notin K$. If an element $x \in F = K(t)$ has no pole except P, then x is of the form $x = \sum_{0 \leqslant i \leqslant n} h_i \varphi^{-i}$, where the $h_i \in K[t]$ are *linear* polynomials. This shows that x is representable as a polynomial in φ^{-1} and $t\varphi^{-1}$. Hence $R = K[\varphi^{-1}, t\varphi^{-1}]$.

Case 2a reduces to Case 2 over the quadratic extension $K' = K(\sqrt{a})$. For, the prime P splits in $F' = FK'$ into two functional primes P_1' and P_2', each of degree 1, which are the pole and the zero of the generator

$$t' = \frac{t + \sqrt{a}}{t - \sqrt{a}}$$

of $F' \mid K'$. Therefore, R is contained in the ring $R' = K'[t', t'^{-1}]$, and $R = F \cap R'$. Let

$$\tau\colon \sqrt{a} \mapsto -\sqrt{a}$$

denote the nontrivial automorphism of $K' \mid K$. Extended to an automorphism of $F' \mid F$, we see that τ sends t' into t'^{-1}. Also, R is the ring of fixed elements under τ. This gives the description

$$R = \mathrm{Fix}^{(\tau)}\, K'[t', t'^{-1}],$$

where

$$K' = K(\sqrt{a}) \qquad \text{and} \qquad \tau\colon \begin{Bmatrix} \sqrt{a} \mapsto -\sqrt{a} \\ t' \mapsto t'^{-1} \end{Bmatrix}.$$

The foregoing discussion gives a *classification* of the subfields $F \subset {}^*K$ of genus zero, with respect to their exceptional behavior. It is complemented by the following *existence statement*: Each of the four cases 0–2a is realized by some subfield $F \subset {}^*K$ of genus zero. In Case 2a we may prescribe the quadratic irrationality $\sqrt{a}$ entering into its description.

To prove this, one has to exhibit nonstandard elements $t_i \in {}^*K$ such that the function field $F_i = K(t_i)$ satisfies case i. We shall only state the mode of construction of these elements, leaving to the reader the straightforward verification that the conditions in the respective cases are indeed satisfied.

First, let us deal with Case 2. Let $u \in K$ be any element which is neither zero nor a root of unity, so that $u^n \neq u^m$ for all $n, m \in \mathbf{N}$. Then we put $t_2 = u^z$, where z is an *infinitely large* natural number in $^*\mathbf{N}$. It is easily seen that t_2 is nonstandard and that $F_2 = K(t_2)$ satisfies Case 2, the exceptional prime divisors being the pole and the zero of t_2. Now we put

$$t_1 = t_2 + t_2^{-1},$$
$$t_0 = t_1 + t_1^{-1}.$$

The fields $F_1 = K(t_1)$ and $F_0 = K(t_0)$ then satisfy Case 1 and Case 0, respectively.

Finally, in case 2a we work over the quadratic extension $K' = K(\sqrt{a})$. We first put $t_2' = u'^z$, the same construction as t_2 above, but now we require in addition that $\tau u' = u'^{-1}$, τ being the nontrivial automorphism of $K' \mid K$. Such an element $u' \in K'$ can be found, for instance, in the form

$$u' = \frac{u + \sqrt{a}}{u - \sqrt{a}},$$

where $u \in K$ is chosen such that u' is not a root of unity. τ extends to a standard automorphism of $^*K'$ over $^*K$, and we have $\tau t_2' = t_2'^{-1}$.

Finally, we put

$$t_{2a} = \sqrt{a}\,\frac{t_2' + 1}{t_2' - 1}.$$

Then $\tau t_{2a} = t_{2a}$; i.e., $t_{2a} \in {}^*K$. It is readily verified that the field $F_{2a} = K(t_{2a})$ satisfies Case 2a, the exceptional prime of degree 2 being the zero of $\varphi = t_{2a}^2 - a$.

The above results can be used to obtain Siegel's classification of those curves of genus zero, which are exceptional in the sense that the Siegel–Mahler theorem does not hold for them. Namely, let $\Gamma : f(x, y) = 0$ be such a curve over K; there are infinitely many K-rational points on Γ whose denominatorial prime divisors all belong to some finite set $\mathfrak{S}$. As explained in Section 1, we then have a *nonstandard* point (x, y) on Γ whose denominatorial primes belong to $\mathfrak{S}$, hence are standard. That is, the elements x, y are exceptional in the function field $F = K(x, y) \subset {}^*K$. According to the above classification, we therefore have one of the following *birational parametrizations* of our curve Γ.

Case 1. $x = \Phi(t)$, $y = \Psi(t)$, where Φ, Ψ are polynomials over K.

Case 2. $x = \Phi(t)$, $y = \Psi(t)$, where Φ, Ψ are finite Laurent series over K.

Case 2a. $x = \Phi(t')$, $y = \Psi(t')$, where Φ, Ψ are finite Laurent series over the field $K' = K(\sqrt{a})$, satisfying the condition $\Phi(t') = \Phi^{\tau}(t'^{-1})$ and similarly for Ψ. Here, τ denotes the nontrivial automorphism of $K' \mid K$, and Φ^{τ} is obtained from Φ by applying τ to its coefficients. These statements do not refer any more to an embedding of the function field of Γ into $^*K$. They are geometric in nature, showing how the points on Γ are parametrized by one parameter t.

These parametrizations are not only necessary, but also sufficient for Γ to be exceptional. Namely, the substitutions $t = t_1$ (resp. $t = t_2$ resp. $t' = t_2'$) yield a nonstandard point (x, y) of Γ such that x and y are exceptional in $^*K$. Let $\mathfrak{S}$ denote the set of those internal primes of $^*K$ which appear in the denominator of x or of y. We know that $\mathfrak{S}$ is internal and does not contain any nonstandard prime; hence, $\mathfrak{S}$ is a *finite* set of standard primes. Since there is a nonstandard point whose denominatorial primes all belong to $\mathfrak{S}$, there are infinitely many standard points with this property. Hence Γ is exceptional.

EPILOGUE[7]

In the summer of 1973, Abraham Robinson visited Heidelberg where he gave a lecture on algebraic function fields and nonstandard arithmetic [16, 17]. He expounded his ideas on embedding function fields into nonstandard models of their ground fields, or into finite extensions of such models. If the ground field is an algebraic number field, then this yields a representation of algebraic functions by (nonstandard) algebraic numbers. In this way it should be possible to explain the arithmetic structure of function fields directly, using well-established nonstandard principles only, by means of the arithmetic structure carried by the ground field.

In my opinion, these ideas of Abraham Robinson are of far-reaching importance, providing us with a new viewpoint and guideline towards our understanding of diophantine problems. It seems wothwhile to put these ideas to a test in order to verify their usefulness and applicability in connection with explicit diophantine problems. Perhaps a good test in this sense would be the explanation, in nonstandard terms, of Weil's theory of distributions and, closely connected with it, the theorems of Mordell and Weil and of Siegel and Mahler. The preceding paper is meant to be

[7] By Peter Roquette.

just such a test and, as we believe, a successful one. Although we deal explicitly with the Siegel–Mahler theorem only, it will be clear to anyone familiar with the subject that Weil's theory of distributions also can be explained in this context.

The paper was written as a result of several weeks of close collaboration with Abraham Robinson at Yale. He completed the first draft by his own hand in November 1973. His severe illness and tragic death prevented him from participating in the discussion of the following versions. Hence, although I want to make it clear that the basic ideas are Robinson's (compare also [16] and [17]), I have to take full responsibility for the form of presentation of the subject. In particular, this refers to Sections 2 and 3, which are of introductory nature. We had in mind to provide an introduction, however short, for those readers who are not acquainted with nonstandard methods but want to understand the basic ideas of our proof. Therefore, these two sections have been added. I hope they serve their purpose: to interest number theorists in nonstandard arithmetic. (By the way, the use of enlargements is not really necessary; the whole proof can be carried out in an ordinary nonstandard model.)

In Robinson's first draft, there was also a section on effectiveness. There, it was pointed out that his methods yield a "relative" effective procedure, relative to the bounds provided by Roth's theorem. This section has been excluded from the present paper; it is planned to publish it separately under the name of Abraham Robinson.

References

1. E. Artin and G. Whaples, Axiomatic characterization of fields by the product formula for valuations, *Bull. Amer. Math. Soc.* **51** (1945), 469–492.
2. A. Baker, Effective methods in the theory of numbers, *in* "Proceedings of the International Congress of Mathematics: Nice, 1970," Vol. 1, pp. 19–26. Paris, 1971.
3. H. Davenport and K. F. Roth, Rational approximations to algebraic numbers, *Mathematika* **2** (1955), 160–167.
4. H. Hasse, Simultane Approximation algebraischer Zahlen durch algebraische Zahlen, *Monatsh. Math. Phys.* **48** (1939), 205–225.
5. H. Hasse, Rein-arithmetischer Beweis des Siegelschen Endlichkeitssatzes für binäre diophantische Gleichungen im Spezialfall des Geschlechts 1, *Abh. Deut. Akad. Wiss. Berlin Kl. Math. Allg. Naturwiss.* 1951, Nr. 2.
6. H. Hasse, "Zahlentheorie," 2. Aufl., Akademie Verlag, Berlin, 1963.
7. S. Lang, Integral points on curves, *Publ. Math. I.H.E.S. Paris* **6** (1960), 27–43.
8. S. Lang, "Diophantine Geometry." Interscience, New York, 1962.
9. W. J. LeVeque, Rational points on curves of genus greater than 1, *J. Reine Angew. Math.* **206** (1961), 45–52.

10. K. MAHLER, Zur Approximation algebraischer Zahlen I, *Math. Ann.* **107** (1933), 691–730.
11. K. MAHLER, Über die rationalen Punkte auf Kurven vom Geschlecht Eins, *J. Reine Angew. Math.* **170** (1934), 168–178.
12. K. MAHLER, A remark on Siegel's theorem on algebraic curves, *Mathematika* **2** (1955), 116–127.
13. D. RIDOUT, The p-adic generalization of the Thue–Siegel–Roth Theorem, *Mathematika* **5** (1958), 40–48.
14. A. ROBINSON, "Non-standard Analysis," North-Holland, Amsterdam, 1966.
15. A. ROBINSON, Nonstandard arithmetic, *Bull. Amer. Math. Soc.* **73** (1967), 818–843.
16. A. ROBINSON, Algebraic function fields and non-standard arithmetic, *in* "Contributions to Non-Standard Analysis, (W. A. J. Luxemburg and A. Robinson, Eds.), North–Holland, Amsterdam, 1972, pp. 1–14.
17. A. ROBINSON, Nonstandard points on algebraic curves, *J. Number Theory* **5** (1973), 301–327.
18. P. ROQUETTE, Arithmetischer Beweis der Riemannschen Vermutung in Kongruenzfunktionenkörpern beliebigen Geschlechts, *J. Reine Angew. Math.* **191** (1953), 199–252.
19. P. ROQUETTE, Bemerkungen zur Theorie der formal p-adischen Körper, *Beitr. Algebra Geometrie* **1** (1971), 177–193.
20. P. ROQUETTE, Zur Theorie des n-Teilungskörpers eines algebraischen Funktionenkörpers, to appear in *Archiv. d. Math.*
21. K. F. ROTH, Rational approximation to algebraic numbers, *Mathematika* **2** (1955), 1–20.
22. C. L. SIEGEL, Über einige Anwendungen diophantischer Approximationen, *Abh. Preuss. Akad. Wiss. Phys.-Math. Kl.* 1929, Nr. 1. Gesammelte Abhandlungen vol. I, pp. 209–226.
23. A. WEIL, L'arithmétique sur les courbes algébriques, *Acta Math.* **52** (1928), 281–315.
24. A. WEIL, Arithmétique et géométrie sur les variétés algébriques, *Actual. Sci. Ind.* no. 206, Hermann & Cie, Paris, 1935.
25. A. WEIL, Arithmetic on algebraic varieties, *Ann. Math.* **53** (1951), 412–444.

STANDARD AND NONSTANDARD NUMBER SYSTEMS

The Brouwer memorial lecture 1973
Leiden, April 26, 1973

ABRAHAM ROBINSON

Ladies and Gentlemen,

Three years ago, RENE THOM gave the first in this series of lectures, and his address recalled the achievements of BROUWER, the topologist. You have asked a logician to present the second Brouwer lecture, no doubt in order to honor BROUWER as a philosopher of mathematics and as the founder of intuitionism. Topology and the foundations of mathematics are the two disciplines to which BROUWER made unforgettable contributions. No one will query the lasting importance of BROUWER's work in topology since much of the modern theory of the subject can be traced back to his pioneer effort. But even those who, like myself, do not adhere to the mathematical practices recommended by BROUWER, the philosopher, will readily acknowledge that his strongly held and profoundly argued views on the foundations of mathematics were of major importance for the subsequent development of this area also. BROUWER's intuitionism is closely related to his conception of mathematics as a dynamic activity of the human intellect rather than the discovery of an immutable abstract universe. This is a conception for which I have some sympathy and which, I believe, is acceptable to many mathematicians who are not intuitionists. It shall be the underlying theme of this lecture that the dynamic evolution of mathematics is an ongoing process not only at the summit, where intricate new results are piling up with great rapidity but also at the more basic level of our number systems, which may seem as eternal to a new generation as yesterday's technological innovations are in the eyes of a child of today. I shall carry on my discussion within the framework of classical mathematics although I may recall that the nature of our numbers was also one of BROUWER's main preoccupations.

1. By the beginning of the nineteenth century the complex numbers, after several hundred years of marginal existence, had achieved full citizenship in the realm of mathematics. The following century saw a series of penetrating and largely successful attempts to clarify and secure the foundations of the standard number systems as they were known about the year 1800, i.e. the natural numbers, the integers, the rational numbers and the real and complex numbers. But during the same period, and partly in connection with this foundational research, there also emerged new number systems and met with varying degrees of acceptance. Of these, I shall mention here (i) GALOIS' imaginaries, (ii) HAMILTON's quaternions, (iii) CANTOR's infinite cardinals and ordinals and (iv) HENSEL's p-adic numbers. It is, inevitably, a matter of convention, which algebraic or arithmetical structures are to be honored by the name of *number systems*. Generally speaking, the more naturally a new system seems to grow out of a *number system* of acknowledged standing, and the more similar to such a system its features, the more likely it is that the new entities also will be regarded as numbers. However, this is not a universal test, as shown by CANTOR's cardinals and ordinals which are regarded as numbers because they represent a natural and farreaching generalization of the positive integers although they lack many of the arithmetical properties of the latter. The degree of acceptance met with by a new number system may sometimes remind us of the fickleness of fate. Thus, HAMILTON's quaternions which, soon after their introduction, came to be regarded as a major tool in geometry and applied mathematics did not realize early expectations. They now are just one of many systems of arrays of real numbers in which certain arithmetical operations have been defined and which may or may not be called *numbers*. As far as the applications are concerned they are overshadowed by vectors and tensors, which are not regarded as numbers at all. I need hardly remind you of the struggle for recognition of the infinite cardinals and ordinals within the framework of general set theory which ended in triumph for these concepts and in tragedy for their creator. And even the p-adic numbers had to fight hard for recognition during the first forty years of their existence, and yet today they are, at least in number theory, regarded as coequal with the real numbers themselves. The p-adic numbers were introduced by HENSEL around 1890 and we may ask whether, except for the occasional use of the term for certain matrices and operators, the process that I have described has finally come to a halt. I shall endeavor to show that this is not the case by

presenting two new types of number systems that have emerged only in recent years.

2. In order to develop the differential and integral calculus with the aid of infinitely small and infinitely large quantities, LEIBNIZ assumed that we may extend the system of real numbers to a larger system, including infinitely small and infinitely large numbers and possessing the same properties as the reals. This assumption was used, explicitly or surreptitiously, until the middle of the nineteenth century. Despite his own denial, the assumption was employed again and again by CAUCHY, who made copious use of infinitesimals, many misleading statements in treatises on the history of mathematics notwithstanding. Taken literally, the assumption contains a glaring inconsistency since the extended system *must differ* from the real numbers at least in its essential feature of containing infinitely large and infinitely small quantities. That the method could nevertheless survived several generations of use is an interesting phenomenon whose explanation, however, does not concern us here. At any rate, it succumbed soon after the middle of the nineteenth century and was replaced by the ε,δ-method of WEIERSTRASS which had been foreshadowed by BOLZANO. Since then there have been numerous attempts to revive the method of infinitesimals. I shall now show how the use of some simple tools from logic can resolve the contradiction which seems to be inherent in the infinitesimal approach. The theory to be described is called *nonstandard analysis*.

Let R be the field of real numbers. Let K be the set of all sentences which are true in R and which are formulated in the first order predicate calculus in a vocabulary that consists of symbols for all the numbers of R and for all the n-ary relations on R, n = 1,2,3,... (including for example equality, $[x = y]$, order $[x < y]$, and the property of being a natural number N(x)). Functions can be represented by relations in the usual way.

Now let H_0 be the set of sentences $r < b$ where r ranges over all the real numbers and b is a new individual constant. Then every finite subset H' of the set $H = K \cup H_0$ has a model. This can be seen by interpreting the symbols of H other than b as in K while interpreting b as some real number greater than all real numbers which (or, more precisely, whose names) occur in H'. It then follows from the so-called *compactness theorem* of the lower predicate calculus that the set H has a model, $^*R$ say. Since K includes all basic sentences (atomic sentences or their negations) which hold in R, e.g.

S(3,5,8) and ¬ P(3,5,4) - where S(x,y,z) and P(x,y,z) stand for $x + y = z$ and $xy = z$ respectively - it follows that R can be injected into $^*R$. Thus, we may choose $^*R$ as an extension of R. $^*R$ is an ordered field since the axioms for an ordered field belong to K. Also, $^*R$ is nonarchimedean for we have $r < b$ for all $r \in R$ so that, in particular, b denotes a number greater than all natural numbers.

Every function in R, e.g. for one variable $y = f(x)$, has a canonical extension to $^*R$. For if F(x,y) denotes the relation in R which holds if and only if $y = f(x)$ then the same F(x,y) denotes in $^*R$ a function which extends f(x) and which has the same properties as f(x) but only as far as these can be expressed in K. Similarly, the set of natural numbers, N say, is extended to a subset $^*N$ of $^*R$. From now on, we shall call all elements of $^*R$ *real numbers*, distinguishing the elements of R as *standard real numbers*. A number $a \in {}^*R$ is *finite* if $|a| < r$ for some standard real r. A real number which is not finite is *infinite*. A number $a \in {}^*R$ is *infinitely small* or *infinitesimal* if $|a| < r$ for all standard positive r. The set of finite numbers in R is a ring called R_0 and the set of infinitesimal numbers is a maximal ideal in R_0 and is called R_1. The quotient ring R_0/R_1 is isomorphic to R under the isomorphism which assigns to every $a \in R_0$ the *unique* standard number $^\circ a$ which is *infinitely close* to a, i.e. $a - {}^\circ a \in R_1$. In general, we say that a_1 is infinitely close to a_2 and we write $a_1 \simeq a_2$ if $a_1 - a_2 \in R_1$. For any $r \in R$ the set of $a \in {}^*R$ which are infinitely close to r is called the *monad* of a, $\mu(a)$. Thus $\mu(0) = R_1$. Within this setting, we may now develop the differential and integral calculus. Here I shall give only a few typical examples.

Let $\{s_n\}$ be a sequence with values in R, $\{s_n\} : N \to R$. As we pass from R to $^*R$, $\{s_n\}$ extends to a sequence from $^*N$ into $^*R$ which by abuse of language, will still be denoted by $\{s_n\}$. Let a be any standard real number. Then we are going to prove

$$\lim_{n\to\infty} s_n = a \textit{ if and only if } s_n \simeq a \textit{ for all infinite } n.$$

The condition is necessary. For let r be any standard positive real number. By the definition of a limit $\lim_{n\to\infty} s_n = a$ implies that there exists a standard positive integer n_0 such that $|s_n - a| < r$ for $n > n_0$. This can be formalized as a sentence of K, which is then true also in $^*R$. It follows that if n is finite - and hence, as can be shown, greater than n_0 - then

$|s_n - a| < r$. This proves that $s_n \simeq a$.

The condition is also sufficient. Let $\varepsilon > 0$ be any standard positive number. We have to show that *there exists a natural number* x *such that for any natural* y *greater than* x, $|s_y - a| < \varepsilon$. The assertion in italics can be formalized as a sentence x in the language of K. If it is not true in R then $\neg X$ (not X) belongs in K and, hence, holds in $^*R$. But for all infinite n, $|s_n - a| \simeq 0$ and so, certainly $|s_n - a| < \varepsilon$. Accordingly, we only have to take $x = n_0$ as an infinite natural number to see that X is true in $^*R$. This is a contradiction, which completes the proof.

A similar argument shows that for a to be a limit point (accumulation point) of s_n it is necessary and sufficient that $s_n \simeq a$ for *some* infinite n. Suppose that (s_n) is bounded, $|s_n| < c$ in R, then this must still be true in $^*R$ and so $|s_n| < c$ also for all infinite n. Take any s_n for infinite n. As we have just seen, s_n is finite so it has a standard part $^\circ s_n = a$. Then $s_n \simeq a$, so a is a limit point of the sequence $\{s_n\}$. We have proved the theorem of BOLZANO and WEIERSTRASS:

Every bounded sequence has a limit point.

Going on to functions, let f(x) be an ordinary (*standard*) real function which is defined for $a < x < b$. Let $a < \xi < b$, ξ standard. Then f(x) extends automatically from R to $^*R$, where the extension will still be denoted by f(x). A method similar to that used above shows:

f(x) *is continuous at* ξ *if and only if* $f(x) \simeq f(\xi)$ *for all* $x \simeq \xi$, *i.e. for all* x *in the monad of* ξ.

If f(x) is defined in the closed interval $a \leq x \leq b$ then the condition holds, with the obvious modifications also at a and at b. Suppose then that f(x) is continuous in the closed interval $a \leq x \leq b$. Suppose also that $f(a) < 0$, $f(b) > 0$. We are going to give a nonstandard proof of the intermediate value theorem of BOLZANO and WEIERSTRASS:

There exists a number ξ, $a < \xi < b$, *such that* $f(\xi) = 0$.

Working first in R divide the interval <a,b> into n equal subintervals with end points $x_k = a + (k/n)(b-a)$, $0 \leq k \leq n$, $x_0 = a$, $x_n = b$. Then *given* n, *there exists a smallest integer* λ *such that* $f(x_\lambda) < 0$, $f(x_{\lambda+1}) \geq 0$. Formulating the assertion in italics as a sentence X in the vocabulary of K we then see that $X \in K$, so X holds also in $^*R$. Passing to $^*R$ and taking n *infinite*, we now find that, for the corresponding λ, x_λ and $x_{\lambda+1}$ are infinitely close to each other, so they have the same standard part $\xi = {}^{\circ}x_{\lambda} = {}^{\circ}x_{\lambda+1}$. But then $f(x_{\lambda}) \simeq f(\xi) \simeq f(x_{\lambda+1})$, by the continuity of $f(x)$ at ξ, and so $f(\xi) = 0$ since $f(x_{\lambda})$ is negative and $f(x_{\lambda+1})$ is nonnegative. This proves the theorem. Continuing in this manner, we can develop the elementary differential and integral calculus without difficulty. However, for less elementary applications, in analysis and elsewhere, we have to deepen our approach.

3. So far, the fact that we have been able to avoid the contradiction inherent in LEIBNIZ' approach would seem to have hinged on the fact that we restricted the transfer of properties of R to $^{*}R$ to those properties that are expressible in the lower predicate calculus. However, it turns out that we can discard this limitation. Thus, we now include in our vocabulary not only first order relations on R, i.e. relations between real numbers, but also relations between relations, sets of relations, relations between sets, etc. and, implicitly, functions of all orders. Moreover, we permit also quantification with respect to such higher order entities ("there exists a set", "for all n-place relations between sets", etc.). From now on, let K be the set of all sentences true in R in this extended language. An analysis which is almost transparent when we use the set theoretical approach shows the following.

There exists an extension of R, still to be called $^{*}R$, which includes infinitely large and infinitely small numbers and which, at the same time satisfies all sentences of K, provided we reinterpret the higher order quantifiers as referring not to all entities of the type in question but only to a subclass of these, called *internal*. For our purposes, it is not necessary to know how exactly the totality of internal functions is constituted. Rather, the mere assertion of the existence of internal classes as explained is sufficient to enable us to decide in concrete cases whether or not an entity is internal. We are going to explain this point in some detail. To any entity (number, relation, etc.) in R there corresponds the internal entity in $^{*}R$ which has the same name in the language of K, e.g. the set of natural numbers in $^{*}R$, $^{*}N$. Such entities are called *standard* (by a slight abuse of language, since one also refers to R as the standard system). There are internal entities which are not standard. For example, if ρ is any positive infinite real number then the set of all real numbers greater than ρ is internal. Indeed, the assertion *for any real number* r *there exists a set consisting of all real numbers greater than* x holds in R and therefore also in $^*R$. By contrast the set of infinite natural numbers, $^*N$-N, is *external* i.e. it is not internal. For $^*N$-N is not empty but it does not contain a smallest element (if a is infinite then a-1 must be infinite also). On the other hand there exists an assertion which is true in R and therefore also in $^*R$ and which states that every nonempty set of natural numbers contains a smallest element. It follows that N (and similarly R) are not internal either within $^*R$.

Going on to sequences, all *standard* sequences, i.e. the extensions to $^*R$ of sequences in R, are internal, e.g. the sequence $1, \frac{1}{2}, \frac{1}{3}, \dots$. So is the sequence which, for a specific infinite integer ρ is equal to 0 for $n \le \rho$ and to 1 for $n > 0$. However, the sequence $\{s_n\}$ such that $s_n = 0$ for finite n and $s_n = 1$ for infinite n is external. For if it were internal then the set $^*N$-N which can be defined in terms of this sequence as the set of all natural numbers n for which $s_n = 1$ would be internal also and we have just seen that this is not the case. Instead of regarding the elimination of certain entities from our theory as a drawback we may interpret it as the imposition of greater regularity on the entities that remain. This is expressed in a particularly interesting way in the following lemma, which turns up with remarkable frequency in the applications.

Let $\{s_n\}$ *be an internal sequence such that* $s_n \simeq 0$ *for all finite* n. *Then there exists an infinite* ρ *such that* $s_n \simeq 0$ *for all* $n \le \rho$.

__Proof__. Let $t_n = (n+1)|s_n|$. Then $\{t_n\}$ is internal and $t_n \simeq 0$ for all finite n and so $t_n \le 1$ for all finite n. If this relation is true for *all* n, choose any infinite $n \le \rho$, $|s_n| < \frac{1}{n+1}$ so s_n is still infinitely small. In the alternative case there is a first ρ such that $t_n > 1$. For all infinite $n \le \rho$, we still have $|s_n| \le \frac{1}{n+1}$ proving the lemma.

We note that the success of nonstandard analysis is based on the complementary use of (i) the facts which can be transferred from R to $^*R$ because they can be formulated in the language of K (internal facts) and (ii) the introduction of the notion of *infinitely large* which is external and cannot be formulated in the language of K as well as of allied notions such as the notion of a monad (infinitesimal neighborhood of a point).

4. Both intuition and the formal description given so far seem to imply that the idea of nonstandard analysis is tied up inextricably with the

metric character of the real numbers. However, this impression is erroneous and a better understanding of the underlying principle is obtained if one adopts a more general point of view.

So let M be a mathematical structure of any kind, e.g. R, or a topological space, or a topological group. Let K be the set of all sentences which are true in M, in a general language, as in section 3. A binary relation Q(x,y) of any type, e.g. between individuals, between sets and ternary relations, etc., will be called *concurrent* if the following property is satisfied. For any finite set of entities in the domain of the first argument of Q, $a_1,\dots,a_n$, there exists a b such that $Q(a_k,b)$ is satisfied by M, $k = 1,\dots,n$. We call an extension of M, $^*M$ an *enlargement* of M if the following conditions are satisfied.

(i) All sentences of K are satisfied by M in the sense of the preceding section.

(ii) For any concurrent relation Q(x,y) in M, there exists an element of $^*M$, to be denoted by b_Q such that $Q(a,b_Q)$ holds in $^*M$, where a now ranges over *all* elements of the first domain of Q in M. b_Q is called a *bound* for Q.

An example of a concurrent relation is provided by the order relation < in R. Any positive infinite b in $^*R$ can serve as a bound for Q. In a concrete application we usually make use only of a small number of bounds for concurrent relations. The notion of a concurrent relation is very convenient but condition (ii) can also be reformulated so as to make it apparent that an enlargement of M is a compact extension of M in a suitable topology.

So far we have not said anything specific concerning the construction of nonstandard models and, in particular, of enlargements. On this topic, we shall observe here only that they can be obtained as *ultrapowers*. However, not every ultrapower is an enlargement. On the other hand, all ultrapowers have an interesting property which we mention here because it involves a question which is of obvious general interest. Suppose that we are given a sequence $\{q_n\}$ of objects of a certain type (numbers, points, relations, etc.) where the term *sequence* here shall indicate that the subscript ranges over N. Does there exist an *internal* sequence $\{s_n\}$ from $^*N$ into the class of entities in question such that $s_n = q_n$ for all finite n? A nonstandard model which satisfies this condition is said to be sequentially *comprehensive.* It turns out that all ultrapowers are sequentially comprehensive.

As a first step in the development of the nonstandard theory of a structure M as above, we may again see how to reformulate its most important notions in nonstandard terms. For example, let M be a topological space and let $^*M$ be an enlargement of M. Let p be a point in M so that p may be regarded also as a point of $^*M$. We define the *monad* of p, $\mu(p)$ as the intersection of all $^*U_p$ where U_p ranges over the open neighborhood of p in M and, for every such U_p, $^*U_p$ is its extension to $^*M$. It will be seen that if M is the field of real numbers, R, then the topological notion of a monad introduced here coincides with the metric concept of the same name defined in section 2. We have the following two theorems.

For M *to be separated (a Hausdorff space) it is necessary and sufficient that for any two distinct points* p,q *in* M *the monads of* p,q *be disjoint.*

A point p in $^*M$ is called *nearstandard* if it belongs to the monad of a standard point. Then

for M *to be compact it is necessary and sufficient that all points in* $^*M$ *be nearstandard.*

Evidently, our theory now applies also to the field of complex numbers. However, this case can be handled also by the methods of the previous sections by regarding a complex number as a pair of real numbers.

5. I shall now sketch the basic nonstandard theory of Hilbert space and shall indicate how it has been used in order to solve a standard problem concerning the existence of invariant subspaces for a certain class of linear operators.

We consider infinite separable Hilbert space, H, and assume that H is represented as the space ℓ_2 of complex valued sequences $\sigma = \{s_1, s_2, s_3, \ldots\}$ such that $\Sigma|s_n|^2 = \|\sigma\|^2 < \infty$. We then have an orthonormal base $B = \{\beta_1, \beta_2, \ldots\}$ where $\beta_n = \{\delta_{n1}, \delta_{n2}, \ldots\}$, ($\delta_{nk} = 0$ for $n \neq k$, $\delta_{kk} = 1$). We now pass to an enlargement of H, $^*H$. The points of $^*H$ are given by internal sequences $\sigma = \{s_1, s_2, \ldots\}$, $\|\sigma\| < \infty$, where the subscripts now range over $^*N$. The set of base points $^*B$ now contains points β_n also for infinite n. In other words, on passing from H to $^*H$ we have enlarged the space not only by adding points σ which are infinitely far from the origin but such that the corresponding unit vector $\sigma/\|\sigma\|$ is standard or nearstandard but we have also, as it were, added new dimensions, i.e. those corresponding to unit vectors β_n for infinite n. For although such a β_n is still at distance ℓ from the origin it is not nearstandard. This can be surmised intuitively from the fact that β_n is orthogonal to all the standard base points β_k (k finite). It also follows from the following test for nearstandardness.

A point $\sigma = \{s_1, s_2, \ldots\}$ *is nearstandard if and only if* (i) $\|\sigma\|$ *is finite and* (ii) *for every infinite* k, *the number* $\sum_{n=k}^{\infty} |s_n|^2$ *is infinitesimal.*

If only the first condition is satisfied σ is said to be *normfinite*. In standard language, a compact linear operator T on H is a linear operator from H into H which transforms every bounded set into a set with compact closure. Remembering the test for compactness which is given at the end of the preceding section, it is then not difficult to prove:

The linear operator T *is compact if and only if it maps every normfinite point on a nearstandard point.*

Pictorially, we may say that T is compact if it swings any point σ that is at a finite distance from the origin (or from any other standard point) into a position that is infinitely close to a point of H.
Suppose now that T is given by a matrix representation with respect to the given base, so $T = (a_{kn})$, $k,n = 1,2,\ldots$ where k stands for the rows and n for the columns. On passing to $^*H$ the matrix, still to be denoted by the same symbol obtains values also for infinite k and n. We now divide (a_{kn}) in $^*H$ into four *quarters* according as (i) k and n are both finite, (ii) k is finite and n is infinite, (iii) k is infinite and n is finite, (iv) k and n are both infinite. For *any* bounded operator on a Hilbert space, the convergence of the sequences $\sum_n |a_{kn}|^2$ and $\sum_k |a_{kn}|^2$ in H shows that in $^*H$ the a_{kn} in the second and third quarter must be infinitesimal. As the identity operator shows this need not be true for the fourth quarter. However, translating the conditions for the compactness of an operator into matrix language one shows that if T is compact then $a_{kn} \simeq 0$ in the fourth quarter also. For let $\sigma = \beta_n = \{0,0,\ldots,1,0,\ldots\}$ where n is infinite; then $\tau = T\sigma = \{a_{1n}, a_{2n}, \ldots, a_{kn}, \ldots\}$. Since σ is normfinite, τ must be nearstandard. And this, by our condition for nearstandardness, implies that $a_{kn} \simeq 0$.

A (nontrivial) *invariant subspace* for an operator T is a closed linear subspace A of H such that $TA \subset A$, where $A \neq \{0\}$, H. It was proved by VON NEUMANN, and later by K.T. SMITH and ARONSZAJN that a compact operator has an invariant subspace. For general bounded linear operators the question is still unsolved. K.T. SMITH and HALMOS posed the question whether any operator whose square is compact possesses an invariant subspace. This question was resolved by A.R. BERNSTEIN and myself by showing that the answer is affirmative for any operator T such that for some complex polynomial p(z) of strictly positive degree, p(T) is compact. I shall now indicate the proof for $p(z) = z^2$ which is the original problem of SMITH and HALMOS.

We start off in the standard domain, by representing T in *almost superdiagonal form*, i.e. $T = (a_{kn})$ is such that $a_{kn} = 0$ for $k > n + 1$. Discarding trivial cases, this can be done by choosing as generators of the space the points $\sigma, T\sigma, T^2\sigma, \ldots$ where σ is an arbitrary point of norm ℓ and by orthogonalizing by means of the GRAM-SCHMIDT procedure. This yields a set of basis vectors $\{\beta_1, \beta_2, \ldots\}$. Then $T\beta_n$ depends linearly on $\beta_1, \ldots, \beta_{n+1}$, so

$$T\beta_n = a_{1n}\beta_1 + a_{2n}\beta_2 + \ldots + a_{n+1,n}\,\beta_{n+1}.$$

This shows that the matrix representation of T by (a_{kn}) is almost superdiagonal. If the matrix (a_{kn}) were actually *superdiagonal* then all linear spaces spanned by $\beta_1, \ldots, \beta_n$ would be invariant subspaces of T. Since this is not necessarily the case, we have to proceed more circuitously. By assumption, T^2 is compact. By direct squaring of the matrix a_{kn} we find that $T^2 = (b_{kn})$ where (b_{kn}) may have nonzero elements, not only in the first diagonal below the main diagonal but also in the second diagonal. More particularly, $b_{k+2,k} = a_{k+2,k+1}\, a_{k+1,k}$ ($= \sum_j a_{k+2,j}\, a_{j,k+1}$, where the remaining terms drop out). But then $b_{k+2,k} \simeq 0$ for infinite k by the property of an infinite matrix stated earlier and so at least one of the two factors $a_{k+2,k+1}$ and $a_{k+1,k}$ must be infinitesimal. *Thus, scanning the diagonal below the main diagonal in the fourth quarter of* (a_{kn}) *from top left to bottom right, we find that at least one of any two consecutive entries must be infinitesimal.*

Now let ω be an infinite natural number and consider the action of the projection operator P_ω which transforms any $\sigma \in {}^*H$, $\sigma = \{s_1, s_2, \ldots, s_\omega, s_{\omega+1}, \ldots\}$ into $\sigma_\omega = \{s_1, s_2, \ldots, s_\omega, 0, 0, \ldots\}$. Write $H_\omega = P_\omega H$ and $T_\omega = P_\omega T P_\omega$. Then T_ω maps H_ω into H_ω. Moreover H_ω is a finite dimensional space in the sense of nonstandard analysis, or as one says, the number of dimensions of H is *starfinite*. On the other hand, for any standard or nearstandard point σ, $P_\omega\sigma$ is infinitely close to σ since $\|P_\omega\sigma - \sigma\|^2 = \sum_{j=1}^{\infty}|s'_{\omega j}|^2$ and, as we have seen, this is infinitesimal. Thus, intuitively speaking, all points of H are approximated up to infinitesimals by points of H_ω. On the other hand, since T_ω is a linear operator in H_ω, the entire theory of operators in finite dimensional spaces applies to it in the sense of nonstandard analysis. It follows that T_ω possesses many invariant subspaces in H_ω. To every such space, E say, there corresponds a *closed* linear subspace ${}^\circ E$ in H which consists of all points of H that are infinitely close to points of E. It is then not difficult to prove that $T\,{}^\circ E \subset {}^\circ E$ leaving only the problem of showing that ${}^\circ E$ is different from $\{0\}$ and H for at least one of the E in question. The proof of this fact is the most subtle part of the argument and involves a choice of ω such that $a_{\omega+1,\omega}$ is infinitesimal. We have indicated that all points of H are infinitely close to points of a subset of H_ω. Although this does not make any essential difference, we shall now show that we may actually choose H_ω in such a way that it *contains* H. At any rate, according to our construction, the set of points of H is to be regarded as a subset of the set of points of ${}^*H$. Consider the following relation in H to be called $Q(x,y)$: x is a point of H and y is a finite dimensional subspace of H that contains x. The relation Q is concurrent for if $a_1, \ldots, a_n$ are points of H then we can find a finite dimensional subspace of H that contains them. We now pass to ${}^*H$. Since we have assumed that ${}^*H$ is an enlargement of H, there is a starfinite dimensional subspace H' of ${}^*H$ such that H' contains all points of H, $H \subset H'$. Now let $\{H_n\}$ be the sequence of subspaces spanned by $\{\beta_1, \ldots, \beta_n\}$, n = 1,2,... in the proof of the above theorem. Since $\cup H_n = H$ we have, passing to ${}^*H$ that, for sufficiently high (infinite) ω, H_ω contains H'. Thus, we have an injection $H \to H_\omega$.

6. Having shown that nonstandard analysis can handle a concrete problem effectively, we now go back to some more basic questions concerning the construction of number systems. We recall that R is isomorphic to R_0/R_1,

where R_0 is the set of finite numbers of $^*R$ and R_1 is the set of infinitesimal numbers. The equivalence classes of R_0 modulo R_1 are just the monads of the standard real numbers. It is not difficult to see that for any standard real number there is a (standard or nonstandard) rational number which is infinitely close to it, i.e. every monad contains such a rational number. It follows that if Q is the field of rational numbers, $^*Q$ an enlargement of Q and Q_0 and Q_1 the sets of finite and infinitesimal numbers in $^*Q$ respectively then Q_1 is still a maximal ideal in Q_0 and $R \simeq Q_0/Q_1$. Thus, *we obtain the real completion of* Q *as a homomorphic image of the subring* Q_0 *of* Q.

The same procedure can be applied also in order to obtain the p-adic completions of Q. A p-adic valuation induces a metric on Q and so we also have a metric on $^*Q$ in the nonstandard sense, $v_p(a) = n$ if $a = p^n r$ where the numerator and denominator of r are prime to q, and $|a|_p = p^{-n}$. Accordingly, we may again introduce the subring of $^*Q$, Q_0^p say, whose elements have finite norm in the metric just defined and the ideal Q_1^p which consists of the infinitesimal elements in Q_0^p. Then the field of p-adic numbers, Q_p, is isomorphic to Q_0^p/Q_1^p. Thus, the p-adic completions of Q also are homomorphic images of subrings of $^*Q$. However, for the p-adic numbers, Q_1^p consists of all the infinite powers of elements of the valuation ideal J_p, while there is no counterpart to J_p in the archimedean case. Exactly analogous constructions can be carried out for an arbitrary algebraic number field if we wish to obtain the various archimedean and nonarchimedean completions of the field. If we confine ourselves to the p-adic integers, Z_p, for a given prime rational p then Z_p can be obtained as the quotient ring $^*Z/J_1^p$ where J_1^p is the (external) ideal in $^*Z$ which consists of all integers divisible by infinite powers of p. This replaces the classical procedure in which Z_p is constructed as the completion of the integers in the p-metric. In his original approach, HENSEL used a more pictorial procedure, as follows. Consider the representation of the positive integers in Z for base p, so $a = b_k \ldots b_2 b_1 b_0$, $0 \le b_j < p$. Now extend this domain by permitting that the sequence continues indefinitely to the left, while still carrying excesses in addition from right to left, as usual. This yields Z_p. In the corresponding nonstandard approach, we take from the outset all internal expressions $b_k \ldots b_2 b_1 b_0$ as above where, however, k may now be finite or infinite but we identify two such expressions if they differ only at infinite subscripts.

7. In the previous section we made use of infinite powers of standard primes. The following reformulation of a famous theorem of AX and KOCHEN involves the use of infinite primes (in the nonstandard sense). This metamathematical result, which led to the solution of an arithmetical problem of ARTIN's, is as follows. Let Q_p be the field of p-adic numbers as before and let $R_p((t))$ be the field of formal power series (Laurent series) over the prime field of characteristic p. Let X be any sentence which is formulated in the lower predicate calculus in terms of equality, addition, and multiplication (and, if we so desire, involving also addition and order in the valuation group and the valuation function $v_p(x)$). Then there exists a positive integer p_0 such that for all $p > p_0$, X holds in $R_p((t))$ if and only if it holds in Q_p. Suppose now that we have passed to an enlargement which includes Q_p and $R_p((t))$ for both finite and infinite p. (The power series of $R_p((t))$ now range from some finite or infinite negative integer and over all positive integers, finite or infinite.) Then the result just quoted shows immediately that for any infinite p, Q_p satisfies X if and only if $R_p((t))$ satisfies X. That is to say, *Q_p and $R_p((t))$ are elementarily equivalent.*

8. As our last example in the nonstandard way of thought, we shall consider the theory of rational points on curves. Let k be an algebraic number field and let f(x,y) be an absolutely irreducible polynomial with coefficients in k, which determines a set of zeros Γ in k^2-space. We imagine that k is embedded in C, the field of complex numbers. Let $^*C$ be a higher order nonstandard extension of C so that in particular $^*C$ may be an enlargement of C. Then $^*k$ is a subfield of $^*C$ and Γ is extended to a set $^*\Gamma$, in $^*k^2$. Now let $A=(\xi,\eta)$ be a *nonstandard* point of $^*\Gamma$, by which we mean that at least one of the coordinates ξ or η is nonstandard. Such a point exists if and only if the set Γ is infinite. Then we claim that A is a generic point (over k) of the curve G determined by $f(x,y) = 0$. To see this suppose that A is not generic. Then there exists a polynomial $g(x,y) \in k[x,y]$ which is satisfied by A, $g(\xi,\eta) = 0$ such that g is not divisible by f. But g and f have only a finite number of distinct points in common in k^2 so there is a statement to that effect in the language of C. It follows that this number cannot increase on passing from C to $^*C$, so all points of $^*\Gamma$ are standard. Thus, $k(\xi,\eta)$ is the function field of G. At the same time $k(\xi,\eta) \subset {}^*k$ so we have an embedding of the function field of G into a nonstandard number

field. This embedding is external since the only internal subfields of $^*k$ are the nonstandard extensions of the algebraic number fields which are included in k.

Now let P be an internal prime ideal in $^*k$ and suppose first that P is nonstandard. There is a valuation $v_P(a)$ of the elements of $^*k$ expressing divisibility by P, with value group $^*Z$. We have $v_P(a) = 0$ for all $a \in k$, $a \neq 0$ since the set prime divisors of a number in k is finite and, again, cannot increase on passing from k to $^*k$. Suppose now that $v_P(a) \neq 0$ for some $a \in k(\xi,\eta)$. Then v_P induces a nontrivial valuation on the function field which is trivial on k and so it must be equivalent to one of the canonical valuations of the function field and hence, as can also be shown directly, it must be a discrete valuation of rank 1.

Suppose next that P is a standard prime ideal. In that case, $v_P(a)$ is finite for all $a \in k$, $a \neq 0$ but may be different from zero for such a. Suppose that $v_P(a)$ is infinite for some $a \in {}^*k$. It is then not difficult to show that $v_P(a)$ still induces a valuation of $k(\xi,\eta)$ which is trivial on k. For this purpose, we choose some $a \neq 0$ in $^*k$ such that $v_P(a)$ is positive infinite, and we then put, for any other $b \neq 0$ in $^*k$, $V_P(b) = {}^\circ(v_P(b)/v_P(a))$ where the expression in parentheses is in fact finite and we have taken its standard part. Since Z is an isolated subgroup of $^*Z$, the map $v_P(b) \to V_P(a)$ can also be obtained by a classical procedure of valuation theory.

Suppose finally that we are given an archimedean valuation of $^*k$ as expressed by an absolute value $|a|$. Suppose that $|a|$ is infinitesimal for some $a \neq 0$ in $^*k$. It is remarkable that even in this case we obtain a rank one valuation of $k(\xi,\eta)$ by putting $V(b) = {}^\circ(\ln|b|/\ln|a|)$, where the $V(b)$ are rational integers for a suitable choice of a.

It is entirely possible that different valuations of $^*k$ yield the same valuation of the function field $k(\xi,\eta)$. The question arises whether all valuations of $k(\xi,\eta)$ can be obtained in this way. This is indeed the case, although the two available proofs of this assertion are not trivial. One of these proceeds in a relatively straightforward manner on the basis of the theorem of RIEMANN-ROCH. The second, and more instructive, proof involves a rather sophisticated analogue of the theorem of ARTIN and WHAPLES on the uniqueness of systems of valuations that satisfy the product formula. It involves a concept that is related to the classical notion of the height of a number and can be used to prove some basic results concerning that notion.

To conclude this account of nonstandard analysis, including nonstandard arithmetic, let me emphasize that it does not present us with a single number system which extends the real numbers, but with many related systems. Indeed, there seems to be no natural way to give preference to just one among them. This contrasts with the classical approach to the real numbers, which are supposed to constitute a unique or, more precisely, categorical totality. However, as I have stated elsewhere, I belong to those who consider that it is in the realm of possibility that at some stage even the established number systems will, perhaps under the influence of developments in set theory, *bifurcate* so that, for example, future generations will be faced with several coequal systems of *real numbers* in place of just one.

9. Nonstandard analysis is now over twelve years old. While the question of its use is and may remain a matter that is up to the choice of the individual mathematician one can, by now, hardly doubt its effectiveness. I shall now describe yet another class of systems, which have emerged only within the last three years. While these systems have not yet been applied to classical mathematics, I shall endeavor to convince you, by the natural way in which they present themselves, that they also deserve our attention.

Let Σ be the class of commutative fields and, within it, let G be the class of algebraically closed fields. Then the following conditions are satisfied:

(i) Every $M \in \Sigma$ can be embedded in an $M' \in G$.

(ii) Let M and M' be two elements of G such that $M \subset M'$. Then M is an *elementary substructure* of M', in symbols $M < M'$. That is to say, if X is any sentence of the lower predicate calculus formulated in terms of equality, addition, multiplication and names of any of the elements of M such that M satisfies X, then M' also satisfies X.

(iii) Let $M \in \Sigma$, $M' \in G$ such that $M < M'$. Then $M \in G$.

Of these, (i) and (iii) are obvious facts of field theory while (ii) is a classical result of applied model theory.

Informally speaking, the construction of an algebraically closed extension to a given field M rounds off, or completes, M. At any rate, the great importance of algebraically closed fields in the theory of commutative fields leads us to ask the question whether we can find an analogous concept for other classes of algebraic structures. For ordered or formally real fields, the concept of a real closed field was put forward by ARTIN and SCHREIER with the obvious intention of producing such an analogue. But it turns out that the same can be done for many other classes Σ under conditions of striking generality. Thus, let Σ be the class of all structures which are models of a set of axioms K in the first order predicate calculus such that Σ is closed under union (inductive limit of) chains. This is the case if K is, or is equivalent to, a set of $\forall\exists$-sentences, i.e. sentences of the forms $(\forall x_1)\ldots(\forall x_m)(\exists y_1)\ldots(\exists y_m)\ Q(x_1,\ldots,x_n,\ y_1,\ldots,y_m)$ where Q does not contains any further quantifiers. Then it can be shown that there exists one and only one subclass G of Σ which satisfies conditions (i), (ii) and (iii), where the vocabulary specified in (ii) is to be replaced by the vocabulary of the set K under consideration. For the case of ordered fields or of formally real fields, the class G turns out indeed to consist of all real closed fields, confirming the intuition of ARTIN and SCHREIER. However, in many interesting cases, G had not been defined previously. The elements of G are called *generic structures* because the existence of G was first proved by a model theoretic analogue of the forcing procedure by means of which PAUL COHEN introduced the notion of a generic set into axiomatic set theory. If Σ has the joint embedding property - any two structures that belong to Σ can be injected into one and the same element of Σ - then all structures in G are elementarily equivalent. This is the case, for example, if Σ is the class of algebraically closed fields of specified characteristic.

There is a related concept whose definition involves less metamathematical machinery. An *existential sentence* is a sentence that can be written in the form $X = (\exists x_1)\ldots(\exists x_n)Q(x_1,\ldots,x_n)$ where Q is free of quantifiers. With K and Σ as before, Σ closed under union of chains a structure $M \in \Sigma$ is called *existentially complete* if it satisfies the following condition. If X is any existential sentence in the vocabulary of M (i.e. with names for any of the elements of M) such that X is satisfied by some extension M' of M which belongs to Σ then X is satisfied also by M. The class of existentially complete structures is denoted by E. It turns out that E includes G. Moreover, if one of these two classes is axiomatizable, i.e. is the class of models of a set of axioms in the lower predicate calculus then it coincides with the other. This is the case, for example, if Σ is the class of commutative fields or the class of (commutative) ordered fields. It is not the case if Σ is the class of groups.

Now let Z be the ring of integers and let H be the set of all sentences which are formulated in terms of the relation of equality, of the operations of addition, subtraction and multiplication, and of the constants 0 and 1. Let Σ_H be the class of all models of H. Then Σ_H is not closed under union of chains and so the theory just sketched does not apply to it. The situation can be remedied by considering instead the class Σ of all subrings of structures of Σ_H that contain the element 1. Then Σ is the class of all models of the set K of all universal sentences - i.e. sentences of the form $(\forall x_1)\ldots(\forall x_n)Q(x_1,\ldots,x_n)$ where Q is free of quantifiers - which are contained in H. This shows that Σ *is* closed under union of chains. Moreover, Σ has the joint embedding property, so all structures of G - the class of generic structures in Σ - are elementary equivalent. It follows that the theory K^F of G, i.e. the set of all sentences true in the structures of G is *complete* (for any sentence X in the stated vocabulary, either X or $\neg X$ belongs to K^F). It can be shown that K^F is different from H although both K^F and H are complete (maximal consistent) extensions of K. It can also be shown that the class E of all existentially complete structures in Σ is strictly larger than G. In particular, E includes Z so that not all structures of E are elementarily equivalent. As is to be expected, the structures of G have many properties in common with the ring of integers **Z** . In particular, they satisfy a limited form of the axiom of induction. On the other hand, in view of the provenance of G, it is also to be expected that its elements have at least some of the desirable properties that algebraically closed fields possess within the class of fields and this is indeed the case. Thus, we may regard the structures of G and of E as new number systems which are closely related to arithmetic and whose further investigation would be of obvious interest.

10. The invention or discovery of new number systems should not be regarded as an end in itself. In fact, each of the two developments that I have described occurred in response to a natural demand. We may ask whether there still exists in mathematics a situation which may give rise to a system of numbers as yet unknown. The answer is that such a situation not only exists but that it is close to some of the most exciting problems of contemporary mathematics. I am thinking of the wellknown analogy between algebraic function fields and their Riemann surfaces on one hand and algebraic number fields and their class of divisors or valuations, and - within the latter domain - of the similarities and differences between archi-

medean and nonarchimedean divisors. In a recent address on metamathematical problems I have suggested that the elucidation of this situation may be regarded as a problem in the foundations of mathematics no less, or perhaps more, than some of the more special questions concerning infinite cardinals or ordinals. Here I wish to add only that perhaps this problem also will receive a satisfactory solution in the discovery of a system of numbers that is unimaginable today and will be commonplace tomorrow.

I thank you for your attention.

References

[1] Robinson, A., *Nonstandard analysis*, Studies in Logic and the Foundations of Mathematics, North Holland Publ. Cy., Amsterdam, (1966).

[2] Robinson, A., *Nonstandard arithmetic and generic arithmetic*, Proceedings of the International Congress for Logic, Philosophy and Methodology of Science, Bucharest 1971, to be published.

[3] Robinson, A., *Metamathematical problems*, to be published in the Journal of Symbolic Logic.

(Received, April 27, 1973)

Yale University
Department of Mathematics

ON THE INTEGRATION OF HYPERBOLIC DIFFERENTIAL EQUATIONS

A. Robinson*.

[*Extracted from the Journal of the London Mathematical Society, Vol.* 25, 1950.]

§1. *Introduction*

In the present paper, we shall be concerned with the integration of linear partial differential equations in two independent variables and of arbitrary order, viz.,

$$\sum_{k=0}^{n} \sum_{l=0}^{k} a_{kl} \frac{\partial^k z}{\partial x^{k-l} \partial y^l} = a_0, \tag{1}$$

where x and y are the two independent variables, z is the dependent variable, n is a positive integer and the a_{kl} and a_0 are functions of x and y.

Our main object will be the development of a new method for the integration of a hyperbolic equation of type (1) for given initial conditions. We shall confine ourselves to the standard case in which the function z and its first $(n-1)$ derivatives with respect to x are given over a range of

* Received 27 May, 1949; read 16 June, 1949.

y for $x = 0$. For analytical coefficients and analytical initial conditions, the existence of a solution is ensured by the fundamental theorem of Cauchy-Kowalewski. However, the requirement of analyticity is unnaturally stringent for problems of this type. More recent work has established the existence and uniqueness of the solution under weaker conditions (see ref. 1, where other references are given). F. Rellich (ref. 2) has generalized Riemann's method to cope with essentially the same problem as that considered in the present note. The case of constant coefficients had been treated earlier by G. Herglotz by a Fourier integral method (ref. 3).

The present method establishes the existence and uniqueness of the solution under considerably less stringent conditions than postulated hitherto. Also, it leads to procedures which, with some modifications, should be suitable for numerical purposes. It can moreover be adapted to deal with some cases—discontinuity of coefficients, etc.—which may be important for the applications, but which are outside the scope of other methods. However, in the present paper we shall confine ourselves to the solution of the standard problem mentioned above. The central idea of the method first arose in connection with some work on stress propagation in beams (refs. 4, 5).

It will be shown in §2 of the present paper that the integration of a hyperbolic equation of type (1) can always be reduced to the successive integration of systems of linear equations of the type

$$\frac{\partial f_i}{\partial x}+c_i\frac{\partial f_i}{\partial y}=b_{i1}f_1+b_{i2}f_2+\ldots+b_{im}f_m+b_i, \quad i=1,\,2,\,\ldots,\,m, \quad m\leqslant n, \tag{2}$$

where the coefficients c_i, b_i and b_{ik} are functions of x and y which depend on the coefficients of the differential equation (1), but not on the initial conditions. However, in the process of determining the coefficients b_{ik}, we shall have to find particular solutions of systems of partial differential equations of the more general type

$$\frac{\partial f_i}{\partial x}+c_i\frac{\partial f_i}{\partial y}=F_i(x,\,y,\,f_1,\,f_2,\,\ldots,\,f_m), \quad i=1,\,2,\,\ldots,\,m. \tag{3}$$

The characteristic feature of both (2) and (3) is that their left hand sides represent the derivatives of the unknown functions along the characteristic curves of the system, while the right-hand sides only involve the unknown functions but not their derivatives. Two methods of integration for systems of this type are indicated in §3.

§2. *The replacement of an equation of order n by special systems of first order equations.*

We proceed to establish a system of equations of type (2) which is related to the given differential equation

$$\sum_{k=0}^{n} \sum_{l=0}^{k} a_{kl}(x, y) \frac{\partial^k z}{\partial x^{k-l} \partial y^l} = a_0 \tag{1}$$

on the assumption that the roots of the characteristic equation

$$a_{n0}\gamma^n - a_{n1}\gamma^{n-1} + \ldots + (-1)^{n-1} a_{n,n-1}\gamma + (-1)^n a_{nn} = 0 \tag{4}$$

are all real and distinct in a given region of the (x, y) plane. We may then assume that $a_{n0} = 1$. We shall suppose, in the first instance, only that the coefficients a_{kl} and a_0 possess continuous derivatives of the first order. Then the roots γ_k, $k = 1, 2, \ldots, n$ of (4) are continuous and distinct functions of x and y in the region in question.

It will be seen that the roots of the equation

$$\gamma^n + a_{n1}\gamma^{n-1} + \ldots + a_{n,n-1}\gamma + a_{nn} = 0$$

are $-\gamma_1, -\gamma_2, \ldots, -\gamma_n$. We may therefore write

$$\gamma^n + a_{n1}\gamma^{n-1} + \ldots + a_{n,n-1}\gamma + a_{nn}$$

$$= (\gamma + \gamma_m)(\gamma^{n-1} + a_1{}^{(m)}\gamma^{n-2} + \ldots + a^{m}_{n-1}), \quad m = 1, 2, \ldots, n, \tag{5}$$

where the coefficients $a_l^{(m)}$ are functions of x and y.

Then

$$\left(\frac{\partial}{\partial x} + \gamma_m \frac{\partial}{\partial y}\right)\left(\frac{\partial^{n-1}}{\partial x^{n-1}} + a_1^{(m)} \frac{\partial^{n-1} z}{\partial x^{n-2}\partial y} + a_2^{(m)} \frac{\partial^{n-1} z}{\partial x^{n-3}\partial y^2} + \ldots + a_{n-1}^{(m)} \frac{\partial^{n-1} z}{\partial y^{n-1}}\right)$$

$$= \frac{\partial^n z}{\partial x^n} + a_{n1} \frac{\partial^n z}{\partial x^{n-1}\partial y} + a_{n2} \frac{\partial^n z}{\partial x^{n-2}\partial y^2} + \ldots + a_{nn} \frac{\partial^n z}{\partial y^n}$$

$$+ \left(\frac{\partial a_1^{(m)}}{\partial x} + \gamma_m \frac{\partial a_1^{(m)}}{\partial y}\right) \frac{\partial^{n-1} z}{\partial x^{n-2}\partial y} + \left(\frac{\partial a_2^{(m)}}{\partial x} + \gamma_m \frac{\partial a_2^{(m)}}{\partial y}\right) \frac{\partial^{n-1} z}{\partial x^{n-3}\partial y^2}$$

$$+ \ldots + \left(\frac{\partial a_{n-1}^{(m)}}{\partial x} + \gamma_m \frac{\partial a_{n-1}^{(m)}}{\partial y}\right) \frac{\partial^{n-1} z}{\partial y^{n-1}}. \tag{6}$$

We put

$$z_m(x, y) = \frac{\partial^{n-1} z}{\partial x^{n-1}} + a_1^{(m)} \frac{\partial^{n-1} z}{\partial x^{n-2} \partial y} + \ldots + a_{n-1}^{(m)} \frac{\partial^{n-1} z}{\partial y^{n-1}} \quad (m = 1, 2, \ldots, n) \tag{7}$$

and define functions $f_m(x, y)$, $m = 1, 2, \ldots, n$, by

$$f_m(x, y) = z_m(x, y) + \sum_{k=0}^{n-2} \sum_{l=0}^{k} b_{kl}^{(m)} \frac{\partial^k z}{\partial x^{k-l} \partial y^l}, \tag{8}$$

where the coefficients $b_{kl}^{(m)} = b_{kl}^{(m)}(x, y)$ will be determined presently.

From (7) and (8)

$$\frac{\partial^{n-1} z}{\partial x^{n-1}} + a_1^{(m)} \frac{\partial^{n-1} z}{\partial x^{n-2} \partial y} + \ldots + a_{n-1}^{(m)} \frac{\partial^{n-1} z}{\partial y^{n-1}} = f_m(x, y) - \sum_{k=0}^{n-2} \sum_{l=0}^{k} b_{kl}^{(m)} \frac{\partial^k z}{\partial x^{k-l} \partial y^l} \quad (m = 1, 2, \ldots, n). \tag{9}$$

It can be shown that the determinant of (9), looked upon as a system of equations for the derivatives

$$\frac{\partial^{n-1} z}{\partial x^{n-1}}, \frac{\partial^{n-1} z}{\partial x^{n-2} \partial y}, \ldots, \frac{\partial^{n-1} z}{\partial y^{n-1}},$$

does not vanish. Its square is in fact equal to the discriminant of the characteristic equation.

Writing $[c_{im}]$ for the inverse of the matrix

$$[1, a_1^{(m)} a_2^{(m)}, \ldots, a_{n-1}^{(m)}] \quad (m = 1, 2, \ldots, n),$$

we then have

$$\frac{\partial^{n-1} z}{\partial x^{n-1-i} \partial y^i} = \sum_{m=1}^{n} c_{im} \left(f_m(x, y) - \sum_{k=0}^{n-2} \sum_{l=1}^{k} b_{kl}^{(m)} \frac{\partial^k z}{\partial x^{k-l} \partial y^l} \right) \quad (i = 0, 1, \ldots, n-1). \tag{10}$$

Also, from (7)

$$\frac{\partial z_m}{\partial x} + \gamma_m \frac{\partial z_m}{\partial y} = a_0 - \sum_{k=0}^{n-2} \sum_{i=0}^{k} a_{ki} \frac{\partial^k z}{\partial x^{k-i} \partial y^i} + \sum_{i=0}^{n-1} \left(\frac{\partial a_i^{(m)}}{\partial x} + \gamma_m \frac{\partial a_i^{(m)}}{\partial y} - a_{n-1,i} \right) \frac{\partial^{n-1} z}{\partial x^{n-1-i} \partial y^i} \quad (m = 1, 2, \ldots, n), \tag{11}$$

where we define the functions $a_0^{(m)}(x, y)$ which occur here for the first time for convenience by $a_0^{(m)}(x, y) = 0$.

Hence, from (8), (10) and (11),

$$\frac{\partial f_m}{\partial x}+\gamma_m\frac{\partial f_m}{\partial y}=\frac{\partial z_m}{\partial x}+\gamma_m\frac{\partial z_m}{\partial y}+\left(\frac{\partial}{\partial x}+\gamma_m\frac{\partial}{\partial y}\right)\left(\sum_{k=0}^{n-2}\sum_{l=0}^{k} b_{kl}^{(m)}\frac{\partial^k z}{\partial x^{k-l}\partial y^l}\right)$$

$$=\frac{\partial z_m}{\partial x}+\gamma_m\frac{\partial z_m}{\partial y}+\sum_{k=0}^{n-2}\sum_{l=0}^{k}\left[\left(\frac{\partial b_{kl}^{(m)}}{\partial x}+\gamma_m\frac{\partial b_{kl}^{(m)}}{\partial y}\right)\frac{\partial^k z}{\partial x^{k-l}\partial y^l}\right.$$
$$\left.+b_{kl}^{(m)}\left(\frac{\partial^{k+1}z}{\partial x^{k+1-l}\partial y^l}+\gamma_m\frac{\partial^{k+1}z}{\partial x^{k-l}\partial y^{l+1}}\right)\right].$$

$$=\frac{\partial z_m}{\partial x}+\gamma_m\frac{\partial z_m}{\partial y}+\sum_{k=0}^{n-1}\sum_{l=0}^{k}\left[\left(\frac{\partial b_{kl}^{(m)}}{\partial x}+\gamma_m\frac{\partial b_{kl}^{(m)}}{\partial y}\right)\right.$$
$$\left.+\left(b_{k-1,l}^{(m)}+\gamma_m b_{k-1,l-1}^{(m)}\right)\right]\frac{\partial^k z}{\partial x^{k-l}\partial y^l}, \quad (12)$$

where we have put $b_{-1,l}^{(m)}=b_{n-1,l}^{(m)}=b_{k,k+1}^{(m)}=b_{k-1,-1}^{(m)}=0$, $m=1, 2, \ldots, n$, $k=0, 1, \ldots, n-1$, $l=0, 1, 2, \ldots, n-1$, for convenience.

Substituting the values of the derivatives

$$\frac{\partial^{n-1}z}{\partial x^{n-1-i}\partial y^i} \quad (i=0, 1, \ldots, n-1),$$

from (10) we obtain

$$\frac{\partial f_m}{\partial x}+\gamma_m\frac{\partial f_m}{\partial y}=\frac{\partial z_m}{\partial x}+\gamma_m\frac{\partial z_m}{\partial y}$$
$$+\sum_{k=0}^{n-2}\sum_{i=0}^{k}\left[\left(\frac{\partial b_{ki}^{(m)}}{\partial x}+\gamma_m\frac{\partial b_{ki}^{(m)}}{\partial y}\right)+\left(b_{k-1,i}^{(m)}+\gamma_m b_{k-1,i-1}^{(m)}\right)\right]\frac{\partial^k z}{\partial x^{k-i}\partial y^i}$$
$$+\sum_{i=0}^{n-1}\sum_{p=1}^{n}\left(b_{n-2,i}^{(m)}+\gamma_m b_{n-2,i-1}^{(m)}\right)c_{ip}\left(f_p-\sum_{k=0}^{n-2}\sum_{l=0}^{k} b_{kl}^{(p)}\frac{\partial^k z}{\partial x^{k-l}\partial y^l}\right). \quad (13)$$

Again, substituting the expressions for $\dfrac{\partial z_m}{\partial x}+\gamma_m\dfrac{\partial z_m}{\partial y}$ as given by (11) and using (10), we have after some modification,

$$\frac{\partial f_m}{\partial x}+\gamma_m\frac{\partial f_m}{\partial y}$$
$$=a_0+\sum_{p=1}^{k}\left[\sum_{i=0}^{n-1}\left(b_{n-2,i}^{(m)}+\gamma_m b_{n-2,i-1}^{(m)}+\frac{\partial a_i^{(m)}}{\partial x}+\gamma_m\frac{\partial a_i^{(m)}}{\partial y}-a_{n-1,i}\right)c_{ip}\right]f_p(x,y)$$
$$+\sum_{k=0}^{n-2}\sum_{l=0}^{n}\left[\left(\frac{b_{kl}^{(m)}}{\partial x}+\gamma_m\frac{b_{kl}^{(m)}}{\partial y}\right)+\left(b_{k-1,l}^{(m)}+\gamma_m b_{k-1,l-1}^{(m)}\right)-a_{kl}\right.$$
$$\left.-\sum_{i=0}^{n-1}\sum_{p=1}^{n}\left(b_{n-2,i}^{(m)}+\gamma_m b_{n-2,i-1}^{(m)}+\frac{\partial a_i^{(m)}}{\partial x}+\gamma_m\frac{\partial a_i^{(m)}}{\partial y}-a_{n-1,i}\right)c_{ip}b_{kl}^{(p)}\right]\frac{\partial^k z}{\partial x^{k-l}\partial y^l}.$$

(14)

It follows that if we define the functions $b_{ki}^{(m)}$ in such a way that the coefficients of $\frac{\partial^k z}{\partial x^{k-l}\partial y^l}$ in (14) all vanish, then (14) becomes

$$\frac{\partial f_m}{\partial x}+\gamma_m\frac{\partial f_m}{\partial y}=b_{m1}f_1+b_{m2}f_2+\ldots+b_{mn}f_n+a_0 \quad (m=1, 2, \ldots, n), \tag{15}$$

where

$$b_{mk}(x, y)=\sum_{i=0}^{n-1}\left(b_{n-2,i}^{(m)}+\gamma_m b_{n-1,i-1}^{(m)}+\frac{\partial a_i^{(m)}}{\partial x}+\gamma_m\frac{\partial a_i^{(m)}}{\partial y}-a_{n-1,i}\right)c_{ik}$$

$$(m, k=1, 2, \ldots, n). \tag{16}$$

Referring to (14), we see that the condition that all the terms involving $\frac{\partial^k z}{\partial x^{k-l}\partial y^l}$ vanish will be satisfied if the functions $b_{kl}^{(m)}$, $m=1, 2, \ldots, n$, $k=0, 1, \ldots, n-2$, $l=0, \ldots, k$, are solutions of the system of partial differential equations

$$\frac{\partial b_{kl}^{(m)}}{\partial x}+\gamma_m\frac{\partial b_{kl}^{(m)}}{\partial y}=a_{kl}-(b_{k-1,l}^{(m)}+\gamma_m b_{k-1,l-1}^{(m)})$$
$$+\sum_{i=0}^{n-1}\sum_{p=1}^{n}c_{ip}\left(b_{n-2,i}^{(m)}+\gamma_m b_{n-2,i-1}^{(m)}+\frac{\partial a_i^{(m)}}{\partial x}+\gamma_m\frac{\partial a_i^{(m)}}{\partial y}-a_{n-1,i}\right)b_{kl}^{(p)}. \tag{17}$$

The number of equations in (17) is $\frac{1}{2}n^2(n-1)$.

Assume now that the values of z and of its first $(n-1)$ derivatives with respect to x are specified for a range of values of y, for $x=0$,

$$z(0, y)=g_0(y), \quad \left(\frac{\partial z}{\partial x}\right)_{x=0}=g_1(y), \quad \ldots, \quad \left(\frac{\partial^{n-1}z}{\partial x^{n-1}}\right)_{x=0}=g_{n-1}(y), \tag{18}$$

where $g_i(y)$, $i=0, \ldots, n-1$ can be differentiated $n-i-1$ times with respect to y. Having solved (17) for arbitrary but sufficiently regular initial conditions (see ref. 6), we obtain the initial conditions for the functions $f_m(x, y)$ from (7), (8) and (18),

$$f_m(0, y)=g_{n-1}(y)+a_1^{(m)}\frac{\partial g_{n-2}}{\partial y}+\ldots+a_{n-1}^{(m)}\frac{\partial^{n-1}g_0}{\partial y^{n-1}}+\sum_{k=0}^{n-2}\sum_{l=0}^{k}b_{kl}^{(m)}\frac{\partial^l g_{k-l}}{\partial y^l}. \tag{19}$$

Having solved (15) for these initial conditions, we may then look upon any one of the equations (9) as a linear partial differential equation of order $(n-1)$ for z. All these equations are of hyperbolic type, the roots of their characteristic equations consisting of $(n-1)$ of the n roots of (4). By the successive application of this procedure, we finally obtain a first order equation for z, which is itself an example of (2).

3. *Integration.*

We shall now discuss the integration of the system of equations

$$\frac{\partial f_i}{\partial x}+c_i\frac{\partial f_i}{\partial y}=F_i(x, y, f_1, f_2, \ldots, f_m) \quad (i=1, 2, \ldots, m) \tag{3}$$

for given initial conditions

$$f_1(0, y)=h_1(y), f_2(0, y)=h_2(y), \ldots, f_m(0, y)=h_m(y). \tag{20}$$

It is assumed throughout that the $c_i=c_i(x, y)$ possess first derivatives in the region under consideration. With each one of the equations in (3) we associate a one-parametric family of characteristic curves $y=\phi_i(x)$ given by

$$\frac{dy}{dx}=c_i(x, y) \quad (i=1, 2, \ldots, m). \tag{21}$$

We denote differentiation along a characteristic curve $y=\phi_i(x)$ by $\frac{D_i}{D_i x}$, so that for any arbitrary function $f(x, y)$, $\frac{D_i f}{D_i x}$ is defined by

$$\frac{D_i f(x, y)}{D_i x}=\lim_{h\to 0}\frac{f\big(x+h, \phi_i(x+h)\big)-f\big(x, \phi_i(x)\big)}{h}, \quad y=\phi_i(x) \tag{22}$$

whenever that limit exists. Thus, if f possesses continuous first derivatives,

$$\frac{D_i f}{D_i x}=\frac{\partial f}{\partial x}+\frac{d\phi_i}{dx}\frac{\partial f}{\partial y}=\frac{\partial f}{\partial x}+c_i(x, y)\frac{\partial f}{\partial y}. \tag{23}$$

Equations (3) can now be written as

$$\frac{D_i f_i}{D_i x}=F_i(x, y, f_1, f_2, \ldots, f_m) \quad (i=1, 2, \ldots, m). \tag{24}$$

Integrating between two values x' and x'' along a characteristic curve $\phi_i(x)$, we obtain

$$f_i\big(x'', \phi_i(x'')\big)-f_i\big(x', \phi_1(x')\big)$$
$$=\int_{x'}^{x''} F_i\Big\{\xi, \phi_i(\xi), f_1\big(\xi, \phi_i(\xi)\big), \ldots, f_m\big(\xi, \phi_i(\xi)\big)\Big\}\, d\xi. \tag{25}$$

In particular, taking $x'=0$ and writing x for x'', we have

$$f_i\big(x, \phi_i(x)\big)=f_i\big(0, \phi_i(0)\big)$$
$$+\int_0^x F_i\Big\{\xi, \phi_i(\xi), f_1\big(\xi, \phi_i(\xi)\big), \ldots, f_m\big(\xi, \phi_i(\xi)\big)\Big\}\, d\xi \quad (i=1, 2, \ldots, m), \tag{26}$$

where ϕ_i is any particular characteristic curve which meets the y axis at some point of the region under consideration.

Equations (26) suggest a method of successive approximation by which the functions $f_i(x, y)$ can be determined for given values. The functions $f_{i0}(x, y)$ $(i = 1, 2, \ldots, m)$ are first defined by

$$f_{1,0}(x, y) = h_1(\phi_1(0)),\ f_{2,0}(x, y) = h_2(\phi_2(0)),\ \ldots,\ f_{m,0}(x, y) = h_m(\phi_m(0)), \quad (27)$$

where the characteristic curves in question pass through the point (x, y),

$$y = \phi_1(x) = \phi_2(x) = \ldots = \phi_m(x).$$

The functions $f_{i\mu}(x, y)$ $(i = 1, 2, \ldots, m,\ \mu \geqslant 1)$ are then determined successively by

$$f_{i\mu}(x, y) = h_i(\phi_i(0)) + \int_0^x F_i\{\xi,\ \phi_i(\xi),\ f_{1,\mu-1}(\xi, \phi_i(\xi)),\ \ldots,$$
$$f_{m,\mu-1}(\xi, \phi_i(\xi))\}\, d\xi, \quad (28)$$

where the characteristic curves are chosen as above.

A set of conditions under which the functions $f_{i\mu}(x, y)$ can be constructed such that the limits $f_i(x, y) = \lim_{\mu\to\infty} f_{i\mu}(x, y)$ exist and form a solution of (3) for the initial values (20) is given in ref. 6.

In addition to giving rise to a method of successive approximation, equations (24) also suggest a step-by-step method of integration, by which the functions $f_i(x, y)$ are determined approximately for $x+\delta x$ and for all y if they are known for x and for all y. The formulae of regression are

$$f_i(x+\delta x, y) = f_i(x, y-c_i\delta x) + F_i\delta x \quad (i = 1, 2, \ldots, m). \quad (29)$$

However, in this form the method is not as yet very suitable for numerical work, since the formulae of regression require a knowledge of $f_i(x, y)$ for arbitrary values of y, whereas in practice it will be known only at a number of isolated points. Hence, in general, an additional operation of interpolation will be required at each step. The convergence of this process is a matter for further investigation.

In conclusion, it may be said that the methods outlined in the present paper appear to have distinct prospects for numerical application. Their main advantage is that since integration is carried out along the characteristic curves, there is no possibility of a failure, such as may occur in the application of some lattice methods (see refs. 7.8). On the other hand the total number of integrations required for the solution of an equation of fairly high order is very large. However, even in these cases the procedure may still form a suitable basis for work on a modern calculating machine.

References.

1. K. Friedrichs, " Non-linear hyperbolic differential equations for functions of two independent variables ", *American Journal of Math.*, 52 (1948).
2. F. Rellich, " Verallgemeinerung der Riemannschen Integrationsmethode, etc.", *Math. Annalen*, 103 (1930).
3. G. Herglotz, " Über die Integration linearer partieller Differenzialgleichungen, etc., I, II, III ", *Ber. Sächs. Akad.*, 1926, 1928.
4. A. Robinson, " Shock transmission in beams", *R.A.E. Report No. S.M.E.* 3319, 1945. To be published as part of *Reports and Memoranda of the Aeronautical Research Council*, No. 2265.
5. A. Robinson, " Shock transmission in beams of variable characteristics ", *R.A.E. Report No. S.M.E.* 3340, 1945. To be published together with ref. **4.**
6. A. Robinson, " On the integration of hyperbolic differential equations ", *College of Aeronautics Report No.* 18, July, 1948.
7. R. Courant, K. Friedrichs and H. Lewy, " Über die partiellen Differenzengleichungen der mathematischen Physik ", *Math. Annalen*, 100 (1928).
8. L. Collatz, " Über das Differenzenverfahren bei Anfangswertproblemen partieller Differenzialgleichungen ", *Z.A.M.M.*, 16 (1936).

The College of Aeronautics,
Cranfield.

ON FUNCTIONAL TRANSFORMATIONS AND SUMMABILITY.

By A. ROBINSON.

[Received 29 July, 1947.—Read 16 October, 1947.]

[*Extracted from the Proceedings of the London Mathematical Society, Ser.* 2, *Vol.* 52, 1950.]

1. *Introduction.*

The present paper is concerned with a generalization of the concept of summability.

Given an infinite matrix $A = \|a_{k,n}\|$, $k = 1, 2, \ldots, n = 1, 2, \ldots)$, $a_{k,n}$ complex, and an infinite sequence (s_n) $(n = 1, 2, \ldots)$, s_n complex, the transformed sequence of (s_n) by A, $(t_k) = A\ (s_n)$ is defined by:

(1) $$t_k = \sum_{n=1}^{\infty} a_{k,n} s_n \quad (k = 1, 2, 3, \ldots);$$

(1) is called a sequence-to-sequence transformation.

In order that this transformation shall have a meaning, it is required that the sum on the right-hand side of (1) converge for all k. In that case A is said to "apply to (s_n)". In order that A should apply to all convergent sequences it is necessary and sufficient that

(2) $$\sum_{n=1}^{\infty} |a_{k,n}| < \infty \quad (k = 1, 2, \ldots).$$

A necessary and sufficient condition for A to transform every convergent sequence into a convergent sequence with the same limit is given by the well-known theorem of Toeplitz and Silverman†.

THEOREM I. *A necessary and sufficient condition for the matrix A to transform any convergent sequence into a convergent sequence with the same*

† Compare, for example, P. Dienes, *The Taylor series* (Oxford, 1931), 389.

limit is:

(3) $$\lim_{k\to\infty} a_{k,n} = 0 \quad (n = 1, 2, \ldots),$$

(4) $$\lim_{k\to\infty} \sum_{n=1}^{\infty} a_{k,n} = 1,$$

(5) $$\sum_{n=1}^{\infty} |a_{k,n}| \leqslant M, \ (k = 1, 2, \ldots), \ M \textit{ independent of } k.$$

Similarly, given an infinite matrix $B = \|b_{k,n}\|$, the series-to-sequence transformation by B of an infinite series $a_1+a_2+a_3+\ldots$ to an infinite sequence (t_k) is given by

(6) $$t_k = \sum_{n=1}^{\infty} b_{k,n} a_n \quad (k = 1, 2, 3, \ldots).$$

As before, in order that this transformation shall have a meaning, it is required that the sum on the right-hand side of (6) converge for all k. In that case B is said to apply to $\sum_{n=1}^{\infty} a_n$. In order that B should apply to all convergent series it is necessary and sufficient that

(7) $$\sum_{n=1}^{\infty} |b_{k,n} - b_{k,n+1}| < \infty \quad (k = 1, 2, \ldots).$$

Corresponding to Theorem I we have Theorem II, due to Carmichael, Bosanquet, and others†.

THEOREM II. *A necessary and sufficient condition for the matrix B to transform any convergent series into a convergent series whose limit equals the sum of the series is:*

(8) $$\sum_{n=1}^{\infty} |b_{k,n} - b_{k,n+1}| \leqslant M \quad (k = 1, 2, \ldots),$$

M *independent of* k,

(9) $$\lim_{k\to\infty} b_{kn} = 1 \quad (n = 1, 2, \ldots).$$

† P. Dienes, *loc. cit.* 396.

Matrices satisfying the conditions of Theorems I and II respectively can be used in order to define the limit of (t_k), wherever it exists, as the generalized limit, or as the generalized sum of (s_n) and $\sum_{n=1}^{\infty} a_n$ respectively. There are similar definitions and conditions for "Semi-continuous matrices" where k is replaced by a real variable w.

The above definitions relate to sequences, or series, of constant terms. When applying the theory to sequences or series of functions, we concentrate on a specified value of the argument, so that the sequence (or series) becomes essentially one of constant terms, and then proceed as before. There are, however, methods of summation which effectively involve the functions which make up the sequence (or series) as a whole. To illustrate this possibility we are going to discuss Borel's (integral) procedure of summation.

Given a power series $\sum_{n=0}^{\infty} a_n z^n$ with a positive radius of convergence, the associated series of $\sum_{n=0}^{\infty} a_n z^n$ is defined as

$$\sum_{n=0}^{\infty} \frac{a_n t^n z^n}{n!}.$$

This series represents an integral function of tz, which we denote by $\phi(tz)$. The generalized Borel-sum of the original series in a point z is then defined as $\int_0^{\infty} e^{-t}\phi(tz)\,dt$ whenever this integral exists. More precisely,

$$B \sum_{n=0}^{\infty} a_n z^n = \lim_{\omega \to \infty} \int_0^{\omega} e^{-t}\phi(tz)\,dt$$

We modify the procedure by putting $p = 1/z$, $x = tz = t/p$. Then the original series becomes

$$\sum_{n=0}^{\infty} \frac{a_n}{p^n} \tag{10}$$

and is now the Laurent series of a function regular at infinity. The associated series is

$$\sum_{n=0}^{\infty} \frac{a_n x^n}{n!} \tag{11}$$

representing an integral function $\phi(x)$. Finally the Borel-integral is

$$B \sum_{n=0}^{\infty} \frac{a_n}{p^n} = p \int_0^{\infty} e^{-px}\phi(x)\,dx = \lim_{\omega \to \infty} p \int_0^{\omega} e^{-px}\phi(x)\,dx. \tag{12}$$

In particular (accepting the regularity of Borel's procedure)

$$\frac{1}{p^n} = p \int_0^\infty e^{-px} \left(\frac{x^n}{n!}\right) dx,$$

which is in fact a well-known result of the operational calculus.

Given a function $f(x)$ defined for real positive x, the integral

$$\bar{f}(p) = \int_0^\infty e^{-px} f(x)\, dx$$

is the Laplace transform of $f(x)$, $\bar{f}(p) = Lf(x)$. The integral

$$f^*(p) = p \int_0^\infty e^{-px} f(x)\, dx$$

is sometimes known as Carson's integral†, and we shall call $f^*(p)$ the Carson transform of $f(x)$, $f^*(p) = Cf(x)$.

Without going deeply into the theory of the Laplace transform in this introduction, it is known that under certain conditions the Laplace transformation possesses an inverse operation L^{-1} (given by the formula of Fourier-Mellin,

$$f(x) = \frac{1}{2i\pi} \int_{\gamma - i\infty}^{\gamma + i\infty} e^{xp} \bar{f}(p)\, dp.$$

Since $f^*(p) = p\bar{f}(p)$, it follows that Carson's transformation also possesses an inverse operation C^{-1}. We may then say that the associated series (11) is obtained from (10) by subjecting it term by term to the transformation C^{-1}; thus

$$\phi(x) = \sum_{n=0}^{\infty} \phi_n(x) = \sum_{n=0}^{\infty} C^{-1} f_n(p)$$

where

$$\phi_n(x) = \frac{a_n x^n}{n!} \text{ and } f_n(p) = \frac{a_n}{p^n}.$$

Again, denoting by C_ω the transformation $p \int_0^\omega e^{-px} f(x)\, dx$, we may

† See H. S. Carslaw and J. C. Jaeger, *Operational methods in applied mathematics* (Oxford, 1941), XV.

rewrite (12)

$$B \sum_{n=0}^{\infty} \frac{a_n}{p^n} = \lim_{\omega\to\infty} C_\omega \phi(x).$$

We notice that C_ω is continuous, so that

$$(13) \qquad B \sum_{n=0}^{\infty} \frac{a_n}{p^n} = \lim_{\omega\to\infty} C_\omega \Big(\sum_{n=0}^{\infty} C^{-1} f_n(p) \Big) = \lim_{\omega\to\infty} \sum_{n=0}^{\infty} C_\omega C^{-1} f_n(p).$$

Thus the generalized Borel-limit is given by

$$B \sum_{n=0}^{\infty} f_n = \lim_{\omega\to\infty} \sum_{n=0}^{\infty} C_\omega C^{-1} f_n.$$

Putting $C_{\omega,0} = C_{\omega,1} = C_{\omega,2} = \ldots = C_\omega C^{-1}$, this may be written

$$(14) \qquad B \sum_{n=0}^{\infty} f_n = \lim_{\omega\to\infty} \sum_{n=0}^{\infty} C_{\omega,n} f_n.$$

We therefore see that Borel's method is essentially the transformation of a series of functions by a (semi-continuous) matrix of linear operators. (The fact that Borel's integral can in fact be replaced by a matrix of constant terms does not hold for general series of functions.)

Instead of assuming explicitly that the elements of the series or sequence are functions of one variable, we may assume that they are more generally elements of a metric linear space S, while the elements of the transforming matrix are defined in an abstract fashion as linear operators, *i.e.* as linear operations in S. We are going to develop the theory for ordinary (discrete) matrices, the treatment of semi-continuous matrices being quite similar.

In addition to its own inherent interest, this approach provides us with the right perspective on the transformation of numerical sequences or series by numerical matrices. In fact, the individual terms $a_{kn} s_n$ in the series $t_k = \sum_{n=0}^{\infty} a_{kn} s_n$ are not essentially the products of two numbers respectively, a_{kn} and s_n, but each one of them is a linear (homogeneous) function of the respective s_n.

2. *Linear transformations of point sequences in abstract spaces.*

Let S be a complete metric linear space, *i.e.* a system of elements $a, b, \ldots$, which form an Abelian group with respect to addition with the system of all complex numbers as operator ring and for which norms $|a|$, $|b|$, ..., are defined with the following properties.

(i) For every $a \varepsilon s$, $|a|$ is a real non-negative number, positive if $a \neq 0$ and 0, if $a = 0$, where 0 is the neutral element in S.

(ii) For every $a \varepsilon s$ and complex λ, $|\lambda a| = |\lambda||a|$.

(iii) For all $a \varepsilon s$, $b \varepsilon s$, $|a+b| \leqslant |a|+|b|$.

(iv) Given an infinite sequence (s_k), $s_k \varepsilon S$ $(k = 1, 2, \ldots)$, the limit of (s_k), wherever it exists, $s = \lim_{k\to\infty} s_k$ is defined in the usual way by $\lim_{k\to\infty} |s-s_k| = 0$, and if, for a given sequence (s_k), $\lim_{m,n\to\infty} |s_m - s_n| = 0$, then (s_k) has a limit.

A (bounded) linear operator T in S is a function defined for all $a \varepsilon S$, taking values c in S, $c = Ta$, and possessing the following properties:

(i) $Ta+Tb = T(a+b)$ for all $a, b \varepsilon S$.

(ii) $\lambda(Ta) = T(\lambda a)$ for all $a \varepsilon S$ and complex λ.

(iii) T is bounded, *i.e.* $\lim \sup_{|x| \leqslant 1} |Tx| < \infty$.

$|T| = \lim \sup |Tx|$ for $|x| \leqslant 1$ is called the bound of T. Given two linear operators T_1 and T_2, their sum T_1+T_2 is defined as the linear operator which transforms any $a \varepsilon S$ into $T_1 a + T_2 a$. A linear operator T is said to be the limit of an infinite sequence (T_p) of linear operators if, for all $a \varepsilon S$, $\lim_{p\to\infty} T_p a = Ta$. If $\lim_{p\to\infty} |T_p - T| = 0$ then $\lim_{p\to\infty} T_p = T$. The sum of an infinite series of linear operators is defined as the limit of the partial sums of the series wherever it exists.

Given a sequence (s_n), $(n = 1, 2, \ldots)$, of points in S, consider the transformation of (s_n) into a sequence (t_k) of points in the same space, where

$$t_k = \sum_{n=1}^{\infty} A_{k,n} s_n, \tag{15}$$

the $A_{k,n}$ $(k = 1, 2, \ldots, n = 1, 2, \ldots)$ being specified (bounded) linear operators in S.

Under what conditions does the matrix $A = \|A_{k,n}\|$ "apply" to every convergent sequence (s_n) in S so that the limit of the transformed sequence

(t_k) coincides with the limit of the original sequence. To solve this problem we introduce the concept of the bound of a finite or infinite sequence of linear operators.

Given a finite sequence of linear operators in S, $(T_1, T_2, \ldots, T_m)$, we define the bound of that sequence, $|T_1, \ldots, T_m|$ by

$$|T_1, \ldots, T_m| = \limsup_{|x_p| \leqslant 1} \left| \sum_p T_p x_p \right|, \quad x_p \in S.$$

If the sequence contains only one term, $T = T_1$, then the bound of (T) so defined coincides with the "ordinary" bound of T as defined previously. It will be convenient to speak of ∞ as a bound, if $\limsup_{|x_p| \leqslant 1} \left| \sum_p T_p x_p \right| = \infty$.

The bound $|T_1, T_2, \ldots|$ of an infinite sequence of linear operators (T_n) will be defined as the upper limit of the bounds of the finite subsequences of (T_n), (finite or infinite as the case may be).

It is easy to show that the bound of a finite or infinite sequence has the following properties:

$$\limsup_{n\to\infty} \left| \sum_{p=1}^{n} T_p \right| \leqslant |T_1, T_2, \ldots, T_p, \ldots|, \tag{16}$$

$$|T_1, T_2, \ldots, T_p, \ldots| \leqslant \sum_{p=1}^{\infty} |T_p|, \tag{17}$$

$$\limsup_{n\to\infty} \left| \sum_{p=1}^{n} T_p x_p \right| \leqslant |T_1, T_2, \ldots, T_p, \ldots| \limsup |x_p| \tag{18}$$

for arbitrary $x_p \in S$,

$$|T_1, \ldots, T_p, \ldots| + |\bar{T}_1, \ldots, \bar{T}_p, \ldots| \geqslant |T_1 + \bar{T}_1, \ldots, T_p + \bar{T}_p, \ldots| \tag{19}$$

where $(T_1, \ldots, T_p, \ldots)$ and $(\bar{T}_1, \ldots, \bar{T}_p, \ldots)$ are arbitrary sequences of linear operators in S.

We are going to prove

THEOREM III. *Given an infinite sequence of (bounded) linear operators* (T_p), *a necessary and sufficient condition for* $\sum_{p=1}^{\infty} T_p s_p$ *to converge whenever* (s_p) *converges is*

$$|T_1, T_2, \ldots, T_p, \ldots| = M < \infty \tag{20}$$

and

$$\sum_{p=1}^{\infty} T_p = T \text{ exists.} \tag{21}$$

The condition is sufficient. Let $t_k = \sum_{p=1}^{k} T_p s_p$, and assume first that $\lim_{p\to\infty} s_p = 0$. Then for sufficiently large p, $p > p_0$, $|s_p| < \epsilon$, where ϵ is any positive quantity given in advance. Hence, for $k > p_0$ and arbitrary (positive, integral) m,

$$|t_{k+m} - t_k| = |\sum_{p=k+1}^{k+m} T_p s_p| \leqslant |T_{k+1}, \ldots, T_{k+m}| \epsilon \leqslant |T_1, T_2, \ldots| \epsilon = M \epsilon.$$

This shows that (t_k) converges.

Again, for a general convergent sequence (s_p), $\lim_{p\to\infty} s_p = s$, we have

$$t_k = \sum_{p=1}^{k} T_p s + \sum_{p=1}^{k} T_p (s_p - s) = (\sum_{p=1}^{k} T_p) s + \sum_{p=1}^{k} T_p (s_p - s) \tag{22}$$

and $s_p - s \to 0$ as $p \to \infty$. Hence, as $k \to \infty$, $(\sum_{p=1}^{k} T_p) s \to Ts$ by (21), while $\sum_{p=1}^{k} T_p\,(s_p - s)$ converges, as shown above.

The necessity of (21) is shown by taking $s_p = s$, $(p = 1, 2, \ldots)$, s arbitrary. To prove that condition (20) is necessary, we show first that all the T_p are collectively bounded. Otherwise there exists a monotonically increasing sequence of positive integers (p_k) so that $\lim_{x\to\infty} |T_{p_k} s_{p_k}| = \infty$, where (s_{p_k}) is a sequence of $s_{p_k} \varepsilon S$, $0 < |s_{p_k}| \leqslant 1$. Hence, defining

$$(\bar{s}_{p_k}) \text{ by } \bar{s}_{p_k} = \frac{1}{|s_{p_k}|^{\frac{1}{2}}} s_{p_k},$$

we have $\lim_{k\to\infty} \bar{s}_{p_k} = 0$ and $\lim_{k\to\infty} |T_{p_k} \bar{s}_{p_k}| = \infty$. We may even select a subsequence of (p_k), say (p_{k_i}), so that

$$|T_{p_{k_i}} \bar{s}_{p_{k_i}}| - \sum_{j=1}^{i-1} |T_{p_{k_i}} \bar{s}_{p_{k_i}}| \geqslant i.$$

Now we define a sequence $(\tilde{s}_n)$ by $\tilde{s}_n = 0$ except when $n = p_{k_i}$ for some i, in which case we define $\tilde{s}_n = \bar{s}_{p_{k_i}}$. Obviously $\lim_{n\to\infty} \tilde{s}_n = 0$. On the other hand,

$$\left|\sum_{n=1}^{p_{k_i}} T_n \tilde{s}_n\right| \geqslant |T_{p_{k_i}} \tilde{s}_{p_{k_i}}| - \sum_{n=1}^{p_{k_i}-1} |T_n \tilde{s}_n| = |T_{p_{k_i}} \tilde{s}_{p_{k_i}}| - \sum_{j=1}^{i-1} |T_{p_{k_j}} s_{p_{k_j}}| \geqslant i,$$

so that the transformed sequence is unbounded. This is contrary to assumption, so that all the T_p are in fact collectively bounded.

In a very similar way it can be shown that all the finite subsequences of (T_p) are collectively bounded, *i.e.* the sequence itself has a bound.

It should be observed that conditions (20) and (21) are independent of each other. In fact, let S be the space of complex numbers. Then the effect of any given linear operator T_p on an element s of S is that of multiplying s by a certain complex number τ_p. And it is easily seen that in that case $|T_1, T_2, \ldots, T_p, \ldots| = \sum_{p=1}^{\infty} |\tau_p|$. On the other hand it is well-known that the ordinary convergence of a series of complex numbers does not imply its absolute convergence. This shows that (20) is independent of (21).

To show that (21) is independent of (20) we consider the space S whose elements are all the bounded infinite sequences of complex numbers $c = (c_1, c_2, \ldots)$. We define $|c| = \max_k |c_k|$. It is easy to verify that S so defined is a complete metric linear space. Moreover if we define a sequence (T_p) of linear operators in S by $T_p = \|\tau^{(p)}_{k,n}\|$ where $\tau^{(p)}_{p,1} = 1$, and $\tau^{(p)}_{k,n} = 0$ for all other suffices, then $|T_1, T_2, \ldots, T_p, \ldots| = 1$. On the other hand the transforms of the sequence $(1, 0, 0, \ldots)$ by the partial sums of the series $\sum_{p=1}^{\infty} T_p$ are:

$$(1, 0, 0, 0, \ldots),\ (1, 1, 0, 0, \ldots),\ (1, 1, 1, 0, \ldots)$$

and it is easy to see that this sequence does not converge to a limit. It follows that $\sum_{p=1}^{\infty} T_p$ does not converge, showing that (21) is independent of (20).

We now come to the main theorem of this section.

THEOREM IV. *In order that* (t_k), *as given by* (15), *exist and converge to the same limit as* (s_n) *whenever* (s_n) *converges,* $\|A_{k,n}\|$ *being a matrix of (bounded) linear operators, it is necessary and sufficient that*

(23) $$\lim_{k\to\infty} A_{k,n} = O, \quad n = 1, 2, \ldots,$$

(24) $$\sum_{n=1}^{\infty} A_{k,n} = A_k \text{ exists for } k = 1, 2, \ldots,$$

(25) $$\lim_{k\to\infty}{}' A_k = I,$$

(26) $$|A_{k,1}, A_{k,2}, \ldots| = a_k \leqslant M < \infty,$$

where M is a constant independent of k.

In the above, O is the zero operator and I the identity operator ($Os = o$ and $Is = s$ for all $s \varepsilon S$).

The condition is sufficient.

Let (s_n) be the original sequence which is supposed to converge to a limit s, and (t_k) the transformed sequence, $t_k = \sum_{n=1}^{\infty} A_{k,n} s_n$ $(k = 1, 2, \ldots)$. For arbitrary given $\epsilon > 0$, choose N_0 so that $|s - s_n| < \epsilon$ for $n > N_0$, and then K_0 so that $|(I - \sum_{n=1}^{\infty} A_{k,n})s| < \epsilon$ for all $k > K_0$, while at the same time $|A_{k,n}(s_n - s)| < \epsilon / N_0$ for all $n \leqslant N_0$, $k > K_0$. The existence of such a K_0 is ensured by (23)-(25). Then, for any $k > K_0$,

$$|t_k - s| = \left| \sum_{n=1}^{N_0} A_{k,n}(s_n - s) + \sum_{n=N_0+1}^{\infty} A_{k,n}(s_n - s) - (I - \sum_{n=1}^{\infty} A_{kn})s \right|$$

$$\leqslant \sum_{n=1}^{N_0} |A_{k,n}(s_n - s)| + \left| \sum_{n=N_0+1}^{\infty} A_{k,n}(s_n - s) \right| + \left| (I - \sum_{n=1}^{\infty} A_{k,n})s \right|$$

$$\leqslant N_0 \frac{\epsilon}{N_0} + M\epsilon + \epsilon = (2 + M)\epsilon.$$

As ϵ is arbitrarily small and positive, we infer that $\lim_{k\to\infty} t_k = s$, showing that the condition is sufficient.

It is easy to see that condition (23) is necessary. In fact, for given n_0, the sequence $(A_{k,n_0} x)$ is the A-transform of the sequence (s_n) defined by $s_n = x$ for $n = k_0$, $s_n = 0$ for $n \neq k_0$, $x \varepsilon S$ otherwise arbitrary. Now $s_n \to 0$ as $n \to \infty$, and so, by assumption $A_{k,n_0} x \to 0$ as $k \to \infty$. This being true for every $x \varepsilon s$, it follows that $\lim_{k\to\infty} A_{k,n_0} = O$.

Condition (24) is necessary to ensure that the matrix A applies to every convergent sequence, by Theorem III. To see the necessity of (25) consider the sequence (s_n) defined by $s_n = x$ $(n = 1, 2, \ldots)$, $x \varepsilon S$, otherwise arbitrary. Its A-transform is the sequence $(\sum_{n=1}^{\infty} A_{k,n} x)$. Obviously, $s_n \to x_1$, *as* $n \to \infty$ and so $\sum_{n=1}^{\infty} A_{k,n} x = (\sum_{n=1}^{\infty} A_{k,n}) x \to x$ as $k \to \infty$. Hence $\sum_{n=1}^{\infty} A_{k,n} \to I$ as $k \to \infty$, in accordance with (24).

In order to show that condition (26) is necessary, we prove as a preliminary that, if a sequence of (bounded) linear operators (T_p) tends to O, then the sequence is collectively bounded.

Let b_p be the bounds of the respective T_p, so that for arbitrary $\epsilon > 0$, there are $x_p \varepsilon S$, $|x_p| \leqslant 1$ $(p = 1, 2, \ldots)$ satisfying $|T_p x_p| \geqslant b_p - \epsilon$, while $|T_p x| \leqslant b_p$ for all $x \varepsilon S$, provided $|x| \leqslant 1$. Assuming now contrary to our assertion that (b_p) is unbounded, we may select a suitable subsequence of (T_p), $(\bar{T}_p)$, say, with the following properties:

$$\text{(i)} \quad \lim_{p \to \infty} \bar{T}_p = O,$$

$$\text{(ii)} \quad |\bar{T}_p x| \leqslant \tfrac{3}{2} c_p \text{ for all } |x| \leqslant 1,$$

$$\text{(iii)} \quad |\bar{T}_p x_p| \geqslant c_p,$$

where $(\bar{x}_p)$ is a sequence of elements in S, $|\bar{x}_p| \leqslant 1$, and (c_p) is a sequence of positive integers so that $c_1 \geqslant 1$ and $C_{p+1} \geqslant 8c_p$ for all p. Moreover, in view of the fact that T_p tends to 0, we may in fact choose the subsequence $(\bar{T}_p)$ in such a way that

$$\text{(iv)} \quad |\bar{T}_p \bar{x}_q| \leqslant \tfrac{1}{2} c_q \quad \text{for } p = 2, 3, \ldots, q = 1, \ldots, p-1.$$

Consider now the infinite series of elements of S,

$$\frac{\bar{x}_1}{c_1} + \frac{2\bar{x}_2}{c_2} + \frac{4\bar{x}_3}{c_3} + \ldots + \frac{2^{p-1}\bar{x}_p}{c_p} + \ldots.$$

This series is majorized by the geometrical series

$$1 + \frac{1}{4} + \frac{1}{4^2} + \ldots,$$

since $c_1 \geqslant 1$ and $c_{p+1} \geqslant 8c_p$, and therefore converges to an element of S, $\bar{x}$, say. We are going to show that $(T_p x)$ does not tend to 0 contrary to (i).

In fact

$$|\bar{T}_p \bar{x}| = \left| \bar{T}_p \left(\sum_{q=1}^{\infty} \frac{2^{q-1}\bar{x}_q}{c_q} \right) \right| = \left| \bar{T}_p \left(\sum_{q=1}^{p-1} \frac{2^{q-1}\bar{x}_q}{c_q} + \sum_{q=p+1}^{\infty} \frac{2^{q-1}\bar{x}_q}{c_q} + \frac{2^{p-1}\bar{x}_p}{c_p} \right) \right|$$

$$\geqslant \left| \bar{T}_p \frac{2^{p-1}\bar{x}_p}{c_p} \right| - \sum_{q=1}^{p-1} \left| \bar{T}_p \frac{2^{q-1}\bar{x}_q}{c_q} \right| - \sum_{q=p+1}^{\infty} \left| \bar{T}_p \frac{2^{q-1}\bar{x}_q}{c_q} \right|$$

$$\geqslant 2^{p-1} - \tfrac{1}{2} \sum_{q=1}^{p-1} 2^{q-1} - \tfrac{3}{2}\, 2^{p-1}(\tfrac{1}{4} + \tfrac{1}{16} + \ldots),$$

using (ii), (iii) and (iv).

Hence

$$|\bar{T}_p \bar{x}| \geqslant 2^{p-1} - \tfrac{1}{2}(2^{p-1} - 1) - \tfrac{1}{2}\, 2^{p-1} \geqslant \tfrac{1}{2}.$$

This shows that $|\bar{T}_p \bar{x}|$ cannot tend to 0, and hence that, if a sequence of bounded linear operators (T_p) tends to 0, then the sequence is collectively bounded.

It follows from (23), which has already been shown to apply, that the $A_{k,n}$ are collectively bounded for very n, *i.e.* there are positive numbers a_n $(n = 1, 2, \ldots)$, so that $|A_{k,n}| \leqslant a_n$ $(k = 1, 2, \ldots)$.

We are now in a position to show that condition (25) is necessary. For this purpose we assume that it is not in fact satisfied, and construct a sequence (s_n) tending to 0 whose transform by A does not tend to 0.

By hypothesis, there is a sequence of positive integers (k_r) $(r = 1, 2, \ldots, k_1 < k_2 < \ldots)$, such that $\lim_{r\to\infty} a_{k_r} = \infty$, where $a_k = |A_{k,1}, A_{k,2}, \ldots|$. This implies the existence of sequences $(x_n^{(r)})$ $(r = 1, 2, \ldots)$, such that

$$\lim_{r\to\infty} \left| \sum_{n=1}^{\infty} A_{k_r, n} x_n^{(r)} \right| = \infty,$$

where the terms $x_n^{(r)}$ for given r differ from 0 only for a finite number of suffices n, and where $|x_n^{(r)}| \leqslant 1$ for all r and n. Thus, there is a sequence of positive integers (N_r) so that $x_n^{(r)} = 0$ for $n > N_r$. Obviously we may choose the sequence (N_r) so that $N_{r+1} \geqslant N_r$ for all $r \geqslant 1$. We now put $\bar{a}_{k_r} = \left| \sum_{n=1}^{\infty} A_{k_r, n} x_n^{(r)} \right|$, and we may assume (if necessary by the omission of a number of the k_r) that $\bar{a}_{k_r} > 0$ for all r and that $\bar{a}_{k_r} \uparrow \infty$ monotonically.

If now we put $\bar{x}_n^{(r)} = \dfrac{1}{(\sqrt{\bar{a}_{k_r}})} x_n^{(r)}$, then $\bar{x}_n^{(r)} \to 0$ as $r \to \infty$ for every given n, and moreover, since $|x_n^{(r)}| \leqslant 1$ for all n and r, it follows that $\bar{x}_n^{(r)} \to 0$ as $r \to \infty$ uniformly (*i.e.* independently of n).

In order to define (s_n) we first determine s_n for $1 \leqslant n \leqslant n_1$ by $s_n = x_n^{(1)}$ where $n_1 = N_1$. We also put $\bar{k}_1 = k_1$ and $r_1 = 1$.

We next choose $r_2 > r_1$ so that, if $\bar{k}_2 = k_{r_2}$, then $\bar{a}_{\bar{k}_2} \geqslant \left[4\left(\sum_{n=1}^{n_1} a_n\right)\right]^2$ and $\bar{a}_{\bar{k}_2} \geqslant 16\bar{a}_{\bar{k}_1}$. We define s_n by $s_n = x_n^{(r_1)}$ for $n_1 < n \leqslant n_2$, where $n_2 = N_{r_2}$.

We next choose $r_3 > r_2$ so that, if $\bar{k}_3 = k_{r_3}$, then $\bar{a}_{\bar{k}_3} \geqslant \left[4\left(\sum_{n=1}^{n_2} a_n\right)\right]^2$ and $\bar{a}_{\bar{k}_3} \geqslant 16\bar{a}_{\bar{k}_2}$. We define s_n by $s_n = x_n^{(r_3)}$ for $n_2 < n \leqslant n_3$, where $n_3 = N_{r_3}$, etc.

Having defined $r_1, \ldots, r_{p-1}$, and thereby $\bar{k}_1, \ldots, \bar{k}_{p-1}, n_1, \ldots, n_{p-1}$, and s_n for $1 \leqslant n \leqslant n_{p-1}$, we choose $r_p > r_{p-1}$ so that, if $\bar{k}_p = k_{r_p}$, then

$$\bar{a}_{\bar{k}_p} \geqslant \left[4\left(\sum_{n=1}^{n_{p-1}} a_n\right)\right]^2 \quad \text{and} \quad \bar{a}_{\bar{k}_p} \geqslant 16\bar{a}_{\bar{k}_{p-1}}.$$

We then define s_n by $s_n = x_n^{(r_p)}$ for $n_{p-1} < n \leqslant n_p$, where $n_p = N_{r_p}$, etc.

The resulting sequence (s_n) tends to 0, and so the matrix A applies to it. On the other hand we are going to show that the transformed sequence (t_k) does not tend to 0. We have in fact, for integral $p > 1$,

$$t_{\bar{k}_p} = \sum_{n=1}^{\infty} A_{\bar{k}_p, n} s_n = \sum_{n=1}^{n_{p-1}} A_{\bar{k}_p, n} s_n + \sum_{n=n_{p-1}}^{n_p} A_{\bar{k}_p, n} s_n + \sum_{n=n_p+1}^{\infty} A_{\bar{k}_p, n} s_n$$

$$= \sum_{n=1}^{n_{p-1}} A_{\bar{k}_p, n} s_n + \sum_{n=n_{p-1}}^{n_p} A_{\bar{k}_p, n} \bar{x}_n^{(r_p)} + \sum_{n=n_p+1}^{\infty} A_{\bar{k}_p, n} s_n$$

$$= \sum_{n=1}^{n_{p-1}} A_{\bar{k}_p, n} (s_n - \bar{x}_n^{(r_p)}) + \sum_{n=1}^{\infty} A_{\bar{k}_{r_p}, n} \bar{x}_n^{(r_p)} + \sum_{n=n_p+1}^{\infty} A_{\bar{k}_p, n} s_n$$

by the definition of (s_n) and taking into account that $\bar{x}_n^{(r_p)} = 0$ for $n > n_p$. Hence

$$|t_{\bar{k}_p}| \geqslant \left| \sum_{n=1}^{\infty} A_{k_{r_p}, n} x_n^{(r_p)} \right| - \left| \sum_{n=1}^{n_{p-1}} A_{\bar{k}_p, n} (s_n - \bar{x}_n^{(r_p)}) \right| - \left| \sum_{n=n_p+1}^{\infty} A_{\bar{k}_p, n} s_n \right|$$

$$\geqslant \frac{1}{\sqrt{(\bar{a}_{k_{r_p}})}} \left| \sum_{n=1}^{\infty} A_{k_{r_p}, n} x_n^{(r_p)} \right| - \left(\sum_{n=1}^{n_{p-1}} |A_{\bar{k}_p, n}| \right) (|s_n| + |\bar{x}_n^{(r_p)}|)$$

$$\frac{1}{\sqrt{(\bar{a}_{k_{p+1}})}} \left| \sum_{n=n_p+1}^{\infty} A_{\bar{k}_p, n} \{\sqrt{(\bar{a}_{k_{p+1}})}\, s_n\} \right|.$$

Now, by our definitions, $|s_n| \leqslant 1$, $|\bar{x}_n^{(r_p)}| \leqslant 1$, $\sum_{n=1}^{n_p-1} |A_{\bar{k}_p, n}| \leqslant \frac{1}{4}\sqrt{(\bar{a}_{\bar{k}_p})}$, $|\sqrt{(\bar{a}_{k_p+1})}\, s_n| \leqslant 1$ for $n > n_p$ and $\sqrt{(\bar{a}_{\bar{k}_{p+1}})} \geqslant 4\sqrt{(\bar{a}_{\bar{k}_p})}$. Hence

$$|t_{\bar{k}_p}| \geqslant \frac{1}{\sqrt{(\bar{a}_{\bar{k}_p})}}\,\bar{a}_{\bar{k}_p} - 2\cdot\tfrac{1}{4}\sqrt{(\bar{a}_{\bar{k}_p})} - \frac{1}{4\sqrt{(\bar{a}_{\bar{k}_p})}}\,\bar{a}_{\bar{k}_p} = \tfrac{1}{4}\sqrt{(\bar{a}_{\bar{k}_p})}.$$

This shows that a subsequence of (t_k) is unbounded. We conclude that condition (26) is in fact necessary.

Theorem IV is the counterpart of the theorem of Toeplitz and Silverman (Theorem I in this paper) for numerical sequences. Theorem IV is not only more general, but also somewhat deeper than Theorem I, mainly because for abstract spaces $\lim_{p\to\infty} T_p = 0$ does not entail $\lim_{p\to\infty} |T_p| = 0$. In fact, let a be the space of all complex sequences that tend to 0, $x = (x_1, x_2, \ldots)$, $\lim_{n\to\infty} x_n = 0$. Define the metric of the space by $|x| = \max_{n=1,2,\ldots} |x_n|$. Now consider the sequence of linear operators (T_p) given by the matrices $T_p = \|\tau_{k,n}^{(p)}\|$, where $\tau_{p,p}^{(p)} = 1$, and $\tau_{k,n}^{(p)} = 0$ in all other cases. It will be readily seen that $|T_p| = 1$ for all p so that $\lim_{p\to\infty} |T_p| \neq 0$. On the other hand, given any $x \varepsilon S$, $x = (x_1, x_2, \ldots)$, we have $x^{(p)} = T_p x = (x_1^{(p)}, x_2^{(p)}, x_3^{(p)}, \ldots)$, where $x_q^{(p)} = x_p$ for $p = q$ and $x_q^{(p)} = 0$ for $p \neq q$. Hence $|x^{(p)}| = |x_p|$ and so $\lim_{p\to\infty} |x^{(p)}| = 0$, *i.e.* $\lim_{p\to\infty} T_p = 0$.

The above example shows that, while a matrix which transforms every convergent sequence in S into a convergent sequence with the same limit must, by Theorem IV, satisfy condition (23), it need not satisfy the stronger condition $\lim_{k\to\infty} |A_{k,n}| = 0$.

3. *Series-to-sequence transformations in abstract spaces.*

The main object of this section is the proof of Theorem V, corresponding to Theorem II for numerical sequences.

For any infinite series $\sum_{n=1}^{\infty} a_n$ of elements of a complete linear metric space S, consider the sequence (t_k) given by

$$t_k = \sum_{n=1}^{\infty} B_{k,n} a_n \quad (k = 1, 2, \ldots), \tag{27}$$

where the $B_{k,n}$ are linear operators in S, taking values in S. We write $B = \| B_{k,n} \|$.

THEOREM V. *A necessary and sufficient condition for the matrix of B of (bounded) linear operators to transform any convergent series into a convergent sequence whose limit equals the sum of the series is*

$$|B_{k,1}-B_{k,2},\ B_{k\,2}-B_{k,3},\ \ldots| = b_k \leqslant M < \infty, \tag{28}$$

where M is a constant independent of k,

$$\lim_{k\to\infty} B_{k,n} = I \quad (n = 1, 2, \ldots). \tag{29}$$

The conditions are sufficient.

We first show that all the elements of B are uniformly bounded. In fact, for any k and n

$$b_{k,n} = |B_{k,n}| = |(B_{k,n}-B_{k,n-1})+(B_{k,n-1}-B_{k,n-2})+\cdots$$
$$+(B_{k,2}-B_{k,1})+B_{k,1}|$$
$$\leqslant |B_{k,1}-B_{k,2},\ B_{k,2}-B_{k,3},\ \ldots,\ B_{k,n-1}-B_{k,n}|+|B_{k,1}| \leqslant b_k+b_{k,1}$$
$$\leqslant M+b_{k,1}.$$

On the other hand, by (29), $\lim_{k\to\infty}(B_{k,1}-I) = 0$, and it follows, as in the proof of Theorem IV, that the operators $B_{k,1}-I$, and thence the operators $B_{k,1}$, are in fact uniformly bounded, $b_{k,1} \leqslant b < \infty$ $(k = 1, 2, \ldots)$. Hence

$$b_{k,n} \leqslant b+M = P \quad \text{for} \quad k = 1, 2, \ldots, n = 1, 2, \ldots.$$

Now let $\sum_{n=1}^{\infty} a_n$ be a convergent series of elements of S so that $\sum_{n=1}^{\infty} a_n = 0$. We write $\sum_{p=1}^{n} a_p = s_n$ $(a = 1, 2, \ldots)$, and we then have, for arbitrary k and m, "Abel's transformation"

$$\sum_{n=1}^{m} B_{k,n} a_n = \sum_{n=1}^{m-1} (B_{k,n}-B_{k,n+1}) s_n + B_{k,m} s_m, \tag{30}$$

and similarly

$$\sum_{n=m}^{m+p} B_{k,n} a_n = \sum_{n=m}^{m+p-1} (B_{k,n}-B_{k,n+1}) s_n + B_{k,m+p} s_{m+p} - B_{k,m-1} s_{m-1}.$$

Now $s_n \to 0$ by assumption so that, for sufficiently large n, $n > N_0(\epsilon)$, say, $|s_n| < \epsilon$ where ϵ is a positive quantity specified in advance. Hence for $m > N(\epsilon)$

$$\left|\sum_{n=m}^{m+p} B_{k,n} a_n\right| \leqslant \left|\sum_{n=m}^{m+p-1} (B_{k,n} - B_{k,n+1}) s_n\right| + |B_{k,m+p} s_{m+p}| + |B_{k,m-1} s_{m-1}|$$

$$\leqslant b_k \epsilon + 2P\epsilon = (b_k + 2P)\epsilon.$$

This shows that $\sum_{n=1}^{\infty} B_{k,n} a_n$ converges, so that B does in fact apply to $\sum_{n=1}^{\infty} a_n$. It also follows from (30) that

$$t_k = \sum_{n=1}^{\infty} B_{k,n} a_n = \sum_{n=1}^{\infty} (B_{k,n} - B_{k,n+1}) s_n,$$

and from (29) that

$$\lim_{k\to\infty} (B_{k,n} - B_{k,n+1}) = 0 \quad (n = 1, 2, \ldots).$$

Now for given $\epsilon > 0$ choose $N_0 = N_0(\epsilon)$ as before, so that $|s_n| < \epsilon$ for $n > N_0$, and then K_0 so that $|(B_{k,n} - B_{k,n+1}) s_n| < \epsilon/N_0$ for all $n \leqslant N_0$, $k > K_0$. Then, for any $k > K_0$,

$$|t_k| = \left|\sum_{n=1}^{N_0} (B_{k,n} - B_{k,n+1}) s_n + \sum_{n=N_0+1}^{\infty} (B_{k,n} - B_{k,n+1}) s_n\right|$$

$$\leqslant \sum_{n=1}^{N_0} |(B_{k,n} - B_{k,n+1}) s_n| + \left|\sum_{n=N_0+1}^{\infty} (B_{k,n} - B_{k,n+1}) s_n\right| \leqslant N_0 \frac{\epsilon}{N_0} + M\epsilon$$

$$= (1+M)\epsilon.$$

Hence $\lim_{k\to\infty} t_k = 0$ as required.

Next, assume $\sum_{n=1}^{\infty} a_n = s \neq 0$, and put $a_1' = a_1 - s$, $an' = a_n$ for $n \neq 1$. Then $\sum_{n=1}^{\infty} a_n' = 0$, and $\sum_{n=1}^{m} B_{k,n} a_n = \sum_{n=1}^{m} B_{k,n} a_n' + B_{k,1} s$ so that $\sum_{n=1}^{\infty} B_{k,n} a_n$ converges, and $t_k = \sum_{n=1}^{\infty} B_{k,n} a_n = \sum_{n=1}^{\infty} B_{k,n} a_n' + B_{k,1} s$.

Now $\lim_{k\to\infty} \sum_{n=1}^{\infty} B_{k,n} a_n' = 0$ as shown above, while $\lim_{k\to\infty} B_{k,1} s = s$, by (29). Hence $\lim_{k\to\infty} t_k = s$, showing that the conditions are sufficient.

The conditions are necessary.

To see the necessity of (29), consider the convergent infinite series $\sum_{n=1}^{\infty} a_n$ where $a_n = 0$ except for a specified n, $n = n_0$ for which $a_n = x$, $x \varepsilon S$ arbitrary, then $\sum_{n=1}^{\infty} a_n = x$, and $t_k = B_{k, n_0} x$, so that we must have $\lim_{k\to\infty} B_{k, n_0} x = x$, *i.e.* $\lim_{k\to\infty} B_{k, n_0} = I$.

To prove that condition (28) is necessary, we show at first, in a way similar to that used in the proof of Theorem III, that

$$|B_{k,1}-B_{k,2},\ B_{k,2}-B_{k,3},\ \ldots| = b_k < \infty$$

for all k ($k = 1, 2, \ldots$). As the first part of the proof of the present theorem shows, this in itself is sufficient to ensure that the matrix B applies to every convergent series, *i.e.* that (t_k) exists. It also follows, as before, that

$$t_k = \sum_{n=1}^{\infty} B_{k,n} a_n = \sum_{n=1}^{\infty} (B_{k,n}-B_{k,n+1}) s_n$$

provided $\sum_{n=1}^{\infty} a_n = 0$. Putting $B_{k,n}-B_{k,n+1} = A_{k,n}$, the matrix $A_{k,n}$ then has the property to transform every sequence (s_n) that converges to 0 into a sequence (t_k), $t_k = \sum_{n=1}^{\infty} A_{k,n} s_n$, with the same limit. On the other hand since, as already shown, B satisfies (29), $A = \|A_{k,n}\|$ must satisfy (23). Assuming now that B does not satisfy (28), it follows that A does not satisfy (26), and on this assumption we may, as in the proof of the necessity of condition (26) in Theorem IV, construct a sequence (s_n) so that $\lim s_n = 0$ but not $\lim_{k\to\infty} t_k = 0$. This is against the assumption, showing that (28) must be satisfied.

4. *Unbounded linear matrices. Semi-continuous matrices.*

So far the discussion has been confined to linear operators in the restricted sense, viz., to bounded linear operators. We may discard this assumption, and consider operators which are not a priori bounded. We then have, in place of Theorem III,

Theorem VI. *Given an infinite sequence of (bounded or unbounded) linear operators* (T_p), *a necessary and sufficient condition for* $\sum_{p=1}^{\infty} T_p s_p$ *to con-*

verge whenever (s_p) *converges is*

$$(31) \qquad |T_{n_0}, T_{n_0+1}, \ldots, T_p, \ldots| = M < \infty$$

for some positive integer n, and

$$(32) \qquad \sum_{p=1}^{\infty} T_p = T \text{ exists.}$$

The proof is similar to that of Theorem III. Theorem IV is now replaced by

THEOREM VII. *In order that* (t_k) *as given by* (15) (*where the* $A_{k,n}$ *may be bounded or unbounded*) *exist and converge to the same limit as* (s_n) *whenever* (s_n) *converges, it is necessary and sufficient that*

$$(33) \qquad \lim_{k\to\infty} A_{kn} = 0 \quad (n = 1, 2, \ldots),$$

$$(34) \qquad \sum_{n=1}^{\infty} A_{k,n} = A_k \quad \text{exists for} \quad k = 1, 2, \ldots,$$

$$(35) \qquad \lim_{k\to\infty} A_k = I,$$

$$(36) \qquad |A_{k,n_0}, A_{k,n_0+1}, A_{k,n_0+2}, \ldots| = a_k \leqq M < \infty,$$

where M and n_0 *are independent of k.*

These conditions are identical with the conditions of Theorem IV except for (36) which replaces (26). The only novel feature in the proof is the argument showing that (36) is necessary. The argument is divided into two steps; the existence of constants $n_0 = n_0(k)$ so that

$$(36') \qquad |A_{k,n_0(k)}, A_{k,n_0(k)+1}, \ldots| = a_k < \infty$$

is first established as being due solely to the fact that A applies to (s_n). The fact that n_0 can be chosen independently of k is then proved by the construction of a suitable convergent sequence whose transform does not converge to the same limit, on the assumption that (36′) is, but (36) is not, satisfied.

The corresponding problem for series-to-sequence transformations by matrices of general (bounded or unbounded) linear operators is slightly more involved. Before coming to the generalization of Theorem V we are going to prove in some detail

THEOREM VIII. *Given an infinite sequence of (bounded or unbounded) linear operators,* (T_p), *a necessary and sufficient condition for* $\sum_{p=1}^{\infty} T_p a_p$ *to converge whenever* $\sum_{p=1}^{\infty} a_p$ *converges is*

$$|T_{p_0}| = t_{p_0} < \infty \text{ for some positive integer } p_0 \tag{37}$$

and

$$|T_{p_0} - T_{p_0+1}, T_{p_0+1} - T_{p_0+2}, \ldots| = t < \infty. \tag{38}$$

To prove sufficiency, we show first of all that $|T_p| = t_p < \infty$ for all $p > p_0$. In fact, for any positive integer m,

$$\begin{aligned} t_{p_0+m} = |T_{p_0+m}| = |T_{p_0} - [(T_{p_0} - T_{p_0+1}) + (T_{p_0+1} - T_{p_0+2}) + \ldots \\ + (T_{p_0+m-1} - T_{p_0+m})]| \\ \leqslant |T_{p_0}| + |T_{p_0} - T_{p_0+1}, T_{p_0+1} - T_{p_0+2}, \ldots, T_{p_0+m-1} - T_{p_0+m}, \\ T_{p_0+m} - T_{p_0+m+1}, \ldots| = t_{p_0} + t. \end{aligned}$$

Given a convergent series $\sum_{p=1}^{\infty} a_p = s$ we have to show that, for sufficiently large q and for any positive integers m, $\left|\sum_{p=q}^{q+m} T_p a_p\right| < \epsilon$ where ϵ is a positive quantity specified in advance. We assume first $\sum_{p=1}^{\infty} a_p = s = 0$; then, as in §3 above,

$$\sum_{p=q}^{q+m} T_p a_p = \sum_{p=q}^{q+m} (T_p - T_{p+1}) s_n + T_{q+m} s_{q+m} - T_{q-1} s_{q-1},$$

and so

$$\left|\sum_{p=q}^{q+m} T_p a_p\right| \leqslant \left|\sum_{p=q}^{q+m} (T_p - T_{p+1}) s_n\right| + |T_{q+m}||s_{q+m}| + |T_{q-1}||s_{q-1}|.$$

Given any positive η, there is a positive integer p_1, so that $|s_p| < \eta$ for $p > p_1$. Then, for all $q > \max(p_0, p_1)$ and for all positive integers m,

$$\left| \sum_{p=q}^{q+m} T_p a_p \right| \leqslant t\eta + t_{q+m}\eta + t_{q-1}\eta = (t + t_{q+m} + t_{q-1})\eta \leqslant (3t + 2t_{p_0})\eta.$$

This shows that $\sum_{p=1}^{\infty} T_p a_p$ converges. Again, for general convergent $\sum_{p=1}^{\infty} a_p = s$ we put $a_1' = a_1 - s$, $a_p' = a_p$, $p > 1$, and hence

$$\sum_{p=1}^{\infty} T_p a_p = \sum_{p=1}^{\infty} T_p a_p' + T_1 s.$$

To prove necessity, we are going to show that only a finite number of the T_p can possibly be unbounded. Assuming the contrary, there would be a sequence of positive integers, (p_k), $(k = 1, 2, 3, \ldots, p_1 < p_2 < p_3 < \ldots)$, and a sequence (x_k) of elements of S so that $|T_{p_k} x_k| > 2^k$ while $|x_k| \leqslant 1$ $(k = 1, 2, \ldots)$. Then the series $\sum_{p=1}^{\infty} a_p$ defined by $a_p = 0$ unless $p = p_k$ for some k, in which case $a_{p_k} = \frac{1}{2^k} x_k$, would be a convergent series $\left(\text{since } |a_{p_k}| \leqslant \frac{1}{2^k}\right)$ while $\sum_{p=1}^{\infty} T_p a_p$ could not possibly converge since $|T_{p_k} a_{p_k}| \geqslant 1$. This shows that only a finite number of T_p can be unbounded and that there is a positive integer p_0 so that $|T_p| = t_p < \infty$ for $p \geqslant p_0$. In particular $|T_{p_0}| = t_{p_0} < \infty$ in accordance with (37).

An exactly similar method can be used in order to show that the operators T_p $(p \geqslant p_0)$ are uniformly bounded so that there is a positive constant $\bar{t}$ for which $|T_p| \leqslant \bar{t}$ $(p = p_0, p_0+1, p_0+2, \ldots)$. It follows that $|T_p - T_{p+1}| \leqslant 2\bar{t}$ $(p = p_0, p_0+1, \ldots)$. Finally, we are going to show that condition (38) must also be satisfied for the same constant p_0.

Assume on the contrary that $|T_{p_0} - T_{p_0+1}, T_{p_0+1} - T_{p_0+2}, \ldots| = \infty$. In that case there exist infinite sequences of positive (p_k) and (m_k) $(k = 1, 2, 3, \ldots, p_0 \leqslant p_1 < p_2 < \ldots, m_k > 0)$ and elements of S, $x_n^{(k)}$ $(n = 0, \ldots, m_k, |x_n^{(k)}| \leqslant 1)$ for all n and k so that

$$\left| \sum_{n=0}^{m_k} (T_{p_k+n} - T_{p_k+n+1}) x_n^{(k)} \right| \geqslant 2^k, \tag{39}$$

and since the operators $(T_{p_k+n} - T_{p_k+n+1})$ $(k = 1, 2, \ldots, n = 0, 1, \ldots, m_k)$

are all bounded, we may even assume that the p_k have been selected in such a way that $p_{k+1} > p_k + m_k$ ($k = 1, 2; \ldots$).

We now define a sequence (s_q) by $s_q = 0$ except when $q = p_k + n$, $0 \leqslant k \leqslant m_k$ for some k ($k = 1, 2, \ldots$), in which case we define s_q by

$$s_q = \frac{1}{2^k} x_n^{(k)}.$$

We also define the infinite series $\sum_{p=1}^{\infty} a_p$ by $a_1 = s_1, a_p = s_p - s_{p-1}$ ($p = 2, 3, \ldots$), so that the s_q are the partial sums of the series, $s_q = \sum_{p=1}^{q} a_p$. It follows from the definition of the sequence (s_q) that $\lim_{q\to\infty} s_q = 0$ so that $\sum_{p=1}^{\infty} a_p = 0$. Also, for sufficiently high q, $q \geqslant \bar{p}$, say, $|s_q| \leqslant 1/(4\bar{t})$. Hence for any $p_k > \bar{p}$,

$$\left|\sum_{p=p_k}^{p_k+m_k} T_p a_p\right| = \left|\sum_{p=p_k}^{p_k+m_k} (T_p - T_{p+1}) s_p + T_{p_k+m_k} s_{p_k+m_k} - T_{p_k-1} s_{p_k-1}\right|$$

$$\geqslant \left|\sum_{p=p_k}^{p_k+m_k} (T_p - T_{p+1}) \frac{1}{2^k} x_{p-p_k}^{(k)}\right| - |T_{p_k+m_k}||s_{p_k+m_k}| - |T_{p_k-1}||s_{p_k-1}|$$

$$\geqslant \frac{2^k}{2^k} - 2\bar{t}\frac{1}{4\bar{t}} = \tfrac{1}{2}.$$

It follows that $\sum_{p=1}^{\infty} T_p a_p$ does not converge.

By similar arguments we may prove, in amplification of Theorem V:

THEOREM IX. *A necessary and sufficient condition for a matrix B of (bounded or unbounded) linear operators to transform any convergent series into a convergent sequence whose limit equals the sum of the series is*

$$(40)\qquad |B_{k,n_0}| = b_{k,n_0} \leqslant P \textit{ for some } n_0 \textit{ and } P.$$

where n_0 and P are constants independent of k,

$$(41)\qquad |B_{k,n_0} - B_{k,n_0+1}, B_{k,n_0+1} - B_{k,n_0+2}, \ldots| = b_k \leqslant M < \infty,$$

where M is a constant independent of k,

$$(42)\qquad \lim_{k\to\infty} B_{k,n} = I \quad (n = 1, 2, \ldots).$$

It is logically obvious, and can easily be verified, that matrices of bounded linear operators, which satisfy the conditions of Theorems IV and V respectively, also satisfy the conditions of Theorems VII and IX respectively. As in the theory of linear transformations of numerical sequences, a matrix satisfying the conditions of Theorem VII will be said to be a T-matrix, and a matrix satisfying the conditions of Theorem IV will be said to be a Γ-matrix.

A considerable proportion of the general theory of numerical T- or Γ-matrices appears to extend (with some modifications if we admit unbounded operators) to matrices of linear operators. For example, the product to two T-matrices exists and is a T-matrix. And if two T-matrices are commutative then they are consistent at least for bounded sequences. Again, a necessary and sufficient condition for two T-matrices

$$\| A_{k,n} \| \text{ and } \| B_{k,n} \|$$

of (bounded or unbounded) linear operators to be absolutely equivalent for bounded sequences† is the existence of a positive integer n_0 independent of k so that

$$\lim_{k \to \infty} | A_{k,n_0} - B_{k,n_0}, A_{k,n_0+1} - B_{k,n_0+1}, \ldots | = 0.$$

More generally, we might consider matrices of linear operators from one space to another, and develop the subject in conformity with the theory of Kojima and Schur for numerical matrices.

As stated in the introduction, there is an exactly similar theory for semi-continuous matrices. In fact some of the most interesting applications of the whole idea underlying the present paper are most naturally formulated in terms of semi-continuous matrices of bounded operators. For such matrices, the transform of a series $\sum_{n=1}^{\infty} a_n$ is defined by

$$t(\omega) = \sum_{n=1}^{\infty} B_n(\omega) a_n, \tag{43}$$

where the operators $B_n(\omega)$ are specified for $n = 1, 2, \ldots$ and (i) for all real ω ($\omega > \omega_0$) or (ii) for all real ω ($0 < |\omega| < \omega_0$), where ω_0 is some positive quantity independent of n in both cases. For case (i) the counterpart

† R. G. Cooke, "On mutual consistency and regular T-limits", *Proc. London Math Soc.* (2), 41 (1936) 113-125.

of Theorem V is

Theorem X. *A necessary and sufficient condition for the semi-continuous matrix of (bounded) linear operators* $\| B_n(\omega) \|$ *to transform every convergent series into a function* $t(\omega)$ *as given by* (43), *so that* $\lim_{\omega \to \infty} t(\omega) = \sum_{n=1}^{\infty} a_n$, *is*

$$(44) \qquad | B_1(\omega) - B_2(\omega),\ B_2(\omega) - B_3(\omega),\ \ldots | = b(\omega) \leqslant M < \infty$$

for all $\omega > \omega_0$ *where* M *is a constant independent of* ω,

$$(45) \qquad \lim_{\omega \to \infty} B_n(\omega) = I \quad (n = 1, 2, \ldots).$$

The corresponding theorem for case (ii) is

Theorem XI. *A necessary and sufficient condition for the semi-continuous matrix of (bounded) linear operators* $\| B_n(\omega) \|$ *to transform every convergent series into a function* $t(\omega)$ *as given by* (43) *so that* $\lim_{\omega \to 0} t(\omega) = \sum_{n=1}^{\infty} a_n$ *is*

$$(46) \qquad | B_1(\omega) - B_2(\omega),\ B_2(\omega) - B_3(\omega),\ \ldots | = b(\omega) \leqslant M < \infty$$

for all $0 < |\omega| < \omega$, *where* M *is a constant independent of* ω,

$$(47) \qquad \lim_{\omega \to 0} B_n(\omega) = I \quad (n = 1, 2, \ldots).$$

Various applications of these theorems will be discussed in the following section.

5. *Applications.*

A very simple yet far-reaching method of producing semi-continuous Γ-matrices of (bounded) linear operators in a complete metric linear space is as follows:

Let T be a (bounded or unbounded) linear operator defined for the elements of a complete metric linear space S' and taking all the elements of S as values once and only once. In these circumstances T has an inverse linear operator, T^{-1} and we shall assume that T^{-1} is bounded. Further, let $T(\omega)$ be defined for (i) $\omega > \omega_1$, or for (ii) $0 < |\omega| < \omega_0$, where ω_0 is a positive constant, as a function of bounded linear operators in S, and such that

(i) $\lim_{\omega\to\infty} T_\omega = T$ or (ii) $\lim_{\omega\to 0} T(\omega) = T$. Then the matrix $\| B_n(\omega) \|$ defined by

$$B_1(\omega) = B_2(\omega) = \ldots = T(\omega)\, T^{-1}$$

satisfies the conditions (i) of Theorem X, or (ii) of Theorem XI respectively. In fact, $B_n(\omega) - B_{n+1}(\omega) = 0$ $(n = 1, 2, \ldots)$, so that condition (44) [or (46)] is satisfied implicitly. Also

$$\lim B_n(\omega) = \lim [T(\omega)\, T^{-1}] = [\lim T(\omega)]\, T^{-1} = T \,.\, T^{-1} = I,$$

in agreement with condition (45) [or (47)].

Alternatively, we may assume that S' is not *a priori* a metric space, but a general linear convergence space, *i.e.* a space whose elements form an Abelian group with respect to addition, and for which a concept of "limit" is defined so that for some sequences (s_n') of elements of S' there is an element $s' \varepsilon S'$, which is the limit of (s_n'), $\lim_{n\to\infty} s_n' = s$. The requirement of boundedness in respect of T^{-1} and $T(\omega)$ is now replaced by the requirement of continuity, *e.g.* $\lim_{n\to\infty} T^{-1} s_n = T^{-1} s$ if $\lim_{n=\infty} s_n = s$. However, the existence of the one-one continuous transformation T^{-1} implies that S can in fact be metricized by putting $|s'| = |Ts'|$, so that the limits defined by this metric coincide with the limits given *a priori*.

It will be seen that an example for the procedure described above is provided by Borel's method as interpreted in the introduction to the present paper. However, before we can apply the results of the preceding section directly to Borel's method, we have to define the spaces S and S' and the metrics imposed on them.

We define S as the space of all functions $f(p)$ of a complex variable which are regular and bounded on and outside a specified circle $|p| = R$. In particular, the functions $f(p)$ are supposed to be regular at infinity. We define the norm of f by $|f| = \max_{|p|=R} |f(p)|$. With this definition S becomes a complete metric linear space.

Now consider the function

$$\phi(x) = \frac{1}{2\pi i} \int e^{xp} \frac{f(p)}{p}\, dp, \tag{48}$$

where f is an arbitrary element of S and the path of integration is any circle of radius $r \geqslant R$. $\phi(x)$ is an integral function of order 1 and of normal type smaller than R. Conversely, every integral function of order 1 and of normal type smaller than R can be obtained in that way†. The system of all these functions constitutes a linear space S', and (48) establishes a one-one correspondence between S and S'. The inverse transformation is given by

$$(49) \qquad f(p) = p\int_0^\infty e^{-px}\phi(x)\,dx, \quad \text{valid for } \Re p > R,$$

i.e. by Carson's integral. We write this transformation $f = C\phi$, and so (48) may be written symbolically $\phi = C^{-1}f$.

The metric defined in S leads to a convergence concept, $\lim_{n\to\infty} f_n = f$, which is equivalent to $\lim_{n\to\infty} f_n(p) = f(p)$ for all $|p| \geqslant R$. Similarly, if we define a metric in S' by $|\phi| = |C\phi|$, then $\lim_{n\to\infty} \phi_n = \phi$ is equivalent to $\lim_{n\to\infty} \phi_n(x) = \phi(x)$ for all x, and S' becomes a complete metric linear space. It follows from the definition of the metric in S that both C and C^{-1} are bounded, $|C| = |C^{-1}| = 1$. Also, considering the linear transformations $C(\omega)$ defined by

$$(50) \qquad f = C(\omega)\phi = p\int_0^\omega e^{-px}\phi(x)\,dx \quad (\omega > 0),$$

we see that the operators $C(\omega)$ are continuous and therefore bounded, and that $\lim_{\omega\to\infty} C(\omega) = C$. Hence, the semi-continuous matrix $B = \| B_n(\omega) \|$, given by $B_1(\omega) = B_2(\omega) = \ldots = C(\omega)C^{-1}$, is a Γ-matrix and so, by Theorem X, whenever the infinite series $\sum_{n=1}^{\infty} f_n$ converges, $\sum_{n=1}^{\infty} f_n = f$, $f_n \in S$ $(n = 1, 2, \ldots)$, then $\lim_{\omega\to\infty} \sum_{n=1}^{\infty} B_n(\omega) f_n = f$. In this statement "convergence" is equivalent to ordinary convergence everywhere outside $|p| = R$. In particular, if $f_n = \frac{a_{n-1}}{p^{n-1}}$ $(n = 1, 2, \ldots)$, where the a_n are arbitrary constants, then

$$\sum_{n=1}^{\infty} B_n(\omega) f_n = C(\omega)\left(\sum_{n=1}^{\infty} C^{-1} f_n\right) = C(\omega)\left(\sum_{n=0}^{\infty} \frac{a_n x^n}{n!}\right) = C(\omega)\phi(x).$$

And so

$$f(p) = \lim_{\omega\to\infty} C(\omega)\phi = p\int_0^\infty e^{-px}\phi(x)\,dx,$$

† See G. Doetsch, *Theorie und Anwendung der Laplace Transformation* (Berlin, 1937). 77.

where the integral on the right-hand side converges for $\Re p > R$. Substituting $p = \frac{1}{z}$, $px = t$, we obtain

$$f\left(\frac{1}{z}\right) = \int_0 e^{-t}\phi(tz)\,dt \quad \left[\Re\left(\frac{1}{z}\right) > R\right],$$

where the path of integration on the right-hand side is however, in general, complex. If instead we take the positive real axis as the path of integration, we obtain Borel's integral, but the precise range of efficiency of that integral is outside the scope of the general method presented in this paper.

Another interesting application can be made by taking S as the space of all functions $f(p)$ that are regular for $\left|1-\frac{1}{p}\right| \leqslant R$, where R is a positive constant greater than one. That is to say, the functions $f(p)$ are regular on and outside a circle of radius $\frac{R}{R^2-1}$, whose centre is on the negative part of the real axis at a distance $\frac{1}{R^2-1}$ from the origin. We define the modulus of f by $|f| = \max_{|1-(1/p)|=R} |f(p)|$. The space S so defined is again a complete metric linear space. The space S' is taken to be the space of all functions $\phi(x)$ obtained from the elements of S by applying the transformation (48). We again write $\phi = C^{-1}f$ so that $f = C(\phi)$ is given by Carson's integral (49), valid for $|p| > \frac{1}{R^2+1}$. The functions of S are all integral functions of order 1 and of normal type smaller than

$$\frac{1}{R^2-1} + \frac{R}{R^2-1} = \frac{1}{R-1}.$$

Defining $|\phi|$ by $|\phi| = |C\phi|$, S' becomes a complete metric linear space. Thus, defining $C(\omega)$ as by (50) and $B = \| B_n(\omega) \|$ by

$$B_1(\omega) = B_2(\omega) = \ldots = C(\omega)\,C^{-1},$$

B is a Γ-matrix as before, and whenever the infinite series $\sum_{n=1}^{\infty} f_n(p)$ $\left(f_n \varepsilon\, s, \left|1-\frac{1}{p}\right| \leqslant R\right)$, converges, to $f(p)$ say, then $\lim_{\omega\to\infty} \sum_{n=1}^{\infty} B_n(\omega) f_n(p) = f(p)$.

In particular, we may assume $f_n = a_{n-1}\left(1-\frac{1}{p}\right)^{n-1}$ $(n = 1, 2, \ldots)$, where the a_{n-1} are constant s. Then, provided the radius of convergence of the power series $\sum_{n=0}^{\infty} a_n r^n$ is greater than $R > 1$, we have, for $\left|1-\frac{1}{p}\right| \leqslant R$,

$$\sum_{n=1}^{\infty} B_n(\omega) f_n = C(\omega) \sum_{n=1}^{\infty} C^{-1} f_n.$$

Now

$$C^{-1} f_n = \frac{a_{n-1}}{2\pi i} \int e^{xp} \left(1-\frac{1}{p}\right)^{n-1} \frac{dp}{p} = L_{n-1}(x),$$

where $L_{n-1}(x)$ is Laguerre's polynomial of order $n-1$ and so, if we define the associated Laguerre series of $f(p)$ by $\phi(x) = \sum_{n=0}^{\infty} a_n L_n(x)$, the series converges everywhere and represents an integral function, and

$$f(p) = p \int_0^{\infty} e^{-px} \phi(x)\, dx, \text{ valid for } \Re p > \frac{1}{R+1}. \tag{51}$$

Putting $1-\frac{1}{p} = z$, *i.e.* $p = \frac{1}{1-z}$, and substituting for the variable of integration x the variable $t = xp = \frac{x}{1-z}$, equation (51) becomes

$$\sum_{n=1}^{\infty} a_n z^n = \int_0 e^{-t} \phi\big((1-z)t\big)\, dt, \tag{52}$$

where, however, the path of integration on the right-hand side is, in general, complex. If, instead, we take the path of integration as the positive real axis, we obtain a counterpart of Borel's method. The precise domain of convergence of the integral, however, is beyond the scope of the present theory and presumably can best be determined by the resources of the general theory of the Laplace transform†.

Another example is provided by Riemann's method for the summation of general trigonometrical series. If we denote by R the linear operation of double differentiation, and by $R(\omega)$ the operation

$$R(\omega)\,\phi = \frac{\phi(x+2\omega)+\phi(x-2\omega)-2\phi(x)}{4w^2},$$

† G. Doetsch, *loc. cit.*

then we may consider the matrix $R = \| R_n(\omega) \|$, where $R_n(\omega) = R(\omega) R^{-1}$ ($\omega = 1, 2, \ldots$), defined for $0 < |\omega| < \omega_0$. Defining the complete metric linear spaces S and S' conveniently, we may then show by the use of Theorem XI that, if $\Sigma f_n(x)$ is a (trigonometrical) series which converges in the mean to a function of L_2, $f(x)$, say, then the transform of

$$\Sigma f_n(x),\ R - \Sigma f_n(x)$$

converges in the mean to $f(x)$, $\operatorname{l.i.m.}_{\omega\to\infty} \sum_{n=1}^{\infty} R_n(\omega) f_n = f$. Here again it would be difficult to prove the precise results of Riemann's theory by the methods of the present paper.

The examples mentioned above are very simple indeed compared with the scope of the general theory. An example of quite a different type will now be given.

Let S be the space of all real functions $f(x)$ defined for $0 \leqslant x \leqslant 1$, and which satisfy Lipschitz's condition $|f(x_1) - f(x_2)| \leqslant k |x_1 - x_2|$ for all $0 \leqslant x_1 \leqslant 1$, $0 \leqslant x_2 \leqslant 1$, where k is a constant depending only on f but not on x_1 or x_2. We define the metric of S by

$$|f| = |f(0)| + \max_{0 \leqslant x_1,\, x_2 \leqslant 1} \left| \frac{f(x_1) - f(x_2)}{x_1 - x_2} \right|.$$

On this definition S becomes a complete metric linear space. Further let $\| b_{k,n} \|$ ($k = 1, 2, \ldots, n = 1, 2, \ldots$) be a numerical Γ-matrix so that $0 \leqslant b_{k,n} \leqslant 1$ for all k and n. Then the transformations $\phi(x) = f(b_{k,n} x)$ define bounded linear operators $B_{k,n}$ in S, $\phi = B_{k,n} f$. Also $\lim_{k\to\infty} B_{k,n} = I$ for all so that the matrix $B = \| B_{n,k} \|$ satisfies condition (29). Since $\| b_{k,n} \|$ is a Γ-matrix, it satisfies condition (8) for a certain positive constant M. Now

$$|(B_{k,n} - B_{k,n+1}) f(x)| = |f(b_{k,n} x) - f(b_{k,n+1} x)| \leqslant |b_{k,n} - b_{k,n+1}| |x| |f|,$$

so that $|B_{k,n} - B_{k,n+1}| \leqslant |b_{k,n} - b_{k,n+1}|$. Hence $\sum_{n=1}^{\infty} |B_{k,n} - B_{k,n+1}| \leqslant M$ for all k, and so, by (17), $|B_{k,1} - B_{k,2},\ B_{k,2} - B_{k,3},\ \ldots| \leqslant M$ for all k, showing that B satisfies condition (28) and so is a Γ-matrix. It follows, by Theorem V, that whenever $\sum_{n=1}^{\infty} f_n = f$ in S, then $\lim_{k\to\infty} \sum_{n=1}^{\infty} B_{k,n} f_n = f$.

Now convergence in S is readily seen to be equivalent to ordinary convergence in every point $0 \leqslant x \leqslant 1$ ($\lim_{n\to\infty} g_n = g$ in S implies, and is implied by, $\lim_{n\to\infty} g_n(x) = g(x)$, $0 \leqslant x \leqslant 1\big)$. Hence, if $\sum_{n=1}^{\infty} f_n(x) = f(x)$ for all $0 \leqslant x \leqslant 1$ and $\|b_{k,n}\|$ is a numerical Γ-matrix, $0 \leqslant b_{k,n} \leqslant 1$ ($k = 1, 2, \ldots$, $n = 1, 2, \ldots$), then

$$\lim_{k\to\infty} \sum_{n=1}^{\infty} f_n(b_{k,n}\, x) = f(x)$$

provided the $f_n(x)$ and $f(x)$ are functions satisfying Lipschitz's condition in $0 \leqslant x \leqslant 1$.

College of Aeronautics,
Cranfield,
Bletchley, Bucks.

CORE-CONSISTENCY AND TOTAL INCLUSION FOR METHODS OF SUMMABILITY

G. G. LORENTZ AND A. ROBINSON

1. **Introduction.** We shall consider methods of summation A, B, . . . defined by matrices of real elements (a_{mn}), (b_{mn}), $(m, n = 1, 2, \ldots)$ which are regular, that is, have the three well-known properties of Toeplitz (**4**, p. 43). A method A is said to be *core-consistent with the method B for bounded sequences* if the A-core (**3**, p. 137; and **4**, p. 55) of each real bounded sequence is contained in its B-core. B is *totally included* in A, $B \ll A$, if each real sequence which is B-summable to a definite limit (this limit may be finite or infinite of a definite sign) is also A-summable to the same limit. It will be shown in the present paper that if the matrix A is core-consistent with the positive matrix B, then A is "almost" divisible by B on the right. This statement is made precise in Theorem 1 below. The proof (§2) involves some elementary properties of convex sets in Banach spaces. In §3, the same method is used to prove a similar result for the relation $B \ll A$ (Theorem 2). Some simple corollaries are given in §4.

Let l_1 be the Banach space of elements $\mathbf{x} = (x_n)$, with norm

$$||\mathbf{x}|| = \sum_{n=1}^{\infty} |x_n|,$$

so that the rows of the matrices A, B are elements $\mathbf{a}_m$, $\mathbf{b}_m$ of l_1. Elements $\mathbf{x}, \mathbf{y} \in l_1$ are called disjoint if $x_n y_n = 0$ $(n = 1, 2, \ldots)$; an element $\mathbf{x} \in l_1$ is positive, $\mathbf{x} \geqslant 0$, if $x_n \geqslant 0$ $(n = 1, 2, \ldots)$. If $\mathbf{x} = (x_1, x_2, \ldots, x_n, \ldots) \in l_1$, we shall write

$$\mathbf{x}^q = (x_1, \ldots, x_q, 0, 0, \ldots), \qquad \mathbf{x}_p = (0, \ldots, 0, x_p, x_{p+1}, \ldots),$$

$$\mathbf{x}_p^{\,q} = (0, \ldots, 0, x_p, \ldots, x_q, 0, \ldots), \qquad p \leqslant q.$$

We also use the same notation for sets $E \subset l_1$, for instance $E_p^{\,q}$ is the set of all $\mathbf{x}_p^{\,q}$ with $\mathbf{x} \in E$. A *cone* $K \subset l_1$ is a set such that

$$\sum_{1}^{n} c_k \mathbf{x}_k \in K$$

whenever $c_k \geqslant 0$, $\mathbf{x}_k \in K$. For instance, the set of all positive elements is a cone in l_1.

We shall prove the following theorems:

THEOREM 1. *Let A, B be regular matrices and let A be core-consistent with B. If B is positive, that is if $\mathbf{b}_m \geqslant 0$ $(m = 1, 2, \ldots)$, there is a positive regular matrix C such that the norm of the mth row of $CB - A$ tends to zero for $m \to \infty$.*

Received December 22, 1952; in revised form May 8, 1953.

The case where the elements of the sequences, or of the matrices, are complex is not essentially different as will be shown in §2.

If $A = (a_{mn})$, we shall write A_p for the matrix obtained from A by replacing all a_{mn} with $n < p$ by zeros.

THEOREM 2. *If A, B are regular row-finite matrices, B positive and*

(i) $$B \ll A,$$

there is an integer p and a regular positive row-finite matrix C such that

(1) $$CB_p = A_p;$$

this remains true if (i) *is replaced by the (formally weaker) hypothesis that*

(ii) $\tau_n \to +\infty$ *always implies* $|\sigma_n| \to +\infty$, *where σ_n and τ_n are the A- and the B- transforms of a sequence s_n, respectively.*

If B is the unit matrix I, these results were known before; for the case of Theorem 1 see Agnew **(1)**, also **(3**, p. 149**)**; for Theorem 2, Hurwitz **(5)** or **(4**, p. 53**)**.

2. Core-consistency. If Theorem 1 is true for a given pair of matrices A, B, it is also true for any two matrices A', B' with rows $\mathbf{a}'_m$, $\mathbf{b}'_m$ satisfying

$$||\mathbf{a}_m - \mathbf{a}'_m|| \to 0, \quad ||\mathbf{b}_m - \mathbf{b}'_m|| \to 0.$$

This and the regularity of A, B imply that we may assume A, B to be row-finite, and such that there is a sequence $n(m)$ increasing to $+\infty$ with $a_{mn} = b_{mn} = 0$ for $n < n(m)$.

LEMMA. *In the above conditions there exist two sequences $p = p(m) < q(m)$ such that $p(m) \to \infty$ for $m \to \infty$ and that*

(2) $$\rho(\mathbf{a}_m, K) = \rho(\mathbf{a}_m, K_p{}^q);$$

here $\rho(\mathbf{a}_m, K)$ is the distance from $\mathbf{a}_m$ to the cone K generated by the $\mathbf{b}_\lambda (\lambda = 1, 2, \ldots)$.

Proof. For a given m, let $m_1 \leqslant m_2$ be such that $\mathbf{b}_\mu$ is disjoint with $\mathbf{a}_m$ if μ does not satisfy $m_1 \leqslant \mu \leqslant m_2$; we may assume that $m_1 \to \infty$ for $m \to \infty$. Let K' be the cone generated by the $\mathbf{b}_\mu$, $m_1 \leqslant \mu \leqslant m_2$, let $p(m) = n(m_1)$ and let q be so large that $b_{\mu n} = 0$, $m_1 \leqslant \mu \leqslant m_2$, $a_{mn} = 0$ for $n > q$. Then $\mathbf{a}_{mp}{}^q = \mathbf{a}_m$, $K'_p{}^q = K'$, and therefore

(3) $$\rho(\mathbf{a}_m, K) \leqslant \rho(\mathbf{a}_m, K') = \rho(\mathbf{a}_m, K'_p{}^q).$$

On the other hand, let $\mathbf{x} \in K$, then $\mathbf{x}$ is a linear combination, with positive coefficients, of some of the $\mathbf{b}_\lambda$. If we omit from it all those $\mathbf{b}_\lambda$ which are not $\mathbf{b}_\mu$, we shall obtain another element $\mathbf{x}' \in K'$. The omitted $\mathbf{b}_\lambda$ are disjoint with $\mathbf{a}_m$ and all $b_{\lambda n}$ satisfy $b_{\lambda n} \geqslant 0$. This implies

$$||\mathbf{a}_m - \mathbf{x}_p{}^q|| \geqslant ||\mathbf{a}_m - \mathbf{x}'_p{}^q||.$$

Since $K'_p{}^q \subset K_p{}^q$, it follows that $\rho(\mathbf{a}_m, K_p{}^q) = \rho(\mathbf{a}_m, K'_p{}^q)$ and using (3) we obtain $\rho(\mathbf{a}_m, K) \leqslant \rho(\mathbf{a}_m, K_p{}^q)$. The inverse inequality is obvious, and (2) follows.

Proof of Theorem 1. We shall show that

$$\rho(\mathbf{a}_m, K) \to 0. \tag{4}$$

If this is not true, there exists by the Lemma an $\epsilon > 0$, a sequence of disjoint $\mathbf{a}_{m_i}$ and a sequence of disjoint intervals $[p_i, q_i]$ with

$$\rho(\mathbf{a}_{m_i}, K_{p_i}^{q_i}) > \epsilon.$$

If

$$\mathbf{y} = \sum c_i \mathbf{a}_{m_i}$$

is a linear combination of the $\mathbf{a}_{m_i}$ with $c_i > 0$, $\sum c_i = 1$ and if $\mathbf{x} \in K$, we can put

$$\mathbf{z}_i = c_i^{-1} \mathbf{x}_{p_i}^{q_i} \in K_{p_i}^{q_i}$$

and have

$$||\mathbf{y} - \mathbf{x}|| = ||\sum c_i \mathbf{a}_{m_i} - \sum c_i \mathbf{z}_i|| = \sum c_i ||\mathbf{a}_{m_i} - \mathbf{z}_i|| > \epsilon.$$

This shows that the convex set E generated by the $\mathbf{a}_{m_i}$ is at a distance $\geqslant \epsilon$ from E, hence the ϵ-neighbourhood E_ϵ of E is disjoint with K. If K_ϵ is the cone generated by E_ϵ, K and K_ϵ are disjoint except for the origin. By a well-known theorem (**7**, Theorem 1.2), there is in l_1 a bounded linear functional $f(\mathbf{x})$ of norm one which is positive on K and negative on K_ϵ. Hence $f(\mathbf{y}) \leqslant -\epsilon$ on E (**7**, Lemma 1.2). This means that there is a bounded sequence s_ν with

$$\sum b_{m\nu} s_\nu \geqslant 0 \qquad (m = 1, 2, \ldots),$$

$$\sum a_{m_i,\nu} s_\nu \leqslant -\epsilon \qquad (i = 1, 2, \ldots),$$

and contradicts the hypothesis of Theorem 1.

From (4) it follows that for some row-finite positive matrix $C = (c_{mn})$,

$$||\mathbf{a}_m - \sum_n c_{mn} \mathbf{b}_n|| \to 0, \qquad m \to \infty.$$

Finally, this C will be necessarily regular, provided we agree to take $c_{mn} = 0$ whenever $\mathbf{b}_n = 0$. For

$$c_{mn} b_{n\nu} \leqslant \sum_{n=1}^{\infty} c_{mn} b_{n\nu} = a_{m\nu} + o(1) = o(1), \qquad m \to \infty$$

implies that $c_{mn} \to 0$ for $m \to \infty$ and each n. On the other hand,

$$\sum_{\nu=1}^{\infty} a_{m\nu} = \sum_\nu \sum_n c_{mn} b_{n\nu} + o(1)$$

$$= \sum_n c_{mn} \sum_\nu b_{n\nu} + o(1)$$

together with

$$\sum_\nu a_{m\nu} = 1 + o(1), \quad \sum_\nu b_{n\nu} = 1 + o(1)$$

imply that $\sum_n c_{mn} \to 1$ for $m \to \infty$. This completes the proof.

The concept of the core is defined also for sequences of complex numbers (**3**, p. 137). Accordingly, we may introduce the concept of core-consistency as well for matrices and sequences with complex elements. With this new definition, Theorem 1 holds literally as before.

For the proof assume that A is a regular matrix with complex elements, B is positive and that A is core-consistent with B for bounded sequences. Hence, by Knopp's core theorem (**6**, p. 115) or (**4**, p. 55), the core of the B-transform of any bounded sequence s_n is included in the core of s_n. But A is core-consistent with B, and so the core of the A-transform of s_n also is included in the core of s_n. This implies (**3**, p. 149) that $A = A' + V$ where A is a positive regular matrix, and the norm of the mth row of the matrix V tends to zero as $m \to \infty$. Clearly, A' also is core-consistent with B, for complex or more particularly, for real sequences. It then follows from the original Theorem 1 that there exists a positive regular matrix C such that the norm of the mth row of $CB - A'$ tends to zero from $m \to \infty$. Consequently, the norm of the mth row of

$$CB - A = CB - A' - V$$

also tends to zero for $m \to \infty$. This proves our assertion.

The converse of this (as well as the converse of Theorem 1) is a direct consequence of Knopp's core theorem. Thus, let A, B, C, be three regular matrices, C positive, such that the norm of the mth row of $CB - A$ tends to zero for $m \to \infty$. Then the core of the transform of any bounded complex sequence s_n by CB coincides with the core of the transform of s_n by A. The transform of s_n by CB is the transform by C of the transform of s_n by B. Hence the core of the transform of s_n by CB is included in the core of the transform of s_n by B, by virtue of Knopp's core theorem. In other words, CB, and hence A, are core-consistent with B for bounded sequences.

3. Total inclusion. We shall now prove Theorem 2, deducing (1) from the hypothesis (ii). Let $\rho_{mp} = \rho(\mathbf{a}_{mp}, K_p)$; we first show that

$$\rho_{mp} = 0 \text{ for all } p \text{ sufficiently large and } m = 1, 2, \ldots. \tag{5}$$

Let (5) be false. Since the ρ_{mp} decrease for m fixed and increasing p and finally become zero, we deduce that for each p, $\rho_{mp} > 0$ for an infinity of m. Now $\rho_{mp} > 0$ implies the existence of $\delta > 0$, $\epsilon > 0$ such that the sphere S in l_1 with center $\mathbf{a}_{mp}$ and radius δ does not have common points with the cone K' generated by the points $\mathbf{b}_{\lambda p}$ ($\lambda = 1, 2, \ldots$), and by the spheres with radii ϵ around those of the $\mathbf{b}_{\mu p}$ ($\mu = 1, 2, \ldots, m$) which are not zero. Hence, there is a functional

$$f(\mathbf{x}) = \sum x_n s_n, \ ||f|| = 1 \text{ in } l_1,$$

generated by a bounded sequence s_n with $s_n = 0$ for $n < p$, such that the hyperplane $f(\mathbf{x}) = 0$ separates S and K' and supports S (by Eidelheit's theorem, (**7**, Theorem 1.6)). If $f(\mathbf{x}) \geqslant 0$ on K', we have

$$\tau_\mu = f(\mathbf{b}_{\mu p}) \geqslant \epsilon \quad \text{for } \mathbf{b}_{\mu p} \neq 0 \qquad (\mu = 1, 2, \ldots, m)$$

(**7**, Lemma 1.2) and

$$0 > \sigma_m = f(\mathbf{a}_{mp}) \geqslant -||f||\delta = -\delta.$$

By fixing $\epsilon > 0$, taking $\delta > 0$ sufficiently small, and then multiplying the s_n with a sufficiently large positive number, we obtain the following statement:

(*) For each m, p with $\rho_{mp} > 0$ and for any two positive numbers M, η, there is a bounded sequence s_n with $s_n = 0$ for $n < p$ such that

$$\sigma_m = \sum_n a_{mn} s_n = -\eta, \quad \tau_\lambda = \sum_n b_{\lambda n} s_n \geqslant 0 \qquad (\lambda = 1, 2, \ldots),$$

$$\tau_\mu > M \text{ if } 1 \leqslant \mu \leqslant m \text{ and } \mathbf{b}_{\mu p} \neq 0.$$

We now define inductively increasing sequences of integers $p_1, p_2, \ldots, m_1, m_2, \ldots$ and bounded sequences $\mathbf{s}^{(i)}$ satisfying $s_n^{(i)} = 0$ for $n < p_i$. If

$$p_1, \ldots, p_{i-i}; m_1, \ldots, m_{i-i}; \mathbf{s}^{(1)}, \ldots, \mathbf{s}^{(i-1)}$$

are already defined, take p_i so large that $a_{\mu n} = 0$ for $n \geqslant p_i$, $\mu = m_1, \ldots, m_{i-1}$, then find an $m_i > m_{i-1}$ with

$$\left|\sum a_{m_i,n}(s_n^{(1)} + \ldots + s_n^{(i-1)})\right| < \tfrac{1}{2}, \quad \rho(\mathbf{a}_{m_i p_i}, K_{p_i}) > 0.$$

By (*), there is a bounded sequence $\mathbf{s}^{(i)}$ with $s_n^{(i)} = 0$ for $n < p_i$ such that

$$\sum_{n=1}^{\infty} a_{m_i,n}(s_n^{(1)} + \ldots + s_n^{(i)}) = -1, \tag{6}$$

$$\sum_n b_{\lambda n} s_n^{(i)} \geqslant 0 \qquad (\lambda = 1, 2, \ldots), \tag{7}$$

$$\sum_n b_{\mu n} s_n^{(i)} > i \text{ if } b_{\mu p_i} \neq 0 \qquad (\mu = 1, \ldots, m). \tag{8}$$

Let $\mathbf{s} = (s_n)$ be the sequence defined by $s_n = \sum_i s_n^{(i)}$; for each n this sum has only a finite number of terms. Since $\mathbf{a}_{m_i}$ and $\mathbf{s}^{(j)}$ are disjoint for $j > i$, we have by (6), $\sigma_{m_i} = -1$, and by (7) and (8), $\tau_\lambda \to \infty$, which contradicts the hypothesis and proves (5).

Fixing a p for which (5) holds, we consider an arbitrary m. For each $\epsilon > 0$ there is an $\mathbf{x}$ in K such that

$$||\mathbf{a}_{mp} - \mathbf{x}_p|| < \epsilon, \qquad \mathbf{x} = \sum c_\mu \mathbf{b}_\mu,\ c_\mu \geqslant 0. \tag{9}$$

Let q be the last index n with $a_{mn} \neq 0$. If we omit from the last sum all $\mathbf{b}_\mu$ for which

$$\sum_{n>q} b_{\mu n} > \sum_{n \leqslant q} b_{\mu n},$$

we shall obtain an element $\mathbf{x}' \in K$ with

$$||\mathbf{a}_{mp} - \mathbf{x}'_p|| \leqslant ||\mathbf{a}_{mp} - \mathbf{x}_p||.$$

It follows that μ in (9) may be assumed bounded for all ϵ. Then we must have

$$\mathbf{a}_{mp} = \sum_{n=1}^{N} c_{mn} \mathbf{b}_{np}, \qquad c_{mn} \geqslant 0. \tag{10}$$

This proves the theorem, for the argument used in the proof of Theorem 1 shows that $C = (c_{mn})$ is regular, provided in (10) we take $c_{mn} = 0$ whenever $\mathbf{b}_{np} = 0$.

We give some corollaries to Theorem 2, assuming that the matrices A, B are regular and row-finite and that B is positive. We compare the following relations (for the definition of the core of a possibly unbounded sequence see (**4**, p. 55)):

(i) $B \ll A$.

(ii) For each sequence s_n, $\tau_n \to +\infty$ implies that $|\sigma_n| \to +\infty$.

(iii) $A_p = CB_p$ for some p with $C \geqslant 0$.

(iv) A is core-consistent with B for all real sequences.

(v) A is core-consistent with B for all complex sequences.

Then we have:

THEOREM 3. *Conditions* (i)-(v) *are equivalent.*

Proof. Clearly, (i) → (ii). Theorem 2 shows that (ii) implies (iii) and it is easy to see that (iii) → (i). From the definitions of the properties concerned we have (v) → (iv) → (ii). Finally, Knopp's core theorem states that (iii) → (v). This completes the proof.

4. Applications. For further illustration of Theorems 1 and 2 we shall give some applications to totally equivalent and core equivalent methods. Two methods A, B are *totally equivalent*, if $A \ll B$ and $B \ll A$; they are *core-equivalent for bounded sequences* if the A-core of each bounded sequence coincides with its B-core. In what follows, V is a matrix such that the norm of the mth row tend to zero for $m \to \infty$, and I is the unit matrix.

THEOREM 4. (i) *A method A is core-equivalent with I for bounded sequences if and only if A has a representation*

$$A = A' + V \tag{11}$$

with positive A', where A' contains a sequence of rows of the form

$$\mathbf{a}'_{m_n} = (0, \ldots, 0, a_{m_n,n}, 0, \ldots), \qquad n = 1, 2, \ldots. \tag{12}$$

(then necessarily $m_n \to \infty$, $a_{m_n,n} \to 1$ for $n \to \infty$.)

(ii) *A regular row-finite method A is totally equivalent with I if and only if for some p, A_p is positive and contains a sequence of rows of the form* (12).

Proof. (i) The conditions are clearly sufficient. It follows from Theorem 1 that (11) with a positive A' is necessary. Again by Theorem 1, there is a positive regular matrix C and a V' with $CA' = I + V'$. For each n we have

$$\sum_{m=1}^{\infty} c_{nm}\mathbf{a}'_m = \mathbf{e}_n \tag{13}$$

with $e_{nl} \geqslant 0$, $\mathbf{e}_n = (e_{nl})$, $e_{nn} \to 1$ and

$$\sum_{l \neq n} e_{nl} \to 0$$

for $n \to \infty$. Let

$$\epsilon_n = \sum_{l \neq n} \frac{e_{nl}}{e_{nn}}.$$

Then $\epsilon_n \to 0$ for $n \to \infty$. Since the c_{nm} are all positive, it follows from (13) that there is at least one $m = m_n$ such that

$$\sum_{l \neq n} \frac{a'_{ml}}{a'_{mn}} \leqslant \epsilon_n.$$

For otherwise, multiplying the relations

$$\sum_{l \neq n} a'_{ml} > \epsilon_n a'_{mn} \qquad (m = 1, 2, \ldots)$$

with c_{nm} and adding we would obtain by means of (13) that

$$\sum_{l \neq n} e_{nl} > \epsilon_n e_{nn},$$

which contradicts the definition of ϵ_n.

We now replace by zero the elements a_{ml} of the rows of A' with $m = m_n$, $l \neq n (n = 1, 2, \ldots)$. Denoting the matrix thus obtained again by A', we see that (11) and (12) are satisfied. This proves (i); the proof of (ii) is similar.

Theorem 4(i) may serve to show, for instance, that if a regular Hausdorff method H_g is core-equivalent with I for bounded sequences, then H_g is identical with I.

A method A is *normal* if $a_{mn} = 0$ for $n > m$ and $a_{nn} \neq 0$ $(n = 1, 2, \ldots)$. In this case A has an inverse A^{-1}. If A, B are normal, there is a triangular matrix C with $A = CB$.

THEOREM 5. *Let the regular normal methods A, B be totally equivalent. Then there exists a sequence $c_m \to 1$ such that for some p,*

$$a_{mn} = c_m b_{mn}, \qquad m = 1, 2, \ldots; n = p, p + 1, \ldots. \tag{14}$$

Proof. Let $A = CB$, $B = DA$, then the matrices C, D are triangular, regular and totally equivalent with I. We have

$$a_{mm} = c_{mm} b_{mm}, \quad b_{mm} = d_{mm} a_{mm},$$

hence

$$c_{mm} d_{mm} = 1,$$

and we obtain $c_{mm} \to 1$. From Theorem 4 (ii) it follows that for all sufficiently large n, $c_{mn} = 0$ if $n \neq m$. Putting $c_m = c_{mm}$, we obtain (14).

It should be added that sometimes it is even possible to prove that A, B are identical if they are totally equivalent. Let $A = H_g$, $B = H_{g_1}$ be two regular and normal Hausdorff methods. Then

$$\sum_{n=0}^{m} |a_{mn}|$$

converges for $m \to \infty$ to the "essential" total variation of $g(x)$. From (14) it follows that

$$\sum_{n=0}^{m} |a_{mn} - b_{mn}| \to 0,$$

hence g and g_1 are essentially identical. Thus we obtain a remark of Bosanquet (**2**, p. 452) that H_g, H_{g_1} are identical if they are totally equivalent.

References

1. R. P. Agnew, *Cores of complex sequences and of their transforms*, Amer. J. Math. *61* (1939), 178–186.
2. S. K. Basu, *On the total relative strength of the Hölder and Cesàro methods*, Proc. London Math. Soc. (2), *50* (1949), 447–462.
3. R. G. Cooke, *Infinite matrices and sequence spaces* (London, 1950).
4. G. H. Hardy, *Divergent series* (Oxford, 1949).
5. W. A. Hurwitz, *Some properties of methods of evaluation of divergent sequences*, Proc. London Math. Soc. (2), *26* (1926), 231–248.
6. K. Knopp, *Zur Theorie der Limitierungsverfahren*. Math. Zeitschrift, *31* (1929–30), pp. 97–127, 276–305.
7. M. G. Krein and M. A. Rutman, *Linear operators leaving invariant a cone in a Banach space*, Uspehi Mat. Nauk (N.S.), *3*, no 23 (1948), 3–95; Amer. Math. Soc. Translations no. 26 (1950).

Wayne University
and
University of Toronto

Philosophy

ON CONSTRAINED DENOTATION

Abraham Robinson

1. Both by the choice of its subject and by its mode of discussion, Russell's essay "On Denoting" [ref. 4] has remained one of the conspicuously influential philosophical papers of this century. Yet its main thesis has not found universal acceptance. The details of Russell's solution of the problem of descriptions are well known and need not be repeated. Here I wish to recall only that Russell rejected the notion that a description is a name. The following argument is given in *Principia Mathematica* in support of this view [ref. 6].

> For if that were the meaning of "Scott is the author of Waverley" [i.e., that "Scott" and "the author of Waverley" are two names for the same object], what would be required for its truth would be that Scott should have been *called* the author of Waverley: if he had been so called, the proposition would be true, even if someone else had written Waverley; while if no one called him so, the proposition would be false, even if he had written Waverley.

However, in spite of the apparent force of this line of reasoning, many writers, both before and since, have propounded theories which imply that a description is indeed a name or a sort of name. The fact that they could do so with impunity indicates that there is a flaw in Russell's argument. It seems to me that this flaw can be found in the tacit assumption that the name of a person or object can always be chosen arbitrarily; or, to put it the other way round, that the interpretation of a name is, initially, a matter of choice. I can see no compelling reason for this assumption. In the present paper we shall be concerned with several formal frameworks which are based on a contrary point of view.

Once we grant that the interpretation of a name may be affected, or *constrained*, by circumstances, my own preference goes to a setting in which a description has a denotation only if there exists a unique object which satisfies its defining condition. This is, roughly, the point of view adopted by Hilbert and Bernays [ref. 1]. Their procedure has been criticized [ref. 5] because it implies that a formula involving a description can be regarded as well-formed only if it has been proved that the description in question has a denotation. However, this objection is met if—as is natural in a model-theoretic approach—we distinguish clearly between well-formedness on one hand and interpretability in a structure on the other hand. Some of the formal details of a program based on this distinction are carried out below. However, it will be our main purpose to discuss possibilities rather than to work out all the consequences of a particular point of view.

2. We may ask whether any particular solution of the problem of descriptions is "best possible." However, I do not think that this question can be answered with finality. Indeed, given any particular feature of a formal language, we first have to

make up our minds whether it is our aim (i) to make the language as effective a tool as possible for a particular purpose such as the investigation of the physical world or the formalization of a mathematical theory, or (ii) to give a precise explication of a corresponding feature found in a natural language. In the present instance, contemporary practice seems to show that as far as (i) is concerned, descriptions can be dispensed with altogether, or can be replaced by the successive introduction of new function symbols or individual constants. As for (ii), a complete explication would have to take into account modal contexts and even the distinction between objective situations and subjective knowledge and it seems unlikely that all these can be covered by a single formalization. At any rate, we shall not consider these aspects of the matter in the present paper.

3. We proceed to specify the vocabulary of our formal language, L. It is to consist of (i) individual constants, $a, b, c, \ldots$; (ii) function symbols, $f, g, h, \ldots$; (iii) relation symbols, $R, S, T, \ldots$; (iv) the identity, $=$; (v) connectives $\neg, \vee, \wedge, \subset$; (vi) quantifiers $(\exists)$, $(\forall)$; (vii) the descriptor (ι); (viii) variables, $x, y, z, \ldots$; and (ix) brackets [,]. In (ii) and (iii), the number of places may or may not be specified for each symbol, as we please.

On the basis of the above vocabulary, we may now compose well-formed formulae (wff). The laws of formation of wff are as usual in the lower predicate calculus, except that the rules for the formation of terms are extended by the stipulation that $[(\iota x)Q(x)]$ is a *term* if Q is any well-formed formula which contains x free—i.e. not under the sign of a quantifier or the descriptor. More generally, we choose to rule out empty quantifications or descriptions (e.g. $[(\iota x)Q]$ where Q does not contain x). Also, in principle, we introduce brackets in order to obtain a wff from an atomic formula and after each introduction of a connective or of the identity or of a quantifier or of the descriptor. For example, $[R(f(a), [(\iota x)[(\exists y)S(x, y)]]]$ and $[y = [y = [(\iota x)T(x)]]]$ are wff. We measure the *complexity* of a wff by the number of pairs of (square) brackets in it. However, in actual fact, we shall omit brackets if no misunderstanding is likely.

A *description* is a term of the form $t = [(\iota x)Q(x)]$. Q is called the *scope* of t. The description $t' = [(\iota x')Q'(x')]$ is said to be an *immediate component* of the description $t = [(\iota x)Q(x)]$ if t' is contained in $Q(x)$ and if no description which is contained in $Q(x)$ has t' in its scope. Let $t_1, \ldots, t_k$, $k \geq 0$ be the immediate components of t. Then t is a description *of order 1* if $k = 0$, and t is of order $n \geq 2$ if all its immediate components are at most of order $n-1$ and at least one of them is exactly of order $n-1$.

4. Now let M be a first order structure and suppose that we are given a (many-one) map C from a set V of individual constants and of relation and function symbols in L onto individuals and functions and relations with the corresponding number of places in M. Let $W_0(V)$ be the set of wff formed from V without the use of the descriptor (but including the use of $=$) and let $S_0(V)$ be the set of sentences (wff without free variables) within $W_0(V)$. Assuming that we have enough individual constants in V to denote all individuals of M we then proceed (using the so-called substitution approach to quantification) to determine for all $X \in S_0(V)$ whether or not X *holds in M* (*is true in M, is satisfied by M*): in symbols $M \models X$. The procedure is too well known to require repetition here. Let us recall only that $M \models \neg X$ by definition if and only if X is not

satisfied by M *but is meaningful* for M; i.e., the extralogical constants (individual constants and relation and function symbols) which occur in it belong to the domain of the map C.

Thus, the (meta-) relation $\models$ depends not only on M and on X but also on C although this is not apparent in our notation. In particular, C induces a correspondence or map also from the set of terms which occur in $S_0(V)$, $T_0(V)$ say, onto the set of individuals of M; in other words it defines a *denotation* in M for any $t \in T_0(V)$.

So far, the map C from the extralogical constants of V onto entities of M has been entirely arbitrary (except for the condition that individual constants are mapped on individuals, function symbols on functions with the same number of places, and relation symbols on relations with the same number of places). The map of the remaining terms of C into M is then determined uniquely, and without the intervention of the relation $\models$.

5. We still have to clarify the role of the identity. One correct definition of the identity from the point of view of first order model theory is undoubtedly to conceive of it as the set of diagonal elements of $M \times M$, i.e., as the set of ordered pairs from M whose first and second elements coincide. The symbol " $=$ " then denotes this relation and it is correct that $M \models a = b$ if "a" and "b" are individual constants which denote the same individual in M or, more generally, that $M \models s = t$ if "s" and "t" are terms which denote the same individual in M. But the identity may also be *introduced* by this condition so that $M \models s = t$, *by definition* if "s" and "t" denote the same individual [e.g., refs. 2, 3] under the correspondence C, which is again assumed implicitly, and this seems more apposite in connection with the discussion of sentences which involve both descriptions and identity. It should be pointed out that this definition has the unusual, and, at first sight, disturbing feature that it involves a proposition *about symbols*. This is masked in the notation "$M \models s = t$", which seems to indicate that the formula after the double turnstile, "$s = t$" is, like any other wff, a statement about M. However, it appears that, once we have appreciated the situation, no harm will come of it. The purist who would write "a" $=$ "b" will perceive his error by looking at the result. And if we interpret the "is" in "Scott is the author of Waverley" by the identity sign in the sense of our second definition, then we obtain just what we want, i.e. that "Scott" and "the author of Waverley" *denote* the same individual.

6. Let $[(\iota x)Q(x)]$ be a description of order 1 such that $Q(x) \in W_0(V)$. (We use x *typically*, i.e., we may have any other variable in place of x.) Suppose that $M \models (\exists x)Q(x)$ and $M \models (\forall x)\ (\forall y)\ [Q(x) \wedge Q(y) \supset x = y]$, where y is some variable not in Q. Then there exists an individual constant a such that $M \models Q(a)$ and *we extend* C by mapping $[(\iota x)\ Q(x)]$ on the individual of x which is the image of a under C. This definition is independent of the particular choice of a. Applying it to all descriptions of order 1 such that $Q \in W_0(V)$, we obtain an extension C_1 of C. Next, we extend $T_0(V)$, to the set $T_1(V)$ consisting of the terms which can be formed from V together with the descriptions of order 1 in the domain of C_1, but without further introductions of the descriptor. (For example, $F([(\iota x)S(x)])$ belongs to $T_1(V)$ if $[(\iota x)S(x)]$ is a description of order 1 such that $S \in W_0(V)$, but $(\iota y)R(f([(\iota x)S(x)]), y)$ does not.) Then C_1 induces a map from $T_1(V)$ into M. Without fear of confusion, we shall still denote this map by C_1.

We extend $W_0(V)$ and $S_0(V)$ correspondingly to sets $W_1(V)$ and $S_1(V)$. Since we have now defined a denotation in M for all terms which occur in $S_1(V)$, we may extend our truth definition in M from $S_1(V)$ to $S_1(V)$, i.e., we may define the relation $M \models X$ for all $X \in S_1(V)$ in the usual way.

Suppose next that $Q(x)$ belongs to $W_1(V)$ but not to $W_0(V)$ so that $[(\iota x)Q(x)]$ is a description of order 2. Suppose that $M \models [(\exists x)Q(x)]$ and $M \models (\forall x)\ (\forall y)\ [Q(x) \wedge Q(y) \supset x = y]$. Then we extend C_1 to a map C_2 by mapping $[(\iota x)Q(x)]$ into an element of M which is denoted by some a for which $M \models Q(a)$. Proceeding as before, we then define sets $T_2(V)$, $W_2(V)$, $S_2(V)$ such that all terms of $T_2(V)$ have a denotation in M and there is a truth definition in M for all sentences of $S_2(V)$.

Continuing in this way through the natural numbers, we finally put $T(V) = \bigcup_n T_n(V)$, $W(V) = \bigcup_n W_n(V)$ and $S(V) = \bigcup_n S_n(V)$ so that every term that belongs to $T(V)$ has a denotation in M and there is a step-by-step (but not a constructive) procedure for determining for any $X \in S(V)$ whether or not $M \models X$.

Our definition of the denotation of a description is purely semantical; it does not affect the question of the well-formedness of a string of symbols and, on the other hand, it is not affected by questions of probability. Like the usual truth definition, it is effective (in the sense of being decidable for a finite structure M but not for a general infinite structure). Indeed, in order to determine whether a description t has a denotation in M we may proceed as follows. We list t as well as its immediate components and their components, etc., yielding a finite set of descriptions $\{t_\nu\}$, $t_\nu = [(\iota x_\nu)Q_\nu(x_\nu)]$, which we arrange in increasing order. We then check, one by one, whether or not the conditions $M \models [(\exists x_\nu)Q_\nu(x_\nu)]$ and $M \models (\forall x_\nu)\ (\forall y)\ [Q_\nu(x_\nu) \wedge Q_\nu(y) \supset x_\nu = y]$ are satisfied. The procedure is effective if M is finite.

By analogy with the double turnstile, $\models$, which denotes satisfaction, we introduce the *double wedge* $\lhd\lhd$ to indicate denotation. Thus, for given M and correspondence C (which is again ignored in the notation), we write $M \lhd\lhd t$ and $M \lhd\lhd Q$ if t and Q are contained in $T(V)$ and $W(V)$, respectively. Then $M \lhd\lhd X$ is a consequence of $M \models X$.

7. The framework outlined above permits the *elimination of descriptions* in the following sense. Let X be any sentence in the language L. Then there exists a sentence X' containing the same extralogical constants but not involving any descriptions such that the following is true. For any structure M and for a specified map C from the extralogical constants of X into appropriate entities of M, M satisfies X' if and only if the descriptions which occur in X have a denotation in M and M satisfies X.

The proof is by induction on the highest order of the descriptions contained in X. If X contains no descriptions, put $X' = X$. If X contains descriptions of order 1 but of no higher order, let these be $t_j = [(\iota x_j)Q_j(x_j)]$, $j = 1, \ldots, m$, where the $Q_j(x_j)$ are then free of descriptions. We then put $X' = \bar{X}$ where

$$\bar{X} = (\exists x_1) \ldots (\exists x_m)\,[X \wedge Q_1(x_1) \wedge \ldots \wedge Q_m(x_m) \wedge (\forall y_1) \ldots (\forall y_m)\,[Q_1(y_1) \wedge \ldots \wedge Q_m(y_m) \supset x_1 = y_1 \wedge \ldots \wedge x_m = y_m]].$$

If X contains descriptions up to order $n + 1$, $n \geq 1$, we use the same definition for $\bar{X}$ with respect to the descriptions which are contained in X immediately. $\bar{X}$ then contains descriptions up to order n only. This proves the assertion.

8. We shall now consider the deductive aspects of our approach. Disregarding individual variations, it is the purpose of an adequate set of axioms and rules of deduction to enumerate all elements of a certain subset T of the set of all sentences of the given language. In the case of the ordinary lower predicate calculus without extralogical axioms, where T is the set of all "tautologies," this enumeration is effective. If extralogical axioms are present in finite numbers, the enumeration is still effective and the same applies if the set of extralogical axioms is countable and effectively given.

If descriptions are present, we have the additional task of enumerating all those descriptions t that have denotations in all structures in which the extralogical constants that occur in t have been interpreted by means of a mapping C. For this purpose we introduce a (meta-) relation $\lhd$ ("wedge") which is related to the double wedge in roughly the same way as the turnstile is related to the double turnstile. Thus, the intended meaning of $K \lhd t$ is "t possesses a denotation in, or is interpreted in, K."

Thus, let K be a set of extralogical axioms (wff) of the lower predicate calculus, without descriptions and, for convenience, without free variables. Then, in the usual versions of the calculus we may have $K \vdash X$ even if X contains free variables, and the rules of deduction are then arranged in such a way that $K \vdash X$, X is deducible from K if and only if the sentence X' is deducible from K, where X' is obtained from X by applying universal quantification to the free variables of the latter. We shall retain this feature here. However, the usual calculus also permits the occurence of extralogical symbols in X which do not occur in K, for example, when X is a tautology (theorem of the lower predicate calculus). Here it is appropriate to exclude this possibility since $K \lhd X$ is introduced precisely in order to deal with interpretability in K.

9. After these preliminary remarks, we are going to describe the details of a suitable deductive calculus. We keep K listed and, for brevity, omit it in our notation. Thus, we write $\lhd X$ and $\vdash X$ in place of $K \lhd X$ and $K \vdash X$. We recall that a *term* is a variable or an individual constant or is obtained from variables and individual constants by one or more applications of function symbols or of the descriptor.

(i) $\lhd t$ where t is any individual constant which occurs in K or any variable.

(ii)
$$\frac{\lhd t_1, \lhd t_2, \ldots, \lhd t_n}{\lhd f(t_1, \ldots, t_n)}$$

for any n-place function symbol which occurs in K.

(iii)
$$\frac{\lhd t_1, \lhd t_2, \ldots, \lhd t_n}{\lhd R(t_1, \ldots, t_n)}$$

for any n-place relations symbol R which occurs in K.

Thus, (ii) is a first rule of deduction. The remaining rules of deduction involving only the wedge are

(iv)
$$\frac{\lhd X}{\lhd \neg X}, \quad \frac{\lhd X, \lhd Y}{\lhd X \vee Y}, \quad \frac{\lhd X, \lhd Y}{\lhd X \wedge Y}, \quad \frac{\lhd X, \lhd Y}{\lhd X \supset Y}$$

provided X and Y are wff, according to the rules laid down previously.

(v)
$$\frac{\lhd X}{\lhd (\exists y)\, X}, \quad \frac{\lhd X}{\lhd (\forall y) X}, \quad \frac{\lhd t_1, \lhd t_2}{\lhd t_1 = t_2}$$

provided $(\exists y)\, X$ and $(\forall y)X$ are wff.

These are all the axioms and rules of deduction which involve only the wedge.

(vi)
$$\vdash X \text{ for all } X \in K$$

(vii)
$$\lhd X_1, \lhd X_2, \ldots, \lhd X_n$$

when $\psi\, (p_1, \ldots, p_n)$ is any tautology of the *propositional calculus* which involves only the connectives $\neg, \vee, \wedge, \supset$.

As usual, (vii) may be replaced by a finite set of axioms for the propositional calculus, anticipating the rule of modus ponens (ix) below, first rule).

(viii)
$$\frac{\lhd (\forall y)Q(y), \lhd t}{\vdash [(\forall y)Q(y)] \supset Q(t)} \quad \frac{\lhd (\exists y)Q(y), \lhd t}{\vdash Q(t) \supset [(\exists y)Q(y)]},$$

where t is a term and where y is an arbitary variable, and no variable free in t is bound in Q. For a given $Q(y)$, $Q(t)$ means that we replace any free occurrence of y by t.

(ix)
$$\frac{\vdash X, \vdash X \supset Y}{\vdash Y}, \quad \frac{\vdash X \supset Q(y)}{\vdash X \supset [(\forall y)Q(y)]}, \quad \frac{\vdash Q(y) \supset X}{\vdash [(\exists y)Q(y)] \supset X}$$

provided X does not contain y.

(x) In the formula following $\vdash$ we may replace any variable by a variable not present previously, provided we do so everywhere. However, for a bound variable which appears repeatedly under the sign of a quantifier, it is permissible to carry out the substitution in a single quantifier and in its scope.

Next, we have the following axiom schemes for the identity

(xi)
$$\frac{\lhd t}{\vdash t = t} \quad \frac{\lhd s, \lhd t}{\vdash s = t \supset t = s} \quad \frac{\lhd s, \lhd t, \lhd u}{\vdash x = t \wedge t = u \supset s = u}$$

and
$$\frac{\lhd s, \lhd t}{\vdash s = t \wedge Q(s) \supset Q'},$$

where the last expression is supposed to be a wff and such that Q' is obtained from $Q(s)$ by replacing some or all of the occurrences of s by t.

Finally, we introduce the crucial rules of deduction which relate to descriptions

(xii)
$$\frac{\vdash (\exists x)Q(x), \vdash (\forall x)\, (\forall y)\, [Q(x) \wedge Q(y) \supset x = y]}{\lhd [(\iota z)Q(z)]}$$

(xiii)
$$\frac{\lhd [(\iota z)Q(z)]}{\vdash Q([(\iota z)Q(z)])}.$$

10. Employing the axioms and rules of deduction detailed above, we may now derive *theorems* in the usual way, except that in the present circumstances the theorems may be of the form $\vdash X$, where X is a wff, or $\lhd X$, where X is a wff or a term.

We are going to show that our calculus is *adequate* in the following sense.

(a) Suppose that M is a model of K under a map C whose domain contains individual constants for all individuals of M, among them the individual constants which occur in K, if any. Then for any well-formed formula *or term* $Q(y_1, \ldots, y_n)$ in the vocabulary of K whose free variables are just $y_1, \ldots, y_n$, $n \geq 0$, and for any individual constants $a_1, \ldots, a_n$ in the domain of C, $\lhd Q(a_1, \ldots, a_n)$ entails $M \lhd\lhd Q(a_1, \ldots, a_n)$; and if, in particular $Q(y_1, \ldots, y_n)$ is a wff, then $\vdash Q(y_1, \ldots, y_n)$ entails $M \models Q(a_1, \ldots, a_n)$.

By checking through the axioms and rules of deduction of section 9 it is in fact not difficult to verify that (a) is satisfied. The fact that $\vdash Q$ and $\lhd Q$ may hold even if Q contains free variables while $M \models Q$, $M \lhd\lhd Q$ can apply only in the absence of free variables does not interfere with this conclusion.

(b) Given K as before, i.e. such that it consists of wff without free variables or descriptions, let $Q(y_1, \ldots, y_n)$ be any wff or term in the vocabulary of K for which the following condition is satisfied. For any mapping C as considered previously, whose domain D includes the vocabulary of K, onto a structure M which is a model of K (under C) and for any $a_1, \ldots, a_n$ in D, $M \lhd\lhd Q(a_1, \ldots, a_n)$. Then $\lhd Q(y_1, \ldots, y_n)$. And if , moreover, Q is a wff and $M \models Q(a_1, \ldots, a_n)$ for all such M, then $\vdash Q(y_1, \ldots, y_n)$.

It will be seen that (*b*) expresses the completeness of our calculus for the circumstances considered here. To show that it holds, we refer to the definition of the order of a description (section 3 above), which applies equally if the description includes free variables. We introduce a corresponding notion of order for any wff or term X by defining that the order of X is that of the description of highest order contained in it. If X does not contain any description, then we say that it is of order 0. In order to establish (b) it is now sufficient to show that it is correct for all wff and terms of order up to $n + 1$, $n \geq 1$, provided it is correct for all wff and terms of order up to n. We may disregard the slight verbal modification which is necessary for $n = 0$ and we may rely on the extended completeness theorem of the lower predicate calculus to tell us that (b) is correct for terms and wff which are free of descriptions.

So let $t(y_1, \ldots, y_n)$ be any description of order $n + 1$ in the vocabulary of K, $t = [(\iota x)Q(x, y_1, \ldots, y_n)]$ where Q contains descriptions up to order n only. Whenever $M \lhd\lhd t(a_1, \ldots, a_n)$, we then have, by the definition of a description, that $M \models X_1$ and $M \models X_2$

$$\text{where } X_1 = (\exists x)Q(x, a_1, \ldots, a_n)$$

$$\text{and } X_2 = (\forall x)(\forall y)[Q(x, a_1, \ldots, a_n) \wedge Q(y, a_1, \ldots, a_n) \supset x = y].$$

Since this is true for arbitrary M, C, and $a_1, \ldots, a_n$, subject to the stated conditions, where then have, by our inductive assumption, $\vdash Q_1(y_1, \ldots, y_n)$ and $\vdash Q_2(y_1, \ldots, y_n)$ where $Q_1 = (\exists x)Q(x, y_1, \ldots, y_n)$ and $Q_2 = (\forall x)(\forall y)[Q(x, y_1, \ldots, y_n) \wedge Q(y, y_1, \ldots, y_n) \supset x = y]$. Hence, by rule (xii), $\lhd t$. Rules (i)–(v) are now sufficient to establish $\lhd X$ also for any other term or wff of order $n + 1$ in the vocabulary of K.

Now let $X = Q(y_1, \ldots, y_k)$ be a wff or order $n + 1$ such that in all the stated

circumstances $M \models Q(a_1, \ldots, a_k)$. This implies $M \lhd\!\lhd Q(a_1, \ldots, a_k)$ and hence $M \lhd\!\lhd t_j(a_1, \ldots, a_k)$ for all descriptions t_j which are contained in $Q(y_1, \ldots, y_k)$ (although $y_1, \ldots, y_n$ need not appear effectively in all of them). By what has been proved already, we may conclude that $\lhd t_j(y_1, \ldots, y_k)$. Suppose in particular that $t_1, \ldots, t_n$ are the descriptions of order $n + 1$ which appear in X, $t_j(y_1, \ldots, y_k) = [(\iota x_j)Q_j(x_j, y_1, \ldots, y_k)]$. Since $\lhd t_j(y_1, \ldots, y_k)$ has been derived, this must have been done by means of rule (xii), implying that $\vdash (\exists x_j)Q_j(x_j, y_1, \ldots, y_k)$ and $\vdash (\forall x_j)(\forall z_j)[Q_j(x_j, y_1, \ldots, y_k) \wedge Q_j(z_j, y_1, \ldots, y_k) \supset x_j = z_j]$
Also, by (xiii), $\vdash Q(t_j(y_1, \ldots, y_k), y_1, \ldots, y_k))$, and so

10.1 $$\vdash (\forall x_j)[Q_j(x_j, y_1, \ldots, y_k) \supset x_j = t_j(y_1, \ldots, y_k)].$$

We may write Q in more detail as $Q(t_1, \ldots, t_m, y_1, \ldots, y_k)$ indicating the appearance of the descriptions $t_1, \ldots, t_m$ in Q. Then $Q'(y_1, \ldots, y_k) = (\exists x_1) \ldots (\exists x_m)[Q(x_1, \ldots, x_m, y_1, \ldots, y_k) \wedge Q_1(x_1, y_1, \ldots, y_k) \wedge \ldots \wedge Q_m(x_m, y_1, \ldots, y_k)]$ is a wff of order n, at most. Also $M \models Q'(a_1, \ldots, a_k)$ for all the M and $a_1, \ldots, a_k$ which have to be taken into account and so, by the inductive assumption,

10.2 $$\vdash (\exists y_1) \ldots (\exists x_m)[Q(x_1, \ldots, x_m, y_1, \ldots, y_k) \wedge Q_1(x_1, y_1, \ldots, y_k) \wedge \ldots \wedge Q_m(x_m, y_1, \ldots, y_k)].$$

Using familiar procedures of the lower predicate calculus, we obtain from 10.1:

10.3 $$\vdash (\forall x_1) \ldots (\forall x_m)[Q(x_1, \ldots, x_m) \wedge Q_1(x_1) \wedge \ldots \wedge Q_m(x_m) \supset x_1 = t_1 \wedge \ldots \wedge x_m = t_m],$$

where, for the sake of brevity, we have not displayed $y_1, \ldots, y_m$. Also, from the last axiom scheme of (xi),

10.4 $$Q(x_1, \ldots, x_m) \wedge x_1 = t_1 \wedge \ldots \wedge x_m = t_m \supset Q(t_1, \ldots, t_n).$$

Combining 10.2 with 10.3 and 10.4, we deduce, finally,

10.5 $$\vdash Q(t_1, \ldots, t_m).$$

But $Q(t_1, \ldots, t_m)$ is X_1 in a different notation. This completes the proof of (b).

Observe that we were able to deduce $\vdash (\exists x)\,Q(x)$ and $\vdash (\forall x)(\forall y)[Q(x) \wedge Q(y) \supset x = y]$ from $\lhd [(\iota x)Q(x)]$ because the latter assertion can be obtained only via the first and second. The situation would be different if we had admitted descriptions to K.

As Scott observes [ref. 5] if a formula involving a description is regarded as well formed only if the intended "value" of the description exists, then the class of wff may be undecidable. In our approach, there is a corresponding phenomenon. Thus, for a suitable structure M there may be no decision procedure for establishing whether or not a term t has a denotation in M, i.e. whether or not $M \lhd\!\lhd t$. More particularly, we recall the fact that in arithmetic there is no decision procedure for $(\exists y)T_1(x, x, y)$ where T_1 is Kleene's predicate and x ranges over the natural numbers. But if M is the standard model of arithmetic then, for any numeral $\underline{\mathbf{n}}$, $M \lhd\!\lhd (\iota y)\,T_1(\underline{\mathbf{n}}, \underline{\mathbf{n}}, y)$ if and only if $M \models (\exists y)T_1(n, n, y)$. Accordingly, there is no decision procedure for settling whether a given description is meaningful in the domain of natural numbers. Similarly, we may use the undecidability of Peano arithmetic in order to show that it does not have a decision procedure for establishing whether or not $\lhd t$ for a given term t.

Suppose that it is a sentence or consequence of K that, for a particular binary relation symbol R,

$$(\exists x)\,(\exists y)\,[R(x, y) \wedge (\forall z)\,[R(x, z) \supset y = z]].$$

This is equivalent to saying that in any model M of K there exists an element, to be denoted by a, such that $M \lhd\!\lhd [(\iota y)R(a, y)]$. Yet there is, in general, no theorem of our deductive calculus which would reflect this fact since $\lhd\ [(\iota y)R(x, y)]$ asserts that $[(\iota y)R(x, y)]$ is meaningful in M whatever the choice of a.

11. I called this paper "On constrained denotation" in order to indicate that we are concerned with circumstances in which the denotation of a symbol, or complex of symbols, is *constrained*, or *restricted*, by the context. For the case of a description the constraint is maximal, in the sense that if a description is meaningful at all, its denotation is completely determined by the very sentences which establish that it *is* meaningful [see section 10, rule (xii)]. There is a more trivial situation, for which the constraint is maximal, i.e., the case of a term composed of constants and function symbols only. Thus, let a, b denote individuals in a structure M, and let f be a function symbol denoting a specific function in M. Then the denotation of $f(a, b)$ is determined completely.

By contrast *Hilbert's ϵ-symbol* provides examples of terms whose denotation is constrained only partially. For orientation, $[(\epsilon x)Q(x)]$ is intended to denote some individual which satisfies Q provided such an individual exists [ref. 1]. We shall call (ϵ) the *selector* and we shall call any instance $[(\epsilon x)Q(x)]$ a *selection*, by analogy with our previous terminology. For simplicity, we shall consider a formal language with selector but without descriptor, although the reader should have no difficulty in combining the two.

Our formal language, L, will be as in section 3, except that in (vii) the descriptor (ι) is replaced by the selector (ϵ). With this modification, well-formed formulae are obtained exactly as before. Next we proceed, as in section (iv), to determine the satisfaction relation for a structure M with respect to a map C of extralogical constants of L onto appropriate entities of M, yielding truth values for sentences which are free of selectors. We then define the order of a selection exactly as the order of a description was defined at the end of section 3.

Now let $t = [(\epsilon x)Q(x)]$ be a selection or order 1. If $M \models [(\exists x)Q(x)]$, then we extend C by *choosing* an element of M for which there exists an individual constant a such that $M \models Q(a)$, and we map t on the image of a in M. The simultaneous choice of such interpretations for all first order selections yields an extension C_1 of C but, in contrast with the case of descriptions, the extension is not more unambiguous. We extend C_1 as before to all terms and wff in the vocabulary of M_1 and we define satisfaction in M for all sentences which can now be interpreted in M. We then repeat the procedure for selections of higher order and the terms, wff, and sentences which involve them.

The *elimination of selections* is analogous to the elimination of descriptions in section 7. Instead of the definition given there for $\bar{X}$, we now take

$$\bar{X} = (\exists x_1) \ldots (\exists x_m) [X \wedge Q_1(x_1) \wedge \ldots \wedge Q_m(x_m)].$$

The repeated application of this formula leads from a sentence with selections, X, to a sentence X' free of selections such that for a given structure M and map C, M satisfies X' if and only if the domain of C can be extended to the selections of X so that M satisfies X. Also, in order to obtain an adequate deductive calculus in the sense of (a) and (b) of section 10 we have to replace rules (xii) and (xiii) in that section by

$$\frac{\vdash (\exists x)Q(x)}{\lhd [(\epsilon x)Q(x)]} \quad \textit{and} \quad \frac{\lhd [(\epsilon z)Q(z)]}{\vdash Q([(\epsilon z \iota Q(z)])}$$

respectively.

12. We now return to the case of a language with descriptors in order to discuss certain alternatives to the formalism developed in sections 3–10.

Let us replace the deductive calculus of section 10 by some standard version of the deductive lower predicate calculus supplemented by the axiom scheme.

12.1 $\vdash [(\exists x) [Q(x, y_1, \ldots, y_k) \wedge (\forall y) [Q(y, y_1, \ldots, y_k) \supset x = y]4$
$\supset Q([(\iota z) Q(z, y_1, \ldots, y_n)], y_1, \ldots, y_n)$

for an arbitrary wff Q in which x, y, and z are not quantified or bound by the descriptor. As long as we avoid clashes of bound variables, we may then deduce

12.2.
$\vdash (\forall y_1) (\exists y_2) \ldots (\forall y_k) [(\exists x) [Q(x, y_1, \ldots, y_k) \wedge (\forall y) [Q(y, y_1, \ldots, y_k)$
$\supset x = y]] \supset [(\forall y_1) (\exists y_2) \ldots (\forall y_n) [Q([(\iota z) Q(z, y_1, \ldots, y_n)] y_1, \ldots, y_n)]$

and there are analogous formulae for any other sequence of quantifiers applied to $y_1, \ldots, y_n$. The sentences 12.1 and 12.2, etc., are satisfied in all models of K *in which they are meaningful,* i.e., interpreted in the sense of the relation $\lhd\lhd$ However, it is entirely possible that the implicans of 12.2 is not satisfied by any model of K even though K is consistent. If so, the implicate of 12.2 is not meaningful in any model of K. Thus, we now have the possibility that a sentence X is a theorem of our deductive calculus; yet it is not meaningful in any model of K. Nevertheless the calculus is adequate in the following sense.

(a) $\vdash Q(y_1, \ldots, y_n)$ entails $M \models Q(a_1, \ldots, a_n)$ where $a_1, \ldots, a_n$ denote individuals in a model M of K *provided* $M \lhd\lhd Q(a_1, \ldots, a_n)$.

(b) Given a wff $Q(y_1, \ldots, y_n)$ in the vocabulary of K, suppose that the following condition is satisfied. Suppose that $M \lhd\lhd Q(a_1, \ldots, a_n)$ and $M \models Q(a_1, \ldots, a_n)$ for any map C from extralogical constants of L onto entities of a model M of K, as before, Then $\vdash Q(y_1, \ldots, y_n)$.

In order to confirm (a), it suffices to check out the case that $Q(y_1, \ldots, y_n)$ is the wff of 12.1 (where Q has a different meaning). In order to verify (b), we refer to the proof of the corresponding assertion in section 10 and observe that the present system is at least as strong as that of section 10 as far as the derivation of assertions of the form $\vdash X$ is concerned. In particular, the rule (xii) in section 10 follows from 12.1 by means of modus ponens. However, we now have no deductive procedure for deciding whether a term or wff is meaningful for all models of K.

13. A more radical departure from the formalism adopted so far is as follows. Instead of introducing $\lhd$ as a metalogical sign analogous to the symbol of assertion, $\vdash$, we adjoin to the formal language L a one place relation $\mathbf{D}(x)$ whose intended meaning is "the term x denotes." This may seem like a confusion of levels in the semantic hierarchy, since $\mathbf{D}(x)$ really makes our assertion about the symbol x and not about the individual denoted by the symbols. However, this is entirely analogous to the phenomenon observed previously in the case of the definition of the identity adopted in section 5, for there also $a = b$ was interpreted as a relation between symbols (i.e., that they denote the same individual).

We definie terms and wff as in section 3, combining $\mathbf{D}(x)$ among the one place relations (just as the identity is counted among the two place relations as far as the rules of formation are concerned, whatever its interpretation).

Now let C be a map from a set V of extralogical constants in L onto corresponding entities of a structure M as in section 4, so that all individuals of M are in the range of C. Let $T_0(V)$, $W_0(V)$, $S_0(V)$ be the terms, wff, sentences formed from V together with $\mathbf{D}$ and the identity. We determine for all $X \in S_0(V)$ whether or not $M \models X$ by the usual procedure, supplemented by the stipulation that $M \models \mathbf{D}(t)$ for all $t \in T_0(V)$.

Let $t = [(\iota x)Q(x)]$ be a description of order 1 such that $Q(x) \in W_0(V)$. We then put $M \models \mathbf{D}(t)$ if and only if $M \models (\exists x)Q(x)$ and $M \models (\forall x)(\forall y)[Q(x) \wedge Q(y) \supset x = y]$. More generally, we put $M \models \mathbf{D}(t)$ for any *term* of order 1 if $M \models \mathbf{D}(t_j)$ for all *descriptions* of order 1 which are contained in t. Also, if R is an n-place relation symbol and $t_1, \ldots, t_n$ are terms of order 1, then we put $M \models R(t_1, \ldots, t_n)$ if $M \models \mathbf{D}(t_j)$, $j = 1, \ldots, n$ and $M \models R(a_1, \ldots, a_n)$ for some $a_1, \ldots, a_n$ which denote the same individuals of M as $t_1, \ldots, t_n$, respectively. Similarly $M \models t_1 = t_2$ if t_1 and t_2 denote, i.e., if $M \models \mathbf{D}(t_1)$ and $M \models \mathbf{D}(t_2)$, and, moreover, denote the same individual in M. Having now laid down the rules for the validity of the satisfaction relation $M \models X$ for all atomic sentences $X \in S_0(V)$, we may proceed to determine it for all other $X \in S_0(V)$ by the usual rules. We repeat this procedure for terms and sentences of higher order.

In this way, the satisfaction relation $M \models X$ is determined, ultimately, for all sentences X formulated from the vocabulary of V together with $\mathbf{D}$ and, of course, the identity. This determination is "ambiguous" only in the sense that our assertion such as $M \models \neg R(t_x)$ does not enable us to decide whether it is the case that t_x does not denote or that t denotes, but the corresponding instance of the relation does not hold in M. In order to decide this question we need to know in addition whether or not $M \models D(t)$. But although the information provided by $M \models \neg R(t)$ thus is perhaps less complete than we might want it to be it is, nevertheless, perfectly determinate. However, it is a consequence of our definitions that not all theorems of the lower predicate calculus remain valid here. For example, it is quite possible that $M \models \neg[(\forall x)Q(x) \supset Q(t)]$. This will be the case if $M \models (\forall x)Q(x)$ but t is a description of order 1 which has no denotation in M, $M \models \neg \mathbf{D}(t)$. Even so we still have $M \models D(t) \supset [(\forall x)Q(x) \supset Q(t)]$, in any case.

YALE UNIVERSITY
FEBRUARY 1974

REFERENCES

1. D. Hilbert and P. Bernays, *Grundlagen der Mathematik* (Springer, Berlin, 1934, 1939).
2. D. Kalish and R. Montague, *Logic, Techniques of Formal Reasoning* (Harcourt, Brace and World, New York-Burlingame, 1964).
3. P. Lorenzen, *Formale Logik* (de Gruyter, Berlin, 1962).
4. B. Russell, "On Denoting," *Mind* 14 (1905), 479–493.
5. D. Scott, *Existence and Description in Formal Logic, B. Russell, Essays in his honor,* R. Schoenman, ed. (Little, Brown and Co. Boston and Toronto, 1967), pp. 181–200.
6. A. N. Whitehead and B. Russell, *Principia Mathematica,* vol. 1 (Cambridge 1910, 1925).

FORMALISM 64

ABRAHAM ROBINSON
University of California, Los Angeles, California, U.S.A.

1. As we look back upon the development of the Philosophy of Mathematics, the fifty years between 1890 and 1940 appear to us as a golden age. We get the impression that the principal opinions which still dominate our thinking in the field were developed during that period, and among them there stand out three well-moulded points of view—Logicism, Intuitionism, Formalism. No doubt, this picture involves a measure of oversimplification. In actual fact, many of the ideas which find expression in one or the other of the above mentioned philosophies have roots which go back beyond, sometimes far beyond, the "golden age." It is also true that these philosophies remained uncommitted on certain important points. For example, the tenets of Logicism seem to be compatible with diametrically opposed views on the problem of existence in Mathematics. This is illustrated by the fact that the early logicists evinced a rather solid belief in the existence of mathematical structures, finite or infinite, while this belief is not shared by the logical positivists (logical empiricists), who represent later Logicism. Again, there are now various constructivist positions related to, but not identical with, Intuitionism. There is also the position of the nominalists which cannot be subsumed under any of the philosophies enumerated above. But in spite of all these qualifications there persists the impression of the period from 1890 to 1940 as an age of renaissance in the Philosophy of Mathematics, an age in which fundamental opinions were vigorously stated—and vigorously attacked.

2. This philosophical activity involved and inspired the development of numerous "technical" methods and theories in Mathematical Logic, and that development continued beyond 1940 and, as we all know, is still going on at an ever increasing pace. At the same time, the general interest in the Philosophy of Mathematics as such has flagged and is only just beginning to regain some of its former vigor. To those who believe that on all matters of principle the correct answers have already been given by one existing school of thought or another, this presents no problem. Personally, I have to admit that I cannot share this optimistic opinion. It seems to me that *all* points of view that have been put forward as a philosophical basis for Mathematics

involve serious gaps and difficulties, including the point of view which I now hold and which I propose to expound in this address. As the title indicates my position is, basically, close to that of Hilbert and his school. I have added the year, 64, not only because it is now known (as it was not known in 1925) that Hilbert's program is doomed to failure but also because the present picture in the foundations of Mathematics has been affected by important developments in several other directions. I wish to add that I cannot subscribe to everything that has been stated by Hilbert in this field. In fact, this talk is to be regarded neither as a description of a historical position nor as a manifesto which tries to lay down the law but rather as a confession, as a personal statement of a point of view arrived at over a number of years. For this point of view I claim neither faithful adherence to an existing school of thought nor basic originality. Nevertheless I hope that some of my remarks, particularly those contained in the closing sections, will encourage discussions and developments in new directions. In *choosing* my position, my approach has been empirical. That is to say, I have tried to take into account all the evidence available to me that may have a bearing on the subject, including both basic thinking and "technical" results in Logic and Mathematics. I do not dispose of a general philosophical creed which, among other things, would determine also my Philosophy of Mathematics.

3. Different opinions concerning the foundations of Mathematics may differ in their estimation of the meaning or significance of the established body of Mathematics. Thus, they may differ as to the correct *description* of the contents of a theory, e.g., as regards the basic nature of the number concept, or as to the interpretation of existential statements for which no constructive proof is available. They may also differ with respect to their *deontic* principles, more particularly in their answer to the question what kind of activity is proper and reasonable for a mathematician *as* mathematician. Such principles normally involve value judgements which affirm that certain activities, or results of activities, are more important, or worthwhile, or relevant, than others. In this connection, we should not forget that even mathematicians with the same view (or with no views at all) on the foundations of Mathematics may differ sharply as regards the value of a particular piece of work.

Discussions between different schools of thought in the Philosophy of Mathematics usually involve divergences both on the descriptive and on the deontic aspects of the problem. The respective attitudes concerning these aspects may well be interrelated but nevertheless it will be helpful to distinguish between them.

4. My position concerning the foundations of Mathematics is based on the following two main points or principles.

(i) Infinite totalities do not exist in any sense of the word (i.e., either really or ideally). More precisely, any mention, or purported mention, of infinite totalities is, literally, *meaningless*.

(ii) Nevertheless, we should continue the business of Mathematics "as usual," i.e., we should act *as if* infinite totalities really existed.

Of the two principles just stated, the first is descriptive while the second is deontic or prescriptive. I proceed to discuss the first principle.

The problem of Infinity has been debated since the dawn of Philosophy. It is not my purpose to give a historical survey of this discussion. But I may recall here that much of the criticism that was directed by Aristotle (in the third book of his "Physics") against the notion of actual infinity is still topical.

The problem of Infinity is related to, and is sometimes regarded as part of, the problem of Existence in Mathematics, or of the problem of the existence of abstract notions in general. To a nominalist, the existence of a set of five elements is no less illusory than the existence of the totality of all natural numbers. At the other end of the scale are the so-called platonic realists or platonists who believe in the ideal existence of mathematical entities in general, including the existence of transfinite sets of arbitrarily large cardinal numbers to the extent to which they can be introduced at all by means of suitable axioms. It has been said that the platonists do not believe in the objective existence of mathematical entities but rather in the objective truth of mathematical theorems. However, it seems to me that as a matter of empirical fact the platonists believe in the objective truth of mathematical theorems *because* they believe in the objective existence of mathematical entities. Since in their conception a mathematical structure is rather like a physical object, such as a house or a tree, or like a collection of physical objects, which can be examined at leisure and in detail, they conclude that any meaningful question that can be asked concerning such a structure must by necessity possess an absolute answer.

K. Gödel, who may be regarded as the outstanding platonist of our time, has emphasized the similarity between the investigation of physical objects on one hand and of mathematical objects on the other. He sees no reason why we should affirm the objective existence of the former but deny that of the latter. I am in sympathy with this point of view to a very limited extent. It appears to me that the notion of a particular *class* of five elements, e.g., of five particular chairs, presents itself to my mind as clearly as the notion of a single individual (a particular chair, a particular table). Thus, I do not feel compelled to follow the nominalists, who seem to have little trouble in grasping the notion of an individual but feel incapable of proceeding from there to the notion of a class. At any rate, so far as the theory of finite sets is concerned the position of the nominalists leads only to modifications which are of little depth from the mathematical point of view. (I do not wish to deny that the discussion of the nominalistic thesis is still of interest on a different philosophical plane.)

By contrast, I feel quite unable to grasp the idea of an actual infinite totality. To me there appears to exist an unbridgeable gulf between sets or structures of one, or two, or five elements, on one hand, and infinite structures on the other hand or, more precisely, between terms denoting sets or structures of one, or two, or five elements, and terms purporting to denote sets or structures the number of whose elements is infinite. As stated here, this point of view concerns the notion of infinity in Abstract Set Theory, with particular reference to infinite cardinality. However, other types of infinity may be relevant to our discussion, more particularly ordinal infinity, as in the theory of ordinal numbers, and geometrical infinity, such as infinite length. While these can be reduced to cardinal infinity by mathematical considerations one may well argue that, from a philosophical point of view, ordinal infinity or geometrical infinity are more basic than cardinal infinity and provide access to an understanding of the notion of infinity as a whole. So I must add that I am just as unable to grasp ordinal infinity or geometrical infinity.

It follows that I must regard a theory which refers to an infinite totality as *meaningless* in the sense that its terms and sentences cannot possess the direct interpretation in an actual structure that we should expect them to have by analogy with concrete (e.g., empirical) situations. This is not to say that such a theory is therefore pointless or devoid of significance.

An opponent to my position might put forward the following arguments.

(i) He might say that I am unable to grasp the idea of an actual infinite totality merely because my brain suffers from a peculiar limitation. He might argue that he, on the contrary, has a clear conception of all sorts of infinite totalities or, at the very least, of the totality of natural numbers.

(ii) Alternatively, my opponent may concede that he, also, is unable to grasp the idea of an infinite totality. But he may say that this does not in any way prove that infinite totalities do not exist. In order to show that they do exist he may appeal to the physical world. Or, if he does not wish to, or feels that he cannot, appeal to the physical world, he may affirm the existence of a platonic world which contains infinities of all sorts, or of some sort.

The first argument, which asserts that there exists an immanent appreciation of infinity of which I am incapable does not permit any further direct debate of the issue. The second argument may be discussed on a philosophical, more precisely, epistemological level. Thus, we may question whether it has any meaning to affirm the existence of an entity while admitting that we are incapable of understanding its basic characteristics. This criticism is particularly cogent if such existence is asserted in the platonic sense. However, we shall not pursue this line here. Instead, we shall continue the discussion by examining how the different points of view stand up in the light of present day mathematical knowledge.

We have said that to the platonist the existence of a mathematical structure is primary, and the theorems about the structure are secondary. Nevertheless the degree of (potential or actual) completeness of our picture of a particular mathematical structure, or of the set-theoretic universe as a whole (to use platonic language) may reasonably be said to affect the strength of the case of Platonism against Formalism. That is to say (using platonic terms) if one is faced with a mathematical concept that one believes to be categorical in an absolute sense (i.e., realized by a unique structure, up to isomorphism), or with the entire universe of sets, also absolute, then one will naturally be upset when confronted with a sentence X which is meaningful for the theory under consideration and such that both X and not-X are consistent with all assumptions which seem intuitively correct with regard to the structure or universe in question. As you know, this is precisely the situation in Set Theory today. If X is the continuum hypothesis, $2^{\aleph_0} = \aleph_1$, then we know from the complementary results of K. Gödel and P. Cohen that both X and non-X are compatible with all known "natural" assumptions regarding the universe of sets (to use platonic language). While this suggests to the formalist that the entire notion of the universe of sets is meaningless (in the sense indicated by our first principle) the platonist merely concludes that the basic and commonly accepted properties of the universe of sets which are known to us at present are insufficient to decide the continuum hypothesis one way or the other. He will maintain, in this and similar cases, that at any rate only one of the alternatives that offer themselves is the correct one, i.e., is in agreement with *the truth*. At this point he has to face the question whether in this matter the truth is by necessity *discernible* by the human mind. As far as I know, only a small minority of mathematicians, even of those with platonist views, accept the idea that there may be mathematical facts which are *true* but unknowable. If, on the contrary, a platonist maintains that every mathematical truth, about the universe of sets or about a specific structure, is by necessity discernible then he is still called upon to analyze his own modal manner of speech. In particular, he has to ponder the question whether a truth is discernible only if it will (by necessity) be discerned some day and if so, whether this must involve the discovery of new "natural" assumptions or forms of argument which are acceptable to all or most mathematicians. At any rate, it seems to me that the present situation in Set Theory favors the Formalist.

The situation is different in Number Theory. For although Gödel's theorem shows that any explicitly specified set of axioms for Number Theory is incomplete, the examples which have been produced in order to demonstrate this incompleteness are such as to bias us towards a decision as to their truth or falsehood. Thus, the Gödel sentence which asserts its own improvability at the same time affirms its own truth. To this extent, the present situation in Number Theory favors the Platonist.

The question discussed here may be called the *bifurcation problem* in Set Theory and Number Theory. At this moment in history, the path of Set Theory seems to have bifurcated, but the platonist believes that this bifurcation is illusory since *in reality* only one of the alternatives covering the continuum hypothesis can be true. On the other hand, while the bifurcation of either Set Theory or Number Theory is not an essential part of the point of view of the formalist he regards both as entirely possible. Thus, although the problem of bifurcation is certainly relevant to our problem, its consideration does not lead to a clear-cut decision of the case for or against Platonism.

5. I pass on to the discussion of the second basic principle of my formalist philosophy. This is the prescription, or suggestion, or advice, to continue to do Mathematics in the classical way, i.e., in particular, to use terms which purport to refer to infinite totalities as if they really existed. It is at this point that Formalism is in direct conflict with the various constructivist or operationist schools of thought and it will be illuminating to develop our position by way of contrast with these.

I recall that the starting point of Intuitionism is, like my own, the rejection of the naive notion of an infinite totality. The rejection of the law of the excluded middle which is, perhaps, a more famous peculiarity of the intuitionists is regarded by them explicitly as a consequence of their critique of infinitary mathematical concepts. There are also those philosophers of Mathematics whose opposition to classical Mathematics is directed chiefly against the use of impredicative definitions, i.e., definitions in which the definiens involves totalities that include the definiendum. But here again the problem is a serious one only in relation to infinite totalities. At any rate all these schools of thought reject the development of theories that involve actual infinite totalities as a regrettable aberration which will be superseded in due course by saner methods. Sometimes the adherents of these views look for support to the history of Science and point out, correctly, that the Greeks were well aware of the dangers of infinity and, accordingly, did their best to use constructive methods and terminologies. It seems to them that the formalist who admits frankly that much of existing Mathematics is *meaningless* in the sense explained above but who nevertheless encourages the development of this kind of Mathematics adopts a position which is altogether indefensible.

I believe that this criticism is based on an attitude which, though at first sight quite reasonable, is nevertheless too narrow. Those who adopt this attitude think that a concept, or a sentence, or an entire theory, is acceptable only if it can be *understood* properly and that a concept, or sentence, or a theory, is understood properly only if all terms which occur in it can be interpreted directly, as explained. By contrast, the formalist holds that direct interpretability is not a necessary condition for the acceptability of a mathematical theory. Evidently, this issue can be argued only if the constructivist does not regard the direct interpretability of all terms of a theory as axiomatic. Supposing that he is indeed willing to concede this point, let us consider in turn several other criteria of acceptability which are frequently regarded as basic.

(i) A mathematical theory shall be regarded as acceptable only if it is consistent.

Comment. It is indeed a regrettable fact that no version of classical Mathematics is provably consistent, and that there is no *unique* remedy for the inconsistencies (antinomies) which have arisen. On the other hand, a constructivist may indeed *believe* that his theory is safe from inconsistencies, but so far as I have been able to follow the matter, this cannot be proved by arguments which are conclusive from a finitistic point of view.

(ii) A mathematical theory is acceptable if it can serve as a foundation for the Natural Sciences.

Comment. Again, this criterion does not imply that the terms of the theory should be interpretable directly and in detail. It is sufficient that we should have rules which tell how to apply certain relevant parts of our theory to the empirical world. From this point of view the acceptability of Abstract Set Theory is affected by the question whether the theory of real numbers provided by it can serve as an adequate foundation for physical measurements but is independent of the interpretability of infinitary notions such as the totality of real numbers or the totality of natural numbers.

You will observe that this comment has a positivistic, more particularly an instrumentalist, flavour. However, it does not imply the acceptance of the positivistic point of view as far as the Empirical sciences as such are concerned and remains uncommitted in this respect.

(iii) A mathematical theory is to be judged by aesthetic standards such as its beauty or internal relevance.

Comment. The standards mentioned under (iii) have so far defied any sort of scientific approach. The attempts that have been made by notable mathematicians to give a precise expression to their ideas on the subject are regarded by others merely as monuments to the prejudices of these mathematicians. At the same time, it is a fact that the organized world of Pure Mathematics is regulated to a very large extent by our vague intuitive ideas on mathematical beauty and purely mathematical importance. The situation being what it is I will say only that my own intuitive judgement regarding the beauty of a mathematical theory is not affected by the question of the existence or interpretability of the infinitary notions which occur in it. Moreover, I believe that this is also the attitude of most pure mathematicians. To be sure, there are some whose admiration for Cantor's achievement is due to the impression that he discovered an immense world of totalities of enormous size which had been there waiting since the beginning of time. Personally I must confess that the creation of the theory of transfinite numbers impresses me neither more nor less than other bold mathematical innovations such as the theory of Hilbert Space, or Non-archimedean Geometry, or the Theory of Groups.

To sum up, the direct interpretability of the terms of a mathematical theory is not a necessary condition for its acceptability; a theory which includes infinitary terms is not thereby less acceptable or less rational than a theory which avoids them. To *understand* a theory means to be able to follow its logical development and not, necessarily, to interpret, or give a denotation for, its individual terms. However, we may grant at this point that a constructive theory may well have an interest of its own, irrespective of any criticism of the formalist approach. The direct applicability of the procedures suggested by such a theory may be important in practice and philosophically relevant.

6. In Hilbert's view the formal or uninterpreted part of a theory belonged entirely to Mathematics. At the same time, the Metamathematics of the theory was supposed to be strictly finitistic and directly interpretable, as explained above. However, eventually the use of infinitary modes of expression imposed itself on Metamathematics just as it had imposed itself previously on Mathematics. This is very much in evidence in the contemporary development of theories which include a non-countable number of symbols, or infinitely long well-formed formulae. However, it is important to note that even in a more restrained metamathematical approach the actual infinite may enter the scene. For example, in the classical treatment of recursive functions, the notions of *consistency* and *completeness* of a recursive scheme involve the totality of natural numbers. All such infinitary metamathematical theorems are, from our point of view, subject to the two principles enumerated earlier, i.e., to the extent to which they involve infinite totalities they are *meaningless* (in the sense indicated previously) but there may be good reasons for developing them all the same. These remarks apply, in particular, to Semantics. At first sight, the Theory of Models gives a perfectly satisfactory account of the notion of truth in relation to both finite and infinite structures. However, to the formalist any alleged truth definition relative to an infinite structure merely amounts to the formal application of the rules of interpretation to which we are accustomed from concrete, more particularly, empirical examples, to infinite totalities.

Even within standard syntax there are references to infinite totalities such as the totality of well-formed formulae in a given language, or the totality of provable formulae (theorems). A classical mathematician or logician has no hesitation in referring to these totalities as if such references were meaningful and, among other things, in applying to them the laws of the excluded middle. More particularly, the inductive definitions of these totalities presuppose a system of natural numbers or, alternatively or conjointly, a system of infinitary Set Theory, and these are notions which come under our first basic principle. Here as elsewhere the formalist considers that the logician who indulges in this kind of analysis may well understand what he is doing as long as he does not fall into the trap of believing that his theory is meaningful (directly interpretable). But the question remains how to correlate this abstract logic with the *use* of logical principles in actual fact. We shall return to this problem presently.

7. At the beginning of this address, I stated that all known positions in the Philosophy of Mathematics, including my own, still involve serious gaps and difficulties. Among these, the gap due to the absence of consistency proofs for the major mathematical theories appears to be inevitable and we have learned to live with it. Nevertheless, the fact that the development of certain infinitary branches of Mathematics such as Abstract Set Theory has gone as far as it did with only relatively minor, and remediable difficulties, raises the question whether these theories are not, after all, more significant than the formalist is willing to concede. The view that Abstract Set Theory, for example, is workable because it is obtained by extrapolation from a theory of finite sets, which is interpretable even according to the formalist, is open to criticism. Indeed, in laying down the axioms of Set Theory we accept some properties which hold in the realm of finite sets,

such as the axiom of power sets, and reject others, such as the irreflexivity of sets. Thus, the process of extrapolation is actually selective. A possible but perhaps not convincing explanation of the apparently satisfactory nature of so many infinitary theories is that we have arrived at them by trial and error, covering most deductions of moderate length. At any rate there is no doubt that on this score the platonist occupies a more comfortable position.

However, it is the nature of the language and rules actually *used* in the development of (possibly infinitary) mathematical theories that presents us with our greatest problem. Since the formalist cannot rely on the semantic interpretation of his theories, their complete formalization is essential to him. This is part of the original formalist approach due to Hilbert. In this approach the part of Logic and Arithmetic which appear already in the Metamathematics, i.e., which are actually *used* in the development and analysis of a formal theory are supposed to be entirely finitistic (see section 6 above) and such that their validity is self-evident. A formal theory is then obtained by the adjunction of "ideal" (uninterpreted) elements to the formal version of the interpreted finite theory. Thus, the part of a theory that is actually *used* is supposed to be given explicitly and there is no need for its identification a posteriori.

Logical positivists and formalists are in agreement in adopting a pragmatic approach to the development of a formal theory and my sympathy with the positivistic point of view, as far as infinitary theories are concerned, may have been apparent. However, the logical positivists extend this approach to all of metamathematics and maintain that the syntax of a language is a matter of choice at all levels. It is at this point that I am unable to follow them. It seems to me that the rules of logic and of certain parts of Arithmetic which are *used* in the analysis of a formal theory are by no means arbitrary. For example, even when considering some non-standard, e.g., three-valued, logic, the logician ultimately assumes that a given concrete situation either obtains or does not obtain, and that these possibilities are mutually exclusive. In other words, at this point the logician actually *uses* two-valued propositional logic or, in syntactical terms, he assumes the law of contradiction, not-(X and not-X), and the law of the excluded middle, X or not-X.

Thus, it appears to me that there are certain basic forms of thought and argument which are prior to the development of formal Mathematics. I propose to show in the remaining sections that the further discussion of the scope of this basic Metamathematics beyond the explication given by Hilbert is both vital and fruitful.

8. A formal theory may be presented by means of a *primitive frame* (a term due to Curry) which delimits (atomic) symbols, terms, well-formed formulae, and theorems (provable formulae). An example of a primitive frame is provided by the following set of rules (A) for a rudimentary form of recursive arithmetic, which is adapted from [3].

(A) *Symbols.* 0, ′, =.
Terms. 0 is a term. If a is a term then a' is a term.
Formulae. If a and b are terms then $a = b$ is a (well-formed) formula.
Theorems. $0 = 0$ is a theorem. If $a = b$ is a theorem then $a' = b'$ is a theorem.

What is the significance of a primitive frame such as (A)? The answer to this question is ambiguous. In spite of the grammatical form of the sentences of (A), a mathematician will *in practice* interpret them as permissive instructions which enable him to record or communicate terms, formulae, and theorems of the theory defined by the frame if he so desires. Putting it in the most concrete form, (A) tells us that if we take three sheets of paper and label them "terms", "formulae", theorems", then we may enter "0" on the sheet of terms, and we may enter additional inscriptions on the three sheets in accordance with the following rules. If we find that a is an inscription on the sheet of terms then a' may be entered on the sheet of terms. (Here, "a" denotes an inscription, and "a'" denotes the inscription obtained from the former by affixing " ′ " to it.) Similarly, if we find a and b on the sheet of terms then we may enter $a = b$ on the sheet of formulae. As for the sheet of theorems, we may enter $0=0$ on it, and if we read $a=b$ on this sheet we may enter on it also $a' = b'$. We observe that in this interpretation no rules of Logic are involved at all.

Many existing primitive frames for formal theories are less simple. For example, in some versions of the Predicate Calculus overlapping quantifications with respect to the same variable are not permitted, i.e., do not yield well-formed formulae. Thus, in order to know whether or not certain operations are permissible we have to be able to check whether an inscription already available does or does not contain configurations of certain kinds. It may also occur that the natural numbers make their appearance in a primitive frame. Such is the case in Gödel's representation of the language of type theory which includes variables of all types, 0, 1, 2, .. and so on, for all n. This sort of formulation should be ruled out explicitly if the language of the primitive frame is to remain at a strictly finitistic level. It is in fact implicit in the present approach that the primitive frame for a formal theory should be such as to be interpretable immediately in terms of concrete instructions. As already stated this does not commit us to any particular logic, not does it commit us to any a priori assumptions on Arithmetic or Set Theory.

The question what Logic or Arithmetic or Set Theory we actually *use* intuitively, arises when we try to interpret a primitive frame as it stands, i.e., as a *description*, or purported description of a universe of "terms", "well-formed formulae" and "theorems". If we were called up to describe only the terms, well-formed formulae, and theorems *actually* written on a particular sheet of paper, or all terms, etc., *actually* recorded at a given time, then I would not hesitate to say that the full first order Predicate Logic as well as a certain fragment of Arithmetic (which is appropriate to the number of symbols etc. already recorded) are applicable here. However, in actual fact we are required to deal not only with the expressions that *have been* written down but also with those that *might possibly be* written down. To regard those as actual totalities would be contrary to our second principle. If, nevertheless, we decide to act *as if* they constituted actual totalities, we thereby move our metamathematical theory to the realm of uninterpreted formal systems. How far can we go without undertaking this fateful step?

Both formalists and constructionists do not doubt that at least some of the rules of the propositional and predicate calculi are applicable in the interpreted (material, "inhaltlich") kind of metamathematics now under consideration. More precisely these may be taken to include all axioms and rules of the Predicate Calculus which are accepted by those intuitionists to whom a codification of the laws of reasoning is at all acceptable. But there the unanimity ends. Even among the formalists there are differences of opinion as to which parts of Arithmetic may be said to be intuitive and acceptable as finitistic. Personally, I cannot see at this time how a form of reasoning which attempts to escape the consequence of Gödel's second theorem (such as Gentzen's consistency proof for Arithmetic or any other consistency proof for Arithmetic) can remain strictly finitistic and, hence, interpreted. Thus, there is room for further discussion on the precise delimitation of finitistic metamathematical reasoning, with particular reference to the arithmetical and set-theoretical arguments contained in it.

9. On the other hand, we may also take the "fateful step" of removing Syntax itself to the realm of uninterpreted formal theories. If we adopt this policy then the basic syntactical notions such as connectives, variables, quantifiers, well-formed formulae, are not regarded as inscriptions which are created gradually at the whim of a writer but are supposed to constitute rigid totalities or sets, which are connected by certain operations such as concatenation. Similarly, the model-theoretic (semantic) interpretation of sentences takes place in a wider axiomatic framework which includes in addition also sets of individuals and of relations. The connection between these individuals and relations and the symbols (in the abstract sense, as above) which denote them is again given by metamathematical relations, more particularly by the relation or relations of designation. Accordingly, designation appears here as a particular kind of correspondence between abstract sets. In many contexts it is perfectly legitimate to suppose that this correspondence reduces to the identity, in other words, that the notion is autonymous. The somewhat dogmatic approach to the problem of denotation which requires a rigid distinction between name and object is no doubt appropriate to cities and names of cities (e.g., Jerusalem and "Jerusalem") but is not essential when transferred to mathematical entities within the above framework. Consider for example the assertion that there is a one-to-one correspondence between numerals and natural numbers (or, alternatively, a many-one correspondence). Evidently, the notion of a numeral here does not refer to inscriptions (or tokens) since the number of inscriptions that have been written down is finite and can even be estimated. Accordingly, even a numeral must be an abstract entity and may be, for example, the corresponding number. However, we are still faced with the problem of describing the connection between numbers or numerals and the related inscriptions or tokens.

It is important to appreciate that the formal approach to Metamathematics sketched above has been used by the majority of logicians either consciously or unconsciously, for a long time. It is conspicuous in Gödel's famous paper on the incompleteness of Arithmetic and as we have said already, the present high tide in the development of Model Theory is based on the same approach when applied to Semantics. From the point of view of the formalist, this type of Syntax and Semantics of a given formal system T may be accommodated within a single system T' which is an extension of T. The procedure may be repeated so as to yield an extension T'' of T' within which are accommodated the Syntax and Semantics of T'. Further repetitions of this procedure are equally possible.

But although T and T' (and T'') are formal and, in general, uninterpreted theories it is undeniable that in certain important instances they include fragments which are indeed capable of interpretation, more particularly, empirical interpretation. For example, suppose that T is the axiomatic Set Theory of Zermelo and Fraenkel. According to our formalist point of view this theory is uninterpreted. But it is nevertheless the case that the rules for the addition and multiplication of finite cardinals, which form part of that theory, are applicable to concrete, more particularly, empirical situations. Also, if T is Peano's Arithmetic and X is the (formal) theorem of T which states that $x^3 + y^3 = z^3$ has no solution with $xyz \neq 0$ we conclude that no such integers will be found in practice (or else that Peano's Arithmetic is inconsistent). Or again, at the metamathematical level, if a sentence X has been proved undecidable (unprovable and irrefutable) within a certain theory T—this being a result of the formal theory T' as above—then we are confident that indeed we shall never *find* a proof or either X or not-X within T. Thus, both at the mathematical and at the metamathematical level we are inclined to believe that our formal theories have concrete implications. In Hilbert's scheme, the extraction of these concrete implicati ns from a formal theory amounts to the separation of the material (*inhaltlich*) elements of the theory from its ideal elements. However, the identification and interpretation of these material elements may be quite difficult. Suppose for example that X has been proved, within T', to be unprovable within T, *unless T is inconsistent*. Then we are inclined to conclude that we shall either never find a proof for X or if we do find a proof for X then we may also find a proof for not-X. But so long as we have not circumscribed the possibility of finding a proof for not-X more closely this *possibility* of finding a proof for not-X need not in itself be clearly defined in constructive terms so that the material or concrete applicability of our metamathematical result remains in doubt.

The idea of supplementing abstract mathematical or metamathematical results with concrete procedures and algorithms is, of course, quite common. It has both practical and theoretical aspects, as in the development of numerical methods on one hand and in the theory of recursive procedures on the other. But the precise limits of such an activity for a given formal theory and its significance in philosophical terms remain matters for further discussion. With a slight inflexion the idea fits well into the scheme of the logical positivists, to whom a formal theory is *adequate* if it can be linked with the empirical world by means of rules of correspondence, and who dispense with the material Metamathematics discussed previously. By contrast it appears to me that the material (*inhaltlich*) aspects of Logic and Mathematics appear both in the basic Metamathematics discussed in section 7 and in the concrete elements which can be extracted from a formal theory as indicated just now.

10. Thus, it remains a fact that, in ways which we do not fully understand, theorems which to the formalist come under the heading of uninterpreted Mathematics or Metamathematics yield results which can be used in material thought and for empirical purposes. However, since we are under no illusions concerning the concrete meaning of the infinitary assumptions included in such formal theories, our philosophy directs us to investigate whether or not such an assumption should be retained universally. In particular, I will mention here the assumption that there exists a *standard* or *intended model* of Arithmetic or (alternatively, but relatedly) of Set Theory. Clearly, to the formalist, the entire notion of standardness must be meaningless, in accordance with our first basic principle. Accordingly, it is perfectly reasonable to posit that the system of Arithmetic for a mathematical theory T and for the syntax of T are not isomorphic, that is to say, to drop the categoricity of Arithmetic when applied to a mathematical theory and to its syntax simultaneously. On the contrary, it is interesting and fruitful to consider languages in which, for example, the length of a sentence may be n, where n is a natural number in the arithmetic of the syntax whic may be infinite from the point of view of the arithmetic incorporated in the original theory [14].

10. Earlier I stated my view that there are forms of thought which are prior to any formal mathematical theory and I maintained that the logic which is applicable to a given concrete, hence finite, system includes all of the Lower Predicate Calculus. I was less definite in discussing the logic which applies to systems of unbounded extent, i.e., which are potentially infinite. I now wish to suggest that for these a form of Modal Logic may be appropriate. In the Appendix, I show how the semantics of an infinite structure can be defined by means of a concept of potential truth for a set of finite structures. In particular, truth in Arithmetic can thus be defined in terms of potential truth in initial segments of natural numbers. The argument of the Appendix is infinitary but, in keeping with the remarks of section 8 above, it may point the way to a corresponding approach which is syntactical and finitary. In any case, it is still the abiding task of the Philosophy of Mathematics to gain a deeper understanding of (potential) infinity.

REFERENCES

[1] P. Benacerraf and H. Putnam (eds.), *Philosophy of Mathematics*, Selected Readings, Prentice-Hall, 1964.

[2] P. Cohen, The independence of the continuum hypothesis, *Proceedings of the National Academy of Sciences*, vol. 50, 1963, pp. 1143–1148; vol. 51, 1964, pp. 105–110.

[3] H. B. Curry, *Outline of a Formalist Philosophy of Mathematics*, Studies in Logic and the Foundations of Mathematics, North-Holland Publishing Company, 1951.

[4] A. A. Fraenkel and Y. Bar-Hillel, *Foundations of Set Theory*, Studies in Logic and the Foundations of Mathematics, North-Holland Publishing Company, 1958.

[5] K. Gödel, Über formal unentscheidbare Sätze der Principia Mathematica und verwandter Systeme, *Monatshefte für Mathematik und Physik*, vol. 38, 1931, pp. 173–198.

[6] ———, *The Consistency of the Axiom of Choice and of the Generalized Continuum-Hypothesis with* the *Axioms of Set Theory*, Annals of Mathematics Studies No. 3, Princeton 1940.
[7] ———, What is Cantor's continuum problem? *American Mathematical Monthly*, vol. 54, 1947, pp. 515–525. Revised version in [1], pp. 258–273.
[8] A. HEYTING, *Intuitionism, an Introduction*, Studies in Logic and the Foundations of Mathematics, North-Holland Publishing Company, 1956.
[9] D. HILBERT, *Gesammelte Abhandlungen*, vol. 3, Julius Springer, 1935.
[10] ———, Über das Unendliche, *Mathematische Annalen*, vol. 95, 1925, pp. 161–190, Partial translation in [1], pp. 134–151.
[11] G. KREISEL, Hilbert's programme, in [1], pp. 157–180.
[12] P. LORENZEN, *Einführung in die operative Logik und Mathematik*, Grundlehren der mathematischen Wissenschaften, vol. 78, Springer-Verlag, 1955.
[13] A. ROBINSON, *Introduction to Model Theory and to the Metamathematics of Algebra*, Studies in Logic and the Foundations of Mathematics, North-Holland Publishing Company, 1963.
[14] ———, On languages which are based on non-standard arithmetic, *Nagoya Mathematical Journal*, vol. 22, 1963, pp. 83–117.

APPENDIX

A NOTION OF POTENTIAL TRUTH

1. We shall consider sets or *structures* (or, relational systems) and we shall use the ordinary symbol of inclusion, $A \subset B$, in order to indicate that the structure B is an *extension* of the structure A or, which is the same, that A is a substructure of B (see [13], p. 24).

Let Δ be a set of structures which is *directed* under the relation $\subset$. That is to say, if A and B are elements of Δ, then Δ contains a structure C such that $A \subset C$ and $B \subset C$. It is implicit in this definition that the same relations are defined in all structures of Δ. The *union* of Δ is a structure M which is defined in the following way. The set of individuals of M is the set-theoretic union of the sets of individuals of the elements of Δ. Now let $R(x_1, \ldots, x_n)$ be an n-ary relation which is defined for the elements of Δ, and let $a_1, a_2, \ldots, a_n$ be individuals of M. Then there exist elements $A_1, \ldots, A_n$ of Δ such that a_i belongs to A_i, $i = 1, \ldots, n$. Moreover, since Δ is directed under the relation $\subset$ there exists an element A of Δ such that $A_i \subset A$, $i = 1, \ldots, n$, and hence, such that a_i belongs to A, $i = 1, \ldots, n$. We now claim that if A' and A'' are any elements of Δ such that a_i belongs to A and to A'' for $i = 1, \ldots, n$, then either $R(a_1, \ldots, a_n)$ holds in both A' and A'' or $R(a_1, \ldots, a_n)$ holds neither in A' nor in A''. Indeed, since Δ is directed under the relation $\subset$ there exists a structure A''' in Δ such that $A' \subset A''$ and $A'' \subset A'''$. If $R(a_1, \cdots, a_n)$ holds in A''' then it must hold in both A' and A''; if $R(a_1, \ldots, a_n)$ does not hold in A''' then $R(a_1, \ldots, a_n)$ holds neither in A' nor in A''. This proves our assertion.

We *define* that $R(a_1, ..., a_n)$ holds in M if $R(a_1, ..., a_n)$ holds in *some* structure A of Δ which contains $a_1, ..., a_n$, and hence in *all* structures of this kind. If there is no such structure A then we define that $R(a_1, ..., a_n)$ does not hold in M. This determines the structure M completely.

A sentence X is said to be *defined* in a structure A if the relations and individual constants of X (denote entities that) are contained in A.

2. For given Δ, and for any element A of Δ and any sentence X which is defined in Δ, we now introduce the notion of the *potential truth of X in A* as follows.

If X is an atomic formula, $X = R(a_1, ..., a_n)$, where a_i belongs to A_i, $i = 1, ..., n$, then X is *potentially true* in A if and only if X is true in A in the usual sense.

If $X = \sim Y$ then X is potentially true in A if and only if Y is not potentially true in A; if $X = Y \vee Z$ then X is potentially true in A if and only if at least one of the sentences Y and Z is potentially true in A; and so on, as by the use of truth tables, for the remaining connectives.

If X is obtained by existential quantification, $X = (\exists y)\,Z(y)$, then X is potentially true in A if and only if there exists an element B of Δ such that $A \subset B$ and such that for some individual b in B, $Z(b)$ is potentially true in all elements of Δ which are extensions of B. If X is obtained by universal quantification, $X = (\forall y)\,Z(y)$, then X is potentially true in A. if and only if $(\exists y)\,[\sim Z(y)]$ is not potentially true in A.

These rules determine the notion of potential truth uniquely for all X and A as described.

THEOREM. *Let A and A' be elements of Δ such that A' is an extension of A and let the sentence X be defined in A and hence in A'. Then X is potentially true in A' if and only if it is potentially true in A.*

PROOF. The assertion is evidently correct if X is atomic, for in this case potential truth coincides with truth. For all other X we shall prove our assertion by induction following the construction of X. If $X = \sim Y$, where the assertion of the theorem has been proved already for Y, and X is potentially true in A, then Y is not potentially true in A. Hence, Y is not potentially true in A' and X is potentially true in A', as required. Similarly, if X is not potentially true in A, Y is potentially true in A and A', hence X is not potentially true in A', as required. If $X = Y \vee Z$, where the assertion of the theorem has been proved already for Y and Z, and X is potentially true in A, then one of Y and Z, e.g., Y, is potentially true in A. It follows that Y is potentially true in A' and hence that X is potentially true in A', as required. If X is not potentially true in A then Y and Z are not potentially true in

A and, hence, are not potentially true in A'. It follows that X is not potentially true in A'. A similar procedure applies to the remaining connectives.

Now suppose that $X = (\exists y)\, Z(y)$, where the assertion of the theorem has been proved for all sentences $Z(a)$. Suppose also that X is potentially true in A. Then for some B in Δ which is an extension of A and for some individual b in B, $Z(b)$ is potentially true in all extensions of B which belong to Δ. But Δ is a directed set, so there exists an element B' of Δ such that $A' \subset B'$ and $B \subset B'$. Then b is contained in B' and $Z(b)$ is potentially true in all extensions of B' which belong to Δ. This shows that X is potentially true in A'.

Suppose on the other hand that X is defined in A and is potentially true in A'. Then there exists a $B' \in \Delta$, $A' \subset B'$ such that, for some individual b' in B', $Z(b')$ is potentially true in all extensions of B' which belong to Δ. But B' is also an extension of A, so X is potentially true in A, as required.

Suppose finally that $X = (\forall y)\, Z(y)$ where the assertion of the theorem has been proved for all sentences $Z(a)$. Then, as shown previously, the assertion of the theorem is true also for all sentences $\sim Z(a)$. It then follows from the preceding argument that $(\exists y)\, [\sim Z(y)]$ is potentially true in A if and only if it is potentially true in A', i.e., that $(\forall y)\, Z(y)$ is *not* potentially true in A if and only if it is not potentially true in A'. This completes the proof of the theorem.

COROLLARY. *A sentence X is either potentially true in all structures of Δ in which it is defined or it is not potentially true in any structure of Δ.*

3. Let M be the union of Δ as defined in section 1 of the appendix. The following result provides the link between the ordinary truth definition in M and the above potential truth definition with respect to Δ.

THEOREM. *Let X be any sentence which is defined in M. Then X is true in M if and only if it is potentially true in all elements of Δ in which it is defined.*

PROOF. The assertion of the theorem is evidently correct if X is atomic. For other X the proof proceeds again by induction following the construction of the formula. Omitting the case of connectives, for which the argument is trivial, we suppose that $X = (\exists y)\, Z(y)$ is a sentence which is defined in M and such that the assertion of the theorem has been proved already for all $Z(b)$ where b is (or denotes) an individual in M.

If X is true in M then $Z(b)$ is true for some b in M. Let A be any element of Δ in which X is defined. Since Δ is directed it includes an element B which is an extension of A and contains b. Then $Z(b)$ is potentially true in all extensions of B which belong to Δ (since the theorem is supposed to have been proved already for each $Z(b)$) and so $(\exists y)\, Z(y)$ is potentially true in A.

Conversely, suppose that X is potentially true in some element A of Δ. (Thus, by the result of the preceding section, X is potentially true in all elements of Δ in which it is defined.) Then $Z(b)$ is potentially true in some extension B of A which belongs to Δ. Hence, $Z(b)$ is true in M and X is true in M, as required.

The case where X is obtained by universal quantification can be reduced immediately to that considered just now. This completes the proof of the theorem.

The conditions imposed on Δ are satisfied in the following two examples.

(i) Let N be the set of natural numbers with the relations of addition, multiplication, and equality, and let Δ be the set of structures which are obtained by restricting N to the sets $N_k = \{0, 1, \ldots, k\}$, $k = 0, 1, 2, \ldots$. Then $N = M$ is the union of Δ.

(ii) Let M be any structure and let Δ be the set of structures which are obtained by restricting M in all possible ways to its *finite* subsets. Then M is the union of Δ.

In both cases, we arrive at the usual truth definition for M by means of a notion of potential truth for *finite* substructures of M. It will be seen that the most important part of our notion of potential truth concerns existential quantification. It corresponds to the intuitive idea that an existential sentence is potentially true in a given structure if we can find an element that satisfies it by extending the structure far enough in the right direction.

MODEL THEORY

by

ABRAHAM ROBINSON

Yale University

1. *Early History.* Model theory is the branch of mathematical logic which is concerned with the interrelation between sets of sentences, usually given in a specified formal language, on one hand and sets of structures which *satisfy* these sentences (equivalently: in which these sentences *hold*, or are *true*, or *valid*) on the other hand. The term 'model theory' was introduced by A. Tarski in the early nineteen-fifties but the development of the basic principles of the subject goes back a good deal further.

The first significant result of mathematical logic which may be regarded as part of model theory, and one that still plays an important role in the subject is the theorem of Löwenheim-Skolem (resp. [1] and [2]). In contemporary parlance, the theorem states that any sentence of the lower predicate calculus which is satisfied by some structure is satisfied already by a structure which is at most countable (i.e., is finite or countable). It is apparent that this result is concerned only with the question of satisfaction by certain structures and does not refer to the rules of deduction of the formal language to which the sentence belongs. The principal result which provides a bridge between the deductive ('syntactical') properties of the lower predicate calculus on one hand and model-theoretic concepts on the other hand was published by Gödel [3] in 1930. Gödel's theorem states that if a sentence of the lower predicate calculus holds in all structures for which it has a meaning then it can be obtained by a sequence of applications of the (previously specified) rules of deduction of the calculus. Or, expressing the same result in a way better suited to our present purpose, if a sentence in the lower predicate calculus is *consistent* in the sense that it cannot be refuted by the rules of deduction of the calculus then there exists a structure in which the sentence is true.

Although the property of a sentence to be true in a particular structure, or in all structures, has always played a part in mathematics as well as in mathematical logic, logicians had traditionally put the emphasis on deducibility rather than on truth. The reason for this is to be found not merely in the lack of a certain insight but also in the fact that it was the original design of logic to state and investigate our concrete 'laws of thought'. While the rules of deduction of the lower predicate calculus and of other finitistic calculi are concrete in the sense that they can be realized, for example, on a computer, this is not the case for the definition of truth in a general structure. In particular (compare section 2, below) a formal sentence $(\exists x)Y(x)$ is, by definition, true in a structure M if there exists an object in M, to be denoted by a, such that Y(a) is true in M. However, this definition does not imply the existence of a procedure which would enable us actually to *find* a, and no amount of formal sophistication can eliminate this non-constructive feature, unless we are willing to modify the original concept. But just as in classical mathematics a non-constructive approach has come to be accepted as a matter of course, so the infinitistic nature of model theory is something that logicians have learned to live with.

After Gödel, the next memorable step was taken by Tarski in a long paper entitled "The concept of truth in formalised languages" [4]. The paper contains the first detailed analysis of the interpretation of formal sentences in a structure (or "universe of discourse") and it is this, even more than an important technical result contained in it, which lends the paper its lasting significance. About the same time, Skolem published an article [5] which, in spite of the very special nature of the problem considered in it, was to have a profound influence on later developments. In it Skolem showed, by means of a suitable construction, that there exist proper extensions of the system of natural numbers (now called *non-standard models of arithmetic*) which have the same properties as the natural numbers to the extent to which these properties can be expressed in the lower predicate calculus (in terms of equality, addition, and multiplication). As the title of his paper indicates, it was Skolem's main purpose to show that formal languages are not adequate to the task of giving a complete characterization of the natural numbers. However, in due course structures similar to those considered by Skolem, came to be studied in their own right (compare sections 6 and 7 below).

Another strand in the early development of model theory received its impetus from algebra. Within mathematics, we may draw a broad distinction between disciplines which concentrate on the properties of a structure which is supposedly unique up to isomorphism, such as number theory or the theory of functions of a complex variable, and disciplines which are from the outset interested in the totality of realizations, or *models*, of a given system of axioms, such as group theory or algebraic field theory. The second approach is particularly noticeable in modern algebra although it occurs also elsewhere, e.g. in the general theory of topological spaces. It is important to note here that the axioms of the most common algebraic concepts, such as groups, rings, and fields, are formulated within the lower predicate calculus while the axioms for a general topological space involve concepts which belong to higher order languages or, alternatively, which presuppose a set-theoretical framework.

Bearing in mind the definition of model theory given at the beginning of this section it was thus natural that algebra should serve both as an inspiration to the newer field and as an area of application for it. The first to work in this direction was the great Russian algebraist and logician A. I. Malcev, who died in July, 1967. In a paper published in 1936 [6], Malcev stated the following *finiteness principle* now known, in a slightly different formulation, as the *compactness theorem* and still the basic proposition of the lower predicate calculus. In Malcev's version, the principle asserts that if for every finite subset K′ of a set of sentences K in the lower predicate calculus there exists a structure M′ which satisfies the sentences of K′ then there exists a structure M which satisfies all sentences of K simultaneously. In 1941 Malcev [7] used this principle in order to prove certain new theorems on infinite groups—probably the first effective application of logic to classical algebra.

A. Robinson's dissertation ([8] 1949, pub. 1951, compare [9]) is based entirely on the idea that logic provides the tools for developing a general theory of theories which are formulated in the lower predicate calculus by analogy with the special theories considered in abstract algebra, in particular with algebraic field theory. Sets of models (now called *varieties* or *elementary classes*) for sets of sentences of the lower predicate calculus are regarded as analogous to sets of zeros of polynomials or polynomial ideals. The dissertation contains various results and methods (e.g., the 'method of diagrams') which later became standard in the subject, as well as a number of effective applications to field theory.

Since 1950, model theory has expanded rapidly; logicians from many countries have contributed to its development, Tarski's school at Berkeley being the most active center for the study of the subject. A selective bibliography will be found at the end of this article. The reader may consult Addison *et al.* [10] for a more comprehensive bibliography up to 1963.

In the following sections, we shall describe some typical methods and results obtained in model theory over the last fifteen years. For the convenience of the reader our style will be somewhat informal but we have retained the full technical apparatus in cases where nothing else seems to give a proper picture of the topic under discussion (e.g. in section *6*).

2. Semantics of a formal language. The formal language of the lower predicate calculus is not defined uniquely, but it is not difficult to correlate its several versions. We start with a list of *atomic symbols*, e.g. individual constants, a, b, c, ..., individual variables, x, y, z, ..., relation symbols of any number of variables $R(\ )$, $S(\ ,\ )$; function symbols of any number of variables, f( ), g(,), ... ; connectives, $\neg$ (not), $\wedge$ (and), $\vee$ (or), $\supset$ (implies), $\equiv$ (if and only if); quantifiers $(\forall)$—for all, and $(\exists)$—there exists; parentheses, [and].

Terms and *well-formed formulae* are built up from atomic symbols by applying certain definite rules of formation, whose number is finite. We shall suppose that the reader is familiar with such a set of rules, or at least that he can decide intuitively whether or not a formula is well-formed. Quantification is permitted only with respect to individual variables. We shall assume that no variable is in a quantifier within the scope of a quantifier with the same variable. A sentence is a well-formed formula without free (unquantified) variables. A sentence is said to be in prenex normal form if no connective precedes a quantifier from left to right, i.e. if the formula is built up in such a way that quantifiers are introduced only after all applications of connectives.

Although this is not essential, we may think of a *structure* M as a set-theoretical construct, consisting of a set of individuals, δ; of a number of relations ρ, σ, ...—i.e. sets for unary relations, sets of pairs for binary relations, and more generally sets of n-tuples for n-ary relations—all with domain δ; and of a number of functions φ, ψ, ... from δ into δ—where, set-theoretically, a function of n variables is simply a special kind of $(n+1)$-ary relation.

The *interpretation* of a sentence X in a structure M presupposes a correspondence which assigns to each relation symbol occurring in X a relation of M and to each function symbol of X a function of M. Moreover, we assume that for every individual α of M—i.e., α in δ—there is an individual constant a in the language which *denotes* α. The correspondence $a \to \alpha$ may be one-one or many-one, as preferred. Let D be the set of individual constants which denote elements of δ, so that we have a correspondence $T \to \delta$ where T is the set of terms which are built up from individual constants of D by means of function symbols that denote functions in M. For example, if $a \to \alpha$, $b \to \beta$, $f(\ ,\ ) \to \varphi(\ ,\ )$, then $f(a, b) \to \varphi(\alpha, \beta)$ and $f(f(a, b), b) \to \varphi(\varphi(\alpha, \beta), \beta)$.

Now suppose that $R(\ , \dots ,\ )$ is an n-ary relation symbol which denotes the n-ary relation ρ in M, and let $t_1, \dots, t_n$ be terms of T denoting individuals $\alpha_1, \dots, \alpha_n$ of δ. Then we say that $R(t_1, \dots, t_n)$ is *true* (or *holds*) in M if $\langle \alpha_1, \dots, \alpha_n \rangle$ belongs to ρ and that $R(t_1, \dots, t_n)$ is *false* (or, *does not hold*) in M if $\langle \alpha_1, \dots, \alpha_n \rangle$ does not belong to ρ. We extend these truth definitions by means of the usual truth-functional interpretation of the connectives and by means of the interpretation of existential quantification given in section *1* above with a corresponding definition for universal quantification. This completes our interpretation of a sentence in a structure. A relation of equality may also be introduced—either as an ordinary relation which is substitutive and an equivalence or as the relation of identity. In the latter case, we supplement our list of symbols by the introduction of '=' and we define that $s = t$ holds in M if s and t denote the same element of δ.

Observe that if the set of individuals of M, δ, is uncountable, then the set of symbols D must be uncountable. The introduction of uncountable sets of symbols at this point can actually be avoided but it is characteristic of contemporary model theory that it makes free reference to uncountable sets of symbols and to other infinitistic notions. Apart from this feature, the above rules for the interpretation of a sentence evidently are only a formal explication of our intuitive understanding of this notion.

3. Prefix problems. We shall now consider some typical questions and results of model theory. Let X be a sentence of the lower predicate calculus and suppose that X is in prenex normal form. The sequence of quantifiers of X is called its *prefix*. We may classify X according to the nature of its prefix. For example, X will be called *existential* if its prefix contains existential quantifiers only (or if X does not contain any quantifiers at all). Similarly, X is universal if it contains universal quantifiers only. X is called an $\forall\exists$-sentence if no universal quantifier in its prefix is preceded by an existential quantifier, etc.

We may ask whether the various classes of sentences introduced above can be characterized also by the set-theoretic properties of the sets of structures which satisfy these sentences. However, in this form the question is not well put for if X holds in a structure M, then clearly any sentence Y which is logically equivalent to X and which contains the same 'non-logical constants' (relation symbols, function symbols, and individual constants), also holds in M. Accordingly, we modify our question and ask what are the set-theoretical properties of the models of a sentence which is equivalent to a sentence of one of the above mentioned classes. Now it is easy to see that if a sentence Y is logically equivalent to an existential sentence X then whenever Y holds in a structure M it holds also in any extension M′ of M. It has been shown by Tarski [11] that the converse assertion is also correct; if a sentence Y is such that whenever it holds in a structure M then it holds also in all extensions of M, then Y is logically equivalent to an existential sentence. Similarly, if X is an ∀∃-sentence or is equivalent to an ∀∃-sentence and holds in all the structures M_i of an ascending chain of structures $M_1 \subset M_2 \subset M_3 \subset \ldots$ then it is easy to see that X holds also in the union of these structures $M = \bigcup_n M_n$; while the converse of this assertion is again non-trivial and constitutes a theorem of Łoś and Suszko [12]. We may mention here also an earlier result which does not refer to the particular form of a sentence X but still requires that X be formulated in the lower predicate calculus. It states, roughly, that if X holds in the intersection of any two of its models (supposed not empty) then whenever X holds in the structures M_i of a chain $M_1 \subset M_2 \subset M_3 \subset \ldots$ then X holds also in their union (Robinson [8], Chang [13]). Using the above mentioned theorem of Łoś and Suszko, we may conclude that X is equivalent to an ∀∃ sentence.

4. Completeness. Elementary equivalence. By *the vocabulary* of a sentence or set of sentences we mean the set of non-logical constants contained in it. Let K be a non-empty set of sentences and X be any sentence. We then have the following possibilities. (i) X is deducible from K; (ii) X is refutable by K, i.e. $\neg X$ is deducible from K; (iii) neither X nor $\neg X$ is deducible from K. If the third possibility does not arise for any X whose vocabulary is contained in the vocabulary of K then K is called *complete.* If K is contradictory then it is obviously complete; we shall exclude this trivial case.

From the model-theoretic point of view, it is easy to see that any consistent set of sentences K possesses a complete extension K′. For if K is consistent then it possesses a model M and if K′ is the set of all sentences whose vocabulary is contained in the vocabulary of K and which hold in M, then K′ is clearly complete. However, this does not provide an effective procedure for obtaining K′ even if K is given effectively.

To put the notion of completeness in model-theoretic terms, let us fix a vocabulary, V, and let us assume that it is meaningful for (interpreted in) two structures, M_1 and M_2. Then M_1 and M_2 are called *elementarily equivalent* for the given vocabulary if any sentence X whose vocabulary is contained in V holds in M_2 if and only if it holds in M_1. A set of sentences K is complete if any two models of K are elementarily equivalent for the vocabulary of K.

For an example of a mathematically important set of axioms which is known to be complete, consider the notion of a real closed field. This notion can be axiomatized in the lower predicate calculus in terms of the relations of equality and (redundantly) order, and in terms of the operations (functions) of addition and multiplication. Perhaps the best known example of a real closed field is the field of real numbers. The fact that (a set of axioms for) this notion is complete was first proved by Tarski (Tarski & McKinsey [14]) who actually provided a procedure which enables us to decide for any sentence X formulated in terms of the above mentioned relations whether X holds in any real closed field and hence, in all real closed fields. The same result can be derived by means of a general theory of complete theories (Robinson [15]) which is based on a modified notion of completeness called *model completeness*. The theory has been used in order to establish completeness also in many other cases.

There is an interesting connection between the notions of completeness and decidability. Suppose that K is a finite or a recursively enumerable set of axioms. Then if K is complete it is also decidable. That is to say, if K is complete then there exists a program which enables us to decide for any sentence X whose vocabulary is in the vocabulary of K whether or not X is deducible from K. In fact, such a program is provided by any systematic procedure which enumerates all sentences whose vocabulary is contained in the vocabulary of K and which are deducible from K. Given any X whose vocabulary is contained in the vocabulary of K, such a procedure will, sooner or later, arrive either at X or at $\neg X$, thus telling us *whether or not* X is deducible from K.

To conclude this section, we may mention the notion of an *elementary* (or *arithmetical*) *extension*, which occurs frequently in model-theoretic considerations (Tarski & Vaught [16]). Let M′ be an extension of a structure M (so that all relations and functions which are defined in M extend to M′). Then M′ is called an elementary extension of M if M′ is elementarily equivalent to M for a vocabulary which includes symbols for all relations and function of M *as well as for all individuals of* M. In terms of this notion the concept of model-completeness which was mentioned earlier in this section may be characterized as follows. A set of sentences K is model-complete if and only if for every pair of models M, M′ of K such that M′ is an extension of M, M′ is also an elementary extension of M.

5. Beth's Theorem. Beth's theorem provides an interesting example of a purely syntactical result which can be proved conveniently by model-theoretic methods. Suppose that the vocabulary of a sentence X contains a relation symbol R(, ... ,) of n variables as well as a number of other relation symbols. Displaying R, we write X = Y(R), and we write X′ = Y(R′) for the sentence which is obtained from X by replacing R everywhere by another n-place relation symbol, R′, which did not occur previously. Suppose further that the sentence

$$X \wedge X' \supset [(\forall x_1) \ldots (\forall x_n)[R(x_1, \ldots, x_n) \equiv R'(x_1, \ldots, x_n)]]$$

is provable in the lower predicate calculus. Then Beth's theorem states that R can be defined explicitly in terms of the remaining vocabulary of X. That is to say, there exists a well-formed formula with the free variables $x_1, \ldots, x_n$, $Q(x_1, \ldots, x_n)$ say, whose vocabulary is contained in the vocabulary of X but which does not contain R, such that

$$X \supset [(\forall x_1) \ldots (\forall x_2)[R(x_1, \ldots, x_n) \equiv Q(x_1, \ldots, x_n)]]$$

is provable in the lower predicate calculus.

In his original paper, Beth [17] indicated how to prove the theorem by means of his *method of semantic tableaux*, which is intermediate between model theory and proof theory and which is related to Herbrand's method. A model-theoretic proof was given by A. Robinson [18] by means of the *consistency lemma*. Going in the opposite direction, Craig [19] established Beth's theorem by a purely proof-theoretic argument depending on his *interpolation lemma*. Although they differ in their basic character, the consistency lemma and the interpolation lemma can be derived from one another. By virtue of Gödel's completeness theorem, Beth's theorem itself can be restated in purely model-theoretic terms.

6. Ultraproducts. So far, we have not discussed the methods by which model theory arrives at the structures whose existence is affirmed by its theorems. During the earlier stages, the detailed features of these structures were in fact not relevant to the discussion and it was sufficient to know that they could be obtained, usually by appealing to some strongly infinitistic principle such as the axiom of choice, or the well-ordering theorem, or Zorn's lemma, or the maximal ideal theorem for Boolean algebras. However, more recently, the following construction, which was foreshadowed to some extent by Skolem's method for obtaining non-standard models of arithmetic (see section *1* above) has proved very fruitful (Frayne *et al.* [20], Keisler [21], Kochen [22]). Let $\{M_\nu\}$ be a set of structures indexed on an infinite index set $I = \{\nu\}$. Suppose that 'the same' relations and functions (i.e. relations and functions denoted by the same symbols in our formal language) are given in the several M_ν. This will be the case, in particular, if all the M_ν are copies of one and the same structure M. We now define a new structure M′ whose 'individuals' are the functions f(x) with domain I such that $f(\nu) \varepsilon M_\nu$ for all $\nu \varepsilon I$. In order to define the relations which hold between these 'individuals', we must be given a free ultrafilter D on I. That is to say, D is a set of subsets of I with the following properties. (i) If $A \varepsilon D$ and $A \subset B \subset I$ then $B \varepsilon D$. (ii) If $A \varepsilon D$ and $B \varepsilon D$ then $A \cap B \varepsilon D$. (iii) For any $A \subset I$, either $A \varepsilon D$ or $I - A \varepsilon D$. (iv) The empty set is not contained in D. (v) No element of I is contained in all elements of D (this rules out a trivial case).

Now let R(x, y, z) denote a relation (assumed ternary by way of example) which is defined in the structures M_ν, and let f, g, h denote 'individuals' of M′ as introduced above. Then R(f, g, h) shall hold in M′, by definition, if and only if the set

$$\{\nu \mid R(f(\nu), g(\nu), h(\nu)) \text{ holds in } M_\nu\}$$

belongs to D. The definition of functions in M′ corresponding to the functions defined in the M_ν is carried out component-wise.

The structure M′ obtained in this way is called an *ultraproduct*. If all the M_ν are copies of a single M then M′ is called an *ultrapower* of M. A basic result on ultraproducts states that in this case M′ is an elementary extension of M. More generally, the ultraproduct construction can be used in order to prove the compactness theorem (section *1*, above). Beyond that, certain properties of ultraproducts can be used to prove theorems about the lower predicate calculus (e.g. Beth's theorem) even though these properties cannot be formulated in the language of the lower predicate calculus.

7. Applications to algebra and analysis. The compactness theorem provides a ready tool for proving embedding theorems in algebra, such as: if every finitely generated subring of a noncommutative ring R can be embedded in a skew field then R can be embedded in a skew field. Another kind of application is related to completeness (section *4* above). For example, there exist results about the field of real numbers which can be formulated in the lower predicate calculus, but which have been proved by topological methods. Tarski's theorem of section *4* shows that such results are true in all real closed fields irrespective of their topological properties, if any. A related method has been used to give a new proof of Artin's theorem concerning Hilbert's seventeenth problem and even to establish the existence of certain numerical bounds whose existence had previously been only conjectured (Robinson [9]). But the most spectacular result in this area is due to Ax & Kochen [23], who used model-theoretic methods in order to dispose of a famous conjecture due to Artin.

There is a far-reaching application of model theory to analysis which is based on the existence of proper (hence non-archimedean) extensions of the real numbers R which share many of the formal properties of R. Such structures, called *non-standard models of analysis* can be obtained, for example, as ultrapowers of R. They provide a satisfactory framework for the development of the differential and integral calculus and of other branches of analysis by means of infinitely large and infinitely small (infinitesimal) quantities as envisaged by Leibniz (Robinson [24]).

8. Model theory and set theory. As mentioned in section *1*, a question relating to the possible cardinality of a model of an axiomatic system was considered at the very beginning of the history of our subject. In recent years, there has been a considerable amount of work on a class of far more refined problems of the same nature (see Vaught [25]). A difficult problem of another type, but still involving cardinalities, was solved by M. D. Morley [26].

As a matter of long tradition, independence and relative consistency results for axiomatic theories are established by the construction of suitable models. However, the methods used for this purpose may well depend on the special nature of the axiomatic theory under consideration. Thus, in set theory, the existential nature of most of the axioms and the availability of the ordinals for the ranking of sets has led to special methods of great depth and power (Cohen [27], Gödel [28]). Quite recently, these methods have been linked with some of the ideas of general model theory and this has led to exciting developments which are still in progress (*Lecture-Notes* [29]).

BIBLIOGRAPHY

ADDISON, J. W., HENKIN, L., & TARSKI, A., (eds.) [10] *The Theory of Models* (Proceedings of the 1963 International Symposium at Berkeley), Amsterdam 1965.

AX, J., & KOCHEN, S., [23] *Diophantine Problems over Local Fields*, American Journal of Mathematics **87**, 605-630, 631-648 (1965).

BETH, E. W., [17] *On Padoa's Method in the Theory of Definition*, Proceedings of the Academy of Sciences of the Netherlands, ser. A **56**, 330-339 (1953).

CHANG, C. C., [13] *On Unions of Chains of Models*, Proceedings of the American Mathematical Society **10**, 120-127 (1959).

CHANG, C. C., & KEISLER, H. J., *Continuous Model Theory*, Princeton (N.J.) 1966.

COHEN, Paul J., [27] *Set Theory and the Continuum Hypothesis*, New York & Amsterdam 1966.

CRAIG, W., [19] *Three Uses of the Herbrand-Gentzen Theorem in Relating Model Theory and Proof Theory*, Journal of Symbolic Logic **22**, 269-285 (1957).

EHRENFEUCHT, A., & MOSTOWSKI, A., *Models of Axiomatic Theories Admitting Automorphisms*, Fundamenta Mathematicae **43**, 50-68 (1956).

FRAÏSSÉ, R. J., *Sur quelques classifications des systèmes de relations* (Publications scientifiques de l'Université d'Algers, sér. A., vol. 1), Algers 1954, pp. 35-182.

FRAYNE, T. E., MOREL, A. C., & SCOTT, D. S., [20] *Reduced Direct Products*, Fundamenta Mathematicae **51**, 195-228 (1962).

FUHRKEN, E. G., *Languages with Added Quantifier "There Exist at Least* $\aleph$", in: *The Theory of Models (Berkeley 1963)*, Amsterdam 1965, pp. 121-131.

GÖDEL, K., [3] *Die Vollständigkeit der Axiome des Logischen Funktionenkalküls*, Monatshefte für Mathematik und Physik **37**, 349-360 (1930).

——— [28] *The Consistency of the Axiom of Choice and of the Generalized Continuum Hypothesis with the Axioms of Set Theory*, Princeton (N.J.) 1940.

HANF, W., *Model-Theoretic Methods in the Study of Elementary Logic*, in: *The Theory of Models (Berkeley 1963)*, Amsterdam 1965, pp. 132-145.

HENKIN, L. A., *The Completeness of the First Order Functional Calculus*, Journal of Symbolic Logic **14**, 159-166 (1949).

——— *Some Interconnections between Modern Algebra and Mathematical Logic*, Transactions of the American Mathematical Society **74**, 410-427 (1953).

HORN, A., *On Sentences which are True of Direct Unions of Algebras*, Journal of Symbolic Logic **16**, 14-21 (1951).

KARP, C., *Languages with Expressions of Infinite Length*, Amsterdam 1964.

KEISLER, H. J., [21] *Ultraproducts and Elementary Classes*, Proceedings of the Academy of Sciences of the Netherlands, ser. A **64**, 477-495 (1961).

——— *Some Applications of the Theory of Models to Set Theory*, in: *Logic, Methodology and Philosophy of Science (Stanford 1960)*, Stanford 1962.

KOCHEN, S., [22] *Ultraproducts in the Theory of Models*, Annals of Mathematics, ser. 2, **74**, 221-261 (1961).

KREISEL, G., *Model-Theoretic Invariants: Applications to Recursive and Hyperarithmetic Operations*, in: *The Theory of Models (Berkeley 1963)*, Amsterdam 1965.

Kripke, S. A., *A Completeness Theorem in Modal Logic,* Journal of Symbolic Logic **24**, 1-14 (1957).
Lecture-Notes [29] prepared in connection with the Summer Institute on Axiomatic Set Theory, Los Angeles 1967.
Lévy, A., *On the Principles of Reflection in Axiomatic Set Theory,* in: *Logic, Methodology and Philosophy of Science (Stanford 1960),* Stanford 1962.
Łoś, J., *On the Extending of Models,* I, Fundamenta Mathematicae **42**, 38-54 (1955).
Łoś, J., & Suszko, R., [12] *On the Extending of Models,* IV. *Infinite Sums of Models,* Fundamenta Mathematicae **44**, 52-60 (1957).
Löwenheim, L., [1] *Über Möglichkeiten in Relativkalkül,* Mathematische Annalen **76**, 447-470 (1915).
Lyndon, R., *Properties Preserved under Homomorphism,* Pacific Journal of Mathematics **9**, 143-154 (1959).
MacDowell, R., & Specker, E., *Modelle der Arithmetik,* in: *Infinitistic Methods (Warsaw 1959),* New York, Oxford, London, Paris & Warsaw 1961.
Malcev, A. I., [6] *Untersuchungen aus dem Gebiete der mathematischen Logik,* Matematičeskij Sbornik **1**, 323-336 (1936).
——— [7] *Ob odnom obščem metode polučenija localni teorem teorij grupp,* Proceedings of the Pedagogical Institute of Ivanovo **1**, 3-9 (1941).
Mendelson, E., *On Non-Standard Models for Number Theory,* in: *Essays on the Foundations of Mathematics,* Jerusalem 1961, pp. 259-268.
Montague, R. M., *Semantical Closure and Non-Finite Axiomatizability,* in: *Infinitistic Methods (Warsaw 1959),* New York, Oxford, London, Paris & Warsaw 1961.
Morley, M. D., [26] *Categoricity in Power,* Transactions of the American Mathematical Society **114**, 514-538 (1965).
Morley, M. D., & Vaught, R. L., *Homogeneous Universal Models,* Mathematica Scandinavica **11**, 37-57 (1962).
Mostowski, A., *On Absolute Properties of Relations,* Journal of Symbolic Logic **12**, 33-42 (1947).
——— *On Recursive Models of Formalized Arithmetic,* Bulletin de l'Académie Polonaise des Sciences, Mathematical-Astronautical-Physical Series **5**, 401-404 (1957).
Rabin, M. O., *Arithmetical Extensions with Prescribed Cardinality,* Proceedings of the Academy of Sciences of the Netherlands, ser. A **62**, 439-446 (1959).
——— *Universal Groups of Automorphisms of Models,* in: *The Theory of*
Robinson, A., [8] *On the Metamathematics of Algebra (London 1949),* Amsterdam 1951.
——— [9] *Introduction to Model Theory and to the Metamathematics of Algebra,* Amsterdam 1963.
——— [15] *Complete Theories,* Amsterdam 1956.
——— [18] *A Result on Consistency and its Application to the Theory of Definition,* Proceedings of the Academy of Sciences of the Netherlands, ser. A. **59**, 42-58 (1956).
——— [24] *Non-Standard Analysis,* Amsterdam 1966.
Scott, D. S., *Measurable Cardinals and Constructive Sets,* Bulletin de l'Académie Polonaise des Sciences, Mathematical-Astronomical-Physical Series **9**, 521-524 (1961).
——— *On Constructing Models for Arithmetic,* in: *Infinitistic Methods (War-*

saw 1959), New York, Oxford, London, Paris & Warsaw 1961, pp. 235-255.

SHEPHERDSON, J. C., *Inner Models for Set Theory*, Journal of Symbolic Logic **16**, 161-190 (1951); **17**, 225-237 (1952); **18**, 145-167 (1953).

SKOLEM, T., [2] *Logisch-kombinatorische Untersuchungen über die Erfüllbarkeit oder Beweisbarkeit mathematischer Sätze nebst einem Theorem über dichte Mengen* (Skrifter Videnskaps-selskapet Kristiania, Mathematical-Scientific Class, no. 4), Kristiania 1920.

——— [5] *Über die Nichtcharakterisierbarkeit der Zahlenreihe mittels endlich oder abzählbar unendlich vieler Aussagen mit ausschliesslich Zahlenvariablen*, Fundamenta Mathematicae **23**, 150-161 (1934).

TARSKI, A., [4] *Der Wahrheitsbegriff in den formalisierten Sprachen* (Polish 1933), Studia Philosophica **1**, 261-405 (1935).

——— [11] *Contributions to the Theory of Models*, Proceedings of the Academy of Sciences of the Netherlands, ser. A **57**, 572-588 (1954); **58**, 56-64 (1955).

TARSKI, A., & MCKINSEY, J. C. C., [14] *A Decision Method for Elementary Algebra and Geometry* (1948), 2nd ed., Berkeley & Los Angeles 1951.

TARSKI, A., & VAUGHT, R. L., [16] *Arithmetical Extensions of Relational Systems*, Compositio Mathematica **13**, 81-102 (1957).

VAUGHT, R. L., [25] *A Löwenheim-Skolem Theorem for Cardinals Far Apart*, in: *The Theory of Models (Berkeley 1963)*, Amsterdam 1965, pp. 390-401.

THE METAPHYSICS OF THE CALCULUS

ABRAHAM ROBINSON

University of California, Los Angeles

1. From the end of the seventeenth century until the middle of the nineteenth, the foundations of the Differential and Integral Calculus were a matter of controversy. While most students of Mathematics are aware of this fact they tend to regard the discussions which raged during that period entirely as arguments over technical details, proceeding from logically vague (Newton) or untenable (Leibniz) ideas to the methods of Cauchy and Weierstrass which meet modern standards of rigor. However, a closer study of the history of the subject reveals that those who actually took part in this dialogue were motivated or influenced quite frequently by basic philosophical attitudes. To them the problem of the foundations of the Calculus was largely a philosophical question, just as the problem of the foundations of Set Theory is regarded in our time as philosophical no less than technical. Thus, d'Alembert states in a passage from which I have taken the title of this address ([2]):

'La théorie des limites est la base de la vraie Métaphysique du calcul différentiel.'

It will be my purpose today to describe and analyse the interplay of philosophical and technical ideas during several significant phases in the development of the Calculus. I shall carry out this task against the background of Non-standard Analysis as a viable Calculus of Infinitesimals. This will enable me to give a more precise assessment of certain historical theories than has been possible hitherto.

The basic ideas of Non-standard Analysis are sketched in the next two sections. A comprehensive development of that theory will be found in [10]. The last chapter of that reference also contains a more detailed discussion of the historical issues raised in the present talk.

2. Let R be the field of real numbers. We introduce a formal language L in order to express within it statements about R. The precise scope of the language depends on the purpose in hand. We shall suppose here that we have chosen L as a very rich language. Thus, L shall include symbols for all individual real numbers, for all sets of real numbers, for all binary, ternary, quaternary, etc., relations between real numbers, and also for all sets and relations of higher order, e.g. the set of all binary relations between real numbers. In addition L shall include the connectives of negation, disjunction, conjunction and implication and also variables and quantifiers. Quantification will be permitted at all levels, but we may suppose, for the sake of familiarity, that type restrictions have been imposed in the usual way. Thus, L is the language of a 'type theory of order ω'. Within it one can express all facts of Real (or of Complex) Analysis. There is no need to introduce function symbols explicitly for to every function of n variables $y=f(x_1, \ldots, x_n)$ there corresponds an $n+1$-ary relation $F(x_1, \ldots, x_n, y)$ which holds if and only if $y=f(x_1, \ldots, x_n)$.

Let K be the set of all sentences in L which hold (are true) in the field of real numbers, R. It follows from standard results of Predicate Logic that there exists a proper extension $^*R$ of R which is a model of K, i.e. such that all sentences of K are true also in $^*R$. However, the statement just made is correct only if the sentences of K are interpreted in $^*R$ 'in Henkin's sense'. That is to say, when interpreting phrases such as 'for all relations' (of a certain type, universal quantification) or 'for some relation' (of a certain type, existential quantification) we take into account not the totality of all relations (or sets) of the given type but only a subset of these, the so-called *internal* or *admissible* relations (or sets). In particular, if S is a set or relation in R then there is a corresponding internal set or relation $^*S$ in $^*R$, where S and $^*S$ are denoted by the same symbol in L. However not all internal entities of $^*R$ are of this kind.

The *Non-standard model of Analysis* $^*R$ is by no means unique. However, once it has been chosen, the totality of its internal entities is given with it. Thus, corresponding to the set of natural numbers N in R, there is an internal set $^*N$ in $^*R$ such that $^*N$

is a proper extension of N. And $^*N$ has 'the same' properties as N, i.e. it satisfies the same sentences of L just as $^*R$ has 'the same' properties as R. N is said to be a *Non-standard model of Arithmetic*. From now on all elements (individuals) of $^*R$ will be regarded as 'real numbers', while the particular elements of R will be said to be *standard*.

$^*R$ is a non-archimedean ordered field. Thus $^*R$ contains non-trivial infinitely small (*infinitesimal*) numbers, i.e. numbers $a \neq 0$ such that $|a| < r$ for all standard positive r. (0 is counted as infinitesimal, trivially.) A number is *finite* if $|a| < r$ for some standard r, otherwise a is *infinite*. The elements of $^*N - N$ are the *infinite natural numbers*. An infinite number is greater than any finite number. If a is any finite real number then there exists a uniquely determined standard real number r, called the *standard part* of a such that $r - a$ is infinitesimal or, as we shall say also, such that r is infinitely close to a, write $r \simeq a$.

3. Let $f(x)$ be an ordinary ('standard') real-valued function of a real variable, defined for $a < x < b$, where a, b are standard real numbers, $a < b$. As we pass from R to $^*R$, $f(x)$ is extended automatically so as to be defined for all x in the open interval (a, b) *in* $^*R$. As customary in Analysis, we shall denote the extended function also by $f(x)$, but we may refer to it, by way of distinction, as '$f(x)$ in $^*R$', as opposed to the original '$f(x)$ in R'.

The properties of $f(x)$ in $^*R$ are closely linked to the properties of $f(x)$ in R by the fact that R and $^*R$ satisfy the same set of sentences, K. A single but relevant example of this interconnection is as follows.

Let $f(x)$ be defined in R, as above, and let x_0 be a standard number such that $a < x_0 < b$. Suppose that $f(x_0 + \xi) \simeq f(x_0)$, i.e. that $f(x_0 + \xi) - f(x_0)$ is infinitesimal, for all infinitesimal ξ, where $f(x)$ is now considered in $^*R$. Then we claim that for every standard $\varepsilon > 0$ there exists a standard $\delta > 0$ such that $|f(x_0 + \xi) - f(x_0)| < \varepsilon$ for all ξ such that $|\xi| < \delta$. – Indeed if ε is any standard positive real number then the statement,

'There exists an $\eta > 0$ such that for all ξ, $|\xi| < \eta$ implies $|f(x_0 + \xi) - f(x_0)| < \varepsilon$', can be formulated as a sentence X within L. Thus, either X or not-X holds in R. But if not-X held in R then it would belong to K and hence, would hold also in $^*R$. Since X holds in $^*R$, by its definition, we conclude that actually X holds also in R. And any f which realizes η in R must be standard since there are no other numbers in R. This proves our assertion.

We may also prove the converse, i.e. if for every standard $\varepsilon > 0$ there exists a standard $\delta > 0$ such that $|f(x_0+\xi) - f(x_0)| < \varepsilon$ for all ξ such that $|\xi| < \delta$ in R, then $f(x_0+\xi) \simeq f(x_0)$ for all infinitesimal ξ in $^*R$. *This shows that $f(x)$ is continuous at x_0 in R if and only if $f(x_0+\xi)$ is infinitely close to $f(x_0)$ in $^*R$.*

Similarly, it can be shown that $f(x)$ is differentiable at x_0 if and only if the ratios $(f(x_0+\xi) - f(x_0))/\xi$ have the same standard part, d, for all infinitesimal $\xi \neq 0$, and d is then the derivative of $f(x)$ at x_0 in the ordinary sense.

For a last example, let $\{s_n\}$ be an infinite sequence of real numbers in R. On passing from R to $^*R$, $\{s_n\}$ is extended so as to be defined also for infinite natural numbers n. Let s be a standard real number. It can then be proved that s is the limit of $\{s_n\}$ in the ordinary sense, $\lim_{n\to\infty} s_n = s$ if and only if s_n is infinitely close to s (or, which is the same, if s is the standard part of s_n) for all infinite natural numbers n.

The above examples may suffice in order to give a hint how the Differential and Integral Calculus can be developed within the framework of Non-standard Analysis.

4. It appears that Newton's views concerning the foundations of the Calculus were somewhat ambiguous. He referred sometimes to infinitesimals, sometimes to moments, sometimes to limits and sometimes, and perhaps preferentially, to physical notions. But although he and his successors remained vague on the cardinal points of the subject, he did envisage the notion of the limit which, ultimately, became the cornerstone of Analysis. By contrast, Leibniz and his successors wished to base the Calculus, clearly and unambiguously, on a system which includes infinitely small quantities. This approach is crystallized in the first sentence of the '*Analyse des infiniment petits pour l'intelligence des lignes courbes*' by the Marquis de l'Hospital. We mention in passing that de l'Hospital, who was a pupil of Leibniz and John Bernoulli, acknowledged his indebtedness to his two great teachers.

De l'Hospital's begins with a number of definitions and axioms. We quote (translated from [7]):

'Definition I. A quantity is *variable* if it increases or decreases continuously; and, on the contrary, a quantity is *constant* if it remains the same while other quantities change. Thus, for a parabola, the ordinates and abscissae are variable quantities while the parameter is a constant quantity.'

'Definition II. The infinitely small portion by which a variable increases or decreases continuously is called its difference . . .'

For *difference* read *differential*. There follows an example with reference to a diagram and a corollary in which it is stated as evident that the differential of a constant quantity is zero. Next, de l'Hospital introduces the differential notation and then goes on –

'First requirement or supposition. One requires that one may substitute for one another [*prendre indifféremment l'une pour l'autre*] two quantities which differ only by an infinitely small quantity: or (which is the same) that a quantity which is increased or decreased only by a quantity which is infinitely smaller than itself may be considered to have remained the same . . .'

'Second requirement or supposition. One requires that a curve may be regarded as the totality of an infinity of straight segments, each infinitely small: or (which is the same) as a polygon with an infinite number of sides which determine by the angle at which they meet, the curvature of the curve . . .'

Here again we have omitted references to a diagram.

In order to appreciate the significance of these lines we have to remember that, when they were written, mathematical axioms still were regarded, in the tradition of Euclid and Archimedes, as empirical facts from which other empirical facts could be obtained by deductive procedures; while the definitions were intended to endow the terms used in the theory with an empirical meaning. Thus (contrary to what a scheme of this kind would signify in our time) de l'Hospital's formulation implies a belief in the reality of the infinitely small quantities with which it is concerned. And the same conclusion can be drawn from the preface to the book –

'Ordinary Analysis deals only with finite quantities: this one [i.e. the Analysis of the present work] penetrates as far as infinity itself. It compares the infinitely small differences of finite quantities; it discovers the relations between these differences; and in this way makes known the relations between finite quantities, which are, as it were, infinite compared with the infinitely small quantities. One may even say that this Analysis extends beyond infinity: For it does not confine itself to the infinitely small differences but discovers the relations between the differences of these differences, . . .'

It is this robust belief in the reality of infinitely small quantities which held sway on the continent of Europe through most of the eighteenth century. And it is this point of view which is commonly believed to have been that of Leibniz. However, although Leibniz was indeed responsible for the technique and notation of this Calculus of Infinitesimals his ideas on the foundations of the subject were quite different and considerably more subtle. In fact, we know from Leibniz' correspondence that he was critical of de l'Hospital's belief in the reality of infinitesimals and even more critical of Fontenelle's emphatic affirmation of this opinion.

Leibniz' own view, as published in 1689 [8] and as repeated and elaborated subsequently in a number of letters, may be summarized as follows. While approving of the introduction of infinitely small and infinitely large quantities, Leibniz did not consider them as real, like the ordinary 'real' numbers, but thought of them as ideal or fictitious, rather like the imaginary numbers. However, by virtue of a general principle of continuity, these ideal numbers were supposed to be governed by the same laws as the ordinary numbers. Moreover, Leibniz maintained that his procedure differed from 'the style of Archimedes' only in its language [*dans les expressions*]. And in describing 'the style of Archimedes' i.e. the Greek method of exhaustion, he used the following perfectly appropriate, yet strikingly modern, phrase (translated from [9]):

'One takes quantities which are as large or as small as is necessary in order that the error be smaller than a given error [*pour que l'erreur soit moindre que l'erreur donnée*] . . .'

However, Leibniz, like de l'Hospital after him, stated that two quantities may be accounted equal if they differ only by an amount which is infinitely small relative to them. And on the other hand, although he did not state this explicitly within his axiomatic framework, de l'Hospital, like Leibniz, assumed that the arithmetical laws which hold for finite quantities are equally valid for infinitesimals. It is evident, and was evident at the time, that these two assumptions cannot be accommodated simultaneously within a consistent framework. They were widely accepted nevertheless, and maintained themselves for a considerable length of time since it was found that their judicious and selective use was so very fruitful. However, Non-standard Analysis shows how a relatively slight modification of these ideas leads to a consistent theory or, at least, to a theory which is consistent relative to classical Mathematics. Thus, instead of claiming that two quantities which differ only by an infinitesimal amount, e.g. x and $x+\mathrm{d}x$, are actually equal, we find only that they are equivalent in a well-defined sense, $x+\mathrm{d}x \simeq x$ and can thus be substituted for one another in some relations but not in others. At the same time, the assertion that finite and infinitary quantities have 'the same' properties is explicated by the statement that both R and $^*R$ satisfy the set of sentences K. And if we ask, for example, whether $^*R$ (like R) satisfies Archimedes' axiom then the answer depends on our interpretation of the question. If by Archimedes' axiom we mean the statement that from every positive number a we can obtain a number greater than 1 by repeated addition –

$$a+a+\ldots+a \text{ (n times)} > 1,$$

where n is an ordinary natural number, then $^*R$ does not satisfy the axiom. But if we mean by it that for any $a>0$ there exists a natural number n (which may be infinite) and that $n\cdot a>1$, then Archimedes' axiom does hold in $^*R$.

5. In the view of many, including the author, the problem of the nature of infinitary notions is still of central importance in the Philosophy of Mathematics. To a logical positivist, the entire argument over the reality of a mathematical structure may seem pointless but even he will have to acknowledge the historical importance of the issue. To de l'Hospital, the infinitely small and large quantities (which were still thought of as geometrical entities) represented the actual infinite. On the other hand, Leibniz stated specifically that although he believed in the actual infinite in other spheres of Philosophy, he did not assume its existence in Mathematics. He also said that he accepted the potential (or as he put it, referring to the schoolmen, 'syncategorematic') infinite as exemplified, in his view, in the number of terms of an infinite series. To sum up, Leibniz accepted the ideal, or fictitious, infinite; accepted the potential infinite; and within Mathematics, rejected or at least dispensed with, the actual infinite.

Like the proponents of the new theory, its critics also were motivated by a combination of technical and philosophical considerations. Berkeley's 'Analyst' ([3]; compare [11]) constitutes a brilliant attack on the logical inadequacies both of the Newtonian Theory of Fluxions and of the Leibnizian Differential Calculus. In discrediting these theories, Berkeley wished to discredit also the views of the scientists on theological matters. But beyond that, and more to the point, Berkeley's distaste for the Calculus was related to the fact that he had no place for the infinitesimals in a philosophy dominated by perception.

6. The second half of the eighteenth century saw several attempts to put the Calculus on a firm footing. However, apart from d'Alembert's affirmation of the importance of the limit concept (and, possibly, some of L. N. M. Carnot's ideas, which may have influenced Cauchy), none of these made a contribution of lasting value. Lagrange's attempt to base the entire subject on the Taylor series expansion was doomed to failure although, indirectly, it may have had a positive influence on the development of the idea of a formal power series.

It is generally believed that it was Cauchy who finally put the Calculus on rigorous foundations. And it may therefore come as a surprise to learn that infinitesimals still played a vital role in his system. I quote from Cauchy's *Cours d'Analyse* (translated from [6]):

'When speaking of the continuity of functions, I was obliged to discuss the principal properties of the infinitesimal quantities, properties which constitute the foundation of the infinitesimal calculus . . .'

However, Cauchy did not regard these entities as basic but tried to derive them from the notion of a variable:

'A variable is a quantity which is thought to receive successively different values . . .'

'When the successive numerical values of a variable decrease indefinitely so as to become smaller than any given number, this variable becomes what is called an *infinitesimal* [*infiniment petit*] or an infinitely small quantity.'

At the same time the *limit* of a variable (when it exists) is defined as a fixed value which is approached by the variable so as to differ from it finally as little as one pleases. It follows that a variable which becomes infinitesimal has zero as limit.

Cauchy did not wish to regard the infinitesimals as numbers. And the assumption that they satisfy the same laws as the ordinary numbers, which had been stated explicitly by Leibniz, was rejected by Cauchy as unwarranted. Moreover, Cauchy stated, on a later occasion, that while infinitesimals might legitimately be used in an argument they had no place in the final conclusion.

However, Cauchy's professed opinions in these matters notwithstanding, he did in fact treat infinitesimals habitually as if they were ordinary numbers and satisfied the familiar rules of Arithmetic. And, as it happens, this procedure led him to the correct result in most cases although there is a famous and much discussed situation in the theory of series of functions in which he was led to the wrong conclusion. Here again, Non-standard Analysis, in spite of its different background, provides a remarkably appropriate tool for the discussion of Cauchy's successes and failures.

For example, the fact that a function $f(x)$ is continuous at a point x_0 if the difference $f(x_0+\xi)-f(x_0)$ is infinitesimal for infinitesimal ξ, which is a *theorem* of Non-standard Analysis (see section 3 above), is also a precise explication of Cauchy's notion of continuity. On the other hand, in arriving at the wrong conclusion that the sum of a series of continuous functions is continuous provided it exists, Cauchy used the unwarranted argument that if

$\lim_{n\to\infty} s_n(x) = s(x)$ over an interval then $s_n(x_0) - s(x_0)$ is, *for all* x_0 *in the interval*, infinitesimal for infinite n. In Non-standard Analysis, it turns out that this is true for standard (ordinary) $s_n(x)$, $s(x)$ and x_0, but not in general for non-standard x_0, e.g. not if $x_0 = x_1 + \xi$ where x_1 is standard and ξ is infinitesimal.

In order to appreciate to what extent Cauchy regarded the infinitesimals as an integral part of his system, it is instructive to consider his definition of a derivative. To him, $f'(x)$, wherever it exists, is the limit of the ratio

$$\frac{\Delta y}{\Delta x} = \frac{f(x+\xi) - f(x)}{\xi}$$

where ξ *is infinitesimal*. In the standard modern approach the assumption that ξ is infinitesimal is completely redundant or, more precisely, meaningless. The fact that it was nevertheless introduced explicitly by Cauchy shows that his mental image of the situation was fundamentally different from ours. Thus, it would appear that, to his mind, a variable does not attain the limit zero directly but only after travelling through a region of infinitesimals.

We have to add that in our 'classical' framework the entire notion of a variable in Cauchy's sense, as a mathematical entity *sui generis*, has no place. We might describe a variable, in a jocular mood, as a function which has lost its argument, while Cauchy's infinitesimals still are, to use Berkeley's famous phrase, the ghosts of departed quantities. But such carping criticism does not help us to understand the just recognition accorded to Cauchy's achievement, which is still thought by many to have resolved the fundamental difficulties that had beset the Calculus previously.

If we wish to find the reasons for Cauchy's success we have to consider, once again, both the technical-mathematical and the basic philosophical aspects of the situation. Cauchy established the central position of the limit concept for good. It is true that d'Alembert, who had emphasized the importance of this concept some decades earlier, in a sense went further than Cauchy by stating (translated from [1]):

'We say that in the *Differential* Calculus there are no infinitely small quantities at all . . .'

But apparently d'Alembert did not work out the consequences of his general principles; while the vast scope and the subtlety of Cauchy's mathematical achievement showed to the world that his tools enabled him to go farther and deeper than his predecessors. He introduced these tools at a time when the great achievements of the earlier and technically more primitive method of infinitesimals had become commonplace. Thus, the momentum which had enabled that method to disregard earlier attacks such as Berkeley's was exhausted before the end of the eighteenth century and due attention was again given to its logical weaknesses (which had been there, for all to see, all the time). These weaknesses had been associated throughout with the introduction of entities which were commonly regarded as denizens of the world of actual infinity. It now appeared that Cauchy was able to remove them from that domain and to base Analysis on the potential infinite (compare [4] and [5]). He did this by choosing as basic the notion of a variable which, intuitively, suggests potentiality rather than actuality. And so it happened that a grateful public was willing to overlook the fact that, from a strictly logical point of view, the new method shared some of the weaknesses of its predecessors and, indeed, introduced new weaknesses of its own.

7. When Weierstrass (who had been anticipated to some extent by Bolzano) introduced the δ,ε-method about the middle of the nineteenth century he maintained the limit concept in its central place. At the same time, Weierstrass' approach is perhaps closer than Cauchy's to the Greek method of exhaustion or at least to the feature of that method which was described by Leibniz ('*pour que l'erreur soit moindre que l'erreur donnée*', see section 4 above). On the issue of the actual infinite versus the potential infinite, the δ,ε-method did not, as such, force its proponents into a definite position. To us, who are trained in the set-theoretic tradition, a phrase such as 'for every positive ε, there exists a positive δ . . .' does in fact seem to contain a clear reference to a well-defined infinite totality, i.e. the totality of positive real numbers. On the other hand, already Kronecker made it clear, in his lectures, that to him the phrase meant that one could *compute* for, every *specified* positive ε, a positive δ with the required property. However, it was not then known that the abstract and the constructive approaches actually lead to different theories of Analysis, so that a mathematician's inability to provide a procedure for computing a function whose existence he has proved by abstract arguments is not necessarily due to his personal inadequacy.

At the same time it is rather natural that Set Theory should have arisen, as it did, from the consideration of certain problems of Analysis which required the further clarification of basic concepts. And its creator, Georg Cantor, argued forcefully and in great detail that Set Theory deals with the actual infinite. Nevertheless, Cantor's attitude towards the theory of infinitely small quantities was entirely negative, in fact he went so far as to claim that he could disprove their existence by means of Set Theory. I quote (translated from [4]):

'*The fact of* [the existence of] *actually-infinitely large numbers is not a reason for the existence of actually-infinitely small quantities*; *on the contrary, the impossibility of the latter can be proved precisely by means of the former.*

'Nor do I think that this result can be obtained in any other way *fully* and *rigorously*.'

The misguided attempt which is summed up in this quotation was concerned not only with the past but was directed against P. du Bois-Reymond and O. Stolz who had just re-established a modest but rigorous theory of non-Archimedean systems. It may be recalled that, at the time, Cantor was fighting hard in order to obtain recognition for his own theory.

Cantor's belief in the actual existence of the infinities of Set Theory still predominates in the mathematical world today. His basic philosophy may be likened to that of de l'Hospital and Fontenelle although their infinite quantities were thought to be concrete and geometrical while Cantor's infinities are abstract and divorced from the physical world. Similarly, the intuitionists and other constructivists of our time may be regarded as the heirs to the Aristotelian traditions of basing Mathematics on the potential infinite. Finally, Leibniz' approach is akin to Hilbert's original formalism, for Leibniz, like Hilbert, regarded infinitary entities as ideal, or fictitious, additions to concrete Mathematics. Thus, we may conclude this talk with the observation that although the very subject matter of foundational research has changed radically over the last two hundred years, there is a remarkable permanency in the concern with the infinite in Mathematics and in the various philosophical attitudes which have been adopted towards this notion.

References

[1] J. Le R. d'Alembert, article 'Différentiel' in *Encyclopédie méthodique ou par ordre de matières* (Mathématiques) 3 vols., Paris–Liège 1784–1789.

[2] J. Le R. d'Alembert, article 'Limite' in *Encyclopédie méthodique ou par ordre de matières* (Mathématiques) 3 vols., Paris–Liège 1784–1789.

[3] G. Berkeley, *The Analyst*, 1734, Collected works, vol. 4 (ed. A. A. Luce and T. E. Jessop) London 1951.

[4] G. Cantor, *Mitteilungen zur Lehre vom Transfiniten*, 1887–1888, Gesammelte Abhandlungen ed. E. Zermelo, Berlin 1932, pp. 378–439.

[5] E. Carruccio, I Fondamenti dell'Analisi matematica nel pensiero di Agostino Cauchy, *Bolletino dell'Unione Matematica Italiana* ser. 3, vol. 12, 1957, pp. 298–307.

[6] A. Cauchy, *Cours d'Analyse de l'Ecole Royale Polytechnique*, 1re partie, Analyse Algébrique, 1821 (Oeuvres complètes ser. 2, vol. 3).

[7] G. F. A. de l'Hospital, *Analyse des infiniment petites pour l'intelligence des lignes courbes*, Paris (1st ed. 1696) 2nd ed. 1715.

[8] G. W. Leibniz, Tentamen de motuum coelestium causis, *Acta Eruditorum*, 1689, Mathematische Schriften (ed. C. I. Gerhardt) vol. 5, 1858, pp. 320–328.

[9] G. W. Leibniz, Mémoire de M. G. G. Leibniz touchant son sentiment sur le calcul différentiel, *Journal de Trévoux*, 1701, Mathematische Schriften (ed. C. I. Gerhardt) vol. 5, 1858, p. 350.

[10] A. Robinson, *Non-standard Analysis*, Studies in Logic and the Foundations of Mathematics, Amsterdam, 1966.

[11] E. W. Strong, Mathematical reasoning and its objects, *George Berkeley lectures*, University of California Publications in Philosophy, vol. 29, Berkeley and Los Angeles, 1957, pp. 65–88.

DISCUSSION

Peter Geach: *Infinity in scholastic philosophy.*

Leibniz, as Robinson's quotation shows, assimilated the distinction of actual and potential infinity to the distinction of categorematic and syncategorematic infinity. This was common form in the scholasticism of his own day (it is to be found already in Suarez [1]); and yet by the standards of an older scholasticism it was a confusion so gross as might excuse an enemy of scholasticism for echoing Lord Chesterfield's remark to one of the College of Heralds, that the foolish man did not even understand his own foolish business.

The distinctions are not even the same *sort* of distinction. The distinction between actual and potential infinity is a distinction between two ways in which outside things, *res extra*, could be said to be infinite. 'Categorematic' and 'syncategorematic' on the other hand are words used to describe (uses of) words in a language; an infinite multitude, say, can no more be syncategorematic than it can be pronominal or adverbial. To be sure, the confusion is explicable. A *categorematic* use of a word is a use of it so that it can be understood as a logical subject or predicate; and just those things are *actually* infinite of which the word 'infinite' taken *categorematically* can with truth be predicated *sans phrase* (*simpliciter*). But this does not make the confusion excusable – especially as there is no such close connexion between the potentially infinite and the syncategorematic use of 'infinite'.

I shall give an example of a sentence calling for the categorematic-syncategorematic distinction and then go on to the medieval rationale of the distinction. When we read in Spinoza that there are infinite Divine attributes, we need to know whether he meant that each attribute is an infinite attribute, or, that there are infinitely many attributes; in fact the latter was his meaning. A medieval scholastic would have said that 'infinite' is taken categore-

[1] Suarez, *De Incarnatione*, disp. 26, s. 4, n. 5.

matically in the first case and syncategorematically in the second case. The example may serve to stop us from thinking in terms of actuality and potentiality; for of course nothing was further from Spinoza's mind than that the infinity of Divine attributes was only potential; all the same, if 'infinite' in 'There are infinite Divine attributes' means 'infinitely many', there is no choice but to parse it as a syncategorematic use of 'infinite'.

Syncategorematic words, for medieval logicians, were words that give form to propositions, such as the copula, negation, quantifiers, and connectives [1]. (Later scholastics also count e.g. adverbs like 'badly' and possessives like 'Cicero's' as syncategorematic; this extended application seems to me misleading.) 'Infinite' in the syncategorematic sense is explicitly assimilated to the quantifiers; and rightly so – 'there are infinitely many' is plainly an expression of the same semantical category as 'there are some' or 'there are none' [2].

The closest connexion that can be made between the syncategorematic use of 'infinite' and the potentially infinite is this: the scholastic account of the syncategorematic word 'infinite' makes it licit to use it without (at least obviously) introducing actually infinite numbers. For 'there are infinite –s' was expounded as meaning 'there are not so many –s that there are no more' (*non sunt tot quin sint plura*); and in modern quantificational terms this comes out as 'for no n are there no more than n –s', where the range of the variable 'n' is restricted to *finite* cardinals.

Hans Freudenthal: *Technique versus metaphysics in the calculus.*

There is more metaphysics in Leibniz' speculations on Calculus than is usually known, e.g. attempts to understand the relation of body and soul by an analogy with that between a magnitude

[1] J. Reginald O'Donnell C.S.B. 'The Syncategoremata of William of Sherwood', *Mediaeval Studies* III, 1941, pp. 48, 54 f.

[2] Walter Burleigh, *De puritate artis logicae*, Tractatus brevior. See also the edition of the *Tractatus longior* by P. Bochner, O.S.B., St. Bonaventure, N.Y., 1955: pp. 259f.

and its differential. The genetic theory of preformation which asserted that the new creature has been preformed in its progenitors and particularly the whole of mankind in Adam, led to the idea of the differential of a genus, from which the genus developed as its integral.

Differentials were well-known in antiquity (atomic lines). Archimedes used them as a heuristic tool. In modern times they were reintroduced by Kepler, Cavalieri (indivisibles), and others, and systematically used by Newton's and Leibniz' predecessors. Their methods were more or less geometrical. This is particularly true of Pascal who behaved idiosyncratically towards Cartesian methods. Leibniz' starting point was an integral transformation he found in Pascal's work and stripped of its geometrical clothing. The gist of Leibniz' efforts was the thorough algebraisation of calculus. The result was an easy and prolific formalism, more practical than Newton's, and rapidly accepted by most creative mathematicians.

A. Heyting: *Technique versus metaphysics in the calculus.*

One of the main points of the lecture was, that questions which originally were considered as metaphysical can later on be considered as merely technical questions. I would like to relativize a little further the difference between metaphysical and technical questions, and perhaps also from another point of view to make it more absolute. To begin with the latter, it is clear that the things which we write on the blackboard, simply considered as signs, are purely technical; on the other side it is clear that the theological considerations by which Cantor motivated his notion of the actual infinite, were metaphysical in nature. But there is quite a gradual scale of notions between the purely technical and the almost purely metaphysical. As soon as we give any interpretation to the signs, we introduce metaphysical or at least philosophical notions. If you consider non-standard analysis as technical, at the same time you consider it as an interpreted set theory, and from that point of view it contains some metaphysics. I agree that what now are considered as questions of philosophical nature, for instance the opposition between constructivism and Cantorism or Platonism, can from a certain point of view be considered as technical, because all these lines of thought are expressed in developments which can be considered from a purely formal point of view. But on the other hand, what now are considered as technicalities can be considered later on as more philosophical; the philosophical implications can become more relevant in the future.

Y. Bar-Hillel: *The irrelevance of ontology to mathematics.*

I don't think we have to wait for the future in order to make up our minds on the ontological (or metaphysical, or philosophical) status of the main mathematical entities. In particular, I don't think we have to wait much longer before realizing that the current practice of many philosophers of mathematics who use such terms as 'real', 'ideal' or 'fictional' to qualify mathematical entities is of little help towards the clarification of the methodological issues involved (in complete analogy to the situation in the empirical sciences). In my view, this is just another instance of the confusions created by using the material mode of speech on an inappropriate occasion, and this seems to me to have been definitely the case in the Hilbert–Bernays way of talking about ideal mathematical entities. (I am aware of the history of this usage.) The sooner we stop being concerned with the 'ontological' problems of recent set theory, which are nothing but the product of an unhappy mode of speech, the sooner will we get down to discussing the real issues raised by recent developments.

M. Bunge: *Non-standard analysis and the conscience of the physicist.*

One century elapsed between the execution and burial of infinitesimals by the ε-δ revolution, and their resurrection in non-standard analysis. The historian may rejoice upon finding that some of the intuitions of Leibniz, Newton, Euler and their followers, though coarse, were after all not stupid. And the physicist may feel relieved. Indeed, he has never ceased to use infinitesimals, e.g. in setting up differential equations representing physical processes. But he has done it with a bad conscience ever since rumours of the Dedekind–Weierstrass revolution reached his ear. He can now refer to non-standard analysis for the rigorous justification of his intuitive infinitesimals, just as he refers to the theory of distributions for the legalization of the various delta 'functions' which his physical intuition led him to introduce. These two cases illustrate the thesis that intuition is not to be rejected provided one can control it rationally. (See the commentator's *Intuition and Science*, 1962.) The only thing to be feared, in connection with the modern heir to the old infinitesimal, is that the standard calculus teacher may feel entitled to teach analysis the easy way (as it was still done at the turn of the century) by identifying modern infinitesimals with old infinitesimals and the latter with differentials. Happily the news of this Robinsonian resurrection will not spread that quickly.

ABRAHAM ROBINSON: *Reply.*

Commenting first on the philosophical points raised by Bar-Hillel and Heyting, it will be evident that in my paper I have dealt with questions of reality in Mathematics from the detached point of view of a historian. However, I am willing to go further and commit myself to the point of stating that in my view these problems should not be discussed in the cavalier fashion advocated by Bar-Hillel. As to what is technical and what is essential, I certainly did not want to suggest that the very differences of opinion between constructivists and platonists are merely technical. But it seems to me likely that questions of detail within transfinite set theory such as the correctness of the continuum hypothesis or the existence of very large cardinals, will come to be regarded as philosophically irrelevant, although I yield to no one in my admiration for the ingenious methods which have been devised to cope with these problems.

In reply to Bunge and Freudenthal, I wish to emphasize that

Leibniz' infinitesimals, like my own, are not in-divisible and in this respect should be distinguished from indivisibles. However, it is true that, historically, the distinction is blurred. I may add that Pascal (writing as 'Monsieur Dettonville') anticipated Leibniz in claiming that the method of the ancients and the method of indivisibles differed only in their expression (*manière de parler*).

I gather that Professor Geach accepts my suggestion that in Leibniz' mind syncategorematic infinity and potential infinity were the same. On my part, I have enjoyed his exposition of the original medieval point of view. However, as he himself points out in his closing remarks, some connection can be made between the two concepts so that their identification is at least not entirely fortuitous.

CONCERNING PROGRESS IN THE PHILOSOPHY OF MATHEMATICS

Abraham ROBINSON[†]

Yale University, New Haven, Conn. U.S.A.

1. The advances made on the technical side of mathematical logic and of the foundations of mathematics over the last one hundred years have been spectacular, and it is hard to imagine any future progress in these areas that will not have to make use of at least some of these developments. While their importance may to some extent remain a matter of opinion, especially if we take into account the judgment of the most zealous constructivists, their correctness or acceptability is no more open to doubt than that of the established results of the classical branches of mathematics.

By comparison, the evolution of our understanding of the essential nature of mathematics has been hesitant and ambiguous and, in any case, the conclusions that have been reached by one school of thought have been rejected by another. One might have expected that some of the major technical advances referred to above would have swayed all or most philosophers of mathematics in one and the same direction, but this has not been the case even in the most outstanding instances, some of which are under consideration here today.

2. Nevertheless, as this Symposium shows, the discussion of the philosophical foundations of mathematics remains full of fascination and challenge. Although I personally have, for the most part, concerned myself with the technical aspects of metamathematics and mathematical logic, I have not remained immune to this challenge. The particular problem in the philosophy of mathematics that has the greatest fascination for me–and I maintain that it is a genuine problem–is that of the existence, or reality, or intelligibility, or objectivity, of infinite totalities.

In an address that I gave a number of years ago [4], my two main points were (i) that mathematical theories which, allegedly, deal with infinite totalities do not have any detailed meaning, i.e. reference, and (ii) that this has no bearing on the question whether or not such theories should be developed and that, indeed, there are good reasons why we should continue to do mathematics in the classical fashion nevertheless. I developed these, by no means new, theses in some detail, emphasizing among other things that (i) applies also to the infinitistic parts of logic itself and, in particular, to model theory. I also discussed some of the weaknesses of my position. Subsequent exchanges, both oral and in published writings, have not induced me to change my views. Moreover, I believed then and I still believe that the well-known recent developments in set theory represent evidence favoring these views. However, on the present occasion, I wish to concentrate on the following more general or, if you like, preliminary question. *What is the kind of progress that has been achieved, over the years, in the philosophy of mathematics and what further progress may we expect of it in the future?*

3. We shall be interested not only in the philosophy of mathematics as such but also in other material that seems relevant to the philosophy of mathematics. As might be expected, it is not easy to delimit either one of these areas and, in particular, where to draw the line of demarcation between them may be a matter of opinion. I interpret the notion of 'material that is relevant to the philosophy of mathematics' rather widely including, on the one hand, topics from mathematics and physics whose significance for the philosophy of mathematics may be adjudged small by most contemporary philosophers and, on the other hand, results which are sometimes regarded as lying within the domain of the philosophy of mathematics as such. Among the former, I may mention the irrationality of $\sqrt{2}$ and the transcendentality of π, the existence of non-euclidean geometries, the foundations of the differential and integral calculus, the ubiquitous nature of topological and categorical concepts, the (approximate) applicability of Kepler's laws to the physical world, and the identification of observables with operators in quantum mechanics. Among the latter I include the existence of uncountable sets, the well ordering theorem, Gödel's theorem on the incompleteness of Peano arithmetic and the existence of nonstandard models of number theory. More generally, while several notions and results of set theory, e.g., in addition to those already mentioned, the theorems of Gödel and Cohen on

the continuum conjecture and on the axiom of choice, or the reducibility of all of mathematics to set theory are highly relevant to the philosophy of mathematics, others including many theorems of cardinal and ordinal arithmetic, in spite of their great intrinsic interest, seem to me no more fundamental than ordinary number theory or the behavior of a function on its circle of convergence.

Be this as it may, no one will challenge the assertion that since the dawn of Greek mathematics and in particular in recent times progress in the vast area that I have staked out with the aid of examples, have been most impressive. Assuming the continuing development of our civilization, I see no reason why similar progress should not be made also in the future.

To facilitate further discussion, we shall subsume the material of this section under *class I*.

4. We now come to the philosophy of mathematics as such within the range that I have accorded to this area. Certainly, if we compare the present situation with that existing one hundred years ago we cannot fail to perceive that here also there has been great progress. I may conveniently divide the several directions in which this progress has been achieved into two groups, to be called *class* IIa and *class* IIb.

In class IIa I include advances in the philosophy of mathematics (or, if you like, in the philosophical foundations of mathematics) which are concerned with specific problems or other points of detail in this area. Here are some important items under this heading.

(i) The Cantor–Frege–Russell concept of a cardinal number;

(ii) the compilation of a set of logical axioms and rules of deduction which is adequate at least for the lower predicate calculus;

(iii) the explication or explications of the notion of computability; and

(iv) the explicit formulation of the truth concept.

As it happens, (i)–(iv) correspond to the routine subdivision of 'Mathematical logic and the foundations of mathematics' into set theory, proof theory, recursion theory, and model theory. The development of an adequate calculus of deduction has the longest history while the emergence of a precise notion of computability represents perhaps one of the most impressive intellectual achievements of our time. Although opinions may differ as to the relative depth and merits of the four topics, each one of them involves results that have met with universal acceptance no less than the most classical theorems of pure mathematics.

In class IIb I shall include work in the philosophy of mathematics that concerns the essential nature of mathematics and of mathematical entities. One hundred years ago, none of the specifically mathematical philosophies that are of central importance in our time had as yet emerged. There was no intuitionism and no predicativism, no logicism and no formalism. Traces of constructivism can in fact be found already in Euclid but the kind of abstract mathematics that was to provoke a constructivistic reaction was only beginning to take shape in 1870. And not only the various constructivistic schools of thought but also logicism and formalism owe their existence largely to this development. Among the great philosophical systems that had been put forward earlier there are indeed several which include more or less detailed discussions of the concepts of mathematics. Some of these are still interesting and influenced later opinions (e.g., nominalism and Platonism) while others are, by our standards, inadequate in detail and in some case outright nonsensical. A general philosophical system which occupies a special position here is logical positivism since it is not only relevant to the philosophy of mathematics but also owes its existence to it.

At any rate the creation of logicism, of formalism, and of the several constructivisms must, in my view, be regarded as important advances, separately and jointly. I have qualified this statement by the phrase 'in my view' because I am aware of the fact that the more fanatical adherents of one or the other of the schools of thought just mentioned will not agree with me. Thus, an engaged constructivist may well believe that the development of nonconstructive mathematics is an aberration which is only compounded by the attempts of logicists or formalists to provide a philosophical basis for it. Conversely, there may be logicists and formalists who, like most uncommitted mathematicians, believe that the constructivists are wasting their time by tying their hands with unnecessary restrictions. Or again, a logical positivist may feel that the very problems whose solutions divide the several mathematical creeds are imaginary or meaningless (Scheinprobleme) and, for that reason, not worth discussing. I cannot share these attitudes. The elaboration of a particular viable point of view in the philosophy of mathematics as such must be regarded as progress in this area. To take a recent development, the very precise delimitation of predicativism of Poincaré. Its clarification of the predicativistic method lends more substance to this approach and helps us to gauge its strength.

Thus, the last one hundred years have yielded much progress in the formulation and elaboration of tenable views in the basic philosophy of mathematics. None of this progress has been such as to eliminate one or the other of the previously mentioned schools of thought altogether. Experience has shown (as in the case of logicism and Russell's paradox) that in this area the discovery of a specific antinomy in a particular approach can be countered by an emendation of the technical details. This is not to say that *any* philosophy of mathematics, once formulated, should be regarded as co-equal with all others. For example, staying within the selected period of one hundred years up to the present, one must say that Hermann Cohen's philosophy of mathematics [1] whatever its historical interest, is deservedly unknown at the present time. It is also true that an individual cannot *commit* himself to two incompatible philosophies simultaneously. Thus, he cannot commit himself to both platonism and formalism since these are incompatible in their beliefs and he cannot commit himself to both formalism and some form of constructivism since these are incompatible in their practices. But at the level of our present discussion all these points of view are equally admissible.

One might have thought that as more and more results became available of the kind that we have called 'relevant to the philosophy of mathematics' (class I) or of the kind enumerated in class IIa, at least some of them would have had sufficient persuasive power to decide the issue in favor (or against) one or the other of the basic philosophical points of view. However, the only instances of such events that I can recall have no bearing on the debates of the present. Thus, the discovery of irrationals may have induced the Greek mathematicians of the classical period to regard geometry rather than number theory as the basis of all mathematics, and the failure of Leibniz' followers to establish a viable calculus of infinitesimals brought about a temporary eclipse of the actual infinite in mathematics. But, by contrast, the incompleteness of Peano arithmetic and the undecidability of the continuum conjecture have not led to a general abandonment of Platonism although it would seem to some, including myself, that they provide evidence against it. To sum up, papers in class IIb must be identified with the emergence and elaboration of differing opinions. There is no evidence that a more trenchant kind of progress in this area is at all likely.

5. In the preceding section I divided the advances that have occurred in the philosophy of mathematics into two classes, IIa and IIb. Under IIa I mentioned the explications of the notions (i) cardinal number, (ii) deduction or inference, (iii) computability, (iv) truth. Class IIb consisted of the philosophical theories that have been developed concerning the essential nature of mathematics. The standard results produced under IIa command universal or nearly universal acceptance. Nothing like it can be said of the various general philosophies of class IIb. Let us therefore look more closely at the topics listed under IIa in order to obtain some guidance concerning the kind of problems in the philosophy of mathematics to which we may expect a definitive answer.

At first sight, each of the four topics has its own character which is generically distinct from that of the others. However, upon further inspection, we notice that they all go beyond the consideration of mathematics as an abstract universe. As far as the definition of the notion *cardinal number* is concerned this is not immediately obvious since we might regard it as a problem of abstract set theory. However, as such it is not a problem in the philosophy of mathematics at all. Rather, its great philosophical importance is due to its immediate relation to the external world, at least as far as our perception of concrete (and hence, finite) collections is concerned. Again, the analysis of *inference* is the analysis of an activity or our mind (or, of the mind) and as such is at a different level from the contemplation of an abstract universe that our mind perceives. The same applies to the notion of computability even if we take it with its most abstract meaning. Moreover, if I consider either inference or computability in relation to someone else's mind then it becomes unquestionably a phenomenon of the external world, which can also be imitated by mechanical procedures (i.e., by a machine). Finally, the notion of truth is concerned with our statements about the external world and so its analysis, also, is at a level distinct from that at which we do pure mathematics. Here also, it is possible and in fact customary, in model theory, to incorporate the truth definition in an extension of classical pure mathematics and to correlate it with deducibility, which has first been adapted in a similar way (as in the completeness theorem of the lower predicate calculus). However, this again deprives these concepts of their basic philosophical significance.

The correctness of the four concepts in question (when taken in conjunction with commonly accepted interpretations and procedures) is subject to test by confirmation or, to use Popper's term, by falsification. The same does not apply to the philosophical theories that make up the last group. This is clear as far as their deontic components are concerned. Thus, a constructivist may condemn the formalists (as well as most uncommitted mathematicians) for adopting nonconstructive procedures, but the adoption of a particular practice as such is not subject to the empirical test just mentioned, which applies only to descriptive theories. As far as the reality of infinite totalities is concerned, which is the main point at issue between platonists and formalists, the situation is more complicated and I shall return to it later. At any rate, a consistent logical positivist may be expected to discard class IIb altogether and to retain only the problems and answers that we have included in classes I and IIa. However, at the level of our present discussion we should not adopt this attitude.

6. Commenting on another point of the preceding argument, we have distinguished between mathematical results which are relevant to the philosophy of mathematics (class I) and explications of specific foundational concepts (class IIa). As I have indicated already in one instance, the line of demarcation between these two classes depends on which aspect of a particular topic is emphasized in preference to others. I certainly do not wish to imply that the topics that I put in the first class are less important for the general issue of the present paper than those mentioned later. Thus I put Gödel's incompleteness theorem into the first class because, once the concept of deducibility has been formalized it may be regarded as purely mathematical. This does not detract in any way from its immense importance. Indeed, it also has an empirical significance since it implies that no one will ever write down a proof of Gödel's sentence on the basis of Peano's axioms. However, the same applies to many results which are unquestionably part of pure mathematics.

To take another case in which we have opted for a possibly controversial assignment, consider the reduction of mathematics to set theory, which we put into class I. One might say that this is inconsistent with the (widely held) judgment, implicit in a remark in Section 4., that the replacement of number theory by geometry as the foundation of Greek mathematics represented a philosophical revolution. In fact, this judgment is only acceptable from the point of view that prevailed in the civilization in which this event occurred. Moreover, confining myself to the present, I must emphasize that I do not regard the classification in question as absolute. In particular, it seems to me that it cannot always be disassociated from the philosophical point of view of the classifier. In the present instance, my choice is related to the fact that I do not believe in the primacy of set theory over all other branches of mathematics and this in turn may be due to my formalistic point of view.

7. After this survey of past developments, we turn to the consideration of things to come. As I have said already, there is every reason to believe that the mathematics of the future, like that of the past, will include developments which are relevant to the philosophy of mathematics. Such developments may occur again in set theory, as they have occurred in the recent past. They may occur in the theory of categories where we see, once again, a largely successful attempt to reduce all of pure mathematics to a single discipline. But, presumably, the future, like the past, will also bring unexpected mathematical innovations which will add new dimensions to the philosophy of mathematics. And it would seem to me that such an event is likely even in a highly technical subject like algebraic number theory.

Continued progress in the remaining two classes (IIa and IIb) is equally likely, of the kind that has characterized developments in these areas in the past. As far as specific topics, in particular explications of intuitive notions, are concerned, I expect that these will have to do with the relationship of mathematics with the external world, including consideration of the active mind (or, according to one's point of view, of thinking individuals). Thus, we may expect a much more detailed elucidation of the relations between pure and applied mathematics than is available at present. I accept the description of this relationship which is given by Körner [3] as '(i) the replacement of empirical concepts and propositions by mathematical [ones], (ii) the deduction of consequences from the mathematical premises so provided, and (iii) the replacement of some of the deduced mathematical propositions by empirical [ones].' I believe that there is room for applying this outline with greater precision to individual mathematical and physical theories.

At this point, we should also consider the possibility of future developments in the empirical sciences which will affect the areas with which we are concerned. I can think of one such development, which will surely occur in the fullness of time, although I have no means of judging when. This is the possibility of analyzing in detail the neurophysiological processes in the brain which correspond to its mathematical activity. As far as I know there is at present no reliable evidence on this subject and the theoretical models that have been produced are highly hypothetical and, in any case, are not sufficiently specific to distinguish details. The material produced by the work on artificial intelligence, in particular in connection with theorem proving by machine, leads to a picture which is, most probably, still far removed from actuality. I believe that when the empirical discoveries that are necessary here are made, this will lead to further advances concerning some of the topics in class IIa that have been mentioned and also to unexpected new developments. In particular, proof theory, in the wider sense of a general theory of (mathematical) proofs, and the problems of pure and applied mathematics, are two of the areas that are likely to be affected.

There is a standard philosophical argument against bringing in the kind of empirical evidence just mentioned. It is that in obtaining and describing them we are necessarily making use of the very mathematical notions that we are trying to explain. However, one of the lessons that we have been taught by the metamathematical approach is that a circularity of this kind is neither avoidable nor does it imply that an investigation involving it must be sterile. A linear development of our philosophical understanding might be more tidy but we cannot impose it on the universe of which we are a part.

8. Finally, we have to discuss future prospects in the debate concerning the essential nature of mathematics (class IIb). I do not expect that the definitive advances achieved in other areas will ever lead to a decision in favor of one school of thought or the other, just as they have failed to do in the past. To illustrate, consider the basic dispute between platonists and Fregean objectivists on one hand and formalists on the other hand. Let us listen to a conversation between P, a platonist of Fregean objectivist, and R, a formalist. P believes that the universe of sets and the universe of natural numbers are in some sense real or objective and hence, that their structure and properties are unique. R does not believe in the reality or objectivity of these infinite universes. He thinks that it is not absolutely meaningful to talk of the structure of these universes and expects that sooner or later the mathematicians will accept several co-equal theories of sets and even of arithmetic. They agree to test their views against the well-known incompleteness results of arithmetic and set theory. R points out that Gödel's incompleteness theorem applies not only to first order Peano arithmetic but also (indeed, in the original version) to the higher order theory of Principia Mathematica. P replies that the Gödel sentence (or any similar sentence) though unprovable and irrefutable within the given formal framework, can nevertheless be shown to be true by informal but rigorous arguments (or by a generally acceptable widening of the formal framework). R concedes that this is correct and suggests that they move on to set theory. P is agreeable and R brings up the continuum hypothesis, pointing out that it is neither provable nor refutable in accepted axiomatic set theory and that, in this case, the question cannot be settled either by an intuitive argument. P retorts that he is aware of this fact but that the discovery of a generally acceptable axiom or intuitive argument may be expected, sooner or later, to decide this problem also. R rejoins that he cannot exclude this but insists that if P wishes to enter the realm of hypothetical developments he should also include the possibility that, sooner or later, someone will discover an absolutely undecidable sentence of elementary arithmetic. Both realize that the conversation has arrived at an impass and they consider the possibility of meeting again one thousand years hence, in order to continue the debate in the light of developments that will occur in the meantime. However, after further discussion they drop this idea because they conclude that, in addition to certain practical difficulties that would be involved in such an arrangement, each of them would still have the same escape argument available, to be applied to either arithmetic or set theory, come what may.

I might add that the attitudes of P and R towards the kind of incompleteness phenomena discussed above are not strictly symmetrical. R will maintain his basic point of view even if, in some way which is not foreseeable at present, he is persuaded that each individual statement of set theory or arithmetic will ultimately be decided. Indeed, even now the infinite structures whose truth sets are recursive are no more real to him than all others. By contrast, *if* some day P is persuaded that there are absolutely undecidable sentences in either set theory or arithmetic, he will realize that he must either give up his philosophy or accept the existence of unknowable truths and he may well opt for the former.

At any rate, while future developments may well lead to an increase or decrease in the relative strengths of the various schools of thought, I do not foresee the final elimination of any one of those now current. So, here again, I expect future advances to be similar in kind to past ones. Thus, several of the existing creeds will be developed further so as to enhance their persuasiveness. Also, new schools of thought or new variants of old ones will perhaps be put on the map. For example, I can well imagine that a serious mathematical philosophy based on the dialectical approach will make its appearance. It seems to be that this approach has already shown its great value in connection with our understanding of the evolution of scientific theories and of their heuristic aspect. As far as the detailed analysis of mathematics or of a mathematical theory (e.g., the calculus) by the dialectical method is concerned, my reading, beginning with Hegel's work in this area, has not led me to find anything that can stand up to serious criticism. It is quite possible that this situation will be remedied in the future.

Logical positivism is another general philosophy whose detailed views on the philosophy of mathematics require further clarification. At first sight, the 'official' position of this school of thought is perfectly clear–mathematics is part of language and, as such is largely arbitrary. However, in actual fact, the acceptance of Frege's logicism has led some positivists to accept also his objectivism, at least in relation to arithmetic while others, more consistently I believe, reject this point of view. Another general philosophy whose application to mathematics will perhaps be worked out in greater detail is nominalism (see [2]).

We may also expect advances which are, in a way, complementary to the prospective developments just described. These will consist of a widening of the *general* philosophical basis of those philosophies of mathematics that have up till now concentrated on the consideration of mathematics alone. For example, I expect that future work on formalism may well include general epistemological and even ontological considerations. Indeed, I think that there is a real need, in formalism and elsewhere, to link our understanding of mathematics with our understanding of the physical world. The notions of objectivity, existence, infinity, are all relevant to the latter as they are to the former (although this again may be contested by a logical positivist) and a discussion of these notions in a purely mathematical context is, for that reason, incomplete.

This brings me to the end of my presentation. I am aware that I have dealt with the proposed subject of this symposium (the relevance to philosophy of recent results in set theory) only in passing but I hope that I have at least been able to indicate a perspective for its discussion.

References

[1] H. Cohen, *Das Prinzip der Infinitesimalmethode und seine Geschichte*, 1st Ed. 1883 (new Ed. Suhrkamp, Frankfurt, 1968).
[2] L. Henkin, Nominalistic analysis of mathematical language, in *Proceedings of the International Congress for Logic, Methodology and Philosophy of Science*, Stanford 1960 (pub. 1962) pp. 187–193.
[3] S. Körner, *The Philosophy of Mathematics* (Hutchinson, London, 1960).
[4] A. Robinson, Formalism 64, in Y. Bar-Hillel, ed., *Logic, Methodology and Philosophy of Science* (North-Holland, Amsterdam, 1965) pp. 228–246.

Some thoughts on the history of mathematics

Dedicated to A. Heyting on the occasion of his 70th birthday

by

Abraham Robinson

1.

The achievements of Mathematics over the centuries cannot fail to arouse the deepest admiration. There are but few mathematicians who feel impelled to reject any of the major results of Algebra, or of Analysis, or of Geometry and it seems likely that this will remain true also in future. Yet, paradoxically, this iron-clad edifice is built on shifting sands. And if it is hard, and perhaps even impossible, to present a satisfactory viewpoint on the foundations of Mathematics today, it is equally hard to give an accurate description of the conceptual bases on which the mathematicians of the past constructed their theories. Some of the suggestions that we shall offer here on this topic are frankly speculative. Some may have been arrived at by comparing similar situations at different times in history, a procedure which is open to challenge and certainly should be used with great caution. Another preliminary remark which is appropriate here concerns the use of the word "real" with reference to mathematical objects. This term is ambiguous and has been stigmatised by some as meaningless in the present context. But the fundamental controversies on the significance of this word should not inhibit its use in a historical study, whose purpose it is to describe and analyze attitudes and not to justify them.

2.

It is commonly accepted that the beginnings of Mathematics as a deductive science go back to the Greek world in the fifth and fourth centuries B.C. It is even more certain that in the course of many hundreds of years before that time people in Egypt and Mesopotamia had accumulated an impressive body of mathematical knowledge, both in Geometry and in Arithmetic. Since this knowledge was recorded in the form of numerical problems and answers it is frequently asserted that pre-Greek Mathematics was purely "empirical". However, unless this expression is meant to indicate merely that pre-Greek Mathematics was not deductive and if it is to be taken literally, we are asked to believe, e.g., that the Mesopotamian mathematicians arrived at Pythagoras' theorem by measuring a large number of right triangles and by inspecting the numbers obtained as the squares of their side lengths. Is it not much more likely that these mathematicians, like their Greek successors, were already familiar with one of the arguments leading to a proof of Pythagoras' theorem by a decomposition of areas, but that no such proof was recorded by them since they regarded the reasoning as intuitively clear? To put it facetiously and anachronistically, if a Sumerian mathematician had been asked for his opinion of Euclid he might have replied that he was interested in *real* Mathematics and not in useless generalizations and abstractions. However, some major advances in Mathematics consisted not in the discovery of new results or in the invention of ingenious new methods but in *the codification of elements of accepted mathematical thought*, i.e. *in making explicit arguments, notions, assumptions, rules, which had been used intuitively for a long time previously*. It is in this light that we should look upon the contributions of the Greek mathematicians and philosophers to the foundations of Mathematics.

3.

For our present discussion, the question whether the major contribution to the system of Geometry recorded in Euclid's Elements was due to Hippocrates or to Eudoxus or to Euclid himself is of no importance (except insofar as it may affect the following problem, for chronological reasons). However, it would be important to know to what extent the emergence of deductive Mathematics was due to the lead given by one of the Greek philosophers or philosophical schools of the fifth and fourth centuries. Is it true, as has been asserted by some, that the creation of the axiomatic method was due to the direct influence of Plato or of Aristotle or, as has been suggested recently by Á. Szabó, that it was a response to the teachings of the Eleatic school? In our time, the immediate influence of philosophers on the foundations of Mathematics is confined to those who are willing to handle technical-mathematical details. But even now, a general philosophical doctrine may, almost imperceptibly, affect the direction taken by foundational research in Mathematics in the long run. In classical Greece, the differentiation between Philosophy and Mathematics was less pronounced, but nevertheless, with the possible exception of Democritus, we do not know of any leading philosopher of that period who originated an important contribution to Mathematics as such. When Plato singled out Theaetetus in order to emphasize the generality of mathematical arguments he was, after all, referring to a real person who had died only a few years earlier, and he wished to take no credit for the achievement described by him. Nevertheless, by laying bare some important characteristics of mathematical thought, both he and Aristotle exerted considerable influence on later generations. Thus Aristotle, having studied the mathematics of the day, established standards of rigor and completeness for mathematical reasoning which went far beyond the level actually reached at that time. And although we may assume that Euclid and his successors were aware of the teachings of Plato and Aristotle, their own aims in the development of Geometry as a deductive science were less ambitious than Aristotle's program from a purely logical point of view. It is in fact well known that even in the domain of purely mathematical postulates Euclid left a number of glaring gaps. And as far as the laws of logic are concerned, Euclid confined himself to axioms of equality (and inequality) and did not include the rules of deduction which had already been made available by Aristotle. Thus Euclid, like Archimedes after him, was content to single out those axioms which could not be taken for granted or which deserved special mention for other reasons and then derived his theorems from those axioms *in conjunction with other assumptions whose truth seemed obvious, by means of rules of deduction whose legitimacy seemed equally obvious.* It would be out of place to ask whether Euclid would have been able to include in his list of postulates this or that assumption if he had wanted just as even today it would, in most cases, be futile to ask a working mathematician to specify the rules of deduction that he uses in his arguments. The chances are that the typical working mathematician would reply that he is willing to leave this task to the logicians and that, by contrast, his own intuition is sound enough to get along spontaneously. For example, when proving that any composite number has a prime divisor (Elements, Book VII Proposition 31), Euclid appealed explicitly to the principle of infinite descent (which is a variant of the "axiom of induction") yet he did not include that principle among his axioms. By contrast, the axiom of parallels was included by Euclid (Elements, Book I, Postulate 5) because though apparently true, it was not intuitively obvious. Similarly "Archimedes axiom" was included by Archimedes (On the sphere and cylinder, Book I, Postulate 5) because although required for developing the method of exhaustion, it was not intuitively obvious either. In fact, Euclid did not accept this axiom at all explicitly but instead introduced a definition (Elements, Book V, Definition 5) which implies that he did not wish to exclude the possibility that magnitudes which are non-archimedean relative to one another actually exist, but that he deliberately confined himself to archimedean systems of magnitudes in order to be able to develop the theory of proportions and, to some extent, the method of exhaustion.

4.

From the beginnings of the axiomatic method until the nineteenth century A.D. axioms were regarded as statements of fact from which other statements of fact could be deduced (by means of legitimate procedures and relying on other obvious facts, see above). However, there is in Euclid an element of "constructivism" which, on one hand, seems to hark back to pre-Greek Mathematics and, on the other hand, should strike a chord in the hearts of those who believe that Mathematics has been pushed too far in a formal-deductive direction and who advocate a more constructive approach to the foundations of Mathematics. And although the first three postulates of the Elements, Book I can be interpreted as purely existential statements, the "constructivist" tenor of their actual style is unmistakable. Moreover, the cautious formulations of the second and fifth postulates seems to show a trace of the distaste for infinity that we find already in Aristotle. In addition, there are, of course, scattered through the Elements many "propositions" which are actually constructions.

5.

Euclid's geometry was supposed to deal with real objects, whether in the physical world or in some ideal world. The definitions which preface several books in the Elements are supposed to communicate what object the author is talking about even though, like the famous definition of the point and the line, they may not be required in the sequel. The fundamental importance of the advent of non-Euclidean geometry is that by contradicting the axiom of parallels it denied the uniqueness of geometrical concepts and hence, their reality. By the end of the nineteenth century, the interpretation of the basic concepts of Geometry had become irrelevant. This was the more important since Geometry had been regarded for a long time as the ultimate foundation of all Mathematics. However, it is likely that the independent development of the foundations of the number system which was sparked by the intricacies of Analysis would have deprived Geometry of its predominant position anyhow.

An ironic fate decreed that only after Geometry had lost its standing as the basis of all Mathematics its axiomatic foundations finally reached the degree of perfection which in the public estimation they had possessed ever since Euclid. Soon after, the codification of the laws of deductive thinking advanced to a point which, for the first time, permitted the satisfactory formalization of axiomatic theories.

6.

In the twentieth century, Set Theory achieved the position, once occupied by Geometry, of being regarded as the basic discipline of Mathematics in which all other branches of Mathematics can be embedded. And, within quite a short time, the foundations of Set Theory went through an evolution which is remarkably similar to the earlier evolution of the foundations of Geometry. First the initial assumptions of Set Theory were held to be intuitively clear being based on natural laws of thought for whose codification Cantor, at least, saw no need. Then Set Theory was put on a postulational basis, beginning with the explicit formulation of the least intuitive among them, the axiom of choice. However, at that point the axioms were still supposed to describe "reality", albeit the reality of an ideal, or Platonic, world. And finally, the realization that it is equally consistent either to affirm or to deny some major assertions of Set Theory such as the continuum hypothesis led, in the mid-sixties, to a situation in which the belief that Set Theory describes an objective reality was dropped by many mathematicians.

The evolution of the foundations of Set Theory is closely linked to the development of Mathematical Logic. And here also we can see how, in our own time, advances have been made through the codification of notions (such as the truth concept) which were used intuitively for a long time previously. And again it may be left open whether the postulates of a system deal with real objects or with idealizations (e.g. the rules of formation and deduction of a formal language). And there is every reason to believe that the codification of intuitive concepts and the reinterpretation of accepted principles will continue also in future and will bring new advances, into territory still uncharted.

Added March 20, 1968:

In an article published since the above lines were written (Non-Cantorian Set Theory, Scientific American, vol. 217, December 1967, pp. 104—116) Paul J. Cohen and Reuben Hersh compare the development of geometry and set theory and anticipate some of the points made here.

(Oblatum 3-1-'68)

Yale University

BIBLIOGRAPHY

Books

On the Metamathematics of Algebra. North-Holland Publ. Co., Amsterdam, 1951. ix + 195 pp.

Théorie Metamathématique des Ideaux. Gauthier-Villars, Paris, 1955. ix + 186 pp.

Wing Theory (with J. A. Laurmann). Cambridge Univ. Press, Cambridge, England, 1956. ix + 569 pp.

Complete Theories. North-Holland Publ. Co., Amsterdam, 1956. ix + 129 pp.

Introduction to Model Theory and to the Metamathematics of Algebra. North-Holland Publ. Co., Amsterdam, 1963. ix + 284 pp. Translated into Russian 1967; into Italian 1974; Second Edition 1974.

Numbers and Ideals. Holden-Day, San Francisco, 1965. ix + 106 pp.

Non-Standard Analysis. North-Holland Publ. Co., Amsterdam, 1966. Second Edition 1974. ix + 293 pp.

Contributions to Non-Standard Analysis (coeditor with W. A. J. Luxemburg). North-Holland Publ. Co., Amsterdam, 1972. ix + 289 pp.

Nonarchimedean Fields and Asymptotic Expansions (with A. H. Lightstone). North-Holland Publ. Co., Amsterdam, 1975. ix + 204 pp.

Algebra—A Model Theoretic Viewpoint (with V. B. Weispfenning). Springer-Verlag, to appear.

Papers

The boldface numbers following entries refer to their location in the Selected Papers of Abraham Robinson, *Volumes* **1, 2,** or **3**. *Entries that are not followed by a boldface number have not been reprinted in these volumes.*

[1] On the independence of the axioms of definiteness. J. Symbolic Logic 4 (1939), 69–72.

[2] On nil-ideals in general rings. Galley proofs recovered posthumously. **1**

[3] On a certain variation of the distributive law for a commutative algebraic field. Proc. Roy. Soc. Edinburgh Sect. A. 61 (1941), 93–101. M.R. 3 #101. **1**

[4] Shock transmission in beams of variable characteristics. Report No. S.M.E. 3340 Roy. Aircraft Establishment, Farnborough (1945), 60 pp.

[5] A minimum energy theorem in aerodynamics. Techn. Note No. S.M.E. 298 (1945), 7 pp.

[6] A note on the interpretation of V-g records (with P. E. Montagnon and S. V. Fagg). Ministry of Supply (London), Aeronaut. Res. Council, Rep. and Memoranda No. 2097 (9460), (1945), 8 pp.

[7] The aerodynamic loading of wings with endplates. Minstry of Supply (London), Aeronaut. Res. Council, Rep. and Memoranda No. 2343 (8615), 1945 (1950), 13 pp.

[8] The wave drag of diamond-shaped aerofoils at zero incidence. Ministry of Supply (London), Aeronaut. Res. Council, Rep. and Memoranda No. 2394 (9780), 1946 (1950), 6 pp. **3**

[9] Acoustic analyses in calculating supersonic resistance. Manuscript (1947), 4 pp.

[10] Shock transmission in beams. Ministry of Supply (London), Aeronaut. Res. Council, Rep. and Memoranda No. 2265 (8769, 9306 and 9344), 1945 (1950), 68 pp. **3**

[11] Aerofoil theory of a flat delta wing at supersonic speeds. Ministry of Supply (London), Aeronaut. Res. Council, Rep. and Memoranda No. 2548 (10,222), 1946 (1952), 21 pp. M.R. 14 #699. **3**

[12] Flutter derivatives of a wing-tailplane combination. College of Aeronautics, Cranfield (1947), 21 pp.

[13] The characterization of algebraic plane curves (with Th. Motzkin). Duke Math. J. 14 (1947), 837–853. M.R. 9 #373. **1**

[14] Note on the application of the linearised theory for compressible flow to transonic speeds (with A. D. Young). Ministry of Supply (London), Aeronaut. Res. Council, Rep. and Memoranda No. 2399 (10,474), 1947 (1951), 6 pp.

[15] The effect of the sweepback of delta wings on the performance of an aircraft at supersonic speeds

(with F. T. Davies). Ministry of Supply (London), Aeronaut. Res. Council, Rep. and Memoranda No. 2476 (10,594), 1947 (1951), 6 pp.

[16] Interference on a wing due to a body at supersonic speeds (with S. Kirkby). Ministry of Supply (London), Aeronaut. Res. Council, Rep. and Memoranda No. 2500 (10,631), 1947 (1952), 10 pp. M.R. 9 #479, M.R. 13 #882.

[17] Rotary derivatives of a delta wing at supersonic speeds. J. Royal Aeronaut. Soc. 52 (1948), 735–752.

[18] On some problems of unsteady supersonic aerofoil theory. Proc. 7th Internat. Congress Appl. Mech. Vol. 2 (1948), 500–514. M.R. 11 #477. **3**

[19] Bound and trailing vortices in the linearised theory of supersonic flow, and the downwash in the wake of a delta wing (with J. H. Hunter-Tod). Ministry of Supply (London), Aeronaut. Res. Council, Rep. and Memoranda No. 2409 (11,296), 1947 (1952), 14 pp. M.R. 9 #479. **3**

[20] On source and vortex distributions in the linearized theory of steady supersonic flow. Quart. J. Mech. Appl. Math. 1, 1947 (1948), 408–432. M.R. 10 #74. M.R. 10 #410. **3**

[21] The aerodynamic derivatives with respect to sideslip for a delta wing with small dihedral at zero incidence at supersonic speeds (with J. H. Hunter-Tod). Ministry of Supply (London), Aeronaut. Res. Council, Rep. and Memoranda No. 2410 (11,322), 1947 (1952), 14 pp. **3**

[22] Numerical solution of integral equations (with S. Kirkby). Note on Computational Methods No. 7, College of Aeronautics, Cranfield, (1949), 67 pp. M.R. 9 #479.

[23] On non-associative systems. Proc. Edinburgh Math. Soc. (2) 8 (1949), 111–118. M.R. 12 #5. **1**

[24] On the integration of hyperbolic differential equations. J. London Math. Soc. 25 (1950), 209–217. M.R. 12 #338. (College of Aeronautics, Cranfield, preprint (1948) reviewed in M.R. 10 #303.) **2**

[25] On functional transformations and summability. Proc. London Math. Soc. (2) 52 (1950), 132–160. M.R. 12 #253. **2**

[26] Les rapports entre le calcul déductif et l'interprétation sémantique d'un systéme axiomatique. Colloques Internationaux du Centre National de la Recherche Scientifique, Paris, No. 36 (1950). 35–52. Centre National de la Recherche Scientifique, Paris, 1953. M.R. 15 #190.

[27] On the application of symbolic logic to algebra. Proc. Internat. Congress of Mathematicians, Cambridge, Mass., 1950, Vol. 1, 686–694. Amer. Math. Soc., Providence, R.I., 1952. M.R. 13 #716. **1**

[28] Wave reflexion near a wall. Proc. Cambridge Philos. Soc. 47 (1951), 528–544. M.R. 12 #875. (College of Aeronautics, Cranfield, Report No. 17 (1950) reviewed in M.R. 12 #454.) **3**

[29] On axiomatic systems which posses finite models. "Methodos" Editrice La Fiaccola, Milano (1951), 140–149. **1**

[30] Aerofoil theory for swallow tail wings of small aspect ratio. Aeronaut. Quart. 4 (1952), 69–82. M.R. 14 #219. (College of Aeronautics, Cranfield, Report No. 41 (1950) reviewed in M.R. 12 #452.)

[31] L'application de la logique formelle aux mathématiques. Applications scientifiques de la logique mathématique (Actes du 2e Colloque International de Logique Mathématique, Paris, 1952), 51–64. Gauthier-Villars, Paris, 1954. M.R. 16 #782.

[32] Non-uniform supersonic flow. Quart. Appl. Math. 10 (1953), 307–319. M.R. 14 #511. **3**

[33] Flow round compound lifting units. Proc. Symposium on High Speed Aerodynamics, Ottawa, Canada 1953. High Aerodynamics Lab. National Aeronaut. Establishment, Ottawa (1953), 26–29.

[34] Core-consistency and total inclusion for methods of summability (with G. G. Lorentz). Canad. J. Math. 6 (1954), 27–34. M.R. 15 #618. **2**

[35] On some problems of unsteady aerofoil theory, Proc. of the Second Canadian Symposium on Aerodynamics, Toronto, 1954. The Institute of Aerophysics, University of Toronto, 1954, 106–122. M.R. 17 #310. **3**

[36] On predicates in algebraically closed fields. J. Symbolic Logic 19 (1954), 103–114. M.R. 15 #925. **1**

[37] Note on an embedding theorem for algebraic systems. J. London Math. Soc. 30 (1955), 249–252. M.R. 17 #449. **1**

[38] Mixed problems for hyperbolic partial differential equations (with L. L. Campbell). Proc. London Math. Soc. (2) 5 (1955), 129–147. M.R. 16 #1116.

[39] Metamathematical considerations on the relative irreducibility of polynomials (with P. C. Gilmore). Canad. J. Math. 7 (1955), 483–489. M.R. 17 #226. **1**

[40] On ordered fields and definite functions. Math. Ann. 130 (1955), 257–271. M.R. 17 #822. **1**

[41] Apercu metamathématique sur les nombres réels. Deux Conférences Prononcées a L'Université de Montréal, Février (1956), 14 pp.

[42] Further remarks on ordered fields and definite functions. Math. Ann. 130 (1956), 405–409. M.R. 17 #118. **1**

[43] A result on consistency and its application to the theory of definition. Nederl. Akad. Wetensch. Proc. Ser. A 59, and Indag. Math. 18 (1956), 47–58. M.R. 17 #1122. **1**

[44] Note on a problem of L.Henkin. J. Symbolic Logic 21 (1956), 33–35. M.R. 17 #817. **1**

[45] Ordered structures and related concepts. In Mathematical Interpretation of Formal Systems. North-Holland Publ. Co., Amsterdam, 1955, 51–56. M.R. 17 #700. **1**

[46] Completeness and persistence in the theory of models. Z. Math. Logik und Grundlagen der Math. 2 (1956), 15–26. M.R. 17 #1173. **1**

[47] Solution of a problem by Erdös–Gillman–Henriksen. Proc. Amer. Math. Soc. 7 (1956), 908–909. M.R. 18 #37. **1**

[48] On the motion of small particles in a potential field of flow. Comm. Pure Appl. Math. 9 (1956), 69–84. M.R. 17 #1147. **3**

[49] Wave propagation in a heterogeneous elastic medium. J. Math. Phys. 36 (1957), 210–222. M.R. 19 #903. **3**

[50] Transient stresses in beams of variable characteristics. Quart. J. Mech. Appl. Math. 10 (1957), 148–159. M.R. 19 #106. **3**

[51] Some problems of definability in the lower predicate calculus. Fund. Math. 44 (1957), 309–329. M.R. 19 #1032. **1**

[52] Syntactical transforms (with A. H. Lightstone). Trans. Amer. Math. Soc. 86 (1957), 220–245. M.R. 19 #934. **1**

[53] On the representation of Herbrand functions in algebraically closed fields (with A. H. Lightstone). J. Symbolic Logic 22 (1957), 187–204. M.R. 20 #4555. **1**

[54] Relative model-completeness and the elimination of quantifiers. Dialectica 12 (1958), 394–407. M.R. 21 #1265. **1**

[55] Relative model-completeness and the elimination of quantifiers. In Summaries of Talks, Summer Institute for Symbolic Logic, Cornell Univ. 1957. Communications Research Division, Institute for Defense Analyses, Princeton, N.J. 1958 (2d. ed., 1960), 155–159.

[56] Proving a theorem (as done by man, logician, or machine). In Summaries of Talks, Summer Institute for Symbolic Logic, Cornell Univ. 1957. Communications Research Division, Institute for Defense Analyses (1958), 350–352.

[57] Applications to field theory. In Summaries of Talks, Summer Institute for Symbolic Logic, Cornell Univ. 1957. Communications Research Division, Institute for Defense Analyses (1958), 326–331.

[58] Outline of an introduction to mathematical logic. Parts I, II, III. Canad. Math. Bull. 1 (1958), 41–54, 113–136, 193–208; Part IV. Canad. Math. Bull. 2 (1959), 33–42. M.R. 20 #5123, 20, #5124; Parts III and IV. M.R. 21 #6321.

[59] Solution of a problem of Tarski. Fund. Math. 47 (1959), 179–204. M.R. 22 #3690. **1**

[60] On the concept of a differentially closed field. Bull. Res. Council Israel 8F (1959), 113–128. M.R. 23 #2323. **1**

[61] Obstructions to arithmetical extension and the theorem of Łoś and Suszko. Nederl. Akad. Wetensch. Proc. Ser. A 62 and Indag. Math. 21 (1959), 489–495. M.R. 22 #2544. **1**

[62] Algèbre différentielle à valeurs locales. Atti del VI Congresso dell'Unione Matematica Italiana, Napoli 1959. Edizioni Cremonese Roma 1959, 1 p.

[63] Local differential algebra. Trans. Amer. Math. Soc. 97 (1960), 427–456. M.R. 23 #A148. **1**
[64] Local differential algebra—the analytic case (with S. Halfin). Technical (Scientific) Note No. 9, Contract No. AF 61 (052)–187 US Airforce, Air Research and Development Command, European Office, Brussels, Belgium (1960), 9 pp.
[65] On the mechanization of the theory of equations. Bull. Research Council of Israel 9F (1960), 47–70. M.R. 26 #4910.
[66] Elementary properties of ordered Abelian groups (with E. Zakon). Trans. Amer. Math. Soc. 96 (1960), 222–236. M.R. 22 #5673. **1**
[67] Recent developments in model theory. Proc. of the 1960 Internat. Congress for Logic, Methodology and Philosophy of Science. Stanford Univ. Press, Stanford, Calif. (1962), 60–79. M.R. 29 #4668. **1**
[68] Model theory and non-standard arithmetic. Infinitistic Methods. In Proc. Sympos. Foundations of Math. Warsaw, 1959. Pergamon Press, N.Y. 1961, 265–302. M.R. 26 #32. **1**
[69] On the construction of models. In Essays on the Foundations of Mathematics. (Fraenkel anniversary volume), Magnes Press, Hebrew Univ., Jerusalem (1961), 207–217. M.R. 29 #23. **1**
[70] On the D-calculus for linear differential equations with constant coefficients. Math. Gazette 45 (1961), 202–206.
[71] Non-standard analysis. Nederl. Akad. Wetensch. Proc. Ser. A 64, and Indag. Math. 23 (1961), 432–440. M.R. 26 #33. **2**
[72] A note on embedding problems. Fund. Math. 50 (1961/62), 455–461. M.R. 25 #2946. **1**
[73] Airfoil theory, Chapter 72 in *Handbook of Engineering Mechanics* (W. Flügge, ed.), McGraw-Hill, New York, 1962, 24 pp.
[74] Modern mathematics and secondary schools. International Review of Education 8 (1962), 34–40.
[75] A basis for the mechanization of the theory of equations. In Computer Programming and Formal Systems. North-Holland Publ. Co., Amsterdam, 1962, 95–99. M.R. 26 #4911.
[76] On the mechanization of the theory of numbers (with M. Machover). Technical Report No. 9, Information Systems Branch Contract No. 62558–2214. U.S. Office of Naval Research (1962), 37 pp.
[77] Complex function theory over non-archimedean fields. Contract No. AF 61(052)–187. US Airforce, Office of Aerospace Research, European Office, Brussels, Belgium (Technical Report No. 30), (1962), 127 pp.
[78] Local partial differential algebra (with S. Halfin). Trans. Amer. Math. Soc. 109 (1963), 165–180. **1**
[79] On languages which are based on non-standard arithmetic. Nagoya Math. J. 22 (1963), 83–117. M.R. 27 #3532. **2**
[80] On symmetric bimatrix games (with J. H. Griesmer and A. J. Hoffman). IBM Research Paper RC-959 (1963), 24 pp.
[81] Some remarks on threshold functions. IBM Research Note NC-291, Thomas J. Watson Research Center, Yorktown Heights, New York (1963), 15 pp. **1**
[82] On generalized limits and linear functionals. Pacific J. Math. 14 (1964), 269–283. M.R. 29 #1534. **2**
[83] Random-access stored-program machines, an approach to programming languages (with C. C. Elgot). J. Assoc. Comp. Math. 11 (1964), 365–399. M.R. 30 #4400. **1**
[84] Between logic and mathematics. I.C.S.U. Review 6, North-Holland Publ. Co. (1964), 218–226. Polish translation in Wiadomosci Mat. 9 (1966), 89–96.
[85] Formalism 64. In Proc. Internat. Congress for Logic, Methodology and Philos. Sci., Jerusalem 1964, North-Holland Publ. Co., Amsterdam 1965, 228–246. M.R. 35 #5281. **2**
[86] On the theory of normal families. Acta Philos. Fenn. 18 (1965), 159–184. M.R. 32 #5845. **2**
[87] Topics in non-archimedean mathematics. In the Theory of Models. Proc. 1963 Internat. Sympos. Berkeley, Calif., North-Holland Publ. Co., Amsterdam, 1965, 285–298. M.R. 33 #5489. **2**
[88] Solution of an invariant subspace problem of K. T. Smith and P. R. Halmos (with A. R. Bernstein). Pacific J. Math. 16 (1966), 421–431. M.R. 33 #1724. **2**
[89] A new approach to the theory of algebraic numbers. I. II. Atti Accad. Naz. Lincei Rend. Cl.

Sci. fis. mat. natur. (8) 40 (1966), 222–225. M.R. 35 #5305 and 770–774 M.R. 35 #5306. **2**

[90] On some applications of model theory to algebra and analysis. Rend. Mat. e Appl. (Roma) (5) 25 (1966), 562–592. M.R. 36 #2489. **2**

[91] Non-standard theory of Dedekind rings. Nederl. Akad. Wetensch. Proc. Ser. A 70 and Indag. Math. 29 (1967), 444–452. M.R. 38 #4723. **2**

[92] Nonstandard arithmetic. Bull. Amer. Math. Soc. 73 (1967), 818–843. M.R. 36 #1319. **2**

[93] Multiple control computer models (with C. C. Elgot and J. D. Rutledge). In Systems and Computer Science (Proc. Conf. London, Ont. 1965), Univ. Toronto Press, Toronto, Ont. (1967), 60–76. M.R. 38 #4073. **1**

[94] The metaphysics of the calculus. In Problems in the Philosophy of Mathematics, North-Holland Publ. Co., Amsterdam, 1967, 28–46. **2**

[95] Some thoughts on the history of mathematics. Compositio Math. 20 (1968), 188–193. M.R. 37 #21. Also in Logic and the Foundations of Mathematics, dedicated to A. Heyting on his 70th birthday. Groningen, Walters-Noordhoff 1968. **2**

[96] Model theory. In Contemporary Philosophy: A Survey. Vol. I: Logic and the Foundations of Mathematics, edited by R. Klibansky. Firenze, La Nuova Italia Editrice 1968, 61–73. **2**

[97] On flexural wave propagation in nonhomogeneous elastic plates (with A. E. Hurd). SIAM J. Appl. Math. 16 (1968), 1081–1089. **3**

[98] Topics in nonstandard algebraic number theory. In Proc. Of Internat. Sympos. on Applications of Model Theory to Algebra, Analysis and Probability. Pasadena, Calif., 1967. Holt, Rinehart and Winston, New York, 1969, 1–17. M.R. 42 #3054. **2**

[99] A set-theoretical characterization of enlargements (with E. Zakon). In Proc. of the Internat. Sympos. on Applications of Model Theory to Algebra, Analysis and Probability. Pasadena, Calif. 1967, Holt, Rinehart and Winston, New York, 1969, 109–122. M.R. 39 #1319. **2**

[100] Germs. In Proc. of the Internat. Sympos. on Applications of Model Theory to Algebra, Analysis and Probability. Pasadena, Calif., 1967. Holt, Rinehart and Winston, New York, 1969, 138–149. M.R. 38 #4723. **2**

[101] Problems and methods of model theory. In Aspects of Math. Logic, C.I.M.E. Corsi 3° Ciclo, Varenna, 1968. Edizioni Cremonese, Roma, 1969, 181–266. M.R. 41 #5208.

[102] Compactification of groups and rings and nonstandard analysis. J. Symbolic Logic 34 (1969), 576–588. M.R. 44 #1765. **2**

[103] Completing theories by forcing (with J. Barwise). Ann. Math. Logic 2 (1970), 119–142. M.R. 42 #7494. **1**

[104] From a formalist's point of view. Dialectica, Vol. 23 (1970), 45–49.

[105] Elementary embeddings of fields of power series. J. Number Theory 2 (1970), 237–247. M.R. 41 #3451. **2**

[106] Forcing in model theory. Symposia Mathematica, Vol. 5, Academic Press, New York, 1971, 69–82. M.R. 43 #4651. See paper 108. **1**

[107] Infinite forcing in model theory. Proc. 2d Scandinavian Logic Sympos. in Oslo, 1970. North-Holland Publ. Co., Amsterdam, 1971, 317–340. M.R. 50 #9574. **1**

[108] Forcing in model theory. Proc. Internat. Congress of Mathematicians, Nice, 1970. Gauthier-Villars, Paris, 1971, 245–250. See paper 106.

[109] On the notion of algebraic closedness for noncommutative groups and fields. J. Symbolic Logic 36 (1971), 441–444. M.R. 45 #43. **1**

[110] Application of logic to pure mathematics. A brief survey, Yale (1971), 6 pp.

[111] Algebraic function fields and non-standard arithmetic. In Contributions to Nonstandard Analysis, North-Holland Publ. Co., Amsterdam (1972), 1–14. **2**

[112] Inductive theories and their forcing companions (with E. R. Fisher). Israel J. Math. 12 (1972), 95–107. M.R. 47 #3163. **1**

[113] The nonstandard $\lambda{:}\phi_2^4(x)$: model I. The technique of nonstandard analysis in theoretical physics (with P. Kelemen). J. Math. Phys. 13 (1972), 1870–1874. Model II. The standard model from a

nonstandard point of view (with P. Kelemen), 1875–1878. **2**

[114] A limit theorem on the cores of large standard exchange economies (with D. J. Brown). Proc. Nat. Acad. Sci. 69 (1972), 1258–1260. **2**

[115] Generic categories. Lecture presented at the Logic Symposium in Orleans, France (1972), 20 pp. **1**

[116] On the real closure of a Hardy field. In Theory of sets and topology (in honor of Felix Hausdorff). VEB. Deutsch. Verlag Wissensch., Berlin, 1972, 427–433. M.R. 40 #4980. **1**

[117] On bounds in the theory of polynomial ideals. Selected questions of algebra and logic. (A collection dedicated to the memory of A. I. Mal'cev), Izdat. "Nauka" Sibirsk. Otdel., Novosibirsk, 1973, 245–252. M.R. 49 #2364. **1**

[118] Model theory as a framework for algebra. In Studies in Model Theory, MAA Studies in Math., Vol. 8, Math. Assoc. Amer. Washington, D.C. (1973), 134–157. M.R. 49 #2365. **1**

[119] Function theory on some nonarchimedean fields. In Papers in the Foundations of Mathematics. Slaught Memorial Papers No. 13. Math. Assoc. Amer. (1973), 87–109. M.R. 48 #8464. **2**

[120] Ordered differential fields. J. Combinatorial Theory Ser. A, 14 (1973), 324–333. M.R. 48 #2124. **1**

[121] Nonstandard points on algebraic curves. J. Number Theory 5 (1973), 301–327. M.R. 53 #443. **2**

[122] Metamathematical problems. J. Symbolic Logic 38 (1973), 500–516. M.R. 49 #2240. **1**

[123] Standard and nonstandard number systems. (The Brouwer Memorial Lecture, Nieuw Arch. Wisk. (3) 21 (1973), 115–133. M.R. 55 #7767. **2**

[124] Numbers—What are they and what are they good for? Yale Scientific Magazine, Vol. 47 (1973), 14–16.

[125] Nonstandard arithmetic and generic arithmetic. In Proc. of the Fourth Internat. Congress for Logic, Methodology and Philosophy of Science, Bucharest, 1971. North-Holland Publ. Co., 1973, 137–154. **1**

[126] A note on topological model theory. Dedicated to Andrzej Mostowski on his 60th Birthday, Fund. Math. 81 (1973/74), 159–171. M.R. 49 #7135. **1**

[127] A decision method for elementary algebra and geometry—revisited. Proc. of the Tarski Symposium, Berkeley, Calif., 1971. Proc. Sympos. Pure Math. Vol. 25, 139–152. Amer. Math. Soc., 1974. M.R. 51 #2902. **1**

[128] Enlarged Sheaves. In Proc. Victoria Symposium on Nonstandard Analysis 1972. Lecture Notes in Mathematics Vol. 369, Springer-Verlag, N.Y., 1974, 249–260. **2**

[129] The Cores of Large Standard Exchange Economies (with D. J. Brown). J. Economic Theory, 9 (1974), 245–254. **2**

[130] Nonstandard Exchange Economies (with D. J. Brown). Econometrica 43 (1975), 41–55. **2**

[131] Concerning progress in the philosophy of mathematics. In Proc. Logic Colloquium at Bristol, 1973. North-Holland Publ. Co., Amsterdam 1975, 41–52. M.R. 52 #13309 **2**

[132] On the finiteness theorem of Siegel and Mahler Concerning Diophantine Equations (with P. Roquette). J. Number Theory 7 (1975), 121–176. M.R. 51 #10222. **2**

[133] Algorithms in algebra. In Model Theory and Algebra. A Memorial Tribute to Abraham Robinson. Lecture Notes in Mathematics, Vol. 498, Springer-Verlag, N.Y. 1975, 15–40. M.R. 52 #7880 and 53 #5298. **1**

[134] On constrained denotation. **2**

Film

Non-standard Analysis—a filmed 1 hour lecture presented by the Mathematical Association of America, 1970.

Articles Published in Encyclopedias

Dictionary of Scientific Biography, New York, Charles Scribner's Sons, 14 Vols. 1970–1976.; L'Hôpital, Guillaume-Francios-Antoine de, Vol. 8, pp. 304–305 (1973); Mittag-Leffler, Magnus Gustav, Vol. 9, pp. 426–427 (1974); Méray, Hughes Charles Robert, Vol. 9, pp. 307–308 (1974); Toeplitz, Otto, Vol. 13, p. 42 (1976); Stoltz, Otto, Vol. 13, p. 81 (1976).

Ha-Entsikedyah Ha'Ivrit (in Hebrew). Vol. 24, article, ''Matematikah,'' section containing discussion on the foundation of mathematics with emphasis on historical development, pp. 756–763 (1972). Vol. 14, article, ''Hidrodinamikah,'' pp. 83–89 (1960).

Enciclopedia del Novecento, Mathematical Logic and the Foundations of Mathematics (in Italian), 67 pp. Rome, Instituto Della Enciclopedia Italiana, Vol. 3, 1978.

Doctoral Students of Abraham Robinson

UNIVERSITY OF TORONTO (1952–56)

L. L. Campbell, Queens University, Kingston, Ontario
A. H. Lightstone, Queens University (deceased)
J. A. Steketee, Technological University of Delft, the Netherlands
R. A. Ross, University of Toronto

HEBREW UNIVERSITY OF JERUSALEM (1957–62)

A. Levy, Hebrew University of Jerusalem (coadviser with A. Fraenkel)
P. Katz, Hebrew University of Jerusalem (coadviser with A. Dvoretzky)
S. Halfin, Bell Laboratories, Murray Hill, New Jersey
A. Meir, University of Alberta, Edmonton, Canada (coadviser with A. Jakimovsky)

UNIVERSITY OF CALIFORNIA AT LOS ANGELES (1962–68)

A. R. Bernstein, University of Maryland
D. L. Dubrovsky, Montreal, Quebec
E. M. Gold
L. D. Kugler, University of Michigan at Flint
W. M. Lambert, Universidad de Costa Rica
Eugene Madison, University of Iowa
R. G. Phillips, University of South Carolina
P. Tripodes, Venice, California

YALE UNIVERSITY (1968–75)

G. L. Cherlin, Rutgers University
C. Wood, Wesleyan University
D. R. Johnson, Jr., University of Pittsburgh
J. Hirschfeld, Tel-Aviv University
W. H. Wheeler, Indiana University
E. E. Kra, Elizabeth, New Jersey
L. M. Manevitz, Hebrew University of Jerusalem (Robinson succeeded by A. Macintrye as adviser)
P. M. Winkler, Stanford University (Robinson succeeded by A. Macintyre as adviser)